Molecular Evolution

Molecular Evolution
A Statistical Approach

ZIHENG YANG

UNIVERSITY PRESS

Molecular Evolution: A Statistical Approach. Ziheng Yang. © Ziheng Yang 2014.
Published 2014 by Oxford University Press.

OXFORD
UNIVERSITY PRESS

Great Clarendon Street, Oxford, OX2 6DP,
United Kingdom

Oxford University Press is a department of the University of Oxford.
It furthers the University's objective of excellence in research, scholarship,
and education by publishing worldwide. Oxford is a registered trade mark of
Oxford University Press in the UK and in certain other countries

© Ziheng Yang 2014

The moral rights of the authors have been asserted

First Edition published in 2014

All rights reserved. No part of this publication may be reproduced, stored in
a retrieval system, or transmitted, in any form or by any means, without the
prior permission in writing of Oxford University Press, or as expressly permitted
by law, by licence or under terms agreed with the appropriate reprographics
rights organization. Enquiries concerning reproduction outside the scope of the
above should be sent to the Rights Department, Oxford University Press, at the
address above

You must not circulate this work in any other form
and you must impose this same condition on any acquirer

Published in the United States of America by Oxford University Press
198 Madison Avenue, New York, NY 10016, United States of America

British Library Cataloguing in Publication Data
Data available

Library of Congress Control Number: 2013956540

ISBN 978-0-19-960261-2

Links to third party websites are provided by Oxford in good faith and
for information only. Oxford disclaims any responsibility for the materials
contained in any third party website referenced in this work.

Foreword

Over the last two decades, Ziheng Yang has been a leading architect of the emergent field of computational molecular evolution. His first book, *Computational Molecular Evolution*, was published in 2006 and became an instant classic. The book broke new ground both in terms of its subject matter and expository style. It presented an up-to-date, detailed, and comprehensive account of computational and statistical aspects of molecular evolutionary analysis, while retaining an informal style and pragmatic perspective that made it highly accessible. The book targeted a readership that included both biologists and applied mathematicians, yet it did not oversimplify in catering to biologists by avoiding advanced calculus or linear algebra, or pandering to mathematicians with the usual theorem-proof format. Somehow, this middle-of-the-road approach seems to have worked. Furthermore, despite the book's graduate textbook flavour the chapters were peppered with Yang's original interpretations and suggestions making it part textbook and part research monograph. Even individuals who were already experienced in computational evolutionary analysis will have gained new insights.

Yang's knowledge and practical experience are evident on every page of his new book, *Molecular Evolution: A Statistical Approach*. What is particularly remarkable is his ability to translate for non-specialists the key developments of this rapidly changing field so effectively. The content represents a significant expansion of his previous book; in particular, the treatment of Bayesian inference is much more extensive. Bayesian inference has become a cornerstone of phylogenetic inference over the last decade, as many programs such as MRBAYES and BEAST are now available which implement Markov chain Monte Carlo (MCMC) simulation methods for this purpose. The book devotes new chapters to the fundamentals of Bayesian inference and MCMC methodologies. Biologists using MCMC programs for molecular evolutionary analyses will benefit from the ground-up approach of these chapters, which introduce the basic principles using motivating examples based on evolutionary processes of obvious practical importance that will be familiar to molecular evolutionists. In this way, remarkably clear explanations are provided for such notoriously difficult concepts as reversible-jump MCMC, Dirichlet processes, Bayes factor calculations for model comparison, and so on. Several excellent books exist on phylogenetic inference, written from either an applied statistical perspective (Felsenstein 2004) or a more rigorous mathematical one (Semple and Steel 2003). However, I am unaware of any book that contains the extensive details found in Yang's book concerning the MCMC implementations (proposal moves, prior distributions, etc.) underlying currently available programs for Bayesian phylogenetic inference.

In this era of cheap next-generation sequencing, multi-locus genomic data are the new norm and therefore the distinction between inference of locus-specific gene trees and multi-locus species trees has become key. *Molecular Evolution: A Statistical Approach* thus contains a new chapter that covers the multi-species coalescent, species tree inference, and species delimitation methods. Yang has been a key contributor to the development of this theory during the last decade and provides one of the clearest explanations of the multi-species coalescent that I have read. For persons whose research interests include computational molecular evolution and molecular phylogenetics this new book from Ziheng Yang is essential reading.

Bruce Rannala

Davis, California
April 2014

Preface

The main objective of this book is to present and explain the statistical methods and computational algorithms developed in molecular evolution, phylogenetics, and phylogeography for the comparative analysis of genetic sequence data. Reconstruction of molecular phylogeny and inference of the molecular evolutionary process are considered problems of statistical inference, and likelihood and Bayesian methods are treated in depth as standard methods of data analysis. Heuristic and approximate methods are discussed from such a viewpoint as well and are often used to introduce the central concepts, because of their simplicity and intuitive appeal. However, the book does not dwell on proofs or mathematical niceties; it emphasizes care but not rigour.

Molecular Evolution: A Statistical Approach represents an expanded and updated treatment of my earlier research monograph *Computational Molecular Evolution*, published by Oxford University Press in 2006. The major change has been the far more comprehensive and extensive coverage of Bayesian methods, while the target audience has been expanded to include upper level undergraduate as well as graduate students. It can also be read by researchers working in such diverse fields as evolutionary biology, molecular systematics, population genetics, statistical phylogeography, bioinformatics and computational biology, computer science, and computational statistics. It is hoped that biologists who have used software programs to analyse their own data will find the book particularly useful in helping them understand the principles of the methods. For applied mathematicians, molecular studies of evolution are 'a source of novel statistical problems' (Neyman 1971), and this book will provide an accessible summary of the exciting and often unconventional inference problems in the field, some of which are yet unsolved.

Although this new book is written at a similar level of mathematical sophistication as my 2006 work, I have taken care to assist the biologist readers who may find the mathematical arguments challenging. First, every important mathematical result is followed by a verbal rendering, and it is reportedly possible to read the book while skipping the equations, at least at first reading. Second, I have included numerous examples of real data analysis and numerical calculations to illustrate the theory, in addition to the working problems at the end of each chapter. Many biologists find numerical calculations less intimidating than abstract formulae. Example datasets and small C and R programs that implement computational algorithms discussed in the book are posted on the web site for the book: http://abacus.gene.ucl.ac.uk/MESA/. Third, I have prepared a primer on probability and statistics, with an overview of mathematical results used in this book, for biologists who would like to grapple with the mathematical details in the book. This has been used as the pre-course reading material for an advanced workshop on Computational Molecular Evolution (CoME) that runs annually in Hinxton, Cambridge, and Heraklion, Crete, co-organized by Aidan Budd, Nick Goldman, Alexandros Stamatakis, and me. It is available at: http://abacus.gene.ucl.ac.uk/PPS/PrimerProbabilityStatistics.pdf.

The 2006 book was used as a textbook for graduate courses on bioinformatics and computational genomics in Peking University (2010) and in ETH Zurich (2011). I thank the students in those courses for their useful feedback. For instructors, I have found an early coverage of the simulation chapter to be useful, as afterwards simulation projects can be assigned as homework when other chapters are taught.

I am grateful to a number of colleagues who read earlier drafts of chapters of this book and provided constructive comments and criticisms: Konstantinos Angelis, Mario dos Reis, Ed Susko, Chi Zhang, and Tianqi Zhu. The following colleagues read and commented on Chapter 9: Daniel Dalquen, Adam Leaché, Liang Liu, and Jim Mallet. Needless to say, all errors that remain are mine. (Please report errors and typos you discover to me at z.yang@ucl.ac.uk. Errata will be posted on the book's web site.) Thanks are also due to Helen Eaton, Lucy Nash, and Ian Sherman at Oxford University Press for their support and patience throughout the project.

Ziheng Yang

London
April 2014

Contents

1 Models of nucleotide substitution   1
   1.1 Introduction   1
   1.2 Markov models of nucleotide substitution and distance estimation   4
     1.2.1 The JC69 model   4
     1.2.2 The K80 model   7
     1.2.3 HKY85, F84, TN93, etc.   9
     1.2.4 The transition/transversion rate ratio   13
   1.3 Variable substitution rates across sites   15
   1.4 Maximum likelihood estimation of distance   17
     1.4.1 The JC69 model   18
     1.4.2 The K80 model   22
     1.4.3 Likelihood ratio test of substitution models   22
     *1.4.4 Profile and integrated likelihood methods   24
   1.5 Markov chains and distance estimation under general models   26
     1.5.1 Markov chains   26
     *1.5.2 Distance under the unrestricted (UNREST) model   27
     *1.5.3 Distance under the general time-reversible model   29
   1.6 Discussions   32
     1.6.1 Distance estimation under different substitution models   32
     1.6.2 Limitations of pairwise comparison   32
   1.7 Problems   33

2 Models of amino acid and codon substitution   35
   2.1 Introduction   35
   2.2 Models of amino acid replacement   35
     2.2.1 Empirical models   35
     2.2.2 Mechanistic models   39
     2.2.3 Among-site heterogeneity   39
   2.3 Estimation of distance between two protein sequences   40
     2.3.1 The Poisson model   40
     2.3.2 Empirical models   41
     2.3.3 Gamma distances   41
   2.4 Models of codon substitution   42
     2.4.1 The basic model   42
     2.4.2 Variations and extensions   44
   2.5 Estimation of d_S and d_N 47
     2.5.1 Counting methods   47
     2.5.2 Maximum likelihood method   55

	2.5.3	Comparison of methods	57
	2.5.4	More distances and interpretation of the d_N/d_S ratio	58
	2.5.5	Estimation of d_S and d_N in comparative genomics	61
	*2.5.6	Distances based on the physical-site definition	63
	*2.5.7	Utility of the distance measures	65
*2.6		Numerical calculation of the transition probability matrix	65
2.7		Problems	68

3 Phylogeny reconstruction: overview 70

3.1		Tree concepts	70
	3.1.1	Terminology	70
	3.1.2	Species trees and gene trees	79
	3.1.3	Classification of tree reconstruction methods	81
3.2		Exhaustive and heuristic tree search	82
	3.2.1	Exhaustive tree search	82
	3.2.2	Heuristic tree search	82
	3.2.3	Branch swapping	84
	3.2.4	Local peaks in the tree space	86
	3.2.5	Stochastic tree search	88
3.3		Distance matrix methods	88
	3.3.1	Least-squares method	89
	3.3.2	Minimum evolution method	91
	3.3.3	Neighbour-joining method	91
3.4		Maximum parsimony	95
	3.4.1	Brief history	95
	3.4.2	Counting the minimum number of changes on a tree	95
	3.4.3	Weighted parsimony and dynamic programming	96
	3.4.4	Probabilities of ancestral states	99
	3.4.5	Long-branch attraction	99
	3.4.6	Assumptions of parsimony	100
3.5		Problems	101

4 Maximum likelihood methods 102

4.1		Introduction	102
4.2		Likelihood calculation on tree	102
	4.2.1	Data, model, tree, and likelihood	102
	4.2.2	The pruning algorithm	103
	4.2.3	Time reversibility, the root of the tree, and the molecular clock	107
	4.2.4	A numerical example: phylogeny of apes	108
	4.2.5	Amino acid, codon, and RNA models	110
	*4.2.6	Missing data, sequence errors, and alignment gaps	110
4.3		Likelihood calculation under more complex models	114
	4.3.1	Mixture models for variable rates among sites	114
	4.3.2	Mixture models for pattern heterogeneity among sites	122
	4.3.3	Partition models for combined analysis of multiple datasets	123
	4.3.4	Nonhomogeneous and nonstationary models	125

4.4	Reconstruction of ancestral states		125
	4.4.1	Overview	125
	4.4.2	Empirical and hierarchical Bayesian reconstruction	127
	*4.4.3	Discrete morphological characters	130
	4.4.4	Systematic biases in ancestral reconstruction	131
*4.5	Numerical algorithms for maximum likelihood estimation		133
	*4.5.1	Univariate optimization	134
	*4.5.2	Multivariate optimization	136
4.6	ML optimization in phylogenetics		138
	4.6.1	Optimization on a fixed tree	138
	4.6.2	Multiple local peaks on the likelihood surface for a fixed tree	139
	4.6.3	Search in the tree space	140
	4.6.4	Approximate likelihood method	143
4.7	Model selection and robustness		144
	4.7.1	Likelihood ratio test applied to rbcL dataset	144
	4.7.2	Test of goodness of fit and parametric bootstrap	146
	*4.7.3	Diagnostic tests to detect model violations	147
	4.7.4	Akaike information criterion (AIC and AIC_c)	148
	4.7.5	Bayesian information criterion	149
	4.7.6	Model adequacy and robustness	150
4.8	Problems		151

5 Comparison of phylogenetic methods and tests on trees 153

5.1	Statistical performance of tree reconstruction methods		153
	5.1.1	Criteria	154
	5.1.2	Performance	156
5.2	Likelihood		157
	5.2.1	Contrast with conventional parameter estimation	157
	5.2.2	Consistency	158
	5.2.3	Efficiency	159
	5.2.4	Robustness	163
5.3	Parsimony		165
	5.3.1	Equivalence with misbehaved likelihood models	165
	5.3.2	Equivalence with well-behaved likelihood models	168
	5.3.3	Assumptions and justifications	169
5.4	Testing hypotheses concerning trees		171
	5.4.1	Bootstrap	172
	5.4.2	Interior-branch test	177
	5.4.3	K-H test and related tests	178
	5.4.4	Example: phylogeny of apes	179
	5.4.5	Indexes used in parsimony analysis	180
5.5	Problems		181

6 Bayesian theory 182

6.1	Overview	182
6.2	The Bayesian paradigm	183

		6.2.1	The Bayes theorem	183
		6.2.2	The Bayes theorem in Bayesian statistics	184
		*6.2.3	Classical versus Bayesian statistics	189
	6.3	Prior		197
		6.3.1	Methods of prior specification	197
		6.3.2	Conjugate priors	198
		6.3.3	Flat or uniform priors	199
		*6.3.4	The Jeffreys priors	200
		*6.3.5	The reference priors	202
	6.4	Methods of integration		203
		*6.4.1	Laplace approximation	203
		6.4.2	Mid-point and trapezoid methods	204
		6.4.3	Gaussian quadrature	205
		6.4.4	Marginal likelihood calculation for JC69 distance estimation	206
		6.4.5	Monte Carlo integration	210
		6.4.6	Importance sampling	210
	6.5	Problems		212

7 Bayesian computation (MCMC) — 214

	7.1	Markov chain Monte Carlo		214
		7.1.1	Metropolis algorithm	214
		7.1.2	Asymmetrical moves and proposal ratio	218
		7.1.3	The transition kernel	219
		7.1.4	Single-component Metropolis–Hastings algorithm	220
		7.1.5	Gibbs sampler	221
	7.2	Simple moves and their proposal ratios		221
		7.2.1	Sliding window using the uniform proposal	222
		7.2.2	Sliding window using the normal proposal	223
		7.2.3	Bactrian proposal	223
		7.2.4	Sliding window using the multivariate normal proposal	224
		7.2.5	Proportional scaling	225
		7.2.6	Proportional scaling with bounds	226
	7.3	Convergence, mixing, and summary of MCMC		226
		7.3.1	Convergence and tail behaviour	226
		7.3.2	Mixing efficiency, jump probability, and step length	230
		7.3.3	Validating and diagnosing MCMC algorithms	241
		7.3.4	Potential scale reduction statistic	242
		7.3.5	Summary of MCMC output	243
	7.4	Advanced Monte Carlo methods		244
		7.4.1	Parallel tempering (MC^3)	245
		7.4.2	Trans-model and trans-dimensional MCMC	247
		7.4.3	Bayes factor and marginal likelihood	256
	7.5	Problems		260

8 Bayesian phylogenetics — 263

	8.1	Overview		263
		8.1.1	Historical background	263

8.1.2	A sketch MCMC algorithm	264
8.1.3	The statistical nature of phylogeny estimation	264

8.2 Models and priors in Bayesian phylogenetics 266
    8.2.1 Priors on branch lengths 266
    8.2.2 Priors on parameters in substitution models 269
    8.2.3 Priors on tree topology 276

8.3 MCMC proposals in Bayesian phylogenetics 279
    8.3.1 Within-tree moves 279
    8.3.2 Cross-tree moves 281
    8.3.3 NNI for unrooted trees 284
    8.3.4 SPR for unrooted trees 287
    8.3.5 TBR for unrooted trees 289
    8.3.6 Subtree swapping 291
    8.3.7 NNI for rooted trees 292
    8.3.8 SPR on rooted trees 293
    8.3.9 Node slider 294

8.4 Summarizing MCMC output 295

8.5 High posterior probabilities for trees 296
    8.5.1 High posterior probabilities for trees or splits 296
    8.5.2 Star tree paradox 298
    *8.5.3 Fair coin paradox, fair balance paradox, and Bayesian model selection 300
    8.5.4 Conservative Bayesian phylogenetics 305

8.6 Problems 306

9 Coalescent theory and species trees 308

9.1 Overview 308

9.2 The coalescent model for a single species 309
    9.2.1 The backward time machine 309
    9.2.2 Fisher–Wright model and the neutral coalescent 309
    9.2.3 A sample of n genes 312
    9.2.4 Simulating the coalescent 315
    9.2.5 Estimation of θ from a sample of DNA sequences 316

9.3 Population demographic process 320
    9.3.1 Homogeneous and nonhomogeneous Poisson processes 321
    9.3.2 Deterministic population size change 322
    9.3.3 Nonparametric population demographic models 323

9.4 Multispecies coalescent, species trees and gene trees 325
    9.4.1 Multispecies coalescent 325
    9.4.2 Species tree–gene tree conflict 331
    9.4.3 Estimation of species trees 335
    9.4.4 Migration 343

9.5 Species delimitation 349
    9.5.1 Species concept and species delimitation 349
    9.5.2 Simple methods for analysing genetic data 351
    9.5.3 Bayesian species delimitation 352

	9.5.4 The impact of guide tree, prior, and migration	355
	9.5.5 Pros and cons of Bayesian species delimitation	358
9.6	Problems	359

10 Molecular clock and estimation of species divergence times — 361

10.1 Overview — 361
10.2 Tests of the molecular clock — 363
    10.2.1 Relative-rate tests — 363
    10.2.2 Likelihood ratio test — 364
    10.2.3 Limitations of molecular clock tests — 365
    10.2.4 Index of dispersion — 366
10.3 Likelihood estimation of divergence times — 366
    10.3.1 Global clock model — 366
    10.3.2 Local clock model — 367
    10.3.3 Heuristic rate-smoothing methods — 368
    10.3.4 Uncertainties in calibrations — 370
    10.3.5 Dating viral divergences — 372
    10.3.6 Dating primate divergences — 373
10.4 Bayesian estimation of divergence times — 375
    10.4.1 General framework — 375
    10.4.2 Approximate calculation of likelihood — 376
    10.4.3 Prior on evolutionary rates — 377
    10.4.4 Prior on divergence times and fossil calibrations — 378
    10.4.5 Uncertainties in time estimates — 382
    10.4.6 Dating viral divergences — 384
    10.4.7 Application to primate and mammalian divergences — 385
10.5 Perspectives — 388
10.6 Problems — 389

11 Neutral and adaptive protein evolution — 390

11.1 Introduction — 390
11.2 The neutral theory and tests of neutrality — 391
    11.2.1 The neutral and nearly neutral theories — 391
    11.2.2 Tajima's D statistic — 393
    11.2.3 Fu and Li's D, and Fay and Wu's H statistics — 394
    11.2.4 McDonald–Kreitman test and estimation of selective strength — 395
    11.2.5 Hudson–Kreitman–Aquade test — 397
11.3 Lineages undergoing adaptive evolution — 398
    11.3.1 Heuristic methods — 398
    11.3.2 Likelihood method — 399
11.4 Amino acid sites undergoing adaptive evolution — 400
    11.4.1 Three strategies — 400
    11.4.2 Likelihood ratio test of positive selection under random-site models — 402
    11.4.3 Identification of sites under positive selection — 405
    11.4.4 Positive selection at the human MHC — 406

11.5	Adaptive evolution affecting particular sites and lineages	408
	11.5.1 Branch-site test of positive selection	408
	11.5.2 Other similar models	409
	11.5.3 Adaptive evolution in angiosperm phytochromes	410
11.6	Assumptions, limitations, and comparisons	411
	11.6.1 Assumptions and limitations of current methods	412
	11.6.2 Comparison of methods for detecting positive selection	413
11.7	Adaptively evolving genes	414
11.8	Problems	416

12 Simulating molecular evolution — 418

12.1	Introduction	418
12.2	Random number generator	418
12.3	Generation of discrete random variables	420
	12.3.1 Inversion method for sampling from a general discrete distribution	420
	12.3.2 The alias method for sampling from a discrete distribution	421
	12.3.3 Discrete uniform distribution	422
	12.3.4 Binomial distribution	423
	12.3.5 The multinomial distribution	423
	12.3.6 The Poisson distribution	423
	12.3.7 The composition method for mixture distributions	424
12.4	Generation of continuous random variables	424
	12.4.1 The inversion method	425
	12.4.2 The transformation method	425
	12.4.3 The rejection method	425
	12.4.4 Generation of a standard normal variate using the polar method	428
	12.4.5 Gamma, beta, and Dirichlet variables	430
12.5	Simulation of Markov processes	430
	12.5.1 Simulation of the Poisson process	430
	12.5.2 Simulation of the nonhomogeneous Poisson process	431
	12.5.3 Simulation of discrete-time Markov chains	433
	12.5.4 Simulation of continuous-time Markov chains	435
12.6	Simulating molecular evolution	436
	12.6.1 Simulation of sequences on a fixed tree	436
	12.6.2 Simulation of random trees	439
12.7	Validation of the simulation program	439
12.8	Problems	440

Appendices — 442

Appendix A. Functions of random variables	442
Appendix B. The delta technique	446
Appendix C. Phylogenetic software	448

References — 450
Index — 488

Reader note: The asterisk next to a heading indicates a more difficult or technical section/problem.

CHAPTER 1

Models of nucleotide substitution

1.1 Introduction

Calculation of the distance between two sequences is perhaps the simplest phylogenetic analysis, yet it is important for two reasons. First, calculation of pairwise distances is the first step in distance matrix methods of phylogeny reconstruction, which use cluster algorithms to convert a distance matrix into a phylogenetic tree. Second, Markov process models of nucleotide substitution used in distance calculation form the basis of likelihood and Bayesian methods of phylogeny reconstruction. Indeed, joint analysis of multiple sequences can be viewed as a natural extension of pairwise distance calculation. Thus, besides discussing distance estimation, this chapter introduces the theory of Markov chains used in modelling nucleotide substitutions in a DNA sequence. It also introduces the method of maximum likelihood (ML). Bayesian estimation of pairwise distances and Bayesian phylogenetics are introduced in Chapters 6–8.

The distance between two sequences is defined as the expected number of nucleotide substitutions per site. If the evolutionary rate is constant over time, the distance will increase linearly with the time of divergence. A simplistic distance measure is the proportion of different sites, sometimes called the p distance. If 10 sites are different between two sequences, each 100 nucleotides long, then $p = 10\% = 0.1$. This raw proportion works fine for very closely related sequences but is otherwise a clear underestimate of the number of substitutions that have occurred. A variable site may result from more than one substitution, and even a constant site, with the same nucleotide observed in the two sequences, may harbour back or parallel substitutions (Figure 1.1). Multiple substitutions at the same site or *multiple hits* cause some changes to be hidden. As a result, p is not a linear function of evolutionary time. Thus the raw proportion p is usable only for highly similar sequences, with $p < 5\%$, say.

To estimate the number of substitutions, we need a probabilistic model to describe changes between nucleotides over evolutionary time. Continuous-time Markov chains are commonly used for this purpose. The nucleotide sites in the sequence are assumed to be evolving independently of each other. Substitutions at any particular site are described by a Markov chain, with the four nucleotides to be the *states* of the chain. The main feature of a Markov chain is that it has no memory: 'given the present, the future does not depend on the past'. In other words, the probability with which the chain jumps into other nucleotide states depends on the current state, but not on how the current state is reached. This is known as the *Markovian property*. Besides this basic assumption, we often place further constraints on substitution rates between nucleotides, leading to

Molecular Evolution: A Statistical Approach. Ziheng Yang. © Ziheng Yang 2014.
Published 2014 by Oxford University Press.

1 MODELS OF NUCLEOTIDE SUBSTITUTION

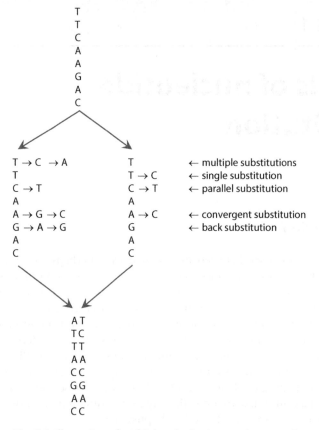

Fig. 1.1 Illustration of multiple substitutions at the same site or multiple hits. An ancestral sequence has diverged into two sequences and has since accumulated nucleotide substitutions independently along the two lineages. Only two *differences* are observed between the two present-day sequences, so that the proportion of different sites is $\hat{p} = 2/8 = 0.25$, while in fact as many as 10 *substitutions* (seven on the left lineage and three on the right lineage) occurred so that the true distance is $10/8 = 1.25$ substitutions per site. Constructed following Graur and Li (2000).

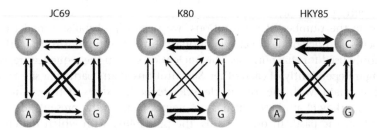

Fig. 1.2 Relative substitution rates between nucleotides under three Markov chain models of nucleotide substitution: JC69, K80, and HKY85. The thickness of the lines represents the substitution rates, while the sizes of the circles represent the steady-state distribution.

different models of nucleotide substitution. A few commonly used models are summarized in Table 1.1 and illustrated in Figure 1.2. These are discussed below.

Table 1.1 Substitution rate matrices for commonly used Markov models of nucleotide substitution

	p	From	To T	C	A	G
JC69 (Jukes and Cantor 1969)	1	T	·	λ	λ	λ
		C	λ	·	λ	λ
		A	λ	λ	·	λ
		G	λ	λ	λ	·
K80 (Kimura 1980)	2	T	·	α	β	β
		C	α	·	β	β
		A	β	β	·	α
		G	β	β	α	·
F81 (Felsenstein 1981)	4	T	·	π_C	π_A	π_G
		C	π_T	·	π_A	π_G
		A	π_T	π_C	·	π_G
		G	π_T	π_C	π_A	·
HKY85 (Hasegawa et al. 1984, 1985)	5	T	·	$\alpha\pi_C$	$\beta\pi_A$	$\beta\pi_G$
		C	$\alpha\pi_T$	·	$\beta\pi_A$	$\beta\pi_G$
		A	$\beta\pi_T$	$\beta\pi_C$	·	$\alpha\pi_G$
		G	$\beta\pi_T$	$\beta\pi_C$	$\alpha\pi_A$	·
F84 (Felsenstein, DNAML program since 1984)	5	T	·	$(1 + \kappa/\pi_Y)\beta\pi_C$	$\beta\pi_A$	$\beta\pi_G$
		C	$(1 + \kappa/\pi_Y)\beta\pi_T$	·	$\beta\pi_A$	$\beta\pi_G$
		A	$\beta\pi_T$	$\beta\pi_T$	·	$(1 + \kappa/\pi_R)\beta\pi_G$
		G	$\beta\pi_T$	$\beta\pi_C$	$(1 + \kappa/\pi_R)\beta\pi_A$	·
TN93 (Tamura and Nei 1993)	6	T	·	$\alpha_1\pi_C$	$\beta\pi_A$	$\beta\pi_G$
		C	$\alpha_1\pi_T$	·	$\beta\pi_A$	$\beta\pi_G$
		A	$\beta\pi_T$	$\beta\pi_C$	·	$\alpha_2\pi_G$
		G	$\beta\pi_T$	$\beta\pi_C$	$\alpha_2\pi_A$	·
GTR (REV) (Tavaré 1986; Yang 1994b; Zharkikh 1994)	9	T	·	$a\pi_C$	$b\pi_A$	$c\pi_G$
		C	$a\pi_T$	·	$d\pi_A$	$e\pi_G$
		A	$b\pi_T$	$d\pi_C$	·	$f\pi_G$
		G	$c\pi_T$	$e\pi_C$	$f\pi_A$	·
UNREST (Yang 1994b)	12	T	·	a	b	c
		C	d	·	e	f
		A	g	h	·	i
		G	j	k	l	·

Note: The diagonals of the matrix are determined by the requirement that each row sums to 0. p is the number of free parameters in the model. If only relative rates are considered (as in a typical likelihood analysis), the number should be reduced by 1. In F84, $\pi_Y = \pi_T + \pi_C$ and $\pi_R = \pi_A + \pi_G$. The equilibrium distribution is $\pi = (\frac{1}{4}, \frac{1}{4}, \frac{1}{4}, \frac{1}{4})$ under JC69 and K80, and $\pi = (\pi_T, \pi_C, \pi_A, \pi_G)$ under F81, F84, HKY85, TN93, and GTR. Under the general unrestricted (UNREST) model, it is given by equation (1.61).

1.2 Markov models of nucleotide substitution and distance estimation

1.2.1 The JC69 model

The JC69 model (Jukes and Cantor 1969) assumes that every nucleotide has the same instantaneous rate λ of changing into every other nucleotide. We use q_{ij} to denote the substitution rate from nucleotides i to j, with i, j = T, C, A, or G. Thus the *substitution rate matrix* is

$$Q = \{q_{ij}\} = \begin{bmatrix} -3\lambda & \lambda & \lambda & \lambda \\ \lambda & -3\lambda & \lambda & \lambda \\ \lambda & \lambda & -3\lambda & \lambda \\ \lambda & \lambda & \lambda & -3\lambda \end{bmatrix}, \qquad (1.1)$$

where the nucleotides are ordered T, C, A, and G. The diagonals are determined by the mathematical requirement that each row of the matrix sums to 0. The total rate of substitution of any nucleotide i is 3λ, which is $-q_{ii}$.

To relate the Markov chain model to sequence data, we need calculate the probability that given the nucleotide i at a site now, it will become nucleotide j time t later. This is known as the *transition probability*, denoted $p_{ij}(t)$. If time t is very small, we have $p_{ij}(t) \approx q_{ij}t$ for $i \neq j$, and $p_{ii}(t) \approx 1 - t\sum_{j \neq i} q_{ij}$. In other words, the *matrix of transition probabilities* is

$$P(t) = \{p_{ij}(t)\} \approx I + Qt = \begin{bmatrix} 1 - 3\lambda t & \lambda t & \lambda t & \lambda t \\ \lambda t & 1 - 3\lambda t & \lambda t & \lambda t \\ \lambda t & \lambda t & 1 - 3\lambda t & \lambda t \\ \lambda t & \lambda t & \lambda t & 1 - 3\lambda t \end{bmatrix}, \text{ for small } t. \qquad (1.2)$$

Suppose a random region of the human genome evolves according to the JC69 model, at the rate of $3\lambda = 2.2 \times 10^{-9}$ substitutions/site/year (Kumar and Subramanian 2002) (Table 1.2). Consider a site occupied by a T right now. The probability that $t = 10^6$ years later this site will have a C will be $\lambda t = 0.00073$, and the probability that it remains to be T will be $1 - 3\lambda t = 0.9978$.

Equation (1.2) does not work well if t is not small. In general,

$$P(t) = e^{Qt} = I + Qt + \frac{1}{2!}(Qt)^2 + \frac{1}{3!}(Qt)^3 + \cdots. \qquad (1.3)$$

We will discuss the calculation of this matrix exponential later. For the moment, we simply give the solution for the JC69 model as

$$P(t) = e^{Qt} = \begin{bmatrix} p_0(t) & p_1(t) & p_1(t) & p_1(t) \\ p_1(t) & p_0(t) & p_1(t) & p_1(t) \\ p_1(t) & p_1(t) & p_0(t) & p_1(t) \\ p_1(t) & p_1(t) & p_1(t) & p_0(t) \end{bmatrix}, \text{ with } \begin{cases} p_0(t) = \frac{1}{4} + \frac{3}{4}e^{-4\lambda t}, \\ p_1(t) = \frac{1}{4} - \frac{1}{4}e^{-4\lambda t}. \end{cases} \qquad (1.4)$$

Imagine a long sequence with nucleotide i at every site, and let every site evolve for a time period t. Then the proportion of nucleotide j in the sequence will be $p_{ij}(t)$, for j = T, C, A, G. The two different elements of the transition probability matrix, $p_0(t)$ and $p_1(t)$, are plotted in Figure 1.3. A few features of the matrix $P(t)$ are worth noting. First, every row of $P(t)$ sums to 1, because at any time t the chain has to be in one of the four nucleotide states. Second, $P(0) = I$, the identity matrix, reflecting the case of no evolution ($t = 0$). Third, rate λ and time t occur in the transition probabilities only in the form of the product λt. Thus if we are given a source sequence and a target sequence,

1.2 MARKOV MODELS OF NUCLEOTIDE SUBSTITUTION

Table 1.2 A sample of estimated mutation/substitution rates

Taxa	Genes/genomes	Mutation/substitution rate	Source
Placental mammals	Genomic mutation rate at four-fold degenerate sites	2.2×10^{-9} per site per year	Kumar & Subramanian (2002)
Primates	12 protein-coding genes in the mitochondrial genome	7.9×10^{-9} per site per year for all codon positions, or 2.2, 0.1, 4.2×10^{-9} per site per year for positions 1, 2, and 3, respectively.	Yang & Yoder (2003)
Human	Family-based genome sequencing	$1.1-1.2 \times 10^{-8}$ per site per generation	Roach et al. (2010), Kong et al. (2012)
Plants (rice and maize)	Nuclear genome	6×10^{-9}/site/year for synonymous 9×10^{-11}/site/year for nonsynonymous	Gaut (1998)
Plants (rice and maize)	Mitochondrial genome	0.3×10^{-9}/site/year for synonymous 1.3×10^{-11}/site/year for nonsynonymous	Gaut (1998)
Plants (rice and maize)	Chloraplast genome	1.1×10^{-9}/site/year for synonymous 1.8×10^{-11}/site/year for nonsynonymous	Gaut (1998)
HIV virus	HIV-1 *env* V3 region	$2-17 \times 10^{-3}$/site/year	Berry et al. (2007)

it will be impossible to tell whether the source has evolved into the target at rate λ over time t or at rate 2λ over time $t/2$. In fact, the sequences will look the same for any combination of λ and t as long as λt is fixed. With no external information about either the time or the rate, we can estimate only the distance, but not time and rate individually.

Lastly, when $t \to \infty$, $p_{ij}(t) = 1/4$, for all i and j. This represents the case where so many substitutions have occurred at every site that the target nucleotide is random, with probability

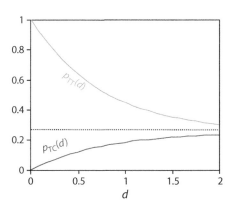

Fig. 1.3 Transition probabilities under the JC69 model (equation (1.4)) plotted against distance $d = 3\lambda t$, measured in the expected number of substitutions per site.

1/4 for every nucleotide, irrespective of the starting nucleotide. The probability that the chain is in state j when $t \to \infty$ is represented by π_j and the distribution $(\pi_T, \pi_C, \pi_A, \pi_G)$ is known as the *limiting distribution* of the chain. For the JC69 model, $\pi_j = 1/4$ for every nucleotide j. If the states of the chain are already in the limiting distribution, the chain will stay in that distribution, so the limiting distribution is also the *steady-state distribution* or *stationary distribution*. In other words, if a long sequence starts with T at every site, the proportions of the four nucleotides T, C, A, and G will drift away from $(1, 0, 0, 0)$ and approach $(1/4, 1/4, 1/4, 1/4)$, as the sequence evolves. If the sequence starts with equal proportions of the four nucleotides, it will continue to have equal proportions of the four nucleotides as the sequence evolves. The Markov chain is said to be stationary, or nucleotide substitutions are said to be in equilibrium. This is an assumption made in almost all models used in phylogenetic analysis, and is violated if the sequences in the data have different base compositions.

How does the Markov chain model correct for multiple hits and recover the hidden changes illustrated in Figure 1.1? This is achieved through the calculation of the transition probabilities using equation (1.3), which accommodates all possible paths the evolutionary process might have taken. In particular, the transition probabilities for a Markov chain satisfy the following equation, known as the Chapman–Kolmogorov equation (e.g. Grimmett and Stirzaker 1992, p. 239):

$$p_{ij}(t_1 + t_2) = \sum_k p_{ik}(t_1) p_{kj}(t_2). \tag{1.5}$$

This is a direct application of the *law of total probability*: the probability that nucleotide i will become nucleotide j time $t_1 + t_2$ later is a sum over all possible states k at any intermediate time point t_1 (Figure 1.4).

We now consider estimation of the distance between two sequences. From equation (1.1), the total substitution rate for any nucleotide is 3λ. If the two sequences are separated by time t (for example, if they diverged from a common ancestor time $t/2$ ago), the distance between the two sequences will be $d = 3\lambda t$. Suppose x out of n sites are different between the two sequences, so that the proportion of different sites is $\hat{p} = x/n$. (The hat or caret is used to indicate that the proportion is an estimate from the data.) To derive the expected probability p of different sites, consider one sequence as the ancestor of the other. By the symmetry of the model (equation (1.4)), this is equivalent to considering the two sequences as descendants of an extinct common ancestor. From equation (1.4), the probability that the nucleotide in the descendant sequence is different from the nucleotide in the ancestral sequence is

$$p(d) = 3p_1(t) = \frac{3}{4} - \frac{3}{4}e^{-4\lambda t} = \frac{3}{4} - \frac{3}{4}e^{-4d/3}. \tag{1.6}$$

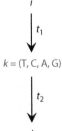

Fig. 1.4 Illustration of the Chapman–Kolmogorov theorem. The transition probability from any nucleotide i to any nucleotide j over time $t_1 + t_2$ is a sum over all possible states k at any intermediate time point t_1.

By equating this to the observed proportion $\hat{p}$, we obtain an estimate of distance as

$$\hat{d} = -\frac{3}{4}\log\left(1 - \frac{4}{3}\hat{p}\right), \qquad (1.7)$$

where the logarithm has base e (sometimes written as ln instead of log). If $\hat{p} \geq \frac{3}{4}$, the distance formula will be inapplicable; two random sequences should have about 75% different sites, and if $\hat{p} \geq \frac{3}{4}$, the distance estimate is infinite. To derive the variance of $\hat{d}$, note that $\hat{p}$ is a binomial proportion with variance $\hat{p}(1-\hat{p})/n$. Considering $\hat{d}$ as a function of $\hat{p}$ and using the so-called delta technique (see Appendix B), we obtain

$$\text{var}(\hat{d}) = \text{var}(\hat{p}) \times \left|\frac{d\hat{d}}{d\hat{p}}\right|^2 = \hat{p}(1-\hat{p})/n \times \frac{1}{(1-4\hat{p}/3)^2} \qquad (1.8)$$

(Kimura and Ohta 1972).

Example 1.1. The observed sequences of human and orangutan 12s rRNA genes from the mitochondrial genome are summarized in Table 1.3. From the table, $x = 90$ out of the $n = 948$ sites are different, so that $\hat{p} = x/n = 0.09494$. By equation (1.7), $\hat{d} = 0.1015$. Equation (1.8) gives the variance of $\hat{d}$ as 0.0001188 and standard error 0.0109. The approximate 95% confidence interval is thus $\hat{d} \pm 1.96 \times \text{SE} = 0.1015 \pm 1.96 \times 0.0109$ or $(0.0801, 0.1229)$. □

1.2.2 *The K80 model*

Substitutions between the two pyrimidines (T ↔ C) or between the two purines (A ↔ G) are called *transitions*, while those between a pyrimidine and a purine (T, C ↔ A, G) are called *transversions*. In real data, transitions often occur at higher rates than transversions. Thus Kimura (1980) proposed a model that accounts for different transition and transversion rates. Note that the biologist's use of the term transition (as opposed to transversion) has nothing to do with the probabilist's use of the same term (as in transition probability). Typically the usage is clear from the context and there is little risk of confusion.

Let the substitution rates be α for transitions and β for transversions. The model is referred to as K80, also known as Kimura's two-parameter model. The rate matrix is as follows (see also Figure 1.2):

Table 1.3 Numbers and frequencies (in parentheses) of sites for the 16 site configurations (patterns) in human and orangutan mitochondrial 12s rRNA genes

Orang	Human				Sum (π_i)
	T	C	A	G	
T	179 (0.188819)	23 (0.024262)	1 (0.001055)	0 (0)	0.2141
C	30 (0.031646)	219 (0.231013)	2 (0.002110)	0 (0)	0.2648
A	2 (0.002110)	1 (0.001055)	291 (0.306962)	10 (0.010549)	0.3207
G	0 (0)	0 (0)	21 (0.022152)	169 (0.178270)	0.2004
Sum (π_j)	0.2226	0.2563	0.3323	0.1888	1

Note: Genbank accession numbers for the human and orangutan sequences are D38112 and NC_001646, respectively (Horai et al. 1995). There are 954 sites in the alignment, but six sites involve alignment gaps and are removed, leaving 948 sites in each sequence. The average base frequencies in the two sequences are 0.2184 (T), 0.2605 (C), 0.3265 (A), and 0.1946 (G).

$$Q = \begin{bmatrix} -(\alpha + 2\beta) & \alpha & \beta & \beta \\ \alpha & -(\alpha + 2\beta) & \beta & \beta \\ \beta & \beta & -(\alpha + 2\beta) & \alpha \\ \beta & \beta & \alpha & -(\alpha + 2\beta) \end{bmatrix}. \quad (1.9)$$

The total substitution rate for any nucleotide is $\alpha + 2\beta$, and the distance between two sequences separated by time t is $d = (\alpha + 2\beta)t$. Note that αt is the expected number of transitions per site and $2\beta t$ is the expected number of transversions per site. One can use αt and βt as the two parameters in the model, but it is often more convenient to use the distance d and the transition/transversion rate ratio $\kappa = \alpha/\beta$. The matrix of transition probabilities is given as

$$P(t) = e^{Qt} = \begin{bmatrix} p_0(t) & p_1(t) & p_2(t) & p_2(t) \\ p_1(t) & p_0(t) & p_2(t) & p_2(t) \\ p_2(t) & p_2(t) & p_0(t) & p_1(t) \\ p_2(t) & p_2(t) & p_1(t) & p_0(t) \end{bmatrix}, \quad (1.10)$$

where the three distinct elements of the matrix are

$$p_0(t) = \tfrac{1}{4} + \tfrac{1}{4}e^{-4\beta t} + \tfrac{1}{2}e^{-2(\alpha+\beta)t} = \tfrac{1}{4} + \tfrac{1}{4}e^{-4d/(\kappa+2)} + \tfrac{1}{2}e^{-2d(\kappa+1)/(\kappa+2)},$$
$$p_1(t) = \tfrac{1}{4} + \tfrac{1}{4}e^{-4\beta t} - \tfrac{1}{2}e^{-2(\alpha+\beta)t} = \tfrac{1}{4} + \tfrac{1}{4}e^{-4d/(\kappa+2)} - \tfrac{1}{2}e^{-2d(\kappa+1)/(\kappa+2)}, \quad (1.11)$$
$$p_2(t) = \tfrac{1}{4} - \tfrac{1}{4}e^{-4\beta t} = \tfrac{1}{4} - \tfrac{1}{4}e^{-4d/(\kappa+2)}$$

(Kimura 1980; Li 1986). Note that $p_0(t) + p_1(t) + 2p_2(t) = 1$.

The sequence data can be summarized as the proportions of sites with transitional and transversional differences. Let these be S and V, respectively. Again, by the symmetry of the model (equation (1.10)), the probability that a site is occupied by nucleotides with a transitional difference is $E(S) = p_1(t)$. Similarly $E(V) = 2p_2(t)$. Equating these to the observed proportions S and V leads to two simultaneous equations in two unknowns, which are easily solved to give

$$\hat{d} = -\tfrac{1}{2}\log(1 - 2S - V) - \tfrac{1}{4}\log(1 - 2V),$$
$$\hat{\kappa} = \frac{2 \times \log(1 - 2S - V)}{\log(1 - 2V)} - 1 \quad (1.12)$$

(Kimura 1980; Jukes 1987). Equivalently the transition distance αt and the transversion distance $2\beta t$ are estimated as

$$\widehat{\alpha t} = -\tfrac{1}{2}\log(1 - 2S - V) + \tfrac{1}{4}\log(1 - 2V),$$
$$\widehat{2\beta t} = -\tfrac{1}{2}\log(1 - 2V), \quad (1.13)$$

The distance formula is applicable only if $1 - 2S - V > 0$ and $1 - 2V > 0$. As S and V are multinomial proportions with $\text{var}(S) = S(1-S)/n$, $\text{var}(V) = V(1-V)/n$, and $\text{cov}(S, V) = -SV/n$, we can use the delta technique to derive the variance–covariance matrix of $\hat{d}$ and $\hat{\kappa}$ (see Appendix B). In particular, the variance of $\hat{d}$ is

$$\text{var}(\hat{d}) = [a^2 S + b^2 V - (aS + bV)^2]/n, \quad (1.14)$$

where

$$a = (1 - 2S - V)^{-1},$$
$$b = \tfrac{1}{2}[(1 - 2S - V)^{-1} + (1 - 2V)^{-1}]. \quad (1.15)$$

Example 1.2. For the 12s rRNA data of Table 1.3, the proportions of transitional and transversional differences are $S = (23 + 30 + 10 + 21)/948 = 0.08861$ and $V = (1 + 0 + 2 + 0 + 2 + 1 + 0 + 0)/948 = 0.00633$. Thus equations (1.12) and (1.14) give the distance and standard error as 0.1046 ± 0.0116 (Table 1.4). The estimate $\hat{\kappa} = 30.836$ indicates that the transition rate is ~30 times higher than the transversion rate. □

1.2.3 HKY85, F84, TN93, etc.

1.2.3.1 TN93

The JC69 and K80 models have symmetrical substitution rates, with $q_{ij} = q_{ji}$ for all $i \neq j$. Such a Markov chain has $\pi_i = \frac{1}{4}$ for all i as the stationary distribution; that is, when the substitution process reaches equilibrium, the sequence will have equal proportions of the four nucleotides. This assumption is unrealistic for most datasets. Here we consider a few models that accommodate unequal base compositions. The model of Tamura and Nei (1993), referred to as TN93, has most of the commonly used models as special cases. Thus we present detailed results for this model, which also apply to its special cases. The substitution rate matrix under the TN93 model is

$$Q = \begin{bmatrix} -(\alpha_1 \pi_C + \beta \pi_R) & \alpha_1 \pi_C & \beta \pi_A & \beta \pi_G \\ \alpha_1 \pi_T & -(\alpha_1 \pi_T + \beta \pi_R) & \beta \pi_A & \beta \pi_G \\ \beta \pi_T & \beta \pi_C & -(\alpha_2 \pi_G + \beta \pi_Y) & \alpha_2 \pi_G \\ \beta \pi_T & \beta \pi_C & \alpha_2 \pi_A & -(\alpha_2 \pi_A + \beta \pi_Y) \end{bmatrix}. \quad (1.16)$$

Table 1.4 Estimates of distance between the human and orangutan 12s rRNA genes

Model and method	$\hat{d}$	Estimates of other parameters	ℓ
Distance formulae			
JC69	0.1015 ± 0.0109		
K80	0.1046 ± 0.0116	$\hat{\kappa} = 30.83 \pm 13.12$	
F81	0.1016		
F84	0.1050	$\hat{\kappa} = 15.548$	
TN93	0.1078	$\hat{\kappa}_1 = 44.228, \hat{\kappa}_2 = 21.789$	
Maximum likelihood			
JC69	0.1015 ± 0.0109		-1710.58
K80	0.1046 ± 0.0116	$\hat{\kappa} = 30.83 \pm 13.12$	-1637.90
F81	0.1017 ± 0.0109	$\hat{\pi} = (0.2251, 0.2648, 0.3188, 0.1913)$	-1691.97
F84	0.1048 ± 0.0117	$\hat{\kappa} = 15.640,$ $\hat{\pi} = (0.2191, 0.2602, 0.3286, 0.1921)$	-1616.60
HKY85	0.1048 ± 0.0117	$\hat{\kappa} = 32.137,$ $\hat{\pi} = (0.2248, 0.2668, 0.3209, 0.1875)$	-1617.27
TN93	0.1048 ± 0.0117	$\hat{\kappa}_1 = 44.229, \hat{\kappa}_2 = 21.781$ $\hat{\pi} = (0.2185, 0.2604, 0.3275, 0.1936)$	-1613.03
GTR (REV)	0.1057 ± 0.0119	$\hat{a} = 2.0431, \hat{b} = 0.0821, \hat{c} = 0.0000, \hat{d} = 0.0670,$ $\hat{e} = 0.0000,$ $\hat{\pi} = (0.2184, 0.2606, 0.3265, 0.1946)$	-1610.36
UNREST	0.1057 ± 0.0120	See equation (1.66) for the estimated Q; $\hat{\pi} = (0.2184, 0.2606, 0.3265, 0.1946)$	-1610.36

Note: ℓ is the log likelihood under the model.

1 MODELS OF NUCLEOTIDE SUBSTITUTION

While parameters $\pi_T, \pi_C, \pi_A, \pi_G$ are used to specify the substitution rates, they also give the stationary (equilibrium) distribution, with $\pi_Y = \pi_T + \pi_C$ and $\pi_R = \pi_A + \pi_G$ to be the frequencies of pyrimidines and purines, respectively.

The matrix of transition probabilities over time t is $P(t) = \{p_{ij}(t)\} = e^{Qt}$. A standard approach to calculating an algebraic function, such as the exponential, of a matrix Q, is to *diagonalize* Q (e.g. Schott 1997, Chapter 3). Suppose Q can be written in the form

$$Q = U\Lambda U^{-1}, \tag{1.17}$$

where U is a nonsingular matrix and U^{-1} is its inverse, and Λ is a diagonal matrix $\Lambda = \text{diag}\{\lambda_1, \lambda_2, \lambda_3, \lambda_4\}$. The λs are the eigenvalues (or latent roots) of Q, and columns of U and rows of U^{-1} are the corresponding right and left eigenvectors of Q, respectively. Equation (1.17) is also known as the *spectral decomposition* of Q. The reader should consult a textbook on linear algebra for calculation of eigenvalues and eigenvectors of a matrix (e.g. Schott 1997, Chapter 3).

From equation (1.17), we have $Q^2 = (U\Lambda U^{-1})(U\Lambda U^{-1}) = U\Lambda^2 U^{-1} = U\,\text{diag}\{\lambda_1^2, \lambda_2^2, \lambda_3^2, \lambda_4^2\}\,U^{-1}$. Similarly $Q^m = U\,\text{diag}\{\lambda_1^m, \lambda_2^m, \lambda_3^m, \lambda_4^m\}\,U^{-1}$ for any integer m. In general, any algebraic function h of matrix Q can be calculated as $h(Q) = U\,\text{diag}\{h(\lambda_1), h(\lambda_2), h(\lambda_3), h(\lambda_4)\}\,U^{-1}$ as long as $h(Q)$ exists. Thus

$$P(t) = e^{Qt} = U\,\text{diag}\left\{e^{\lambda_1 t}, e^{\lambda_2 t}, e^{\lambda_3 t}, e^{\lambda_4 t}\right\} U^{-1}. \tag{1.18}$$

For the TN93 model, the spectral decomposition of Q is analytical. We have $\lambda_1 = 0$, $\lambda_2 = -\beta$, $\lambda_3 = -(\pi_R \alpha_2 + \pi_Y \beta)$, and $\lambda_4 = -(\pi_Y \alpha_1 + \pi_R \beta)$, and

$$U = \begin{bmatrix} 1 & 1/\pi_Y & 0 & \pi_C/\pi_Y \\ 1 & 1/\pi_Y & 0 & -\pi_T/\pi_Y \\ 1 & -1/\pi_R & \pi_G/\pi_R & 0 \\ 1 & -1/\pi_R & -\pi_A/\pi_R & 0 \end{bmatrix}, \tag{1.19}$$

$$U^{-1} = \begin{bmatrix} \pi_T & \pi_C & \pi_A & \pi_G \\ \pi_T \pi_R & \pi_C \pi_R & -\pi_A \pi_Y & -\pi_G \pi_Y \\ 0 & 0 & 1 & -1 \\ 1 & -1 & 0 & 0 \end{bmatrix}. \tag{1.20}$$

Substituting Λ, U, and U^{-1} into equation (1.18) gives

$$P(t) = \begin{bmatrix} \pi_T + \frac{\pi_T \pi_R}{\pi_Y}e_2 + \frac{\pi_C}{\pi_Y}e_4 & \pi_C + \frac{\pi_C \pi_R}{\pi_Y}e_2 - \frac{\pi_C}{\pi_Y}e_4 & \pi_A(1-e_2) & \pi_G(1-e_2) \\ \pi_T + \frac{\pi_T \pi_R}{\pi_Y}e_2 - \frac{\pi_T}{\pi_Y}e_4 & \pi_C + \frac{\pi_C \pi_R}{\pi_Y}e_2 + \frac{\pi_T}{\pi_Y}e_4 & \pi_A(1-e_2) & \pi_G(1-e_2) \\ \pi_T(1-e_2) & \pi_C(1-e_2) & \pi_A + \frac{\pi_A \pi_Y}{\pi_R}e_2 + \frac{\pi_G}{\pi_R}e_3 & \pi_G + \frac{\pi_G \pi_Y}{\pi_R}e_2 - \frac{\pi_G}{\pi_R}e_3 \\ \pi_T(1-e_2) & \pi_C(1-e_2) & \pi_A + \frac{\pi_A \pi_Y}{\pi_R}e_2 - \frac{\pi_A}{\pi_R}e_3 & \pi_G + \frac{\pi_G \pi_Y}{\pi_R}e_2 + \frac{\pi_A}{\pi_R}e_3 \end{bmatrix}, \tag{1.21}$$

where $e_2 = \exp(\lambda_2 t) = \exp(-\beta t)$, $e_3 = \exp(\lambda_3 t) = \exp\{-(\pi_R \alpha_2 + \pi_Y \beta)t\}$, $e_4 = \exp(\lambda_4 t) = \exp\{-(\pi_Y \alpha_1 + \pi_R \beta)t\}$.

When t increases from 0 to ∞, the diagonal element $p_{jj}(t)$ decreases from 1 to π_j, while the off-diagonal element $p_{ij}(t)$ increases from 0 to π_j, with $p_{ij}(\infty) = \pi_j$, irrespective of the starting nucleotide i. The limiting distribution $(\pi_T, \pi_C, \pi_A, \pi_G)$ is also the stationary distribution. Also the rate of convergence to the stationary distribution, that is, the rate at which $p_{ij}(t) - \pi_j$ approaches zero, is determined by the largest nonzero eigenvalue.

We now consider estimation of the sequence distance under the model. First, we will look at the definition of distance. The substitution rate of nucleotide i is $-q_{ii} = \sum_{j \neq i} q_{ij}$, and differs among the four nucleotides. When the substitution process is in equilibrium, the amount of time the Markov chain spends in the four states T, C, A, and G is proportional to the equilibrium frequencies π_T, π_C, π_A and π_G, respectively. Similarly, if we consider a long DNA sequence in substitution equilibrium, the proportions of sites occupied by nucleotides T, C, A, and G are π_T, π_C, π_A and π_G, respectively. Thus the average substitution rate, either defined as an average over a long time for one site or as an average over many sites in a long sequence at one time point, is

$$\lambda = -\sum_i \pi_i q_{ii} = 2\pi_T \pi_C \alpha_1 + 2\pi_A \pi_G \alpha_2 + 2\pi_Y \pi_R \beta. \tag{1.22}$$

The distance between two sequences separated by time t is $d = \lambda t$.

To derive a distance estimate, we use the same strategy as for the K80 model discussed above. We call the nucleotides across sequences at a site as a *site configuration* or *site pattern*. Our strategy is to equate the observed proportions of sites with certain site patterns to their expected probabilities. Let S_1 be the proportion of sites occupied by two different pyrimidines (i.e. sites with patterns TC or CT), S_2 the proportion of sites with two different purines (i.e. sites with patterns AG or GA), and V the proportion of sites with a transversional difference.

Next, we need to derive the expected probabilities for those sites: $E(S_1), E(S_2)$, and $E(V)$. We cannot use the symmetry argument as for JC69 and K80 since Q is not symmetrical. However, Q satisfies the following condition:

$$\pi_i q_{ij} = \pi_j q_{ji}, \text{ for all } i \neq j. \tag{1.23}$$

Equivalently, $\pi_i p_{ij}(t) = \pi_j p_{ji}(t)$, for all t and for all $i \neq j$. Markov chains satisfying such conditions are said to be *time-reversible*. Reversibility means that the process will look the same whether time runs forward or backward; that is, whether we view the substitution process from the present into the future or from the present back into the past. As a result, given two sequences, the probability of data at a site is the same whether one sequence is ancestral to the other or both are descendants of an ancestral sequence. Equivalently, equation (1.23) means that the expected amount of change from i to j is equal to the expected amount of change in the opposite direction. Note that the *rates* of change may be different in the two directions: $q_{ij} \neq q_{ji}$. Now consider sequence 1 to be the ancestor of sequence 2, separated by time t. Then

$$E(S_1) = \pi_T p_{TC}(t) + \pi_C p_{CT}(t) = 2\pi_T p_{TC}(t). \tag{1.24}$$

The first term in the sum, $\pi_T p_{TC}(t)$, is the probability that a site has nucleotide T in sequence 1 and C in sequence 2. This equals the probability of having T in sequence 1, given by π_T, times the transition probability $p_{TC}(t)$ that T will become C in sequence 2 time t later. Thus $\pi_T p_{TC}(t)$, is the probability of observing site pattern TC. The second term in the sum, $\pi_C p_{CT}(t)$, is the probability for site pattern CT. Similarly $E(S_2) = 2\pi_A p_{AG}(t)$ and $E(V) = 2\pi_T p_{TA}(t) + 2\pi_T p_{TG}(t) + 2\pi_C p_{CA}(t) + 2\pi_C p_{CG}(t)$. Equating the observed proportions S_1, S_2, and V to their expected probabilities leads to three simultaneous equations in three unknowns: e_2, e_3, and e_4 in the transition probability matrix (1.21) or equivalently, $d, \kappa_1 = \alpha_1/\beta$, and $\kappa_2 = \alpha_2/\beta$. Note that the nucleotide frequency parameters π_T, π_C, π_A, and π_G can be estimated using the average observed frequencies. Solving the system of equations gives the following estimates:

$$\hat{d} = \frac{2\pi_T \pi_C}{\pi_Y}(a_1 - \pi_R b) + \frac{2\pi_A \pi_G}{\pi_R}(a_2 - \pi_Y b) + 2\pi_Y \pi_R b,$$
$$\hat{\kappa}_1 = \frac{a_1 - \pi_R b}{\pi_Y b}, \tag{1.25}$$
$$\hat{\kappa}_2 = \frac{a_2 - \pi_Y b}{\pi_R b},$$

where

$$a_1 = -\log\left(1 - \frac{\pi_Y S_1}{2\pi_T \pi_C} - \frac{V}{2\pi_Y}\right),$$
$$a_2 = -\log\left(1 - \frac{\pi_R S_2}{2\pi_A \pi_G} - \frac{V}{2\pi_R}\right), \quad (1.26)$$
$$b = -\log\left(1 - \frac{V}{2\pi_Y \pi_R}\right)$$

(Tamura and Nei 1993).

The formulae are inapplicable whenever π_Y or π_R is 0 or any of the arguments to the logarithm functions are ≤ 0, as may happen when the sequences are divergent. The variance of the estimated distance $\hat{d}$ can be obtained by using the delta technique, ignoring errors in the estimates of nucleotide frequencies and noting that $S_1, S_2,$ and V are multinomial proportions. This is similar to the calculation under the K80 model (equation (1.14)); see Tamura and Nei (1993).

Example 1.3. For the 12s rRNA data of Table 1.3, we have the observed proportions $S_1 = (23 + 30)/948 = 0.05591$, $S_2 = (10 + 21)/948 = 0.03270$, and $V = 6/948 = 0.00633$. Equation (1.25) gives the estimates as $\hat{d} = 0.1078$, $\hat{\kappa}_1 = 44.228$, and $\hat{\kappa}_2 = 21.789$. □

1.2.3.2 HKY85 and F84 models

Two models that are commonly used in likelihood and Bayesian phylogenetics are special cases of TN93. The first is due to Hasegawa and colleagues (Hasegawa et al. 1984, 1985). This is now commonly known as HKY85, instead of HYK84, apparently due to my misnaming (Yang 1994b). The model is obtained by setting $\alpha_1 = \alpha_2 = \alpha$ or $\kappa_1 = \kappa_2 = \kappa$ in the TN93 model (Table 1.1). The transition probability matrix is given by equation (1.21), with α_1 and α_2 replaced by α. It is not straightforward to derive a distance formula under this model (Yang 1994b), although Rzhetsky and Nei (1994) suggested a few possibilities.

The second special case of the TN93 model was implemented by Joseph Felsenstein in his DNAML program since Version 2.6 (1984) of the PHYLIP package. This is now known as the F84 model. The rate matrix was first published by Hasegawa and Kishino (1989) and Kishino and Hasegawa (1989). It is obtained by setting $\alpha_1 = (1 + \kappa/\pi_Y)\beta$ and $\alpha_2 = (1 + \kappa/\pi_R)\beta$ in the TN93 model, requiring one fewer parameter (Table 1.1). Under this model, the eigenvalues of the Q matrix become $\lambda_1 = 0$, $\lambda_2 = -\beta$, $\lambda_3 = \lambda_4 = -(1 + \kappa)\beta$. There are only three distinct eigenvalues, as for the K80 model, and thus it is possible to derive a distance formula.

From equation (1.22), the sequence distance is $d = \lambda t = 2(\pi_T \pi_C + \pi_A \pi_G + \pi_Y \pi_R)\beta t + 2(\pi_T \pi_C/\pi_Y + \pi_A \pi_G/\pi_R)\kappa\beta t$. The expected probabilities of sites with transitional and transversional differences are

$$E(S) = 2(\pi_T \pi_C + \pi_A \pi_G) + 2\left(\frac{\pi_T \pi_C \pi_R}{\pi_Y} + \frac{\pi_A \pi_G \pi_Y}{\pi_R}\right) e^{-\beta t} - 2\left(\frac{\pi_T \pi_C}{\pi_Y} + \frac{\pi_A \pi_G}{\pi_R}\right) e^{-(\kappa+1)\beta t},$$
$$E(V) = 2\pi_Y \pi_R (1 - e^{-\beta t}). \quad (1.27)$$

By equating the observed proportions S and V to their expectations, one can obtain a system of two equations in two unknowns, which can be solved to give

$$\hat{d} = 2\left(\frac{\pi_T \pi_C}{\pi_Y} + \frac{\pi_A \pi_G}{\pi_R}\right) a - 2\left(\frac{\pi_T \pi_C \pi_R}{\pi_Y} + \frac{\pi_A \pi_G \pi_Y}{\pi_R} - \pi_Y \pi_R\right) b,$$
$$\hat{\kappa} = a/b - 1, \quad (1.28)$$

where

$$a = \overline{(\kappa+1)\beta t} = -\log\left\{1 - \frac{S}{2\left(\frac{\pi_T\pi_C}{\pi_Y} + \frac{\pi_A\pi_G}{\pi_R}\right)} - \frac{\left(\frac{\pi_T\pi_C\pi_R}{\pi_Y} + \frac{\pi_A\pi_G\pi_Y}{\pi_R}\right)V}{2(\pi_T\pi_C\pi_R + \pi_A\pi_G\pi_Y)}\right\}, \quad (1.29)$$

$$b = \overline{\beta t} = -\log\left\{1 - \frac{V}{2\pi_Y\pi_R}\right\}$$

(Tateno et al. 1994; Yang 1994a). The approximate variance of $\hat{d}$ can be obtained similarly to that under K80 (Tateno et al. 1994). The estimated distance under F84 for the 12s rRNA genes is shown in Table 1.4.

If we assume $\alpha_1 = \alpha_2 = \beta$ in the TN93 model, we obtain the F81 model (Felsenstein 1981) (Table 1.1). A distance formula was derived by Tajima and Nei (1982). Estimates under this and some other models for the 12s rRNA dataset of Table 1.3 are listed in Table 1.4. It may be mentioned that the matrices Λ, U, U^{-1} and $P(t)$ derived for the TN93 model hold for its special cases, such as JC69 (Jukes and Cantor 1969), K80 (Kimura 1980), F81 (Felsenstein 1981), HKY85 (Hasegawa et al. 1984, 1985), and F84. Under some of those simpler models, simplifications are possible (see Problem 1.2).

1.2.4 The transition/transversion rate ratio

Unfortunately at least three definitions of the 'transition/transversion rate ratio' are in use in the literature. The first is the ratio of the numbers (or proportions) of transitional and transversional differences between the two sequences, without correcting for multiple hits (e.g. Wakeley 1994). This is $E(S)/E(V) = p_1(t)/(2p_2(t))$ under the K80 model (see equation (1.10)). For highly similar sequences, this is close to $\alpha/(2\beta)$ under K80. At intermediate levels of sequence divergence, $E(S)/E(V)$ increases with $\alpha/(2\beta)$, but the pattern is complex. When the sequences are very different, $E(S)/E(V)$ approaches $1/2$ irrespective of $\alpha/(2\beta)$. Figure 1.5 plots the ratio $E(S)/E(V)$ against the sequence divergence. Thus the ratio is meaningful only for closely related sequences. In real datasets, however, highly similar sequences may not contain much information and the estimate may involve large sampling errors. In general, the $E(S)/E(V)$ ratio is a poor measure of the transition–transversion rate difference and should be avoided.

The second measure is $\kappa = \alpha/\beta$ in the models of Kimura (1980) and Hasegawa et al. (1985), with $\kappa = 1$ meaning no rate difference between transitions and transversions. A third measure may be called the average transition/transversion ratio, and is the ratio

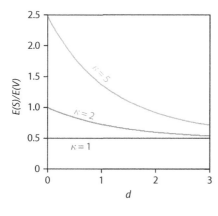

Fig. 1.5 The transition/transversion ratio $E(S)/E(V)$ under the K80 model plotted against sequence divergence d. This is $p_1/(2p_2)$ in equation (1.11) and corresponds to infinitely long sequences.

Table 1.5 Average transition/transversion ratio R

Model	Average transition/transversion rate ratio (R)
JC69	$\dfrac{1}{2}$
K80	$\dfrac{\kappa}{2}$
F81	$\dfrac{\pi_T\pi_C + \pi_A\pi_G}{\pi_Y\pi_R}$
F84	$\dfrac{\pi_T\pi_C(1+\kappa/\pi_Y) + \pi_A\pi_G(1+\kappa/\pi_R)}{\pi_Y\pi_R}$
HKY85	$\dfrac{(\pi_T\pi_C + \pi_A\pi_G)\kappa}{\pi_Y\pi_R}$
TN93	$\dfrac{\pi_T\pi_C\kappa_1 + \pi_A\pi_G\kappa_2}{\pi_Y\pi_R}$
REV (GTR)	$\dfrac{\pi_T\pi_C a + \pi_A\pi_G f}{\pi_T\pi_A b + \pi_T\pi_G c + \pi_C\pi_A d + \pi_C\pi_G e}$
UNREST	See equation (1.30) in text

of the expected numbers of transitional and transversional substitutions between the two sequences. This is the same measure as the first one, except that it corrects for multiple hits. For a general substitution rate matrix (the UNREST model in Table 1.1 but note that $q_{TC} = a, q_{TA} = b$, etc.), this is

$$R = \frac{\pi_T q_{TC} + \pi_C q_{CT} + \pi_A q_{AG} + \pi_G q_{GA}}{\pi_T q_{TA} + \pi_T q_{TG} + \pi_C q_{CA} + \pi_C q_{CG} + \pi_A q_{AT} + \pi_A q_{AC} + \pi_G q_{GT} + \pi_G q_{GC}}. \quad (1.30)$$

Note that the Markov chain spends a proportion π_T of time in state T, while q_{TC} is the rate that T changes to C. Thus $\pi_T q_{TC}$ is the amount of 'flow' from T to C. The numerator in equation (1.30) is the average amount of transitional change, while the denominator gives the amount of transversional change. Table 1.5 gives R for commonly used simple models. Under the model of Kimura (1980), $R = \alpha/(2\beta)$ and equals $1/2$ when there is no transition–transversion rate difference. As from each nucleotide, one change is a transition and two changes are transversions, we expect to see twice as many transversions as transitions, hence the ratio $1/2$.

Note that parameter κ has different definitions under the F84 and HKY85 models (Table 1.1). Without the transition–transversion rate difference, $\kappa_{F84} = 0$ and $\kappa_{HKY85} = 1$. Roughly, $\kappa_{HKY85} \simeq 1 + 2\kappa_{F84}$. By forcing the average ratio R to be identical under the two models (Table 1.5), one can derive a more accurate approximation (Goldman 1993):

$$\kappa_{HKY85} \simeq 1 + \frac{\pi_T\pi_C/\pi_Y + \pi_A\pi_G/\pi_R}{\pi_T\pi_C + \pi_A\pi_G} \kappa_{F84}. \quad (1.31)$$

Overall, R is more convenient to use for comparing estimates under different models while κ is more suitable for formulating the null hypothesis of no transition–transversion rate difference.

Some authors (e.g. Singh et al. 2009) used the measure

$$R^* = \frac{q_{TC} + q_{CT} + q_{AG} + q_{GA}}{q_{TA} + q_{TG} + q_{CA} + q_{CG} + q_{AT} + q_{AC} + q_{GT} + q_{GC}}. \quad (1.32)$$

Under the K80 model this has the ratio $1/2$ if there is no transition–transversion rate difference. For other models, this is hard to interpret and should be avoided.

1.3 Variable substitution rates across sites

All models discussed in §1.2 assume that different sites in the sequence evolve at the same rate. This assumption may be unrealistic in real data. First, the mutation rate may vary among sites (Hodgkinson and Eyre-Walker 2011). Second, different sites may play different roles in the structure and function of the gene and are thus under different selective pressures. Mutations at different sites may thus be fixed in the population at different rates. When the substitution rates vary, the hotspots may accumulate many changes, while the conserved sites remain unchanged. Thus, for the same sequence distance or the same amount of evolutionary change, we will observe fewer differences than if the rate is constant. In other words, ignoring variable rates among sites leads to underestimation of the sequence distance.

One can accommodate the rate variation by assuming that the rate r for any site is a random variable drawn from a statistical distribution. The most commonly used distribution is the gamma distribution. The resulting models are represented by a suffix '+Γ', such as JC69 + Γ, K80 + Γ, etc., and the distances are sometimes called *gamma distances*. The density function of the gamma distribution is

$$g(r; \alpha, \beta) = \frac{\beta^\alpha}{\Gamma(\alpha)} e^{-\beta r} r^{\alpha-1}, r > 0, \quad (1.33)$$

where $\alpha > 0$ and $\beta > 0$ are the shape and rate parameters. Here

$$\Gamma(\alpha) = \int_0^\infty e^{-t} t^{\alpha-1} dt \quad (1.34)$$

is the gamma function. For integer n, $\Gamma(n) = (n-1)!$. The mean and variance of the gamma distribution are $E(r) = \alpha/\beta$ and $\text{var}(r) = \alpha/\beta^2$. To avoid using too many parameters, we set $\beta = \alpha$ so that the mean of the distribution is 1, with variance $1/\alpha$. The shape parameter α is then inversely related to the extent of rate variation at sites (Figure. 1.6). If $\alpha > 1$, the distribution is bell-shaped, meaning that most sites have intermediate rates around 1, while few sites have either very low or very high rates. In particular, when $\alpha \to \infty$, the distribution degenerates into the model of a single rate for all sites. If $\alpha \leq 1$, the distribution has a highly skewed L-shape, meaning that most sites have very low rates or are nearly 'invariable', but there are some substitution hot spots with high rates. Estimation of α from real data requires joint comparison of multiple sequences as it is virtually impossible to do so using only two sequences. We will discuss estimation of α later, in §4.3.1. Here we assume that α is given.

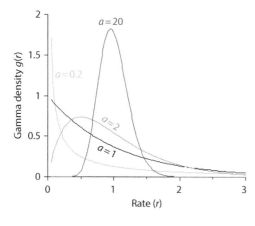

Fig. 1.6 Probability density function of the gamma distribution for variable rates among sites. The rate parameter of the distribution is fixed so that the mean is one; as a result, the density involves only the shape parameter a. The x-axis is the substitution rate, while the y-axis is proportional to the number of sites with that rate.

With variable rates among sites, the sequence distance is defined as the expected number of substitutions per site, averaged over all sites. Here we will derive gamma distances under the JC69 and K80 models, and comment on similar derivations under other models.

First we consider the JC69 + Γ model. Since the mean rate in the gamma distribution is one, the distance averaged across all sites is still d. As in the case of the JC69 model with one rate for all sites, we equate the observed proportion of different sites to the expected probability under JC69 + Γ to derive a distance formula. If a site has rate r, the distance between the sequences at that site is $d \cdot r$, and the probability of observing a difference at the site is $p(d \cdot r)$, with p given in equation (1.6). However, r is an unknown random variable, so we have to consider contributions from sites with different rates. In other words, we average over the distribution of r to calculate the unconditional probability of different sites:

$$p = \int_0^\infty \left(\frac{3}{4} - \frac{3}{4}e^{-4d \cdot r/3}\right) g(r) \, dr = \frac{3}{4} - \frac{3}{4}\left(1 + \frac{4d}{3\beta}\right)^{-\alpha}. \tag{1.35}$$

Here we made use of the following result for the gamma distribution:

$$\int_0^\infty e^{-cr} g(r) \, dr = \int_0^\infty e^{-cr} \frac{\beta^\alpha}{\Gamma(\alpha)} e^{-\beta r} r^{\alpha-1} \, dr = \left(\frac{\beta+c}{\beta}\right)^{-\alpha} \int_0^\infty \left[\frac{(\beta+c)^\alpha}{\Gamma(\alpha)} e^{-(\beta+c)r} r^{\alpha-1}\right] dr = \left(1 + \frac{c}{\beta}\right)^{-\alpha}, \tag{1.36}$$

for any constant $c > 0$. Note that the quantity in the square brackets is the probability density function (PDF) for the gamma distribution with parameters α and $\beta + c$ and thus the integral is 1.

By equating p of equation (1.35) to the observed $\hat{p}$, we obtain the JC69 + Γ distance as

$$\hat{d} = \frac{3}{4}\alpha\left[\left(1 - \frac{4}{3}\hat{p}\right)^{-1/\alpha} - 1\right] \tag{1.37}$$

(Golding 1983), with variance

$$\text{var}(\hat{d}) = \text{var}(\hat{p}) \times \left|\frac{dd}{dp}\right|^2 = \frac{\hat{p}(1-\hat{p})}{n} \times \left(1 - \frac{4}{3}\hat{p}\right)^{-2/\alpha-2}. \tag{1.38}$$

Now we consider the K80 + Γ model. To avoid confusion about the notation, we use d and κ as parameters under the K80 model and α and β as parameters of the gamma distribution. The distance is defined as an average across all sites. As in the case of K80 with one rate for all sites, we equate the observed proportions of transitional and transversional differences to their expected probabilities to derive a distance formula. If a site has rate r, both transition and transversion rates at the site are multiplied by r, with the same transition/transversion rate ratio κ assumed across sites. As in the case of JC69 + Γ, we derive the probability for a transitional difference by averaging $p_1(d \cdot r)$, with p_1 given in equation (1.11), over the gamma distribution:

$$E(S) = \int_0^\infty p_1(d \cdot r) g(r) \, dr$$
$$= \int_0^\infty \left[\frac{1}{4} + \frac{1}{4}\exp\left(\frac{-4d \cdot r}{\kappa+2}\right) - \frac{1}{2}\exp\left(\frac{-2(\kappa+1)d \cdot r}{\kappa+2}\right)\right] g(r) \, dr \tag{1.39}$$
$$= \frac{1}{4} + \frac{1}{4}\left(1 + \frac{4d}{(\kappa+2)\alpha}\right)^{-\alpha} - \frac{1}{2}\left(1 + \frac{2(\kappa+1)d}{(\kappa+2)\alpha}\right)^{-\alpha},$$

making use of equation (1.36). Similarly the probability that we observe a transversional difference is

$$E(V) = \int_0^\infty 2p_2(d \cdot r) g(r) \, dr = \frac{1}{2} - \frac{1}{2}\left(1 + \frac{4d}{(\kappa+2)\alpha}\right)^{-\alpha}, \tag{1.40}$$

where p_2 is given in equation (1.11). Equating the above to the observed proportions S and V leads to

$$\hat{d} = \frac{\alpha}{2}[(1-2S-V)^{-1/\alpha}-1] + \frac{\alpha}{4}[(1-2V)^{-1/\alpha}-1],$$
$$\hat{\kappa} = \frac{2[(1-2S-V)^{-1/\alpha}-1]}{[(1-2V)^{-1/\alpha}-1]} - 1 \qquad (1.41)$$

(Jin and Nei 1990). Compared with equation (1.12) for the one-rate model, the only change is that the logarithm function $\log(y)$ becomes $-\alpha(y^{-1/\alpha}-1)$. This is a general feature of gamma distances. The large-sample variance of $\hat{d}$ is given by equation (1.14) except that now

$$a = (1-2S-V)^{-1/\alpha-1},$$
$$b = \frac{1}{2}[(1-2S-V)^{-1/\alpha-1} + (1-2V)^{-1/\alpha-1}]. \qquad (1.42)$$

In general, note that equation (1.18) can be written equivalently as

$$p_{ij}(t) = \sum_{k=1}^{4} c_{ijk} e^{\lambda_k t} = \sum_{k=1}^{4} u_{ik} u_{kj}^{-1} e^{\lambda_k t}, \qquad (1.43)$$

where λ_k is the kth eigenvalue of the rate matrix Q, u_{ik} is the ikth element of U, and u_{kj}^{-1} is the kjth element of U^{-1} in equation (1.18). Thus the probability of observing nucleotides i and j in the two sequences at a site is

$$f_{ij}(t) = \int_0^\infty \pi_i p_{ij}(t \cdot r) g(r) \, \mathrm{d}r = \pi_i \sum_{k=1}^{4} c_{ijk} (1 - \lambda_k t/\alpha)^{-\alpha}, \qquad (1.44)$$

by equation (1.36). The exponential functions under the one-rate model are replaced by the power functions under the gamma model. Under the one-rate model, we can view the exponential functions as unknowns to solve the equations, and now we can view those power functions as unknowns. Thus, one can derive a gamma distance under virtually every model for which a one-rate distance formula is available. Those include the F84 model (Yang 1994a) and the TN93 model (Tamura and Nei 1993), among others.

Example 1.4. We calculate the sequence distance between the two mitochondrial 12s rRNA genes under the K80 + Γ model, with $\alpha = 0.5$ fixed. The estimates of the distance and the transition/transversion rate ratio κ are $\hat{d} \pm$ SE $= 0.1283 \pm 0.01726$ and $\hat{\kappa} \pm$ SE $= 37.76 \pm 16.34$. Both estimates are much larger than under the one-rate model (Table 1.4). It is well known that ignoring rate variation among sites leads to underestimation of both the sequence distance and the transition/transversion rate ratio (Wakeley 1994; Yang 1996a). The underestimation is more serious at larger distances and with more variable rates (that is, smaller α). □

1.4 Maximum likelihood estimation of distance

In this section, we discuss the ML method for estimating sequence distances. ML is a general methodology for estimating parameters in a model and for testing hypotheses concerning the parameters. It plays a central role in statistics and is widely used in molecular phylogenetics. It forms the basis of much material covered later in this book. We will focus mainly on the JC69 and K80 models, re-deriving the distance formulae discussed earlier. While discovering what we already know may not be very exciting, it may be effective in helping us understand the workings of the likelihood method. Note that ML is an 'automatic' method, as it tells us how to proceed even when the estimation problem is difficult and our intuition fails. Interested readers should consult a statistics textbook, for example, DeGroot and Schervish (2002), Kalbfleisch (1985), and Edwards (1992) at elementary levels, or Cox and Hinkley (1974) and Stuart et al. (1999) at more advanced levels.

1.4.1 The JC69 model

Let X be the observed data and θ the parameter we hope to estimate. The probability of observing data X, when viewed as a function of the unknown parameter θ with the data given, is called the *likelihood function*: $L(\theta; X) = f(X|\theta)$. Probability and likelihood are fundamentally different concepts; see Box 1.1 for a summary. According to the *likelihood principle*, the likelihood function contains all information in the data about θ. The value of θ that maximizes the likelihood, say $\hat{\theta}$, is our best point estimate, called the *maximum likelihood estimate* (MLE). Furthermore, the likelihood curve around $\hat{\theta}$ provides information about the uncertainty in the point estimate. The theory applies to problems with either a single parameter or with multiple parameters. In the later case θ is a vector.

Box 1.1 PROBABILITY VERSUS LIKELIHOOD

Probability and likelihood are fundamentally different concepts.

- Likelihood is defined up to a proportionality constant. Likelihood does not integrate (over the parameter space) to 1. Probability sums (over the sample space) to 1.
- The probability curve should be interpreted by area under the curve, while its height is not meaningful. The likelihood should be compared using the height and point by point, e.g. $L(\theta_1) > L(\theta_2)$ for two points θ_1 and θ_2. The area under the likelihood curve is not meaningful.
- Likelihood is invariant to reparametrization. Suppose we are interested in the size of a crater lake, which is a circle. We can use either α (the area of the circle) or β (the radius) as the parameter in the model, with $\alpha = \pi\beta^2$. Then if $L(\alpha_1) > L(\alpha_2)$, we have $L(\beta_1) > L(\beta_2)$. In general, if α and β are alternative parametrizations, with $\beta = h(\alpha)$ where h is a one-to-one monotonic function, then the MLEs are invariant to reparametrization: $\hat{\beta} = \hat{h}(\alpha) = h(\hat{\alpha})$. In contrast, variable transformation (if it is nonlinear) in general changes the shape of the probability density. For example, Appendix A includes an example in which x has a two-moded distribution while $y = h(x)$ has only one mode.

Here we apply the theory to estimation of the distance between two sequences under the JC69 model. The single parameter is the distance d. The data are two aligned sequences, each n sites long, with x differences. From equation (1.6), the probability of observing different nucleotides at a site between two sequences separated by distance d is

$$p = 3p_1 = \frac{3}{4} - \frac{3}{4}e^{-4d/3}. \tag{1.45}$$

Thus, the probability of observing the data, that is, x differences out of n sites, is given by the binomial probability

$$L(d; x) = f(x|d) = Cp^x(1-p)^{n-x} = C\left(\frac{3}{4} - \frac{3}{4}e^{-4d/3}\right)^x \left(\frac{1}{4} + \frac{3}{4}e^{-4d/3}\right)^{n-x}. \tag{1.46}$$

As the data x are observed, this probability is now considered a function of the parameter d. Values of d with higher L are better supported by the data than values of d with lower L. As multiplying the likelihood by any function of the data that is independent of the parameter θ will not change our inference about θ, the likelihood is defined up to a proportionality constant. We will use this property to introduce two changes to the likelihood of equation (1.46). First, the binomial coefficient $C = n!/[x!(n-x)!]$ is a constant and will be dropped. Second, to use the same definition of likelihood for all substitution models, we will distinguish 16 possible data outcomes at a site (the 16 possible site patterns) instead of just two outcomes (that is, difference with probability p and identity with probability $1-p$) as in equation (1.46). Under JC69, the four constant patterns (TT, CC, AA, GG) have the same probability of occurrence, as do the 12 variable site patterns (TC, TA, TG etc.). This will not be the case for other models. Thus the redefined likelihood is given by the multinomial probability with 16 cells:

$$L(d; x) = \left(\frac{1}{4}p_1\right)^x \left(\frac{1}{4}p_0\right)^{n-x} = \left(\frac{1}{16} - \frac{1}{16}e^{-4d/3}\right)^x \left(\frac{1}{16} + \frac{3}{16}e^{-4d/3}\right)^{n-x}, \quad (1.47)$$

where p_0 and p_1 are from equation (1.4). Each of the 12 variable site patterns has probability $\frac{1}{4} p_1$ or $\frac{1}{12} p$. For example, the probability for site pattern TC is equal to $\frac{1}{4}$, the probability that the starting nucleotide is T, times the transition probability $p_{TC}(t) = p_1$ from equation (1.4). Similarly, each of the four constant site patterns (TT, CC, AA, GG) has probability $\frac{1}{4} p_0$ or $(1-p)/4$. The reader can verify that equations (1.46) and (1.47) differ only by a proportionality constant (Problem 1.4).

Furthermore, the likelihood L is typically extremely small and awkward to work with. Thus its logarithm $\ell(d) = \log\{L(d)\}$ is commonly used instead. As the logarithm function is monotonic, we achieve the same result; that is, $L(d_1) > L(d_2)$ if and only if $\ell(d_1) > \ell(d_2)$. The *log likelihood function* is thus

$$\ell(d; x) = \log\{L(d; x)\} = x \log\left(\frac{1}{16} - \frac{1}{16}e^{-4d/3}\right) + (n-x) \log\left(\frac{1}{16} + \frac{3}{16}e^{-4d/3}\right). \quad (1.48)$$

To estimate d, we maximize L or equivalently its logarithm ℓ. By setting $d\ell/dd = 0$, we can determine that ℓ is maximized at

$$\hat{d} = -\frac{3}{4} \log\left(1 - \frac{4}{3} \times \frac{x}{n}\right). \quad (1.49)$$

This is the MLE of d. It is the distance formula, equation (1.7), which we derived earlier.

We now discuss some statistical properties of MLEs. Under quite mild regularity conditions we will not go into, the MLEs have nice asymptotic (large-sample) properties (see, e.g. Stuart et al. 1999, pp. 46–116). For example, they are consistent, asymptotically unbiased and efficient. Consistency means that the estimate $\hat{\theta}$ converges to the true value θ when the sample size $n \to \infty$. Unbiasedness means that the expectation of the estimate equals the true parameter value: $E(\hat{\theta}) = \theta$. Efficiency means that no other unbiased estimate can have a smaller variance than the MLE. Furthermore, the MLEs are asymptotically normally distributed. These properties are known to hold in large samples. How large the sample size has to be for the approximation to be reliable depends on the particular problem.

Another important property of MLEs is that they are invariant to transformations of parameters or reparametrizations. The MLE of a function of parameters is the same function of the MLEs of the parameters: $\hat{h}(\theta) = h(\hat{\theta})$. Thus if the same model can be formulated using either parameters θ_1 or θ_2, with θ_1 and θ_2 constituting a one-to-one mapping, use of either parameter leads to the same inference. For example, we can use the probability of a difference between the two sequences p as the parameter in the JC69 model instead of the distance d. The two form a one-to-one mapping through equation (1.45). The log likelihood function for p corresponding to equation (1.47) is $L(p; x) = \left(\frac{p}{12}\right)^x \left(\frac{1-p}{4}\right)^{n-x}$, from which we get the MLE of p: $\hat{p} = x/n$. We can then view d as a function of p and obtain its MLE $\hat{d}$, as given by equation (1.49). Whether we use p or d as the parameter, the same inference is made, and the same log likelihood is achieved: $\ell(\hat{d}) = \ell(\hat{p}) = x \log \frac{x}{12n} + (n-x) \log \frac{n-x}{4n}$.

Two approaches can be used to calculate a confidence interval for the MLE. The first relies on the theory that the MLE $\hat{\theta}$ is asymptotically normally distributed around the true value θ when the sample size $n \to \infty$. The asymptotic variance can be calculated using either the observed information $-\frac{d^2\ell}{d\theta^2}$ or the expected (Fisher) information $-E\left(\frac{d^2\ell}{d\theta^2}\right)$. While both are reliable in large samples, the observed information is preferred in real data analysis (e.g. Efron and Hinkley 1978). This is equivalent to using a quadratic polynomial to approximate the log likelihood around the MLE. Here we state the result for the multivariate case, with k parameters in the model:

$$\hat{\theta} \sim N_k(\theta, -H^{-1}), \text{ with } H = \left\{\frac{\partial^2 \ell}{\partial \theta_i \partial \theta_j}\right\}. \tag{1.50}$$

In other words, the MLEs $\hat{\theta}$ have an asymptotic k-variate normal distribution, with the mean to be the true values θ, and the variance–covariance matrix to be $-H^{-1}$, where H is the matrix of second derivatives, also known as the Hessian matrix (Stuart et al. 1999, pp. 73–74).

In our example, the asymptotic variance for $\hat{d}$ is

$$\text{var}(\hat{d}) = -\left(\frac{d^2\ell}{dd^2}\right)^{-1} = \frac{\hat{p}(1-\hat{p})}{(1-4\hat{p}/3)^2 n}. \tag{1.51}$$

This is equation (1.8). An approximate 95% confidence interval for d can be constructed as $\hat{d} \pm 1.96\sqrt{\text{var}(\hat{d})}$.

The normal approximation has a few drawbacks. First, if the log likelihood curve is not symmetrical around the MLE, the normal approximation will be unreliable. For example, if the parameter is a probability, which ranges from 0 to 1, and the MLE is close to 0 or 1, the normal approximation may be very poor. Second, the confidence interval constructed this way includes parameter values that have lower likelihood than values outside the interval. Third, even though the MLEs are invariant to reparametrizations, the confidence intervals constructed using the normal approximation are not.

The second approach is called the *likelihood interval*, which avoids all three problems associated with the normal approximation. It is based on the likelihood ratio test (LRT). In large samples, the LRT statistic, $2[\ell(\hat{\theta}) - \ell(\theta)]$, where θ is the true parameter value and $\hat{\theta}$ is the MLE, has a χ^2_k distribution with the degree of freedom k equal to the number of parameters. Thus one can lower the log likelihood from the peak $\ell(\hat{\theta})$ by $\frac{1}{2}\chi^2_{k,\,5\%}$ to construct a 95% confidence (likelihood) region. Here $\chi^2_{k,\,5\%}$ is the 5% critical value of the χ^2 distribution with k degrees of freedom. The likelihood region contains parameter values with the highest likelihood, values that cannot be rejected by an LRT at the 5% level when compared against $\hat{\theta}$. This likelihood ratio approach is known to give more reliable intervals than the normal approximation. The normal approximation works well for some parametrizations but not for others; the likelihood interval automatically uses the best parametrization.

Example 1.5. For the 12s rRNA data of Table 1.3, we have $\hat{p} = x/n = 90/948 = 0.09494$, and $\hat{d} = 0.1015$. The variance of $\hat{d}$ is 0.0001188, so that the 95% confidence interval based on the normal approximation is $(0.0801, 0.1229)$. If we use p as the parameter instead, we have $\text{var}(\hat{p}) = \hat{p}(1-\hat{p})/n = 0.00009064$, so that the 95% confidence interval for p is $(0.0763, 0.1136)$. These two intervals do not match; for example, if we use the lower bound for p to calculate the lower bound for d, the result will be different. The log likelihood curves are shown in Figure 1.7, with the peak at $\ell(\hat{d}) = \ell(\hat{p}) = -1710.577$. By lowering the log likelihood ℓ by $\frac{1}{2}\chi^2_{1,\,5\%} = 3.841/2 = 1.921$ from its peak, we obtain the 95% likelihood intervals $(0.0817, 0.1245)$ for d and $(0.0774, 0.1147)$ for p. Compared with the intervals based on the normal approximation, the likelihood intervals are asymmetrical and are shifted to the right, reflecting the steeper drop of log likelihood and thus

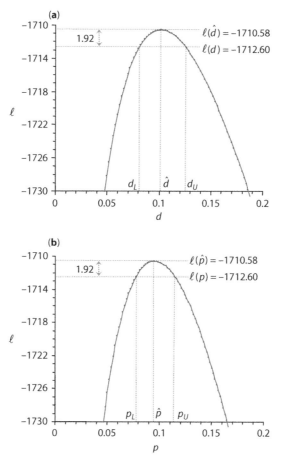

Fig. 1.7 Log likelihood curves and construction of confidence (likelihood) intervals under the JC69 model. The parameter in the model is the sequence distance d in (**a**) and the probability of different sites p in (**b**). The mitochondrial 12s rRNA genes of Table 1.3 are analysed.

more information on the left side of the MLE. Also the likelihood intervals for p and d match each other. The likelihood interval is invariant to reparametrization. □

1.4.2 The K80 model

The likelihood theory applies to models with multiple parameters. We apply the method to estimation of the sequence distance d and the transition/transversion rate ratio κ under the K80 model (Kimura 1980). The data are the numbers of sites with transitional (n_S) and transversional (n_V) differences, with the number of constant sites to be $n - n_S - n_V$. In deriving the probabilities of observing such sites, we again consider all 16 site patterns, as for the JC69 model. Thus the probability is $\frac{1}{4} p_0$ for any constant site (e.g. TT), $\frac{1}{4} p_1$ for any site with a transitional difference (e.g. TC), and $\frac{1}{4} p_2$ for any site with a transversional difference (e.g. TA), with p_0, p_1, p_2 given in equation (1.11). The log likelihood is

$$\ell(d, \kappa; n_S, n_V) = \log\{f(n_S, n_V | d, \kappa)\}$$
$$= (n - n_S - n_V)\log(p_0/4) + n_S \log(p_1/4) + n_V \log(p_2/4). \tag{1.52}$$

MLEs of d and κ can be derived from the likelihood equation $\partial\ell/\partial d = 0$, $\partial\ell/\partial\kappa = 0$. The solution can be shown to be equation (1.12), with $S = n_S/n$ and $V = n_V/n$. A simpler argument relies on the invariance property of the MLEs. Suppose we consider the probabilities of transitional and transversional differences $E(S) = p_1$ and $E(V) = 2p_2$ as parameters in the model instead of d and κ. From the log likelihood (equation (1.52)), the MLEs of $E(S)$ and $E(V)$ are simply S and V. The MLEs of d and κ can be obtained through the one-to-one mapping between the two sets of parameters, which involves the same step taken when we derived equation (1.12) in §1.2.2 by equating the observed proportions S and V to their expected probabilities.

Example 1.6. For the 12s rRNA data of Table 1.3, we have $S = 0.08861$ and $V = 0.00633$. The MLEs are thus $\hat{d} = 0.1046$ for the sequence distance and $\hat{\kappa} = 30.83$ for the transition/transversion rate ratio. These are the same as calculated in Example 1.2. The maximized log likelihood is $\ell(\hat{d}, \hat{\kappa}) = -1637.905$. Application of equation (1.50) leads to the variance–covariance matrix (see Appendix B):

$$\text{var}\begin{pmatrix} \hat{d} \\ \hat{\kappa} \end{pmatrix} = \begin{pmatrix} 0.0001345 & 0.007253 \\ 0.007253 & 172.096 \end{pmatrix}. \tag{1.53}$$

From this, one can get the approximate SEs to be 0.0116 for $\hat{d}$ and 13.12 for $\hat{\kappa}$. The log likelihood surface contour is shown in Figure 1.8, which indicates that the data are much more informative about d than about κ. One can lower the log likelihood from its peak by $\frac{1}{2}\chi^2_{2,5\%} = 5.991/2 = 2.996$, to construct a 95% confidence (likelihood) region for the two parameters (Figure 1.8). □

1.4.3 Likelihood ratio test of substitution models

We may ask whether the transition and transversion rates are indeed different or whether K80 fits the data much better than JC69. The LRT provides a general framework for testing hypotheses concerning model parameters in the likelihood framework. Most of the tests we have learned in a biostatistics course are its special cases or approximations, such as the t test comparing two population means, the t test of the linear regression coefficient, and χ^2 test of association in a 2 × 2 contingency table. The two hypotheses under comparison are nested, with one being a special case of the other. Suppose the simpler hypothesis (the

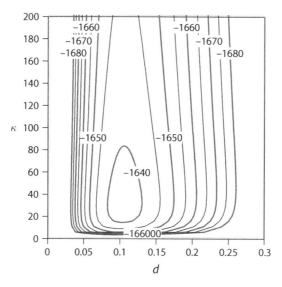

Fig. 1.8 Log likelihood contours for the sequence distance d and the transition/transversion rate ratio κ under the K80 model. The mitochondrial 12s rRNA genes of Table 1.3 are analysed. The peak of the surface is at the MLEs $\hat{d} = 0.1046$, $\hat{\kappa} = 30.83$, with $\ell = -1637.905$. The 95% likelihood region is surrounded by the contour line at $\ell = -1637.905 - 2.996 = -1640.901$ (not shown).

null hypothesis H_0) involves q parameters and the more general hypothesis (the *alternative hypothesis* H_1) has p parameters. Let the maximum log likelihood values under the two models be $\ell_0 = \log\{L(\hat{\theta}_0)\} = \ell(\hat{\theta}_0)$ and $\ell_1 = \log\{L(\hat{\theta}_1)\} = \ell(\hat{\theta}_1)$, where $\hat{\theta}_0$ and $\hat{\theta}_1$ are the MLEs under the two models, respectively. Then under certain regularity conditions, the LRT statistic

$$2\Delta\ell = 2\log\left(\frac{L_1}{L_0}\right) = 2(\ell_1 - \ell_0) \tag{1.54}$$

is asymptotically distributed as χ^2_{p-q} if H_0 is true. In other words, if the null model is true, twice the log likelihood difference between the null and alternative models is approximately χ^2 distributed with the degree of freedom equal to the difference in the number of parameters between the two models. The approximation applies to large samples.

Example 1.7. We use the LRT to compare JC69 and K80 using the 12s rRNA data of Table 1.3. The (maximized) log likelihood under H_0 (JC69) is $\ell_0 = -1710.58$ and that under H_1 (K80) is $\ell_1 = -1637.90$ (Table 1.4). The LRT statistic is $2\Delta\ell = 2(\ell_1 - \ell_0) = 2[-1637.90 - (-1710.58)] = 145.36$. Note that JC69 is equivalent to K80 with parameter $\kappa = 1$ fixed, so that JC69 is nested within K80. Thus $2\Delta\ell$ should be compared with the χ^2 distribution with one degree of freedom (df = 1), with the significance values to be 3.84 at 5% and 6.63 at 1%. K80 fits the dataset much better. It has been observed that we can often easily reject simpler models using molecular sequence data, possibly because the datasets are typically large. □

*1.4.4 Profile and integrated likelihood methods

Suppose we are interested in the sequence distance d under the K80 model (Kimura 1980) but not in the transition/transversion rate ratio κ. However we want to consider κ in the model as transition and transversion rates are known to differ and the rate difference may affect our estimation of d. Parameter κ is thus appropriately called a *nuisance parameter*, while d is our parameter of interest. Dealing with nuisance parameters is commonly considered a weakness of the likelihood method. The approach we described above, estimating both d and κ with ML and using $\hat{d}$ while ignoring $\hat{\kappa}$, is known variously as the *relative likelihood*, *pseudo likelihood* or *estimated likelihood*, since the nuisance parameters are replaced by their estimates.

A more respected approach is the *profile likelihood*. This defines a log likelihood for the parameters of interest only, which is calculated by optimizing the nuisance parameters at fixed values of the parameters of interest. In other words, the profile log likelihood for d is $\ell(d) = \ell(d, \hat{\kappa}_d)$, where $\hat{\kappa}_d$ is the MLE of κ for the given d. This is a pragmatic approach that most often leads to reasonable answers. The likelihood interval for $\hat{d}$ is constructed from the profile likelihood in the usual way.

Example 1.8. For the 12s rRNA genes, the highest likelihood $\ell(\hat{d}) = -1637.905$ is achieved at $\hat{d} = 0.1046$ and $\hat{\kappa} = 30.83$. Thus the point estimate of d is the same as before. We fix d at different values. For each fixed d, the log likelihood (1.52) is a function of the nuisance parameter κ, and is maximized to estimate κ. Let the estimate be $\hat{\kappa}_d$, with the subscript indicating it is a function of d. It does not seem possible to derive $\hat{\kappa}_d$ analytically, so we use a numerical optimization algorithm instead (as discussed later in §4.5). The optimized likelihood is the profile likelihood for d: $\ell(d) = \ell(d, \hat{\kappa}_d)$. This is plotted against d in Figure 1.9a, together with the estimate $\hat{\kappa}_d$. We lower the log likelihood by $\frac{1}{2}\chi^2_{1,\,5\%} = 1.921$ to construct the profile likelihood interval for d: (0.0836, 0.1293). □

If the model involves many parameters, and in particular, if the number of parameters increases without bound with the increase of the size of the data, the likelihood method may run into deep trouble, so deep that the MLEs may not even be consistent (e.g. Kalbfleisch and Sprott 1970; Kalbfleisch 1985, pp. 92–96). A useful strategy in this case is to assign a statistical distribution to describe the variation or uncertainties in the *nuisance parameters*, and integrate them out in the likelihood. This is known as *integrated likelihood* or *marginal likelihood* and has the flavour of a Bayesian approach.

Here we apply the idea to deal with the nuisance parameter κ. Let $f(\kappa)$ be the distribution assigned to κ, also known as a prior. Then the integrated likelihood is

$$L(d) = \int_0^\infty f(\kappa)\, f(n_S, n_V \mid d, \kappa)\, d\kappa$$
$$= \int_0^\infty f(\kappa) \times \left(\frac{p_0}{4}\right)^{n-n_S-n_V} \left(\frac{p_1}{4}\right)^{n_S} \left(\frac{p_2}{4}\right)^{n_V} d\kappa, \qquad (1.55)$$

where p_0, p_1, and p_2 are from equation (1.11). For the present problem, it is possible to use an improper prior: $f(\kappa) = 1$, $0 < \kappa < \infty$. The prior is *improper* as it does not integrate to 1 and is not a proper probability density. The integrated likelihood is then

$$L(d) = \int_0^\infty f(n_S, n_V \mid d, \kappa)\, d\kappa = \int_0^\infty \left(\frac{p_0}{4}\right)^{n-n_S-n_V} \left(\frac{p_1}{4}\right)^{n_S} \left(\frac{p_2}{4}\right)^{n_V} d\kappa. \qquad (1.56)$$

---

* indicates a more difficult or technical section.

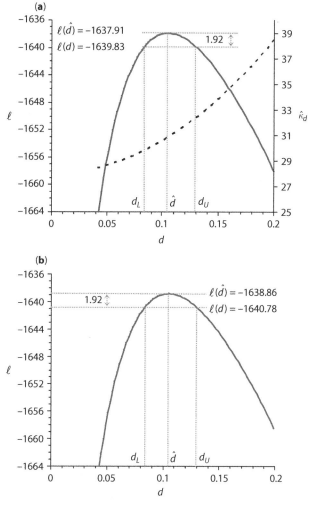

Fig. 1.9 Profile (**a**) and integrated (**b**) log likelihood for distance d under the K80 model. The mitochondrial 12s rRNA genes of Table 1.3 are analysed. (**a**) The profile likelihood $\ell(d) = \ell(d, \hat{\kappa}_d)$ is plotted against d. The estimated nuisance parameter $\hat{\kappa}_d$ at fixed d is also shown. The profile log likelihood is lowered from the peak by 1.921 to construct a likelihood interval for parameter d. (**b**) The integrated likelihood for d is calculated by integrating over the nuisance parameter κ using equation (1.55), with a uniform prior $\kappa \sim U(0, 99)$.

Example 1.9. We apply equation (1.55) to the 12s rRNA data of Table 1.3. We use a uniform prior $\kappa \sim U(0, c)$, with $c = 99$ so that $f(\kappa) = 1/c$ for $0 < \kappa < c$. Analytical calculation of the integral appears awkward, so a numerical method is used instead. The log integrated likelihood $\ell(d) = \log\{L(d)\}$, with $L(d)$ given by equation (1.55), is plotted in Figure 1.9b. This is always lower than the profile log likelihood (Figure 1.9a). The MLE of d is obtained numerically as $\hat{d} = 0.1048$, with the maximum log likelihood $\ell(\hat{d}) = -1638.86$. By lowering ℓ by 1.921, we construct the likelihood interval for d to be (0.0837, 0.1295). For this example, the profile and integrated likelihood methods produced very similar MLEs and likelihood intervals. □

1.5 Markov chains and distance estimation under general models

We have discussed most of the important properties of continuous-time Markov chains useful in this book. In this section we provide a more systematic overview, and also discuss two general Markov chain models: the general time-reversible (GTR) model and the general unconstrained model. The theory will be applied in a straightforward manner to model substitutions between amino acids and between codons in Chapter 2. Note that Markov chains (processes) are classified according to whether time and state are discrete or continuous. In the Markov chains we consider in this chapter, the states (the four nucleotides) are discrete while time is continuous. In Chapter 7, we will encounter Markov chains with discrete time and either discrete or continuous states. Interested readers should consult a textbook on Markov chains and stochastic processes (e.g. Karlin and Taylor 1975; Grimmett and Stirzaker 1992; Ross 1996; Norris 1997). Note that some authors use the term Markov chain if time is discrete, and Markov process if time is continuous.

1.5.1 *Markov chains*

Let the state of the chain at time t be $X(t)$. This is one of the four nucleotides T, C, A, or G. We assume that different sites in a DNA sequence evolve independently, and the Markov chain model is used to describe nucleotide substitutions at any site. The Markov chain is characterized by its *generator* or the substitution rate matrix $Q = \{q_{ij}\}$, where $q_{ij}, i \neq j$, is the instantaneous rate of change from i to j; that is, $\Pr\{X(t + \Delta t) = j | X(t) = i\} = q_{ij} \Delta t$, for any $j \neq i$. If q_{ij} does not depend on time, as we assume here, the process is said to be *time-homogeneous*. The diagonals q_{ii} are specified by the requirement that each row of Q sums to zero, that is, $q_{ii} = -\sum_{j \neq i} q_{ij}$. Thus $-q_{ii}$ is the substitution rate of nucleotide i, the rate at which the Markov chain leaves state i. The general model without any constraint on the structure of Q will have 12 free parameters.

The dynamics of a Markov chain with only a finite number of states is fully determined by its Q matrix. For example, the Q matrix specifies the *transition probability matrix* over any time $t > 0$: $P(t) = \{p_{ij}(t)\}$, where $p_{ij}(t) = \Pr\{X(t) = j | X(0) = i\}$. Indeed $P(t)$ is the solution to the following differential equation:

$$\frac{dP(t)}{dt} = P(t)Q, \qquad (1.57)$$

with the boundary condition $P(0) = I$, the identity matrix (e.g. Grimmett and Stirzaker 1992, p. 242). This has the solution

$$P(t) = e^{Qt}. \qquad (1.58)$$

(e.g. Lang 1987, Chapter 8).

As Q and t occur only in the form of a product, it is conventional to multiply Q by a scale factor so that the average rate is 1. Time t will then be measured by distance, that is, the expected number of substitutions per site. Thus we use Q to define the relative substitution rates only.

If the Markov chain $X(t)$ has the initial distribution $\pi^{(0)} = (\pi_T^{(0)}, \pi_C^{(0)}, \pi_A^{(0)}, \pi_G^{(0)})$, then time t later the distribution $\pi^{(t)} = (\pi_T^{(t)}, \pi_C^{(t)}, \pi_A^{(t)}, \pi_G^{(t)})$ will be given by

$$\pi^{(t)} = \pi^{(0)} P(t). \quad (1.59)$$

If a long sequence initially has the four nucleotides in proportions $\pi_T^{(0)}, \pi_C^{(0)}, \pi_A^{(0)}, \pi_G^{(0)}$, then time t later the proportions will become $\pi^{(t)}$. For example, consider the frequency of nucleotide T in the target sequence: $\pi_T^{(t)}$. Such a T can result from any nucleotide in the source sequence at time 0. Thus $\pi_T^{(t)} = \pi_T^{(0)} p_{TT}(t) + \pi_C^{(0)} p_{CT}(t) + \pi_A^{(0)} p_{AT}(t) + \pi_G^{(0)} p_{GT}(t)$. Written in matrix notation, this is equation (1.59).

If the initial and target distributions are the same, $\pi^{(0)} = \pi^{(t)}$, the chain will stay in that distribution forever. The chain is then said to be stationary or at equilibrium, and the distribution (let it be π) is called the *stationary* or *steady-state distribution*. Our Markov chain can move from any state to any other state in finite time with positive probability. Such a chain is called *irreducible* and has a unique stationary distribution, which is also the *limiting distribution* when time $t \to \infty$. As indicated above, the stationary distribution is given by

$$\pi P(t) = \pi. \quad (1.60)$$

This is equivalent to

$$\pi Q = 0 \quad (1.61)$$

(e.g. Grimmett and Stirzaker 1992, p. 244). This can also be written as $\sum_i \pi_i q_{ij} = 0$ or $\sum_{i \neq j} \pi_i q_{ij} = -\pi_j q_{jj}$ for any j. The total amount of flow into any state j is $\sum_{i \neq j} \pi_i q_{ij}$, while the total amount of flow out of state j is $-\pi_j q_{jj}$. Equation (1.61) states that the two are equal when π is the stationary distribution. Equation (1.61), together with the obvious constraints $\pi_j \geq 0$ and $\sum_j \pi_j = 1$, allows us to determine the stationary distribution from Q for any Markov chain.

*1.5.2 Distance under the unrestricted (UNREST) model

In the most general model of nucleotide substitution, all the non-diagonal elements of the rate matrix Q are free parameters. This model, referred to as UNREST, was implemented by Yang (1994b) for estimating the pattern of nucleotide substitution, in comparison with the GTR (REV) model. The rate matrix Q is shown in Table 1.1. Note that the equilibrium nucleotide frequencies $\{\pi_T, \pi_C, \pi_A, \pi_G\}$ are given by equation (1.61), as functions of the rate parameters in Q; they should not be counted as additional parameters in the model. In this regard, note that the rate matrix for this model is often given incorrectly in the phylogenetics literature (e.g. Swofford et al. 1996, eq. 3). In general Q may have complex eigenvalues and eigenvectors, so its implementation requires care.

We mention here an interesting special case of the UNREST model, the strand-symmetry model, proposed by Sueka (1995) and implemented by Bielawski and Gold (2002; see also Singh et al. 2009). This assumes that the mutation rates are the same on the two strands of the DNA, so that in the comparison of homologous sequences on the same strand, we have, say, $q_{TC} = q_{AG}$. The rate matrix involves six parameters:

---

* indicates a more difficult or technical section.

28 1 MODELS OF NUCLEOTIDE SUBSTITUTION

$$Q = \begin{bmatrix} -(b+c+e) & b & c & e \\ a & -(a+d+f) & d & f \\ c & e & -(b+c+e) & b \\ d & f & a & -(a+d+f) \end{bmatrix}. \quad (1.62)$$

Perhaps by coincidence, all eigenvalues of this matrix are real: $\lambda_0 = 0$, $\lambda_1 = -(a+b+d+e)$, and $\lambda_{2,3} = -\frac{1}{2}[a+b+d+e+2c+2f \pm [(a+b+d+e+2c+2f)^2 - 8(ac+ae+cd+bd+bf+2cf)]^{1/2}]$. The equilibrium distribution is given analytically by Singh et al. (2009) as

$$(\pi_T, \pi_C, \pi_A, \pi_G) = \left(\frac{a+d}{2(a+b+d+e)}, \frac{b+c}{2(a+b+d+e)}, \frac{a+d}{2(a+b+d+e)}, \frac{b+c}{2(a+b+d+e)} \right). \quad (1.63)$$

To estimate the sequence distance under the UNREST model, note that the model can in theory identify the root of the two-sequence tree (Figure. 1.10a), so that two branch lengths (t_1 and t_2) are involved. In addition there are 11 relative-rate parameters in the Q matrix (suppose we fix $q_{GA} = l = 1$ in Q, Table 1.1). The likelihood is given by the multinomial probability with 16 cells, corresponding to the 16 possible site patterns. Let $f_{ij}(t_1, t_2, Q)$ be the probability for the ijth cell, that is, the probability that any site has nucleotide i in sequence 1 and j in sequence 2. Since such a site can result from all four possible nucleotides in the ancestor, we have to average over them:

$$f_{ij}(t_1, t_2, Q) = \sum_k \pi_k p_{ki}(t_1) p_{kj}(t_2). \quad (1.64)$$

Here π_k is the equilibrium frequency of nucleotide k, given by equation (1.61), together with the constraint $\sum_j \pi_j = 1$, as a function of Q. Let n_{ij} be the number of sites in the ijth cell. The log likelihood is then

$$\ell(t_1, t_2, Q) = \sum_{i,j} n_{ij} \log\{f_{ij}(t_1, t_2, Q)\}. \quad (1.65)$$

The model involves 13 parameters: 11 relative rates in Q plus two branch lengths t_1 and t_2. There are two problems with this unconstrained model. First, numerical methods are necessary to find the MLEs of parameters as no analytical solution seems possible. The eigenvalues of Q may be complex numbers. Second, and more importantly, typical datasets may not have enough information to estimate so many parameters. In particular, even though t_1 and t_2 are identifiable, their estimates are highly correlated. For this reason the model is not advisable for use in distance calculations.

Example 1.10. For the 12s rRNA data of Table 1.3, the log likelihood appears flat when t_1 and t_2 are estimated as separate parameters. We thus force $t_1 = t_2$ during the numerical

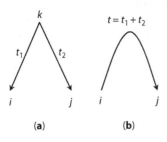

Fig. 1.10 A tree for two sequences, showing the observed nucleotides i and j at one site and the direction of evolution. (**a**) Two sequences diverged from a common ancestor (root of the tree) t_1 and t_2 time units ago; time is measured by the distance or the amount of sequence change. (**b**) Sequence 1 is ancestral to sequence 2. Under time-reversible models, we cannot identify the root of the tree, as the data will look the same whether both sequences were descendants of a common ancestor (as in **a**), or one sequence is ancestral to the other (as in **b**), or wherever the root of the tree is along the single branch connecting the two sequences.

maximization of the log likelihood. The estimate of the sequence distance $t = (t_1 + t_2)$ is 0.1057, very close to estimates under other models (Table 1.4). The MLE of rate matrix Q is

$$Q = \begin{pmatrix} -1.46 & 1.40 & 0.06 & 0.00 \\ 1.16 & -1.22 & 0.06 & 0.00 \\ 0.05 & 0.04 & -0.60 & 0.51 \\ 0.00 & 0.00 & 0.85 & -0.85 \end{pmatrix}, \quad (1.66)$$

scaled so that the average rate is $-\sum_i \pi_i q_{ii} = 1$. The steady-state distribution is calculated from equation (1.61) to be $\hat{\pi} = (0.2184, 0.2606, 0.3265, 0.1946)$, virtually identical to the observed frequencies (Table 1.3). The log likelihood is -1610.36. □

*1.5.3 Distance under the general time-reversible model

A Markov chain is said to be time-reversible if and only if

$$\pi_i q_{ij} = \pi_j q_{ji}, \text{ for all } i \neq j. \quad (1.67)$$

Note that π_i is the proportion of time the Markov chain spends in state i, and $\pi_i q_{ij}$ is the amount of 'flow' from states i to j, while $\pi_i q_{ij}$ is the flow in the opposite direction. Equation (1.67) states that the flow between any two states is the same in the opposite directions and is known as the *detailed-balance* condition. There does not appear to be any biological reason to expect the substitution process to be reversible, so reversibility is a mathematical convenience. Models discussed in this chapter, including JC69 (Jukes and Cantor 1969), K80 (Kimura 1980), F84, HKY85 (Hasegawa et al. 1985), and TN93 (Tamura and Nei 1993), are all time-reversible. Equation (1.67) is equivalent to

$$\pi_i p_{ij}(t) = \pi_j p_{ji}(t), \text{ for all } i \neq j \text{ and for any t.} \quad (1.68)$$

Another equivalent condition for reversibility is that the rate matrix can be written as a product of a symmetrical matrix multiplied by a diagonal matrix; the diagonal elements in the diagonal matrix will then specify the equilibrium frequencies. Thus the rate matrix for the GTR model of nucleotide substitution is

$$Q = \{q_{ij}\} = \begin{bmatrix} \cdot & a\pi_C & b\pi_A & c\pi_G \\ a\pi_T & \cdot & d\pi_A & e\pi_G \\ b\pi_T & d\pi_C & \cdot & f\pi_G \\ c\pi_T & e\pi_C & f\pi_A & \cdot \end{bmatrix} = \begin{bmatrix} \cdot & a & b & c \\ a & \cdot & d & e \\ b & d & \cdot & f \\ c & e & f & \cdot \end{bmatrix} \begin{bmatrix} \pi_T & 0 & 0 & 0 \\ 0 & \pi_C & 0 & 0 \\ 0 & 0 & \pi_A & 0 \\ 0 & 0 & 0 & \pi_G \end{bmatrix}, \quad (1.69)$$

with the diagonals of Q given by the requirement that each row of Q sums to 0. This matrix involves nine free parameters: the rates a, b, c, d, e, and f and three frequency parameters. The model was first applied by Tavaré (1986) to sequence distance calculation and by Yang (1994b) to estimation of relative substitution rates (substitution pattern) between nucleotides using ML. It is commonly known as GTR or REV.

Keilson (1979) discussed a number of nice mathematical properties of reversible Markov chains. One of them is that all eigenvalues of the rate matrix Q are real (see §2.6). Thus efficient and stable numerical algorithms can be used to calculate the eigenvalues of Q. Alternatively, it appears possible to diagonalize Q of equation (1.69) analytically: one eigenvalue is 0, so that the characteristic equation that the eigenvalues should satisfy is a cubic equation (e.g. Lang 1987, Chapter 8), which is solvable. Even so, analytical calculation appears too tedious to be practical.

---

* indicates a more difficult or technical section.

In the phylogenetic analysis of sequence data, reversibility leads to an important simplification to the likelihood function. The probability of observing site pattern ij in equation (1.64) becomes

$$\begin{aligned} f_{ij}(t_1, t_2) &= \sum_k \pi_k p_{ki}(t_1) p_{kj}(t_2) \\ &= \sum_k \pi_i p_{ik}(t_1) p_{kj}(t_2) \\ &= \pi_i p_{ij}(t_1 + t_2). \end{aligned} \quad (1.70)$$

The second equality is because of the reversibility condition $\pi_k p_{ki}(t_1) = \pi_i p_{ik}(t_1)$, while the third equality is due to the Chapman–Kolmogorov theorem (equation (1.5)).

Two remarks are in order. First, f_{ij} depends on $t_1 + t_2$ but not on t_1 and t_2 individually; thus we can estimate $t = t_1 + t_2$ but not t_1 and t_2 separately. Equation (1.70) thus becomes

$$f_{ij}(t) = \pi_i p_{ij}(t). \quad (1.71)$$

Second, while we defined f_{ij} as the probability of a site when both sequences are descendants of a common ancestor (Figure 1.10a), $\pi_i p_{ij}(t)$ is the probability of the site if sequence 1 is ancestral to sequence 2 (Figure 1.10b). The probability is the same if we consider sequence 2 as the ancestor of sequence 1, or wherever we place the root along the single branch linking the two sequences. Thus under the model, the log likelihood (1.65) becomes

$$\ell(t, a, b, c, d, e, \pi_T, \pi_C, \pi_A) = \sum_i \sum_j n_{ij} \log\{f_{ij}(t)\} = \sum_i \sum_j n_{ij} \log\{\pi_i p_{ij}(t)\}. \quad (1.72)$$

We use Q to represent the relative rates, with $f = 1$ fixed in Q, and multiply the whole matrix by a scale factor so that the average rate is $-\sum_i \pi_i q_{ii} = 1$. Time t is then the distance: $d = -t \sum_i \pi_i q_{ii} = t$. The model thus involves nine parameters, which can be estimated numerically by solving a nine-dimensional optimization problem. Sometimes the base frequency parameters are estimated using the average observed frequencies, in which case the dimension is reduced to six.

Note that the log likelihood functions under the JC69 and K80 models, that is, equations (1.48) and (1.52), are special cases of equation (1.72). Under these two models, the likelihood equation is analytically tractable, so that numerical optimization is not needed. Equation (1.72) also gives the log likelihood for other reversible models such as F81, HKY85, F84, and TN93. MLEs under those models were obtained through numerical optimization for the 12s rRNA genes of Table 1.3 and listed in Table 1.4. Note that the distance formulae under F81, F84, TN93 etc., discussed in §1.2, are not MLEs, despite claims to the contrary. First, the observed base frequencies are in general not MLEs of the base frequency parameters. Second, all 16 site patterns have distinct probabilities under those models and are not collapsed in the likelihood function (1.72), but collapsed site patterns are used in the distance formulae (such as the constant patterns TT, CC, AA, GG). Nevertheless, it is expected that the distance formulae will give estimates very close to the MLEs (see, e.g. Table 1.4).

Under GTR + Γ, with gamma-distributed rates among sites, the log likelihood is still given by equation (1.72) but with $f_{ij}(t)$ given by equation (1.44). Thus the distance can be estimated by maximizing the likelihood. This is the ML method for distance estimation under the GTR + Γ model described by Gu and Li (1996) and Yang and Kumar (1996).

Besides the ML estimation, a few distance formulae have been suggested in the literature for the GTR and even the UNREST models. We consider the GTR model first. Note that in matrix notation, equation (1.71) becomes

1.5 MARKOV CHAINS AND DISTANCE ESTIMATION UNDER GENERAL MODELS

$$F(t) = \{f_{ij}(t)\} = \Pi P(t), \tag{1.73}$$

where $\Pi = \text{diag}\{\pi_T, \pi_C, \pi_A, \pi_G\}$. As $P(t) = e^{Qt}$, we can estimate Qt by

$$\overline{Qt} = \log\{\hat{P}\} = \log\{\hat{\Pi}^{-1}\hat{F}\}. \tag{1.74}$$

where we use the average observed frequencies to estimate Π and use $\hat{f}_{ij} = \hat{f}_{ji} = (n_{ij} + n_{ji})/n$ to estimate the F matrix. The logarithm of $\hat{P}$ is computed by diagonalizing $\hat{P}$. When Q is defined as the relative substitution rates with the average rate to be 1, both t and Q can be recovered from the estimate of Qt. Note that the sequence distance can be defined as $d = -\sum_i \pi_i q_{ii} t = -\text{trace}\{\Pi Qt\}$, where trace $\{A\}$ is the sum of the diagonal elements of matrix A. An estimate is thus:

$$\hat{d} = -\text{trace}\{\hat{\Pi} \log(\hat{\Pi}^{-1}\hat{F})\}. \tag{1.75}$$

This approach was first suggested by Tavaré (1986, Equation 3.12), although Rodriguez et al. (1990) were the first to publish equation (1.75). A number of authors (e.g. Gu and Li 1996; Yang and Kumar 1996; Waddell and Steel 1997) apparently rediscovered the distance formula, and also extended the distance to the case of gamma-distributed rates among sites, using the same idea for deriving gamma distances under JC69 and K80 (see §1.3).

The distance (1.75) is inapplicable when any of the eigenvalues of $\hat{P}$ is ≤ 0, which can occur often at high sequence divergences. This is similar to the inapplicability of the JC69 distance when more than 75% of sites are different. As there are nine free parameters in the model and also nine free observables in the symmetrical matrix $\hat{F}$, the invariance property of MLEs suggests that equation (1.75), if applicable, should give the MLEs.

Next we describe a distance suggested by Barry and Hartigan (1987a), which works without the reversibility assumption and even without assuming a stationary model:

$$\hat{d} = -\frac{1}{4} \log\{\text{Det}(\hat{\Pi}^{-1}\hat{F})\}, \tag{1.76}$$

where $\text{Det}(A)$ is the determinant of matrix A, which is equal to the product of the eigenvalues of A. The distance is inapplicable when the determinant is ≤ 0 or when any of the eigenvalues of $\hat{\Pi}^{-1}\hat{F}$ is ≤ 0. Barry and Hartigan (1987a) referred to equation (1.76) as the *asynchronous distance*. It is now commonly known as the *Log-Det distance*.

Let us consider the behaviour of the distance under simpler stationary models in very long sequences. In such a case, $\hat{\Pi}^{-1}\hat{F}$ will approach the transition probability matrix $P(t)$, and its determinant will approach $\exp(\sum_k \lambda_k t)$, where the λ_ks are the eigenvalue of the rate matrix Q (see equation (1.18)). Thus $\hat{d}$ in equation (1.76) will approach $-\frac{1}{4}\sum_k \lambda_k t$. For the K80 model, the eigenvalues of the rate matrix (1.9) are $\lambda_1 = 0, \lambda_2 = -4\beta, \lambda_3 = \lambda_4 = -2(\alpha + \beta)$, so that $\hat{d}$ approaches $(\alpha + 2\beta)t$, which is the correct sequence distance. Obviously this will hold true for the simpler JC69 model as well. However, for more complex models with unequal base frequencies, $\hat{d}$ of equation (1.76) does not estimate the correct distance, even though it grows linearly with time. For example, under the TN93 model, $\hat{d}$ approaches $\frac{1}{4}(\pi_Y \alpha_1 + \pi_R \alpha_2 + 2\beta)t$.

Barry and Hartigan (1987a) defined $\hat{f}_{ij} = n_{ij}/n$, so that $\hat{F}$ is not symmetrical, and interpreted $\hat{\Pi}^{-1}\hat{F}$ as an estimate of $P_{12}(t)$, the matrix of transition probabilities from sequences 1 to 2. The authors argued that the distance should work even if the substitution process is not homogeneous or stationary, that is, if there is systematic drift in base compositions during the evolutionary process. Evidence for the performance of the distance when different sequences have different base compositions is mixed. The distance appears to have acquired a paranormal status when it was rediscovered or modified by Lake (1994), Steel

(1994b), Zharkikh (1994), among others. For a more recent discussion of the distance, see Massingham and Goldman (2007).

1.6 Discussions

1.6.1 *Distance estimation under different substitution models*

One might expect more complex models to be more realistic and to produce more reliable distance estimates. However, the situation is more complex. At small distances, the different assumptions about the structure of the Q matrix do not make much difference, and simple models such as JC69 and K80 produce very similar estimates to those under more complex models. The two 12s rRNA genes analysed in this chapter are different at about 10% of the sites. The different distance formulae produced virtually identical estimates, all between 0.10 and 0.11 (Table 1.4). This is the case despite the fact that simple models like JC69 are grossly wrong, judged by the log likelihood values achieved by the models (see §1.4.3 and Problem 1.7). The rate variation among sites has much more impact, as seen in §1.3.

At intermediate distances, for example, when the sequences are about 20% or 30% different, model assumptions become more important. It may be favourable to use realistic models for distance estimation, especially if the sequences are not short. At large distances, for example, when the sequences are >40% different, the different methods often produce very different estimates, and the estimates, especially those under more complex models, involve large sampling errors. Sometimes the distance estimates become infinite or the distance formulae become inapplicable. This happens far more often under more complex models than under simpler models. In such cases, a useful approach is to add more sequences to break down the long distances and to use a likelihood-based approach to compare all sequences jointly on a phylogeny.

1.6.2 *Limitations of pairwise comparison*

If there are only two sequences in the whole dataset, pairwise comparison is all we can do. If we have multiple sequences, however, pairwise comparison may be hampered as it ignores the other sequences, which should also provide information about the relatedness of the two sequences being compared. Here, brief comment will be made on two obvious limitations of the pairwise approach. The first is the lack of internal consistency. Suppose we use the K80 model for pairwise comparison of three sequences: a, b, and c. Let $\hat{\kappa}_{ab}$, $\hat{\kappa}_{bc}$, and $\hat{\kappa}_{ca}$ be the estimates of the transition/transversion rate ratio κ in the three comparisons. Considering that the three sequences are related by a phylogenetic tree, we see that we estimated κ for the branch leading to sequence a as $\hat{\kappa}_{ab}$ in one comparison but as $\hat{\kappa}_{ca}$ in another. This inconsistency is problematic when complex models involving unknown parameters are used, and when information about model parameters is visible only when multiple sequences are compared simultaneously. An example is the variation of evolutionary rates among sites. With only two sequences, it is virtually impossible to decide whether a site has a difference because the rate at the site is high or because the overall divergence between the two sequences is high. Even if the parameters in the rate distribution (such as the shape parameter α of the gamma distribution) are fixed, the pairwise approach does not guarantee that a high-rate site in one comparison is also a high-rate site in another.

A second limitation is important in analysis of highly divergent sequences, in which substitutions have nearly reached *saturation*. The distance between two sequences is the sum of branch lengths on the phylogeny along the path linking the two sequences. By adding branch lengths along the tree, the pairwise distance can become large even if all branch lengths on the tree are small or moderate. As discussed above, large distances involve large sampling errors in the estimates or even cause the distance formulae to be inapplicable. By summing up branch lengths, the pairwise approach exacerbates the problem of saturation and may be expected to be less tolerant of high sequence divergences than likelihood or Bayesian methods, which compare all sequences simultaneously.

1.7 Problems

1.1 Use the transition probabilities under the JC69 model (equation (1.4)) to confirm the Chapman–Kolmogorov theorem (equation (1.5)). It is sufficient to consider two cases: (**a**) $i = T, j = T$; and (**b**) $i = T, j = C$. For example, in case (**a**), confirm that $p_{TT}(t_1 + t_2) = p_{TT}(t_1)p_{TT}(t_2) + p_{TC}(t_1)p_{CT}(t_2) + p_{TA}(t_1)p_{AT}(t_2) + p_{TG}(t_1)p_{GT}(t_2)$.

1.2 Derive the transition probability matrix $P(t) = e^{Qt}$ for the JC69 model. Set $\pi_T = \pi_C = \pi_A = \pi_G = 1/4$ and $\alpha_1 = \alpha_2 = \beta$ in the rate matrix (1.16) for the TN93 model to obtain the eigenvalues and eigenvectors of Q under JC69, using results of §1.2.3. Alternatively you can derive the eigenvalues and eigenvectors from equation (1.1) directly. Then apply equation (1.18).

1.3 Derive the transition probability matrix $P(t)$ for the Markov chain with two states 0 and 1 and rate matrix $Q = \begin{bmatrix} -u & u \\ v & -v \end{bmatrix}$. Confirm that the spectral decomposition of Q is given as

$$Q = U\Lambda U^{-1} = \begin{bmatrix} 1 & -u \\ 1 & v \end{bmatrix} \begin{bmatrix} 0 & 0 \\ 0 & -u-v \end{bmatrix} \begin{bmatrix} \frac{v}{u+v} & \frac{u}{u+v} \\ -\frac{1}{u+v} & \frac{1}{u+v} \end{bmatrix}, \qquad (1.77)$$

so that

$$P(t) = e^{Qt} = Ue^{\Lambda t}U^{-1} = \frac{1}{u+v}\begin{bmatrix} v + ue^{-(u+v)t} & u - ue^{-(u+v)t} \\ v - ve^{-(u+v)t} & u + ve^{-(u+v)t} \end{bmatrix}. \qquad (1.78)$$

Note that the stationary distribution of the chain is given by the first row of U^{-1}, as $(\frac{v}{u+v}, \frac{u}{u+v})$, which can also be obtained from $P(t)$ by letting $t \to \infty$. A special case is $u = v = 1$, when we have

$$P(t) = \begin{bmatrix} \frac{1}{2} + \frac{1}{2}e^{-2t} & \frac{1}{2} - \frac{1}{2}e^{-2t} \\ \frac{1}{2} - \frac{1}{2}e^{-2t} & \frac{1}{2} + \frac{1}{2}e^{-2t} \end{bmatrix}. \qquad (1.79)$$

This is the binary equivalent of the JC69 model.

1.4 Confirm that the two likelihood functions for the JC69 model, equations (1.46) and (1.47), are proportional and the proportionality factor is a function of n and x but not of d. Confirm that the likelihood equation, $\frac{d\ell}{dd} = \frac{d \log\{L(d)\}}{dd} = 0$, is the same whichever of the two likelihood functions is used.

1.5 Derive the equilibrium nucleotide frequencies for the K80 model. Solve the system of linear equations generated by equation (1.61) and the constraint $\sum_j \pi_j = 1$.

1.6 A large genomic region evolves neutrally according to the JC69 model, at the rate 2×10^{-8} substitutions/site/year (this is roughly the rate in the mitochondria in mammals). (**a**) Suppose initially the sequence consists of Ts only. What will be the proportions of T, C, A, and G in the sequence 10^6 and 10^8 years later? (**b**) Do the same calculation assuming that the sequence initially had Cs only. (**c**) Do the same calculation if the initial proportions of T, C, A, and G are $\pi_0 = (0.4, 0.3, 0.2, 0.1)$.

1.7 Use the 12s rRNA data of Table 1.3 to conduct the LRT to compare K80 against HKY85, and HKY85 against GTR. The numbers of parameters under the models are listed in Table 1.1, and the log likelihood values are listed in Table 1.4, but you may prefer running a program (such as BASEML in the PAML package, Yang 2007b) to do the calculation yourself.

1.8 Use the protein-coding DNA sequences from the human and gibbon mitochondrial genomes to calculate the sequence distance at the three codon positions. Download the sequences from GenBank (accession numbers X93334 for *Homo sapiens* and X99256 for *Hylobates lar*) and concatenate the 12 protein-coding genes encoded on the same H strand of the genome (that is, excluding NADH6, which is coded on the other strand with very different base compositions). Align the sequences using an alignment program such as CLUSTAL or PRANK and make manual adjustments if necessary. Separate the three codon positions into three independent datasets, and calculate the distance between the two species under various substitution models: JC69, K80, GTR, and JC69 + Γ_5, K80 + Γ_5, and GTR + Γ_5 (with $\alpha = 0.5$ fixed for the gamma models). Discuss the impact of model assumptions on distance estimation.

1.9* Suppose $x = 9$ heads and $r = 3$ tails are observed in $n = 12$ independent tosses of a coin. Derive the MLE of the probability of heads (θ). Consider two mechanisms by which the data are generated.

(**a**) *Binomial*. The number $n = 12$ tosses was fixed beforehand. In $n = 12$ tosses, $x = 9$ heads were observed. Then the number of heads x has a binomial distribution, with probability

$$f(x|\theta) = \binom{n}{x} \theta^x (1-\theta)^{n-x}. \qquad (1.80)$$

(**b**) *Negative binomial*. The number of tails $r = 3$ was fixed beforehand, and the coin was tossed until $r = 3$ tails were observed, at which point it was noted that $x = 9$ heads were observed. Then x has a negative binomial distribution, with probability

$$f(x|\theta) = \binom{r+x-1}{x} \theta^x (1-\theta)^{n-x}. \qquad (1.81)$$

Confirm that under both models, the MLE of θ is x/n.

---

* indicates a more difficult or technical problem.

CHAPTER 2

Models of amino acid and codon substitution

2.1 Introduction

In Chapter 1 we discussed continuous-time Markov chain models of nucleotide substitution and their application to estimation of the distance between two nucleotide sequences. This chapter discusses similar Markov chain models to describe substitutions (replacements) between amino acids in proteins, or between codons in protein-coding genes. We make a straightforward use of the Markov chain theory introduced in Chapter 1, except that the states of the chain are now the 20 amino acids or the 61 sense codons (in the universal genetic code), instead of the four nucleotides.

With protein-coding genes, we have the advantage of being able to distinguish the *synonymous* or *silent* substitutions (nucleotide substitutions that do not change the encoded amino acid) from the *nonsynonymous* or *replacement* substitutions (those that do change the amino acid). As natural selection operates mainly at the protein level, synonymous and nonsynonymous mutations are under very different selective pressures and are fixed at very different rates. Thus comparison of synonymous and nonsynonymous substitution rates provides a means to understand the effect of natural selection on the protein, as pointed out by pioneers of molecular evolution as soon as DNA sequencing techniques became available (e.g. Kafatos et al. 1977; Kimura 1977; Jukes and King 1979; Miyata and Yasunaga 1980). This comparison does not require knowledge of absolute substitution rates or species divergence times. Chapter 11 provides a detailed discussion of models developed to detect natural selection through phylogenetic comparison of multiple sequences. In this chapter, we consider comparison of only two sequences, to calculate a distance for synonymous substitutions (d_S) and another for nonsynonymous substitutions (d_N).

2.2 Models of amino acid replacement

2.2.1 *Empirical models*

A distinction can be made between empirical and mechanistic models of amino acid substitution. *Empirical* models attempt to describe the relative rates of substitution between amino acids without explicitly considering factors that influence the amino acid replacement process. They are often constructed by analysing large quantities of sequence data, as compiled from sequence databases. *Mechanistic models*, on the other hand, consider the biological process involved in amino acid substitution, such as mutational biases in

Molecular Evolution: A Statistical Approach. Ziheng Yang. © Ziheng Yang 2014.
Published 2014 by Oxford University Press.

the DNA, translation of the codons into amino acids, and acceptance or rejection of the resulting amino acid after filtering by natural selection. Mechanistic models have more interpretative power and are particularly useful for studying the forces and mechanisms of gene sequence evolution. For phylogenetic tree reconstruction, empirical models appear at least equally efficient.

Empirical models of amino acid substitution are all constructed by estimating relative substitution rates between amino acids under the general time-reversible model. The rate q_{ij} from amino acids i to j is assumed to satisfy the detailed balance condition

$$\pi_i q_{ij} = \pi_j q_{ji}, \text{ for any } i \neq j. \tag{2.1}$$

This is equivalent to the requirement that the rate matrix can be written as the product of a symmetrical matrix and a diagonal matrix:

$$Q = S\Pi, \tag{2.2}$$

where $S = \{s_{ij}\}$ with $s_{ij} = s_{ji}$ for all $i \neq j$, and $\Pi = \text{diag}\{\pi_1, \pi_2, \ldots, \pi_{20}\}$, with π_j to be the equilibrium frequency of amino acid j (see §1.5.3). Whelan and Goldman (2001) referred to the s_{ij} as the amino acid *exchangeabilities*.

The first empirical amino acid substitution matrix was constructed by Dayhoff and colleagues (Dayhoff et al. 1978). They compiled and analysed protein sequences available at the time, using a parsimony argument to reconstruct ancestral protein sequences and tabulating amino acid changes along branches on the phylogeny. To reduce the impact of multiple hits, only similar sequences that are different from one another at < 15% of sites were used. Inferred changes were merged across all branches without regard for their different lengths. Dayhoff et al. (1978) approximated the transition probability matrix for an expected distance of 0.01 changes per site, called 1 PAM (for point-accepted mutations). This is $P(0.01)$ in our notation, from which the instantaneous rate matrix Q can be constructed. (See Kosiol and Goldman (2005) for a discussion of this construction.) The resulting rate matrix is known as the Dayhoff matrix.

The Dayhoff matrix was updated by Jones et al. (1992), who analysed a much larger collection of protein sequences, using the same approach as did Dayhoff et al. (1978). The updated matrix is known as the JTT matrix.

A variation of these empirical models is to replace the equilibrium amino acid frequencies (the π_js) in the empirical matrices by the frequencies observed in the data being analysed, while using the amino acid exchangeabilities from the empirical model (Cao et al. 1994). This strategy adds 19 free frequency parameters but is often found to improve the fit of the model considerably. The models are then known as Dayhoff + F, JTT + F, etc., with a suffix '+F'.

It is also straightforward to estimate the rate matrix Q under the general time-reversible model from the dataset by using maximum likelihood (ML) (Adachi and Hasegawa 1996a). This is the same procedure as is used to estimate the pattern of nucleotide substitution (Yang 1994b). We will describe the details of likelihood calculation on a phylogeny in Chapter 4. The reversible rate matrix involves $20 \times 19/2 - 1 = 189$ relative rate parameters in the symmetrical matrix S of exchangeabilities as well as 19 free amino acid frequency parameters, with a total of 208 parameters (see equation (2.2)). Usually the amino acid frequency parameters are estimated using the observed frequencies, reducing the dimension of the optimization problem by 19. The dataset should be relatively large to allow estimation of so many parameters; 50 or 100 reasonably divergent protein sequences appear sufficient to provide good estimates. Examples of such matrices include the MTREV (Adachi and Hasegawa 1996a) and MTMAM (Yang et al. 1998) models for

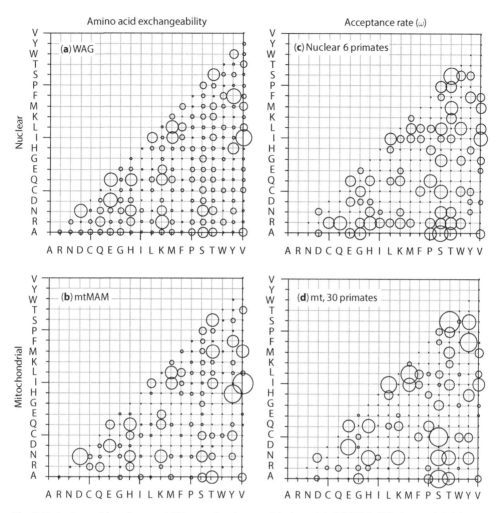

Fig. 2.1 Amino acid exchangeabilities under the empirical models (**a**) WAG (Whelan and Goldman 2001) and (**b**) MTMAM (Yang et al. 1998), and nonsynonymous acceptance rates (ωs) estimated from primate genomic data (**c** and **d**). The rate is represented by the size (area) of the bubble. The exchangeability between two amino acids i and j is s_{ij} in equation (2.2). The five highest exchangeabilities are for amino acid pairs Ile-Val, Phe-Tyr, Asp-Glu, Glu-Gln, Asp-Asn under WAG and are Ile-Val, His-Tyr, Asp-Asn, Met-Thr, Ser-Tyr under MTMAM. The nonsynonymous acceptance rates (**c** and **d**) are ω in equation (2.7), except that a separate ω parameter is estimated for every pair of amino acids that can interchange by one nucleotide substitution. Data used for (**c**) are 857 nuclear genes from six primates (human, chimpanzee, gorilla, orangutan, macaque, and marmoset), and those for (**d**) are 12 protein-coding genes in the mitochondrial genomes of 30 primate species. Both datasets are from dos Reis et al. (2012), which shows the phylogeny in Figure S1. Branch lengths in the tree were estimated under the F3 × 4 model with one single ω (the one-ratio model M0). Branch lengths and κ are then fixed to estimate 75 ω parameters for the nuclear genome (**c**) and 70 ω parameters for the mitochondrial genome (**d**). The five highest acceptance rates are $\hat{\omega}_{ST} = 0.592$, $\hat{\omega}_{AS} = 0.564$, $\hat{\omega}_{IV} = 0.424$, $\hat{\omega}_{RH} = 0.419$, and $\hat{\omega}_{CS} = 0.390$ for the nuclear genes, and are $\hat{\omega}_{ST} = 0.268$, $\hat{\omega}_{CS} = 0.219$, $\hat{\omega}_{FY} = 0.219$, $\hat{\omega}_{AS} = 0.201$, and $\hat{\omega}_{LM} = 0.196$ for the mitochondrial genome. The estimates under the simple codon model M0 (one-ratio) is $\hat{\kappa} = 3.65$ and $\hat{\omega} = 0.212$ for the nuclear genes and $\hat{\kappa} = 10.83$ and $\hat{\omega} = 0.062$, for the mitochondrial genome.

vertebrate or mammalian mitochondrial proteins, and the cpREV model for chloroplast proteins (Adachi et al. 2000). Whelan and Goldman (2001) used a similar approach to estimate a combined substitution rate matrix from 182 alignments of nuclear proteins. This is known as the WAG matrix, and is an update of the Dayhoff and JTT matrices. An even more recent update is the LG matrix (Le and Gascuel 2008). Special empirical matrices have also been estimated, for example, for retroviruses (RTREV, Dimmic et al. 2002), and for arthropod (MTART, Abascal et al. 2007) and metazoan (MTZOA, Rota-Stabelli et al. 2009) mitochondrial proteins.

Figure 2.1a and b presents the symmetrical matrix S of exchangeabilities under the empirical models WAG (Whelan and Goldman 2001) and MTMAM (Yang et al. 1998). The Dayhoff (Dayhoff et al. 1978), JTT (Jones et al. 1992), and LG (Le and Gascuel 2008) models are all derived from a large number of nuclear proteins and are similar to WAG. The MTMAM model is derived from comparisons of the 12 proteins encoded by the H-strand of the mitochondrial genome from 20 mammals and is very similar to MTREV24 (Adachi and Hasegawa 1996a), derived from comparison of 24 vertebrates.

Several features of these matrices are worth noting. First, amino acids with similar physico-chemical properties tend to exchange with each other at higher rates than dissimilar amino acids (Zuckerkandl and Pauling 1965; Clark 1970; Dayhoff et al. 1972, 1978; Grantham 1974). This pattern is particularly conspicuous when we compare rates between amino acids separated by only one codon position difference (Miyata et al. 1979). For example, aspartic acid (D) and glutamic acid (E) have high rates of exchange, as do isoleucine (I) and valine (V); in each pair the amino acids are similar. Cysteine, however, has low rates of exchange with all other amino acids. Second, the 'mutational distance' between amino acids determined by the structure of the genetic code has a major impact on the exchange rates, with amino acids separated by two or three codon position differences having lower rates than amino acids separated by one difference (Table 2.1). For example, empirical models for nuclear proteins (WAG) and for mitochondrial proteins (MTMAM) have very different exchange rates between arginine (R) and lysine (K). These two amino acids are chemically similar and exchange frequently in nuclear proteins, as they can

Table 2.1 The universal genetic code

Phe F	TTT	Ser S	TCT	Tyr Y	TAT	Cys C	TGT
	TTC		TCC		TAC		TGC
Leu L	TTA		TCA	End *	TAA	End *	TGA
	TTG		TCG		TAG	Trp W	TGG
Leu L	CTT	Pro P	CCT	His H	CAT	Arg R	CGT
	CTC		CCC		CAC		CGC
	CTA		CCA	Gln Q	CAA		CGA
	CTG		CCG		CAG		CGG
Ile I	ATT	Thr T	ACT	Asn N	AAT	Ser S	AGT
	ATC		ACC		AAC		AGC
	ATA		ACA	Lys K	AAA	Arg R	AGA
Met M	ATG		ACG		AAG		AGG
Val V	GTT	Ala A	GCT	Asp D	GAT	Gly G	GGT
	GTC		GCC		GAC		GGC
	GTA		GCA	Glu E	GAA		GGA
	GTG		GCG		GAG		GGG

reach each other through one codon position change. However, they rarely exchange in mitochondrial proteins as the codons differ at two or three positions in the mitochondrial code (Adachi and Hasegawa 1996a). Similarly, arginine (R) exchanges, albeit at low rates, with methionine (M), isoleucine (I), and threonine (T) in nuclear proteins but such exchanges are virtually absent in mitochondrial proteins. Furthermore, the two factors may be operating at the same time. When codons were assigned to amino acids during the origin and evolution of the genetic code, error minimization appears to have played a role so that amino acids with similar chemical properties tend to be assigned codons close to one another in the code (e.g. Osawa and Jukes 1989; Freeland and Hurst 1998).

Empirical amino acid substitution matrices are also used in alignment of multiple protein sequences. Cost (weight) matrices are used to penalize mismatches, with heavier penalties applied to rarer changes. The penalty for a mismatch between amino acids i and j is usually defined as $-\log\{p_{ij}(t)\}$, where $p_{ij}(t)$ is the transition probability between i and j, and t measures the sequence divergence or branch length. After Dayhoff et al. calculated PAM matrices for different sequence distances, such as PAM_{100} for $t = 1$ substitution per site, and PAM_{250} for $t = 2.5$, a number of score matrices have been published, including the BLOSSUM matrix (Henikoff and Henikoff 1992), the VT matrix (Müller and Vingron 2000), and the Gonnet matrix (Gonnet et al. 1992). Such matrices are useful for multiple sequence alignment but are in general too crude for use in phylogenetic analysis.

2.2.2 Mechanistic models

Yang et al. (1998) implemented a few mechanistic models of amino acid substitution. They are formulated at the level of codons and explicitly model the biological processes involved, i.e. different mutation rates between nucleotides, translation of the codon triplet into amino acid, and acceptance or rejection of the amino acid due to selective pressure on the protein. We will discuss models of codon substitution later in this chapter. Such codon-based models of amino acid substitution may be called *mechanistic*. Yang et al. (1998) presented an approach for constructing a Markov process model of amino acid replacement from a model of codon substitution, by aggregating synonymous codons into one state (the encoded amino acid). Analysis of the mitochondrial genomes of 20 mammalian species suggests that the mechanistic models fit the data better than the empirical models such as those of Dayhoff (Dayhoff et al. 1978) and JTT (Jones et al. 1992). Some of the mechanistic models implemented incorporate physico-chemical properties of amino acids (such as size and polarity), by assuming that dissimilar amino acids have lower exchange rates. While the use of such chemical properties improved the fit of the models, the improvement was not extraordinary, perhaps reflecting our poor understanding of which of the many chemical properties are most important and how they affect amino acid substitution rates. As Zuckerkandl and Pauling (1965) remarked, 'apparently chemists and protein molecules do not share the same opinions regarding the definition of the most prominent properties of a residue'.

2.2.3 Among-site heterogeneity

The selective pressure is expected to vary across sites or regions of a protein as they perform different roles in the protein structure and function. The simplest example of such among-site heterogeneity is variable substitution rates. Empirical models of amino acid substitution can be combined with the gamma model of rates among sites (Yang 1993, 1994a), leading to models such as Dayhoff + Γ, JTT + Γ, etc., with a suffix '+Γ'.

Table 2.2 Maximum likelihood estimates of the gamma shape parameter α from a few datasets

Data	$\hat{\alpha}$	Refs
DNA sequences		
1063 human and chimpanzee Mitochondrial D-loop HVI sequences	0.42–0.45	(Excoffier and Yang 1999)
SSU rRNA from 40 species (Archaea, Bacteria, and Eukaryotes)	0.60	(Galtier 2001)
LSU rRNA from 40 species (Archaea, Bacteria, and Eukaryotes)	0.65	(Galtier 2001)
13 hepatitis B viral genomes	0.26	(Yang et al. 1995b)
Protein sequences		
6 nuclear proteins (APO3, ATP7, BDNF, CNR1, EDG1, ZFY) from 46 mammalian species	0.12–0.93	(Pupko et al. 2002b)
4 nuclear proteins (A2AB, BRCA1, IRBP, vmF) from 28 mammalian species	0.29–3.0	(Pupko et al. 2002b)
12 mitochondrial proteins concatenated from 18 mammalian species	0.29	(Cao et al. 1999)
45 chloroplast proteins concatenated from 10 plants and cynobacteria species	0.56	(Adachi et al. 2000)

Note: For nucleotide sequences, the substitution models HKY85 + Γ or GTR + Γ were assumed, which account for different transition and transversion rates and different nucleotide frequencies. For amino acid models, JTT + Γ, MTREV + Γ, or CPREV + Γ were assumed. Estimates under the 'G+I' models, which incorporates a proportion of invariable sites in addition to the gamma distribution of rates for sites, are not included in this table, as parameter α has different interpretations in the 'Γ' and 'I+Γ' models.

This is the same gamma model as discussed in §1.3, and assumes that the *pattern* of amino acid substitution is the same among all sites in the protein, but the rate is variable. The rate matrix for a site with rate r is rQ, with the same Q shared for all sites. The shape parameter α of the gamma distribution measures how variable the rates are, with small αs indicating strong rate variation. Table 2.2 provides a small sample of estimates of α from real data. While estimates from much larger datasets now exist, those in the table are still representative. For example, in nearly every functional protein, the shape parameter α is less than 1, indicating strong rate variation among sites.

Besides the rates, the pattern of amino acid substitution represented by the matrix of relative substitution rates may also vary among sites. For example, different sites in the protein may prefer different amino acids. Bruno (1996) described an amino acid model in which a set of amino acid frequency parameters are used for each site in the sequence. This model involves too many parameters for use in the ML analysis but may be useful in Bayesian inference (see §8.2 in Chapter 8 later). Thorne et al. (1996) and Goldman et al. (1998) described models that allow for a few classes of sites to evolve under different Markov chain models with different rate matrices. Such site classes may correspond to secondary structural categories in the protein. Similar models were described by Koshi and Goldstein (1996b) and Koshi et al. (1999). Fitting such models requires joint analysis of multiple sequences on a phylogeny.

2.3 Estimation of distance between two protein sequences

2.3.1 *The Poisson model*

If every amino acid has the same rate λ of changing into any other amino acid, the number of substitutions over time t will be a Poisson-distributed variable (Bishop and Friday

1985, 1987). This is the amino acid equivalent of the Jukes and Cantor (1969) model for nucleotide substitution. The sequence distance, defined as the expected number of amino acid substitutions per site, is then $d = 19\lambda t$, where t is the total time separating the two sequences (i.e. twice the time of divergence). We let $\lambda = 1/19$ so that the substitution rate of each amino acid is 1 and then time t is measured by distance. We thus use t and d interchangeably. Suppose x out of n sites are different between the two protein sequences, with the proportion $\hat{p} = x/n$. The maximum likelihood estimate (MLE) of distance t is then

$$\hat{t} = -\frac{19}{20} \log\left(1 - \frac{20}{19}\hat{p}\right). \tag{2.3}$$

The variance of $\hat{t}$ can be derived similarly to that under the JC69 model for nucleotides (see §1.2.1).

2.3.2 Empirical models

Under empirical models such as Dayhoff, JTT, or WAG, different amino acids have different substitution rates. The Q matrix representing the relative substitution rates is assumed to be known, so that the only parameter to be estimated is the sequence distance d. It is a common practice to multiply Q by a scale constant so that the average rate is $-\sum_i \pi_i q_{ii} = 1$. Then $t = d$. Under empirical models Dayhoff + F, JTT + F, etc., amino acid frequencies in the observed data are used to replace the equilibrium frequencies in the empirical model. Then again only the distance t needs to be estimated.

It is straightforward to use ML to estimate t, using the same procedure for calculating the distance under the GTR model for nucleotides (see §1.5.3). Let n_{ij} be the number of sites occupied by amino acids i and j in the two sequences. The log likelihood function is

$$\ell(t) = \sum_i \sum_j n_{ij} \log\{f_{ij}(t)\} = \sum_i \sum_j n_{ij} \log\{\pi_i p_{ij}(t)\}, \tag{2.4}$$

where $f_{ij}(t)$ is the probability of observing a site with amino acids i and j in the two sequences, π_i is the equilibrium frequency of amino acid i, and $p_{ij}(t)$ is the transition probability. This is equation (1.72). The one-dimensional optimization problem can be easily managed numerically. Calculation of the transition probability matrix $P(t) = \{p_{ij}(t)\}$ is discussed in §2.6.

An alternative approach is to estimate t by matching the observed proportion of different sites between the two sequences with the expected proportion under the model

$$p = \sum_{i \neq j} f_{ij}(t) = \sum_{i \neq j} \pi_i p_{ij}(t) = \sum_i \pi_i (1 - p_{ii}(t)) = 1 - \sum_i \pi_i p_{ii}(t). \tag{2.5}$$

Figure 2.2 shows the relationship between t and p for several commonly used models. As $f_{ij}(t)$ is not the same for all $i \neq j$, this approach differs from the MLE. They should nevertheless be very similar if the empirical model is not too wrong.

2.3.3 Gamma distances

If the rates vary according to the gamma distribution with given shape parameter α, and the relative rates are the same between any two amino acids, the sequence distance becomes, instead of equation (2.3),

$$\hat{t} = \frac{19}{20}\alpha\left[\left(1 - \frac{20}{19}\hat{p}\right)^{-1/\alpha} - 1\right], \tag{2.6}$$

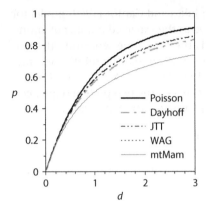

Fig. 2.2 The expected proportion of different sites (p) between two sequences separated by time or distance d under different models. The models are, from top to bottom, Poisson, WAG (Whelan and Goldman 2001), JTT (Jones et al. 1992), Dayhoff (Dayhoff et al. 1978), and MTMAM (Yang et al. 1998). Note that the results for WAG, JTT, and Dayhoff are almost identical.

Table 2.3 Distance estimates between the cat and rabbit p53 proteins

model	One rate ($\alpha = \infty$)	Gamma rates ($\alpha = 0.5$)
Poisson	0.191 ± 0.024	0.234 ± 0.035
WAG	0.191 ± 0.024	0.237 ± 0.037

where $\hat{p}$ is the proportion of different sites. This gamma distance under the Poisson model is very similar to the gamma distance under the JC69 model for nucleotides (equation (1.37)).

For empirical models such as Dayhoff, one can use ML to estimate the sequence distance under the gamma model. The theory discussed in §1.3 for nucleotides can be implemented for amino acids in a straightforward manner.

Example 2.1. Distance between cat and rabbit p53 protein sequences. We calculate the distance between the amino acid sequences of the tumour suppressor protein p53 from the cat (*Felis catus*, GenBank accession number D26608) and the rabbit (*Oryctolagus cuniculus*, X90592). There are 386 and 391 amino acids in the cat and rabbit sequences, respectively. We delete alignment gaps, leaving 382 sites in the sequence. Of these, 66 sites ($\hat{p} = 17.3\%$ of all sites) are different. The distance estimates under the Poisson and WAG models are summarized in Table 2.3, assuming either the same rate for all sites or gamma-distributed rates among sites with shape parameter $\alpha = 0.5$. It is obvious that the among-site rate variation has much greater impact on distance estimation than the empirical substitution rate matrix. □

2.4 Models of codon substitution

2.4.1 *The basic model*

Markov chain models of codon substitution were proposed by Goldman and Yang (1994) and Muse and Gaut (1994). The codon triplet is considered the unit of evolution, and a Markov chain is used to describe substitutions from one codon to another. The state space of the chain is the sense codons in the genetic code (i.e. 61 sense codons in the universal code or 60 in the vertebrate mitochondrial code). Stop codons are not allowed inside a functional protein and are not considered in the chain. The model in common

2.4 MODELS OF CODON SUBSTITUTION

use is a simplified version of the model of Goldman and Yang (1994), and incorporates three features of sequence evolution: the transition/transversion rate ratio κ, the nonsynonymous/synonymous rate ratio ω, and different codon frequencies (π_J for codon J). Here and in the next subsection, we use the upper-case letters I and J to refer to the codon triplets and the lower-case letters i and j to the nucleotides. The instantaneous rate of substitution from codons I to J is specified as

$$q_{IJ} = \begin{cases} 0, & \text{if } I \text{ and } J \text{ differ at two or three codon positions,} \\ \pi_J, & \text{if } I \text{ and } J \text{ differ by a synonymous transversion,} \\ \kappa\pi_J, & \text{if } I \text{ and } J \text{ differ by a synonymous transition,} \\ \omega\pi_J, & \text{if } I \text{ and } J \text{ differ by a nonsynonymous transversion,} \\ \omega\kappa\pi_J, & \text{if } I \text{ and } J \text{ differ by a nonsynonymous transition} \end{cases} \quad (2.7)$$

(Nielsen and Yang 1998; Yang and Nielsen 1998). Mutations are assumed to occur independently at the three codon positions, so that simultaneous changes at two or three positions are expected to occur at negligible rates. Parameters κ and π_J characterize processes (such as mutational biases and selective constraints) at the DNA level, while ω characterizes natural selection on the protein level. Simply, $\omega = 1$ represents neutral protein evolution in which synonymous and nonsynonymous mutations are fixed at the same rate, $\omega < 1$ represents purifying selection removing nonsynonymous mutations, and $\omega > 1$ represents positive selection accelerating the fixation of nonsynonymous mutations.

Different assumptions can be made concerning the equilibrium codon frequency π_J. The simplest model assumes that each codon has the same frequency (called Fequal). The F1 × 4 model calculates the codon frequencies using the frequencies of the four nucleotides, with three free parameters used. Suppose the frequency of nucleotide j is π_j^*, the equilibrium frequency of codon $J = j_1 j_2 j_3$ is then

$$\pi_J = \frac{1}{C} \pi_{j_1}^* \pi_{j_2}^* \pi_{j_3}^*, \quad (2.8)$$

where the scale factor C is to ensure that π_J sum over the sense codons to 1. The F3 × 4 model uses three sets of nucleotide frequencies for the three codon positions, with nine free parameters (F3 × 4). The most parameter-rich model (F61) uses all codon frequencies as parameters with the constraint that their sum is 1.

Table 2.4 Substitution rates to the same target codon CTA (Leu)

Substitution	Relative rate	Substitution type
TTA (Leu) → CTA (Leu)	$q_{TTA,CTA} = \kappa\pi_{CTA}$	Synonymous transition
ATA (Ile) → CTA (Leu)	$q_{ATA,CTA} = \omega\pi_{CTA}$	Nonsynonymous transversion
GTA (Val) → CTA (Leu)	$q_{GTA,CTA} = \omega\pi_{CTA}$	Nonsynonymous transversion
CTT (Leu) → CTA (Leu)	$q_{CTT,CTA} = \pi_{CTA}$	Synonymous transversion
CTC (Leu) → CTA (Leu)	$q_{CTC,CTA} = \pi_{CTA}$	Synonymous transversion
CTG (Leu) → CTA (Leu)	$q_{CTG,CTA} = \kappa\pi_{CTA}$	Synonymous transition
CCA (Pro) → CTA (Leu)	$q_{CCA,CTA} = \kappa\omega\pi_{CTA}$	Nonsynonymous transition
CAA (Gln) → CTA (Leu)	$q_{CAA,CTA} = \omega\pi_{CTA}$	Nonsynonymous transversion
CGA (Arg) → CTA (Leu)	$q_{CGA,CTA} = \omega\pi_{CTA}$	Nonsynonymous transversion

Note: Instantaneous rates from all other codons to CTA are 0.

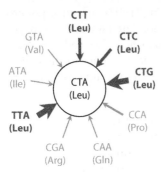

Fig. 2.3 Substitution rates to the same target codon CTA from its nine neighbours. Neighbours are codons that differ from the concerned codon at only one position. Some codons have fewer than nine neighbours as changes to and from stop codons are disallowed. The thicknesses of the arrows represent different rates. The diagram is drawn using $\kappa = 2$ and $\omega = 1/3$, so that there are four different rates to codon CTA, in proportions $1:2:3:6$, for nonsynonymous transversion, nonsynonymous transition, synonymous transversion and synonymous transition, respectively (see equation (2.7)).

As an example, the substitution rates to the same target codon J = CTA (Leu) are given in Table 2.4 and illustrated in Figure 2.3 (see the genetic code in Table 2.1).

It is easy to confirm that the Markov chain specified by the rate matrix $Q = \{q_{IJ}\}$ (equation (2.7)) satisfies the detailed-balance condition (equation (2.1)) for time-reversible chains. To relate the model to sequence data, we need calculate the transition probability matrix $P(t) = \{p_{IJ}(t)\} = e^{Qt}$. Numerical algorithms for this calculation are discussed in §2.6.

2.4.2 Variations and extensions

The basic codon models (Goldman and Yang 1994; Muse and Gaut 1994) have been improved and extended in numerous ways in the last two decades. The book edited by Cannarozzi and Schneider (2012) summarizes recent developments. Here, a summary of the major extensions in the formulation of the Q matrix is provided. The model of Muse and Gaut (1994) and the mutation-selection model of Yang and Nielsen (2008) are described, and a few other variations are discussed. Extensions of the codon models to detect positive selection during protein evolution are discussed in Chapter 11. Note also that those models are almost always fitted to multiple sequences on a phylogeny (rather than a pair of sequences), and methods for fitting such models are not discussed until §4.2 later in Chapter 4.

The Muse–Gaut model. We focus on the rate between two codons $I = i_1 i_2 i_3$ and $J = j_1 j_2 j_3$ that differ at only one position, say position k, with $i_k \neq j_k$. Muse and Gaut (1994) specify the substitution rate as

$$q_{IJ} = \begin{cases} 0, & \text{if } I \text{ and } J \text{ differ at two or three codon positions,} \\ \pi^*_{j_k}, & \text{if } i_k \text{ and } j_k \text{ is a synonymous difference,} \\ \omega \pi^*_{j_k}, & \text{if } i_k \text{ and } j_k \text{ is a nonsynonymous difference,} \end{cases} \quad (2.9)$$

where $\pi^*_{j_k}$ is the frequency of target nucleotide j_k at position k. For instance, the rate from I = TCA to J = TCG is $q_{IJ} = \pi^*_G$. This can be rewritten as $q_{IJ} = \left[\frac{1}{\pi^*_T \pi^*_C}\right] \times (\pi^*_T \pi^*_C \pi^*_G)$. Note that the quantity in square brackets is identical between q_{IJ} and q_{JI} and the quantity in parentheses depends on J but not I. The model thus satisfies the detailed-balance condition (equation (2.2)) and is time-reversible, with the equilibrium distribution

$$\pi_J \propto \pi^*_{j_1} \pi^*_{j_2} \pi^*_{j_3}. \quad (2.10)$$

The mutation-selection (FMutSel) model. The FMutSel model (Yang and Nielsen 2008) explicitly accounts for mutational bias and selection on codon usage. The mutation rate per generation from nucleotides i to j is $\mu_{ij} = a_{ij}\pi_j^*$, with $a_{ij} = a_{ji}$ for all $i \neq j$. Here π_j^* is not really the frequency of nucleotide j, but instead reflects mutation bias; if π_T^* is large, mutations are biased towards T. We scale those parameters so that $\sum \pi_j^* = 1$. This is the time-reversible (GTR or REV) mutation model. For the special case of the HKY85 model, we have $a_{ij} = \kappa$ if i and j differ by a transition, and $a_{ij} = 1$ if i and j differ by a transversion.

We then model selection on codon usage by introducing a fitness parameter f_I for codon I (Akashi 1994, 1995; Xia 1998). The selection coefficient for the mutation that changes the 'wildtype' codon I into a new mutant codon J is thus $s_{IJ} = f_J - f_I$. The probability of fixation of the mutation is $\frac{2s_{IJ}}{1-e^{-2Ns_{IJ}}}$, where N is the effective chromosomal population size (Fisher 1930b; Wright 1931; Kimura 1957). Let $F_I = 2Nf_I$ be the scaled fitness of codon I, and $2Ns_{IJ} = 2N(f_J - f_I) = F_J - F_I$ will be the scaled selection coefficient for the $I \to J$ mutation. Note that the substitution rate from codons I to J is the number of $I \to J$ mutations per generation $(N\mu_{i_k j_k})$, multiplied by the fixation probability of the mutation. Thus

$$q_{IJ} = \begin{cases} 0, & \text{if } I \text{ and } J \text{ differ at two or three codon positions,} \\ a_{i_k j_k} \pi_{j_k}^* \frac{F_J - F_I}{1 - e^{F_I - F_J}}, & \text{if } i_k \text{ and } j_k \text{ differ by a synonymous difference,} \\ \omega a_{i_k j_k} \pi_{j_k}^* \frac{F_J - F_I}{1 - e^{F_I - F_J}}, & \text{if } i_k \text{ and } j_k \text{ differ by a nonsynonymous difference.} \end{cases} \quad (2.11)$$

One of the F_Js is redundant and is set to 0.

As an example, under the HKY85 mutation model, the rate from $I = \text{TCA}$ to $J = \text{TCG}$ is $q_{IJ} = \kappa \pi_G^* \frac{F_{TCG} - F_{TCA}}{1 - e^{F_{TCA} - F_{TCG}}}$. This can be rewritten as $q_{IJ} = \left[\kappa \times \frac{1}{\pi_T^* \pi_C^*} \times \frac{F_{TCG} - F_{TCA}}{e^{F_{TCG}} - e^{F_{TCA}}} \right] \times (\pi_T^* \pi_C^* \pi_G^* \times e^{F_{TCG}})$. Note that the quantity in square brackets is identical between q_{IJ} and q_{JI} and the quantity in parentheses depends on J but not I. Thus the model is time-reversible, with the equilibrium distribution

$$\pi_J \propto \pi_{j_1}^* \pi_{j_2}^* \pi_{j_3}^* \times e^{F_J}. \quad (2.12)$$

This clearly reflects the joint effect of mutational bias $(\pi_{j_1}^* \pi_{j_2}^* \pi_{j_3}^*)$ and natural selection (e^{F_J}).

If the codon frequencies (π_J) are estimated using the observed frequencies in the sequences, this model will add three parameters $(\pi_T^*, \pi_A^*, \pi_G^*)$, and the estimation is stable.

The Muse and Gaut model (equation (2.9)) is a special case of equation (2.11), with all codons having the same fitness so that $s_{IJ} = 0$ for every $I \neq J$. Note that $x/(1 - e^{-x}) \to 1$ when $x \to 0$.

Yap et al. (2010) emphasized the importance of the context effect of DNA sequence evolution (i.e. the frequency of dinucleotide TC differs from the predicted frequency based on mono-nucleotide frequencies π_T and π_C). The model constructed by Yap et al. has some similarities to the FMutSel model but lacks its interpretability.

The ω ratio and selection on the protein. It is clearly too simplistic to use a single ω ratio to characterize selection at the protein level. Goldman and Yang (1994) and Yang et al. (1998) used physico-chemical distances between amino acids to modify the nonsynonymous substitution rates, with the expectation that chemically dissimilar amino acids exchange with each other less often than similar amino acids. Two such distances are well known, based on physico-chemical properties such as composition, polarity, volume, etc. (Grantham 1974; Miyata et al. 1979). Yang et al. (1998) used the form

$$\omega_{IJ} = ae^{-bd_{IJ}}, \qquad (2.13)$$

where d_{IJ} is the chemical distance between the amino acids encoded by codons I and J, and where a and b are parameters to be estimated. Goldman and Yang (1994) assumed $a = 1$, but this was found to be a very poor assumption: estimates of a from real data are much less than 1, indicating that the amino acid change may have other impacts on the folding and function of the protein, even if the concerned amino acids have similar chemical and physical properties (Yang et al. 1998). While use of the amino acid chemical distances leads to significant improvement in the model's fit to the data, the improvement is not extraordinary. It is possible that the selective pressure and the preference for different amino acids vary widely among different domains of the protein so that use of one relationship between ω and d for the whole protein is unrealistic. In this regard, one should note that it is not logically sound to use empirical amino acid substitution matrices such as the Dayhoff matrix to modify the codon substitution rates in a codon model (as in Doron-Faigenboim and Pupko 2007). The empirical matrix reflects the mutational distances between amino acids in the genetic code, whereas mutational distances are already explicitly accommodated in the codon model.

One can also estimate different ω ratios for a few pre-specified types of nonsynonymous substitutions, for example, for those involving conserved or radical amino acid changes (Zhang 2000). In the extreme, one can estimate a separate ω parameter for every pair of amino acids that can be reached by one codon position change, with 75 or 70 ωs estimated for the universal code and the vertebrate mitochondrial code, respectively (Yang et al. 1998). Figure 2.1c and d plot estimates from such an analysis using nuclear and mitochondrial genomic data from primates (dos Reis et al. 2012, 2013a). Note that the ω ratios (the acceptance rates) are very different from the amino acid exchangeabilities (Figure 2.1a and b). Two amino acids can have a high exchangeability because they can reach each other by one codon position change, because the involved nucleotide mutation is a transition, or because the nonsynonymous mutation has a high acceptance rate.

Empirical codon models and double and triple mutations. The models of Goldman and Yang (1994) and Muse and Gaut (1994) assume that the substitution rate between codons I and J that differ at two or three positions is zero. Note that this means that two or three positions should not change instantaneously, but for any positive time t, such changes are allowed, as $p_{IJ}(t) > 0$. There may be two scenarios in which this assumption is unrealistic. First, some complex mutations may involve two or more nucleotide sites. Second, the fixation of a mutation at one codon position may be followed quickly by the occurrence and fixation of another mutation at a neighbouring position of the same codon, perhaps aided by selection fixing compensatory mutations (Savill et al. 2001). In between-species comparisons, the successive fixations of distinct mutations will appear as simultaneous substitutions. How often such compensatory substitutions occur is unclear. In this regard, 'synonymous' differences between the two sets of serine codons (TCN and AGY) which cannot exchange by a single nucleotide mutation appear quite common in real gene sequence alignments, and it is unclear whether they are the product of complex mutations involving two or three nucleotides or of two nonsynonymous point mutations (via either TGY for cysteine or ACN for threonine). Brenner (1988) found that both sets of serine codons are used to code for serine in an otherwise highly conserved segment of serine protease. Based on the exon–intron structure of the encoding genes, he argued that the different serine codons are the result of separate lines of descent (possibly having evolved from precursor cysteine codons) rather than double or triple mutations.

Yang et al. (1998) used amino acid sequence data to fit the general time-reversible (REV/GTR) model in comparison with a simpler model (REV0) that allows one-step amino acid changes only (i.e. between amino acids that can interchange by a single nucleotide mutation). While REV fitted the data significantly better, the log likelihood difference was greatly reduced when the among-site rate variation was taken into account. It seemed as if some other aspects of the model were violated.

Schneider et al. (2005) fitted an empirical matrix of codon substitution to many alignments of vertebrate genes, using a similar counting method to that Dayhoff et al. (1978) used to derive empirical amino acid substitution matrices. Kosiol et al. (2007) used ML to fit an empirical codon model, assuming the REV/GTR model for the 61 states, with 1890 parameters in the rate matrix. The estimation was achieved using the expectation–maximization algorithm of Siepel and Haussler (2004).

Zoller and Schneider (2010) performed a principal component analysis of 3,666 empirical codon substitution rate matrices estimated from gene families. The two most significant factors they found to influence substitution rates are the rate difference between synonymous and nonsynonymous substitutions and the rate difference between single and multiple substitutions. The results suggest that the initial choices made by Goldman and Yang (1994) and Muse and Gaut (1994) to accommodate the nonsynonymous–synonymous rate ratio and to disallow double or triple substitutions were largely sound. A drawback of the empirical codon models is that they are hard to interpret. They do not appear to teach us much about the process of gene sequence evolution even if they achieve better fit to the sequence data than parametric models with only a few parameters.

2.5 Estimation of d_S and d_N

Two distances are usually calculated between protein-coding DNA sequences, for synonymous and nonsynonymous substitutions, respectively. They are defined as the number of synonymous substitutions per synonymous site (d_S or K_S) and the number of nonsynonymous substitutions per nonsynonymous site (d_N or K_A). Two classes of methods have been developed for estimating d_S and d_N: the heuristic counting method and the ML method.

2.5.1 Counting methods

The counting method proceeds similarly to distance calculation under nucleotide substitution models such as JC69. The only difference is that now a distinction is made between the synonymous and nonsynonymous types when sites and differences are counted. The methods involve three steps:

i. Count synonymous and nonsynonymous *sites*;
ii. Count synonymous and nonsynonymous *differences*;
iii. Calculate the proportions of differences and correct for multiple hits.

The first counting methods were developed in the early 1980s shortly after DNA sequencing techniques were invented (Miyata and Yasunaga 1980; Perler et al. 1980). Miyata and Yasunaga (1980) assumed a simple mutation model with equal rates between nucleotides (as in JC69) when counting sites and differences, and used amino acid chemical distances developed by Miyata et al. (1979) to weight evolutionary pathways when counting differences between codons that differ at two or three positions (see below). The method was simplified by Nei and Gojobori (1986), who abandoned the weighting scheme and used equal weighting instead. Li et al. (1985) pointed out the importance of the transition and transversion rate difference, and dealt with it by partitioning codon positions into different degeneracy classes. Their procedure was then improved by Li (1993), Pamilo and Bianchi (1993), Comeron (1995), and Ina (1995). Moriyama and Powell (1997) discussed the influence of unequal codon usage, which implies that the substitution rates are not symmetrical between codons, as assumed by the methods mentioned above. Yang and Nielsen (2000) accommodated both the transition–transversion rate difference and unequal codon frequencies in an iterative algorithm.

Here we describe the method of Nei and Gojobori (1986), referred to as NG86, to illustrate the basic concepts. This is a simplified version of the method of Miyata and Yasunaga (1980) and is similar to the JC69 model of nucleotide substitution. We then discuss the effects of complicating factors such as the unequal transition and transversion rates and unequal codon usage. It is worth noting that estimation of d_S and d_N is a complicated exercise, as manifested by the fact that even the simple transition–transversion rate difference is nontrivial to deal with. Unfortunately, different methods often produce very different estimates.

2.5.1.1 *Counting sites*

Each codon has three nucleotide sites, which are divided into synonymous and nonsynonymous categories. Take the codon TTT (Phe) as an example. As each of the three codon positions can change into three other nucleotides, the codon has nine immediate neighbours: TTC (Phe), TTA (Leu), TTG (Leu), TCT (Ser), TAT (Tyr), TGT (Cys), CTT (Leu), ATT (Ile), and GTT (Val). Out of these, codon TTC codes for the same amino acid as the original codon (TTT). Thus there are $3 \times 1/9 = 1/3$ synonymous sites and $3 \times 8/9 = 8/3$ nonsynonymous sites in codon TTT (Table 2.5). Mutations into stop codons are disallowed during the counting. We apply the procedure to all codons in sequence 1 and sum up

Table 2.5 Counting sites in codon TTT (Phe)

Target codon	Mutation type	Rate ($\kappa = 1$)	Rate ($\kappa = 2$)
TTC (Phe)	Synonymous	1	2
TTA (Leu)	Nonsynonymous	1	1
TTG (Leu)	Nonsynonymous	1	1
TCT (Ser)	Nonsynonymous	1	2
TAT (Tyr)	Nonsynonymous	1	1
TGT (Cys)	Nonsynonymous	1	1
CTT (Leu)	Nonsynonymous	1	2
ATT (Ile)	Nonsynonymous	1	1
GTT (Val)	Nonsynonymous	1	1
Sum		9	12
# syn. Sites		1/3	1/2
# nonsyn. sites		8/3	5/2

Note: κ is the transition/transversion rate ratio.

the counts to obtain the total numbers of synonymous and nonsynonymous sites in the whole sequence. We then repeat the process for sequence 2, and average the numbers of sites between the two sequences. Let those be S and N, with $S + N = 3 \times L_c$, where L_c is the number of codons in the sequence.

2.5.1.2 Counting differences

The second step is to count the numbers of synonymous and nonsynonymous *differences* between the two sequences. In other words, the observed differences between the two sequences are partitioned into the synonymous and nonsynonymous categories. We proceed again codon by codon. This is straightforward if the two compared codons are identical (e.g. TTT vs. TTT), in which case the numbers of synonymous and nonsynonymous differences are 0, or if they differ at one codon position only (e.g. TTC vs. TTA), in which case it is obvious whether the single difference is synonymous or nonsynonymous. However, when the two codons differ at two or three positions (e.g. CCT vs. CAG or GTC vs. ACT), there exist two or six evolutionary pathways from one codon to the other. The multiple pathways may involve different numbers of synonymous and nonsynonymous differences. Most counting methods give equal weights to the different pathways.

For example, there are two pathways between codons CCT and CAG (Table 2.6). The first goes through the intermediate codon CAT and involves two nonsynonymous differences, while the second goes through CCG and involves one synonymous and one nonsynonymous difference. If we apply equal weights to the two pathways, there are 0.5 synonymous differences and 1.5 nonsynonymous differences between the two codons. If the synonymous rate is higher than the nonsynonymous rate, as is the case in almost all genes, the second pathway should be more likely than the first. However, without knowing the d_N/d_S ratio and the sequence divergence beforehand, it is difficult to weight the pathways appropriately. Nevertheless, weighting is expected to have minor effects on the estimates if the sequences are not very divergent (Nei and Gojobori 1986).

The counting is done codon by codon across the sequence, and the numbers of differences are summed to produce the total numbers of synonymous and nonsynonymous differences between the two sequences. Let these be S_d and N_d, respectively.

2.5.1.3 Correcting for multiple hits

We now have

$$p_S = S_d/S,$$
$$p_N = N_d/N \tag{2.14}$$

to be the proportions of differences at the synonymous and nonsynonymous sites, respectively. These are equivalent to the proportion of differences under the JC69 model for nucleotides. Thus we apply the JC69 correction for multiple hits

Table 2.6 Two pathways between codons CCT and CAG

Pathway	Differences	
	Synonymous	Nonsynonymous
CCT (Pro) ↔ CAT (His) ↔ CAG (Gln)	0	2
CCT (Pro) ↔ CCG (Pro) ↔ CAG (Gln)	1	1
Average	0.5	1.5

2 MODELS OF AMINO ACID AND CODON SUBSTITUTION

$$d_S = -\frac{3}{4} \log\left(1 - \frac{4}{3} p_S\right),$$
$$d_N = -\frac{3}{4} \log\left(1 - \frac{4}{3} p_N\right). \quad (2.15)$$

As pointed out by Lewontin (1989), this step is logically flawed. The JC69 formula is suitable for noncoding regions and assumes that any nucleotide can change into *three* other nucleotides with equal rates. When we focus on synonymous sites and differences only, each nucleotide does not have *three* other nucleotides to change into. In practice, the effect of multiple-hit correction is minor, at least at low sequence divergences, so that the bias introduced by the correction formula is not very important.

2.5.1.4 *Application to the rbcL genes*

We apply the NG86 method to estimate d_S and d_N between the cucumber and tobacco genes for the plastid protein ribulose-1,5-bisphosphate carboxylase/oxygenase large subunit (*rbcL*). The GenBank accession numbers are NC_007144 for the cucumber (*Cucumis sativus*) and Z00044 for the tobacco (*Nicotiana tabacum*). There are 476 and 477 codons in the cucumber and tobacco genes, respectively, with 481 codons in the alignment. We delete codons that are alignment gaps in either species, leaving 472 codons in the sequence.

A few basic statistics from the data are listed in Table 2.7, obtained by applying the nucleotide-based model HKY85 (Hasegawa et al. 1985) to analyse the three codon positions separately. The base compositions are unequal, and the third codon positions are AT-rich. The estimates of the transition/transversion rate ratio for the three codon positions are in the order $\hat{\kappa}_3 > \hat{\kappa}_1 > \hat{\kappa}_2$. Estimates of the sequence distance are in the same order $\hat{d}_3 > \hat{d}_1 > \hat{d}_2$. Such patterns are common for protein-coding genes, and reflect the structure of the genetic code and the fact that essentially all proteins are under selective constraint, with higher synonymous than nonsynonymous substitution rates. When the genes are examined codon by codon, 345 codons are identical between the two species, while 115 differ at one position, 95 of which are synonymous and 20 nonsynonymous. Ten codons differ at two positions, and two codons differ at all three positions.

We then apply the NG86 method. The 1,416 nucleotide sites are partitioned into $S = 343.5$ synonymous sites and $N = 1072.5$ nonsynonymous sites. There are 141 differences observed between the two sequences, which are partitioned into $S_d = 103.0$ synonymous differences and $N_d = 38.0$ nonsynonymous differences. The proportions of differences at the synonymous and nonsynonymous sites are thus $p_S = S_d/S = 0.300$ and $p_N = N_d/N = 0.035$. Application of the JC69 correction gives $d_S = 0.383$ and $d_N = 0.036$, with the ratio

Table 2.7 Basic statistics for the cucumber and tobacco *rbcL* genes

Position	Sites	π_T	π_C	π_A	π_G	$\hat{\kappa}$	$\hat{d}$
1	472	0.179	0.196	0.239	0.386	2.202	0.057
2	472	0.270	0.226	0.299	0.206	2.063	0.026
3	472	0.423	0.145	0.293	0.139	6.901	0.282
All	1416	0.291	0.189	0.277	0.243	3.973	0.108

Note: Base frequencies are observed frequencies at the three codon positions, averaged over the two sequences. These are very close to MLEs under the HKY85 model (Hasegawa et al. 1985). The transition/transversion rate ratio κ and sequence distance *d* are estimated under the HKY85 model.

$\hat{\omega} = d_N/d_S = 0.095$. According to this estimate, the protein is under strong selective constraint, and a nonsynonymous mutation has only 9.5% the chance of a synonymous mutation of spreading through the population.

2.5.1.5 Transition–transversion rate difference and codon usage

From the structure of the genetic code, we can see that transitions at the third codon positions are more likely to be synonymous than transversions are. Thus, one may observe many synonymous substitutions, not because natural selection has removed nonsynonymous mutations but because transitions occur at higher rates than transversions, producing many synonymous mutations. Ignoring the transition–transversion rate difference thus leads to underestimation of the number of synonymous sites (S) and overestimation of the number of nonsynonymous sites (N), resulting in overestimation of d_S and underestimation of d_N (Li et al. 1985).

To accommodate the different transition and transversion rates, Li et al. (1985) classified each nucleotide site, i.e. each position in a codon, into *nondegenerate, two-fold degenerate*, and *four-fold degenerate* classes. The degeneracy of a codon position is determined by how many of the three possible mutations are synonymous. At a four-fold degenerate site (e.g. the third position of CCT, for Pro), every change is synonymous, while at the nondegenerate site (e.g. the first position of GCT, for Ala), every change is nonsynonymous. At a two-fold site (e.g. the third position of TTT, for Phe), only one change (transition) is synonymous. The irregular nature of the genetic code causes some problems with this classification strategy. For example, in the universal code the third positions of the three isoleucine codons are three-fold degenerate so that they do not fall naturally into any of the three categories. The first positions of arginine codons (such as AGA and AGG) are also problematic: they are two-fold degenerate but with transversions to be synonymous. These issues are typically dealt with differently by different computer programs.

Let the total numbers of sites in the three degeneracy classes, averaged over the two sequences, be L_0, L_2, and L_4. Similarly the numbers of transitional and transversional differences within each degeneracy class are counted. Li et al. (1985) then used the K80 model (Kimura 1980) to estimate the number of transitions and the number of transversions per site within each degeneracy class. Let these be A_i and B_i, $i = 0, 2, 4$, with $d_i = A_i + B_i$ to be the total distance. A_i and B_i are estimates of αt and $2\beta t$ in the K80 model, given by equation (1.12) in Chapter 1. Thus $L_2 A_2 + L_4 d_4$ and $L_2 B_2 + L_0 d_0$ are the total numbers of synonymous and nonsynonymous substitutions between the two sequences, respectively.

To estimate d_S and d_N, we also need the numbers of synonymous and nonsynonymous sites. Each four-fold degenerate site is a synonymous site and each nondegenerate site is a nonsynonymous site. The case with two-fold degenerate sites is less clear. Li et al. (1985) counted each two-fold site as 1/3 synonymous and 2/3 nonsynonymous, based on the assumption of equal mutation rates. The numbers of synonymous and nonsynonymous sites are thus $L_2/3 + L_4$ and $2L_2/3 + L_0$, respectively. The distances then become

$$d_S = \frac{L_2 A_2 + L_4 d_4}{L_2/3 + L_4},$$
$$d_N = \frac{L_2 B_2 + L_0 d_0}{2L_2/3 + L_0} \qquad (2.16)$$

(Li et al. 1985). This method is referred to as LWL85.

Li (1993) and Pamilo and Bianchi (1993) pointed out that the rule of counting a two-fold site as 1/3 synonymous and 2/3 nonsynonymous ignores the transition–transversion rate difference, and causes underestimation of the number of synonymous sites and overestimation of d_S (and underestimation of d_N). Instead, these authors suggested the following formulae

$$d_S = \frac{L_2 A_2 + L_4 A_4}{L_2 + L_4} + B_4,$$

$$d_N = A_0 + \frac{L_0 B_0 + L_2 B_2}{L_0 + L_2}. \qquad (2.17)$$

This method, known as LPB93, effectively uses the distance at the four-fold sites, $A_4 + B_4$, to estimate the synonymous distance d_S, but replaces the transition distance A_4 with an average between two-fold and four-fold sites: $(L_2 A_2 + L_4 A_4)/(L_2 + L_4)$, assuming that the transition rate is the same at the two-fold and four-fold sites. Similarly d_N may be considered an estimate of the distance at the nondegenerate site, $(A_0 + B_0)$, but with the transversion distance B_0 replaced by the average over the two-fold degenerate and nondegenerate sites, $(L_0 B_0 + L_2 B_2)/(L_0 + L_2)$.

An alternative approach is to calculate each of d_S and d_N as the number of substitutions divided by the number of sites, as in the LWL85 method, but replace 1/3 with an estimate of the proportion ρ of two-fold sites that are synonymous:

$$d_S = \frac{L_2 A_2 + L_4 d_4}{\rho L_2 + L_4},$$

$$d_N = \frac{L_2 B_2 + L_0 d_0}{(1-\rho) L_2 + L_0}. \qquad (2.18)$$

We can use the distances A_4 and B_4 at the four-fold sites to estimate the transition/transversion rate ratio and to partition two-fold sites into synonymous and nonsynonymous categories, as suggested by Pamilo and Bianchi (1993). In other words, $\hat{\kappa} = 2A_4/B_4$ is an estimate of κ in the K80 model and

$$\hat{\rho} = \hat{\kappa}/(\hat{\kappa} + 2) = A_4/(A_4 + B_4). \qquad (2.19)$$

We will refer to equations (2.18) and (2.19) as the LWL85m method. Similar but slightly more complicated distances were defined by Tzeng et al. (2004) in their modification of the LWL85 method.

It is not essential to partition sites according to codon degeneracy. One can take into account the transition–transversion rate difference by counting synonymous and nonsynonymous sites in proportion to synonymous and nonsynonymous 'mutations'. This is the approach of Ina (1995), and it has the advantage of not having to deal with the irregularities of the genetic code, such as the existence of three-fold sites and the fact that not all transitions at two-fold sites are synonymous. Table 2.5 explains the counting of sites in codon TTT when the transition/transversion rate ratio $\kappa = 2$ is given. One simply uses the mutation rates to the nine neighbouring codons as weights when partitioning sites, with each transitional change (to codons TTC, TCT, and CTT) having twice the rate of each transversional change (to the six other codons). Thus there are $3 \times 2/12 = 1/2$ synonymous sites and $3 \times 10/12 = 5/2$ nonsynonymous sites in codon TTT, compared with 1/3 and 8/3 when $\kappa = 1$. The percentage of synonymous sites as a function of κ under the universal code is plotted in Figure 2.4. Ina (1995) described two methods for estimating κ.

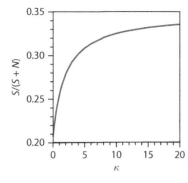

Fig. 2.4 Proportion of synonymous sites $S/(S + N)$ as a function of the transition/transversion rate ratio κ. Codon frequencies are assumed to be the same (1/61). Redrawn from Yang and Nielsen (1998).

The first uses the third codon positions and tends to overestimate κ. The second uses an iterative algorithm. Another commonly used approach is to use the four-fold degenerate sites, as mentioned above (Pamilo and Bianchi 1993; Yang and Nielsen 2000).

Besides the transition–transversion rate difference, another major complicating factor in the estimation of d_S and d_N is the unequal codon frequencies. Recall that the JC69 and K80 models of nucleotide substitution assume symmetrical rates and predict uniform base frequencies. Similarly, symmetrical codon substitution rates predict equal frequencies of all sense codons, and the fact that the observed codon frequencies are not equal means that the rates are not symmetrical. Overall, rates to common codons are expected to be higher than rates to rare codons. Such rate differences affect our counting of sites and differences. Yang and Nielsen (2000) accommodated unequal codon frequencies as well as the transition–transversion rate difference in an iterative procedure for estimating d_S and d_N.

Example 2.2. We apply the methods discussed above to the cucumber and tobacco *rbcL* genes (Table 2.8). The numbers of nondegenerate, two-fold degenerate, and four-fold degenerate sites are $L_0 = 916.5, L_2 = 267.5$, and $L_4 = 232.0$. The 141 differences between the two sequences are partitioned into 15.0 transitions and 18.0 transversions at the nondegenerate sites, 44.0 transitions and 8.5 transversions at the two-fold degenerate sites, and 32.0 transitions and 23.5 transversions at the four-fold degenerate sites. Application of the K80 correction formula leads to $A_0 = 0.0169$ and $B_0 = 0.0200$ as the transitional and transversional distances at the nondegenerate sites. Similarly $A_2 = 0.2073$ and $B_2 = 0.0328$ at the two-fold sites and $A_4 = 0.1801$ and $B_4 = 0.1132$ at the four-fold sites. The LWL85 method calculates the numbers of synonymous and nonsynonymous sites as $S = L_2/3 + L_4 = 321.2$ and $N = L_2 \times 2/3 + L_0 = 1094.8$, so that $d_S = 0.385$ and $d_N = 0.039$, with $d_N/d_S = 0.101$.

The LPB93 method gives $d_S = 0.308$ and $d_N = 0.040$, with $d_N/d_S = 0.129$. If we use $\hat{\rho} = A_4/(A_4 + B_4) = 0.6141$ to estimate the percentage of two-fold sites that are synonymous, and apply equation (2.18), we get $S = 396.3$ and $N = 1019.7$ for the LWL85m method. The distances are then $d_S = 0.312$ and $d_N = 0.042$, with $d_N/d_S = 0.134$. Note that S estimates here are much larger than S calculated by the NG86 method, which ignores the transition and transversion rate difference. The YN00 method incorporates unequal codon frequencies as well as the transition and transversion rate difference. This gives $S = 308.4$, even smaller than S calculated by the NG86 method. The biased codon usage had the opposite effect to the transition–transversion rate difference, and its effect counter-balanced the effect of the transition–transversion rate difference. The distance estimates then become $d_S = 0.498$ and $d_N = 0.035$, with $d_N/d_S = 0.071$. □

Table 2.8 Estimates of d_S and d_N between the cucumber and tobacco $rbcL$ genes

Model	$\hat{\kappa}$	$\hat{S}$	$\hat{N}$	$\hat{d}_S$	$\hat{d}_N$	$\hat{d}_N/\hat{d}_S$ (ω)	$\hat{t}$	ℓ
Counting methods								
NG86 (Nei and Gojobori 1986)	1	343.5	1072.5	0.383	0.036	0.095		
LWL85 (Li et al. 1985)	N/A	321.2	1094.8	0.385	0.039	0.101		
LWL85m (equation (2.18))	3.18	396.3	1019.7	0.312	0.042	0.134		
LPB93 (Li 1993; Pamilo and Bianchi 1993)	N/A	N/A	N/A	0.308	0.040	0.129		
Ina95 (Ina 1995)	5.16	418.9	951.3	0.313	0.041	0.131		
YN00 (Yang and Nielsen 2000)	2.48	308.4	1107.6	0.498	0.035	0.071		
ML method (Goldman and Yang 1994)								
(A) Fequal, $\kappa = 1$	1	360.7	1055.3	0.371	0.036	0.096	0.363	−2466.33
(B) Fequal, κ estimated	2.59	407.1	1008.9	0.322	0.037	0.117	0.358	−2454.26
(C) F1 × 4, $\kappa = 1$ fixed	1	318.9	1097.1	0.417	0.034	0.081	0.361	−2436.17
(D) F1 × 4, κ estimated	2.53	375.8	1040.2	0.362	0.036	0.099	0.367	−2424.98
(E) F3 × 4, $\kappa = 1$ fixed	1	296.6	1119.4	0.515	0.034	0.066	0.405	−2388.35
(F) F3 × 4, κ estimated	3.13	331.0	1085.0	0.455	0.036	0.078	0.401	−2371.86
(G) F61, $\kappa = 1$ fixed	1	263.3	1152.7	0.551	0.034	0.061	0.389	−2317.76
(H) F61, κ estimated	2.86	307.4	1108.6	0.473	0.035	0.074	0.390	−2304.47
(I) FMutSel, κ estimated	2.58	328.8	1087.2	0.433	0.035	0.082	0.383	−2287.51

Note: NG86, LWL85, and LPB93 were implemented in CODEML (Yang 2007b). MEGA (Kumar et al. 2005a; Tamura et al. 2011) gave $d_S = 0.300$ and $d_N = 0.035$ for NG86, $d_S = 0.398$ and $d_N = 0.036$ for LWL85, and $d_S = 0.316$ and $d_N = 0.037$ for LPB93. Both programs use equal weighting of pathways when the two compared codons differ at two or three positions. Ina's program was used for Ina's method (Ina 1995), and the YN00 program in the PAML package (Yang 2007b) was used for YN00 (Yang and Nielsen 2000). The likelihood method was implemented in the CODEML program in PAML (Yang 2007b). The models assumed are Fequal, with equal codon frequencies ($\pi_j = 1/61$ for all j); F1 × 4, with four nucleotide frequencies used to calculate the expected codon frequencies (using three free parameters); F3 × 4, with nucleotide frequencies at three codon positions used to calculate codon frequencies (nine free parameters); F61, with all 61 codon frequencies used as free parameters (60 free parameters because the sum is 1); and FMutSel, which uses mutation bias parameters and 60 codon fitness parameters (Yang and Nielsen 2008). '$\kappa = 1$ fixed' means that transition and transversion rates are assumed to be equal, while 'κ estimated' accommodates different transition and transversion rates (see equation (2.7)). ℓ is the log likelihood value under the model.

2.5.2 Maximum likelihood method

2.5.2.1 Likelihood estimation of d_S and d_N

The ML method (Goldman and Yang 1994) fits a Markov model of codon substitution, such as equation (2.7), to data of two sequences to estimate parameters in the model, including t, κ, ω, and π_j. The likelihood function is given by equation (2.4), except that i and j now refer to the 61 sense codons (in the universal genetic code) rather than the 20 amino acids. The codon frequencies (if they are not fixed at 1/61) are usually estimated by using the observed frequencies in the data, while parameters t, κ, and ω will be estimated by numerical maximization of the log likelihood ℓ. Then d_S and d_N are calculated from the estimates of t, κ, ω, and π_j, according to their definitions. Here we describe the definition of d_S and d_N when parameters in the codon model are given, so that the rate matrix Q and sequence distance t are known. In real data analysis, the same calculation, with the parameters replaced by their MLEs, will produce the MLEs of d_S and d_N, according to the invariance property of MLEs.

We define the numbers of sites and substitutions on a per codon basis. First, the expected number of substitutions per codon from codons i to j, $i \neq j$, over any time t is $\pi_i q_{ij} t$. Thus the numbers of synonymous and nonsynonymous substitutions per codon between two sequences separated by time (or distance) t are

$$S_d = t\rho_S = \sum_{i \neq j,\ aa_i = aa_j} \pi_i q_{ij} t,$$
$$N_d = t\rho_N = \sum_{i \neq j,\ aa_i \neq aa_j} \pi_i q_{ij} t. \quad (2.20)$$

In other words, S_d sums over codon pairs that differ by a synonymous difference while N_d sums over codon pairs that differ by a nonsynonymous difference, with aa_i to be the amino acid encoded by codon i. As the rate matrix is scaled so that one nucleotide substitution is expected to occur in one time unit, we have $S_d + N_d = t$, and ρ_S and ρ_N are the proportions of synonymous and nonsynonymous substitutions, respectively.

Next, we count sites. The proportions of synonymous and nonsynonymous 'mutations', ρ_S^1 and ρ_N^1, are calculated in the same way as ρ_S and ρ_N in equation (2.20) except that $\omega = 1$ is fixed in the rate matrix

$$\rho_S^1 = \tfrac{1}{C} \times \sum_{i \neq j,\ aa_i = aa_j} \pi_i q_{ij}^1,$$
$$\rho_N^1 = \tfrac{1}{C} \times \sum_{i \neq j,\ aa_i \neq aa_j} \pi_i q_{ij}^1, \quad (2.21)$$

where q_{ij}^1 is the ijth element of Q with $\omega = 1$ fixed and $C = \sum_{i \neq j} \pi_i q_{ij}^1$ is a scale factor so that ρ_S^1 and ρ_N^1 sum to 1. The numbers of synonymous and nonsynonymous sites per codon are then

$$S = 3\rho_S^1,$$
$$N = 3\rho_N^1. \quad (2.22)$$

The definition of sites by equation (2.22) is equivalent to Ina's (1995, table 1) if codon frequencies are equal ($\pi_j = 1/61$) and the mutation rate is the same at the three codon positions. However, if the three codon positions have different base compositions and different rates, Ina's method counts each codon position as one site and will lead to results which are hard to interpret. In equation (2.22), each codon is counted as three sites, but each codon position will be counted as more or less than one site depending on whether the mutation rate at the position is higher or lower than the average rate at all three positions. Another point worth mentioning here is that the 'mutation rate' referred to here may have been influenced by selection acting on the DNA (but not selection on the protein). In other words, the proportions of sites represent what we would expect to observe if DNA level processes, which cause the transition–transversion rate difference and unequal codon usage, are operating but there is no selection at the protein level (i.e. $\omega = 1$). Definitions of sites are discussed further in §2.5.4 and §2.5.6.

The distances are then given by

$$d_S = S_d/S = t\rho_S/(3\rho_S^1),$$
$$d_N = N_d/N = t\rho_N/(3\rho_N^1). \qquad (2.23)$$

Note that $\omega = d_N/d_S = (\rho_N/\rho_S)/(\rho_N^1/\rho_S^1)$ is a ratio of two ratios; the numerator ρ_N/ρ_S is the ratio of the numbers of synonymous and nonsynonymous substitutions that are inferred to have occurred, while the denominator ρ_N^1/ρ_S^1 is the corresponding ratio expected if there had been no selection on the protein (so that $\omega = 1$). Thus ω measures the perturbation in the proportions of synonymous and nonsynonymous substitutions caused by natural selection on the protein.

The variances of MLEs of d_S and d_N can be calculated using the delta technique (Example B4 in Appendix B), by using the asymptotic variance–covariance matrix for parameters t, κ, and ω, ignoring the sampling errors in codon frequency parameters (if these are parameters).

Note that the average number of nucleotide substitutions per codon (t) is approximately related to d_N and d_S as

$$t = \frac{3S}{N+S} \times d_S + \frac{3N}{N+S} \times d_N = \frac{3d_S(S + N\omega)}{N+S} \qquad (2.24)$$

(Yang and Nielsen 2000). Here $S/(N + S)$ and $N/(N + S)$ are the proportions of synonymous and nonsynonymous sites, and there are three nucleotide sites in a codon.

2.5.2.2 Estimation of d_S and d_N between the cucumber and tobacco rbcL genes

We use the ML method under different models to estimate d_S and d_N between the cucumber and tobacco *rbcL* genes, in comparison with the counting methods. The results are summarized in Table 2.8.

First, we assume equal transition and transversion rates ($\kappa = 1$) and equal codon frequencies ($\pi_j = 1/61$) (model A in Table 2.8). This is the model underlying the NG86 method. There are two parameters in the model: t and ω (see equation (2.7)). The log likelihood contour is shown in Figure 2.5. The MLEs are found numerically to be $\hat{t} = 0.363$ and $\hat{\omega} = 0.096$. The estimated sequence divergence corresponds to $472 \times 0.363 = 171.4$ nucleotide substitutions over the whole sequence, after correction for multiple hits, compared with 141 raw observed differences. Application of equation (2.20) gives 133.8 synonymous substitutions and 37.6 nonsynonymous substitutions for the whole sequence. We then calculate the proportions of synonymous and nonsynonymous sites to be $\rho_S^1 = 0.243$ and $\rho_N^1 = 0.757$. The numbers of sites for the whole sequence are $S = L_c \times 3\rho_S^1 = 360.7$ and $N = L_c \times 3\rho_N^1 = 1055.3$. Counting sites according to equation (2.22) under this model is equivalent to counting sites in the NG86 method explained in §2.5.1. The slight differences in the counts are due to the fact that the ML method averages over the codon frequencies (1/61) assumed in the model, while the NG86 method averages over the

2.5 ESTIMATION OF D_S AND D_N

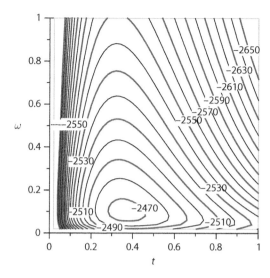

Fig. 2.5 Log likelihood contour as a function of the sequence distance t and the rate ratio ω for the cucumber and tobacco *rbcL* genes. The model assumes equal transition and transversion rates ($\kappa = 1$) and equal codon frequencies ($\pi_j = 1/61$).

observed codons in the two sequences. The distances then become $d_S = 133.8/360.7 = 0.371$ and $d_N = 37.6/1055.3 = 0.036$, very similar to estimates from NG86.

Next, we assume equal codon frequencies but different transition and transversion rates (model B in Table 2.8: Fequal, κ estimated). This is the model underlying the LWL85 (Li et al. 1985), LPB93 (Li 1993; Pamilo and Bianchi 1993), and Ina's (1995) methods. The model involves three parameters: t, κ, and ω. Their MLEs are found numerically to be $\hat{t} = 0.358$, $\hat{\kappa} = 2.59$, and $\hat{\omega} = 0.117$. The ω ratio is 22% larger than the estimate obtained when the transition–transversion rate difference is ignored ($\kappa = 1$). The numbers of synonymous and nonsynonymous substitutions at those parameter values are 131.0 and 37.8, respectively. Counting sites by fixing $\omega = 1$ and using $\hat{\kappa} = 2.59$ gives $\rho_S^1 = 0.288$ and $\rho_N^1 = 0.719$, and $S = 407.1$ and $N = 1008.9$ for the whole sequence. Thus we have $d_S = 131.0/407.1 = 0.322$ and $d_N = 37.8/1008.9 = 0.037$. Note that incorporating the transition–transversion rate difference has had much greater effect on the numbers of sites than on the numbers of substitutions. Also, as $S + N$ is fixed, an increase in S means a decrease in N at the same time, leading to even greater differences in the ω ratio. As S is much smaller than N, the effect on S and d_S is greater than that on N and d_N.

Last, we use models that accommodate both the transition–transversion rate difference and unequal codon frequencies, and the estimates are given in Table 2.8 (models D, F, H). The counting method of Yang and Nielsen (2000) approximates the likelihood method under the F3 × 4 model (model F in Table 2.8), and indeed produced very similar results to the likelihood method. The FMutSel model (I in Table 2.8) uses mutation bias and codon fitness parameters to explain codon usage (equation (2.11)), involving three more parameters than F61. The estimates are similar between the two models.

2.5.3 Comparison of methods

The results in Table 2.8 suggest considerable differences in estimates of d_S and d_N and their ratio ω among the methods. Similar differences have been observed in analyses of many real and simulated datasets (e.g. Muse 1996; Bielawski et al. 2000; Yang and Nielsen 2000). Based on those results, the following observations can be made.

(1) Ignoring the transition and transversion rate difference leads to underestimation of S, overestimation of d_S and underestimation of the ω ratio (Li et al. 1985).

(2) Unequal codon usage often has the opposite effect to the transition–transversion rate difference; ignoring unequal codon usage leads to overestimation of S, underestimation of d_S, and overestimation of ω (Yang and Nielsen 2000). Extreme base or codon frequencies can overwhelm the effect of the transition–transversion rate difference, leading to the ironic result that NG86, which ignores both the transition–transversion rate difference and codon usage, can produce more reliable estimates of d_S and d_N than LPB93, LWL85m or Ina's method, which accommodates the transition–transversion rate difference but not codon usage.

(3) Different methods or model assumptions can produce very different estimates even when the two sequences compared are highly similar. This is in contrast to distance calculation under nucleotide models, where the difference between methods is minimal at low sequence divergences (see, e.g. Table 1.4 in Chapter 1). The reason for this unfortunate sensitivity to model assumptions is that different methods produce different counts of sites.

(4) Assumptions often matter more than methods, at least at low or moderate sequence divergences. On one hand, the counting and likelihood methods often produce similar estimates under the same assumptions. The NG86 method is seen to produce results similar to likelihood under the same model (Muse 1996). Similarly, LPB93 and Ina's method produced results similar to likelihood accommodating the transition–transversion rate difference but not codon usage. On the other hand, different counting methods can produce very different estimates as they are based on different assumptions. So do the likelihood methods under different models. For example, the MLEs of ω in Table 2.8 are almost two-fold different.

Compared with the counting methods, the likelihood method has two advantages. The first is conceptual simplicity. In the counting method, dealing with features of DNA sequence evolution such as different transition and transversion rates and unequal codon frequencies is challenging. For example, some methods take into account the transition/transversion rate ratio κ, but a reliable estimate of κ is hard to obtain. All counting methods use nucleotide-based correction formulae to correct for multiple hits, which are logically flawed. In the likelihood method, we have to specify the instantaneous substitution rates only and leave it to the probability calculus to produce a sensible inference. At the level of instantaneous rates, there are no multiple changes in a codon and it is straightforward to decide whether each change is synonymous or nonsynonymous. The difficult tasks of estimating the transition/transversion rate ratio, weighting evolutionary pathways between codons, correcting for multiple hits at the same site, and dealing with irregularities of the genetic code are achieved automatically in the likelihood calculation. Second, it is much simpler to accommodate more realistic models of codon substitution in the likelihood method than in the counting method. For example, it is straightforward to use a GTR-style mutation model instead of the HKY85-style model in equation (2.7).

2.5.4 More distances and interpretation of the d_N/d_S ratio

2.5.4.1 More distances based on the codon model

The expected number of substitutions from codon i to codon j over time t is given by $\pi_i q_{ij} t$. Thus one can use the rate matrix Q and divergence time t to define various measures to characterize the evolutionary process of a protein-coding gene. For example, one can consider whether the $i \to j$ change is synonymous or nonsynonymous and contrast synonymous and nonsynonymous substitution rates. Similarly one can contrast the transitional and transversional substitutions, substitutions at the three codon positions, substitutions causing conservative or radical amino acid changes, and so on. Such measures are functions of parameters in the codon substitution model (such as κ, ω, and codon frequency π_j). As soon as the MLEs of model parameters are obtained, the rate matrix Q and divergence time t are known, so that the MLEs of such measures can be generated in a straightforward manner according to their definitions. Here we introduce a few more distance measures.

First the distances at the three codon positions can be calculated separately under the codon model: d_{1A}, d_{2A}, d_{3A}. The suffix 'A' in the subscript means 'after selection on the protein', as will become clear later. The expected number of nucleotide substitutions at the first codon position over any time t is the sum of $\pi_i q_{ij} t$ over all pairs of codons i and j that differ at the first position only (see equation (2.7)). As there is one site at the first position in each codon, this is also the number of nucleotide substitutions per site at the first position

$$d_{1A} = \sum_{\{i,j\} \in A_1} \pi_i q_{ij} t, \qquad (2.25)$$

where the summation is over set A_1, which includes all codon pairs i and j that differ at position 1 only. Distances d_{2A} and d_{3A} for the second and third positions can be defined similarly. Here equation (2.25) serves two purposes: (i) to define the distance d_{1A}, as a function of parameters in the codon model (t, κ, ω, π_j), and (ii) to present the ML method for estimating the distance; i.e. the MLE of d_{1A} is given by equation (2.25), with parameters t, κ, ω, and π_j replaced by their MLEs. All distance measures discussed below in this section should be understood in this way.

Example 2.3. We use the F3 × 4 model (model F in Table 2.8) to calculate these distances between the cucumber and tobacco *rbcL* genes. The estimates are $d_{1A} = 0.046$, $d_{2A} = 0.041$, $d_{3A} = 0.314$, with an average of 0.134. These are comparable with the distances calculated by applying the nucleotide substitution model HKY85 to each codon position: $d_1 = 0.057$, $d_2 = 0.026$, and $d_3 = 0.282$ (Table 2.7). In the nucleotide-based analysis, five parameters (d, κ, and three nucleotide frequency parameters) are estimated at each position, and the substitution rate (or sequence distance) is allowed to vary freely among codon positions. In the codon-based analysis, 12 parameters (t, κ, ω, and nine nucleotide frequency parameters) are estimated for the whole dataset of all three positions, and the use of parameter ω to accommodate different synonymous and nonsynonymous rates leads to different substitution rates at the three codon positions, even though the codon model does not explicitly incorporate such rate differences. □

Next we define three distances d_{1B}, d_{2B}, and d_{3B} for the three codon positions, where the suffix 'B' stands for 'before selection on the protein'. If there is no selection at the DNA level, these will be mutational distances at the three positions. Otherwise, they measure distances at the three positions before natural selection at the protein level has affected nucleotide substitution rates. For the first position, we have

$$d_{1B} = \sum_{\{i,j\} \in A_1} \pi_i q_{ij}^1 t, \qquad (2.26)$$

where q_{ij}^1 is q_{ij} but calculated with $\omega = 1$ fixed, as we did when defining the proportion of synonymous sites (see equation (2.22)). Distances d_{2B} and d_{3B} are defined similarly for the second and third positions.

Another distance is d_4, the expected number of substitutions per site at the four-fold degenerate sites of the third codon position, often used as an approximation to the neutral mutation rate. A commonly used heuristic approach to estimating d_4 applies a nucleotide model to data of four-fold degenerate sites at the third codon position. A third position is counted as a four-fold degenerate site if the first two positions are identical across all sequences compared and if the encoded amino acid does not depend on the third position (e.g. Perna and Kocher 1995; Adachi and Hasegawa 1996a; Duret 2002; Kumar and Subramanian 2002; Waterston et al. 2002). This is a conservative definition of four-fold degenerate sites, based on comparison of sequences, and it differs from the definition used in the LWL85 method, which counts four-fold sites along each sequence and then takes an average between the two sequences. For example, if the codons in the two sequences are ACT and GCC, the LWL85 method will count the third position as four-fold degenerate, and the T-C difference at the third position as a four-fold degenerate difference. The heuristic approach does not use such a third position, as its status might have changed during the evolutionary history. The heuristic approach, however, has the drawback that the number of usable four-fold sites decreases with the increase of sequence divergence and with inclusion of more divergent sequences.

Here we describe an ML method for estimating d_4, which overcomes this drawback and which uses a codon model instead of a nucleotide model to correct for multiple hits. We define the expected number of four-fold degenerate sites per codon as the sum of frequencies (π_j) of all four-fold degenerate codons (i.e. codons in which all changes at the third position are synonymous). This number does not decrease with the increase of sequence divergence. The expected number of four-fold degenerate nucleotide substitutions per codon is given by summing $\pi_i q_{ij} t$ over all codon pairs i and j that represent a four-fold degenerate substitution; i.e. codons i and j are both four-fold degenerate and they have a difference at the third position. Then d_4 is given by the number of four-fold degenerate substitutions per codon divided by the number of four-fold degenerate sites per codon. Replacing parameters κ, ω, and π_j in this definition by their MLEs gives the MLE of d_4. It is easy to see that d_4 defined this way converges to that in the heuristic approach when sequence divergence t approaches 0, but the ML method should work for divergent sequences as well. Also note that this method counts four-fold degenerate sites in a similar way to the LWL85 method, although it calculates the number of four-fold degenerate substitutions differently.

Example 2.4. For the cucumber and tobacco *rbcL* genes, application of equation (2.26) under the F3 × 4 model (model F in Table 2.8) gives $d_{1B} = 0.463$, $d_{2B} = 0.518$, and $d_{3B} = 0.383$, with the average over the three codon positions to be 0.455, which is d_S, as will be explained below. The ratios d_{1A}/d_{1B}, d_{2A}/d_{2B}, and d_{3A}/d_{3B} are the proportions of mutations at the three codon positions that are accepted after filtering by selection on the protein, and may be called the acceptance rates (Miyata et al. 1979). These are calculated to be 0.100, 0.078, and 0.820 for the three codon positions, respectively. At the second position, all mutations are nonsynonymous, so that the acceptance rate is exactly ω. At the first and third positions, some changes are synonymous, so that the acceptance rate is $> \omega$. In the heuristic method of estimating d_4, 215 four-fold degenerate sites (i.e. 15.2% of all sites in the sequence) can be used, and application of the HKY85 model to data at these sites produces the estimate $d_4 = 0.344$. Using the ML method, 245.4 four-fold degenerate sites (17.3% of all sites) are used, with the MLE of d_4 being 0.386. This is somewhat larger than the estimate from the heuristic method, but is very close to d_{3B}, which is calculated using 472 third position sites. □

2.5.4.2 Interpretation of the d_N/d_S ratio

Methods for estimating d_S and d_N were historically developed to quantify the effect of natural selection on the protein on nucleotide substitution rates (Miyata and Yasunaga 1980; Gojobori 1983; Li et al. 1985). The proportions of synonymous and nonsynonymous sites are defined as the expected proportions of synonymous and nonsynonymous mutations. Nearly all the methods developed in the literature are of this kind, including the counting methods LWL85 (Li et al. 1985), LPB93 (Li 1993; Pamilo and Bianchi 1993), Ina's (1995) method, and YN00 (Yang and Nielsen 2000), as well as the ML method GY94 (Goldman and Yang 1994).

It is straightforward to show that d_S and d_N, defined in equation (2.23), satisfy the following relationships:

$$\begin{aligned} d_S &= (d_{1B} + d_{2B} + d_{3B})/3, \\ d_N &= d_S \times \omega, \end{aligned} \tag{2.27}$$

where d_{kB} (k = 1, 2, 3) is the 'mutation' distance at the kth codon position, defined in equation (2.26). Thus d_S is the average mutation (substitution) rate over the three codon positions before selection on the protein. If silent site evolution is not neutral and is affected by DNA level selection, d_S will reflect that selection as well. It would appear more appropriate to use symbols d_B (distance *before* selection on the protein) and d_A (distance *after* selection) in place of d_S and d_N. We verified this result in Example 2.4 and will demonstrate it below in a few simple models, but the reader may wish to confirm it (Problem 2.3).

In the literature, it is often taken for granted that use of the ω ratio to detect protein level natural selection requires the assumption of neutral evolution at silent sites. A concern is that if selection acts on codon usage (e.g. Akashi 1994, 1995), or on some other aspects of the DNA sequence (Rubinstein et al. 2011), codon models may be misled to produce an ω ratio greater than one because of reduced synonymous rate and not because of elevated nonsynonymous rate. It should be pointed out that this reasoning is flawed and that use of the ω ratio to detect protein level natural selection does not need the assumption of neutral evolution at synonymous sites. Comparison between d_S and d_N is a contrast between the rates before and after the action of selection on the protein (equation (2.27)), so that the comparison is valid whether evolution at silent sites is driven by mutation or selection. The FMutSel model (Yang and Nielsen 2008), discussed in §2.4.2, uses the codon fitness parameters explicitly to model selection on codon usage.

2.5.5 Estimation of d_S and d_N in comparative genomics

The synonymous and nonsynonymous rates (d_S and d_N) are commonly estimated in comparative analysis of genomes. Figure 2.6 shows the distribution of the estimated d_N, d_S, and ω ratios between the human and orangutan and between the mouse and rat in 857 orthologous genes. The data are from dos Reis et al. (2012). The median values of d_N, d_S, and ω are 0.005, 0.030, and 0.149 for the human–orangutan comparison, and 0.019, 0.195, and 0.101 for the mouse–rat comparison. The divergence times of those two species pairs are similar (dos Reis et al. 2012), but the rodent genes are far more divergent than those of the apes and furthermore the rodent ω is much smaller. The smaller ω in rodents appears to be due to the larger population sizes, and thus there is more effective purifying selection removing deleterious nonsynonymous mutations in rodents.

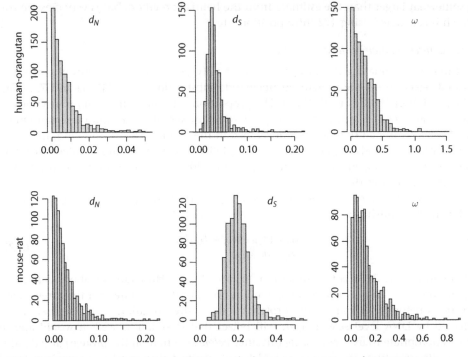

Fig. 2.6 Histograms of d_N, d_S, and ω between the human–orangutan and mouse–rat comparisons of 857 orthologous genes from dos Reis et al. (2012). The median values of d_N, d_S, and ω are 0.005, 0.030, and 0.149 for the human–orangutan comparison, and 0.019, 0.195, 0.101 for the mouse–rat comparison. Rodents have larger population sizes than primates, so that natural selection is more effective in removing deleterious nonsynonymous mutations, resulting in smaller ωs than in primates.

A common problem in comparative analysis of genomes to estimate d_S and d_N is that the time scale or sequence divergence may be inappropriate. Estimation of d_S and d_N requires a time window in which the sequences are neither too similar nor too divergent. If the species are too distantly related or the genomes are too divergent, the synonymous substitutions may have reached saturation, so that it is impossible to obtain reliable estimates of d_S. While any criterion is arbitrary, it appears prudent to treat estimates of d_S greater than 3 with caution. It is virtually impossible to distinguish data with on average five changes per site from data with 50 changes per site, even though as estimates of d_S, those values are very different. When the sequences are too divergent, one useful strategy may be to include other genomes and compare multiple species on a phylogenetic tree, thus breaking the large distance between species into many shorter branches. Such methods are discussed in Chapters 4 and 11.

At the other extreme, if the genomes (from individuals of the same species or from some bacterial strains) are virtually identical, it is impossible to obtain reliable estimates of the d_N/d_S ratio. Furthermore, the interpretation of the d_N/d_S ratio calculated using gene sequences from the same species may be complex. It has been observed in many datasets that the d_N/d_S ratio tends to decrease with the increased sequence divergence. This appears to be partly due to the biases in MLEs and correlations between MLEs of t and ω in small datasets (dos Reis and Yang 2013b). Another factor is the short time scale in within-population comparisons, where deleterious nonsynonymous mutations may not have had enough time to become lost. They are thus more likely to be found in comparison of sequences that separated a short time ago (Rocha et al. 2006; Kryazhimskiy and Plotkin 2008). More research is needed to understand the dynamics of the d_N/d_S estimate from population data and to design more sensitive measures of protein level selection.

*2.5.6 Distances based on the physical-site definition

Bierne and Eyre-Walker (2003) pointed out the existence of an alternative definition based on *physical sites* and argued that distances based on such a definition may sometimes be more appropriate than d_S and d_N, based on the mutation-opportunity definition. Here I define new distance measures d_S^* and d_N^* using a physical-site definition, and explore a few idealized models to illustrate their differences from d_S and d_N. I then briefly discuss the uses of this large collection of distance measures. I will focus on the likelihood method as it is conceptually simpler. However, the discussion should apply to the counting methods as well, and to alternative formulations of the basic codon substitution model, such as equations (2.9) or (2.11). The issue is the conceptual definition of parameters of biological interest, with the sampling errors and estimation inefficiency ignored in the discussion.

Definition of d_S^* and d_N^*. Using a physical-site definition, we count synonymous and nonsynonymous sites S^* and N^* as follows. A nondegenerate site is counted as one nonsynonymous site, a two-fold degenerate site as 1/3 synonymous and 2/3 nonsynonymous, a three-fold degenerate site as 2/3 synonymous and 1/3 nonsynonymous, and a four-fold degenerate site as one synonymous site. The mutation/substitution model is ignored during the counting, as are differences in transition and transversion rates or in codon frequencies, even if such factors are considered in the model. The numbers of synonymous and nonsynonymous sites per codon are averaged over the codon frequencies (π_j) expected in the model. The numbers of synonymous and nonsynonymous substitutions per codon are calculated as before (equation (2.20)). The distances are then defined as

$$d_S^* = S_d/S^*,$$
$$d_N^* = N_d/N^*, \tag{2.28}$$

instead of equation (2.23).

The distances d_S and d_N produced by the methods of Miyata and Yasunaga (1980) and Nei and Gojobori (1986) can be interpreted using either the mutational-opportunity or the physical-site definitions. As these methods assume that the mutation rate is the same between any two nucleotides, there is no numerical difference between the two definitions of sites.

Example 2.5. We apply equations (2.28) to obtain MLEs of d_S^* and d_N^* between the cucumber and tobacco *rbcL* genes, with a subset of the models of Table 2.8 used. The results are summarized in Table 2.9. If we assume equal codon frequencies (models A and B), the counts of (physical) sites are the same whether we consider the transition–transversion rate difference: $S^* = 353.8$ and $N^* = 1062.2$. Indeed the small differences in the counts between the Fequal and F3 × 4 models are due to the different stationary codon frequencies: the counts will be identical if we use the counting methods instead of ML, and count the sites in the codons in the observed sequences. Because of the similar counts of sites, the different models produced similar estimates of d_S^* and d_N^*, in contrast to the estimates of d_S and d_N, which vary considerably among models (Table 2.8). □

---

* indicates a more difficult or technical section.

Table 2.9 MLEs of d_S^* and d_N^* between the cucumber and tobacco *rbcL* genes

Model	$\hat{\kappa}$	$\hat{S}^*$	$\hat{N}^*$	$\hat{d}_S^*$	$\hat{d}_N^*$
(A) Fequal, $\kappa = 1$	1	360.7	1055.3	0.371	0.036
(B) Fequal, κ estimated	2.59	360.7	1055.3	0.363	0.036
(E) F3 × 4, $\kappa = 1$ fixed	1	353.8	1062.2	0.432	0.036
(F) F3 × 4, κ estimated	3.13	353.8	1062.2	0.425	0.036

Note: Calculated using the CODEML program in PAML.

We now consider a few idealized examples to contrast the different distance measures.

The two-fold regular code (Bierne and Eyre-Walker 2003). Imagine a genetic code in which there are no stop codons and every codon is two-fold degenerate. Every first or second codon position is nondegenerate, and every third position is two-fold degenerate, with transitions to be synonymous and transversions nonsynonymous. This code would encode 32 amino acids. Suppose that the 64 codons have equal frequencies, and mutations occur according to the K80 model, with transition rate α and transversion rate β, with $\kappa = \alpha/\beta$. Suppose there is no selection on silent sites, and the proportion of nonsynonymous mutations that are neutral is ω; all other nonsynonymous mutations are lethal and are removed by purifying selection. Thus over any time interval t, the expected number of synonymous substitutions per codon is αt and the expected number of nonsynonymous substitutions per codon is $2(\alpha + 2\beta)\omega t + 2\beta\omega t$.

Using a physical-site definition, we see that every codon has $S^* = 1/3$ synonymous sites and $N^* = 1 + 1 + 2/3 = 8/3$ nonsynonymous sites. The number of synonymous substitutions per (physical) synonymous site is then $d_S^* = \alpha t/(1/3) = 3\alpha t$. This synonymous rate is equal to the nucleotide substitution rate under the JC69 model in which both the transition and transversion rates are α. The number of nonsynonymous substitutions per (physical) nonsynonymous site is $d_N^* = [2(\alpha + 2\beta)\omega t + 2\beta\omega t]/(8/3) = 3(\alpha + 3\beta)\omega t/4$. The ratio $d_N^*/d_S^* = (1 + 3/\kappa)\omega/4$ differs from ω if $\kappa \neq 1$. Consider two genes with the same transversion rate β, but with the transition rate to be $\alpha = \beta$ in the first gene and $\alpha = 5\beta$ in the second. We see that d_S^* is five times as high in the second gene as in the first, as one expects from a physical definition of sites. The ratio $d_N^*/d_S^* = \omega$ in the first gene, but $= 0.4\omega$ in the second. In the second gene, the nonsynonymous rate appears to be reduced relative to the synonymous rate, not because selective constraint on the protein has removed nonsynonymous mutations but because transitions occur at higher rates than transversions.

To use the mutational-opportunity definition of sites, we note that synonymous sites are all at the third position, and that synonymous and nonsynonymous mutations occur in proportions $\alpha : 2\beta$ at the third codon position. Thus the number of synonymous sites in a codon is $S = \alpha/(\alpha + 2\beta)$, and the number of nonsynonymous sites in a codon is $N = 1 + 1 + 2\beta/(\alpha + 2\beta)$, with $S + N = 3$. This is the counting method illustrated in Table 2.5 for LWL85m and Ina's methods (equation (2.18)) and in §2.5.2 for the ML method. Then $d_S = \alpha t/S = (\alpha + 2\beta)t$, and $d_N = [2(\alpha + 2\beta)\omega t + 2\beta\omega t]/N = (\alpha + 2\beta)\omega t = d_S \times \omega$. Note that d_S is the mutation rate at each codon position. It is not the synonymous rate in the usual sense of the word. For the two genes with different transition rates discussed above ($\alpha = \beta$ versus $\alpha = 5\beta$), $d_S = 7\beta t$ in the second gene is only 7/3 times as high as in the first gene ($d_S = 3\beta t$), which would seem strange if one incorrectly uses a physical-site definition to interpret d_S. In both genes, $d_N/d_S = \omega$.

The four-fold regular code. Imagine a genetic code in which all codons are four-fold degenerate, and the 64 sense codons encode 16 amino acids. Again suppose there is no selection on the DNA. Then d_S^* will be the mutation rate at the third position and will be equal to d_4 and d_{3B}, while d_S will be the average mutation rate over the three codon positions. Consider two cases. In the first, suppose mutations occur according to the K80 model, and the 64 codons have the same frequency. Then we have $d_S = d_S^*$, since the mutation rate is the same at the three codon positions, and $d_N = d_N^*$. In the second case, suppose the codon frequencies are unequal and the three codon positions have different base compositions. Then the mutation rate will differ among the three positions and d_S will be different from d_S^*.

*2.5.7 Utility of the distance measures

Both sets of distances (d_S and d_N vs. d_S^* and d_N^*) are valid distance measures. They increase linearly with time and can be used to calculate species divergence times or to reconstruct phylogenetic trees. Both can be used to compare different genes with different codon usage patterns or to test models of molecular evolution. For testing adaptive protein evolution, d_S and d_N can be used while d_S^* and d_N^* are inappropriate. For example, we do not expect $d_N^*/d_S^* = 1$ when there is no selection on the protein. In likelihood-based methods, such tests are usually conducted using parameter ω in the codon model (see Chapter 11), so that estimation of rates of synonymous and nonsynonymous substitutions per site is unnecessary. In the counting methods, the ω ratio is calculated by estimating d_S and d_N first, and the sensitivity of d_S and d_N to model assumptions is a source of concern.

Distance measures $d_{1A}, d_{2A}, d_{3A}, d_{1B}, d_{2B}, d_{3B}$, as well as d_4, discussed in this subsection, all use physical-site definitions of sites. The mutation rate at the third codon position may be estimated using any of d_{3B}, d_4, and d_S^*, but d_{3B} should be preferred to d_4, which should in turn be preferred to d_S^*, based on the numbers of sites used by those distances. The synonymous substitution rate has been used to examine its correlation with codon usage bias (e.g. Bielawski et al. 2000; Bierne and Eyre-Walker 2003). For this purpose, d_{3B}, d_4, and d_S^* appear more appropriate than d_S if the codon usage bias is measured by GC_3, the GC content at the third codon position. It does not make much sense to correlate GC_3 with the average rate over all three codon positions (d_S). If codon usage bias is measured by the *effective number of codons* or ENC (Wright 1990), the situation is less clear, since ENC depends on all positions and codon usage bias is more complex than base composition differences at the third codon position.

It is noted that distances based on physical sites are not so sensitive to model assumptions. In contrast, distances d_S and d_N, which define sites based on mutational opportunities, are much more sensitive. With extreme codon usage or base compositions, different methods or model assumptions can lead to estimates that are several-fold different. In such genes, d_S may be substantially different from d_S^*, d_4, or d_{3B}, and use of different distances may lead to very different conclusions.

*2.6 Numerical calculation of the transition probability matrix

Fitting Markov chain models to sequence data requires calculation of the transition probability matrix $P(t)$, which is given as a matrix exponential through the Taylor expansion

$$P(t) = e^{Qt} = I + Qt + \frac{1}{2!}(Qt)^2 + \frac{1}{3!}(Qt)^3 + \cdots, \quad (2.29)$$

* indicates a more difficult or technical section.

where I is the identity matrix. For nucleotide models, the calculation is either analytically tractable or otherwise inexpensive due to the small size of the matrix. However, for amino acid and codon models, this calculation can be costly and unstable, so that use of a reliable algorithm is more important. One method is to use the first k terms in the expansion of equation (2.29). This should work fine if the matrix Qt does not have large elements, i.e. if $\left|(Qt)_{ij}\right| < 1$. However if some elements in the matrix are large, use of equation (2.29) may require many terms (i.e. large k) and may cause serious cancellations in floating-point arithmetic on the computer. Note that the diagonal elements of Qt are negative while the off-diagonals are positive, so that the terms of equation (2.29) may have elements that are huge but have opposite signs. Here we describe two approaches that are more effective. See Moler and van Loan (2003) for a review of algorithms for calculating matrix exponentials.

Scaling and squaring. The first approach makes use of the relationship

$$e^{Qt} = \left(e^{Qt/m}\right)^m. \tag{2.30}$$

If we choose $m = 2^j$ for some integer j, the mth power of $e^{Qt/m}$ can be calculated by repeated matrix squaring, while $e^{Qt/m}$ can be reliably calculated by using a few terms in the Taylor expansion

$$e^{Qt/m} \approx I + (Qt/m) + \frac{1}{2!}(Qt/m)^2 + \cdots + \frac{1}{k!}(Qt/m)^k. \tag{2.31}$$

To see why the Taylor expansion of equation (2.31) is reliable while that of equation (2.29) is not, note that as long as m or j is large enough, the matrix $(I + Qt/m)$ will have positive elements only. Moler and van Loan (2003: table 1) provides guidelines on the choices of j and k, depending on the norm of matrix Qt (i.e. on how large the elements of the matrix are). With the rate matrix Q scaled to have rate 1, the norm of matrix Qt largely depends on t. Thus reasonable choices include $j = 7$ and $k = 1$ for $t < 1$, and $j = 6$ and $k = 5$ for $t < 10$. For even larger t, larger values of j and k may be necessary. This algorithm works for general rate matrices and is used in the PAML package (Yang 1997a, 2007b) to calculate $P(t)$ for the UNREST model of nucleotide substitution (see Table 1.1 in Chapter 1).

Spectral decomposition. A second approach is numerical computation of the eigenvalues and eigenvectors of the rate matrix Q:

$$Q = U \Lambda U^{-1}, \tag{2.32}$$

where $\Lambda = \text{diag}\{\lambda_1, \lambda_2, \ldots, \lambda_c\}$ is a diagonal matrix, with the eigenvalues of Q on the diagonal, while columns of U are the corresponding right eigenvectors and rows of U^{-1} are the left eigenvectors. All these matrices are of size $c \times c$, with $c = 4$ for nucleotides, 20 for amino acids, and 61 for codons (under the universal code). Then

$$P(t) = e^{Qt} = U e^{\Lambda t} U^{-1} = U \, \text{diag}\{e^{\lambda_1 t}, e^{\lambda_2 t}, \ldots, e^{\lambda_c t}\} U^{-1}. \tag{2.33}$$

A general real matrix can have complex eigenvalues and eigenvectors and their numerical computation may be unstable (Golub and Van Loan 1996). However, a real symmetrical matrix has only real eigenvalues and eigenvectors and their numerical computation is both fast and stable. The rate matrix Q for a time-reversible Markov process is *similar* to a real symmetrical matrix and thus has only real eigenvalues and eigenvectors (e.g. Keilson 1979). Two matrices A and B are said to be *similar* if there exists a nonsingular matrix T so that

2.6 NUMERICAL CALCULATION OF THE TRANSITION PROBABILITY MATRIX

$$A = TBT^{-1}. \tag{2.34}$$

Similar matrices have identical eigenvalues. This fact can be exploited to calculate the eigenvalues and eigenvectors of the Q matrix for a reversible Markov chain (Yang 1995c). Note that Q is similar to

$$B = \Pi^{1/2} Q \Pi^{-1/2}, \tag{2.35}$$

where $\Pi^{1/2} = \text{diag}\{\pi_1^{1/2}, \pi_2^{1/2}, \ldots, \pi_c^{1/2}\}$ and $\Pi^{-1/2}$ is its inverse. From equation (2.2), B is symmetrical if Q is the rate matrix for a reversible Markov chain. Thus B and Q have identical real eigenvalues. We diagonalize B:

$$B = R\Lambda R^{-1}, \tag{2.36}$$

where $R^{-1} = R^T$, the transpose of R. Then

$$Q = \Pi^{-1/2} B \Pi^{1/2} = \left(\Pi^{-1/2} R\right) \Lambda \left(R^{-1} \Pi^{1/2}\right). \tag{2.37}$$

Comparing equations (2.37) with (2.32), we have $U = (\Pi^{-1/2} R)$ and $U^{-1} = R^{-1} \Pi^{1/2}$.

The above algorithm assumes that the frequency $\pi_i > 0$ for every i. If m states are missing (with $\pi_i = 0$), the Markov chain has in effect $c - m$ states. We can rearrange the Q matrix so that the submatrix for the existing states Q_0, of size $(c-m) \times (c-m)$, has the spectral decomposition:

$$Q_0 = U_0 \Lambda_0 V_0. \tag{2.38}$$

Then

$$Q = \begin{bmatrix} Q_0 & 0 \\ 0 & 0 \end{bmatrix} = \begin{bmatrix} U_0 & 0 \\ 0 & I \end{bmatrix} \begin{bmatrix} \Lambda_0 & 0 \\ 0 & 0 \end{bmatrix} \begin{bmatrix} U_0^{-1} & 0 \\ 0 & I \end{bmatrix}, \tag{2.39}$$

with

$$P(t) = \begin{bmatrix} U_0 & 0 \\ 0 & I \end{bmatrix} \begin{bmatrix} \exp\{\Lambda_0 t\} & 0 \\ 0 & I \end{bmatrix} \begin{bmatrix} U_0^{-1} & 0 \\ 0 & I \end{bmatrix} = \begin{bmatrix} U_0 \exp\{\Lambda_0 t\} U_0^{-1} & 0 \\ 0 & I \end{bmatrix}. \tag{2.40}$$

Transition probabilities from and to missing states (with $\pi_i = 0$) are never used in the likelihood calculation so they are not relevant.

For reversible Markov chains, the algorithm of matrix diagonalization (equation (2.37)) may be both faster and more accurate than the algorithm of scaling and squaring (equation (2.31)). The former is particularly efficient when applied to likelihood calculation for multiple sequences on a phylogeny. In that case, we often need to calculate e^{Qt} for fixed Q and different ts. For example, when we fit empirical amino acid substitution models, the rate matrix Q is fixed for the whole tree while the branch length t varies among branches. Then one has to diagonalize Q only once, after which calculation of $P(t)$ for each t involves only matrix multiplications by equation (2.33). The algorithm of scaling and squaring can be useful for any substitution rate matrix, including those for non-reversible models.

2.7 Problems

2.1 Download the human and orangutan NADH6 gene sequences from GenBank (accession numbers X93334 for *Homo sapiens* and D38115 for *Pongo pygmaeus*), align them and apply the methods discussed in this chapter to estimate d_S and d_N. One way of aligning protein-coding DNA sequences is to use CLUSTAL (Thompson et al. 1994) to align the protein sequences first and then construct the DNA alignment based on the protein alignment, using, for example, MEGA (Tamura et al. 2011). Use CODEML to estimate $d_S, d_N, d_{1B}, d_{2B}, d_{3B}, d_S^*, d_N^*$, etc. Assess the impact of allowing for the transition–transversion rate difference on the estimation. Also examine the impact of the model assumption about codon frequencies, by using models such as Fequal, F1 × 4, F3 × 4, F61, and FMutSel.

2.2* Are there really three nucleotide sites in a codon? How many synonymous and non-synonymous sites are in the codon TAT (under the universal code)? Assume the transition/transversion rate ratio $\kappa = 1$.

2.3 Verify $d_S = (d_{1B} + d_{2B} + d_{3B})/3$ (equation (2.27)).

2.4 Conduct a computer simulation to examine the impact of sequence divergence on estimation of ω in comparison of two gene sequences. Assume the Fequal model ($\pi_j = 1/61$ for all j), with $\kappa = 2$, and set $\omega = 0.5$. Assume 500 codons in the gene. Use a few different sequence distances, such as $t = 0.1, 0.3, 0.5, 1.0, 1.5, 2$, and 3 nucleotide substitutions per codon. Use EVOLVER to generate the data and CODEML to analyse them. Use 1,000 replicates to calculate the mean and variance of $\hat{\omega}$ for each sequence distance. What sequence divergence appears to be optimal for estimating ω? Take a guess before you conduct the simulation experiment.

2.5 How large a sample is large enough for the χ^2 approximation to the LRT statistic to be reliable? Conduct a computer simulation to examine the null distribution of the LRT statistic ($2\Delta\ell$) for testing the hypothesis $\omega = 1$, in comparison with χ_1^2 (the χ^2 distribution with df = 1) (see §1.4.3). Assume the Fequal model (with $\pi_j = 1/61$ for codon j), with $t = 0.5, \kappa = 2$, and $\omega = 0.5$. Use different numbers of codons in the gene, such as 50, 100, 200, 300, 500, or 1000. Use EVOLVER to generate the data and CODEML to analyse them. Use 1,000 or 10,000 replicates. Analyse each replicate dataset under H_0 (with $\omega = 1$ fixed) and H_1 (with ω estimated) to calculate the test statistic $2\Delta\ell = 2(\ell_1 - \ell_0)$. Plot the histogram of $2\Delta\ell$ (e.g. using the R command hist) and compare with χ_1^2. Also examine the estimated significance values and compare with the theoretical expectations (3.84 at 5% and 6.63 at 1%).

2.6* *Regular genetic code with mixed two-fold and four-fold codons.* Imagine a genetic code in which a proportion γ of codons are four-fold degenerate while all other codons are two-fold degenerate. (If $\gamma = 48/64$, the code would encode exactly 20 amino acids.) Suppose that neutral mutations occur according to the K80 model, with transition rate α and transversion rate β, with $\alpha/\beta = \kappa$. The proportion of nonsynonymous mutations that are neutral is ω. The numbers of nondegenerate, two-fold and four-fold degenerate sites in a codon are $L_0 = 2, L_2 = 1 - \gamma$, and $L_4 = \gamma$. Over time interval t, the numbers of transitional and transversional substitutions at the three degeneracy

---

* indicates a more difficult or technical problem.

classes are thus $A_0 = \alpha t\omega$, $B_0 = 2\beta t\omega$, $A_2 = \alpha t$, $B_2 = 2\beta t\omega$, $A_4 = \alpha t$, and $B_4 = 2\beta t$. (**a**) Show that the LWL85 method (equation (2.16)) gives

$$d_S = \frac{3(\kappa + 2\gamma)\beta t}{1 + 2\gamma},$$
$$d_N = \frac{3(\kappa + 3 - \gamma)\beta t\omega}{4 - \gamma}, \tag{2.41}$$

with the ratio $d_N/d_S = \omega[(\kappa + 3 - \gamma)(1 + 2\gamma)]/[(4 - \gamma)(\kappa + 2\gamma)]$, which becomes $\omega(\kappa + 3)/(4\kappa)$ if $\gamma = 0$ (so that the code is the two-fold regular code), and ω if $\gamma = 1$ (so that the code is the four-fold regular code). (**b**) Show that both LPB93 (equation (2.17)) and LWL85m (equations (2.18) and (2.19)) give $d_S = (\alpha + 2\beta)t$ and $d_N = d_S\omega$. (Comment: under this model, LWL85 gives d_S^* and d_N^*, distances under the physical-site definition, while LPB93 and LWL85m give d_N and d_S, distances under the mutation-opportunity definition.)

CHAPTER 3

Phylogeny reconstruction: overview

3.1 Tree concepts

This chapter introduces basic concepts related to phylogenetic trees and discusses general features of tree reconstruction methods. Distance and parsimony methods are described in this chapter as well, while likelihood and Bayesian methods are discussed later in Chapters 4 and 6–8.

3.1.1 Terminology

3.1.1.1 Trees, nodes (vertexes), and branches (edges)

A phylogeny or phylogenetic tree is a representation of the genealogical relationships among species, among genes, among populations, or even among individuals. Mathematicians define a *graph* as a set of *vertexes* and a set of *edges* connecting the vertexes, and a tree as a connected graph without loops (see, e.g. Tucker 1995, p. 1). Biologists instead use *nodes* for vertexes and *branches* for edges. Here we consider trees for species, but the description also applies to trees of genes or individuals. The *tips*, *leaves*, or *external nodes* represent present-day species, while the *internal nodes* usually represent extinct ancestors for which no sequence data are available. The ancestor of all sequences is the *root* of the tree.

3.1.1.2 Root of the tree and rooting the tree

A tree with the root specified is called a *rooted tree* (Figure 3.1a), while a tree with the root unknown or unspecified is called an *unrooted tree* (Figure 3.1b). If the evolutionary rate is constant over time, an assumption known as the *molecular clock*, distance matrix, maximum likelihood (ML) and Bayesian methods can identify the root and produce rooted trees. Such use of the clock assumption to determine the root of the tree is known as *molecular clock rooting*.

Another related rooting method, often used in the analysis of population data, is *midpoint rooting*. First, an unrooted tree is inferred without the clock assumption. Then the most distant pair of sequences is identified, with the distance calculated as the sum of branch lengths connecting the two sequences. The root is then placed at the midpoint between the two sequences. Midpoint rooting also relies on the molecular clock assumption.

When the species are distantly related or the sequences are fairly divergent, the clock assumption is most often violated. Incorrectly assuming the clock then can cause serious

Molecular Evolution: A Statistical Approach. Ziheng Yang. © Ziheng Yang 2014.
Published 2014 by Oxford University Press.

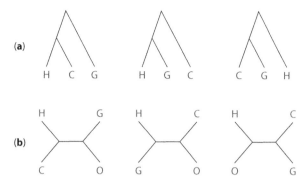

Fig. 3.1 Outgroup rooting. To infer the relationships among human (H), chimpanzee (C), gorilla (G), represented by the three rooted trees in (**a**), we use orangutan (O) as the outgroup. Tree reconstruction methods allow us to estimate an unrooted tree, i.e. one of the trees in (**b**). As the root is along the branch leading to the outgroup, these three unrooted trees for all four species correspond to the three rooted trees for the ingroup species H, C, and G.

errors in phylogeny reconstruction and in rooting. Without the clock and with independent rates for branches on the tree, most tree reconstruction methods are unable to identify the root of the tree and instead produce unrooted trees. Then the commonly used approach to rooting the tree is *outgroup rooting*. Distantly related species, called the *outgroups*, are included in tree reconstruction, while in the reconstructed unrooted tree for all species, the root is placed on the branch leading to the outgroups, so that the subtree for the *ingroups* is rooted. In the example of Figure 3.1, the orangutan is used as the outgroup to root the tree for the ingroup species: human, chimpanzee, and gorilla. In general, outgroups closely related to the ingroup species are better than distantly related outgroups.

In the universal tree of life, no outgroup species exist. Then a strategy is to root the tree using ancient gene duplications that occurred prior to the divergence of all existing life forms (Gogarten et al. 1989; Iwabe et al. 1989). The subunits of ATPase arose through a gene duplication before the divergence of eubacteria, eukaryotes, and archaebacteria. Protein sequences from both paralogues were used to construct a composite unrooted tree, and the root was placed on the branch separating the two duplicates (Gogarten et al. 1989). Elongation factors Tu and G constitute another ancient duplication, and were used in rooting the universal tree of life (Iwabe et al. 1989).

One should note that the output from a tree reconstruction program may look like a rooted tree. The user of the program is expected to know whether the analysis should produce rooted or unrooted trees and to interpret the tree accordingly.

3.1.1.3 *Tree topology, branch lengths, and the parenthesis notation*

The branching pattern of a tree is called the tree structure or *topology*. The length of a branch may represent the amount of sequence divergence or the time period covered by the branch. A tree showing only the topology without the branch length information is

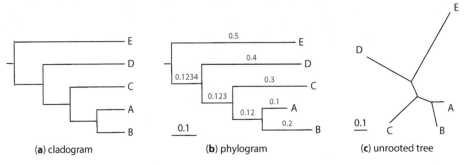

Fig. 3.2 The same tree shown in different styles. (**a**) The cladogram shows the tree topology without branch lengths or with branch lengths ignored. (**b**) In a phylogram, branches are drawn in proportion to their lengths. Here the branch lengths are shown along the branches. (**c**) In an unrooted tree, the location of the root is unknown or ignored.

sometimes called a *cladogram* (Figure 3.2a), while a tree showing both the topology and branch lengths is called a *phylogram* (Figure 3.2b).

For use in computer programs, trees are often represented using the nested parenthesis format or the *Newick format*, named after a lobster restaurant in Dover, New Hampshire, where the format was proposed (Felsenstein 2004, p. 590). For example, the trees in Figure 3.2 may be represented as:

a and **b**: ((((A, B), C), D), E);
b: ((((A: 0.1, B: 0.2): 0.12, C: 0.3): 0.123, D: 0.4): 0.1234, E: 0.5);
c: (((A, B), C), D, E);
c: (((A: 0.1, B: 0.2): 0.12, C: 0.3): 0.123, D: 0.4, E: 0.6234);.

Each internal node is represented by a pair of parentheses, which groups its daughter nodes, while the order of the daughter nodes is arbitrary. The outmost pair of parentheses groups the daughter nodes of the root. Tip nodes are represented by their names. A node can be followed by a semicolon together with a number that is the length of the branch ancestral to the node. Branch lengths here are measured by the expected number of nucleotide substitutions per site, like the sequence distance discussed in Chapter 1.

This format is natural to represent rooted trees. Unrooted trees are represented by placing the root at an arbitrary internal node and by having a trifurcation at the root. The representation is not unique, as the root can be placed anywhere on the tree. For example, the unrooted tree of Figure 3.2c can also be represented as '(A, B, (C, (D, E)));'.

Just as the Newick format does not represent the same tree in a unique way, there is much arbitrariness when a tree is drawn. For example, the root can be on the top, at the bottom, or on the side. To decide whether the different trees are equivalent, think about whether they represent the same evolutionary/genealogical relationships: for example, the three trees shown in Figure 3.3 are identical.

Because of different rates of evolution in different lineages, there may not be direct correspondence between evolutionary *relatedness* and sequence *distance* between two species: two closely related species may not have the smallest sequence distance. The distance is the amount of sequence evolution and is equal to the sum of branch lengths on the paths connecting the two species, while the relatedness is measured by the time of divergence between the two species (see Problem 3.3).

Fig. 3.3 All three trees are identical as they represent the same genealogical relationships among A, B, C, D, and E. This is the same tree of Figure 3.2b: ((((A: 0.1, B: 0.2): 0.12, C: 0.3): 0.123, D: 0.4): 0.1234, E: 0.5).

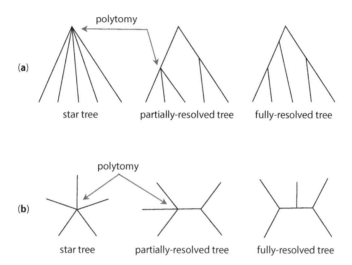

Fig. 3.4 Unresolved and resolved phylogenetic trees. **(a)** Rooted trees. **(b)** Unrooted trees.

3.1.1.4 *Bifurcating and multifurcating trees*

The number of branches connected to a node is called the *degree* of the node. Leaves have a degree of 1. If the root node has a degree greater than 2 or a non-root node has a degree greater than 3, the node represents a *polytomy* or *multifurcation*. A tree with no polytomies is called a *binary tree*, *bifurcating tree*, or *fully resolved tree*. The most extreme unresolved tree is the *star* or *big-bang* tree, in which the root is the only internal node (see Figure 3.4 for example). A polytomy representing truly simultaneous species divergences is sometimes called a *hard polytomy*. It would seem extremely unlikely for one species to diverge into several at exactly the same time, and it may be argued that hard polytomies do not exist. Most often the polytomy represents lack of information in the data to resolve the relationships within a clade (a group of species). Such polytomies are called *soft polytomies*.

3.1.1.5 *The number of trees*

We can work out the total number of unrooted trees by the following *stepwise addition algorithm* (Cavalli-Sforza and Edwards 1967) (Figure 3.5). We start with the single tree for the first three species. This has three branches to which the fourth species can be added. Thus there are three possible trees for the first four species. Each four-species tree has

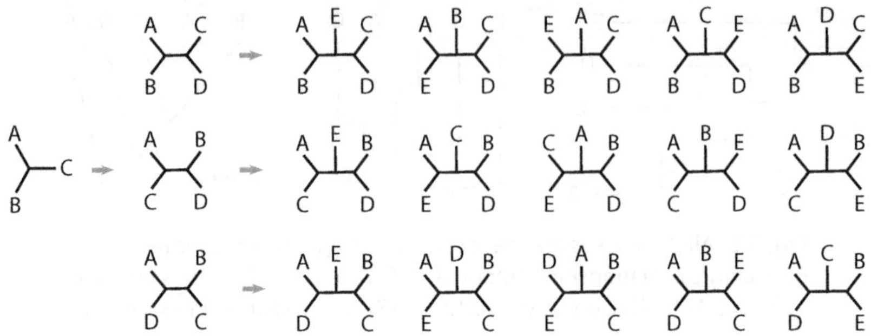

Fig. 3.5 Enumeration of all trees for five taxa A, B, C, D, and E using the stepwise addition algorithm.

five branches, to which the fifth species can be added, resulting in five different five-species trees for each four-species tree. There are thus 3×5 possible trees for five species. In general, a tree of the first $n-1$ species has $(2n-5)$ branches, to which the nth species can be added, so that each of the $(n-1)$-species trees generates $(2n-5)$ distinct n-species trees. Thus the total number of unrooted bifurcating trees for n species is

$$U_n = U_{n-1} \cdot (2n-5) = 3 \cdot 5 \cdot 7 \cdots (2n-5) = \frac{1 \cdot 2 \cdot 3 \cdot 4 \cdot 5 \cdots (2n-5)}{2 \cdot 4 \cdots (2n-6)} = \frac{(2n-5)!}{2^{n-3}(n-3)!}. \tag{3.1}$$

To work out the number of rooted trees for n species, note that each unrooted tree has $(2n-3)$ branches, and the root can be placed on any of those branches, generating $(2n-3)$ rooted trees from each unrooted tree. Thus the number of rooted trees for n species is

$$R_n = U_n \times (2n-3) = U_{n+1} = \frac{(2n-3)!}{2^{n-2}(n-2)!}. \tag{3.2}$$

In certain applications, we also need the concept of *labelled histories*. For example, under the coalescent model, the Yule model of pure birth, or the birth–death process model, all possible labelled histories have the same probability (Aldous 2001). A labelled history is a rooted tree with the internal nodes rank-ordered according to their ages (Edwards 1970). Thus a rooted tree may correspond to several labelled histories. For example, the symmetrical rooted tree for four species $((a, b), (c, d))$ corresponds to two labelled histories, depending on whether or not the common ancestor of a and b is older than the common ancestor of c and d. The asymmetrical rooted tree $(((a, b), c), d)$ corresponds to a single labelled history since there is only one ordering of the internal nodes. The number of possible labelled histories for n sequences is

$$H_n = \frac{n(n-1)}{2} \times \frac{(n-1)(n-2)}{2} \times \cdots \times \frac{2 \cdot 1}{2} = \frac{n!(n-1)!}{2^{n-1}}. \tag{3.3}$$

The counting is done by the so-called coalescent process, which traces the genealogy backwards in time to find common ancestors. Initially there are n lineages, so there are $\binom{n}{2} = \frac{1}{2}n(n-1)$ possible ways of choosing two lineages to join (to coalesce). After the first coalescent event, there will be $n-1$ lineages left so there are $\frac{1}{2}(n-1)(n-2)$ possible ways of choosing two lineages to join, and so on. The last coalescence joins two lineages at the root of the tree. Obviously this coalescent process of joining lineages respects the order of

Table 3.1 The numbers of unrooted trees (U_n), rooted trees (R_n), and labelled histories (H_n) for n species

n	Unrooted trees (U_n)	Rooted trees (R_n)	Labelled histories (H_n)
3	1	3	3
4	3	15	18
5	15	105	180
6	105	945	2,700
7	945	10,395	56,700
8	10,395	135,135	1,587,600
9	135,135	2,027,025	57,153,600
10	2,027,025	34,459,425	2,571,912,000
20	$\sim 2.22 \times 10^{20}$	$\sim 8.20 \times 10^{21}$	$\sim 5.64 \times 10^{29}$
50	$\sim 2.84 \times 10^{74}$	$\sim 2.75 \times 10^{76}$	$\sim 3.29 \times 10^{112}$

coalescent events or the ranking of node ages on the tree; it thus enumerates the labelled histories correctly. The coalescent process is discussed in detail in Chapter 9.

As we can see from Table 3.1, the number of unrooted trees U_n increases explosively with the number of species n. The number of rooted trees R_n and the number of labelled histories H_n rise even faster.

3.1.1.6 Distance between trees

Sometimes we would like to measure how different two trees are. For example, we may be interested in how different the trees estimated from different genes are, or how different the estimated tree is from the true tree in a computer simulation conducted to evaluate a tree reconstruction method. A commonly used measure of topological distance between two trees is the *partition distance* defined by Robinson and Foulds (1981) (see also Penny and Hendy 1985). We give the definition for unrooted trees first. Note that each branch on the tree defines a *bipartition* or *split* of the species; if we chop the branch, the species will fall into two mutually exclusive sets. For example, branch b in tree T_1 of Figure 3.6 partitions the eight species into two sets: (1, 2, 3) and (4, 5, 6, 7, 8). This partition is also present on tree T_2. Partitions defined by terminal branches are in all possible trees and are thus not informative for comparisons between trees. Thus we focus on internal branches only. Partitions defined by branches b, c, d, and e of tree T_1 are the same as partitions defined by branches b', c', d', and e' of tree T_2, respectively. The partition defined by branch a of tree T_1 is not in tree T_2, nor is the partition defined by branch a' of tree T_2 in tree T_1. The partition distance is defined as the total number of bipartitions that exist in one tree but not in the other. Thus T_1 and T_2 have a partition distance of 2. As an unrooted binary tree of n species has $(n-3)$ internal branches, the partition distance ranges from 0 (if the two trees are identical) to $2(n-3)$ (if the two trees do not share any bipartition).

The partition distance can be equivalently defined as the number of contractions and expansions needed to transform one tree into the other. Removing an internal branch by reducing its length to zero is a contraction, while creating an internal branch is an expansion. Trees T_1 and T_2 of Figure 3.6 are separated by a contraction (from T_1 to T_0) and an expansion (from T_0 to T_2), so that their partition distance is 2.

For rooted trees, we use the same definition as for unrooted trees, but imagine the existence of an outgroup species attached to the root. As a rooted binary tree of n species

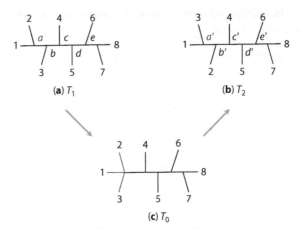

Fig. 3.6 The partition distance between two trees T_1 and T_2 is the total number of bipartitions that are in one tree but not in the other. It is also the number of contractions and expansions needed to change one tree into another. A contraction converts T_1 into T_0 and an expansion converts T_0 into T_2, so the distance between T_0 and T_1 is 1 while the distance between T_1 and T_2 is 2.

has $(n-2)$ internal branches, the partition distance ranges from 0 (if the two trees are identical) to $2(n-2)$ (if the two trees do not share any bipartition).

The partition distance has limitations. First, the distance does not recognize certain similarities between trees. The three trees in Figure 3.7 are identical concerning the relationships among species 2–7 but do not share any bipartitions, so that the partition distance between any two of them is the maximum possible. Indeed, the probability that a random pair of unrooted trees achieve the maximum distance is 70–80% for $n = 5$–10, and is even greater for larger n. Figure 3.8 shows the distribution of partition distance between two random unrooted trees for the case of $n = 10$. Second, the partition distance ignores branch lengths in the tree. Intuitively, two trees that are in conflict around short internal branches are less different than two trees that are in conflict around long internal branches. It is unclear how to incorporate branch lengths in a definition of tree distance. One such measure has been suggested by Kuhner and Felsenstein (1994), and is defined as the sum of squared differences between branch lengths in the two trees

$$B_S = \sum_i (b_i - b'_i)^2, \tag{3.4}$$

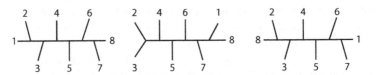

Fig. 3.7 Three trees that do not share any bipartitions and thus achieve the maximum partition distance.

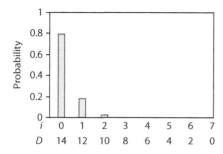

Fig. 3.8 The probability that two random trees from all possible unrooted trees of ten species share i bipartitions or have partition distance D. Note that $D = 2 \times (10 - 3 - i)$.

where b_i and b'_i are branch lengths in the two trees, respectively. If a branch exists in one tree but not in the other, the missing branch has length 0 in the calculation. Third, the partition distance may be misleading if either of the two trees has multifurcations. Suppose we conduct a computer simulation to compare two tree reconstruction methods, using an unrooted binary tree to simulate datasets, and use the partition distance to measure performance: $P = 1 - D/D_{max}$, where $D_{max} = 2(n-3)$ is the maximum distance and D is the distance between the true tree and the estimated tree. When both the true tree and the estimated tree are binary, P is the proportion of bipartitions in the true tree that are recovered in the estimated tree. Suppose that with no information in the data, one method returns the star tree as the estimate while the other method returns an arbitrarily resolved binary tree. Now for the first method, $D = (n-3) = 1/2 D_{max}$, so that $P = 50\%$, which may seem very impressive. The second method has a performance of $P = 1/3$ when $n = 4$ or nearly 0 for large n, since a random tree is very unlikely to share any bipartition with the true tree. However, the two methods clearly have the same performance, and the measure based on the partition distance is unreasonable for the first method.

3.1.1.7 *Consensus trees*

While the partition distance measures how different two trees are, a consensus tree summarizes common features among a collection of trees. Many different consensus trees have been defined; see Bryant (2003) for a comprehensive review. Here we introduce two of them.

The *strict consensus tree* shows only those branches (partitions or splits) that are shared among all trees in the set, with those not supported by all trees collapsed into polytomies. Consider the three trees in Figure 3.9a. The strict consensus tree is shown in Figure 3.9b. The group (A, B) is in the first and third trees but not in the second, while (A, B, C) is in all three trees. Thus the strict consensus tree shows (A, B, C) as a trichotomy, as well as (F, G, H). The strict consensus tree is a conservative way of summarizing the trees and may not be very useful as it often produces the star tree.

The *majority-rule consensus tree* shows branches or splits that are supported by at least half of the trees in the set. It is also common practice to show the percentage of trees that support every node on the consensus tree (Figure 3.9c). For example, the group (A, B) is in two out of the three trees and is thus shown in the majority-rule consensus tree as resolved, with the percentage of support (2/3) shown next to the node. It is known that all groups that occur in more than half of the trees in the set can be shown on the same consensus tree without generating any conflict.

Like the partition distance, the majority-rule consensus tree, as a summary of trees in the set, has limitations. Suppose that there are only three distinct trees in the set, which

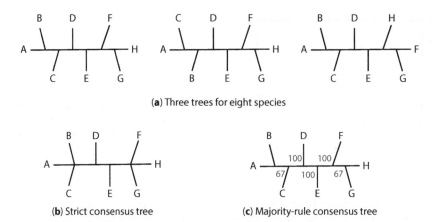

Fig. 3.9 Three trees for eight species (**a**) and their strict consensus tree (**b**) and majority-rule consensus tree (**c**).

are the trees of Figure 3.7, each occurring in proportions around 33%. Then the majority-rule consensus tree will be the star tree. In such cases, it appears more informative to report the first few whole trees with the highest support values.

It may be fitting to emphasize here that polytomies in a consensus tree are a heuristic way of summarizing (or visualizing) phylogenetic uncertainties, and do not represent simultaneous speciation events. The consensus tree may thus be unsuitable for use in downstream phylogenetic analyses, such as molecular clock dating, because used in such an analysis, the consensus tree with polytomies is treated as an exact mathematical model of simultaneous speciation. Instead one should use a fully resolved tree inferred from the data, such as the ML tree or the neighbour-joining (NJ) tree.

3.1.1.8 *Monophyly, paraphyly, clade, and clan*

While phylogenetics is concerned with inference of the phylogeny or reconstruction of the evolutionary relationships of the species, classification or taxonomy is the science of describing, naming, and classifying organisms. It is now widely accepted that phylogeny should be the basis of taxonomic classifications. While classification is beyond the scope of this book, some terms are commonly used in molecular phylogenetic analysis, and will be described here.

A monophyletic group includes all the descendants of a common ancestor. Such a group is also called a *clade*. Taxa in a monophyletic group are more *closely related*; i.e. they have more recent common ancestors than those outside the group. We also use the term *sister species* when two species are each other's closest relatives. A group that includes some descendants of a common ancestor but excludes some others is *non-monophyletic*. Some authors distinguish two types of non-monophyly: paraphyly and polyphyly. A *paraphyletic* group does not include all of the descendants of a single common ancestor. For example, 'apes' include chimpanzees, gorillas, orangutans, and gibbons but exclude humans. Apes are thus a paraphyletic group. Similarly 'reptiles' are a paraphyletic group (Figure 3.10). Reptiles include crocodiles, lizards, and turtles but exclude birds (because birds have novel anatomy and behaviour), but crocodiles are more closely related to birds than they are to other reptiles. A *polyphyletic* group includes species that have multiple common ancestors but excludes some other descendants of those ancestors so that the last common ancestor

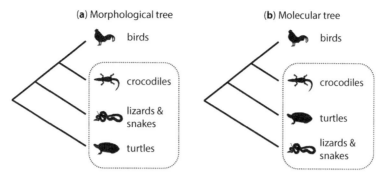

Fig. 3.10 Traditional morphological tree (**a**) and recent molecular tree (**b**) for birds, crocodiles, lizards, and turtles (e.g. Iwabe et al. 2005; Crawford et al. 2012). No matter which of those two trees is correct, 'reptiles' (circled) are a paraphyletic group.

of the group is not a member of the group. A polyphyletic group is often the result of erroneous taxonomic classification based on morphological similarities that are the result of convergent evolution. For example, 'pachyderms' include elephants, rhinoceroses, hippopotamuses, etc. Those mammals all have thick skins but belong to different orders and are a polyphyletic group. 'Vultures' are another polyphyletic group. The Old and New World vultures have striking similarities (such as bald heads) due to convergent evolution. However, the Old World vultures evolved from birds of prey (such as eagles, kites, hawks) while the New World vultures evolved from storks.

Use of terms such as monophyly and clade implies a knowledge of the root of the tree. As most phylogeny reconstruction methods produce unrooted trees, those terms are sometimes applied to unrooted trees as well, with the assumption that the root is in a place such that the use of those terms would be sensible. Wilkinson et al. (2007) recommend the use of the term *clan* (instead of clade) to mean a group of species identified on an unrooted tree. When we cut an internal branch on an unrooted tree, the species will fall into two groups (partitions or splits). These are two clans, and one of them must be a clade.

3.1.2 Species trees and gene trees

The phylogeny representing the relationships among a group of species is called the *species tree* or *organismal* tree. The phylogeny for sequences at a particular gene locus from those species is called the *gene tree*. A number of factors may cause the gene tree to differ from the species tree.

First, phylogeny reconstruction errors may cause the estimated gene tree to be different from the species tree even if the true gene tree agrees with the species tree. The estimation errors may be either random, due to the limited amount of sequence data, or systematic, due to deficiencies of the tree reconstruction method or serious violations of its assumptions. One such case is convergent evolution. For example, the lysozyme has apparently undergone convergent evolution in ruminants (e.g. the cow) and the leaf-eating colobine monkeys (e.g. the langur), as it is recruited as a bacteriolytic enzyme in the fermentative foreguts of those animals (Stewart et al. 1987). As a result, the stomach lysozymes of mammals from those two groups share some physico-chemical and catalytical

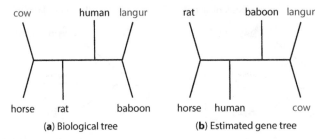

Fig. 3.11 Convergent evolution in the stomach lysozyme of the cow and the langur misleads phylogeny reconstruction methods to infer an incorrect gene tree. The organismal tree for six species of mammals is the one in (**a**), and this should also be the true gene tree for the lysozyme. However, the parsimony (and ML) methods incorrectly infer the gene tree to be the one in (**b**), grouping the cow and the langur together. Drawn following Stewart et al. (1987).

properties as well as certain key amino acids. When the protein sequences are used in tree reconstruction, an incorrect tree is inferred, grouping the cow and the langur together (Figure 3.11).

Second, during the early stages of evolution near the root of the universal tree of life, there appears to have been substantial lateral (horizontal) gene transfer (LGT). As a result, different genes or proteins may have different gene trees, in conflict with the species tree. The LGT appears to be so extensive that some researchers question the concept of a universal tree of life (see, e.g. Doolittle 1998). Third, gene duplications, especially if followed by gene losses, can cause the gene tree to be different from the species tree if paralogous copies of the gene are used for phylogeny reconstruction (Figure 3.12a). Note that

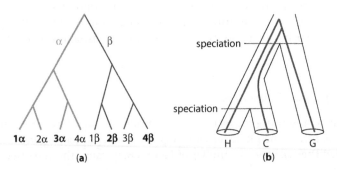

Fig. 3.12 Conflict between species tree and gene tree can be due to gene duplication (**a**) or ancestral polymorphism (**b**). In (**a**), a gene duplicated in the past, creating paralogous copies α and β, followed by divergences of species 1, 2, 3, and 4. If we use gene sequences 1α, 3α, 2β, 4β for phylogeny reconstruction, the true gene tree is ((1α, 3α), (2β, 4β)), different from the species tree ((1, 2), (3, 4)). In (**b**), the species tree is ((human, chimpanzee), gorilla). However, due to ancestral polymorphism and incomplete lineage sorting, the true gene tree is (human, (chimpanzee, gorilla)).

paralogues are genes that originated from gene duplications and may not reflect species relationships, while orthologues are genes that originated from species divergences and thus track speciation events (Fitch 1970). Fourth, *ancestral polymorphism* or polymorphism in ancestral species means that gene sequences in different modern species may be descendants of different ancestral sequences, so that the gene genealogy may fail to track the species phylogeny, a phenomenon called *incomplete lineage sorting*. An example is shown in Figure 3.12b. Here the species tree for human, chimpanzee, and gorilla is ((H, C), G). However, because of sequence variations (polymorphisms) in the extinct ancestral species, the true gene tree is (H, (C, G)). The probability that the gene tree differs from the species tree is greater if the speciation events are closer in time (i.e. if the species tree is almost a star tree) and if the population size of the H-C common ancestor is greater. Such information concerning the gene tree-species tree conflict can be used to estimate the effective population sizes of extinct common ancestors and to infer phylogeographic processes. We will discuss modern computational approaches to such inference using multiple-locus sequence data later, in Chapter 9.

3.1.3 Classification of tree reconstruction methods

Here we consider some overall features of phylogeny reconstruction methods. First, some methods are *distance based*. In those methods, distances are calculated from pairwise comparisons of sequences, and the resulting distance matrix is used in subsequent analysis. A cluster algorithm is often used to convert the distance matrix into a phylogenetic tree (Everitt et al. 2001). The most popular methods in this category include UPGMA (Unweighted Pair-Group Method using Arithmetic Averages, Sneath 1962) and NJ (neighbour-joining, Saitou and Nei 1987). Other methods are *character based*, which attempt to fit the characters (nucleotides or amino acids, say) observed in all species at every site to a tree. Maximum parsimony (Fitch 1971b; Hartigan 1973), ML (Felsenstein 1981), and Bayesian methods (Rannala and Yang 1996; Mau and Newton 1997; Li et al. 2000) are all character based. Distance methods are often computationally faster than character-based methods, and can be easily applied to analyse different kinds of data as long as pairwise distances can be calculated.

Tree reconstruction methods can also be classified as being either *algorithmic* (cluster methods) or *optimality* based (search methods). The former include UPGMA and NJ, which use cluster algorithms to arrive at a single tree from the data as the best estimate of the true tree. Optimality-based methods use an optimality criterion (objective function) to measure a tree's fit to data, and the tree with the optimal score is the estimate of the true tree (Table 3.2). In the maximum parsimony method, the tree score is the minimum number of character changes required for the tree, and the *maximum parsimony tree* or *most parsimonious tree* is the tree with the smallest tree score. The ML method uses the log

Table 3.2 Optimality criteria used for phylogeny reconstruction

Method	Criterion (tree score)
Maximum parsimony	Minimum number of changes, minimized over ancestral states
Maximum likelihood	Log likelihood score, optimized over branch lengths and model parameters
Minimum evolution	Tree length (sum of branch lengths, often estimated by least squares)
Bayesian	Posterior probability, calculated by integrating over branch lengths and substitution parameters

likelihood value of the tree to measure the fit of the tree to the data, and the *maximum likelihood tree* is the tree with the highest log likelihood value. In the Bayesian method, the posterior probability of a tree is the probability that the tree is true given the data. The tree with the maximum posterior probability is the estimate of the true tree, known as the maximum *a posteriori* (MAP) *tree*. In theory, methods based on optimality criteria have to solve two problems: (i) calculation of the criterion (tree score) for a given tree and (ii) search in the space of all trees to identify the tree with the best score. The first problem can be expensive if the tree is large, but the second is much worse when the number of sequences is greater than 20 or 50 because of the huge number of possible trees. As a result, heuristic algorithms are used for tree searches. Optimality-based search methods are usually much slower than algorithmic cluster methods.

Some tree reconstruction methods are model based. Distance methods use nucleotide or amino acid substitution models to calculate pairwise distances. Likelihood and Bayesian methods use substitution models to calculate the likelihood function. These methods are clearly model based. Parsimony does not make explicit assumptions about the evolutionary process. Opinions differ as to whether the method makes any implicit assumptions, and, if so, what they are. We will return to this issue in Chapter 5.

3.2 Exhaustive and heuristic tree search

3.2.1 *Exhaustive tree search*

For parsimony and likelihood methods of tree reconstruction, which evaluate trees according to an optimality criterion, one should in theory calculate the score for every possible tree and then identify the tree having the best score. Such a strategy is known as *exhaustive search* and is guaranteed to find the best tree. As mentioned above, the stepwise addition algorithm provides a way of enumerating all possible trees for a fixed number of species (Figure 3.5).

An exhaustive search is, however, computationally unfeasible except for small datasets with, say, fewer than ten taxa. For the parsimony method, a branch-and-bound algorithm has been developed to speed up the exhaustive search (Hendy and Penny 1982). Even so, the computation is feasible for small datasets only. For the likelihood method, such an algorithm is not available. Thus computer programs use heuristic algorithms to search in the tree space, which are not guaranteed to find the optimal tree.

3.2.2 *Heuristic tree search*

Heuristic search algorithms may be grouped into two categories. The first includes hierarchical *cluster algorithms*. These may be subdivided into *agglomerative* methods, which proceed by successive fusions of the *n* species into groups, and *divisive* methods, which separate the *n* species successively into finer groups (Everitt et al. 2001). Whether each step involves a fusion or fission, the algorithm involves choosing one out of many alternatives, and the optimality criterion is used to make that choice. The second category of heuristic tree search algorithms includes *tree-rearrangement* or *branch-swapping* algorithms. They propose new trees through local perturbations to the current tree, and the optimality criterion is used to decide whether or not to move to a new tree. The procedure is repeated until no improvement can be made in the tree score. We describe two cluster algorithms in this subsection and a few branch-swapping algorithms in the next.

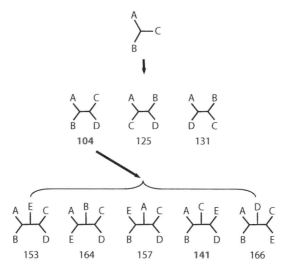

Fig. 3.13 Stepwise addition algorithm under the maximum parsimony criterion. The tree score is the minimum number of changes required by the tree.

Stepwise addition or *sequential addition* is an agglomerative algorithm. It adds sequences one by one, until all sequences are in the tree. When each new sequence is added, all the possible locations are evaluated and the best is chosen using the optimality criterion. Figure 3.13 illustrates the algorithm for the case of five sequences, using the parsimony score as the optimality criterion. Note that this algorithm of heuristic tree search is different from the stepwise addition algorithm for enumerating all possible trees explained in Figure 3.5. In the heuristic search, the locally best subtree is selected at each step, and trees that can be generated from the suboptimal subtrees are ignored. In our example, the ten five-species trees on the second and third rows of Figure 3.5 are never visited in the heuristic search. Thus the algorithm is not guaranteed to find the globally optimal tree. It is less clear whether one should add the most similar sequences or the most divergent sequences first. A common practice is to run the algorithm multiple times, adding sequences in a random order.

Star decomposition is a divisive cluster algorithm. It starts from the star tree of all species, and proceeds to resolve the polytomies by joining two taxa at each step. From the initial star tree of n species, there are $n(n-1)/2$ possible pairs, and the pair that results in the greatest improvement in the tree score is grouped together. The root of the tree then becomes a polytomy with $(n-1)$ taxa. Every step of the algorithm reduces the number of taxa connected to the root by one. The procedure is repeated until the tree is fully resolved. Figure 3.14 shows an example of five sequences, using the log likelihood score for tree selection.

For n species, the stepwise addition algorithm evaluates three trees of four species, five trees of five species, seven trees of six species, with a total of $3+5+7+\cdots+(2n-5) = (n-1)(n-3)$ trees. In contrast, the star decomposition algorithm evaluates $n(n-1)/2 + (n-1)(n-2)/2 + \cdots + 3 = \frac{1}{6}n(n^2-1) - 7$ trees in total, all of which are for n species. Thus for $n > 4$, the star decomposition algorithm evaluates many more and bigger trees than the stepwise addition algorithm and is expected to be much slower. The scores for trees constructed during different stages of the stepwise addition algorithm are not directly

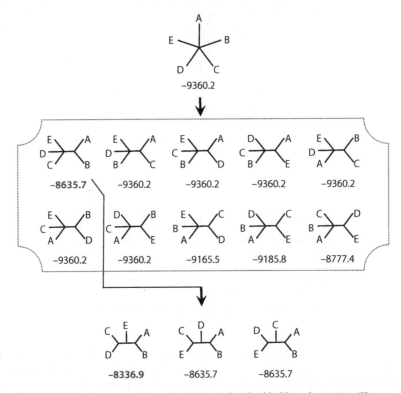

Fig. 3.14 Star decomposition algorithm under the likelihood criterion. The tree score is the log likelihood value calculated by optimizing branch lengths on the tree.

comparable as the trees are of different sizes. Trees evaluated in the star decomposition algorithm are all of the same size and their tree scores are comparable.

Both the stepwise addition and star decomposition algorithms produce resolved trees of all n species. If we stop at the end of either algorithm, we have an algorithmic cluster method for tree reconstruction based on the optimality criterion. However, in most programs, trees generated from these algorithms are treated as starting trees and subjected to local rearrangements. Below are a few such algorithms.

3.2.3 Branch swapping

Branch swapping or tree rearrangements are heuristic algorithms of hill climbing in the tree space. An initial tree is used to start the process. This can be a random tree, or a tree produced by stepwise addition or star decomposition algorithms, or by other faster tree reconstruction methods such as NJ. The branch-swapping algorithm generates a collection of neighbour trees around the current tree. The optimality criterion is then used to decide which neighbour to move to. The branch-swapping algorithm affects our chance of finding the best tree and the amount of computation it takes to do so. If the algorithm generates too many neighbours, each step will require evaluation of too many candidate trees. If the algorithm generates too few neighbours, we do not have to evaluate many trees at each step, but there may be many local peaks in the tree space (see below), and the search can easily get stuck at a local peak.

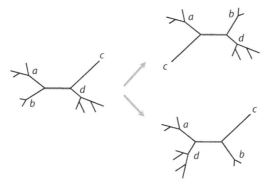

Fig. 3.15 The NNI algorithm. Each internal branch in the tree connects four subtrees or nearest neighbours (*a, b, c, d*). Interchanging a subtree on one side of the branch with another on the other side constitutes an NNI. Two such rearrangements are possible for each internal branch.

Nearest neighbour interchange (NNI). Each internal branch defines a relationship among four subtrees, say, *a*, *b*, *c*, and *d* (Figure 3.15). Suppose the current tree is $((a, b), c, d)$ and the two alternative trees are $((a, c), b, d)$ and $((a, d), b, c)$. The NNI algorithm allows us to move from the current tree to the two alternative trees, by swapping a subtree on one side of the branch with a subtree on the other side. An unrooted tree for *n* species has $n - 3$ internal branches. The NNI algorithm thus generates $2(n - 3)$ immediate neighbours. The neighbourhood relationships among the 15 trees for five species are illustrated in Figure 3.17.

Two other commonly used algorithms are *subtree pruning and regrafting* (SPR) and *tree bisection and reconnection* (TBR) (Swofford et al. 1996). In the former, a subtree is pruned and then reattached to a different location on the tree (Figure 3.16a). In the latter, the tree

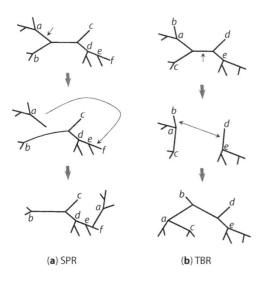

(a) SPR (b) TBR

Fig. 3.16 (a) Branch swapping by SPR. A subtree (for example, the one represented by node *a*) is pruned, and then reattached to a different location on the tree. (b) Branch swapping by TBR. The tree is broken into two subtrees by cutting an internal branch. Two branches, one from each subtree, are then chosen and rejoined to form a new tree.

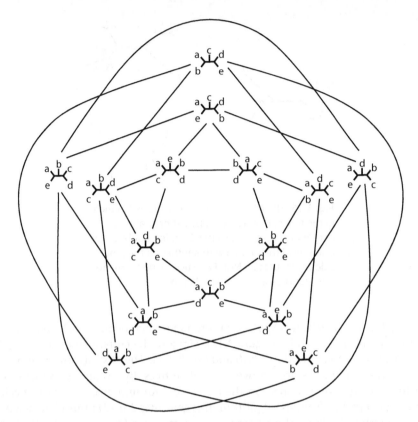

Fig. 3.17 The 15 trees for five species, with neighbourhood relationships defined by the NNI algorithm. Trees that are neighbours under NNI are connected. Note that this visually appealing representation has the drawback that trees close by may not be neighbours. Drawn following Felsenstein (2004).

is cut into two parts by chopping an internal branch and then two branches, one from each subtree, are chosen and rejoined to form a new tree (Figure 3.16b). TBR generates more neighbours than SPR, which in turn generates more neighbours than NNI.

3.2.4 Local peaks in the tree space

Maddison (1991) and Charleston (1995) discussed local peaks or tree islands in the tree space. Figure 3.18 shows an example for five species and 15 trees. The neighbourhood relationship is defined using the NNI algorithm (see Figure 3.17). Each tree has four neighbours, while the ten other trees are two NNI steps away. The parsimony tree lengths for the two trees on the top of the graph, T_1 and T_2, are 1366 and 1362. T_1 is the best tree by the likelihood and Bayesian methods, while T_2 is the most parsimonious tree. Other trees are much worse than those two trees by both the likelihood and parsimony criteria. The eight trees that are neighbours of T_1 or T_2 have tree lengths ranging from 1406 to 1438, while the five trees that are two steps away from T_1 and T_2, have tree lengths ranging from 1488 to 1500. Trees T_1 and T_2 are separated from each other by other trees of much poorer scores and are thus local peaks. They are local peaks for the SPR and TBR algorithms as well. Also T_1 and T_2 are local peaks when the data are analysed under

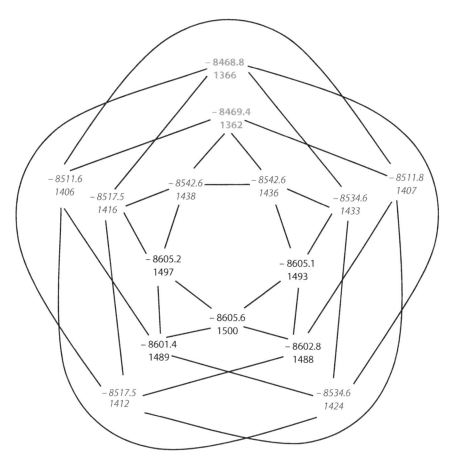

Fig. 3.18 Local peaks in the tree space. The log likelihood values (above) and parsimony scores (below) for the 15 trees of Figure 3.17, shown in the same locations. The dataset was simulated following the construction of Mossel and Vigoda (2005). It consists of 2,000 nucleotide sites simulated under JC69 using the top two trees in Figure 3.17: T_1: ((a, b), c, (d, e)) and T_2: ((a, e), c, (d, b)), with 1,000 sites from each tree. All branch lengths are fixed at 0.1. Trees T_1 and T_2 are two local optima under both parsimony and likelihood criteria. The posterior probabilities for T_1 and T_2 are ~0.64 and ~0.36, respectively.

ML. Indeed for this dataset, the rank order of the 15 trees is almost identical under the likelihood and parsimony criteria. Similarly the dataset may pose serious computational problems for Bayesian Markov chain Monte Carlo algorithms, as discussed by Mossel and Vigoda (2005).

One can design a branch-swapping algorithm under which trees T_1 and T_2 are neighbours. However, such an algorithm will define a different neighbourhood relationship among trees, and may have different local peaks or may have local peaks for different datasets. The problem should be more serious for larger trees with more species, as the tree space is much larger. Similarly, in larger sequence datasets with more sites, the peaks tend to be higher and the valleys deeper, making it very difficult to traverse between peaks (Salter 2001).

3.2.5 *Stochastic tree search*

An optimization algorithm that always goes uphill may get stuck at a local peak. Some algorithms attempt to overcome the problem of local peaks by allowing downhill moves. They can work under either parsimony or likelihood criteria.

The first such algorithm is *simulated annealing* (Metropolis et al. 1953; Kirkpatrick et al. 1983). This is inspired by annealing in metallurgy, a technique involving heating and controlled cooling of a metal to reduce defects. The heat causes the atoms to move at random, exploring various configurations, while the slow cooling allows them to find configurations with low internal energy. In a simulated annealing algorithm of optimization, the objective function is modified to have a flattened surface during the early (heating) stage of the search, making it easy for the algorithm to move between peaks. At this stage downhill moves may be accepted nearly as often as uphill moves. The 'temperature' is gradually reduced as the simulation proceeds, according to some 'annealing schedule'. At the final stage of the algorithm, only uphill moves are accepted, as in a greedy algorithm. Simulated annealing algorithms are highly specific to the problem, and their implementation is more art than science. The efficiency of the algorithm is affected by the neighbourhood function (branch-swapping algorithms) and the annealing schedule. Implementations in phylogenetics include Goloboff (1999) and Barker (2004) for parsimony, and Salter and Pearl (2001) for likelihood. Fleissner et al. (2005) used simulated annealing for simultaneous sequence alignment and phylogeny reconstruction.

A second stochastic tree search algorithm is the *genetic algorithm*. A 'population' of trees is kept in every generation; these are allowed to 'breed' to produce trees of the next generation. The algorithm uses operations that are similar to mutation and recombination in genetics to generate new trees from the current ones. The 'survival' of each tree into the next generation depends on its 'fitness', which is the optimality criterion. Lewis (1998), Katoh et al. (2001), and Lemmon and Milinkovitch (2002), among others, have implemented genetic algorithms to search for the ML tree.

A third stochastic tree search algorithm is the Bayesian Markov chain Monte Carlo (MCMC) algorithm. This is a statistical approach and produces not only a point estimate (the tree with the highest likelihood or posterior probability) but also a measure of uncertainty in the point estimate through posterior probabilities estimated during the search. While MCMC algorithms allow downhill as well as uphill moves, high peaks and deep valleys in the search space can cause serious computational problems. In this regard, we note that both simulated annealing and genetic algorithms have been used to design advanced MCMC algorithms for Bayesian computation. We will discuss Bayesian phylogenetic methods in Chapters 7 and 8.

3.3 Distance matrix methods

Distance methods of phylogeny reconstruction involve two steps: (i) calculation of the distance between every pair of species and (ii) reconstruction of a phylogenetic tree from the distance matrix. The first step has been discussed in Chapters 1 (for nucleotide sequence data) and 2 (for amino acid and codon sequence data). Here we discuss the second step. We describe two optimality-based methods (least-squares and minimum evolution) and one cluster algorithm (neighbour-joining). All distance methods treat the matrix of pairwise distances as observed data. Some of them in addition make use of the variances (and even the covariances) of the estimated distances. After the distance matrix is calculated, the original sequence alignment is no longer used.

3.3.1 Least-squares method

The least-squares (LS) method takes the pairwise distances as observed data and estimates branch lengths on any given tree by trying to match those distances as closely as possible, i.e. by minimizing the sum of squared differences between the observed and expected distances. The expected distance between two species is calculated as the sum of branch lengths along the path on the tree connecting the two species. The minimum sum of squared differences achieved on the tree then measures the fit of the tree to the distance data and is used as the tree score. The tree with the best (least) score is the LS tree, which is the estimate of the true tree. This method was proposed by Cavalli-Sforza and Edwards (1967; see also Edwards and Cavalli-Sforza 1963b), who called it the *additive-tree method*.

More formally, let the observed (calculated) distance between species i and j be d_{ij} and the expected distance be δ_{ij}, which is the sum of branch lengths along the path from species i to j on the tree. Their difference is the error $e_{ij} = d_{ij} - \delta_{ij}$. The closer the errors are to zero, the better the tree and branch lengths fit the data. The LS method estimates the branch lengths by minimizing the sum of the squared errors:

$$S = \sum_{i<j} (d_{ij} - \delta_{ij})^2. \qquad (3.5)$$

For example, the pairwise distances (d_{ij}) calculated under the K80 model for the mitochondrial data of Brown et al. (1982) are shown in Table 3.3. These are taken as observed data. Now consider the tree ((human, chimpanzee), gorilla, orangutan), with its five branch lengths t_0, t_1, t_2, t_3, and t_4 (Figure 3.19). The expected distances in the tree are thus $\delta_{12} = t_1 + t_2$ between the human and the chimpanzee, $\delta_{13} = t_1 + t_0 + t_3$ between the human and the gorilla, and so on. The sum of squared differences is then

$$S = (d_{12} - \delta_{12})^2 + (d_{13} - \delta_{13})^2 + (d_{14} - \delta_{14})^2 + (d_{23} - \delta_{23})^2 + (d_{24} - \delta_{24})^2 + (d_{34} - \delta_{34})^2. \qquad (3.6)$$

In this setup, the distances (d_{ij}) are the observed data and the δs (or more precisely, the five branch lengths t_0, t_1, t_2, t_3, and t_4) are the unknown parameters to be estimated. The

Table 3.3 Pairwise distances for the mitochondrial DNA sequences

1. Human				
2. Chimpanzee	0.0965			
3. Gorilla	0.1140	0.1180		
4. Orangutan	0.1849	0.2009	0.1947	
	1. Human	2. Chimpanzee	3. Gorilla	4. Orangutan

Note: The distance matrix is symmetrical so that only the lower-triangular part is shown. The diagonals are zero.

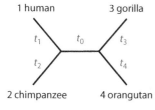

Fig. 3.19 A tree to demonstrate the LS criterion for estimating branch lengths.

Table 3.4 LS branch lengths under K80 for the distance matrix of Table 3.3

Tree	t_0 for internal branch	t_1 for H	t_2 for C	t_3 for G	t_4 for O	S_j
T_1: ((H, C), G, O)	0.008840	0.043266	0.053280	0.058908	0.135795	0.000035
T_2: ((H, G), C, O)	0.000000	0.046212	0.056227	0.061854	0.138742	0.000140
T_3: ((C, G), H, O)						
T_0: (H, C, G, O)						

values of branch lengths that minimize S are the LS estimates. These can be found numerically to be $\hat{t}_0 = 0.008840$, $\hat{t}_1 = 0.043266$, $\hat{t}_2 = 0.053280$, $\hat{t}_3 = 0.058908$, and $\hat{t}_4 = 0.135795$, with the corresponding tree score $S = 0.00003547$. Similar calculations can be done for the other two trees. Indeed, the other two binary trees both converge to the star tree, with the internal branch length estimated to be 0; see Table 3.4. Here we assumed that the branch lengths are estimated under the nonnegative constraint. The tree ((human, chimpanzee), gorilla, orangutan) has the smallest S and is called the LS tree. It is the LS estimate of the true phylogeny. Note that two optimizations are involved: the optimization of branch lengths to calculate the tree score S, and the search in the space of trees for the one with the best tree score, i.e. the LS tree.

Estimation of branch lengths on a fixed tree by the LS criterion uses the same principle as calculating the line of best fit $y = a + bx$ on a scatter plot. If there are no constraints on the branch lengths, the solution is analytical and can be obtained by solving a set of linear equations (Cavalli-Sforza and Edwards 1967). Efficient algorithms that require less computation and less space have also been developed by Rzhetsky and Nei (1993) and Bryant and Waddell (1998). Those algorithms may produce negative branch lengths, which are not meaningful biologically. If the branch lengths are constrained to be nonnegative (as in the above example), the problem becomes one of constrained optimization, which is expensive. However, if we ignore the interpretation of branch lengths, the unconstrained LS is at least consistent: when more and more data are available and the distances approach their true values, the LS tree converges to the true tree. Simulation studies suggest that constraining branch lengths to be nonnegative leads to improved performance in tree reconstruction (e.g. Kuhner and Felsenstein 1994; Gascuel 1997). However, most computer programs implement the LS method without the constraint. It is noted that when the estimated branch lengths are negative, they are most often close to zero.

The LS method described above (i.e. the criterion S of equation (3.5)) uses equal weights for the different pairwise distances and is known as the ordinary least squares (OLS). As in the case of fitting a straight line to a scatter plot, OLS is based on the assumptions that the errors are independent and have equal variance, or equivalently that the (observed) distances are independent and have equal variance. These assumptions are incorrect in the case of pairwise distances. First, larger distances tend to have larger variances. Second, the distances may be correlated because they share branch lengths on the tree. For example, in the tree of Figure 3.19 the distances d_{12} and d_{13} involve the same branch length t_1 so that they both tend to be larger if t_1 is larger: indeed d_{12} and d_{13} have a positive covariance that is equal to the variance of branch length t_1 (Nei and Jin 1989).

The standard approach to dealing with unequal variances is weighted least squares (WLS), which minimizes

$$S = \sum_{i<j} w_{ij}(d_{ij} - \delta_{ij})^2, \tag{3.7}$$

where the weight $w_{ij} = 1/\text{var}(d_{ij})$ (Bulmer 1990). In the method of Fitch and Margoliash (1967), $w_{ij} = 1/d_{ij}^2$ is used. Note that OLS is a special case of WLS with $w_{ij} = 1$. A further extension to WLS that accommodates the correlations (covariances) between the distances, as well as the unequal variances, is the generalized least squares (GLS). While computer simulations suggest that WLS works better than OLS in tree reconstruction, WLS, and especially GLS, involve more computation and are not commonly used.

3.3.2 *Minimum evolution method*

In the LS method discussed above, the LS criterion is used both to estimate the branch lengths on a given tree and to search for the best tree in the tree space. The minimum S for a tree achieved by optimizing its branch lengths is the score for that tree, and at least in theory all possible trees should be compared to find the one with the best score, the LS tree.

In the minimum evolution (ME) method, the LS criterion is usually used to estimate the branch lengths, but tree selection relies on the sum of branch lengths (the tree length). This is based on the plausible but heuristic idea that the true tree is most likely to be the one that involves the minimum amount of evolutionary change. A number of researchers had the same idea at about the same time, including Edwards and Cavalli-Sforza (1963a), Camin and Sokal (1965), and Eck and Dayhoff (1966) (see Edwards 1996, 2009a; Felsenstein 2004). In their analysis of blood group allele frequencies to reconstruct the human population relationships, Edwards and Cavalli-Sforza (1963a) arrived at the *principle of minimum evolution*, which states that 'The most plausible estimate of the evolutionary tree is that which invokes the minimum net amount of evolution'. While the word 'principle' was used, it was intended from the start to be an approximation to the ML method (Edwards 1996). Also it was intended to apply to both continuous and discrete characters. For discrete characters, the amount of evolutionary change should be the minimum number of character changes; so this ME method is now known as parsimony (Camin and Sokal 1965) (see §3.4). For distance data, the amount of evolutionary change is the sum of branch lengths on the tree. Phylogeny reconstruction under the ME criterion based on distances is studied in detail by Kidd and Sgaramella-Zonta (1971) and Rzhetsky and Nei (1993). Gascuel et al. (2001) and Desper and Gascuel (2005) provided excellent reviews of this class of methods.

Variations exist in the practical implementation of the ME principle. First, branch lengths are usually estimated using LS, but as discussed above, variations exist concerning whether the variances and covariances of the observed distances are taken into account and whether the branch lengths are optimized under the nonnegative constraint. Second, several definitions of the tree length exist, differing in their treatment of negative branch lengths. Gascuel et al. (2001) analysed the consistency properties of those variations, and the results are summarized in Table 3.5. A further definition (or estimation method) of tree length is described by Pauplin (2000). This will be described in the next subsection in our discussion of the NJ method.

3.3.3 *Neighbour-joining method*

The simplest distance method is perhaps UPGMA (Sneath 1962). This is a cluster algorithm based on the molecular clock assumption and generates rooted trees. It is thus applicable to population data or closely related species but is not suitable for inferring species phylogenies in general, as the clock is often violated when the sequences are divergent.

3 PHYLOGENY RECONSTRUCTION: OVERVIEW

Table 3.5 Consistency status of minimum evolution method for phylogeny reconstruction

Method	Tree length			
	All-BL	Positive-BL	Absolute-BL	Nonnegative-BL
Ordinary LS	Consistent	Consistent	Consistent	Unknown
Weighted LS	Inconsistent	Inconsistent	Inconsistent	Inconsistent
Generalized LS	Inconsistent	Inconsistent	Inconsistent	Inconsistent

Note: Branch lengths are estimated using Ordinary LS, Weighted LS, or Generalized LS. They are then summed to give the tree length, which is minimized according to the ME criterion. All-BL means that the tree length is calculated as the sum of all branch lengths (both positive and negative) (Rzhetsky and Nei 1993); Positive-BL means the sum of the positive branch lengths, ignoring the negative ones (Swofford and Olsen 1990); Absolute-BL means the sum of the absolute values of the branch lengths (Kidd and Sgaramella-Zonta 1971); and Nonnegative-BL means the sum of the (nonnegative) branch lengths estimated under the nonnegative constraint. For any of the method for estimating branch lengths and for calculating tree length, the ME method selects the shortest tree as being the estimate of the true phylogeny. From Gascuel et al. (2001).

Here we discuss NJ, which is a divisive cluster algorithm proposed by Saitou and Nei (1987). See §3.2.2 for a discussion of divisive and agglomerative cluster algorithms. NJ does not require the clock assumption and produces unrooted trees. It is widely used because it is computationally fast, produces reasonable trees, and has easy-to-use software implementations (Tamura et al. 2011). It starts with a star tree and then chooses a pair of nodes (neighbours) to join (Figure 3.20). A new node is then created to replace the two joined nodes, reducing the number of nodes connected to the root by one and reducing the dimension of the distance matrix by one. The procedure is repeated until the tree is fully resolved. The branch lengths are updated during every step of the algorithm.

Suppose at the current stage of the algorithm, there are r nodes connected to the root (node o). Out of the $r(r-1)/2$ possible pairs of nodes, the pair that minimizes the following Q criterion is chosen to be neighbours for joining:

$$Q_{ij} = (r-2)d_{ij} - \sum_{k=1}^{r}(d_{ik} + d_{jk}), \text{ for } i < j \leq r. \quad (3.8)$$

Suppose nodes i and j are the selected nodes and they are joined to form node u. NJ estimates the length of the branch i–u as

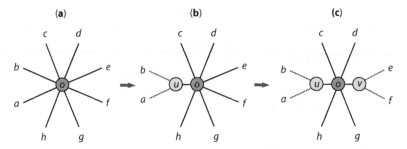

Fig. 3.20 The neighbour-joining method of tree reconstruction is a divisive cluster algorithm. It starts with the star tree (**a**), and chooses a pair of nodes (say, a and b) to join. The two joined nodes are replaced by a new node (node u), reducing the number of nodes connected to the root (node o) by one (**b**). The process is repeated until the tree is fully resolved.

$$d_{iu} = \frac{1}{2}d_{ij} + \frac{1}{2(r-2)}\left[\sum_{k=1}^{r} d_{ik} - \sum_{k=1}^{r} d_{jk}\right]. \tag{3.9}$$

The branch length d_{ju} is calculated similarly. Finally, NJ replaces i and j by u in the distance matrix, using the reduction formula

$$d_{uk} = \frac{1}{2}(d_{ik} - d_{iu}) + \frac{1}{2}(d_{jk} - d_{ju}), \tag{3.10}$$

where k is any node connected to the root (o) other than i and j. Equations (3.8)–(3.10) are due to Studier and Keppler (1988). They are equivalent to and computationally more efficient than those given by Saitou and Nei (1987), according to Gascuel (1994). The two versions always construct the same tree, both in terms of topology and branch lengths.

While good performance of NJ was noted early in computer simulations, its assumptions were not well understood until after mathematical analysis several years later. The discussion below draws heavily on Gascuel and Steel (2006).

Saitou and Nei (1987; see also Nei and Kumar 2000) provided a proof of the consistency of NJ. They showed that pair selection using their equivalent of equation (3.8) minimizes the OLS estimate of the tree length. Accordingly, they considered NJ to be an ME method, minimizing the OLS tree length. However, this proof is not strictly correct because it applies only to the first step of the algorithm, when all the nodes connected to the root are tips, but does not apply to the later steps, when some nodes are interior nodes resulting from joining early neighbours (Gascuel and Steel 2006). Also the consistency of the ME method under OLS tree lengths was not established until Rzhetsky and Nei (1993), later than Saitou and Nei (1987).

Nevertheless, NJ is based on an ME criterion, but the tree length is estimated using a different method from OLS (Gascuel and Steel 2006). Pauplin (2000) studied a method for calculating the tree length directly using the distance matrix. Note that for the example tree of four tips of Figure 3.21a, the tree length is given as

$$l = \frac{1}{2}(d_{ac} + d_{cd} + d_{db} + d_{ba}). \tag{3.11}$$

The rule here is to traverse the tree by visiting pairs of tips in the clockwise direction, i.e. in the order a, c, d, and b, as indicated by the dashed lines. This way each branch on the tree is passed twice, hence the factor $\frac{1}{2}$. If the tree is perfectly additive with the distances to be the true values, equation (3.11) will give the true tree length. Otherwise if the distances are estimates, equation (3.11) will give an estimate of the true tree length. However, the same tree can be drawn in different ways, so that this estimate of tree length is not unique. Then it is natural, as suggested by Pauplin (2000), to average over all possible ways of drawing the same tree. In our example, there is a second way of drawing the same tree, shown in Figure 3.21b, and this gives $l = \frac{1}{2}(d_{ad} + d_{dc} + d_{cb} + d_{ba})$. Averaging over the two ways gives

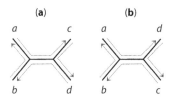

Fig. 3.21 Two different ways of drawing the same tree for four tips to explain Paulin's (2000) direct calculation of tree length from the distance matrix.

$$l = \frac{1}{2}(d_{ab} + d_{cd}) + \frac{1}{4}(d_{ac} + d_{ad} + d_{bc} + d_{bd}). \tag{3.12}$$

In general, Pauplin's (2000) estimate of tree length on a binary tree is

$$l = \sum_{i<j} w_{ij} d_{ij}, \tag{3.13}$$

which averages over all pairwise distances d_{ij}, with the weight w_{ij} to be $\frac{1}{2}$ raised to the power of the number of interior nodes on the path between i and j. This estimate was extended to multifurcating trees by Semple and Steel (2003), in which case the weight w_{ij} is calculated as follows. Consider the directed path from i to j, and for each interior node on the path, count the number of outgoing branches. Multiply those numbers and the reciprocal of the product will be w_{ij}. For example, with the tree of Figure 3.20c, we have $w_{ab} = 1/2$, $w_{cd} = 1/5$, $w_{ae} = 1/(2 \times 5 \times 2)$, and so on. For w_{ae}, note that the path from a to e passes three interior nodes (u, o, v), and the numbers of outgoing branches at those nodes are 2, 5, and 2 respectively. Semple and Steel showed that with this redefinition of w_{ij}, the tree length estimate of equation (3.13) gives exactly the average of estimates of the form of equation (3.11) over all possible ways of drawing the same tree.

Going back to our discussion of NJ, Gascuel and Steel (2006) showed that pair selection in the NJ algorithm by equation (3.8) is equivalent to minimizing the tree length defined in equation (3.13). Of course NJ is a cluster algorithm and does not search for the globally optimal tree under the criterion. One may wonder whether an exhaustive search or a more thorough search than NJ can lead to better performance. The answer to this question is 'Yes'. Indeed this ME method, based on the tree length of equation (3.13), was proposed by Pauplin (2000) and implemented by Desper and Gascuel (2002) as the *balanced ME* method in their FASTME program. Desper and Gascuel's (2002) simulations suggest that FASTME performs better than NJ and other available distance methods (see also Vinh and von Haeseler 2005). Note that equation (3.13) has some flavour of WLS, because a large distance d_{ij} tends to be separated by more interior nodes so that the weight w_{ij} will tend to be smaller (Desper and Gascuel 2004). Desper and Gascuel (2004) also showed that the balanced ME method is consistent.

In summary, NJ is an ME method, but it minimizes the tree length of equation (3.13), not the OLS estimate of tree length. Furthermore the tree length of equation (3.13) is better than the OLS estimate. This explains some counterintuitive results observed in several simulation studies (Gascuel 1997, 2000; Nei et al. 1998). Nei et al. (1998) found that minimizing the OLS tree length leads to poorer performance than NJ. The results prompted the authors to question the optimization principle. The result is unusual, as NJ was justified on the ground that it was based on the ME principle but a more correct implementation (by a more thorough search in the tree space) of the ME principle actually leads to poorer performance than NJ. The optimization principle is justified, but the criterion being optimized is important.

As implied above, a major concern with any distance matrix method is that large distances are poorly estimated, and it is important to take into account their large variances, for example, by using WLS. Besides WLS, Gascuel (1997) modified the formula for updating branch lengths in the NJ algorithm to incorporate approximate variances and covariances of distance estimates. This method, called BIONJ, is close to WLS, and was found to outperform NJ, especially when substitution rates are high and variable among lineages. Another modification is the weighted NJ or WEIGHBOR method of Bruno et al. (2000). This uses an approximate likelihood criterion for joining nodes to accommodate the fact that large distances are poorly estimated. Computer simulations suggest that

WEIGHBOR produces trees similar to ML, and is more robust to the problem of long-branch attraction (see §3.4.5) than NJ (Bruno et al. 2000). Another idea, due to Ranwez and Gascuel (2002), is to improve distance estimates. When calculating the distance between a pair of sequences, those authors used a third sequence to break the long distance into two parts and used ML to estimate three branch lengths on the tree of the three sequences; the pairwise distance is then calculated as the sum of the two branch lengths. Simulations suggest that the improved distance, when combined with the NJ, BIONJ, and WEIGHBOR algorithms, led to improved topological accuracy.

3.4 Maximum parsimony

3.4.1 *Brief history*

Felsenstein (2004) and Edwards (2009a) have published accounts of the early history of phylogeny reconstruction methods. Edwards and Cavalli-Sforza (1963a) suggested the *minimum evolution principle* (later renamed the minimum evolution method) as an approximation to the ML solution. For discrete characters, the amount of evolution should be measured by the minimum number of character changes on the tree. In modern terminology, the method applied to discrete data is known as parsimony, while ME refers to methods minimizing the sum of branch lengths after correcting for multiple hits, as discussed in last section. For discrete morphological characters, Camin and Sokal (1965) suggested the use of the minimum number of changes as a criterion for tree selection, justifying it by arguing that evolution follows the shortest paths, a view sharply criticized by Edwards (1996, 2009a). For molecular data, minimizing changes on the tree to infer ancestral proteins appears most natural and was practised by many pioneers in the field, for example, by Pauling and Zuckerkandl (1963) and Zuckerkandl (1964) as a way of 'restoring' ancestral proteins for 'paleogenetic' studies of their chemical properties, and by Eck and Dayhoff (1966) to construct empirical matrices of amino acid substitution rates. Fitch (1971b) was the first to present a systematic algorithm to enumerate all and only the most parsimonious reconstructions. Fitch's algorithm works on binary trees only. Hartigan (1973) considered multifurcating trees as well and provided a mathematical proof for the algorithm. Since then, much effort has been made to develop fast algorithms for the parsimony analysis of large datasets; see, e.g. Ronquist (1998), Nixon (1999), and Goloboff (1999).

3.4.2 *Counting the minimum number of changes on a tree*

The minimum number of character changes at a site on a given tree is often called the *character length* or *site length*. The sum of character lengths over all sites in the sequence is the minimum number of required changes for the entire sequence and is called the *tree length*, *tree score*, or *parsimony score*. The tree with the smallest tree score is the estimate of the true tree, called the *maximum parsimony tree* or the *most parsimonious tree*. It is common, especially when the sequences are very similar, for multiple trees to be equally best; i.e. they have the same minimum score and are all shortest trees.

Suppose the data for four species at a particular site are AAGG, and consider the minimum number of changes required by the two trees of Figure 3.22. We calculate this number by assigning character states to the extinct ancestral nodes. For the first tree, this is achieved by assigning A and G to the two nodes, and one change (A ↔ G on the internal branch) is required. For the second tree, we can assign either AA (shown) or

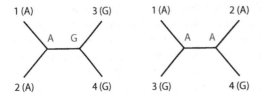

Fig. 3.22 Data AAGG at one site for four species mapped onto two alternative trees ((1, 2), 3, 4) and ((1, 3), 2, 4). The tree on the left requires a minimum of one change while the tree on the right requires two changes to explain the data.

GG (not shown) to the two internal nodes; in either case, a minimum of two changes is required. Note that the set of character states (nucleotides) at a site assigned to ancestral nodes is called an *ancestral reconstruction*. The total number of reconstructions at each site is thus $4^{(n-2)}$ for nucleotides or $20^{(n-2)}$ for amino acids as a binary unrooted tree of n species has $n - 2$ interior nodes. The reconstruction that achieves the minimum number of changes is called the *most parsimonious reconstruction*. Thus, for the first tree, there is one single most parsimonious reconstruction, while for the second tree, two reconstructions are equally parsimonious. The algorithm of Fitch (1971b) and Hartigan (1973) calculates the minimum number of changes and enumerates all the most parsimonious reconstructions at a site. We will not describe this algorithm here. Instead we describe in the next subsection a more general algorithm due to Sankoff (1975), which is very similar to the likelihood algorithm to be discussed in Chapter 4.

Some sites do not contribute to the discrimination of trees and are thus noninformative. For example, a constant site, at which the different species have the same nucleotide, requires no change for any tree. Similarly a *singleton* site, at which two states are observed but one is observed only once (e.g. TTTC or AAGA), requires one change for every tree and is thus not informative. Perhaps more strikingly, a site with data AAATAACAAG (for ten species) is not informative either, as a minimum of three changes are required by any tree, which is also achieved by every tree by assigning A to all ancestral nodes. For a site to be a *parsimony-informative* site, at least two characters have to be observed, each at least twice. Note that the concepts of informative and noninformative sites apply to parsimony only. In distance and likelihood methods, all sites including the constant sites affect the calculation and should be included.

We often refer to the observed character states in all species at a site as a *site configuration* or *site pattern*. The above discussion means that for four species, only three site patterns are informative: *xxyy*, *xyxy*, and *xyyx*, where x and y are any two distinct states. It is obvious that those three site patterns 'support' the three trees T_1: ((1, 2), 3, 4); T_2: ((1, 3), 2, 4); and T_3: ((1, 4), 2, 3), respectively. Suppose the numbers of sites with those site patterns are n_1, n_2, and n_3, respectively. Then T_1, T_2, or T_3 is the most parsimonious tree if n_1, n_2, or n_3 is the greatest among the three.

3.4.3 Weighted parsimony and dynamic programming

The algorithm of Fitch (1971b) and Hartigan (1973) assumes that every change has the same cost. In weighted parsimony, different weights are assigned to different types of character changes. Rare changes are penalized more heavily than frequent changes. For example, transitions are known to occur at a higher rate than transversions and can be assigned a lower cost (weight). Weighted parsimony uses a *step matrix* or *cost matrix* to specify the cost of every type of change. An extreme case is *transversion parsimony*, which gives a penalty of 1 for a transversion but no penalty for a transition. Below we describe Sankoff's (1975) dynamic programming algorithm, which calculates the minimum cost at

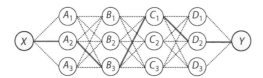

Fig. 3.23 Caravan travelling example used for illustrating the dynamic programming algorithm. It is required to determine the shortest route from X to Y, through four countries A, B, C, and D. Stops between neighbouring countries are connected, with their distances known.

a site and enumerates the reconstructions that achieve this minimum given any arbitrary cost matrix.

We first illustrate the basic idea of dynamic programming algorithms using a fictitious example of a caravan travelling on the silk route. We start from the source city X, Chang-an in central China, to go to the destination Y, Baghdad in Iraq (Figure 3.23). The route goes through four countries A, B, C, and D, and has to pass one of three caravan stops in every country: A_1, A_2, or A_3 in country A; B_1, B_2, or B_3 in country B; and so on. We know the distance between any two stops in two neighbouring countries, such as XA_2 and A_1B_2. We seek to determine the shortest distance and the shortest route from X to Y. An obvious strategy is to evaluate all possible routes, but this can be expensive as the number of routes (3^4 in the example) grows exponentially with the number of countries. A dynamic programming algorithm answers many smaller questions, with the new questions building on answers to the old ones. First we ask for the shortest distances (from X) to stops A_1, A_2, and A_3 in country A. These are just the given distances. Next we ask for the shortest distances to stops in country B, and then the shortest distances to stops in country C, and so on. Note that the questions at every stage are easy given the answers to the previous questions. For example, consider the shortest distance to C_1, when the shortest distances to B_1, B_2, and B_3 are already determined. This is just the smallest among the distances of the three routes going through B_1, B_2, or B_3, with the distance through B_j ($j = 1$, 2, 3) being the shortest distance (from X) to B_j plus the distance between B_j and C_1. After the shortest distances to D_1, D_2, and D_3 are determined, it is easy to determine the shortest distance to Y itself. It is important to note that adding another country to the problem will add another stage in the algorithm, so that the amount of computation grows linearly with the number of countries.

We now describe Sankoff's algorithm. We seek to determine the minimum cost for a site on a given tree as well as the ancestral reconstruction that achieves that minimum. We use the tree of Figure 3.24 as an example. The observed nucleotides at the site at the six tips are CCAGAA. Let $c(x, y)$ denote the cost of change from state x to state y, so $c(x, y) = 1$ for a transitional difference and $c(x, y) = 1.5$ for a transversion (Figure 3.24).

Instead of the minimum cost for the whole tree, we calculate the minimum costs for many subtrees. We refer to a branch on the tree by the node it leads to or by the two nodes it connects. For example, branch 10 is also branch 8–10 in Figure 3.24. We say that each node i on the tree defines a subtree, referred to as subtree i, which consists of branch i, node i, and all its descendant nodes. For example, subtree 3 consists of the single tip branch 10–3 while subtree 10 consists of branch 8–10 and nodes 10, 3, and 4. Define $S_i(x)$ as the minimum cost incurred on subtree i, given that the mother node of node i has

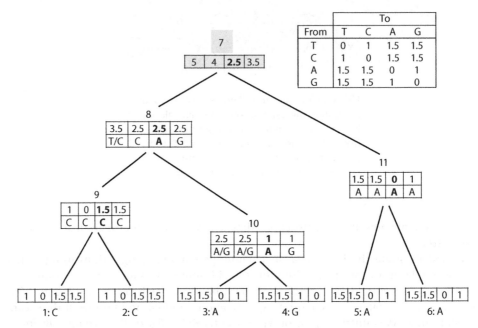

Fig. 3.24 Dynamic programming algorithm for calculating the minimum cost and enumerating the most parsimonious reconstructions using weighted parsimony. The site has observed data CCAGAA. The cost vector at each node gives the minimum cost of the subtree induced by that node (which includes the node itself, its mother branch and all its descendants), given that the mother node has nucleotides T, C, A, or G. The nucleotides at the node that achieved the minimum cost are shown below the cost vector. For example, the minimum cost of the subtree induced by node 3 (including the single branch 10–3) is 1.5, 1.5, 0, or 1, if node 10 has T, C, A, or G, respectively. The minimum cost of the subtree induced by node 10 (including branches 8–10 and nodes 10, 3 and 4) is 2.5, 2.5, 1, or 1, if node 8 has T, C, A, or G, respectively; the said minimum is achieved by node 10 having A/G, A/G, A, or G, respectively. The cost vectors are calculated for every node, starting from the tips and proceeding towards the root. At the root (node 7), the cost vector gives the minimum cost of the whole tree as 5, 4, 2.5, or 3.5, if the root has T, C, A, or G, respectively.

state x. Thus $\{S_i(T), S_i(C), S_i(A), S_i(G)\}$ constitutes a cost vector for subtree i at node i. They are like the shortest distances to stops in a particular country in the caravan example. We calculate the cost vectors for all nodes on the tree, starting with the tips and visiting a node only after we have visited all its descendant nodes. For a tip node i, the subtree is just the tip branch and the cost is simply read from the cost matrix. For example, tip 3 has the cost vector $\{1.5, 1.5, 0, 1\}$, meaning that the (minimum) cost of subtree 3 is 1.5, 1.5, 0, or 1, if mother node 10 has T, C, A, or G, respectively (Figure 3.24). If the nucleotide at the tip is undetermined, the convention is to use the minimum cost among all compatible states (Fitch 1971b). For an interior node i, suppose its two daughter nodes are j and k. Then

$$S_i(x) = \min_y [c(x, y) + S_j(y) + S_k(y)]. \tag{3.14}$$

Note that subtree i consists of branch i plus subtrees j and k. Thus the minimum cost of subtree i is the cost along branch i, $c(x, y)$, plus the minimum costs of subtrees j and k, minimized over the state y at node i. We use $C_i(x)$ to record the state y that achieved the minimum.

Consider node 10 as an example, for which the cost vector is calculated to be $\{S_{10}(T), S_{10}(C), S_{10}(A), S_{10}(G)\} = \{2.5, 2.5, 1, 1\}$. Here the first entry, $S_{10}(T) = 2.5$, means that the minimum cost of subtree 10, given that mother node 8 has T, is 2.5. To see this, consider the four possible states at node 10: $y = $ T, C, A, or G. The (minimum) cost on subtree 10 is $3 = 0 + 1.5 + 1.5, 4, 2.5$, or 2.5, if node 10 has the state $y = $ T, C, A, or G, respectively (and if node 8 has T). Thus the minimum is 2.5, achieved by node 10 having $y = $ A or G; i.e. $S_{10}(T) = 2.5$ and $C_{10}(T) = $ A or G (Figure 3.24). This is the minimization over y in equation (3.14). Similarly, the second entry in the cost vector at node 10, $S_{10}(C) = 2.5$, means that the minimum cost for subtree 10, given that node 8 has C, is 2.5. This minimum is achieved by having $C_{10}(C) = $ A/G at node 10.

Similar calculations can be done for nodes 9 and 11. We now consider node 8, which has daughter nodes 9 and 10. The cost vector is calculated to be $\{3.5, 2.5, 2.5, 2.5\}$, meaning that the minimum cost of subtree 8 is 3.5, 2.5, 2.5, or 2.5, if mother node 7 has T, C, A, or G, respectively. Here we derive the third entry $S_8(A) = 2.5$, with mother node 7 having A. By using the cost vectors for nodes 9 and 10, we calculate the minimum cost on subtree 8 to be $5 = 1.5 + 1 + 2.5, 4, 2.5$, or 4.5, if node 8 has T, C, A, or G, respectively (and if mother node 7 has A). Thus $S_8(A) = 2.5$ is the minimum, achieved by node 8 having $C_8(A) = $ A.

The algorithm is applied successively to all nodes in the tree, starting from the tips and moving towards the root. This upper pass calculates $S_i(x)$ and $C_i(x)$ for all nodes i except the root. Suppose the root has daughter nodes j and k and note that the whole tree consists of subtrees j and k. The minimum cost of the whole tree, given that the root has y, is $S_j(y) + S_k(y)$. This cost vector is $\{5, 4, 2.5, 3.5\}$, for $y = $ T, C, A, G at the root (Figure 3.24). The minimum is 2.5, achieved by having A at the root. In general, if j and k are the daughter nodes of the root, the minimum cost for the whole tree is

$$S = \min_y [S_j(y) + S_k(y)]. \tag{3.15}$$

After calculation of $S_i(x)$ and $C_i(x)$ for all nodes through the upper pass, a down pass reads out the most parsimonious reconstructions. In our example, given A for the root, node 8 achieves the minimum for subtree 8 by having A. Given A at node 8, nodes 9 and 10 should have C and A, respectively. Similarly given A for the root, node 11 should have A. Thus the most parsimonious reconstruction at the site is $y_7 y_8 y_9 y_{10} y_{11} = $ AACAA, with the minimum cost 2.5.

3.4.4 Probabilities of ancestral states

Obviously the ancestral states reconstructed by parsimony may not always be the true states. Many authors thus recognized the desirability of calculating the probability that the parsimony reconstructions are the true states (Fitch 1971b; Maddison and Maddison 1982). This can only be achieved by the use of a character evolution model. Unfortunately most of those calculations do not appear to be correct (e.g. Fitch 1971b; Schluter 1995; Pagel 1999) or relevant (e.g. Maddison 1995). We defer the discussion of such calculations to §4.4, where the correct approach is described.

3.4.5 Long-branch attraction

Felsenstein (1978b) demonstrated that the parsimony method can be statistically inconsistent under certain combinations of branch lengths on a four-species tree. When the

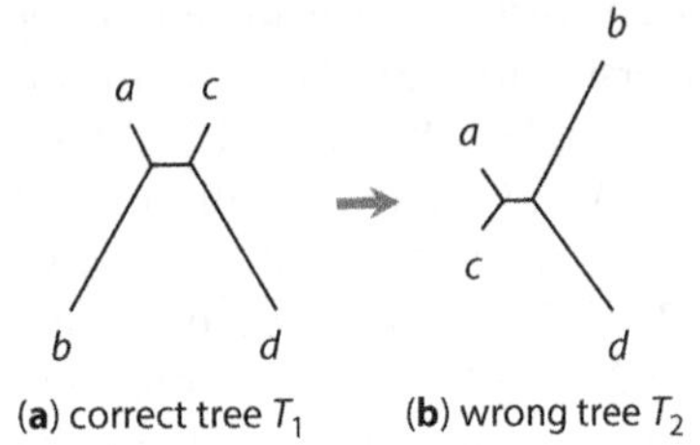

Fig. 3.25 Long-branch attraction. When the correct tree (T_1) has two long branches separated by a short internal branch, parsimony tends to recover a wrong tree (T_2) with the two long branches grouped together.

amount of data (the number of sites) increases to infinity, it becomes more and more certain that the most parsimonious tree is an incorrect tree.

The tree Felsenstein used has the characteristic shape shown in Figure 3.25a, with two long branches separated by a short internal branch. The estimated tree by parsimony, however, tends to group the two long branches together (Figure 3.25b). This phenomenon is now known as 'long-branch attraction'. Using a simple model of character evolution, Felsenstein calculated the probabilities of observing sites with the three site patterns *xxyy*, *xyxy*, *xyyx*, where *x* and *y* are any two distinct characters, and found that Pr(*xyxy*) > Pr(*xxyy*) when the two long branches are much longer than the three short branches. This calculation will be described later in Chapter 4 (see Problem 4.3). Thus with more and more sites in the sequence, it will be increasingly certain that more sites will have pattern *xyxy* than pattern *xxyy*, and that parsimony will recover the wrong tree T_2 instead of the true tree T_1. The phenomenon has been demonstrated in many simulated and real datasets (see, e.g. Huelsenbeck 1998) and is due to the failure of parsimony to correct for parallel changes on the two long branches. Likelihood and distance methods using simplistic and unrealistic evolutionary models show the same behaviour.

3.4.6 *Assumptions of parsimony*

A discussion of the assumptions underlying the parsimony method of phylogeny reconstruction is provided in Chapter 5. Here we comment on a few obvious concerns on the parsimony reconstruction of ancestral states. First, the method ignores branch lengths. Some branches on the tree are longer than others, meaning that they have accumulated more evolutionary changes than other branches. It is thus illogical to assume that a change is as likely to occur on a long branch as on a short one, as parsimony does, when character states are assigned to ancestral nodes on the tree. Second, the simple parsimony criterion ignores different rates of changes between nucleotides. Such rate differences are taken into account by weighted parsimony through the use of a step matrix, although determining the appropriate weights may be nontrivial. In theory, how likely a change is to occur on a particular branch should depend on the length of the branch as well as the relative rate of the change. If one attempts to derive appropriate weights from the observed data, one will naturally be led to the likelihood method, which uses a Markov chain model to describe the nucleotide substitution process, relying on probability theory to accommodate unequal branch lengths, unequal substitution rates between nucleotides, and any other features of the evolutionary process. This is the topic of next chapter.

3.5 Problems

3.1 Draw the tree

(((human: 0.040, chimpanzee: 0.052): 0.016, gorilla: 0.059): 0.047, orangutan: 0.090, gibbon: 0.125);

The branch lengths are the MLEs under JC69 obtained from the mitochondrial data of Brown et al. (1982). Identify the most distant pair of species and use midpoint rooting to root the tree. Draw the resulting rooted tree.

3.2 Write two equivalent Newick representations of the tree in Figure 3.9b.

3.3 The following rooted tree is shown in Figure 3.26:

(a:0.05, (c: 0.07, ((b:0.015, f:0.12) :0.01, (d:0.01, e:0.4) :0.005) :0.03) :0.025);

Which of the following statements are incorrect?
(a) Species d and e are most closely related.
(b) Sequences b and d are most similar.
(c) Species b is more closely related to d than to e.
(d) Species d is more closely related to c than to f.

Fig. 3.26 A tree showing branch lengths for Problem 3.3.

3.4 Calculate the partition distance between the two trees of Figure 3.11.

3.5 Use the three trees of Figure 3.27 to construct the majority-rule consensus tree, and show the support values for the nodes on it.

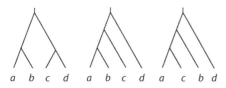

Fig. 3.27 Three rooted trees for constructing the majority-rule consensus tree in Problem 3.5.

CHAPTER 4

Maximum likelihood methods

4.1 Introduction

In this chapter, we will discuss likelihood calculation for multiple sequences on a phylogenetic tree. As indicated at the end of last chapter, this is a natural extension to the parsimony method, when we want to incorporate differences in branch lengths and in substitution rates between nucleotides. Likelihood calculation on a tree is also a natural extension to estimation of the distance between two sequences, discussed in Chapter 1. Indeed Chapter 1 has covered the general principles of Markov chain theory and maximum likelihood (ML) estimation needed in this chapter.

It may be beneficial to distinguish two applications of ML in phylogenetic analysis. The first is estimation of parameters in the evolutionary model and testing of hypotheses concerning the evolutionary process when the tree topology is known or fixed. The likelihood method, with its nice statistical properties, provides a powerful and flexible framework for such analysis (e.g. Stuart et al. 1999). The second is inference of the tree topology. The log likelihood for each tree is maximized by optimizing branch lengths and other substitution parameters, and the optimized log likelihood is used as a tree score for comparing different trees. This second application of ML corresponds to comparison of many statistical models. It involves complexities, which will be discussed in Chapter 5.

4.2 Likelihood calculation on tree

4.2.1 Data, model, tree, and likelihood

The likelihood is defined as the probability of observing the data when the parameters are given, although it is considered to be a function of the parameters. The data consist of s aligned homologous sequences, each n nucleotides long, and can be represented as an $s \times n$ matrix $X = \{x_{jh}\}$, where x_{jh} is the hth nucleotide in the jth sequence. Let $\mathbf{x}_h$ denote the hth column in the data matrix. To define the likelihood, we have to specify the model by which the data are generated. Here we use the K80 nucleotide substitution model (Kimura 1980). We assume that different sites evolve independently of each other and evolution in one lineage is independent of other lineages. We use the tree of five species of Figure 4.1 as an example to illustrate the likelihood calculation. The observed data at a particular site, TCACC, are shown. The ancestral nodes are numbered 0, 6, 7, and 8, with 0 being the root. The length of the branch leading to node i is denoted t_i, defined as the expected number of nucleotide substitutions per site. The parameters in the model include the branch lengths and the transition/transversion rate ratio κ, collectively denoted $\theta = \{t_1, t_2, t_3, t_4, t_5, t_6, t_7, t_8, \kappa\}$.

Molecular Evolution: A Statistical Approach. Ziheng Yang. © Ziheng Yang 2014.
Published 2014 by Oxford University Press.

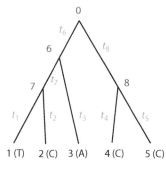

Fig. 4.1 A tree of five species used to demonstrate calculation of the likelihood function. The nucleotides observed at the tips at a site are shown. Branch lengths t_1–t_8 are measured by the expected number of nucleotide substitutions per site.

Because of the assumption of independent evolution among sites, the probability of the whole dataset (the alignment) is the product of the probabilities of data at individual sites:

$$L(\theta) = f(X|\theta) = \prod_{h=1}^{n} f(\mathbf{x}_h|\theta). \tag{4.1}$$

Equivalently the log likelihood is a sum over sites in the sequence

$$\ell = \log\{L(\theta)\} = \sum_{h=1}^{n} \log\{f(\mathbf{x}_h|\theta)\}. \tag{4.2}$$

Here we consider calculation of ℓ when parameters θ are given. We focus on one site, with the data $\mathbf{x}_h$ = TCACC, say. We use x_i to represent the state at ancestral node i, and suppress the subscript h. Since the data at the site can result from any combination of ancestral nucleotides $x_0 x_6 x_7 x_8$, calculation of $f(\mathbf{x}_h)$ has to sum over all possible nucleotide combinations for the extinct ancestors (nodes 0, 6, 7, and 8)

$$f(\mathbf{x}_h|\theta) = \sum_{x_0}\sum_{x_6}\sum_{x_7}\sum_{x_8} \left[\pi_{x_0} p_{x_0 x_6}(t_6) p_{x_6 x_7}(t_7) p_{x_7 T}(t_1) p_{x_7 C}(t_2) p_{x_6 A}(t_3) p_{x_0 x_8}(t_8) p_{x_8 C}(t_4) p_{x_8 C}(t_5)\right]. \tag{4.3}$$

Here the summation over each of x_0, x_6, x_7, x_8 is over the four nucleotides T, C, A, G. The quantity in the square brackets is the probability of data TCACC for the tips and $x_0 x_6 x_7 x_8$ for the ancestral nodes. This is equal to the probability that the root (node 0) has x_0, which is given by $\pi_{x_0} = 1/4$ under K80, multiplied by eight transition probabilities along the eight branches of the tree. We discussed calculation of the transition probabilities in Chapter 1; for example, those under K80 are given in equation (1.10).

Note that given θ, we are able to calculate $f(\mathbf{x}_h|\theta)$ and the log likelihood ℓ. The ML method then estimates θ by maximizing ℓ, often using numerical optimization algorithms (to be discussed in §4.5).

4.2.2 The pruning algorithm

4.2.2.1 Horner's rule and the pruning algorithm

Summing over all combinations of ancestral states is expensive because there are 4^{s-1} possible combinations for $s-1$ interior nodes. The situation is even worse for amino acid or codon sequences as there will then be 20^{s-1} or 61^{s-1} possible combinations. An important technique that is useful in calculating such sums is to identify common factors and calculate them only once. This is known as the *nesting rule* or *Horner's rule*, published by the Irish mathematician William Horner in 1830. The rule was also published in 1820 by a London watchmaker, Theophilus Holdred, and the same principle had been used in 1303 by the Chinese mathematician Zhu Shijie (朱世杰). By this rule, an nth-order polynomial can be calculated with only n multiplications and n additions. For example, a naïve calculation of $1 + 2x + 3x^2 + 4x^3$, as $1 + 2 \cdot x + 3 \cdot x \cdot x + 4 \cdot x \cdot x \cdot x$, requires six multiplications and three additions. However, by writing it as $1 + x \cdot (2 + x \cdot (3 + 4 \cdot x))$, only three multiplications and three additions are needed. As another example, $\sum_{i=1}^{10} \sum_{j=1}^{10} (x_i y_{ij}) = \sum_{i=1}^{10} \left[x_i \left(\sum_{j=1}^{10} y_{ij} \right) \right]$, but the left-hand side involves 100 multiplications and 99 additions while the right-hand side involves only ten multiplications and 99 additions.

If we apply the nesting rule and move the summation signs in equation (4.3) to the right as far as possible, we get

$$f(\mathbf{x}_h|\theta) = \sum_{x_0} \pi_{x_0} \left\{ \sum_{x_6} p_{x_0 x_6}(t_6) \left[\left(\sum_{x_7} p_{x_6 x_7}(t_7) p_{x_7 T}(t_1) p_{x_7 C}(t_2) \right) p_{x_6 A}(t_3) \right] \right\} \\ \times \left[\sum_{x_8} p_{x_0 x_8}(t_8) p_{x_8 C}(t_4) p_{x_8 C}(t_5) \right]. \quad (4.4)$$

Thus we sum over x_7 before x_6, and sum over x_6 and x_8 before x_0. In other words, we sum over ancestral states at a node only after we have done so for all its descendant nodes.

The pattern of parentheses and the occurrences of the tip states in equation (4.4), in the form [(T, C), A], [C, C], match the tree of Figure 4.1. This is no coincidence. Indeed calculation of $f(\mathbf{x}_h|\theta)$ by equation (4.4) constitutes the *pruning algorithm* of Felsenstein (1973b, 1981). This is a variant of the dynamic programming algorithm discussed in §3.4.3. Its essence is to successively calculate probabilities of data at the site on many subtrees. Define $L_i(x_i)$ to be the conditional probability of observing data at the tips that are descendants of node i, given that the nucleotide at node i is x_i. For example, tips 1, 2, 3 are descendants of node 6, so $L_6(T)$ is the probability of observing $x_1 x_2 x_3$ = TCA, given that node 6 has the state x_6 = T. With x_i = T, C, A, G, we calculate a vector of conditional probabilities for each node i. In the literature, the conditional probability $L_i(x_i)$ is often referred to as the 'partial likelihood' or 'conditional likelihood'; these are misnomers since likelihood refers to the probability of the whole dataset and not probability of data at a single site or part of a single site.

If node i is a tip, its descendant tips include tip i itself only, so that $L_i(x_i) = 1$ if x_i is the observed nucleotide, or 0 otherwise. If node i is an interior node with daughter nodes j and k, we have

$$L_i(x_i) = \left[\sum_{x_j} p_{x_i x_j}(t_j) L_j(x_j) \right] \times \left[\sum_{x_k} p_{x_i x_k}(t_k) L_k(x_k) \right]. \quad (4.5)$$

This is a product of two terms, corresponding to the two daughter nodes j and k. Note that tips that are descendants of node i must be descendants of either j or k. Thus the probability $L_i(x_i)$ of observing all descendant tips of node i (given the state x_i at node i) is equal to the probability of observing data at the descendant tips of node j (given x_i) times the probability of observing data at the descendant tips of node k (given x_i). These are the two terms in the two pairs of brackets in equation (4.5), respectively. For example, node i = 6 has daughter nodes j = 7 and k = 3, and descendant tip nodes 1, 2, 3. The probability of observing $x_1 x_2 x_3$ given x_6 is the probability of observing $x_1 x_2$ given x_6, times the probability of observing x_3 given x_6. Given the state x_i at node i, the two parts of the tree down node i are independent. If node i has more than two daughter nodes, $L_i(x_i)$ will be a product of as many terms. Now consider the first term, the term in the first pair of brackets, which is the probability of observing data at descendant tips of node j (given the state x_i at node i). This is the probability $p_{x_i x_j}(t_j)$ that x_i will become x_j over branch length t_j times the probability $L_j(x_j)$ of observing the tips of node j given the state x_j at node j, summed over all possible states x_j.

We calculate the conditional probability vector $L_i(x_i)$ for node i only after the vectors $L_j(x_j)$ and $L_k(x_k)$ for its daughter nodes j and k have been calculated. Thus we calculate the probabilities of data $x_1 x_2$ down node 7, then the probabilities of data $x_1 x_2 x_3$ down node 6, then the probabilities of data $x_4 x_5$ down node 8, and finally the probabilities of the whole data $x_1 x_2 x_3 x_4 x_5$ down node 0. The calculation proceeds from the tips of the tree towards the root, visiting each node only after all its descendant nodes have been visited. In computer science, this way of visiting all nodes on the tree is known as the *post-order tree traversal* (as opposed to *pre-order tree traversal*, in which ancestors are visited before descendants). After visiting all nodes on the tree and calculating the probability vector for the root $L_0(x_0)$, the probability of data at the site is given as

$$f(\mathbf{x}_h|\theta) = \sum_{x_0} \pi_{x_0} L_0(x_0). \tag{4.6}$$

Note that π_{x_0} is the (prior) probability that the nucleotide at the root is x_0, given by the equilibrium frequency of the nucleotide x_0 under the model.

Example 4.1. We use the tree of Figure 4.1 to provide a numerical example of the calculation using the pruning algorithm at one site (Figure 4.2). For definiteness, we fix internal

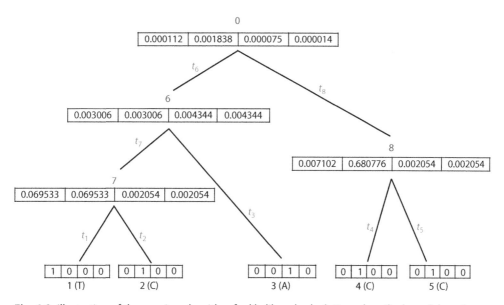

Fig. 4.2 Illustration of the pruning algorithm for likelihood calculation when the branch lengths and other parameters are fixed. The tree of Figure 4.1 is reproduced, showing the vector of conditional probabilities at each node. The four elements in the vector at each node are the probabilities of observing data at the descendant tips, given that the node has T, C, A, or G, respectively. For example, 0.069533 for node 7 is the probability of observing data $x_1 x_2 =$ TC at tips 1 and 2, given that node 7 has T. The K80 model is assumed, with $\kappa = 2$. The branch lengths are fixed at 0.1 for the internal branches and 0.2 for the external branches. The transition probability matrices are shown in the text.

branch lengths at $t_6 = t_7 = t_8 = 0.1$ and external branch lengths at $t_1 = t_2 = t_3 = t_4 = t_5 = 0.2$. We also set $\kappa = 2$. The two transition probability matrices are as follows, in which the ijth element is $p_{ij}(t)$, with the nucleotides ordered T, C, A, and G (see equation (1.10) for K80):

$$P(0.1) = \begin{bmatrix} 0.906563 & 0.045855 & 0.023791 & 0.023791 \\ 0.045855 & 0.906563 & 0.023791 & 0.023791 \\ 0.023791 & 0.023791 & 0.906563 & 0.045855 \\ 0.023791 & 0.023791 & 0.045855 & 0.906563 \end{bmatrix},$$

$$P(0.2) = \begin{bmatrix} 0.825092 & 0.084274 & 0.045317 & 0.045317 \\ 0.084274 & 0.825092 & 0.045317 & 0.045317 \\ 0.045317 & 0.045317 & 0.825092 & 0.084274 \\ 0.045317 & 0.045317 & 0.084274 & 0.825092 \end{bmatrix}.$$

Consider node 7, which has daughter nodes 1 and 2. Using equation (4.5), we obtain the first entry in the probability vector as $L_7(T) = p_{TT}(0.2) \times p_{TC}(0.2) = 0.825092 \times 0.084274 = 0.069533$. This is the probability of observing T and C at tips 1 and 2, given that node 7 has T. The other entries, $L_7(C), L_7(A)$, and $L_7(G)$, can be calculated similarly, as can the vector for node 8. Next the vector at node 6 can be calculated, by using the conditional probability vectors at daughter nodes 7 and 3. Finally, we calculate the vector for node 0, the root. The first entry, $L_0(T) = 0.000112$, is the probability of observing the descendant tips (1, 2, 3, 4, 5) of node 0, given that node 0 has $x_0 = T$. Equation (4.5) gives this as the product of two terms. The first term, $\sum_{x_6} p_{x_0 x_6}(t_6) L_6(x_6)$, sums over x_6 and is the probability of observing data TCA at the tips 1, 2, 3, given that node 0 has T. This is $0.906563 \times 0.003006 + 0.045855 \times 0.003006 + 0.023791 \times 0.004344 + 0.023791 \times 0.004344 = 0.003070$. The second term, $\sum_{x_8} p_{x_0 x_8}(t_8) L_8(x_8)$, is the probability of observing data CC at tips 4 and 5, given that node 0 has T. This is $0.906563 \times 0.007102 + 0.045855 \times 0.680776 + 0.023791 \times 0.002054 + 0.023791 \times 0.002054 = 0.037753$. The product of the two terms gives $L_0(T) = 0.00011237$. Other entries in the vector for node 0 can be similarly calculated. Finally application of equation (4.6) gives the probability of data at the site as $f(\mathbf{x}_h|\theta) = 0.000509843$, with $\log\{f(\mathbf{x}_h|\theta)\} = -7.581408$. □

4.2.2.2 Savings on computation

The pruning algorithm is a major time saver. As in the dynamic programming algorithm discussed in §3.4.3, in the pruning algorithm the amount of computation required by one calculation of the likelihood increases linearly with the number of nodes or the number of species, even though the number of combinations of ancestral states increases exponentially.

Some other obvious savings may be mentioned here as well. First, the same transition probability matrix is used for all sites or site patterns in the sequence and may be calculated only once for each branch. Second, if two sites have the same data, the probabilities of observing them will be the same and need be calculated only once. Collapsing sites into *site patterns* thus leads to a saving in computation, especially if the sequences are highly similar so that many sites have identical patterns. Under JC69, some sites with different data, such as TCAG and TGCA, also have the same probability of occurrence and can be collapsed further (Saitou and Nei 1986). The same applies to K80, although the saving is not as much as under JC69. It is also possible to collapse *partial site patterns* corresponding to subtrees (e.g. Kosakovsky Pond and Muse 2004). For example, consider the tree of Figure 4.1 and two sites with data TCACC and TCACT. The conditional probability vectors for interior nodes 6 and 7 are the same (because the data for species 1, 2, and 3 are the same at the two sites) and can be calculated only once. However, such collapsing of partial site patterns depends on the tree topology and involves an overhead for bookkeeping. Reports vary as to its effectiveness.

4.2.2.3 Hadamard conjugation

It is fitting to mention here an alternative method, called *Hadamard conjugation*, for calculating the site pattern probabilities and thus the likelihood. The Hadamard matrix is a square matrix consisting of –1 and 1 only. With –1 and 1 representing grey and dark grey, respectively, the matrix is useful for designing pavements. Indeed it was invented by the English mathematician James Sylvester (1814–1897) under the name 'anallagmatic pavement' and later studied by the French mathematician Jacques Hadamard (1865–1963). It was introduced to molecular phylogenetics by Hendy and Penny (1989), who used it to transform branch lengths on an unrooted tree to the site pattern probabilities, and vice versa. The transformation or conjugation works for binary characters or under Kimura's (1981) 3ST model of nucleotide substitution, which assumes three substitution types: one rate for transitions and two rates for transversions. It is computationally feasible for small trees with < 20 species, and is sometimes useful in theoretical analysis of phylogenetic methods (Felsenstein 2004; Hendy 2005).

4.2.3 Time reversibility, the root of the tree, and the molecular clock

As discussed in Chapter 1, most substitution models used in molecular phylogenetics describe time-reversible Markov chains. For such chains, the transition probabilities satisfy $\pi_i p_{ij}(t) = \pi_j p_{ji}(t)$ for any i, j, and t. Reversibility means that the chain will look identical probabilistically whether we view the chain with time running forward or backward. An important consequence of reversibility is that the root can be moved arbitrarily along the tree without affecting the likelihood. This is called the *pulley principle* by Felsenstein (1981). For example, substituting $\pi_{x_6} p_{x_6 x_0}(t_6)$ for $\pi_{x_0} p_{x_0 x_6}(t_6)$ in equation (4.3), and noting $\sum_{x_0} p_{x_6 x_0}(t_6) p_{x_0 x_8}(t_8) = p_{x_6 x_8}(t_6 + t_8)$, we have, by the Chapman–Kolmogorov theorem (equation (1.5)),

$$f(\mathbf{x}_h|\theta) = \sum_{x_6} \sum_{x_7} \sum_{x_8} \left[\pi_{x_6} p_{x_6 x_7}(t_7) p_{x_6 x_8}(t_6 + t_8) p_{x_7 T}(t_1) p_{x_7 C}(t_2) p_{x_6 A}(t_3) p_{x_8 C}(t_4) p_{x_8 C}(t_5) \right]. \quad (4.7)$$

This is the probability of the data if the root is at node 6, and the two branches 0–6 and 0–8 are merged into one branch 6–8, of length $t_6 + t_8$. The resulting tree is shown in Figure 4.3, with the root at node 6.

Equation (4.7) also highlights the fact that the model is over-parametrized in Figure 4.1, with one branch length too many. The likelihood is the same for any combinations of t_6

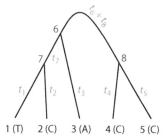

Fig. 4.3 The ensuing unrooted tree when the root is moved from node 0 to node 6 in the tree of Figure 4.1.

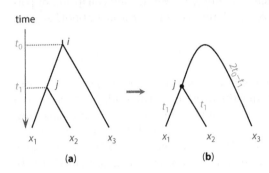

Fig. 4.4 (a) On a rooted tree for three species and under the clock, the model involves two parameters t_0 and t_1, measured by the expected number of substitutions per site from the ancestral node to the present time. The likelihood calculation has to sum over ancestral states i and j at the two ancestral nodes. (b) If the model is reversible, the same calculation can be achieved by moving the root to the ancestor of species 1 and 2, and summing over state j at the new root; the tree then becomes a star tree with three branches of lengths t_1, t_1, and $2t_0 - t_1$.

and t_8 as long as $t_6 + t_8$ is the same. The data do not contain information to estimate t_6 and t_8 separately and only their sum is estimable. Thus with reversibility, only unrooted trees can be identified if the molecular clock (rate constancy over time) is relaxed and every branch has its own rate.

If we assume the molecular clock, however, the root of the tree can indeed be identified. With a single rate throughout the tree, every tip is equidistant from the root, and the natural parameters are the ages of the ancestral nodes, measured by the expected number of substitutions per site. A binary tree of s species has $s - 1$ internal nodes, and thus $s - 1$ branch length parameters under the clock model. An example for three species is shown in Figure 4.4a. The branch length is then given as the difference of the ages of the two nodes at the ends of the branch. Given the branch lengths, likelihood calculation or the pruning algorithm proceed as before.

Even under the clock, the pulley principle may be used to simplify the likelihood calculation in theoretical studies of small trees. For example, likelihood calculation on the tree of Figure 4.4a involves summing over the ancestral states i and j at the two ancestral nodes. However, it is simpler to move the root to the common ancestor of species 1 and 2, so that one has to sum over ancestral states at only one node (Figure 4.4b). The probability of data $x_1 x_2 x_3$ at a site becomes

$$f(x_1 x_2 x_3 | \theta) = \sum_i \sum_j \pi_i p_{ij}(t_0 - t_1) p_{jx_1}(t_1) p_{jx_2}(t_1) p_{ix_3}(t_0)$$
$$= \sum_j \pi_j p_{jx_1}(t_1) p_{jx_2}(t_1) p_{jx_3}(2t_0 - t_1), \qquad (4.8)$$

where $\theta = \{t_0, t_1\}$ are the parameters under the model. Such arbitrary moving of the root is very similar to the case of two sequences discussed in equation (1.70) and Figure 1.10.

4.2.4 A numerical example: phylogeny of apes

We use the sequences from the 12 proteins encoded by the heavy strand of the mitochondrial genome from seven ape species. The data are a subset of the mammalian sequences analysed by Cao et al. (1998). The 12 proteins are concatenated into one long sequence and analysed as one dataset as they appear to have similar substitution patterns. The other protein in the genome, ND6, is not included as it is encoded by the opposite strand of the DNA with quite different base compositions. The species and the GenBank accession numbers for the sequences are human (*Homo sapiens*, D38112), common chimpanzee (*Pan troglodytes*, D38113), bonobo chimpanzee (*Pan paniscus*, D38116), gorilla (*Gorilla gorilla*, D38114), Bornean orangutan (*Pongo pygmaeus pygmaeus*, D38115), Sumatran orangutan

(*Pongo pygmaeus abelii*, X97707), and gibbon (*Hylobates lar*, X99256). Alignment gaps are removed, with 3,331 amino acids in the sequence.

There are 945 binary unrooted trees for seven species, so we evaluate them exhaustively. We assume the empirical MTMAM model for mammalian mitochondrial proteins (Yang et al. 1998). The ML tree is shown in Figure 4.5, which has the log likelihood score $\ell = -14,558.59$. The worst binary tree has the score $-15,769.00$, while the star tree has the score $-15,777.60$. Figure 4.6 shows that the same tree (the one of Figure 4.5) has the highest log likelihood, the shortest tree length by parsimony, and also the shortest likelihood tree length (the sum of maximum likelihood estimates (MLEs) of branch lengths). Thus ML, maximum parsimony, and minimum evolution all selected the same best tree for this dataset. (Note that minimum evolution normally estimates branch lengths by applying

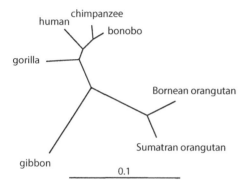

Fig. 4.5 The ML tree for seven ape species estimated from the 12 mitochondrial proteins. Branches are drawn in proportion to their lengths, measured by the number of amino acid substitutions per site. The MTMAM model is assumed.

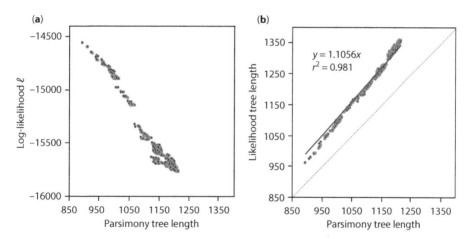

Fig. 4.6 Different criteria for tree selection calculated for all 945 binary unrooted trees for the mitochondrial protein data. (**a**) The log likelihood score ℓ is plotted against the parsimony tree length. (**b**) The likelihood tree length is plotted against the parsimony tree length. The likelihood tree length is measured by the estimated number of amino acid substitutions on the tree, calculated as the sum of estimated branch lengths multiplied by the number of sites. The parsimony tree length is the minimum number of changes and is thus an under-count. The underestimation is slightly more serious for the poor trees (with large tree lengths) but is nearly proportional. All three criteria (parsimony tree length, likelihood tree length, and log likelihood score) select the tree of Figure 4.5 as the best estimate.

least squares to a distance matrix while here I used ML to estimate branch lengths.) Similarly, use of the Poisson model in the likelihood analysis, assuming the same rate between any two amino acids, gives the same tree as the ML tree, with $\ell = -16,566.60$.

4.2.5 Amino acid, codon, and RNA models

The discussions up to now assume models of nucleotide substitution applied to noncoding DNA sequences. The same theory, including the pruning algorithm, can be applied in a straightforward manner to analyse protein sequences under models of amino acid substitution (Bishop and Friday 1985, 1987; Kishino et al. 1990) or protein-coding DNA sequences under models of codon substitution (Goldman and Yang 1994; Muse and Gaut 1994). Discussions of such models are given in Chapter 2. A difference is that the substitution rate and transition probability matrices are of sizes 20×20 for amino acids or 61×61 for codons (as there are 61 sense codons in the universal genetic code), instead of 4×4 for nucleotides. Furthermore, the summation over ancestral states is now over all ancestral amino acids or codons. As a result, likelihood computation under amino acid and codon models is much more expensive than under nucleotide models.

For phylogenetic analysis of ribosomal RNA (rRNA) genes, Markov models with 16 states for dinucleotides have been developed to describe substitutions in the helical (stem) regions where complementary nucleotides tend to change together (Schoeniger and von Haeseler 1994; Rzhetsky 1995; Tillier and Collins 1998; see also Siepel and Haussler 2004). Assumptions are made to reduce the number of parameters in the 16×16 rate matrix. Some studies confirm the benefit of taking RNA secondary structures into account in the model, with improved accuracy and robustness of phylogenetic reconstruction (Telford et al. 2005; Keller et al. 2010). Nevertheless, Letsch and Kjer (2011) pointed out that RNA covariation models often fail to recover reasonable trees because the highly divergent single-stranded loop regions contribute much of the information for the analysis while the covarying sites in the stem regions are effectively down-weighted. They advise caution in the uncritical application of RNA substitution models and suggest that loop regions should be assessed for substitutional saturation and alignment difficulties, and should be removed if they appear to contain too much noise.

*4.2.6 Missing data, sequence errors, and alignment gaps

4.2.6.1 General theory

The likelihood function provides a natural framework for accommodating incompletely determined nucleotides (ambiguities) and nucleotide changes caused by sequencing errors or DNA degradation in ancient DNA. Let X be the observed data, which may involve ambiguities or sequence errors, and Y be the unknown true alignment with fully determined nucleotides, which is the full exact data that we wish to have observed. The likelihood is the probability of X, and not of Y. The two are related by $f(X|Y, \psi)$, the probability of observing X given the full data Y, which is specified by the mechanism by which missing data or sequence errors arise, and may involve unknown parameters ψ. The likelihood is given by the law of total probability

$$L(\theta, \psi) = f(X|\theta, \psi) = \sum_{Y} f(Y|\theta) f(X|Y, \psi), \qquad (4.9)$$

where the summation (integration) is over all possible full data Y and where $f(Y|\theta)$ is the probability of the full data Y, as calculated in equations (4.1)–(4.4).

* indicates a more difficult or technical section.

We assume that the ambiguities or sequencing errors occur independently at different sites in the sequence. This assumption may be plausible for ambiguities or missing data, but appears quite unrealistic for sequence errors, which tend to be related to local sequence features (e.g. Nakamura et al. 2011). Under this independence assumption, $f(X|Y, \psi)$ is a product of probabilities at different sites and so is $f(Y|\theta)$. Thus

$$L(\theta, \psi) = f(X|\theta, \psi) = \prod_{h=1}^{n} f(\mathbf{x}_h|\theta, \psi) = \prod_{h=1}^{n}\left[\sum_{\mathbf{y}_h} f(\mathbf{y}_h|\theta) f(\mathbf{x}_h|\mathbf{y}_h, \psi)\right], \qquad (4.10)$$

where n is the number of sites (alignment columns), and $\mathbf{y}_h$ and $\mathbf{x}_h$ are the full and observed data at site h, respectively,

We describe a general procedure to deal with both missing data and sequence errors, and then discuss the particulars of each. The Nomenclature Committee of the International Union of Biochemistry (NC-IUB) recognizes 15 incompletely specified bases: T (= U), C, A, G, Y (T or C), R (A or G), M (C or A), K (T or G), S (C or G), W (T or A), H (not G), B (not A), V (not U), D (not C), and N (any base). Suppose that the 'rate' of missing data or sequence errors is sequence-specific, and ambiguities and sequence errors occur at random in the sequence, independently of other species. We define a 4×15 matrix $E^{(i)} = \{\varepsilon_{yx}^{(i)}\}$, where $\varepsilon_{yx}^{(i)}$ is the probability that nucleotide or ambiguity symbol $x \in \{T, C, A, G, Y, R, M, K, S, W, H, B, V, D, N\}$ is observed in the sequence at tip i given that the true nucleotide is $y \in \{T, C, A, G\}$. Note that $\sum_x \varepsilon_{yx}^{(i)} = 1$ for every y and every tip i. Parameters involved in the $E^{(i)}$ matrices for the tips are collected into the vector ψ.

Felsenstein's pruning algorithm allows the summation in equation (4.10) over the true nucleotides ($\mathbf{y}_h$) at site h to be achieved automatically, by setting

$$L_i(y) = \varepsilon_{yx_i}^{(i)}, \qquad (4.11)$$

i.e. by setting the conditional probability vector at tip i to $\left\{\varepsilon_{Tx_i}^{(i)}, \varepsilon_{Cx_i}^{(i)}, \varepsilon_{Ax_i}^{(i)}, \varepsilon_{Gx_i}^{(i)}\right\}$. Recall $L_i(y)$ is the probability of observing data (x_i) at tip i at the site given that the true nucleotide is y. The general model involves too many parameters and does not appear to be identifiable. We can constrain the model and reduce the number of parameters to be estimated from the data.

4.2.6.2 Ambiguities and missing data

Now we consider the case of ambiguities and missing data, assuming no sequence errors. In other words, if an observed nucleotide is one of T, C, A, or G, it is assumed to be the true nucleotide. As an example, suppose in the alignment of the three sequences for Figure 4.4, sequences 1 and 3 involve ambiguities at different rates (prevalence). Consider a site with the observed data to be $\mathbf{x}_h = $ YTR. The probability of observing such a site is a sum over all nucleotide configurations ($\mathbf{y}_h$) that are compatible with $\mathbf{x}_h$. From equation (4.10), we have

$$f(\text{YTR}|\theta, \psi) = f(\text{TTA}|\theta) \times \varepsilon_{TY}^{(1)}\varepsilon_{AR}^{(3)} + f(\text{TTG}|\theta) \times \varepsilon_{TY}^{(1)}\varepsilon_{GR}^{(3)}$$
$$+ f(\text{CTA}|\theta) \times \varepsilon_{CY}^{(1)}\varepsilon_{AR}^{(3)} + f(\text{CTG}|\theta) \times \varepsilon_{CY}^{(1)}\varepsilon_{GR}^{(3)}. \qquad (4.12)$$

Note that $f(\mathbf{x}_h|\mathbf{y}_h, \psi) = \varepsilon_{TY}^{(1)}\varepsilon_{AR}^{(3)}$ for $\mathbf{y}_h = $ TTA, or $= \varepsilon_{TY}^{(1)}\varepsilon_{GR}^{(3)}$ for $\mathbf{y}_h = $ TTG, and so on. It appears plausible to assume that $\varepsilon_{TY}^{(1)} = \varepsilon_{CY}^{(1)}$ and $\varepsilon_{AR}^{(3)} = \varepsilon_{GR}^{(3)}$; i.e. the chance for T to be read as Y is the same as that for C to be read as Y, etc. Note that the relevant probability here is the probability of seeing Y when the true base is T (or C), rather than the probability that the observed Y is in fact T (or C). The former informs us about the sequencing technology and error-generating mechanism, while the latter may mainly reflect the base compositions in the sequence. At any rate, under the assumption that the probability $f(\mathbf{x}_h|\mathbf{y}_h, \psi)$ does not depend on the true state $\mathbf{y}_h$, equation (4.12) can be written as

$$f(\mathbf{x}_h|\theta, \psi) = f(\text{YTR}|\theta) = c[f(\text{TTA}|\theta) + f(\text{TTG}|\theta) + f(\text{CTA}|\theta) + f(\text{CTG}|\theta)], \quad (4.13)$$

where $c = \varepsilon_{TY}^{(1)} \varepsilon_{AR}^{(3)}$ is a constant if we are not interested in parameters ψ; i.e. c is independent of the tree topology and parameters θ such as the branch lengths. We can ignore c and the likelihood for θ is still correctly defined.

The sum in equation (4.13) can be achieved by setting $L_i(y)$ for tip i to 1 for any nucleotide y that is compatible with the observed state (x_i) and to 0 otherwise; in other words, the conditional probability vectors are (1, 1, 0, 0) for tip 1 and (0, 0, 1, 1) for tip 3. For a site with exact data such as $\mathbf{x}_h = $ TCA, we have $\mathbf{y}_h = $ TCA, and $f(\mathbf{x}_h|\theta, \psi) = f(\mathbf{y}_h|\theta) \times \varepsilon_{TT}^{(1)} \varepsilon_{AA}^{(3)}$. Again if we are not interested in parameters ψ, we can ignore the constant $\varepsilon_{TT}^{(1)} \varepsilon_{AA}^{(3)}$. This idea of dealing with ambiguities by setting the conditional probabilities at the tips to 0s and 1s is due to Felsenstein (2004, pp. 255–6), and is the version used in ML phylogenetics programs. This has the nice property that the likelihood stays exactly the same if an empty column is added to the alignment which has an 'N' or '?' in every sequence. The discussion above, however, suggests that the strategy works only if the probabilities of seeing the observed ambiguous nucleotide are the same for the different compatible true nucleotides, i.e. if $\varepsilon_{TY} = \varepsilon_{CY}, \varepsilon_{TN} = \varepsilon_{CN} = \varepsilon_{AN} = \varepsilon_{GN}$, and so on. If T were harder to read by the sequencing machine and tended to become an ambiguity Y more often than C, this treatment would not be correct. In general, missing data are easier to accommodate in the model if the probability of missing data does not depend on the true state (Little and Rubin 1987, pp. 88–92).

4.2.6.3 Sequence errors

The sequence errors we consider here are those in a single sequence, and not those in multiple reads of a single sequence or of a mixture of alleles. If data from multiple genomic regions are available, it may be reasonable to assume that the sequence errors are genome-specific, so that the same error model is applied to sequences from multiple loci of the same genome. Suppose sequence/genome i has errors but no ambiguities, we can define a transition matrix $E^{(i)}$, of size 4×4 (instead of 4×15),

$$E = \{\varepsilon_{yx}\} = \begin{bmatrix} \varepsilon_{TT} & \varepsilon_{TC} & \varepsilon_{TA} & \varepsilon_{TG} \\ \varepsilon_{CT} & \varepsilon_{CC} & \varepsilon_{CA} & \varepsilon_{CG} \\ \varepsilon_{AT} & \varepsilon_{AC} & \varepsilon_{AA} & \varepsilon_{AG} \\ \varepsilon_{GT} & \varepsilon_{GC} & \varepsilon_{GA} & \varepsilon_{GG} \end{bmatrix}, \quad (4.14)$$

where ε_{yx} is the probability of observing base x if the true base is y, with the nucleotides ordered T, C, A, and G. Here we suppress the superscript i for genome i. Each row sums to 1, so there are 12 free parameters in the general matrix. The error rate may depend on the true nucleotide, and evolution and sequencing errors may have different patterns, such as different transition/transversion rate ratios. To reduce the number of parameters, we can use the HKY85 model to describe sequence errors, with five parameters. The simplest model will be JC69, with the diagonal to be $\varepsilon_{TT} = \varepsilon_{CC} = \varepsilon_{AA} = \varepsilon_{GG} = 1 - 3\varepsilon$ and the off-diagonals to be ε. The conditional probability vector at the tip of the tree is then set to $\{1 - 3\varepsilon, \varepsilon, \varepsilon, \varepsilon\}$ for an observed T, $\{\varepsilon, 1 - 3\varepsilon, \varepsilon, \varepsilon\}$ for C, and so on.

While the error model can be arbitrarily complex in a computer simulation, for inference, we have to consider the information content in the data and avoid non-identifiability. It is prudent to require at least one genome to be free of sequence errors. In comparison of closely related species, the molecular clock may be assumed as well. Sequence errors have the effect of adding an extra amount of evolution to the tip branches of the tree. Thus the apparent violation of the clock due to the sequence errors and the different patterns between evolution and sequence errors may provide information for estimating the error rate parameters (Burgess and Yang 2008).

4.2.6.4 *Alignment gaps*

Alignment gaps pose greater difficulties to likelihood calculation than ambiguities, sequence errors or DNA degradation. The general theory of equation (4.9) still holds, with X being the observed alignment, and Y the unobserved true alignment. However, both $f(Y|\theta)$ and $f(X|Y, \psi)$ are hard to calculate. Here $f(X|Y, \psi)$ is the probability of the 'observed' alignment generated by the alignment program given the true alignment, while $f(Y|\theta)$ is the probability of the true alignment given the phylogeny, branch lengths, insertion and deletion rates, etc.

In theory, it is advantageous to develop models of insertions and deletions as well as substitutions to align sequences in a probabilistic framework (Bishop and Thompson 1986; Thorne et al. 1991, 1992). Such an approach will generate estimates of the insertion and deletion rates, and also provide a probabilistic measurement of alignment accuracy. However, the early methods are based on simplistic assumptions about insertions and deletions and involve intensive computation even for two sequences. While improvements are being made both to the biological realism of the model and to the computational efficiency (see, e.g. Hein et al. 2000, 2003; Lunter et al. 2005), this modelling approach has not reached the stage of producing a useable method or software program. As a result, almost all multiple sequence alignments are generated using heuristic methods. There is also a keen interest in inferring alignment and phylogeny simultaneously, for example, using the Bayesian framework (Fleissner et al. 2005; Holmes 2005; Redelings and Suchard 2005).

Here we discuss a few *ad hoc* procedures for treating gaps in a given alignment, that is, calculation of $f(Y|\theta)$, with alignment errors ignored. There are three options. The first is to treat an alignment gap as the fifth nucleotide or the 21st amino acid, different from all other character states. This is used in some parsimony algorithms, but is uncommon in likelihood implementations (but see McGuire et al. 2001). One problem with this approach is that it treats a stretch of five gaps as five independent evolutionary events, even though it may well represent one event (an insertion or deletion of five nucleotides). Two worse options are commonly used: (i) to delete all sites at which there are alignment gaps in any sequence and (ii) to treat alignment gaps as undetermined nucleotides (ambiguities). The information loss caused by deleting sites with alignment gaps can be substantial if the sequences are divergent and the alignment includes many columns with gaps. The approach of treating alignment gaps as undetermined nucleotides is problematic as well, since gaps mean that the nucleotides do not exist and not that they exist but are unknown. It is not so clear what effects those two approaches have on phylogenetic tree reconstruction. It seems reasonable to delete a site if it contains alignment gaps in most species and to keep the site if it contains alignment gaps in very few species. In analysis of highly divergent species, it is common practice to remove regions of the protein for which the alignment is unreliable.

4.3 Likelihood calculation under more complex models

The discussion in the section above assumes that all sites in the sequence evolve at the same rate according to the same rate matrix. This may be a very unrealistic assumption for real sequences. Much progress has been made in the last two decades in extending models used in likelihood analysis. A few important extensions are discussed in this section.

4.3.1 Mixture models for variable rates among sites

In real sequences, the substitution rates are often variable across sites. Ignoring rate variation among sites can have a major impact on phylogenetic analysis (e.g. Tateno et al. 1994; Huelsenbeck 1995a; Yang 1996c; Sullivan and Swofford 2001). To accommodate variable rates in a likelihood model, one should not in general use a rate parameter for every site, as otherwise there will be too many parameters to estimate and the likelihood method may misbehave. A sensible approach is to use a statistical distribution to model the rate variation. Both discrete and continuous rate distributions have been used.

4.3.1.1 Discrete-rate model

In this model, sites are assumed to fall into K discrete classes with different rates (Table 4.1). The rate at any site in the sequence takes a value r_k with probability p_k, with $k = 1, 2, \ldots, K$. The rs and ps are parameters, to be estimated by ML from the data. We place two constraints to avoid the use of too many parameters. First the probabilities sum to one: $\sum p_k = 1$. Second, the average rate is fixed at $\sum p_k r_k = 1$, so that the branch length is measured as the expected number of nucleotide substitutions per site averaged over the site or rate classes. The model with K site classes thus involves $2(K-1)$ free parameters. The rates rs are thus relative multiplication factors. Every site is in effect evolving along the same tree topology with proportionally elongated or shrunken branches. In other words, the substitution rate matrix at a site with rate r is rQ, with Q shared across all sites.

As we do not know which site class each site belongs to, the probability of observing data at any site is a weighted average over the site classes

$$f(\mathbf{x}_h|\theta) = \sum_{k=1}^{K} p_k \times f(\mathbf{x}_h|r = r_k; \theta). \tag{4.15}$$

The likelihood is again calculated by multiplying the probabilities across sites. The conditional probability, $f(\mathbf{x}_h|r; \theta)$, of observing data $\mathbf{x}_h$ given the rate r, is just the probability under the one-rate model, with all branch lengths multiplied by r. It can be calculated using the pruning algorithm for each site class. A variable-rate model with K site classes thus takes K times as much computation as the one-rate model.

Table 4.1 The discrete-rate model

Site class	1	2	3	...	K
Probability	p_1	p_2	p_3	...	p_K
Rate	r_1	r_2	r_3	...	r_K

As an example, consider the tree of Figure 4.4b. We have

$$f(x_1 x_2 x_3 | \theta) = \sum_{k=1}^{K} p_k \times f(x_1 x_2 x_3 | r = r_k; \theta), \qquad (4.16)$$

$$f(x_1 x_2 x_3 | r; \theta) = \sum_j \pi_j p_{j x_1}(t_1 r) p_{j x_2}(t_1 r) p_{j x_3}((2 t_0 - t_1) r)$$

(compare with equation (4.8)).

Discrete-rate models are known as *finite-mixture models*, since the sites are a mixture from K classes. The general model of Table 4.1 is implemented by Yang (1995a). As is typical in such finite-mixture models, one can fit only a few rate classes to practical datasets, so K should not exceed 3 or 4. Also the estimates of the rate and proportion parameters from a given dataset tend to change dramatically when K changes, making it hard to interpret those parameters. A special case of the general discrete-rate model is the *invariable-site* model, which assumes two site classes: the class of *invariable sites* with rate $r_0 = 0$ and another class with a constant rate r_1(Hasegawa et al. 1985). As the average rate is $p_0 r_0 + (1 - p_0) r_1 = (1 - p_0) r_1 = 1$, we have $r_1 = 1/(1 - p_0)$, where the proportion of invariable sites p_0 is the only parameter in the model. Note that a variable site in the alignment cannot have rate $r_0 = 0$. Thus the probability of data at a site, i.e. equation (4.15), becomes

$$f(\mathbf{x}_h | \theta) = \begin{cases} p_0 + p_1 \times f(\mathbf{x}_h | r = r_1; \theta), & \text{if the site is constant,} \\ p_1 \times f(\mathbf{x}_h | r = r_1; \theta), & \text{if the site is variable.} \end{cases} \qquad (4.17)$$

Example 4.2 Discrete-rate model for mitochondrial 12S rRNA. We fit the general discrete-rate model (Table 4.1) to a dataset of mitochondrial small subunit (12S) rRNA genes from 30 Old World and New World Monkeys. There are 978 sites in the alignment. The tree topology is shown in Figure 4.7. We assume the HKY85 model and examine the effect of the number of rate classes. The log likelihood (ℓ) is plotted against K in Figure 4.8, while parameter estimates are listed in Table 4.2. Each additional rate class adds two free parameters to the model. The log likelihood improves hugely (by 685.35 units) by the addition of the second rate class (i.e. when K increases from 1 to 2), by 49.35 for the third class, and only by 0.58 for the fourth class. For this dataset, it is not possible to fit more than four classes; models with $K \geq 5$ simply collapse to the model with $K = 4$. Note that use of the constant-rate model leads to underestimation of branch lengths, tree length, and the transition/transversion rate ratio (Wakeley 1994; Yang et al. 1994, 1995c). □

4.3.1.2 Gamma-rate model

A second approach is to use a continuous distribution to approximate variable rates among sites. The most commonly used distribution is the gamma (see Figure 1.6 in Chapter 1). In §1.3 we discussed the use of the same gamma model in calculating pairwise distances. The gamma density is

$$g(r; \alpha, \beta) = \frac{\beta^\alpha r^{\alpha-1} e^{-\beta r}}{\Gamma(\alpha)}, \qquad (4.18)$$

with mean α/β and variance α/β^2. We let $\beta = \alpha$ so that the mean is 1. The shape parameter α is inversely related to the extent of rate variation among sites. As in the discrete-rate model, we do not know the rate at the site, and have to average over the rate distribution

$$f(\mathbf{x}_h | \theta) = \int_0^\infty g(r) f(\mathbf{x}_h | r; \theta) \, \mathrm{d}r. \qquad (4.19)$$

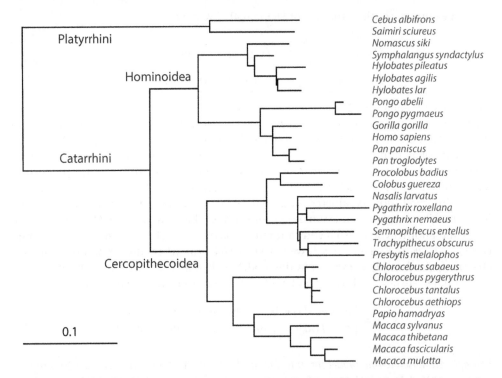

Fig. 4.7 The tree for the mitochondrial 12S rRNA genes from 30 primate species (catarrhini and platyrrhini), with branch lengths estimated under the HKY + Γ_5 model. The unrooted tree is used to fit different models of rate variation among sites in Table 4.2 and Figure 4.8, but here the tree is rooted for clarity.

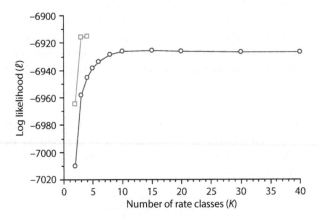

Fig. 4.8 The log likelihood value (ℓ) as a function of the number of rate classes (K) in the discrete-rate model (□) and in the discrete gamma model (○). In the discrete-rate model ℓ stays the same when $K \geq 4$, and in both models, $\ell = -7650.15$ when $K = 1$. The mitochondrial 12S rRNA genes from 30 primates are analysed (see Figure 4.7 and Table 4.2).

Table 4.2 Log likelihood values and MLEs of parameters under variable-rate models with different rate classes

Model	p	ℓ	$\hat{T}$	$\hat{\kappa}$	$\hat{r}_k$ and $\hat{p}_k$ (or $\hat{a}$)
Constant-rate model					
$K = 1$	61	−7650.15	1.445	9.88	
Discrete-rate model					
$K = 2$	63	−6964.80	1.920	12.97	$\hat{r}_k$: 0.096 2.954
					$\hat{p}_k$: 0.684 0.316
$K = 3$	65	−6915.45	2.159	14.30	$\hat{r}_k$: 0.036 1.466 5.860
					$\hat{p}_k$: 0.608 0.301 0.092
$K = 4$	67	−6914.87	2.168	14.35	$\hat{r}_k$: 0.000 0.341 1.667 6.062
					$\hat{p}_k$: 0.486 0.171 0.259 0.084
$K = 5$	69	as for $K = 4$			
Discrete gamma model					$\hat{a}$
$K = 2$	62	−7010.24	2.418	12.55	0.257
$K = 3$	62	−6958.11	2.137	13.11	0.247
$K = 4$	62	−6945.29	2.080	13.50	0.254
$K = 5$	**62**	**−6938.65**	**2.143**	**13.87**	**0.259**
$K = 6$	62	−6933.72	2.220	14.21	0.256
$K = 8$	62	−6928.26	2.250	14.53	0.249
$K = 10$	62	−6926.36	2.221	14.63	0.247
$K = 15$	62	−6925.79	2.187	14.73	0.247
$K = 20$	62	−6926.13	2.194	14.79	0.247
$K = 30$	62	−6926.56	2.226	14.85	0.246
$K = 40$	62	−6926.68	2.251	14.88	0.245
$K = 50$	62	−6926.70	2.266	14.89	0.244
$K = 100$	62	−6926.59	2.288	14.89	0.242

Note: The HKY85 model is assumed, with the observed base frequencies used as estimates. p is the number of parameters in the model and T is the tree length (sum of 57 branch lengths). Rates among sites are assumed to be constant, or modelled using the general discrete-rate model or the discrete gamma model, with K rate classes. The data are mitochondrial 12S rRNA gene sequences from 30 primates. The phylogeny is shown in Figure 4.7, which uses the branch length estimates under HKY85 + Γ_5 (highlighted here in bold).

Here the collection of parameters θ includes α as well as branch lengths and other parameters in the substitution model (such as κ). Note that the sum in equation (4.15) for the discrete model is now replaced by an integral for the continuous model.

An algorithm for calculating the likelihood function of equation (4.19) was described by Yang (1993), Gu et al. (1995), and Kelly and Rice (1996). Note that the conditional probability $f(\mathbf{x}_h|r)$ is a sum over all combinations of ancestral states (equation (4.3)). Each term in the sum is a product of transition probabilities across the branches. Each transition probability has the form $p_{ij}(tr) = \sum_k c_{ijk} e^{\lambda_k tr}$, where t is the branch length, λ_k is the eigenvalue, and c_{ijk} is a function of the eigenvectors (see equation (1.43)). Thus, after expanding all the products, $f(\mathbf{x}_h|r)$ is a sum of many terms of the form ae^{br}, and then the integral over r can be obtained analytically (equation (1.36)). However, this algorithm is very slow because of the huge number of terms in the sum, and is only practical for small trees with fewer than ten sequences.

4.3.1.3 Discrete gamma model

One may use the discrete-rate model as an approximation to the continuous gamma, leading to the *discrete gamma model*. Yang (1994a; see also Waddell et al. 1997) tested this strategy, using K equal probability site classes, with the mean or median for each class used to represent all rates in that class (Figure 4.9). Thus $p_k = 1/K$ while r_k is calculated as a function of the gamma shape parameter α. If the mean rate is used, we have

$$r_k = K \int_a^b r \times g(r; \alpha, \beta) \, dr, \qquad (4.20)$$

where a and b are the boundary points for the kth bin. The probability of data at a site is then given by equation (4.15). The model involves one single parameter α, just like the continuous gamma. The discrete gamma may be considered a crude way of calculating the integral of equation (4.19). Yang's (1994a) test on small datasets suggested that as few as $K = 4$ site classes provided good approximation. In large datasets with hundreds of sequences, more categories may be beneficial (Mayrose et al. 2005). Because of the discretization, the discrete gamma means less rate variation among sites than the continuous gamma for the same parameter α. Thus, the discrete gamma almost always produces a smaller estimate of α than the continuous gamma when the two models are fitted to the same dataset (which has a fixed amount of rate variation).

Felsenstein (2001a) and Mayrose et al. (2005) discussed the use of numerical integration (quadrature) algorithms to calculate the integral of equation (4.19). We will discuss methods of numerical integration later, in §6.4. Note that the integral is the area under the integrand curve. The simplest of algorithms for calculating the integral cut the x-axis

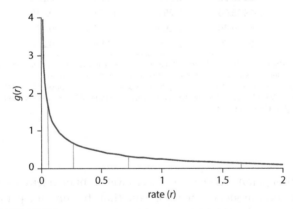

Fig. 4.9 The discrete gamma model of variable rates across sites uses K equal-probability categories to approximate the continuous gamma distribution, with the mean rate in each category used to represent all rates in that category. Shown here is the use of $K = 5$ categories to approximate the gamma density $g(r)$ with $\alpha = 0.5$ (and mean 1). The four vertical lines are at $r = 0.06418, 0.27500, 0.70833,$ and 1.64237. These are the 20%, 40%, 60%, and 80% percentiles of the distribution and cut the density into five categories, each of proportion 1/5. The mean rates in the five categories are 0.02121, 0.15549, 0.46708, 1.10712, and 3.24910.

into equal-sized segments and then approximate the integral by the sum of the areas of the rectangles on the segments. More sophisticated algorithms calculate the values of the integrand for a fixed set of values of r and then approximate the integrand by using simpler functions such as the polynomial. A concern is that the integrand in equation (4.19) may peak at zero for constant sites and at large values for highly variable sites, so that an algorithm using fixed points may not approximate the integral well for all sites. Adaptive algorithms are more reliable as they sample more densely in regions where the integrand is large. However, they are too expensive, as different points are used for different sites, meaning that the transition probability matrix has to be calculated for every site along every branch for every rate class. Given that both the continuous and discrete gamma models are empirical without mechanistic biological justifications, there should be no strong preference for either of them over the other. Thus there is not a great need for very accurate calculation of the integral in the continuous gamma model as the discrete model may work equally well.

Example 4.3 Discrete gamma model for mitochondrial 12S rRNA. The discrete gamma model is fitted to the primate mitochondrial 12S rRNA genes analysed in Example 4.2. The HKY85 model is assumed in combination with the discrete gamma model of variable rates for sites, with different rate categories (K) used (Table 4.2). Note that K is not a parameter, and the models with different $K \geq 2$ are non-nested models with the same number of parameters. One does not optimize K but instead uses a reasonable fixed value (say, 5 or 4) and then the estimates of parameter α will be comparable across datasets. In terms of model adequacy (the log likelihood), the discrete-rate model with $K = 3$ provides a good fit to this dataset, better than the discrete gamma model. However, the former uses four parameters in the rate distribution (r_k and f_k) while the latter uses only one (α). Furthermore, the estimates of r_k and f_k in the discrete-rate model are unstable. Overall, given the high similarity of estimates of parameters such as κ and branch lengths across those variable-rate models, the discrete gamma appears to be a better model than the general discrete-rate model if our objective is to reconstruct the phylogeny or estimate branch lengths, etc. □

Example 4.4 The phylogeny of seven ape species. We use the MTMAM + Γ_5 model to analyse the dataset of mitochondrial protein sequences (see §4.2.4), with a discrete gamma model used to accommodate variable rates among sites, with $K = 5$ site classes used. The shape parameter is fixed at $\alpha = 0.4$, which is the estimate from mammalian species (Yang et al. 1998). The ML tree is the same as under the MTMAM model (Figure 4.5). The log likelihood value for the tree is $\ell = -14,413.38$, much higher than that under the one-rate model ($\ell_0 = -14,558.59$). If α is estimated from the data, the same tree is found to be the ML tree, with $\hat{\alpha} = 0.333$ and $\ell = -14,411.90$. According to the likelihood ratio test (LRT), we should compare $2\Delta\ell = 2(\ell_1 - \ell_0) = 293.38$ with χ_1^2 (the χ^2 distribution with one degree of freedom) or with the 1:1 mixture of 0 and χ_1^2 (Whelan and Goldman 2000). There is no doubt that the discrete gamma model fits the data much better than the one-rate model. □

4.3.1.4 Other rate distributions

In addition to the gamma model, several other statistical distributions have been implemented and tested. Waddell et al. (1997) considered the log-normal model. Kelly and Rice (1996) took a nonparametric approach, assuming a general continuous distribution without specifying any particular distributional form. With such a general model, not much inference is possible.

Gu et al. (1995) added a proportion of invariable sites to the gamma distribution, so that a proportion p_0 of sites are invariable while the other sites (with proportion $p_1 = 1 - p_0$)

have rates drawn from the gamma. The model is known as 'I + Γ'. The mean is fixed at 1: $E(r) = (1 - p_0) \alpha/\beta = 1$ so that $\beta = (1 - p_0) \alpha$. The variance is $V(r) = \frac{1 + \alpha p_0}{\alpha(1 - p_0)}$. This model is somewhat pathological as the gamma distribution with $\alpha \leq 1$ already allows for sites with very low rates; as a result, adding a proportion of invariable sites creates a strong correlation between p_0 and α, making it hard to estimate those parameters reliably (Yang 1993; Sullivan et al. 1999; Mayrose et al. 2005). Another drawback of the model is that the estimate of p_0 is very sensitive to the number and divergences of the sequences included in the data. The proportion p_0 is never larger than the observed proportion of constant sites; with the addition of more and divergent sequences, the proportion of constant sites drops, and the estimate of p_0 tends to go down as well. The I + Γ model is typically the most complex among models implemented in phylogenetic programs and is often the recommended model by automatic model selection procedures such as MODELTEST (Posada and Crandall 1998; Posada 2008), due to the large size of phylogenetic datasets. With the drawbacks mentioned here, one should avoid the I + Γ models and use the gamma models instead.

Mayrose et al. (2005) suggested a gamma mixture model, which assumes that the rate for any site is a random variable from a mixture of two gamma distributions with different parameters. This appears more stable than the I + Γ model. For estimating branch lengths and phylogenies, the different distributions tend to produce similar results, so that the simple gamma should be adequate. For estimating rates at sites from large datasets with many sequences, the mixture of two gammas may be preferable.

Note that under both continuous- and discrete-rate models, data at different sites are independent and identically distributed (i.i.d.). While the model allows different sites to evolve at different rates, it does not specify *a priori* which sites should have which rates. Instead the rate for any site is a random draw from a common distribution. As a result, data at different sites have the same distribution.

4.3.1.5 *Empirical Bayesian estimation of substitution rates at sites*

Large datasets with many sequences may provide opportunities for multiple changes at individual sites, making it possible to estimate the relative substitution rate at each site. In both discrete- and continuous-rate models, the rates for sites are random variables and are integrated out in the likelihood function. Thus to estimate the rate, we use the conditional (posterior) distribution of the rate given the data at the site

$$f(r|\mathbf{x}_h; \theta) = \frac{f(r|\theta) f(\mathbf{x}_h | r; \theta)}{f(\mathbf{x}_h | \theta)}. \tag{4.21}$$

Parameters θ may be replaced by their estimates, such as the MLEs. This is known as the empirical Bayes (EB) approach. Under the continuous-rate model, one can use the posterior mean as the estimate of the rate at the site (Yang and Wang 1995). Under the discrete-rate model, the rate r in equation (4.21) takes one of K possible values, with $f(r|\theta) = p_k$. One may calculate the posterior mean or use the rate with the highest posterior probability as the best estimate (Yang 1994a).

The EB approach ignores sampling errors in the parameter estimates, which may be a source of concern in small datasets. One can assign a prior on the parameters and use a full Bayesian approach to deal with uncertainties in the parameters. Mateiu and Rannala (2006) developed an interesting Markov chain Monte Carlo (MCMC) algorithm for estimating rates at sites, using a uniformized Markov chain (Kao, 1997 pp. 273–277) to implement the continuous gamma model. This strategy allows large datasets with hundreds of sequences to be analysed under the continuous model. The authors' simulation shows that the discrete gamma model provides good estimates of branch lengths, but tends to underestimate high rates unless a very large number of site classes is used (with $K \geq 40$, say).

Another likelihood approach to estimating rates at sites was implemented by Nielsen (1997), who treated the rate at every site as a parameter, estimated by ML. This method suffers from the use of too many parameters, so that the estimates are often zero or infinity.

4.3.1.6 *Correlated rates at adjacent sites*

Yang (1995a) and Felsenstein and Churchill (1996) implemented models that allow rates to be correlated across adjacent sites. In the *auto-discrete gamma* model (Yang 1995a), the rates at two adjacent sites have a bivariate gamma distribution (i.e. the marginal distributions of both rates are gamma), discretized to make the computation feasible. This is an extension to the discrete gamma model and includes a parameter ρ, which measures the strength of autocorrelation. The model is implemented through a hidden Markov chain, which describes the transition from one rate class to another along the sequence. Felsenstein and Churchill (1996) described a similar hidden Markov chain model, in which a segment of nucleotide sites is assumed to have the same rate, with the length of the segment reflecting the strength of the correlation. The hidden Markov model involves about the same amount of computation as the independent discrete-rate model. Tests on real data suggest that substitution rates are indeed highly correlated. Nevertheless, estimation of branch lengths or other parameters in the model did not seem to be affected by ignoring the correlation of rates at adjacent sites, although the variances of the MLEs were underestimated when the correlation is ignored.

4.3.1.7 *Covarion models*

In both the discrete- and continuous-rate models, the rate for a site is applied to all branches in the tree, so that a fast-evolving site is fast-evolving throughout the phylogeny. This assumption is relaxed in the *covarion* (for COncomitantly VARIable codON) models, which are based on Fitch's (1971a) idea of coevolving codons in a protein-coding gene, with substitutions in one codon affecting substitutions in other codons. Such models allow a site to switch from one site class to another. As a result, a site may be fast-evolving along some lineages while slowly evolving along others. Tuffley and Steel (1998) discussed an extension to the invariable-site model, in which a site switches between two states: the 'on' state (+), in which the nucleotide has a constant rate of evolution and the 'off' state (−), in which the nucleotide is invariable. A likelihood implementation of this model is provided by Huelsenbeck (2002). Similarly Galtier (2001) implemented an extension to the discrete gamma model, in which a site switches from one site class to another over evolutionary time, with K site classes from the discrete gamma.

Here we use the simpler model of Tuffley and Steel (1998) and Huelsenbeck (2002) to describe the implementation of such models. A simple approach is to construct a Markov chain with an expanded state space. Instead of the four nucleotides, we consider eight states: T_+, C_+, A_+, G_+, T_-, C_-, A_-, and G_-, with '+' and '−' representing the 'on' and 'off' states, respectively. Nucleotides of the '+' state can change between themselves, according to a substitution model such as JC69 or HKY85. A nucleotide of the '−' state can change to the same nucleotide of the '+' state only.

In the likelihood calculation, one has to sum over all the eight states for each ancestral node to calculate the probability of data at each site. Furthermore, a nucleotide observed at a tip node can be in either '+' or '−' states, so that the probability of data at a site will be a sum over all compatible patterns of the expanded states. Suppose the observed data at a site for three species are TCA. Then Pr(TCA) will be a sum over eight site patterns: $T_+C_+A_+$, $T_+C_+A_-$, $T_+C_-A_+$, ..., and $T_-C_-A_-$. This summation can be achieved efficiently using likelihood treatment of missing data in the pruning algorithm, as discussed in §4.2.6; we let the conditional probability $L_i(x_i)$ for the tip node i be 1 for both states $x_i = T_+$ and $x_i = T_-$ if the observed nucleotide for tip i is T, say. The model of Galtier (2001) can be implemented in the same way, although the Markov chain has $4K$ instead of 8 states.

An interesting use of the same idea is made by Guindon et al. (2004), who implemented a codon-based switching model, which allows a codon to switch between site classes with different ω ratios. The model thus allows the selective pressure on the protein indicated by the ω ratio to vary both among sites and among lineages, and is perhaps close to Fitch's (1971a) original idea of coevolving codons. Chapter 11 provides further discussions of codon-based models. It may be noted that all the covarion models discussed above assume that data at different sites are i.i.d.

4.3.2 Mixture models for pattern heterogeneity among sites

Rate variation among sites is a common feature of genetic sequence evolution. While in nonfunctional and noncoding DNA that is fast evolving without any selective constraint, the mutation rate may be nearly constant over sites, most datasets used in phylogenetic analysis show considerable mutation/substitution rate variation. The dramatic improvement in the fit of the model to data, indicated by the huge increase in the log likelihood, upon adding the single gamma shape parameter seen in Examples 4.3 and 4.4 is typical of many datasets. Furthermore, incorrectly assuming a constant rate for sites when rates vary can have a significant impact on different aspects of phylogenetic analysis such as tree topology reconstruction and branch length estimation (Yang 1996c).

While perhaps not as important as the rate, other aspects of the evolutionary process may vary among sites as well. In theory, the same approaches as discussed above can be taken to accommodate any such among-site heterogeneity in the model. For example, Huelsenbeck and Nielsen (1999) used a gamma distribution to describe variable transition/transversion rate ratios among sites, and another gamma distribution to describe variable rates, both discretized. For protein sequences, the different domains of the protein may have strong preferences for different amino acids (Bruno 1996; Halpern and Bruno 1998). One could use a set of amino acid frequency parameters for each site, but the model would involve too many parameters to implement in an ML method. Treating the amino acid frequencies as random variables (like the gamma-distributed rates for sites) would mean high-dimensional integrals in the likelihood function.

Besides amino acid frequencies, the whole rate matrix Q representing the amino acid substitution pattern may be allowed to vary among sites (Koshi and Goldstein 1996b; Thorne et al. 1996; Goldman et al. 1998; Koshi et al. 1999). Given the plethora of empirical amino acid substitution matrices, a computationally feasible strategy is to have a mixture model of several site classes, with different empirical matrices used for different classes. As the empirical matrices do not involve new parameters, the computational cost is tolerable. Such phylogenetic mixture models were implemented by Le et al. (2008) in the PHYML program (Guindon and Gascuel 2003), combined with gamma-distributed rates among sites. The authors' systematic test using large alignments in a database found that the mixture models of substitution pattern heterogeneity provide improved fit to data.

In general, it is easier to implement parameter-rich mixture models in the Bayesian framework (e.g. Lartillot and Philippe 2004) than in ML, so we will discuss those models in Chapter 8.

4.3.3 Partition models for combined analysis of multiple datasets

Two approaches may be taken to accommodate the heterogeneity of the substitution rate and pattern among sites. If we know *a priori* which sites are likely to be evolving fast and which sites evolving slowly, it is natural to use *partition* models, which assign different parameters to sites in different partitions. If we lack such information, we can assume a random statistical distribution, i.e. a *mixture model* to accommodate the among-site heterogeneity, as in last subsection. It is also possible to combine the two, with the partition model accommodating the large-scale variations among partitions and mixture model accommodating the remaining variation within each partition. In statistical jargon, partition models are *fixed-effect* models and mixture models are *random-effect* models. Models that include both fixed and random effects are called *mixed-effects models*.

Such partition and partition-mixture models have been implemented by Yang, Lauder, and Lin (1995b; see also Yang 1995a, 1996b; Pupko et al. 2002b) and applied to analysis of hominoid mitochondrial DNA sequences. Different rates, transition/transversion rate ratios and base frequencies are assigned to sites at the three codon positions while the gamma model is used to accommodate the remaining rate variation within each codon position. LRTs are used to compare models of different complexity. The codon position-based partition models are also used by Shapiro et al. (2006; see also Ren et al. 2005) to analyse hundreds of alignments of viral and yeast genes. While the nucleotide-based partition models may not fit the protein-coding gene sequences as well as the codon models, they involve much less computation and appear to be effective in phylogenetic analysis of protein-coding genes.

Note that the site partitioning should be done *a priori* (i.e. not based on analysis of the same data) but can be based on any criteria. The rationale is that sites within the same partition have similar evolutionary characteristics, describable using the same parameters, while sites from different partitions have different evolutionary dynamics, requiring different parameters to accommodate the heterogeneity. The different partitions may correspond to different genes, which evolve at different rates, have different base frequencies, and so on. They may also correspond to different codon positions in a protein-coding gene. Also besides the rate, parameters reflecting other features of the evolutionary process, such as the transition/transversion rate ratio or base compositions, and even the tree topology, can be allowed to differ among partitions.

Consider the rate as an example. Suppose there are K partitions (e.g. genes or codon positions), with rates $r_1, r_2, \ldots, r_K$. To avoid the use of too many parameters, we may fix $r_1 = 1$, so that the branch length is measured by the expected number of substitutions for the first partition. Equivalently, we may fix the average rate to 1, so that the branch length is measured by the number of substitutions per site, averaged over all partitions. Let $I(h)$ label the partition that site h belongs to; i.e. $I(h) = 3$ if site h is from the third codon position. The log likelihood is then

$$\ell(\theta, r_1, r_2, \ldots, r_K; X) = \sum_h \log\{f(\mathbf{x}_h | r_{I(h)}; \theta)\}. \tag{4.22}$$

Here the rates for site partitions are parameters in the model, while θ includes branch lengths and other substitution parameters. The probability of data at a site is calculated by using the correct rate parameter $r_{I(h)}$ for the site, in contrast to the random-rate (mixture) model, in which the probability for a site is an average over the site classes (equations (4.15) and (4.19)). Likelihood calculation under the fixed-rates (partition) model therefore takes about the same amount of computation as under the one-rate model. The partition model dealing with among-site rate heterogeneity is sometimes known as the *site-specific rate model*, even though the rates are partition-specific but not site-specific.

The most parameter-rich form of the partition model assumes that all parameters in the substitution model are different among partitions. ML under this model is equivalent to separate analysis of data of different partitions, summing up the log likelihood values. This was used by Yang (1996b) when the tree topology is fixed, and by Hasegawa et al. (1997), who evaluated different tree topologies, referring to the method as the *total evidence* approach.

In the partition model, one knows which partition or site class each site is from, and the rates (and other parameters) for site partitions are parameters, estimated by ML. In the mixture models, one does not know which rate each site has; instead one treats the rate as a random draw from a statistical distribution and estimates parameters of that distribution (such as α for the gamma model) as a measure of the variability in the rates. In the mixture model, data at different sites are i.i.d. In the partition model, data at different sites within the same partition are i.i.d., but sites from different partitions have different distributions. This distinction should be taken into account in statistical tests based on resampling sites, such as the bootstrap (Felsenstein 1985a).

The partition models are useful for analysing multiple heterogeneous loci, to assemble information concerning common features of the evolutionary process among genes while accommodating their heterogeneity. It allows the estimation of gene-specific rates and parameters, and hypothesis testing concerning similarities and differences among genes. Both likelihood (Yang 1996b; Pupko et al. 2002b; Leigh et al. 2008) and Bayesian (Suchard et al. 2003; Nylander et al. 2004; Pagel and Meade 2004) approaches can be taken. A similar use of this strategy is in the estimation of species divergence times under local-clock models, in which the divergence times are shared across loci while the multiple loci may have different evolutionary characteristics (Kishino et al. 2001; Yang and Yoder 2003).

In the literature there has been a debate concerning *combined analysis* versus *separate analysis* (see, e.g. Huelsenbeck et al. 1996). In the former, sequences from multiple loci are concatenated and then treated as one 'super-gene', with possible differences in the evolutionary dynamics among the genes ignored. The latter analyses different genes separately. The separate analysis can reveal differences among genes but does not provide a natural way of assembling information from multiple heterogeneous datasets. It may also over-fit the data. Thus neither approach is ideal. The appropriate approach should be a combined analysis that accommodates the heterogeneity among partitions. Another related debate is between *supermatrix* and *supertree* approaches, especially in the context of analysing genomic datasets in which some genes may be missing in some species (see, e.g. Bininda-Emonds 2004). The supermatrix approach concatenates sequences, patching up missing sequences with question marks, and is equivalent to the combined analysis mentioned above. It fails to accommodate possible differences among loci. The supertree approach reconstructs phylogenies using data from different genes, and then assembles the subtrees into a supertree for all species. It is typically difficult to accommodate the uncertainties in the individual subtrees so that heuristic approaches are used to deal with conflicts among the subtrees (Wilkinson et al. 2005). The supertree approaches may be useful in assembling information concerning phylogenies estimated from different sources (such as molecules and morphology). For analysis of sequence data, the likelihood-based approach to combined analysis, which accommodates the heterogeneity among multiple datasets, should have an advantage (Ren et al. 2009).

4.3.4 *Nonhomogeneous and nonstationary models*

In the analysis of divergent sequences, one often observes considerable variation in nucleotide or amino acid compositions among sequences. The assumption of a homogeneous and stationary Markov chain model is clearly violated. One can test whether base compositions are homogeneous by using a contingency table of nucleotide or amino acid counts in the sequences to construct a X^2 statistic (e.g. Tavaré 1986; Ababneh et al. 2006). However, such a formal test is hardly necessary because typical molecular datasets are large and such a test can reject the null hypothesis with ease. Empirical studies (e.g. Lockhart et al. 1994) suggest that unequal base compositions can mislead tree reconstruction methods, causing them to group sequences according to the base compositions rather than genetic relatedness.

Dealing with the drift of base compositions over time in a likelihood model is difficult. Yang and Roberts (1995) implemented a few nonhomogeneous models, in which every branch in the tree is assigned a separate set of base frequency parameters ($\pi_T, \pi_C, \pi_A, \pi_G$) in the HKY85 model. Thus the sequences drift to different base compositions during the evolutionary process, after they diverged from the root sequence. This model involves many parameters and is useable for small trees with a few species only. A modification to the model allows the user to specify how many sets of frequency parameters should be assumed and which set each branch should be assigned to. Previously, Barry and Hartigan (1987b) described a model with even more parameters, in which a general transition probability matrix P with 12 parameters is estimated for every branch. The model does not appear to have ever been used in data analysis. Galtier and Gouy (1998) implemented a simpler version of the Yang and Roberts model. Instead of the HKY85 model, they used the model of Tamura (1992), which assumes that G and C have the same frequency and A and T have the same frequency, so that only one frequency parameter for the GC content is needed in the substitution rate matrix. Galtier and Gouy (1999; Boussau and Gouy 2006) estimated different GC content parameters for branches on the tree. Because of the reduced number of parameters, this model was successfully used to analyse relatively large datasets. However, in some datasets, the base compositional drift may not be described properly by just GC content change.

The problem of too many parameters may be avoided by constructing a prior on them, using a stochastic process to describe the drift of base compositions over branches or over time. The likelihood calculation then has to integrate over the trajectories of base frequencies. This integral is daunting, but may be calculated using Bayesian MCMC algorithms (Foster 2004; Blanquart and Lartillot 2006). We discuss such models in Chapter 8.

4.4 Reconstruction of ancestral states

4.4.1 *Overview*

Evolutionary biologists have had a long tradition of reconstructing traits in extinct ancestral species and using them to test interesting hypotheses. The MacClade program (Maddison and Maddison 2000) provides a convenient tool for ancestral reconstruction using different variants of the parsimony method. Maddison and Maddison (2000) also provided an excellent review of the many uses (and misuses) of ancestral reconstruction. The *comparative method* (e.g. Felsenstein 1985b; Harvey and Pagel 1991; Schluter 2000) uses reconstructed ancestral states to uncover associated changes between two characters. Although association does not necessarily mean a cause–effect relationship, establishing an evolutionary association is the first step in inferring the adaptive significance of a trait. For example, butterfly larvae may be palatable (P_+) or unpalatable (P_-) and they may be solitary (S_+) or gregarious (S_-). If we can establish a significant association between character states P_+ and S_+, and if in particular, P_+ always appears before S_+ on the phylogeny based on ancestral state reconstructions, a plausible explanation is that palatability drives the evolution of solitary behaviour (Harvey and Pagel 1991). In such analysis, ancestral reconstruction is often the first step.

For molecular data, ancestral reconstruction has been used to estimate the relative rates of substitution between nucleotides or amino acids (e.g. Dayhoff et al. 1978; Gojobori et al. 1982), to count synonymous and nonsynonymous substitutions on the tree to infer adaptive protein evolution (e.g. Messier and Stewart 1997; Suzuki and Gojobori 1999), to infer changes in nucleotide or amino acid compositions (Duret et al. 2002), to detect coevolving nucleotides or amino acids (e.g. Shindyalov et al. 1994; Tuff and Darlu 2000; Dutheil et al. 2005), and to conduct many other analyses. Many of these procedures have been superseded by more rigorous likelihood analyses. A major application of ancestral sequence reconstruction is in the so-called chemical paleogenetic restoration studies envisaged by Pauling and Zuckerkandl (1963; see also Zuckerkandl 1964). Those studies use parsimony or likelihood to infer ancestral proteins and then synthesize them using site-directed mutagenesis and examine their chemical and physiological properties in the laboratory (e.g. Malcolm et al. 1990; Stackhouse et al. 1990; Libertini and Di Donato 1994; Jermann et al. 1995; Thornton et al. 2003; Ugalde et al. 2004; Gaucher et al. 2008). Hypotheses concerning the sequences, functions, and structures of ancient proteins are formulated and tested in this way. A number of reviews have been published on the topic (e.g. Golding and Dean 1998; Chang and Donoghue 2000; Benner 2002; Thornton 2004; Dean and Thornton 2007), as well as an edited book (Liberles 2009).

We discussed a dynamic programming algorithm for ancestral reconstruction under weighted parsimony in §3.4.3, which includes as special cases the parsimony method of Fitch (1971b) and Hartigan (1973). While discussing ancestral reconstruction by parsimony, Fitch (1971b) and Maddison and Maddison (1982) emphasized the advantage of a probabilistic approach to ancestral reconstruction and the importance of quantifying the uncertainty in the reconstruction. Nevertheless, early studies (and some recent ones) failed to calculate the right probabilities. When a Markov chain model is used to describe the evolution of characters, the ancestral states are random variables in the model: they do not appear in the likelihood function, which averages over all possible ancestral states (see §4.2), and cannot be estimated from the likelihood function. To infer ancestral states, one should calculate the conditional (posterior) probabilities of ancestral states given the data. This is the EB approach, proposed by Yang et al. (1995a) and Koshi and Goldstein (1996a). It is empirical, as parameter estimates (such as MLEs) are used in the calculation of posterior probabilities of ancestors. Many statisticians consider EB to be a likelihood approach (as opposed to Bayesian or full Bayesian approach). To avoid confusion, I will refer to the approach as EB (instead of likelihood).

Compared with parsimony reconstruction, the EB approach takes into account different branch lengths and different substitution rates between nucleotides or amino acids. It also provides posterior probabilities as a measure of the accuracy of the reconstruction. The EB approach has the drawback of not accommodating sampling errors in parameter estimates, and may be problematic in small datasets, which lack information to estimate parameters reliably. Huelsenbeck and Ballback (2001) implemented a full (hierarchical) Bayesian approach to ancestral state reconstruction, which assigns priors on parameters and averages over their uncertainties through MCMC algorithms (see Chapters 7 and 8). Another approach, proposed by Nielsen (2002; Huelsenbeck et al. 2003; Bollback 2006; Minin and Suchard 2008) and known as *stochastic mapping*, samples substitutions on branches of the tree. Used correctly, the samples of substitutions on the tree should lead to the same inference as reconstructed states at interior nodes, but the approach may have a computational advantage at low sequence divergence. For divergent sequences, there may be many substitutions at a site and the approach of sampling substitutions on the tree is computationally expensive.

In this section, I will describe the EB approach to ancestral sequence reconstruction, and discuss the modifications in the hierarchical Bayesian approach. I will also discuss the reconstruction of ancestral states of a discrete morphological character, which can be achieved using the same theory, but the inference involves greater uncertainties due to lack of information to estimate model parameters.

4.4.2 Empirical and hierarchical Bayesian reconstruction

A distinction can be made between the *marginal* and *joint reconstructions*. The former assigns a character state to a single node, while the latter assigns a set of character states to all ancestral nodes. For instance, given the observed nucleotides at the tips $x_1 x_2 x_3 x_4 x_5$ = TCACC in Figure 4.2, x_6 = T is a marginal reconstruction for node 6, while $x_0 x_6 x_7 x_8$ = TTTT is a joint reconstruction. Marginal reconstruction is more suitable when one wants the sequence at a particular node, as in the molecular restoration studies. Joint reconstruction is more suitable when one counts changes at each site.

Here we use the example of Figure 4.2 to illustrate the EB approach to ancestral reconstruction. We pretend that the branch lengths and the transition/transversion rate ratio κ, used in calculating the conditional probabilities (i.e. the $L_i(x_i)$'s shown on the tree), are the true values. In real data analysis, the parameters should be replaced by the MLEs from the data, and furthermore, an unrooted tree should be used when the molecular clock is not assumed.

4.4.2.1 Marginal reconstruction

We calculate the posterior probabilities of character states at one ancestral node. Consider node 0, the root. The posterior probability that node 0 has the nucleotide x_0, given the data at the site $\mathbf{x}_h$, is

$$f(x_0|\mathbf{x}_h;\theta) = \frac{f(x_0|\theta)f(\mathbf{x}_h|x_0;\theta)}{f(\mathbf{x}_h|\theta)} = \frac{\pi_{x_0} L_0(x_0)}{\sum_{x_0} \pi_{x_0} L_0(x_0)}. \qquad (4.23)$$

Note that $\pi_{x_0} L_0(x_0)$ is the joint probability of the state x_0 at the root and the states $\mathbf{x}_h$ at the tips. This is calculated by summing over all other ancestral states except x_0 (see §4.2.2, for the definition of $L_0(x_0)$, and equation (4.5)). Figure 4.2 shows $L_0(x_0)$, while the prior probability $f(x_0|\theta) = \pi_{x_0} = 1/4$ for any nucleotide x_0 under the K80 model. The probability of data at the site is $f(\mathbf{x}_h|\theta) = 0.000509843$. Thus the posterior probabilities at node 0 are 0.055 (= 0.25 × 0.00011237/0.000509843), 0.901, 0.037, and 0.007, for T, C, A, and G, respectively. C is the most probable nucleotide at the root, with posterior probability 0.901.

Posterior probabilities at any other interior node can be calculated by moving the root to that node and redoing the calculation using the same algorithm. These are 0.093 (T), 0.829 (C), 0.070 (A), 0.007 (G) for node 6; 0.153 (T), 0.817 (C), 0.026 (A), 0.004 (G) for node 7; and 0.010 (T), 0.985 (C), 0.004 (A), 0.001 (G) for node 8. From these marginal reconstructions, one might guess that the best joint reconstruction is $x_0 x_6 x_7 x_8 = \text{CCCC}$, with posterior probability $0.901 \times 0.829 \times 0.817 \times 0.985 = 0.601$. However, this calculation is incorrect, since the states at different nodes are not independent. For example, given that node 0 has $x_0 = \text{C}$, the probability that nodes 6, 7, and 8 will have C as well will be much higher than when the state at node 0 is unknown.

The above description assumes the same rate for all sites, but applies also to the fixed-rate (partition) models. In a random-rate model, equation (4.23) still gives the correct posterior probability but both $f(\mathbf{x}_h|\theta)$ and $f(\mathbf{x}_h|x_0;\theta)$ are sums over the rate categories and can similarly be calculated using the pruning algorithm.

4.4.2.2 Joint reconstruction

With this approach, we calculate the posterior probability for a set of character states assigned to all interior nodes at a site. Let $\mathbf{y}_h = (x_0, x_6, x_7, x_8)$ be such an assignment or reconstruction.

$$f(\mathbf{y}_h|\mathbf{x}_h;\theta) = \frac{f(\mathbf{x}_h, \mathbf{y}_h|\theta)}{f(\mathbf{x}_h|\theta)} \qquad (4.24)$$

$$= \frac{\pi_{x_0} p_{x_0 x_6}(t_6) p_{x_6 x_7}(t_7) p_{x_7 T}(t_1) p_{x_7 C}(t_2) p_{x_6 A}(t_3) p_{x_0 x_8}(t_8) p_{x_8 C}(t_4) p_{x_8 C}(t_5)}{f(\mathbf{x}_h|\theta)}.$$

The numerator $f(\mathbf{x}_h, \mathbf{y}_h|\theta)$ is the joint probability of the tip states $\mathbf{x}_h$ and the ancestral states $\mathbf{y}_h$, given parameters θ. This is the term in the square brackets in equation (4.3). The probability of data at a site, $f(\mathbf{x}_h|\theta)$, is a sum over all possible reconstructions $\mathbf{y}_h$, while the percentage of contribution from any particular reconstruction $\mathbf{y}_h$ is the posterior probability for that reconstruction.

The difficulty with a naïve use of this formula, as in Yang et al. (1995a) and Koshi and Goldstein (1996a), is the great number of ancestral reconstructions (all combinations of x_0, x_6, x_7, x_8). Note that only the numerator is used to compare the different reconstructions, as $f(\mathbf{x}_h|\theta)$ is fixed. Instead of maximizing the product of the transition probabilities to find the best reconstruction, we can maximize the sum of the logarithms of the transition probabilities. The dynamic programming algorithm for determining the best reconstructions for weighted parsimony, described in §3.4.3, can thus be used after the following minor modifications. First, each branch now has its own cost matrix while a single cost matrix was used for all branches for parsimony. Second, the score for each reconstruction involves an additional term $\log(\pi_{x_0})$. The resulting algorithm is equivalent to that described by Pupko et al. (2000). It works under the one-rate and fixed-rate models (the partition models) but not under the random-rate models (the mixture models). For the latter, Pupko et al. (2002a) implemented a branch-and-bound algorithm.

Application of the dynamic programming algorithm to the problem of Figure 4.2 leads to $x_0 x_6 x_7 x_8 = \text{CCCC}$ as the best reconstruction, with posterior probability 0.784. This agrees with the marginal reconstruction, according to which C is the most probable nucleotide for every node, but the probability is higher than the incorrect value 0.601, mentioned above. The next few best reconstructions can be obtained by a slight extension of the dynamic programming algorithm, as TTTC (0.040), CCTC (0.040), CTTC (0.040), AAAC (0.011), and CAAC (0.011). The above calculations are for illustration only. In real data analysis, we should use the MLEs of branch lengths and other parameters, and also use an unrooted tree since we are not assuming the clock.

The marginal and joint reconstructions use slightly different criteria. They normally produce consistent results, with the most probable joint reconstruction for a site consisting of character states that are also the best in the marginal reconstructions. Conflicting results may arise when the competing reconstructions have similar probabilities, in which case neither reconstruction is very reliable. The significance of the distinction lies mostly in that one should not multiply the probabilities for the marginal reconstructions to calculate the probability for the joint reconstruction.

4.4.2.3 *Comparison with parsimony*

If we assume the JC69 model with symmetrical substitution rates and also equal branch lengths, the EB and parsimony approaches will give exactly the same rankings of the joint reconstructions. Under JC69, the off-diagonal elements of the transition probability matrix are all equal and smaller than the diagonals: $P_{ij}(t) < P_{ii}(t)$, so that a reconstruction requiring fewer changes will have a higher posterior probability than one requiring more changes (see equation (4.24)). When branch lengths are allowed to differ as they are in a likelihood analysis, and the substitution rates are unequal as under more complex substitution models than JC69, parsimony, and EB may produce different results. Even in such cases, the two approaches are expected to produce very similar results, in that the most parsimonious reconstructions most often have the highest posterior probabilities. In both approaches, the main factor influencing the accuracy of ancestral reconstruction is the sequence divergence level. The reconstruction is less reliable at more variable sites or for more divergent sequences (Yang et al. 1995a; Zhang and Nei 1997). The main advantage of EB over parsimony is that EB provides posterior probabilities as a measure of accuracy.

4.4.2.4 *Hierarchical Bayesian approach*

In real data analysis, parameters θ in equations (4.23) and (4.24) are replaced by their estimates, say, the MLEs. In large datasets, this may not be a problem as the parameters are reliably estimated. In small datasets, the parameter estimates may involve large sampling errors, so that the EB approach may suffer from inaccurate parameter estimates. For example, if a branch length is estimated to be zero, no change will be possible along that branch during ancestral reconstruction, even though the zero estimate may not be reliable. In such a case, it is advantageous to use a hierarchical (full) Bayesian approach, which assigns a prior on parameters θ and integrates them out. Such an approach is implemented by Huelsenbeck and Bollback (2001; see also Pagel et al. 2004).

Uncertainty in the phylogeny is a more complex issue. If the purpose of ancestral reconstruction is for use in further analysis, as in comparative methods (Felsenstein 1985b; Harvey and Purvis 1991), one can average over uncertainties in the phylogenies, in substitution parameters, as well as in the ancestral states by sampling from the posterior of these quantities in an MCMC algorithm (Huelsenbeck and Bollback 2001). The most important uncertainty in this regard is probably that of the ancestral states; in other words, use of the most likely ancestral states while ignoring the suboptimal reconstructions may produce different results from averaging over different ancestral reconstructions (see §4.4.4 about biases in ancestral reconstruction). If the purpose is to reconstruct the sequence at a particular ancestral node, the best approach may be to use a fixed phylogeny that is as reliable as possible (e.g. the ML tree). Use of a multifurcating consensus tree is not advisable as the consensus tree is necessarily incorrect.

Averaging over binary trees in a Bayesian algorithm may not have an advantage over using a fixed tree. Hanson-Smith et al. (2010) used simulation to examine whether accommodating uncertainties in the phylogeny improves the accuracy of ancestral sequence reconstruction. They found that use of the ML tree produces robust and accurate reconstructions and that the Bayesian approach of incorporating phylogenetic uncertainties is not necessary or beneficial. When there is phylogenetic uncertainty, the plausible trees produce identical ancestral reconstructions; conversely, when different phylogenies produce different ancestral states, there is little or no ambiguity about the true phylogeny.

*4.4.3 Discrete morphological characters

It may be fitting to discuss here the similar problem of reconstructing morphological characters. Parsimony used to be the predominant method for such analysis. Schluter (1995), Mooers and Schluter (1999), and Pagel (1999) emphasized the importance of quantifying the uncertainty in ancestral reconstruction and championed the likelihood reconstruction. This effort has met with two difficulties. First, their formulation of the likelihood method was not correct. Second, a single morphological character has little information for estimating model parameters, such as branch lengths on the tree and relative substitution rates between characters, and this lack of information makes the analysis highly unreliable. Below I will discuss the first problem but can only lament on the second.

The problem considered by Schluter and Pagel is reconstruction of a binary morphological character evolving under a Markov model with rates q_{01} and q_{10} (see Problem 1.3 in Chapter 1). To reduce the number of parameters to be estimated, all branches on the tree are assumed to have the same length. The correct solution to this problem is the EB approach (equation (4.24)) (Yang et al. 1995a; Koshi and Goldstein 1996a). As discussed above, the likelihood (EB) method of ancestral reconstruction does not directly use the likelihood function. Instead Schluter (1995), Mooers and Schluter (1999), and Pagel (1999) used $f(\mathbf{x}_h, \mathbf{y}_h|\theta)$ of equation (4.24) as the 'likelihood function' for comparing ancestral reconstructions $\mathbf{y}_h$. This is equivalent to the EB approach except that Mooers and Schluter neglected the π_{x_0} term, and Pagel used $1/2$ for π_{x_0}; the prior probabilities at the root should be given by the substitution model as $\pi_0 = q_{10}/(q_{01} + q_{10})$ and $\pi_1 = q_{01}/(q_{01} + q_{10})$. Note that $f(\mathbf{x}_h, \mathbf{y}_h|\theta)$ is the joint probability of $\mathbf{x}_h$ and $\mathbf{y}_h$ and is not the likelihood of $\mathbf{y}_h$. The 'log likelihood ratio' for comparing two states at a node discussed by those authors is the ratio of the posterior probabilities for the two states and cannot be interpreted in the sense of Edwards (1992); thus a 'log likelihood ratio' of 2 means posterior probabilities 0.88 $(= e^2/(e^2 + 1))$ and 0.12 for the two states. Pagel (1999) described the EB calculation as the 'global' approach and preferred an alternative 'local approach', in which both substitution parameters θ and ancestral states $\mathbf{y}_h$ are estimated from the joint probability $f(\mathbf{x}_h, \mathbf{y}_h|\theta)$. This local approach is invalid, as θ should be estimated from the likelihood $f(\mathbf{x}_h|\theta)$, which averages over all possible ancestral states $\mathbf{y}_h$.

If one insisted on estimating ancestral states from the likelihood function, it would be possible to do so only for the marginal reconstruction at the root. Note that the likelihood function is the probability of the observed data, i.e. the states at the tips. The likelihood function would be $L_0(x_0)$, the probability of data at the site given the state x_0 at the root. The placement of the root would then affect the likelihood and the rooted tree should be used. For molecular data, this approach suffers from the problem that the number of parameters grows without bound when the sample size increases. At any rate, it is not

---

* indicates a more difficult or technical section.

possible to use the likelihood function for the joint reconstruction. For the example of Figure 4.2 for nucleotides

$$f(\mathbf{x}_h|\mathbf{y}_h;\theta) = p_{x_7T}(t_1)p_{x_7C}(t_2)p_{x_6A}(t_3)p_{x_8C}(t_4)p_{x_8C}(t_5). \tag{4.25}$$

This is a function of $x_6, x_7,$ and x_8 (for nodes that are connected to the tips), and not of x_0 (for nodes that are not directly connected to the tips): under the Markov model, given the states at the immediate ancestors of the tips, the states at the tips are independent of states at older ancestors. Thus the likelihood function is independent of states at the interior nodes not directly connected to the tips and it is impossible to infer ancestral states from the likelihood function.

Even with the EB approach, the substitution rates q_{01} and q_{10} cannot be estimated reliably from only one character. One might hope to use an LRT to compare the one-rate ($q_{01} = q_{10}$) and two-rate ($q_{01} \neq q_{10}$) models. However, this test is problematic as the asymptotic χ^2 distribution may not be reliable due to the small sample size (which is one) and more importantly, the failure to reject the one-rate model may reflect a lack of power of the test rather than a genuine symmetry of the rates. Analyses by a number of authors (e.g. Mooers and Schluter 1999) suggest that ancestral reconstruction is sensitive to the assumption of rate symmetry. Another worrying assumption is that of equal branch lengths. This means that the expected amount of evolution is the same for every branch on the tree, and is more unreasonable than the clock assumption (rate constancy over time). An alternative may be to use estimates of branch lengths obtained from molecular data, but this is open to the criticism that the molecular branch lengths may not reflect the amount of evolution in the morphological character.

An approach to dealing with uncertainties in the parameters is hierarchical Bayesian, which averages over uncertainties in substitution rates and branch lengths through a prior. Schultz and Churchill (1999) implemented such an algorithm, and found that the posterior probabilities for ancestral states are very sensitive to priors on relative substitution rates, even in seemingly ideal situations where the parsimony reconstruction is unambiguous. Those studies highlight the misleading overconfidence of parsimony as well as the extreme difficulty of reconstructing ancestral states for a single character. The Bayesian approach does offer the consolation that if different inferences are drawn from the same data under the same model, they must be due to differences in the prior.

In general, classical statistical approaches such as ML are not expected to work well in datasets consisting of one sample point, or in problems involving as many parameters as the observed data. When ML is applied to analyse one or two morphological characters (Pagel 1994; Lewis 2001), the asymptotic theory for MLEs and LRTs may not apply.

4.4.4 Systematic biases in ancestral reconstruction

The ancestral character states reconstructed by parsimony or likelihood (EB) are our best guesses under the respective criteria. However, if they are used for further statistical analyses or tests, one should bear in mind that they are inferred pseudo-data rather than real observed data. They involve random errors due to uncertainties in the reconstruction. Worse still, use of only the best reconstructions while ignoring the suboptimal ones can cause systematic biases. A number of authors (Collins et al. 1994; Perna and Kocher 1995; Eyre-Walker 1998) have discussed such biases with parsimony reconstruction. They also exist with likelihood (EB) reconstruction. The problem lies not in the use of parsimony versus likelihood or Bayesian for ancestral reconstruction, but in the use of the optimal reconstructions while ignoring the suboptimal ones. For example,

132   **4 MAXIMUM LIKELIHOOD METHODS**

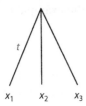

Fig. 4.10 A tree of three species for demonstrating the bias in ancestral reconstruction. The three branch lengths are equal, at $t = 0.2$ substitutions per site.

using simulation, Williams et al. (2006) found that use of the optimal reconstructions by parsimony and likelihood leads to overestimated thermostability of ancestral proteins, while a Bayesian method that sometimes chooses less probable residues from the posterior probability distribution does not.

As an example, consider the star tree of three sequences of Figure 4.10. Suppose that the substitution process has been stationary and followed the model of Felsenstein (1981), with parameters $\pi_T = 0.2263$, $\pi_C = 0.3282$, $\pi_A = 0.3393$, and $\pi_G = 0.1062$ (these frequencies are from the human mitochondrial D-loop hypervariable region I). Suppose that each branch length is 0.2 nucleotide substitutions per site, so that the transition probability matrix is

$$P(0.2) = \begin{bmatrix} 0.811138 & 0.080114 & 0.082824 & 0.025924 \\ 0.055240 & 0.836012 & 0.082824 & 0.025924 \\ 0.055240 & 0.080114 & 0.838722 & 0.025924 \\ 0.055240 & 0.080114 & 0.082824 & 0.781821 \end{bmatrix}. \quad (4.26)$$

We use the correct model and branch lengths to calculate posterior probabilities for ancestral states at the root, and examine the frequencies of A and G in the reconstructed ancestral sequence. This mimics studies which use ancestral reconstruction to detect possible drift in base compositions. At sites with data AAG, AGA, and GAA, the posterior probabilities for the states at the root are 0.006 (T), 0.009 (C), 0.903 (A), and 0.083 (G), so that A is much more likely than G (see Problem 4.4). However, if we use A and ignore G at every such site, we will over-count A and under-count G. Similarly, at sites with data GGA, GAG, and AGG, the posterior probabilities are 0.002 (T), 0.003 (C), 0.034 (A), and 0.960 (G). The best reconstruction is G, so that we over-count G and under-count A at such sites. The greater frequency of A than of G in the data means that there are more sites with data AAG, AGA, and GAA (with probability 0.02057) than sites with data GGA, GAG, and AGG (with probability 0.01680). The net bias is then an over-count of A and under-count of G in the ancestor. Indeed, the base compositions in the reconstructed sequence for the root are 0.212 (T), 0.324 (C), 0.369 (A), and 0.095 (G), more extreme than the frequencies in the observed sequences. Thus ancestral reconstruction indicates an apparent gain of the rare nucleotide G over the time of evolution from the root to the present. As the process is in fact stationary, this apparent drift in base compositions is an artefact of the EB (and parsimony) reconstruction, caused by ignoring the suboptimal reconstructions at every site.

Jordan et al. (2005) used parsimony to reconstruct ancestral protein sequences and observed a systematic gain of rare amino acids (and loss of common ones) over evolutionary time. The trend is the same as discussed above, and appears to be an artefact of ancestral reconstruction (Goldstein and Pollock 2006). Perna and Kocher (1995) studied the use of parsimony to infer ancestral states and then to count changes along branches to estimate the substitution rate matrix. The bias involved in such an analysis appears even greater than in counts of nucleotides discussed above. Clearly, the problem is more serious for more divergent sequences since ancestral reconstruction is poorer, but the bias can be considerable even in datasets of closely related sequences, such as the human mitochondrial D-loop sequences (Perna and Kocher 1995) or population data (Hernandez et al. 2007).

Despite those caveats, the temptation to infer ancestors and use them to perform all sorts of statistical tests appears too great to resist. Ancestral reconstruction is thus used frequently, with many interesting and spurious discoveries being made all the time.

Instead of ancestral reconstruction, one should try to rely on a likelihood-based approach, which sums over all possible ancestral states, weighting them appropriately according to their probabilities of occurrence (see §4.2). For example, the relative rates of substitutions between nucleotides or amino acids can be estimated by using a likelihood model (Yang 1994b; Adachi and Hasegawa 1996a; Yang et al. 1998; Whelan and Goldman 2001; Le and Gascuel 2008). Possible drifts in nucleotide compositions may be tested by implementing models that allow different base frequency parameters for different branches (Yang and Roberts 1995; Galtier and Gouy 1998). Chapter 11 discusses a few more examples in which both ancestral reconstruction and full likelihood-based approaches are used to analyse protein-coding genes to detect positive selection.

If a likelihood analysis under the model is too complex and one has to resort to ancestral reconstruction, a heuristic approach to reducing the bias may be to use the suboptimal as well as the optimal reconstructions in the analysis. One may use a simpler existing likelihood model to calculate posterior probabilities for ancestral states and use them as weights to accommodate both optimal and suboptimal reconstructions. In the example above, the posterior probabilities for the root state at a site with data AAG are 0.006 (T), 0.009 (C), 0.903 (A), and 0.083 (G). Instead of using A for the site and ignoring all other states, one can use both A and G, with weights 0.903 and 0.083 (rescaled so that they sum to one). If we use all four states at every site in this way, we will recover the correct base compositions for the root sequence with no bias at all, since the posterior probabilities are calculated under the correct model. If the likelihood model assumed in ancestral reconstruction is too simplistic and incorrect, the posterior probabilities (weights) will be incorrect as well. Even so this approach may be less biased than ignoring suboptimal reconstructions entirely. Akashi et al. (2007) applied this approach to count changes between preferred (frequently used) and unpreferred (rarely used) codons in protein-coding genes in the *Drosophila melanogaster* species subgroup and found that it helped to reduce the bias in ancestral reconstruction. Similarly Dutheil et al. (2005) used reconstructed ancestral states to detect coevolving nucleotide positions, indicated by an excess of substitutions at two sites that occur along the same branches. They calculated posterior probabilities for ancestral states under an independent-site model and used them as weights to count substitutions along branches at the two sites under test.

*4.5 Numerical algorithms for maximum likelihood estimation

The likelihood method estimates parameters θ by maximizing the log likelihood ℓ. In theory, one may derive the estimates by setting to zero the first derivatives of ℓ with respect to θ and solving the resulting system of equations, called the *likelihood equations*:

$$\frac{\partial \ell}{\partial \theta} = 0. \qquad (4.27)$$

---

* indicates a more difficult or technical section.

This approach leads to analytical solutions to pairwise distance estimation under the JC69 and K80 models, as discussed in §1.4. For three species, the analytical solution is possible only in the simplest case of binary characters evolving under the molecular clock (Yang 2000a). The problem becomes intractable as soon as we consider the case of nucleotides with four states (Problem 4.2) or the case of four species. The latter case, of four species under the clock, was studied by Chor and Snir (2004), who derived analytical estimates of branch lengths for the 'fork' tree ((a, b), (c, d)) but not for the 'comb' tree (((a, b), c), d).

In general, numerical iterative algorithms have to be used to maximize the log likelihood. Developing a reliable and efficient optimization algorithm for practical problems is a complicated task. This section gives only a flavour of such algorithms. Interested readers should consult a textbook on nonlinear programming or numerical optimization, such as Gill et al. (1981) and Fletcher (1987).

Maximization of a function (called the *objective function*) is equivalent to minimization of its negative. Below, we will follow the convention and describe our problem of likelihood maximization as a problem of minimization. The objective function is thus the negative log likelihood: $f(\theta) = -\ell(\theta)$. Note that algorithms that reach the minimum with fewer function evaluations are more efficient, since $f(\theta)$ is expensive to calculate.

*4.5.1 Univariate optimization

If the problem is one-dimensional, the algorithm is called *line search* since the search is along a line. Suppose we determine that the minimum is in the interval $[a, b]$. This is called the *interval of uncertainty*. Most line search algorithms reduce this interval successively, until its width is smaller than a pre-specified small value. Assuming that the function is unimodal in the interval (i.e. there is only one valley between a and b), we can reduce the interval of uncertainty by comparing the function values at two interior points θ_1 and θ_2 (Figure 4.11). Different schemes exist concerning choices of the points. Here we describe the golden section search.

4.5.1.1 Golden section search

Suppose the interval of uncertainty is $[0, 1]$; this can be rescaled to become $[a, b]$. We place two interior points at γ and $(1-\gamma)$, where the golden ratio $\gamma \approx 0.6180$ satisfies $\gamma/(1-\gamma) = 1/\gamma$. The new interval becomes $(1-\gamma, 1)$ if $f(\gamma) < f(1-\gamma)$ or $(0, \gamma)$ otherwise (Figure 4.12). No matter how the interval is reduced, one of the two points will be in

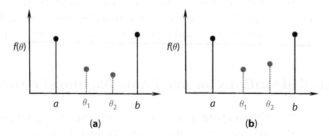

Fig. 4.11 Reduction of the interval of uncertainty (a, b), which contains the minimum. The objective function f is evaluated at two interior points θ_1 and θ_2. (**a**) If $f(\theta_1) \geq f(\theta_2)$, the minimum must lie in the interval (θ_1, b). (**b**) Otherwise if $f(\theta_1) < f(\theta_2)$, the minimum must lie in the interval (a, θ_2).

* indicates a more difficult or technical section.

4.5 NUMERICAL ALGORITHMS FOR MAXIMUM LIKELIHOOD ESTIMATION

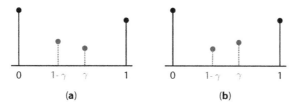

Fig. 4.12 The golden section search. Suppose the minimum is inside the interval (0, 1). Two points are placed at γ and $(1 - \gamma)$ of the interval, where $\gamma = 0.6180$. (a) If $f(\gamma) < f(1 - \gamma)$, the new interval will be $(1 - \gamma, 1)$. (b) If $f(\gamma) \geq f(1 - \gamma)$, the new interval will be $(0, \gamma)$. No matter how the interval is reduced, one of the two points will be in the correct position inside the new interval.

the correct position inside the new interval for the next iteration. With the golden section search, the interval of uncertainty is reduced by γ at each step. The algorithm is said to have a linear convergence rate.

4.5.1.2 Newton's method and polynomial interpolation

For smooth functions, more efficient algorithms can be implemented by approximating f using simple functions whose minimum can be obtained analytically. For example, if we approximate f by a parabola (quadratic), of the form

$$\tilde{f} = a\theta^2 + b\theta + c, \tag{4.28}$$

with $a > 0$, then $\tilde{f}$ has a minimum at $\theta^* = -b/(2a)$. If the function value and its first and second derivatives at the current point θ_k are known, we can use the first three terms of the Taylor expansion to approximate $f(\theta)$:

$$\tilde{f}(\theta) = f(\theta_k) + f'(\theta_k)(\theta - \theta_k) + \frac{1}{2}f''(\theta_k)(\theta - \theta_k)^2. \tag{4.29}$$

This is a quadratic function in θ, in the form of equation (4.28), with $a = f''(\theta_k)/2$ and $b = f'(\theta_k) - f''(\theta_k)\theta_k$. If $f''(\theta_k) > 0$, the quadratic (4.29) achieves its minimum at

$$\theta_{k+1} = -\frac{b}{2a} = \theta_k - \frac{f'(\theta_k)}{f''(\theta_k)}. \tag{4.30}$$

As f may not be a quadratic, θ_{k+1} may not be the minimum of f. We thus use θ_{k+1} as the new current point to repeat the algorithm. This is Newton's method, also known as the Newton–Raphson method.

Newton's method is highly efficient. Its rate of convergence is quadratic, meaning that, roughly speaking, the number of correct figures in θ_k doubles at each step (e.g. Gill et al. 1981, p. 57). A problem, however, is that it requires the first and second derivatives, which may be expensive, troublesome, or even impossible to compute. Without the derivatives, a quadratic approximation can be constructed by using the function values at three points. Similarly a cubic polynomial can be constructed by using the function values and first derivatives (but not the second derivatives) at two points. It is generally not worthwhile fitting high-order polynomials. Another serious problem with Newton's method is that its fast convergence rate is only local, and if the iteration is not close to the minimum, the algorithm may diverge hopelessly. The iteration may also encounter numerical difficulties if $f''(\theta_k)$ is zero or too small. Thus it is important to obtain good starting values for Newton's method, and certain safeguards are necessary to implement the algorithm.

A good strategy is to combine a guaranteed reliable method (such as golden section) with a rapidly convergent method (such as Newton's quadratic interpolation), to yield an algorithm that will converge rapidly if f is well-behaved, but is not much less efficient than the guaranteed method in the worst case. Suppose a point $\tilde{\theta}$ is obtained by quadratic interpolation. We can check to make sure that $\tilde{\theta}$ lies in the interval of uncertainty (a, b) before evaluating $f(\tilde{\theta})$. If $\tilde{\theta}$ is too close to the current point or too close to either end of the interval, one may revert to the golden section search. Another idea of safeguarding Newton's method is to redefine

$$\theta_{k+1} = \theta_k - \alpha f'(\theta_k)/f''(\theta_k), \tag{4.31}$$

where the step length α, which equals 1 initially, is repeatedly halved until the algorithm is non-increasing, that is, until $f(\theta_{k+1}) < f(\theta_k)$.

*4.5.2 Multivariate optimization

Most models used in likelihood analysis in molecular phylogenetics include multiple parameters, and the optimization problem is multidimensional. A naïve approach is to optimize one parameter at a time with all other parameters fixed. However, this is inefficient when the parameters are correlated. Figure 4.13 shows an example in which the two parameters are (positively) correlated. A search algorithm that updates one parameter at a time makes impressive improvements to the objective function initially, but becomes slower and slower when it is close to the minimum. As every search direction is at a 90° angle to the previous search direction, the algorithm zigzags in tiny baby steps. Standard optimization algorithms update all variables simultaneously.

4.5.2.1 Steepest-descent search

Many optimization algorithms use the first derivatives, $g = df(\theta)$, called the *gradient*. The simplest among them is the *steepest-descent* algorithm (or *steepest-ascent* for maximization). It finds the steepest-descent direction, locates the minimum along that direction,

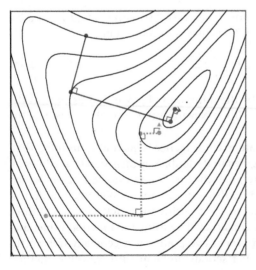

Fig. 4.13 The log likelihood contour when two parameters are positively correlated. A search algorithm changing one parameter at a time (the dotted lines and arrow) is very inefficient. Similarly the steepest-ascent search (the solid lines and arrow) is inefficient because its search direction is always perpendicular to the previous search direction.

* indicates a more difficult or technical section.

and repeats the procedure until convergence. The gradient g is the direction that the function increases the quickest locally and is thus *the steepest-ascent direction*, while $-g$ is the *steepest-descent direction*. Note that the minimum along a search direction occurs when the search direction becomes a tangent line to a contour curve. At that point, the new gradient is perpendicular to the tangent line or the previous search direction (Figure 4.13). The steepest-descent algorithm suffers from the same problem as the naïve algorithm of changing one variable at a time: every search direction forms a 90° angle with the previous search direction. The algorithm descends very quickly initially but becomes slower and slower when it is close to the minimum.

4.5.2.2 Newton's method

The multivariate version of the Newton algorithm relies on quadratic approximation to the objective function. Let $G = d^2 f(\theta)$ be the *Hessian matrix* of second partial derivatives. With p variables, both θ and g are $p \times 1$ vectors while G is a $p \times p$ matrix. A second-order Taylor expansion of f around the current point θ_k gives

$$f(\theta) \approx f(\theta_k) + g_k^T(\theta - \theta_k) + \frac{1}{2}(\theta - \theta_k)^T G_k (\theta - \theta_k), \tag{4.32}$$

where the superscript ^{T} means transpose. By minimizing the right-hand side of equation (4.32), i.e. by setting its gradient to 0, one obtains the next iterate as

$$\theta_{k+1} = \theta_k - G_k^{-1} g_k. \tag{4.33}$$

Note that this is the same as equation (4.30) for the univariate case. Similarly, the multivariate version shares the rapid convergence rate as well as the major drawbacks of the univariate algorithm; i.e. the method requires calculation of the first and second derivatives and may diverge when the iterate is not close enough to the minimum. A common strategy is to take $s_k = G_k^{-1} g_k$ as a search direction, called the *Newton direction*, and to perform a line search to determine how far to go along that direction.

$$\theta_{k+1} = \theta_k + \alpha s_k = \theta_k - \alpha G_k^{-1} g_k. \tag{4.34}$$

Here α is called the *step length*. It is often too expensive to optimize α in this way. A simpler version is to try $\alpha = 1, 1/2, 1/4, \ldots$, until $f(\theta_{k+1}) \leq f(\theta_k)$. This is sometimes known as the *safe-guided Newton algorithm*. When G_k is not *positive definite*, one can reset it to the identity matrix: $G_k = I$.

When the objective function is the negative log likelihood, $f = -\ell$, the Hessian matrix $G = -d^2 \ell(\theta)$ is also called the *observed information matrix*. In some simple statistical problems, the expected information, $-E\{d^2 \ell(\theta)\}$, may be easier to calculate, and can be used in Newton's algorithm. The method is then known as *scoring*. Both Newton's method and scoring have the benefit that the approximate variance–covariance matrix of the MLEs is readily available at the end of the iteration. We used such approximate variances in §1.4.1.

4.5.2.3 Quasi-Newton methods

Quasi-Newton methods include a class of methods that require first derivatives but not second derivatives. While Newton's method calculates the second derivatives G at every current point, quasi-Newton methods build up information about G or its inverse from the calculated values of the objective function f and the first derivatives g during the iteration. If the first derivatives are not available, they may be calculated using the difference approximation. Without the need for second or even first derivatives, quasi-Newton algorithms greatly increased the range of problems that can be solved. The basic algorithm can be sketched as follows:

a. Supply an initial guess θ_0.
b. For $k = 0, 1, 2, \ldots$, until convergence:
   1. Test θ_k for convergence;
   2. Calculate a search direction $s_k = -B_k g_k$;
   3. Perform a line search along s_k to determine the step length $\alpha_k : \theta_{k+1} = \theta_k + \alpha_k s_k$;
   4. Update B_k to give B_{k+1}.

Here B_k is a symmetric positive definite matrix, which can be interpreted as an approximation to G_k^{-1}, the inverse of the Hessian matrix. Note the similarity of this algorithm to Newton's method. In step b3, the scalar $\alpha_k > 0$ is chosen to minimize $f(\theta_{k+1})$, by using a line search algorithm as discussed above. A number of strategies have been developed to update the matrix B. Well-known ones include the Broyden–Fletcher–Goldfard–Shanno (BFGS) and Davidon–Fletcher–Powell (DFP) formulae. See Gill et al. (1981) and Fletcher (1987) for details.

When the first derivatives are impossible or expensive to calculate, an alternative approach is to use a *derivative-free method*. See Brent (1973) for discussions of such methods. According to Gill et al. (1981), quasi-Newton methods, even if the first derivatives are calculated using the difference approximation, are more efficient than derivative-free methods.

4.5.2.4 *Bounds and constraints*

The discussion up to now has assumed that the parameters are unconstrained and can take values over the whole real line. In most practical problems, parameters have bounds. For example, branch lengths should be nonnegative, and the nucleotide frequency parameters π_1, π_2, π_3 should satisfy the following constraints: $\pi_1, \pi_2, \pi_3 > 0$ and $\pi_1 + \pi_2 + \pi_3 < 1$. Even unconstrained optimization can be improved by having rough guesses of the parameter values and applying bounds on the parameters.

Algorithms for constrained optimization are much more complex, with simple lower and upper bounds being easier than general linear inequality constraints (Gill et al. 1981). Variable transform is often an effective approach for dealing with linear inequality constraints. For example, to estimate the nucleotide frequencies $(\pi_1, \pi_2, \pi_3, \pi_4)$, we can use new variables $-\infty < x_1, x_2, x_3 < \infty$, with $x_4 = 0, \pi_1 = x_1/s, \pi_2 = x_2/s, \pi_3 = x_3/s$, and $\pi_4 = 1/s$, where $s = e^{x_1} + e^{x_2} + e^{x_3} + 1$. Another transform is $x_1 = \pi_1/\pi_4, x_2 = \pi_2/\pi_4, x_3 = \pi_3/\pi_4$ so that $0 < x_1, x_2, x_3 < \infty$, but this is often less efficient than the exponential transform. As another example, consider estimation of divergence times (node ages) t_0, t_1, t_2, t_3 on a five-species tree, with the constraints $t_0 > t_1 > t_2 > t_3$. We can define new variables $x_0 = t_0$ (for the root age), $x_1 = t_1/t_0, x_2 = t_2/t_1, x_3 = t_3/t_2$, so that the new variables have simple bounds: $0 < x_0 < \infty, 0 < x_1, x_2, x_3 < 1$. In general the new variable x_i for any non-root node i is defined as the ratio of the node age (t_i) to the age of its mother node.

4.6 ML optimization in phylogenetics

4.6.1 *Optimization on a fixed tree*

In a phylogenetic problem, the parameters to be estimated include branch lengths in the tree and parameters in the substitution model. Given the values of parameters, one can use the pruning algorithm to calculate the log likelihood. In theory, one can then apply any of the general purpose optimization algorithms discussed above to find the MLEs iteratively. However, this will almost certainly produce an inefficient algorithm. Consider the recursive calculation of the conditional probabilities $L_i(x_i)$ in the pruning algorithm. When a branch length changes, $L_i(x_i)$ for only those nodes ancestral to that branch are changed, while those for all other nodes are not affected. Direct application of a general-purpose multivariate optimization algorithm thus leads to many duplicated calculations of the same quantities.

To take advantage of such features of likelihood calculation on a tree, one can optimize one branch length at a time, keeping all other branch lengths and substitution parameters fixed. Suppose one branch connects nodes a and b. By moving the root to coincide with node a, we can rewrite equation (4.6) as

$$f(\mathbf{x}_h|\theta) = \sum_{x_a}\sum_{x_b} \pi_{x_a} p_{x_a x_b}(t_b) L_a(x_a) L_b(x_b). \tag{4.35}$$

The first and second derivatives of ℓ with respect to t_b can then be calculated analytically (Adachi and Hasegawa 1996b; Yang 2000b), so that t_b can be optimized efficiently using Newton's algorithm. One can then estimate the next branch length by moving the root to one of its ends. A change to any substitution parameter, however, typically changes the conditional probabilities for all nodes, so saving is not possible. To estimate substitution parameters, two strategies appear possible. Yang (2000b) tested an algorithm that cycles through two phases. In the first phase, branch lengths are optimized one by one while all substitution parameters are held fixed. Several cycles through the branch lengths are necessary to achieve convergence. In the second phase, the substitution parameters are optimized using a multivariate optimization algorithm such as BFGS, with branch lengths fixed. This algorithm works well when the branch lengths and substitution parameters are not correlated, for example, under the HKY85 or GTR (REV) models, in which the transition/transversion rate ratio κ for HKY85 or the rate ratio parameters in GTR are not strongly correlated with the branch lengths. However, when there is strong correlation, the algorithm can be very inefficient. This is the case with the gamma model of variable rates at sites, in which the branch lengths and the gamma shape parameter α often have strong negative correlations. A second strategy (Swofford 2000) is to embed the first phase of the above algorithm into the second phase. One uses a multivariate optimization algorithm (such as BFGS) to estimate the substitution parameters, with the log likelihood for any given values of the substitution parameters calculated by optimizing the branch lengths. It may be necessary to optimize the branch lengths to a high precision, as inaccurate calculations of the log likelihood (for given substitution parameters) may cause problems for the BFGS algorithm, especially if the first derivatives are calculated numerically using difference approximation.

4.6.2 Multiple local peaks on the likelihood surface for a fixed tree

Numerical optimization algorithms discussed above are all local hill-climbing algorithms. They converge to a local peak, but may not reach the globally highest peak if multiple local peaks exist. Fukami and Tateno (1989) presented a proof that under the F81 model (Felsenstein 1981) and on a tree of any size, the log likelihood curve for one branch length has a single peak when other branch lengths are fixed. Tillier (1994) suggested that the result applies to more general substitution models as well. However, Steel (1994a) pointed out that this result does not guarantee one single peak in the whole parameter space. Consider a two-parameter problem and imagine a peak in the northwest region and another peak in the southeast region of the parameter space. Then there is always one peak if one looks only in the north–south direction or west–east direction, although in fact two local peaks exist. Steel (1994a) and Chor et al. (2000) further constructed counterexamples to demonstrate the existence of multiple local peaks for branch lengths even on small trees with four species.

Nevertheless, local peaks do not appear to be common in real data analysis unless the assumed model is complex and parameter-rich. A symptom of the problem is that different runs of the same analysis starting from different initial values may lead to different results. Rogers and Swofford (1999) used computer simulation to examine the problem, and reported that local peaks were less common for the ML tree than for other poorer trees. It is hard to imagine that the likelihood surfaces are qualitatively different for the different trees, so one possible reason for this finding may be that more local peaks exist at the boundary of the parameter space (say, with zero branch lengths) for the poor trees than for the ML tree. Existence of multiple local peaks is sometimes misinterpreted as an indication of the unrealistic nature of the assumed substitution model. In fact multiple local peaks are much less common under simplistic and unrealistic models like JC69 than under more realistic parameter-rich models.

There is no foolproof remedy to the problem of local peaks. A simple strategy is to run the iteration algorithm multiple times, starting from different initial values. If multiple local peaks exist, one should use the estimates corresponding to the highest peak. Stochastic search algorithms that allow downhill moves, such as simulated annealing and genetic algorithms, are useable as well (see §3.2.5).

4.6.3 Search in the tree space

The above discussion concerns estimation of branch lengths and substitution parameters for a given tree topology. This will be sufficient if our purpose is to estimate parameters in the substitution model or to test hypotheses concerning them, with the tree topology known or fixed. This is a straightforward application of the conventional ML estimation when the likelihood function $L(\theta; X) = f(X|\theta)$ is fully specified.

If our interest is in reconstruction of the phylogeny, we take the optimized log likelihood for the tree as the score for tree selection, and should, at least in theory, solve as many optimization problems as the number of tree topologies. This is far more complex than the conventional ML estimation. We have two levels of optimization: one of optimizing branch lengths (and substitution parameters) on each fixed tree to calculate the tree score and the other of searching for the best tree in the tree space.

An analytical characterization of the tree space is available for the case of rooted trees for three species with binary characters evolving under the molecular clock (Yang 2000a; see also Newton 1996; Steel 2011). This is the simplest problem of tree reconstruction. The transition probability matrix under the model is given in equation (1.79). The three binary rooted trees are $\tau_1 = ((1, 2), 3)$, $\tau_2 = ((2, 3), 1)$, and $\tau_3 = ((3, 1), 2)$, shown in Figure 4.14, where t_{i0} and t_{i1} are the two branch lengths in tree τ_i, for $i = 1, 2, 3$. The star tree τ_0 has only one branch length t_{01}. The data are summarized as the counts $\boldsymbol{n} = (n_0, n_1, n_2, n_3)$ or frequencies (f_0, f_1, f_2, f_3) of the four site patterns: xxx, xxy, yxx, and xyx, where x and y are any two distinct characters. Since $f_0 + f_1 + f_2 + f_3 = 1$, the sample space, i.e. the space of all possible datasets, is thus the tetrahedron OABC of Figure 4.15. Let p_0, p_1, p_2, p_3 be the probabilities of the site patterns given a tree and its branch lengths. The likelihood functions for the three trees are then

4.6 ML OPTIMIZATION IN PHYLOGENETICS

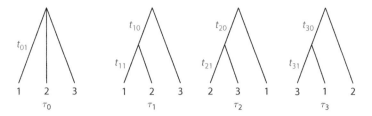

Fig. 4.14 The three rooted trees for three species: τ_1, τ_2, and τ_3. Branch lengths t_{i0} and t_{i1} in each tree τ_i ($i = 1, 2, 3$) are measured by the expected number of character changes per site. The star tree τ_0 is also shown with its branch length t_{01}.

$$f(\mathbf{n}|\tau_1, t_0, t_1) = p_0^{n_0} p_1^{n_1} p_2^{n_2+n_3},$$
$$f(\mathbf{n}|\tau_2, t_0, t_1) = p_0^{n_0} p_1^{n_2} p_2^{n_3+n_1}, \qquad (4.36)$$
$$f(\mathbf{n}|\tau_3, t_0, t_1) = p_0^{n_0} p_1^{n_3} p_2^{n_1+n_2},$$

where the probabilities for the site patterns are

$$p_0(t_0, t_1) = \tfrac{1}{4} + \tfrac{1}{4}e^{-4t_1} + \tfrac{1}{2}e^{-4(t_0+t_1)},$$
$$p_1(t_0, t_1) = \tfrac{1}{4} + \tfrac{1}{4}e^{-4t_1} - \tfrac{1}{2}e^{-4(t_0+t_1)}, \qquad (4.37)$$
$$p_2(t_0, t_1) = \tfrac{1}{4} - \tfrac{1}{4}e^{-4t_1} = p_3(t_0, t_1).$$

One may use the site pattern probabilities as parameters and represent τ_1 as $p_0 > p_1 > p_2 = p_3$, τ_2 as $p_0 > p_2 > p_3 = p_1$, and τ_3 as $p_0 > p_3 > p_1 = p_2$, while the star tree is $p_0 > p_1 = p_2 = p_3$. Those parameter spaces are superimposed on the sample space as well. The (probability) space for tree τ_1 with parameters $0 < t_{10}, t_{11} < \infty$ corresponds to the triangle *OPR*; in other words, p_1 and p_2 in the triangle (with $0 < p_2 = p_3 \leq p_1 \leq p_0$) is a reparametrization of t_{10} and t_{11} (with $0 < t_{10}, t_{11} < \infty$) (Problem 4.1). Similarly the parameter space of τ_2 is the triangle *OPS* and that for τ_3 is *OPT*. The probability spaces of the four trees thus form a 'paper airplane' (Newton 1996) or 'paper dart' (Steel 2011).

For a given dataset or data point (f_1, f_2, f_3), the MLE of t_{10} and t_{11} for τ_1 corresponds to the point in the triangle *OPR* closest to the data point, with the distance measured by the Kullback–Leibler (K-L) divergence. The K-L divergence from distribution p to distribution f is defined as

$$D_{\text{KL}}(f, p) = \sum_i f_i \log \frac{f_i}{p_i} = \sum_i f_i \log f_i - \sum_i f_i \log p_i. \qquad (4.38)$$

Note that $\sum_i f_i \log f_i$ is a constant when the data are observed, while $n \sum_i f_i \log p_i = \sum_i n_i \log p_i$ is the log likelihood, so that minimization of D_{KL} is equivalent to maximization of the log likelihood. Finally, if the data point is closer (by D_{KL}) to triangle *OPR* than it is to *OPS* or *OPT*, τ_1 will be the ML tree. The full likelihood solution is given by Yang (2000a, Table 4) and summarized in Figure 4.15: the ML tree is $\tau_1, \tau_2,$ or τ_3, if the data point is in the region *OPFAD*, *OPDBE*, or *OPECF*, respectively.

Thus search in the space of trees is more complex than optimization in the conventional ML estimation. In the latter, one can define the gradient (the direction of steepest ascent) and local curvature (Figure 4.13) and use a quadratic approximation to predict the peak of

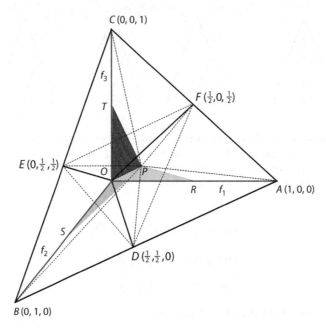

Fig. 4.15 The sample space (the space of all possible datasets) and the parameter space (the space of all trees) for the phylogeny–inference problem of Figure 4.14. Binary characters evolving under the clock with symmetrical substitution rates are used to reconstruct the rooted trees for three species. Each dataset (i.e. the alignment of three sequences) can be summarized as the counts (n_0, n_1, n_2, n_3) or proportions (f_0, f_1, f_2, f_3) of sites with the four site patterns *xxx*, *xxy*, *yxx*, and *xyx*. Since $f_0 = 1 - (f_1 + f_2 + f_3)$, each dataset is a point in the $f_1 - f_2 - f_3$ space or a point inside the tetrahedron *OABC*. The sample space *OABC* is partitioned into four regions, corresponding to the four trees; tree τ_i is the ML tree if and only if the data point falls within region i ($i = 0, 1, 2, 3$). The region for τ_0 is the line segment *OP* plus the tetrahedron *PDEF*. In this region, the three binary trees have the same likelihood as the star tree, so τ_0 is taken as the ML tree. The region for τ_1 is a contiguous block *OPFAD*, consisting of three tetrahedrons *OPAD*, *OPAF*, and *PDAF*. The regions for τ_2 and τ_3 are *OPDBE* and *OPECF*, respectively. The parameter (probability) space for each tree is superimposed onto the sample space. For τ_0 this is line segment *OP*, corresponding to $0 \leq f_1 = f_2 = f_3 < 1/4$ or $0 \leq t_{01} < \infty$. The parameter spaces for τ_1, τ_2, and τ_3 are triangles *OPR*, *OPS*, or *OPT*, respectively. The coordinates for the points are $O(0, 0, 0)$, $P(\frac{1}{4}, \frac{1}{4}, \frac{1}{4})$, $R(1/2, 0, 0)$, $S(0, 1/2, 0)$, and $T(0, 0, 1/2)$. Phylogeny reconstruction thus corresponds to partitioning the sample space and may also be viewed as projecting the observed data point onto the three parameter planes, with the distance measured by the K-L divergence. Redrawn after Yang (2000a).

the surface, as in the Newton or quasi-Newton algorithms. In tree search, such concepts are not meaningful when one moves from one tree (one triangle) to another (another triangle). With more species, the huge number of trees have intricate relationships among themselves, and the landscape may be complex, with local peaks (see Figure 3.18) and plateaus (Charleston 1995).

In practice, tree search by ML does not attempt to determine the best search direction in the tree space, but instead uses tree-rearrangement algorithms (such as NNI or SPR) to move between trees. Because the neighbouring trees generated during branch swapping share subtrees, tree search algorithms should avoid repeated computations of the same quantities. Much effort has been expended to develop fast algorithms for likelihood tree search, such as the FASTDNAML algorithm of Olsen et al. (1994), the genetic algorithms of Lewis (1998) and Lemmon and Milinkovitch (2002), and the parallel likelihood programs PHYML (Guindon and Gascuel 2003) and RAXML (Stamatakis 2006, Stamatakis et al. 2012). For example, it is well recognized that optimizing branch lengths to a high precision may not be a good use of time if the tree is poor. Thus in the algorithm of Guindon and Gascuel (2003), tree topology and branch lengths are adjusted simultaneously. Candidate trees are generated by local rearrangements of the current tree, for example, by using the NNI or SPR algorithms, and in calculating their likelihood scores, only the branch lengths affected by the local rearrangements are optimized, while branch lengths within subtrees unaffected by the rearrangements are not always optimized.

The likelihood algorithms have advanced greatly in the past decade, so that the ML method is now feasible for the analysis of large datasets with thousands of species/sequences (Guindon and Gascuel 2003; Stamatakis 2006; Zwickl 2006). Algorithms that take advantage of new computer hardware with multicore processors and graphical processing units (GPUs) (Suchard and Rambaut 2009; Zierke and Bakos 2010; Stamatakis et al. 2012) are pushing the boundary even further.

4.6.4 *Approximate likelihood method*

The computational burden of the likelihood method prompted the development of approximate methods. One idea is to use other methods to estimate branch lengths on a given tree rather than optimizing branch lengths by ML. For example, Adachi and Hasegawa (1996b) used least-squares estimates of branch lengths calculated from a pairwise distance matrix to calculate approximate likelihood scores. Similarly Rogers and Swofford (1998) used parsimony reconstruction of ancestral states to estimate approximate branch lengths. The approximate branch lengths provide good starting values for a proper likelihood optimization, but the authors suggested that they could also be used to calculate the approximate likelihood values for the tree without further optimization, leading to approximate likelihood methods of tree reconstruction.

Strimmer and von Haeseler (1996), and Schmidt et al. (2002) implemented an approximate likelihood algorithm for tree search called *quartet puzzling*. This uses ML to evaluate the three trees for every possible quartet of species. The full tree for all s species is then constructed by a majority-rule consensus of those quartet trees. This method may not produce the ML tree but is fast. Ranwez and Gascuel (2002) implemented an algorithm that combines features of NJ and ML. It is based on triplets of taxa, and shares the divide-and-conquer strategy of the quartet approach.

With the development of modern likelihood programs such as RAXML, those approximate methods are no longer important. However, they may be useful for generating initial trees, or good candidates or proposals for ML or Bayesian tree search.

4.7 Model selection and robustness

4.7.1 Likelihood ratio test applied to rbcL dataset

We introduced the LRT in §1.4.3. Here we apply it to the data of the plastid *rbcL* genes from 12 plant species. The sequence alignment was kindly provided by Dr Vincent Savolainen. There are 1,428 nucleotide sites in the sequence. The tree topology is shown in Figure 4.16, which will be used to compare models. Table 4.3 shows the log likelihood values and parameter estimates under three sets of nucleotide substitution models. The first set includes JC69, K80, and HKY85, in which the same Markov model is applied to all sites in the sequence. The second set includes the '+Γ_5' models, which use the discrete gamma to accommodate the variable rates among sites, using five site classes (Yang 1994a). The third set includes the partition ('+C') models (Yang 1995a, 1996b). They assume that different codon positions have different substitution rates (r_1, r_2, r_3), and, for K80 and HKY85, different substitution parameters as well. Parameters such as branch lengths in the tree, κ in K80 and HKY85, and the relative rates for codon positions in the '+C' models, are

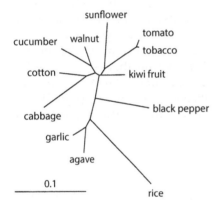

Fig. 4.16 The ML tree for the plastid *rbcL* genes from 12 plant species, estimated under the HKY85+Γ_5 model. The branches are drawn in proportion to their estimated lengths.

Table 4.3 Log likelihood values and MLEs of parameters under different models for the *rbcL* dataset

Model	p	ℓ	MLEs
JC69	21	−6,262.01	
K80	22	−6,113.86	$\hat{\kappa} = 3.561$
HKY85	25	−6,101.76	$\hat{\kappa} = 3.620$
JC69 + Γ_5	22	−5,937.80	$\hat{\alpha} = 0.182$
K80 + Γ_5	23	−5,775.40	$\hat{\kappa} = 4.191, \hat{\alpha} = 0.175$
HKY85 + Γ_5	26	−5,764.26	$\hat{\kappa} = 4.296, \hat{\alpha} = 0.175$
JC69 + C	23	−5,922.76	$r_1 : \hat{r}_2 : \hat{r}_3 = 1 : 0.556 : 5.405$
K80 + C	26	−5,728.76	$\hat{\kappa}_1 = 1.584, \hat{\kappa}_2 = 0.706, \hat{\kappa}_3 = 5.651,$
			$r_1 : \hat{r}_2 : \hat{r}_3 = 1 : 0.556 : 5.611$
HKY85 + C	35	−5,624.70	$\hat{\kappa}_1 = 1.454, \hat{\kappa}_2 = 0.721, \hat{\kappa}_3 = 6.845$
			$r_1 : \hat{r}_2 : \hat{r}_3 = 1 : 0.555 : 5.774$

Note: p is the number of parameters in the model, including 21 branch lengths in the tree of Figure 4.16. The base frequency parameters under the HKY85 models are estimated using the observed frequencies (see Table 4.4).

4.7 MODEL SELECTION AND ROBUSTNESS

Table 4.4 Observed base compositions at the three codon positions for the 12 *rbcL* dataset (Figure 4.16)

Position	π_T	π_C	π_A	π_G
1	0.1829	0.1933	0.2359	0.3878
2	0.2659	0.2280	0.2998	0.2063
3	0.4116	0.1567	0.2906	0.1412
All	0.2867	0.1927	0.2754	0.2452

estimated by ML. The base frequency parameters in the HKY85 models are estimated by using the observed frequencies, averaged over the sequences (Table 4.4).

Here we consider three tests in detail. First, the JC69 and K80 models can be compared using an LRT, to test the null model H_0: $\kappa = 1$. Note that K80 will be reduced to JC69 when parameter $\kappa = 1$ is fixed. Thus JC69 is the null model, and includes $p_0 = 21$ branch lengths as parameters. The (optimized) log likelihood is $\ell_0 = -6{,}262.01$. K80 is the alternative model, with one extra parameter κ. The log likelihood is $\ell_1 = -6{,}113.86$. The test statistic is $2\Delta\ell = 2(\ell_1 - \ell_0) = 296.3$. This is much greater than the critical value $\chi^2_{1,1\%} = 6.63$, indicating that JC69 is rejected by a big margin. The transition and transversion rates are very different, as is clear from the MLE $\hat{\kappa} = 3.56$ under K80. Similarly comparison between K80 and HKY85 using an LRT with three degrees of freedom leads to rejection of the simpler K80 model.

A second test compares the null model JC69 against JC69+Γ_5, to test the hypothesis that different sites in the sequence evolve at the same rate. The one-rate model is a special case of the gamma model when the shape parameter $\alpha = \infty$ is fixed. The test statistic is $2\Delta\ell = 648.42$. In this test, the regularity conditions are not all satisfied, as the value ∞ is at the boundary of the parameter space in the alternative model. As a result, the null distribution is not χ^2_1, but is a 1:1 mixture of the point mass at 0 and χ^2_1 (Chernoff 1954; Self and Liang 1987; Whelan and Goldman 2000). In other words, one expects $2\Delta\ell$ to be 0 in half of the datasets and to be χ^2_1 distributed in the other half when many datasets are simulated under the null model. The critical values are 2.71 at 5% and 5.41 at 1%, rather than 3.84 at 5% and 6.63 at 1% according to χ^2_1. This null mixture distribution may be intuitively understood by considering the MLE of the parameter in the alternative model. If the true value is inside the parameter space, its estimate will have a normal distribution around the true value, and will be smaller than the true value half of the time and greater than the true value half of the time. If the true value is at the boundary, half of the time the estimate would be outside the space if there were no constraint; in such cases, the estimate will be forced to the true value and the log likelihood difference will be 0. Note that the use of χ^2_1 makes the test too conservative; if the test is significant under χ^2_1, it will be significant when the mixture distribution is used. For the *rbcL* dataset, the observed test statistic is huge, so that the null model is rejected whichever null distribution is used. The rates are highly variable among sites, as also indicated by the estimate of α. Similar tests using the K80 and HKY85 models also suggest significant variation in rates among sites.

A third test compares JC69 and JC69+C. In the alternative model JC69+C, the three codon positions are assigned different relative rates $r_1 (= 1), r_2$, and r_3, while the null model JC69 is equivalent to constraining $r_1 = r_2 = r_3$, reducing the number of parameters by 2. The test statistic is $2\Delta\ell = 678.50$, to be compared with χ^2_2. The null model is clearly rejected, suggesting that the rates are very different at the three codon positions. The same conclusion is reached if the K80 or HKY95 model is used in the test.

The two most complex models in Table 4.3, HKY85+Γ$_5$ and HKY85+C, are the only models not rejected in such LRTs. (The two models themselves are not nested and a χ^2 approximation to the LRT is not applicable.) This pattern is typical in the analysis of molecular datasets, especially large datasets; we seem to have no difficulty in rejecting an old simpler model whenever we develop a new model by adding a few extra parameters, sometimes even if the biological justification for the new parameters is dubious. The pattern appears to reflect the fact that most molecular datasets are very large and the LRT tends to favour parameter-rich models in large datasets.

4.7.2 Test of goodness of fit and parametric bootstrap

The LRT we discussed above compares two nested and closely related parametric models and addresses the question whether one model (the general model such as K80) fits the data significantly better than another model (the simpler model such as JC69). Even if K80 fits the data much better than JC69, neither may be adequate. A *test of general adequacy* of a model, also known as the *goodness of fit test*, can be constructed by noting that the data or the site pattern counts follow a multinomial distribution, with each possible site pattern being a category of the multinomial. For a dataset of s sequences, there are 4^s possible site patterns, and thus 4^s categories, with $4^s - 1$ probability parameters. For example, the 64 site patterns for $s = 3$ sequences are TTT, TTC, TTA, TTG, TCT, TCC, ..., GGG. Under the multinomial model, the MLEs of the probabilities for the categories are the observed frequencies, so that the maximum log likelihood is

$$\ell_{\max} = \sum_{i=1}^{4^s} n_i \log \left\{ \frac{n_i}{n} \right\}, \qquad (4.39)$$

where n is the total number of sites (columns) in the alignment and n_i is the number of sites with the ith site pattern. Many of the site patterns may be missing in the data ($n_i = 0$) and they do not contribute to $\ell_{\max}$ (note that the convention is $0^0 = 1$ and $0 \log 0 = 0$).

Substitution models such as JC69, K80, HKY85+Γ$_5$, etc., assume identical and independent distribution among sites. They are all special cases of the general multinomial model. One might expect the χ^2 distribution to apply when we compare the JC69 model, say, against this multinomial by an LRT, with df $= (4^s - 1) - (2s - 3)$. However, this is not so because many of the possible site patterns (categories) have very low counts or are entirely missing. The rule of thumb is that there should be at least five expected counts in each category for the χ^2 to be reliable.

When the χ^2 distribution does not apply, we can use simulation to generate the null distribution. This approach is called a *parametric bootstrap* (e.g. Goldman 1993a; Yang et al. 1994). Here we illustrate it by using the plastid *rbcL* genes from the 12 plant species analysed above to test the goodness of fit of HKY85+Γ$_5$. We do not have a model for alignment gaps, so we remove sites with gaps, with 1,312 sites left.

First we use the two models to analyse the original real data. Under the general multinomial model, we have $\ell_{\max} = -4025.03$. We then fit HKY85+Γ$_5$ to estimate the 21 branch lengths and the parameters in the HKY85+Γ$_5$ model. The latter are $\hat{\kappa} = 4.40165$ and $\hat{\alpha} = 0.196551$. The base frequencies are estimated using the observed frequencies in the data: 0.29071 (T), 0.19372 (C), 0.27274 (A), and 0.24282 (G), although they could be obtained by ML. The log likelihood is $\ell_{\text{HKYG}} = -5242.35$. (The difference from the value in Table 4.3 is due to the removal of sites with alignment gaps.) The test statistic is $\Delta \ell = \ell_{\max} - \ell_{\text{HKYG}} = 1217.32$.

Next we simulate datasets to estimate the null distribution of the test statistic $\Delta \ell$. Use the MLEs of branch lengths for the tree of Figure 4.16 as well as the substitution parameters in

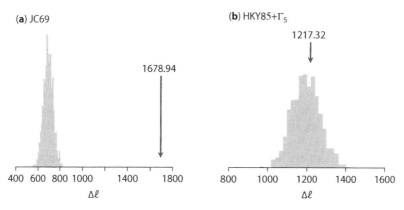

Fig. 4.17 The null distribution estimated using parametric bootstrap for the test of goodness of fit of (**a**) the JC69 model and (**b**) the HKY85+Γ_5 model. The plastid *rbcL* genes from 12 plant species (Figure 4.16) are analysed.

the HKY85+Γ_5 model obtained from the original data to simulate 1,000 replicate datasets. Analyse each replicate dataset in the same way as the original data are analysed. In other words, each replicate dataset i is analysed to calculate $\ell_{\max}$ and ℓ_{HKYG} and their difference Δ_i. Collect the Δ_i values into a histogram (Figure 4.17b), which is an estimate of the null distribution. The observed statistic, $\Delta\ell = 1217.32$, falls well within the null distribution, so HKY85+Γ_5 provides an adequate fit to the data according to this test. The *p*-value is the proportion of the simulated Δ_i values that are greater than the observed $\Delta\ell$ value; this is $p = 38.7\%$.

A similar test of the general adequacy of the JC69 model leads to strong rejection of the model (Figure 4.17a) (Problem 4.5).

The parametric bootstrap is a general method for deriving the null distribution when it is unknown. It can be used to compare two models that are not nested or to compare two nested models when the dataset is small so that the asymptotic χ^2 distribution may not be reliable (Goldman 1993a) (see Problem 2.5). It is nevertheless expensive as each replicate dataset has to be analysed in the same way as the original dataset.

*4.7.3 Diagnostic tests to detect model violations

When the LRT of the goodness of fit of a model rejects the model, one may use diagnostic statistics (quantities that can be calculated using the observed data) to detect which aspects of the model assumptions are violated. Those statistics do not work well for data with alignment gaps or ambiguity sites, which should be removed before applying the tests. For any test statistic, one can use simulation (parametric bootstrap) to generate the distribution expected under the model, against which the observed value of the statistic can be compared. Here we mention a few such statistics (Tavaré 1986; Goldman 1993b; Rzhetsky and Nei 1995). We will consider nucleotide sequences although the test works for amino acid sequences as well.

The first diagnostic statistic is the number of distinct site patterns (Goldman 1993b). Similar statistics include the number of constant sites and the number of highly variable sites, appropriately defined. Early applications of such statistics highlight the importance of accommodating the among-site heterogeneity in the substitution model (Reeves 1992; Goldman 1993b; Yang et al. 1994, 1995c). When substitution rates vary among sites but the model ignores the rate variation, the data tend to show too few distinct

---

* indicates a more difficult or technical section.

site patterns, too many constant or highly variable patterns and too few intermediately variable patterns, relative to the model's expectations. Note that the expectation under the model, or the null distribution of the test statistic under the model can be generated by parametric bootstrap (simulation). One uses the parameter estimates under the model to generate 1,000 replicate datasets and then to calculate the statistic in each to produce a histogram. One can then decide whether the observed number of distinct site patterns is too small or too large relative to the model expectation (Goldman 1993b). Alternatively, one may calculate the mean and variance of the statistic under the model directly if the dataset is very small. As discussed above, the sequence data can be summarized as the counts from a multinomial distribution with 4^s cells for s sequences, corresponding to the 4^s site patterns. The statistic is the number of non-empty cells in the multinomial. If s is very small, one can calculate the probabilities for all site patterns using the MLEs of parameters under the model, and then calculate the mean and variance of the statistic expected under the model. However, it is far simpler to generate the null distribution by simulation.

If the substitution process is stationary, one should expect the different sequences to have the same nucleotide (or amino acid) frequencies, apart from chance fluctuations. Thus one can construct an $s \times 4$ sequence $\times$ nucleotide contingency table and test whether the nucleotide frequencies are homogeneous across sequences (e.g. Rzhetsky and Nei 1995). When the substitution process is reversible and stationary, the probability of observing a site with nucleotides i and j in two sequences equals the probability for a site with j and i. Such expected symmetry can be tested by using the 4×4 contingency table of site pattern counts for two sequences (Tavaré 1986). Let N_{ij} be the number of sites with nucleotide i in sequence 1 and j in sequence 2. Then

$$X^2 = \sum_{i<j} \frac{(N_{ij} - N_{ji})^2}{N_{ij} + N_{ji}} \qquad (4.40)$$

is compared with the asymptotic χ^2 distribution with six degrees of freedom. Jermiin et al. (2008) have explored a few variants of this matched-pairs test. Such tests are in general expected to be very powerful, and are able to detect small violations that may not have a noticeable impact on phylogenetic analysis.

4.7.4 Akaike information criterion (AIC and AIC_c)

The LRT is applicable for comparing two nested models. Although Cox (1961, 1962; see also Atkinson 1970; Lindsey 1974a, 1974b; Sawyer 1984) discussed the use of LRT to compare non-nested models, the idea has not been used much in practical data analysis (but see Goldman 1993a). In the Cox test one of the two non-nested models is designated as the null hypothesis and the other as the alternative hypothesis. Simulation (parametric bootstrap) is often necessary to derive the null distribution of the LRT statistic. With non-nested models, it is often arbitrary which one should be the null hypothesis. To avoid this arbitrariness, two tests are often conducted instead of one, with each of the two models being used as the null hypothesis. The tests may then fail to reject either model, in which case it is unclear how to make inferences about the quantities of interest, especially if the two models lead to different conclusions. In other datasets (especially large ones) the tests may reject both models, in which case we do not have a useful comparison of the two models apart from knowing that neither fits the data well.

The Akaike information criterion (AIC, Akaike 1974) can be used to compare models that are not necessarily nested. The AIC score is calculated for each model, defined as

$$\text{AIC} = -2\ell + 2p, \tag{4.41}$$

where $\ell = \ell(\hat{\theta})$ is the optimum log likelihood under the model, and p is the number of parameters. Models with small AICs are preferred. According to this criterion, an extra parameter is worthwhile if it improves the log likelihood by more than one unit.

The AIC is perceived not to penalize parameter-rich models enough. A correction is thus introduced, which incorporates the data size in the criterion (Sugiura 1978; Hurvich and Tsai 1989)

$$\text{AIC}_c = -2\ell + \frac{2np}{n-p-1} = \text{AIC} + \frac{2p(p+1)}{n-p-1}. \tag{4.42}$$

4.7.5 Bayesian information criterion

In large datasets, both LRT and AIC are known to favour complex parameter-rich models and to reject simpler models too often (Schwarz 1978). The Bayesian information criterion (BIC) is based on a Bayesian argument and penalizes parameter-rich models more severely. It is defined as

$$\text{BIC} = -2\ell + p\log(n), \tag{4.43}$$

where n is the sample size (sequence length) (Schwarz 1978). Again models with small BIC scores are preferred.

Qualitatively, LRT, AIC, and BIC are all mathematical formulations of the *parsimony principle* of model building. Extra parameters are deemed necessary only if they bring about significant or considerable improvements to the fit of the model to data, and otherwise simpler models with fewer parameters are preferred. However, in large datasets, these criteria can differ markedly. For example, if the sample size $n > 8$, BIC penalizes parameter-rich models far more severely than does AIC.

Model selection is an active research area in statistics. Posada and Buckley (2004) provided a nice overview of methods and criteria for model selection in molecular phylogenetics. For automatic model selection, Posada and Crandall (1998; Posada 2008) developed MODELTEST. Well-known substitution models are compared hierarchically using the LRT, AIC, or BIC. The program enables the investigator to avoid making thoughtful decisions concerning the model to be used in phylogeny reconstruction. However, mechanical application of MODELTEST has led to widespread use of the most complex models, such as the pathological 'I + Γ' models, in real data analysis.

Example 4.5. Model selection for the ape mitochondrial protein data. We use the model selection criteria LRT, AIC, AIC$_c$, and BIC to compare a few models applied to the dataset analysed in §4.2.4. The ML tree of Figure 4.5 is assumed. The three empirical amino acid substitution models DAYHOFF (Dayhoff et al. 1978), JTT (Jones et al. 1992), and MTMAM (Yang et al. 1998) are fitted to the data, with either one rate for all sites or gamma rates for sites. In the discrete gamma model, five rate categories are used; the estimates of the shape parameter α range from 0.30 to 0.33 among the three models. The results are shown in Table 4.5. The LRT can be used to compare nested models only, so each empirical model (e.g. DAYHOFF) is compared with its gamma counterpart (e.g. DAYHOFF +Γ$_5$). The LRT statistics are very large, so there is no doubt that the substitution rates are highly variable among sites, whether χ_1^2 or the 50:50 mixture of 0 and χ_1^2 is used for the test. The AIC, AIC$_c$, and BIC scores can be used to compare non-nested models, such as the three empirical models. As they involve the same number of parameters, the ranking using AIC or BIC is the same as using the log likelihood. MTMAM fits the data better than the other two

Table 4.5 Comparison of models for the mitochondrial protein sequences from the apes

Model	p	ℓ	LRT	AIC	AIC$_c$	BIC
DAYHOFF	11	−15,766.72		31,555.44	31,555.52	31,622.66
JTT	11	−15,332.90		30,687.80	30,687.88	30,755.02
MTMAM	11	−14,558.59		29,139.18	29,139.26	29,206.40
DAYHOFF + Γ_5	12	−15,618.32	296.80	31,260.64	31,260.73	31,333.97
JTT + Γ_5	12	−15,192.69	280.42	30,409.38	30,409.47	30,482.71
MTMAM + Γ_5	12	−14,411.90	293.38	28,847.80	28,847.89	28,921.13

Note: p is the number of parameters in the model. The sample size is $n = 3,331$ amino acid sites for the AICc and BIC calculations. The LRT column shows the test statistic $2\Delta\ell$ for comparing each empirical model with the corresponding gamma model.

models, which is expected since the data are mitochondrial proteins while DAYHOFF and JTT were derived from nuclear proteins. The best model for the data according to all three criteria is MTMAM + Γ_5. □

4.7.6 Model adequacy and robustness

All models are wrong but some are useful. (George Box, 1979)

Models are used for different purposes. In some cases, the model is an interesting biological hypothesis we wish to test, so that the model (hypothesis) itself is our focus. For example, the molecular clock (rate constancy over time) is an interesting hypothesis predicted by certain theories of molecular evolution, and it can be examined by using an LRT to compare a clock model and a nonclock model. In other cases, the model, or at least some aspects of the model assumptions, is not our main interest, but has to be dealt with in the analysis. For example, in testing the molecular clock, we need a Markov model of nucleotide substitution (JC69 or HKY85+Γ). We are not interested in the substitution model but we may be concerned about its impact on our test of the molecular clock. In phylogeny reconstruction, our interest is in the tree, but we have to assume an evolutionary model to describe the mechanism by which the data are generated. The model is then a nuisance, but its impact on our analysis cannot be ignored. Some writers distinguish a *hypothesis* from a *model*, and use the term *hypothesis* to refer to the first case (where the model is our focus) and *model* to refer to the second case (where the model is a nuisance). While we do not make such a distinction here, we should bear in mind what the model is used for. Model selection discussed in this section refers to selection of models that are not the focus of our analysis.

It should be stressed that a model's fit to data and its impact on inference are two different things. Often model robustness is even more important than model adequacy. It is neither possible nor necessary for a model to match the biological reality in every detail. The aim of model selection is not to find the 'true model' but to find a model with sufficient parameters to capture the key features of the data (Steel 2005). What features are important will depend on the question being asked, and one has to use knowledge of the subject matter to make the judgement. Structural biologists tend to emphasize the uniqueness of every residue in the protein. Similarly one has every reason to believe that every species is unique. However, by no means should one use one separate parameter for every site and every branch in formulating a statistical model to describe the evolution of the protein sequence. Such a model, saturated with parameters, is not workable.

One should appreciate the power of the i.i.d. models, which assume that the sites in the sequence are independent and identically distributed. A common misconception is that i.i.d. models assume that every site evolves at the same rate and follows the same pattern. It should be noted that almost all models implemented in molecular phylogenetics, such as models of variable rates among sites (Yang 1993, 1994a), models of variable selective pressures among codons (Nielsen and Yang 1998), and the covarion models that allow the rate to vary both among sites and among lineages (Tuffley and Steel 1998; Galtier 2001; Guindon et al. 2004) are i.i.d. models. The i.i.d. assumption is a statistical device which is useful for reducing the number of parameters.

Some features of the process of sequence evolution are both important to the fit of the model to the data and critical to our inference. They should be incorporated in the model. Variable rates among sites appear to be such a factor for phylogeny reconstruction or estimation of branch lengths (Tateno et al. 1994; Huelsenbeck 1995a; Gaut and Lewis 1995; Sullivan et al. 1995; Yang 1996c). Some factors may be important to the model's fit, as judged by the likelihood, but may have little impact on the analysis. For example, adding the transition/transversion rate ratio κ to the JC69 model almost always leads to a huge improvement to the log likelihood, but often has minimal effect on estimation of branch lengths. The difference between HKY85 and GTR (REV) is even less important, even though HKY85 is rejected in most datasets when compared against GTR. The most troublesome factors are those that have little impact on the fit of the model but a huge impact on our inference. For example, in estimation of species divergence times under local molecular clock models, different models for lineage rates appear to fit the data almost equally well, judged by their log likelihood values, but they can produce very different time estimates (see Chapter 10). Such factors have to be carefully assessed even if the statistical test does not indicate their importance.

For phylogeny reconstruction, a number of computer simulations have been conducted to examine the robustness of different methods to violations of model assumptions. Such studies have in general found that model-based methods such as ML are quite robust to the underlying substitution model (e.g. Hasegawa et al. 1991; Gaut and Lewis 1995). However, the importance of model assumptions appears to be dominated by the shape of the tree reflected in the relative branch lengths, which determines the overall level of difficulty of tree reconstruction. 'Easy' trees, with long internal branches or with long external branches clustered together, are successfully reconstructed by all methods and models; indeed wrong simplistic models tend to show even better performance than the more complex true model (we will discuss such counterintuitive results later in §5.2.3). 'Hard' trees, with short internal branches and with long external branches spread over different parts of the tree, are difficult to reconstruct by all methods. For such trees, simplistic models may not even be statistically consistent, and use of complex and realistic models is critical.

4.8 Problems

4.1 Calculate the probabilities of site patterns *xxx*, *xxy*, *yxx*, and *xyx* as a function of the branch lengths t_{10} and t_{11} in the tree τ_1 of Figure 4.14. Assume the symmetrical substitution model for binary characters (equation (1.79)).

4.2* Try to estimate the single branch length under the JC69 model for the star tree of three sequences under the molecular clock (see Saitou 1988; Yang 1994c, 2000a, for

* indicates a more difficult or technical problem.

discussions of likelihood tree reconstruction under this model). The tree is shown in Figure 4.10, where t is the only parameter to be estimated. Note that there are only three site patterns, with one, two, or three distinct nucleotides, respectively. The data are the observed numbers of sites with such patterns: n_0, n_1, and n_2, with the sum to be n. Let the proportions be $f_i = n_i/n$. The log likelihood is $\ell = n \sum_{i=0}^{2} f_i \log(p_i)$, with p_i to be the probability of observing site pattern i. Derive p_i by using the transition probabilities under the JC69 model, given in equation (1.4). You can calculate $p_0 = \Pr(TTT)$, $p_1 = \Pr(TTC)$, and $p_2 = \Pr(TCA)$. Then set $d\ell/dt = 0$. Show that the MLE for the transformed parameter $z = e^{-4t/3}$ is a solution to the following quintic equation:

$$36z^5 + 12(6 - 3f_0 - f_1)z^4 + (45 - 54f_0 - 42f_1)z^3 + (33 - 60f_0 - 36f_1)z^2 + (3 - 30f_0 - 2f_1)z$$
$$+ (3 - 12f_0 - 4f_1) \equiv 0.$$

(4.44)

4.3 *Long-branch attraction for parsimony.* Calculate the probabilities of sites with data *xxyy*, *xyyx*, and *xyxy* in four species for the unrooted tree of Figure 4.18, using two branch lengths p and q under a symmetrical substitution model for binary characters (equation (1.79)). Here it is more convenient to define the branch length as the proportion of different sites at the two ends of the branch. Show that $\Pr(xxyy) < \Pr(xyxy)$ if and only if $q(1-q) < p^2$. With such branch lengths, parsimony for tree reconstruction is inconsistent (Felsenstein 1978a).

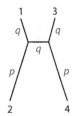

Fig. 4.18 A tree of four species with two branch lengths p and q, defined as the probability that any site is different at the two ends of the branch. For a binary character, this probability is $p = (1 - e^{-2t})/2$, where t is the expected number of character changes per site (see equation (1.79)).

4.4 Bias in ancestral state reconstruction. Calculate the posterior probabilities for T, C, A, and G at the root of the tree of Figure 4.10 when the observed data at the site is $x_1x_2x_3$ = AAG. Assume the F81 substitution model, with base frequency parameters π_T = 0.2263, π_C = 0.3282, π_A = 0.3393, and π_G = 0.1062. Suppose that each branch length is 0.2, and the transition probability matrix is given in equation (4.26). Hint: Use equation (4.23).

4.5 Use the plastid *rbcL* genes from 12 plant species to test the goodness of fit of the JC69 model. Follow the example of §4.7.2. Use BASEML in the PAML package to analyse the original data to generate branch lengths and calculate $\Delta\ell = \ell_{\max} - \ell_{JC}$. Use those branch lengths to simulate 1,000 datasets using the program SEQ-GEN or EVOLVER. Then use a likelihood program (such as BASEML) to analyse the 1,000 replicate datasets to calculate Δ_i to construct a histogram. Your results should be similar to Figure 4.17a.

4.6 *Phylogenetic reconstruction using ML.* Use your own data or find a small dataset of 10–50 species from the literature to infer the ML phylogeny under various substitution models, such as JC69, K80, HKY85, GTR, and the gamma variants JC69+Γ_5, K80+Γ_5, HKY85+Γ_5, and GTR +Γ_5. You can use PHYML to run tree search under those models.

CHAPTER 5

Comparison of phylogenetic methods and tests on trees

5.1 Statistical performance of tree reconstruction methods

This chapter discusses two problems: the evaluation of statistical properties of tree reconstruction methods and test of significance of estimated phylogenies. Both problems are complex and controversial. The view taken in this book is that phylogenetic estimation is a statistical inference problem, but has peculiarities that have to be treated carefully.

The 1980s–1990s saw heated debates concerning the merits and demerits of various tree reconstruction methods, centred mostly on philosophical issues such as whether one can avoid making assumptions about the evolutionary process when inferring phylogenies, or whether one should use model-based methods such as likelihood or 'model-free' methods such as parsimony (e.g. Felsenstein 1973b, 1983; Farris 1983). In the 1990s–2000s, many computer simulation studies were conducted to assess the performance of different methods. Modern phylogenetic analyses typically use more than one method, and discuss the performance of different methods at the same time of data analysis.

Section §5.1 discusses criteria for assessing the statistical properties of tree reconstruction methods and provides a summary of simulation studies conducted to evaluate different methods. Sections §5.2 and §5.3 deal with the likelihood-versus-parsimony controversy from the likelihood and parsimony perspectives, respectively. Even within the statistical framework, tree estimation is much more complex than conventional parameter estimation. Those difficulties are discussed in §5.2. Section §5.3, on parsimony, summarizes previous attempts to identify the assumptions underlying the method or to provide a statistical justification for it by establishing an equivalence to likelihood under a particular model.

Section §5.4 provides an overview of methods for assessing the sampling errors in estimated phylogenies. The reconstructed tree, derived by whatever method, is similar to a point estimate. One would like to attach a measure of its reliability, in the same way that a confidence interval provides a measure of confidence for a point estimate of a conventional parameter. However, the unconventional nature of the problem causes difficulties with this effort. In this chapter we discuss how to attach a measure of confidence for the parsimony, distance, or likelihood trees. Bayesian phylogenetics is discussed in Chapter 8, in which the posterior probability provides the natural measure of confidence.

5.1.1 *Criteria*

In comparing different methods of phylogeny reconstruction, two kinds of error should be distinguished. *Random errors*, also called sampling errors, are due to the finite nature of the dataset. In most models used in molecular phylogenetics, the sample size is the number of sites (nucleotides, amino acids, or codons) in the sequence. When the sequence length approaches infinity, sampling errors will decrease and approach zero. *Systematic errors* are due to incorrect assumptions or some other kind of deficiencies of the method. When the sample size increases, systematic errors will persist and even intensify.

One can judge a tree reconstruction method using a variety of criteria. The computational speed is perhaps the easiest to assess. In general, distance methods are much faster than parsimony, which is in turn faster than likelihood or Bayesian methods. Here we consider the statistical properties of a method.

5.1.1.1 *Identifiability*

If the probability of the data $f(X|\theta)$ is exactly the same for two parameter values θ_1 and θ_2, i.e. $f(X|\theta_1) = f(X|\theta_2)$ for all possible data X, no method will be able to distinguish θ_1 from θ_2 using the observed data. The model is said to be unidentifiable. For example, we will have an identifiability problem if we attempt to estimate both the species divergence time t and the substitution rate r using a pair of sequences from two species under any time-reversible model such as JC69 (see equation (1.70)). The probability of the data or likelihood is exactly the same for $\theta_1 = (t, r)$ and $\theta_2 = (2t, r/2)$, say, and it is impossible to estimate t and r separately. Even if θ is not identifiable, some functions of θ may be identifiable – in this case, the distance $d = t \cdot r$ is. Unidentifiable models are usually due to mistakes in model formulation and should be avoided.

5.1.1.2 *Consistency*

An estimation method or estimator $\hat{\theta}$ is said to be (statistically) consistent if it converges to the true parameter value θ when the sample size n increases. Formally, $\hat{\theta}$ is consistent if

$$\lim_{n \to \infty} \Pr(|\hat{\theta} - \theta| < \varepsilon) = 1, \tag{5.1}$$

for any small number $\varepsilon > 0$. Also $\hat{\theta}$ is said to be *strongly consistent* if

$$\lim_{n \to \infty} \Pr(\hat{\theta} = \theta) = 1. \tag{5.2}$$

Phylogenetic trees are not regular parameters, but we may use the idea of strong consistency and say that a tree reconstruction method is consistent if the probability that the estimated tree is the true tree approaches one when $n \to \infty$. For model-based methods, the definition of consistency assumes the correctness of the model.

There has been much discussion of the consistency of the tree reconstruction method since Felsenstein (1978b) demonstrated that parsimony can be inconsistent under certain combinations of branch lengths on a four-species tree (see §3.4.5). In a conventional estimation problem, consistency is a weak statistical property, easily satisfied by many good and poor estimators. For example, the usual estimator of the probability p of 'success' from a binomial sample with x successes out of n trials is the sample proportion $\hat{p} = x/n$. This is consistent, but so is an arbitrary poor estimator $\tilde{p} = (x - 1000)/n$, which is even negative if $n < 1000$. Sober (1988) considered likelihood to be a more fundamental criterion than consistency, claiming that consistency is not necessary since real datasets are always finite. Such a position appears untenable, as it prefers likelihood for its own sake rather than for its good performance. Statisticians, from both the likelihood/frequentist and Bayesian schools (e.g. Stuart et al. 1999, pp. 3–4; O'Hagan and Forster 2004, pp. 72–74), have never doubted that consistency is a property that any sensible estimator should possess. Fisher (1970, p. 12) considered inconsistent estimators to be 'outside the pale of decent usage'. See Goldman (1990) for more discussions.

5.1.1.3 Efficiency

A consistent estimator with asymptotically the smallest variance is said to be efficient. The variance of a consistent and unbiased estimator cannot be smaller than the Cramér–Rao lower bound; i.e. for any unbiased estimator $\hat{\theta}$ of θ:

$$\text{var}(\hat{\theta}) \geq 1/I, \tag{5.3}$$

where $I = -E\left(\frac{d^2 \log(f(X|\theta))}{d\theta^2}\right)$ is known as the *expected information* or *Fisher information* (Stuart et al. 1999, pp. 9–14, see also §1.4.1). Under quite mild regularity conditions, the maximum likelihood estimate (MLE) has desirable asymptotic properties: i.e. when $n \to \infty$, the MLE is consistent, unbiased, and normally distributed, and attains the minimum variance bound of equation (5.3) (e.g. Stuart et al. 1999, Chapter 18).

If t_1 is an efficient estimator and t_2 is another estimator, one may measure the efficiency of t_2 relative to t_1 as $E_{21} = n_1/n_2$, where n_1 and n_2 are the sample sizes required to give both estimators equal variance, i.e. to make them equally precise (Stuart et al. 1999, p. 22). In large samples, the variance is often proportional to the reciprocal of the sample size, in which case $E_{21} = V_1/V_2$, where V_1 and V_2 are the variances of the two estimators at the same sample size. Suppose we wish to estimate the mean μ of a normal distribution $N(\mu, \sigma^2)$ with variance σ^2 known. The sample mean has variance σ^2/n and is efficient in the sense that no other unbiased estimator can have a smaller variance. The sample median has variance $\pi\sigma^2/(2n)$ for large n. The efficiency of the median relative to the mean is $2/\pi = 0.637$ for large n. Thus the mean achieves the same accuracy as the median with a sample that is 36.3% smaller. The median is nevertheless less sensitive to outliers.

The variance of an estimated tree topology is not a meaningful concept. However, the relative efficiency of two tree reconstruction methods may be measured by

$$E_{21} = n_1(P)/n_2(P), \tag{5.4}$$

where $n_1(P)$ and $n_2(P)$ are the sample sizes required for both methods to recover the true tree with the same probability P (Saitou and Nei 1986; Yang 1996a). As $P(n)$, the probability of recovering the correct tree given the sample size n, can be more easily estimated using computer simulation than $n(P)$, an alternative measure is

$$E_{21}^* = \frac{1 - P_1(n)}{1 - P_2(n)}. \tag{5.5}$$

Thus the efficiency of tree reconstruction method 2 relative to method 1 can be measured by the error in method 1 divided by the error in method 2 for a given sample size n. When both methods are consistent, their errors will approach 0 as $n \to \infty$, but they may do so at very different rates. Method 2 is more efficient than method 1 if $E_{21}^* > 1$. Equations (5.4) and (5.5) are not expected to be identical, but should give the same conclusion concerning the relative performance of the two methods. Later we will use equation (5.5) to analyse the maximum likelihood (ML) method.

5.1.1.4 Robustness

A model-based method is said to be robust if it still performs well when its assumptions are slightly wrong. Clearly some assumptions matter more than others. Robustness is often examined by computer simulation, in which the data are generated under one model but analysed under another wrong and simpler model.

5.1.2 *Performance*

To evaluate different tree reconstruction methods, a number of studies have exploited situations in which the true tree is known or believed to be known. The first strategy is the use of laboratory-generated phylogenies. Hillis et al. (1992) generated a known phylogeny by 'evolving' the bacteriophage T7 in the laboratory. The restriction-site map of the phage was determined at several time points, and the phage was separated and allowed to diverge into different lineages. Thus both the phylogeny and the ancestral states were known from the design of the experiment. Parsimony and four distance-based methods were then applied to analyse restriction-site maps of the terminal lineages to infer the evolutionary history. Very impressively, all methods recovered the true phylogeny! Parsimony also reconstructed the ancestral restriction-site maps with >98% accuracy.

A second approach uses so-called well-established phylogenies, i.e. phylogenetic relationships that are generally accepted on the basis of evidence from fossils, morphology, and previous molecular data. Such phylogenies can be used to assess the performance of tree reconstruction methods as well as the utilities of different gene loci. For example, Cummings et al., (1995), Russo et al., (1996), and Zardoya and Meyer (1996) evaluated the performance of different tree reconstruction methods and of mitochondrial protein-coding genes in recovering mammalian or vertebrate phylogenies. Most empirical phylogenetic studies are of this nature, as researchers use a variety of methods to analyse multiple gene loci, and assess the reconstructed phylogenies against previous estimates (e.g. Cao et al. 1998; Takezaki and Gojobori 1999; Brinkmann et al. 2005).

The third approach is computer simulation. Many replicate datasets are generated under a simulation model and then analysed using various tree reconstruction methods to estimate the true tree, to see how often the true tree is recovered or what percentage of the true splits are recovered. The size and shape of the tree topology, the evolutionary model and the values of its parameters, and the size of the data are all under the control of the investigator and can be varied to see their effects. See Chapter 12 for discussions of simulation techniques. A criticism of simulation studies is that the models used may not reflect the complexity of sequence evolution in the real world. Another is that a simulation study can examine only a very limited set of parameter combinations, and yet the relative performance of tree reconstruction methods may depend on the model and tree shape. It is thus unsafe to extrapolate conclusions drawn from simulations in a small portion of the parameter space to general real data situations. Nevertheless, patterns revealed in simulations are repetitively discovered in real data analysis, so simulation will continue to be relevant for comparing and validating phylogenetic methods, especially as an analytical treatment of those methods is impossible.

Several review articles have been published that summarize previous simulation studies, such as Felsenstein (1988), Huelsenbeck (1995b), and Nei (1996). Conflicting views are often offered concerning the relative performance of different tree reconstruction methods. The following observations appear to be generally accepted.

(1) Methods that assume the molecular clock such as UPGMA perform poorly when the clock is violated, so one should use methods that infer unrooted trees without assuming the clock. UPGMA is nevertheless appropriate for highly similar sequences, such as population data, in which the clock is expected to hold.

(2) Parsimony, as well as distance and likelihood methods under simplistic models, is prone to the problem of long-branch attraction, while likelihood under complex and more-realistic models are more robust (e.g. Kuhner and Felsenstein 1994; Gaut and Lewis 1995; Yang 1995b; Huelsenbeck 1998).

(3) Likelihood methods are often more efficient in recovering the correct tree than parsimony or distance methods (e.g. Hasegawa et al. 1991; Hasegawa and Fujiwara 1993; Kuhner and Felsenstein 1994; Tateno et al. 1994; Gaut and Lewis 1995; Huelsenbeck 1995b). However, counter-examples have been found, some of which will be discussed in the next section.

(4) Distance methods do not perform well when the sequences are highly divergent or contain many alignment gaps, mainly because of difficulties in obtaining reliable distance estimates (e.g. Gascuel 1997; Bruno et al. 2000).

(5) The level of sequence divergence has a great impact on the performance of tree reconstruction methods. Highly similar sequences lack information, so that no method can recover the true tree with any confidence. Highly divergent sequences contain too much noise, as substitutions may have saturated. The amount of information in the data is optimized at intermediate levels of divergence (Goldman 1998). Thus one should ideally sequence fast evolving genes to study closely related species and slowly evolving genes or proteins to study deep phylogenies. In simulation studies, phylogenetic methods appear to be quite tolerant of multiple substitutions at the same site (Yang 1998b; Bjorklund 1999). However, high divergences are often accompanied by other problems, such as difficulty in alignment and unequal base or amino acid compositions among sequences, which indicate a clear violation of the assumption that the substitution process has been stationary.

(6) The shape of the tree as reflected in the relative branch lengths has a huge effect on the success of the reconstruction methods and on their relative performance. 'Hard' trees are characterized by short internal branches and long external branches, with long external branches scattered in different parts of the tree. In such trees, parsimony as well as distance and likelihood methods under simplistic models are prone to errors. 'Easy' trees are characterized by long internal branches. With such trees, every method seems to work fine, and it is even possible for naïvely simplistic likelihood models or parsimony to outperform likelihood under complex models.

5.2 Likelihood

5.2.1 *Contrast with conventional parameter estimation*

Yang (1994c; see also Yang 1996a; Yang et al. 1995c) argued that the problem of tree reconstruction is not one of statistical parameter estimation but instead is one of model selection. In the former case, the probability distribution of the data, $f(X|\theta)$, is fully specified except for the values of parameters θ. The objective is then to estimate θ. In the latter case, there are several competing data-generating models, $f_1(X|\theta_1), f_2(X|\theta_2)$, and $f_3(X|\theta_3)$, each with its own unknown parameters. The objective is then to decide which model is true or closest to the truth. Here, the models correspond to tree topologies while the parameters θ_1, θ_2, or θ_3 correspond to branch lengths in different trees. Tree reconstruction falls into the category of model selection, because the likelihood function and the definition of branch lengths depend on the tree topology (Nei 1987, p. 325). There is a minor difference, however. In a typical model selection situation, our interest is rarely in the model itself; rather we are interested in inference concerning certain parameters and we have to deal with the model because our inference may be unduly influenced by model misspecification. Selection of the substitution model for tree reconstruction is one such case. In contrast, in phylogenetic tree reconstruction, we assume that one of the trees is true, and our main objective is to identify the true tree.

The distinction between parameter estimation and model selection is not a pedantic one. The mathematical theory of ML estimation, such as the consistency and efficiency of MLEs, is developed in the context of parameter estimation and not of model selection. The next two subsections demonstrate that phylogeny reconstruction by ML is consistent but not asymptotically efficient.

5.2.2 Consistency

Early arguments (e.g. Felsenstein 1973b, 1978b; see also Swofford et al. 2001) for the consistency of the likelihood method of tree reconstruction (Felsenstein 1981) referred to the proof of Wald (1949) and did not take full account of the fact that the likelihood function changes among tree topologies. Nevertheless, the consistency of ML under commonly used models is easy to establish. Yang (1994c) provided such a proof, assuming that the model is well-formulated so that the different trees are identifiable. The essence of the proof is that when the number of sites approaches infinity, the true tree will predict probabilities of site patterns that match exactly the observed frequencies, thus achieving the maximum possible likelihood for the data; as a result, the true tree will be chosen as the estimate under the ML criterion.

In the model we assume that different sites evolve independently and according to the same stochastic process. The data at different sites are then independently and identically distributed (i.i.d.) and can be summarized as the counts of 4^s site patterns for s species: n_i for site pattern i, with the sequence length $n = \sum n_i$. Note that in typical datasets, many of the site patterns are not observed so that $n_i = 0$ for some i. The counts of site patterns are random variables from a multinomial distribution, which has 4^s categories corresponding to the 4^s site patterns for s species. Without loss of generality, suppose tree τ_1 is the true tree, while all other trees are incorrect. The probability of the ith category (site pattern) in the multinomial is given by the true tree and true values of parameters (branch lengths and substitution parameters), $\theta_*^{(1)}$, as $p_i = p_i^{(1)}(\theta_*^{(1)})$. Here the superscript $^{(1)}$ indicates that the probabilities and parameters are defined on tree τ_1.

Let $f_i = n_i/n$ be the proportion of sites with the ith pattern. Given the data, the log likelihood for any tree cannot exceed the following upper limit

$$\ell_{\max} = n \sum_i f_i \log(f_i). \tag{5.6}$$

Let $p_i^{(k)}(\theta^{(k)})$ be the probability of site pattern i for tree τ_k with its parameters to be $\theta^{(k)}$. The maximized log likelihood for tree τ_k is thus

$$\ell_k = n \sum_i f_i \log \left\{ p_i^{(k)}(\hat{\theta}^{(k)}) \right\}, \tag{5.7}$$

where $\hat{\theta}^{(k)}$ are the MLEs. Then $\ell_k \leq \ell_{\max}$, and the equality holds if and only if $f_i = p_i^{(k)}$ for all i. Similarly

$$(\ell_{\max} - \ell_k)/n = \sum_i f_i \log \left(\frac{f_i}{p_i^{(k)}(\hat{\theta}^{(k)})} \right) \tag{5.8}$$

is known as the *Kullback–Leibler (K-L) divergence* between the two distributions f_i and p_i. The K-L divergence is nonnegative, and equals zero if and only if the two distributions are identical; i.e. only if $f_i = p_i^{(k)}$ for all i.

Now estimation of $\theta^{(1)}$ on the true tree τ_1 is a conventional parameter estimation problem, and the standard proof (e.g. Wald 1949) applies. When $n \to \infty$, the data frequencies approach the site-pattern probabilities predicted by the true tree, $f_i \to p_i^{(1)}(\theta_*^{(1)})$, the MLEs approach their true values $\hat{\theta}^{(1)} \to \theta_*^{(1)}$, and the maximum log likelihood for the true tree approaches the maximum possible likelihood $\ell_1 \to \ell_{\max}$. The true tree thus provides a perfect match to the data.

A question is whether it is possible for a wrong tree (say, τ_2) to attain the same highest log likelihood $l_{\max}$, or to provide a perfect match to the site pattern probabilities predicted by tree τ_1: $p_i^{(1)}(\theta_*^{(1)}) = p_i^{(2)}(\theta^{(2)})$ for all i for τ_2. If this occurs, tree τ_1 with parameters $\theta_*^{(1)}$ and tree τ_2 with parameters $\theta^{(2)}$ generate datasets that are identical probabilistically, and the model is unidentifiable. Unidentifiable models are usually due to conceptual errors in model formulation. A number of authors have demonstrated that models in common use in phylogenetic analysis are identifiable (e.g. Chang 1996a; Rogers 1997; Allman and Rhodes 2006; Allman et al. 2008).

5.2.3 *Efficiency*

In some simulation studies, distance methods such as NJ (Saitou and Nei 1987) were noted to recover the true tree with a higher probability when a wrong model was used to calculate sequence distances than when the true model was used (Saitou and Nei 1987; Sourdis and Nei 1988; Tateno et al. 1994; Rzhetsky and Sitnikova 1996). In a similar vein, Goldstein and Pollock (1994), and Tajima and Takezaki (1994) constructed distance formulae under deliberately 'wrong' models for tree reconstruction even though the true model was available. Those results are counterintuitive but may not be surprising, since distance methods are not expected to make full use of the information in the data and no theory predicts that they should have optimal performance.

Similar results, however, were observed when the likelihood method was used. Gaut and Lewis (1995) and Yang (1996a) (see also Siddall 1998; Swofford et al. 2001) reported simulation studies where likelihood under the true model had a lower probability of recovering the true tree than parsimony or likelihood under a wrong and simplistic model. Such counterintuitive results might be 'explained away' by suggesting that the datasets may be too small for the asymptotic properties of ML to be applicable.

To see whether small sample sizes are to blame, one can study the asymptotic behaviour of the relative efficiency (measured by equation (5.5)) of the methods when the sequence length $n \to \infty$. A few examples from Yang (1997b) are shown in Figure 5.1. The datasets are generated under the JC69 + Γ_4 model with rates for sites drawn from a gamma distribution with shape parameter $\alpha = 0.2$. The data are analysed using likelihood under JC69 + Γ_4 with α fixed at either 0.2 or ∞; the latter is equivalent to JC69 with one rate for all sites. The two analyses or methods are referred to as *True* and *False*, and both estimate five branch lengths on every tree. Both methods are consistent on the three trees of Figure 5.1, with the probability of recovering the true tree $P \to 1$ when $n \to \infty$. However, on trees *a* and *b*, *False* approaches this limit faster than *True*, as indicated by the relative efficiency of *True* relative to *False*, $E_{\text{TF}}^* = (1 - P_\text{F})/(1 - P_\text{T})$, where P_T and P_F are the probabilities of recovering the true tree by the two methods. Despite the intractability of the likelihood analysis, it appears safe to expect $E_{\text{TF}}^* < 1$ when $n \to \infty$ in trees *a* and *b*. In tree *b*, E_{TF}^* apparently

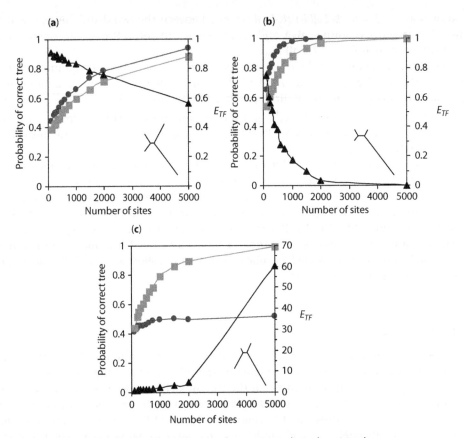

Fig. 5.1 The probability that the ML tree is the true tree plotted against the sequence length. Datasets are simulated under the JC69 + Γ_4 model, with shape parameter $a = 0.2$. The ML analysis assumed either the true JC69 + Γ_4 model with $a = 0.2$ fixed (■, the *True* method) or the false JC69 model, with $a = \infty$ fixed (●, the *False* method). The relative efficiency (▲) of *True* relative to *False* is defined as $\left(E_{TF}^*\right) = (1 - P_F)/(1 - P_T)$. The true trees, shown as insets, have the branch lengths as follows: ((a: 0.5, b: 0.5): 0.1, c: 0.6, d: 1.4) in (a); ((a: 0.05, b: 0.05): 0.05, c: 0.05, d: 0.5) in (b); and ((a: 0.1, b: 0.5) : 0.1, c: 0.2, d: 1.0) in (c). Both *False* and *True* are consistent for those three trees. Redrawn after Figures 1B, D, and C of Yang (1997b).

approaches zero. Additional simulation shows that increasing the value of α fixed in the *False* method from the true value (0.2) to ∞, so that the model becomes progressively more wrong, leads to progressively better performance in recovering this tree. Also use of the continuous gamma model instead of the discrete gamma produces similar results. Similar results may be obtained when parsimony is compared with likelihood under the one-rate method (with $\alpha = \infty$) when the data are generated using the one-rate model; on some trees, parsimony outperforms likelihood, with the efficiency of likelihood relative to parsimony, $E_{\text{ML, MP}} \to 0$ when $n \to \infty$ (Yang 1996a). In sum, the counterintuitive results are not due to small sample sizes.

A number of authors have suggested that the results might be explained by a 'bias' in parsimony or likelihood under a wrong and simplistic model (e.g. Yang 1996a; Huelsenbeck 1998; Bruno and Halpern 1999; Swofford et al. 2001). Swofford et al. (2001) illustrated this bias by the following analogy. Suppose we want to estimate parameter θ, and one method is to ask an oracle that always responds '0.492'. This method cannot be beaten if the true θ is 0.492, and may indeed be a very good method in finite datasets when the true θ is close to 0.492. Similarly, parsimony has a tendency to group long branches together irrespective of the true relationships. If the true tree happens to have the long branches together, the inherent bias of parsimony works in its favour, causing the method to outperform likelihood under the true model. This is demonstrated in Figure 5.2, where the probability of recovering the true tree is estimated by simulating 10,000 replicate datasets. The sequence length is 1,000 sites. In Figure 5.2a, data are generated under the JC69 model and analysed using parsimony as well as ML under JC69. In Figure 5.2b, data are generated under JC69 + Γ_4 with $\alpha = 0.2$ and analysed using parsimony as well as ML under both the true model (JC69 + Γ_4 with $\alpha = 0.2$ fixed) and the wrong JC69 model. The results of Figure 5.2b are rather similar to those of Figure 5.2a, with likelihood under the wrong JC69 model behaving in the same way as parsimony, so we focus on Figure 5.2a. In the right half of the graph, the true tree is $((a, c), b, d)$, with the two long branches a and b separated. Trees of this shape are said to be in the 'Felsenstein zone'. In the left half of the graph, the true tree is $((a, b), c, d)$, with the long branches a and b grouped together. Such trees are said to be in the 'Farris zone'. In both zones, likelihood under the true model recovers the correct tree with reasonable accuracy, with the accuracy improving as the internal branch length t_0 increases. In contrast, parsimony behaves very differently: it recovers the true tree with probability ~100% in the 'Farris zone' but with probability ~0% in the 'Felsenstein zone'; parsimony is indeed inconsistent for trees in the right half of the graphs in Figure 5.2. If the true tree is a star tree with $t_0 = 0$, parsimony produces the tree $((a, b), c, d)$, with probability ~100%, and 0% for the other two trees. Now suppose the true tree is $((a, b), c, d)$, with $t_0 = 10^{-10}$. This branch is so short that not a single change is expected to occur on it at any of the 1,000 sites in any of the 10,000 datasets, as the expected total number of changes is $10,000 \times 1,000 \times 10^{-10} = 0.001$. However, parsimony still recovers the true tree in every dataset. The site patterns supporting the true tree (*xxyy*) are all generated by convergent evolution or *homoplasies*. Swofford et al. argued that in this case the evidence for the correct tree is given too much weight by parsimony and is evaluated correctly by likelihood under the correct model. The analyses of Swofford et al. (2001; see also Bruno and Halpern 1999) provide an intuitive explanation for counterintuitive results such as those of Figures 5.1 and 5.2.

Yet, they are not relevant to the question I posed (Yang 1996a, 1997b): 'Is the likelihood method of tree reconstruction asymptotically efficient, as is a conventional MLE?' The answer to this question has to be 'No'. The asymptotic inefficiency of ML for tree reconstruction is not limited to one set of branch lengths, but applies to a nontrivial region of the parameter space. In the four-species problem, the parameter space is usually defined as a five-dimensional cube, denoted R_+^5, with each of the five branch lengths going from 0 to ∞. Inside R_+^5, one may define a subspace, call it $\aleph$, in which ML is asymptotically less efficient than another method, such as parsimony or ML under a simple and wrong model. For the present, $\aleph$ is not well characterized; we do not know what shape it takes or even whether it consists of one region or several disconnected regions. However, its existence appears to be beyond doubt. Now suppose that datasets are always generated from within $\aleph$. If the statistician applies ML for tree reconstruction, ML will be asymptotically less efficient than parsimony over the whole parameter space ($\aleph$). The 'bias' argument discussed above lacks force, since in this case the 'bias' is always in the direction of the truth. In the case of conventional parameter estimation, one may demand an

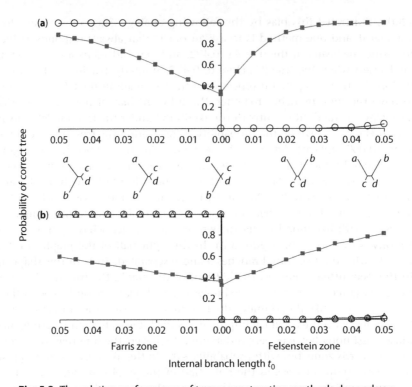

Fig. 5.2 The relative performance of tree reconstruction methods depends on the tree shape. The external branch lengths, in expected number of substitutions per site, are 0.5 for long branches and 0.05 for short branches. The internal branch length (t_0) varies along the x-axis. On the left is a tree in the 'Farris zone', with the internal branch becoming progressively shorter. Then at the origin ($t_0 = 0$) the topology switches to a tree in the 'Felsenstein zone', with the internal branch becoming increasingly longer. Performance is measured by the probability of recovering the true tree, estimated from 10,000 simulated datasets. The sequence length is 1,000 sites. (**a**) The true model is JC69. Each dataset is analysed using likelihood under the true JC69 model (■) and parsimony (○). (**b**) The true model is JC69 + Γ_4 with shape parameter $\alpha = 0.2$. Each dataset is analysed by likelihood under the true JC69 + Γ_4 model, with $\alpha = 0.2$ fixed (■), by likelihood under the false JC69 model (with $\alpha = \infty$) (△), and by parsimony (○). Parsimony in both (**a**) and (**b**) and likelihood under the wrong JC69 model in (**b**) are inconsistent for trees in the Felsenstein zone. At $t_0 = 0$, there is a change in the parameter space and in the definition of the true tree, so that the probability curve for each method is discontinuous. Constructed following Swofford et al. (2001).

estimator that is asymptotically more efficient than ML over a narrow but nonempty interval, say $\theta \in [0.491, 0.493]$. Such an estimator does not exist (e.g. Cox and Hinkley 1974, p. 292).

Here it may be fitting to take a detour into the exotic land of 'super-efficiency'. LeCam (1953) has constructed estimators that are even more efficient than the MLE. Consider t to be a consistent estimator of θ, asymptotically normally distributed with variance proportional to $1/n$. Define a new statistic

$$t' = \begin{cases} t, & \text{if } |t| \geq n^{-1/4}, \\ 0.01t, & \text{if } |t| < n^{-1/4}. \end{cases} \tag{5.9}$$

In other words, if t is very close to 0, make it even closer; otherwise do not change it. Then when $n \to \infty$, we have $\text{var}(t')/\text{var}(t) \to 1$ if $\theta \neq 0$ and $\to 0.01^2$ if $\theta = 0$. Thus at one point ($\theta = 0$), t' is more efficient than t and nowhere is it worse. One can do this for a countably infinite number of parameter values (such as all integers), and thus we have a method that is more efficient than the MLE at infinitely many parameter points and nowhere is it worse. This 'super-efficient' estimator is infinitely smarter than Swofford et al.'s oracle, but even it does not distract from the efficiency of the MLE: the catch is that such 'super-efficiency' can arise only for a set of θ values that collectively have measure zero (LeCam 1953). Here we do not have to bother us with measure-theoretic terminology, and it suffices to take measure to mean the width in one dimension (or area in two dimensions). In other words, 'super-efficiency' can arise only for an isolated set of points that collectively have zero width (such as the collection of all integers or all rational numbers). In one dimension (with one parameter in the model), any nonempty interval $a < \theta < b$ has positive measure, and it is impossible to find a method that is more efficient than the MLE for all θ in the interval. In view of this, 'super-efficiency' is often regarded as an unsuccessful distraction from the efficiency of the MLE.

While there are reports of real data examples in which a simple and wrong model recovered the phylogeny better than more complex and realistic models (e.g. Posada and Crandall 2001), it should be emphasized that the above discussion is not an endorsement of parsimony or likelihood under simplistic models in real data analysis. The discussion here is more of a philosophical nature. The issue is not that ML often performs more poorly than other methods such as parsimony; it is that this can happen in a certain portion of the parameter space at all. It is important to note that under the true model, ML is always consistent, while parsimony and ML under wrong and simplistic models may be inconsistent. For practical data analysis, the biological process is likely to be far more complex than any of the substitution models we use, and likelihood under complex and realistic models may be necessary to avoid the problem of long-branch attraction. It is prudent to use such complex and realistic models in real data analysis, as recommended by Huelsenbeck (1998) and Swofford et al. (2001).

5.2.4 Robustness

Computer simulations have been conducted to examine the performance of likelihood and other model-based methods when some model assumptions are violated. The results are complex, and dependent on a number of factors such as the precise assumptions being violated as well as the shapes of trees assumed in the simulation. Overall the simulations suggest that ML is highly robust to violations of assumptions. ML was found to be more robust to rate variation among sites than distance methods such as neighbour-joining (NJ) (Fukami-Kobayashi and Tateno 1991; Hasegawa et al. 1991; Kuhner and Felsenstein 1994; Tateno et al. 1994; Yang 1994c, 1995b; Gaut and Lewis 1995; Huelsenbeck 1995a). Certain assumptions have a huge impact on the fit of the model to data, but do not appear to have a great effect on tree reconstruction; the transition–transversion rate difference and unequal base compositions appear to fall into this category (e.g. Huelsenbeck 1995a).

In contrast, certain aspects of among-site heterogeneity appear to be far more important. Here we consider two cases. The first is substitution rate variation among sites. Chang

(1996b) generated data using two rates, so that some sites are fast evolving and others are slowly evolving, and found that likelihood assuming one rate can become inconsistent. Even though both sets of sites are generated under the same tree, when they are analysed together a wrong tree is obtained even from an infinite amount of data. Similar results were found when rates vary according to a gamma distribution (Kuhner and Felsenstein 1994; Tateno et al. 1994; Huelsenbeck 1995a; Yang 1995b). Of course, if the model accommodates variable rates among sites, ML is consistent.

A second case of among-site heterogeneity was considered by Kolaczkowski and Thornton (2004). Datasets were generated using two trees, or more precisely, two sets of branch lengths for the same four-species tree $((a, b), (c, d))$, shown in Figure 5.3. Half of the sites evolved on tree 1 with branches a and c short and b and d long, while the other half evolved on tree 2 with branches a and c long and b and d short. The biological rationale for this design is that different sites in a gene sequence may evolve at different rates but the rates for sites may also drift over time (possibly due to changing functional constraints), a process called *heterotachy*. In such data parsimony performs poorly, but likelihood under the homogeneous model, assuming one set of branch lengths for all sites, is even worse. Note that each of the two trees is the long-branch attraction tree (see Figure 3.25), and parsimony can be inconsistent on any of them if the internal branch is short enough (Felsenstein 1978b). However, the probabilities of observing the three parsimony-informative site patterns, $xxyy$, $xyyx$, $xyxy$, are the same on both trees and remain the same if the data are a mixture of sites generated on the two trees. Thus the performance of parsimony is the same whether one tree or two trees are used to generate the data. In contrast, the use of two trees violates the likelihood model, causing ML to become inconsistent and to perform even worse than parsimony. Previously, Chang (1996b) found that a distance method assuming a homogeneous model in distance calculation is inconsistent under such a mixture model, and his explanation appears to apply to ML as well. In datasets generated under the mixture model, species a and c have an intermediate distance since they are very close on tree 1 but far away on tree 2. Similarly, species b and d have an intermediate distance since they are close on tree 2 but far away on tree 1. However, every other pair of species has a large distance since they are far away on both trees. The topology compatible with these requirements is $((a, c), b, d)$, which is different from the true topology.

It is not so clear how common such mixtures of trees are in real datasets. A number of studies examined other combinations of branch lengths between trees and found that likelihood under the homogeneous model performed much better than parsimony (e.g. Gadagkar and Kumar 2005; Gaucher and Miyamoto 2005; Philippe et al. 2005; Spencer et al. 2005; Lockhart et al. 2006). Of course, if two sets of branch lengths are assumed in the model, likelihood will perform well (Spencer et al. 2005; Kolaczkowski and Thornton 2008). Such a mixture model, however, contains many parameters, especially on large trees and thus has questionable utility for practical data analysis. A general approach to dealing with heterotachy is to assume in the model one rate parameter for every site in

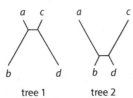

Fig. 5.3 If the data are generated using two trees (or more precisely, two sets of branch lengths on the same topology), it is possible for parsimony to outperform likelihood assuming one tree.

the sequence and every branch on the tree. This is a delicate model and can easily run into the trap of too many parameters. This is discussed in §5.3.1.

5.3 Parsimony

Parsimony was initially developed to analyse discrete morphological characters. The analysis was typically non-statistical, based on intuition and empirical knowledge concerning morphological character evolution. A semi-philosophical argument was developed afterwards to justify the practice. When molecular sequences became available, the method was applied to molecules as well, with each position (nucleotide or amino acid) considered a character. There was a prolonged debate concerning whether phylogeny reconstruction should be viewed as a statistical problem and whether parsimony or likelihood should be the method of choice. See Felsenstein (2001b, 2004) and Albert (2005) for assessments of different views.

Parsimony does not make explicit assumptions about the evolutionary process. Some authors argue that parsimony makes no assumptions at all and that, furthermore, phylogenies should ideally be inferred without invoking any assumptions about the evolutionary process (Wiley 1981). Others point out that it is impossible to make any inference without a model; that a lack of explicit assumptions does not mean that the method is 'assumption-free' as the assumptions may be merely implicit; that the requirement for explicit specification of model assumptions is a strength rather than weakness of the model-based approach, since then the fit of the model to data can be evaluated and improved (e.g. Felsenstein 1973b).

The latter position is taken in this book. Given this position, it is meaningful to ask what assumptions parsimony makes concerning the evolutionary process, and how parsimony can be justified. This section provides an overview of the studies that attempt to establish an equivalence between parsimony and likelihood under particular models, and of the arguments put forward to justify parsimony.

5.3.1 *Equivalence with misbehaved likelihood models*

A number of authors have sought to establish an equivalence between parsimony and likelihood under a particular model. Equivalence here means that the most parsimonious tree and the ML tree under the said model are identical in *every* possible dataset. One can easily show that the two methods produce the same tree in particular datasets, but such results are not useful. Our focus is on the tree topology, so that additional inferences, such as reconstruction of ancestral states or estimation of branch lengths, if provided by the method, are ignored. An established equivalence between parsimony and likelihood may serve two purposes. First, it may indicate that the likelihood model is the evolutionary model assumed by parsimony, thus helping us to understand the implicit assumptions made by parsimony. Second, it may provide a statistical (likelihood) justification for parsimony; if likelihood has nice statistical properties under the model, parsimony will share these properties.

However, both objectives have met with difficulties. First, as Sober (2004) argued, an equivalence between parsimony and likelihood under a particular model means that the said likelihood model is sufficient for the equivalence but may not be necessary. In other words, it is possible for parsimony to be equivalent to likelihood under some other evolutionary models. Second, likelihood comes in many flavours and some likelihood methods are known not to work well. Establishing an equivalence with a misbehaved likelihood method provides little justification for parsimony. Indeed there are about a dozen likelihood methods proposed in the literature, such as profile, integrated, marginal, conditional, partial, relative, estimated, empirical, hierarchical, and penalized likelihood methods. Most of them have been proposed in order to deal with nuisance parameters. (Profile and integrated likelihoods are described in §1.4.4. See Goldman (1990) for a careful discussion of some of those methods.) The well-known asymptotic properties of MLEs, such as consistency and asymptotic efficiency, hold only under certain regularity conditions. One such condition is that the number of parameters in the model should not increase without bound with increase in the sample size. Unfortunately, most likelihood models used to establish an equivalence with parsimony are such pathological ones with infinitely many parameters, under which likelihood may misbehave.

Here I provide a brief review of studies that attempt to establish an equivalence between parsimony and likelihood under such infinite-parameters models. Such likelihood models are rarely used in real data analysis. The next subsection considers equivalence with well-behaved likelihood models, which are in common use.

Felsenstein (1973b, 2004, pp. 97–102) considered a Markov model of character evolution in which a set of branch lengths are applied to all sites in the sequence, but every site has a rate. The model is very similar to models of variable rates across sites discussed in §4.3.1, except that the rates for sites are parameters, so that the number of rate parameters increases without bound as the sequence length increases. Felsenstein showed that the parsimony and likelihood trees coincide when the substitution rates for sites approach zero. The result, though short of a rigorous proof, provides a mathematical justification for the intuitive expectation that parsimony and likelihood are highly similar at low sequence divergences. The proof, however, requires that the MLE of rate for any site $\hat{r}$ approaches zero (with rates at different sites approaching zero proportionally), which is slightly different from letting the rate parameters rs approach zero and showing that likelihood and parsimony trees coincide in every dataset. Even if the true rate r is small, datasets are still possible in which $\hat{r}$ is not very small, in which case parsimony and likelihood may not be equivalent.

Farris (1973) considered a stochastic model of character evolution and attempted to derive MLEs of not only the ancestral sequences at the interior nodes of the tree but also the time points of changes, i.e. a complete history of character evolution through time. He argued that likelihood and parsimony gave the same estimates. Goldman (1990) used a symmetrical model of character evolution to estimate the tree topology, branch lengths, and the ancestral states at the interior nodes of the tree. All branches on the tree are assumed to have the same length, so that the model lacks a time structure (Thompson 1975). Goldman showed that likelihood under the model produces the same tree as parsimony. He argued that estimation of ancestral character states makes the likelihood method inconsistent, and equivalence with such a likelihood method may provide an explanation for the inconsistency of parsimony. This likelihood model is similar to the 'maximum parsimony likelihood' of Barry and Hartigan (1987b), in that both estimate ancestral states together with model parameters such as branch lengths or branch transition probabilities. Strictly speaking, the likelihood formulations in Farris (1973), Barry and Hartigan (1987b), and Goldman (1990) may be hard to justify. For example, the 'likelihood function' used by Goldman (1990, Equation 6) is $f(X, Y|\tau, t)$, where X are sequences at the tips of the tree, Y are the ancestral states, τ is the tree topology, and t is the branch length. Although $f(X, Y|\tau, t)$ may be given a penalized or hierarchical likelihood interpretation (Silverman 1986, pp. 110–119; Lee and Nelder 1996), the approach is not likelihood in the usual sense of the word. The same criticism applies to the 'maximum parsimony likelihood' of Barry and Hartigan (1987b, pp. 200–201). Given the assumed model of character evolution, the ancestral states are random variables, with fully specified distributions. Thus the correct likelihood should be $f(X|\tau, t)$, which is an average over the ancestral states Y, as in Felsenstein (1973b, 1981) (see equation (4.3)); Y should not be treated as parameters.

Tuffley and Steel (1997) presented an interesting analysis of Felsenstein's (1981) likelihood method applied to datasets consisting of one single character (site) (see Yang 2006, §6.5 for a more accessible account of the theory). The principal result is that under a model of equal rate between any two characters, the maximized likelihood on any tree is given by $(1/c)^{1+l}$, where c is the number of character states ($c = 4$ for the JC69 model) and l is the character length or the minimum number of changes according to parsimony. Thus tree 1 is shorter than tree 2 if and only if it has a higher likelihood than tree 2; parsimony and likelihood under the model will always produce the same tree on one character. In analysis of data of multiple characters, the same holds true if a separate set of branch lengths is assumed for every character; likelihood analysis will then be effectively separate analyses of the different characters. The tree with the highest likelihood, calculated by multiplying probabilities across characters, is then always the minimum length or most parsimonious tree, with the character lengths summed over characters. Tuffley and Steel called this model the *no-common mechanism model*. Note that the model assumes the same relative rate between character states so that it does assume certain common mechanism, although it allows the amount of evolution to vary across sites and branches.

As stressed by Tuffley and Steel (1997; see also Steel and Penny 2000), their theory does not apply to data of multiple characters if the same set of branch lengths is applied to all characters, as in Felsenstein's (1981) formulation. Nor does it establish an equivalence between parsimony and likelihood under commonly used likelihood models. The theory is of a philosophical nature. The application of likelihood to models involving infinitely many parameters is detested by statisticians (e.g. Stein 1956; Kalbfleisch and Sprott 1970; see also Felsenstein and Sober 1986; Goldman 1990), and the no-common mechanism model is not an acceptable way of dealing with among-site heterogeneity in the likelihood framework (Steel 2011).

If one is concerned with certain aspects of the evolutionary process being heterogeneous among sites, one may let such quantities be drawn from a parametric or nonparametric distribution, in the same way that models of stochastic rate variation among sites are constructed. One may then use likelihood to estimate the parameters in such a 'super-process'. Such an approach accommodates the among-site heterogeneity while keeping the number of parameters under control. Interestingly, my analysis of such a model with branch lengths varying at random among sites simply leads to Felsenstein's (1981) likelihood, effectively with a redefinition of branch lengths (Yang 2006, p. 201). Similarly Huelsenbeck et al. (2011) implemented the no-common mechanism model in a Bayesian framework, and used the Bayes factor (see §7.4.3 later in Chapter 7) to examine its fit to data in comparison with commonly used models. They found that the no-common mechanism model fits the data far more poorly than other commonly used biologically inspired models; in other words, the cost of using many parameters in the no-common mechanism model considerably outweighs the improvement in the model's fit to data. For inference of deep phylogenies where heterotachy may be a concern, Zhou et al. (2010) implemented the so-called *covarion-mixture* model in the program PhyloBayes, which uses a Dirichlet process prior to assign rate parameters to sites and branches.

5.3.2 Equivalence with well-behaved likelihood models

Models in common use in molecular phylogenetics have the essential features that the model is identifiable and the number of parameters does not increase without bound with the increase of sequence length. Establishing an equivalence between parsimony and such a likelihood model will go a long way to justifying parsimony. Unfortunately, likelihood analysis under such models is most often intractable.

The only tractable phylogenetic likelihood model, which also happens to establish an equivalence between likelihood and parsimony, concerns inference of three rooted trees for three species using binary characters evolving under the molecular clock. This has been discussed in §4.6.3. The three rooted trees are $\tau_1 = ((1,2),3)$, $\tau_2 = ((2,3),1)$, and $\tau_3 = ((3,1),2)$, shown in Figure 4.14, where t_{i0} and t_{i1} are the two branch lengths in tree τ_i, for $i = 1, 2, 3$. There are only four site patterns: xxx, xxy, yxx, and xyx, where x and y are any two distinct characters. Let n_0, n_1, n_2, n_3 be the counts of sites with those patterns. If τ_1 is the true tree and if evolution proceeds at a constant rate, one may expect pattern xxy to occur with higher probability than patterns yxx and xyx, since the former pattern is generated by a change over a longer time period while either of the latter patterns is generated by a change in a shorter time period. Similarly, patterns yxx and xyx 'support' trees τ_2 and τ_3, respectively. Thus one should select tree τ_1, τ_2, or τ_3 as the estimate of the true tree if n_1, n_2, or n_3 is the greatest among the three, respectively. This may be considered a parsimony argument, although it differs from the parsimony method in common use, which minimizes the number of changes and does not distinguish among rooted trees. The likelihood solution is illustrated in Figure 4.15 (Yang 2000a, Table 4). If we ignore extreme datasets in which the sequences are more different than random sequences, the likelihood solution matches the parsimony solution: the ML tree is τ_i if n_i is the greatest among n_1, n_2, and n_3. This model was discussed by Sober (1988), but his likelihood analysis is limited to one character and does not account for the fact that the estimated branch lengths differ across the three trees.

Indeed under this model, all sensible tree reconstruction methods give the same estimate of the tree topology. For example, the least-squares tree based on pairwise distances is the same as the ML or parsimony tree (Yang 2000a). Similarly, if one assigns priors on branch lengths t_0 and t_1 and define the likelihood for tree τ_i as

$$L_i = \int\int f(n_0, n_1, n_2, n_3 | \tau_i, t_{i0}, t_{i1}) f(t_{i0}) f(t_{i1}) dt_{i0} dt_{i1}, \qquad (5.10)$$

the *maximum integrated likelihood tree* is τ_1, τ_2, or τ_3 if n_1, n_2, or n_3 is the greatest, respectively (Yang and Rannala 2005). In addition, if the three binary trees are assigned the same prior probability, the maximum *a posteriori* probability (MAP) tree for the Bayesian approach will also be the ML tree.

A slightly more complex case will be the use of the JC69 model for four nucleotides instead of binary characters (Saitou 1988; Yang 2000a). There is then an additional site pattern xyz, but this pattern does not affect tree selection under either parsimony or likelihood (again except for rare datasets of extremely divergent sequences). The full likelihood solution does not appear tractable (see Problem 4.2), but it is easy to show that if the ML tree is a binary tree, it must correspond to the greatest of n_1, n_2, and n_3 (Yang 2000a, pp. 115–116). The conditions under which the ML tree is the star tree (when none of the binary trees has higher likelihood than the star tree) are not well characterized. Again one may disregard rare extreme datasets of very divergent sequences and consider parsimony and likelihood to be equivalent under the JC69 model for rooted trees with the clock.

More complex models and larger trees with more species are hard to analyse. However, a number of authors have studied cases in which parsimony is inconsistent. In such cases likelihood and parsimony must not be equivalent since likelihood is always consistent. Hendy and Penny (1989) showed that with four species and binary characters evolving under the clock, parsimony is always consistent in recovering the (unrooted) tree, although ML and parsimony do not appear to be equivalent under the model. With five or more species evolving under the clock, it is known that parsimony can be inconsistent in estimating the unrooted trees (Hendy and Penny 1989; Zharkikh and Li 1993; Takezaki and Nei 1994). Thus it is not equivalent to likelihood. For estimating unrooted trees without the clock assumption, parsimony can be inconsistent for trees of four or more species and is not equivalent to likelihood (Felsenstein 1978b; DeBry 1992). Those studies often use small trees with four to six species. The cases for much larger trees are less clear. However, it appears easier to identify cases of inconsistency of parsimony on large trees than on small trees (Kim 1996; Huelsenbeck and Lander 2003), suggesting that likelihood and parsimony are in general not equivalent on large trees.

Some authors (e.g. Yang 1996a) have suggested that parsimony is closer to a simplistic likelihood model, such as JC69, than to a more complex model, such as HKY85 + Γ. Under the assumption of equal rates between any two character states, as in JC69, and under the additional assumption that all branch lengths on the tree are equal (Goldman 1990), likelihood will produce exactly the same ancestral reconstructions as parsimony. However, it is unclear whether the assumptions of the JC69 model and equal branch lengths are sufficient to guarantee that ML and parsimony always produce identical unrooted tree topologies.

5.3.3 *Assumptions and justifications*

5.3.3.1 *Occam's Razor and maximum parsimony*

The *principle of parsimony* is an important general principle in the generation and testing of scientific hypotheses. Also known as Occam's Razor, it states that one should not increase, beyond what is necessary, the number of entities required to explain anything. This principle, used sharply by William Ockham (died *c.* 1349), assumes that simpler explanations are inherently better than complicated ones. In statistical model building, models with fewer parameters are preferred to models with more parameters if they fit the data nearly equally well. likelihood ratio test (LRT), Akaiki information criterion (AIC), and Bayesian information criterion (BIC) are all mathematical exemplifications of this principle; see §4.7.

The parsimony (minimum-step) method of tree reconstruction is sometimes claimed to be based on the parsimony principle in science. The number of character changes on the tree is taken as the number of *ad hoc* assumptions that one has to invoke to explain the data. This correspondence appears superficial, in term rather than in content. For example, both the parsimony and minimum evolution methods minimize the amount of evolution to select the tree, the only difference being that minimum evolution uses the number of changes after a correction for multiple hits while parsimony uses the number of changes without such a correction (see §3.3 and §3.4). It is not reasonable to claim that parsimony, because of its failure to correct for multiple hits, enjoys a philosophical justification that minimum evolution lacks. One may ask how parsimony should proceed and be justified if the number of changes at every site is known and given.

5.3.3.2 *Is parsimony a nonparametric method?*

Some authors argue that parsimony is a nonparametric method (e.g. Sanderson and Kim 2000; Holmes 2003; Kolaczkowski and Thornton 2004). This claim appears to be a misconception. In statistics, some methods make no or weak assumptions about the distribution of the data, so that they may apply even if the distributional assumptions of the parametric methods are violated. For example, for two normally distributed variables, we usually calculate Karl Pearson's product–moment correlation coefficient r and use a t statistic to test its difference from 0. However, this parametric method may not work well if the variables are proportions or ranks, in which case the normal assumption is violated. One may then use Spearman's rank correlation coefficient r_s, which does not rely on the normal assumption and is a nonparametric measure. Nonparametric methods make fewer or less stringent assumptions about the data-generating model. However, they are not assumption-free. They often make the same assumptions of randomness and independence of data samples as parametric methods. A nonparametric method works under situations where the parametric method works, perhaps with a slight loss of power. Parsimony for tree reconstruction is known to break down over a range of parameter values under simple parametric models (Felsenstein 1978b). This failure disqualifies it as a nonparametric method. As pointed out by Spencer et al. (2005), simply not requiring a parametric model is not a sufficient criterion for a satisfactory nonparametric method. A useful nonparametric model should perform well over a wide range of possible evolutionary models, but parsimony does not have this property.

5.3.3.3 *Inconsistency of parsimony*

As mentioned in Chapter 3, parsimony can be inconsistent over some portions of the parameter space; in particular, the method tends to suffer from the problem of long-branch attraction. Felsenstein (1973b, 1978b) conjectured that when the amount of evolution is small and the rate of evolution is more or less constant among lineages, parsimony may be consistent. Later studies have shown that even the existence of a molecular clock combined with a small amount of evolution does not guarantee consistency, and the situation seems to become worse with the increase of the number of sequences in the data (Hendy and Penny 1989; Zharkikh and Li 1993; Takezaki and Nei 1994). Huelsenbeck and Lander (2003) generated random trees using a model of cladogenesis, and found that parsimony is often inconsistent.

Farris (1973, 1977; see also Sober in Felsenstein and Sober 1986) dismissed the argument of Felsenstein (1978b) by claiming that the model Felsenstein used is overly simplistic and thus irrelevant to the performance of parsimony in real data analysis. This response appears to be off the mark. The model is unrealistic because it assumes that evolution occurs independently among sites and lineages, at the same rate across sites and between character states, and so on. A more complex and realistic model relaxing those restrictions will include the simplistic models as special cases: for example, the rates may be allowed to differ between character states but they do not have to. The inconsistency of parsimony under a simplistic model means inconsistency under the more complex and realistic model as well. Furthermore, as pointed out by Kim (1996), the conditions that lead to inconsistency of parsimony are much more general than outlined by Felsenstein (1978b); further relaxation of the assumptions simply exacerbates the problem.

5.3.3.4 *Assumptions of parsimony*

While many authors agree that parsimony makes assumptions about the evolutionary process, it has been difficult to identify them. It appears to be safe to suggest that parsimony assumes independence of the evolutionary process across characters (sites in the sequence) and among evolutionary lineages (branches on the tree). However, much more cannot be said without generating some controversy. Parsimony assigns equal weights to different types of changes, which appears to suggest that it assumes equal rates of change between characters; weighted parsimony is designed to relax this assumption. Similarly, parsimony assigns equal weights to character lengths at different sites, which suggests that it assumes the same (stochastic) evolutionary process at different sites; successive weighting (Farris 1969) attempts to modify this assumption by down-weighting sites with more changes.

Felsenstein (1973a, 1978a) argued that parsimony makes the assumption of low rates; see §5.3.1. As Farris (1977; see also Kluge and Farris 1969) pointed out, the minimum-step criterion does not mean that the method works only if changes are rare. Similarly, the likelihood method maximizes the likelihood (the probability of the data) to estimate parameters, but it is not true that the method works only if the likelihood is high. In large datasets, the likelihood is vanishingly small, but it is valid to compare likelihood for different hypotheses. The effect of the amount of evolution on the performance of parsimony depends on the tree shape. If the tree is hard and parsimony is inconsistent, reducing the evolutionary rate indeed helps to remedy the problem. However, when the tree is easy, parsimony can outperform likelihood, with greater superiority at higher evolutionary rates (e.g. Yang 1996a, Figure 6).

One aspect of the similarity between parsimony and likelihood may be worth emphasizing. In the Markov chain models used in likelihood analysis, the transition probability matrix $P(t) = e^{Qt}$ has the property that the diagonal element is greater than the off-diagonals in the same column:

$$p_{jj}(t) > p_{ij}(t), \tag{5.11}$$

for any fixed j and any t. As a result, an identity tends to have a higher probability than a difference if both can fit the data; ancestral reconstructions and phylogenetic trees requiring fewer changes tend to have higher likelihoods, irrespective of the sequence divergence level. Sober (1988) discussed this implication and referred to equation (5.11) as 'backward inequality'. Based on this observation, one would intuitively expect parsimony to provide a reasonable approximation to likelihood (Edwards and Cavalli-Sforza 1964, p. 64; Cavalli-Sforza and Edwards 1966). Edwards (1996) explicitly states that 'The idea of the method of minimum evolution arose solely from a desire to approximate the maximum likelihood solution'. Perhaps one should be content to consider parsimony as a heuristic method of tree reconstruction that often works well under simple conditions, rather than seeking a philosophical or statistical justification for it. Heuristic methods often involve *ad hoc* treatments of the data that cannot be justified rigorously under any model. As long as its limitations are borne in mind, parsimony is a simple and useful method.

5.4 Testing hypotheses concerning trees

A phylogenetic tree reconstructed using likelihood, parsimony, or distance methods may be viewed as a point estimate. It is desirable to attach a measure of reliability to it. However, the tree represents a complex structure that is quite different from a conventional parameter, making it difficult to apply conventional procedures for constructing confidence intervals or performing significance tests. This section describes several commonly used methods for evaluating the reliability of the estimated tree: bootstrap, test of internal branch lengths, and the likelihood-based tests of Kishino and Hasegawa (1989) and Shimodaira and Hasegawa (1999). Our focus here is on parsimony, distance, and likelihood methods. For the Bayesian method the posterior probability provides the natural measure of accuracy (see Chapter 8).

5.4.1 Bootstrap

5.4.1.1 Bootstrap standard errors and confidence intervals

The bootstrap method is a computationally intensive method of statistical analysis based on simulation. It was initially developed by Efron (1979) to calculate the standard error for a parameter estimator, and has since been used to conduct all sorts of analyses in Frequentist statistics, such as standard errors, confidence intervals, and significance tests. The books of Efron and Tibshirani (1993) and Davison and Hinkley (1997) provide good summaries of the method and its many applications.

Suppose we estimate a parameter θ from a dataset of size n, $\boldsymbol{x} = \{x_1, x_2, ..., x_n\}$ using the statistic $t : \hat{\theta} = t(\boldsymbol{x})$. For example, if θ is the population mean, the sample mean $\hat{\theta} = t(\boldsymbol{x}) = \frac{1}{n}\sum_{i=1}^{n} x_i$ may be a good estimator. The bootstrap method for calculating the standard error for $\hat{\theta}$ involves generating many *bootstrap pseudo-samples*. To generate a bootstrap sample, $\boldsymbol{x}^* = \{x_1^*, x_2^*, \ldots, x_n^*\}$, sample n times with replacement from the n data points in the original dataset $\{x_1, x_2, \ldots, x_n\}$. Note that if u is a random number, $\lfloor nu \rfloor + 1$ will be a random integer between 1 and n. The asterisks here indicate that $\boldsymbol{x}^*$ is not the actual dataset $\boldsymbol{x}$, but a resampled version. For instance, we might have $x_1^* = x_2$, $x_2^* = x_9$, $x_3^* = x_9$, and so on. Some data points in the original dataset are sampled multiple times, while others not at all. We do this B times, to generate B bootstrap datasets, $\boldsymbol{x}_1^*, \boldsymbol{x}_2^*, \ldots, \boldsymbol{x}_B^*$, each of the same size (n) as the original dataset. Typically $B = 100$ or $1,000$ is adequate. We analyse each bootstrap dataset in the same way that we analyse the original dataset, generating an estimate $\hat{\theta}_b^* = t(\boldsymbol{x}_b^*) = \frac{1}{n}\sum_{i=1}^{n} x_i^*$ from bootstrap dataset b. The standard error for $\hat{\theta}$ is then estimated by

$$\text{SE} = \frac{1}{B}\sum_{b=1}^{B} (\hat{\theta}_b^* - \bar{\theta}^*)^2, \quad (5.12)$$

where $\bar{\theta}^*$ is the average of $\hat{\theta}_b^*$ over the B bootstrap replicates.

The central idea of bootstrap resampling is that the distribution of the bootstrap estimate around the estimate from the original data, $\hat{\theta}_b^* - \hat{\theta}$, is a good approximation to the distribution of the original estimate around the true value, $\hat{\theta} - \theta$. The bootstrap can be applied automatically whatever the complexity of the estimation procedure (or the complexity of the statistic $t(\boldsymbol{x})$) or whatever the parametric model is; one simply plugs in the same formula that is used in the analysis of the original dataset. This *plug-in principle* is a big advantage of the method and can be used to calculate the SE or to construct the confidence interval of a parameter estimate (Efron and Tibshirani 1993). We illustrate the latter using an example.

Example 5.1: Bootstrap confidence interval for sequence distance under JC69. We generate 1,000 bootstrap pseudo-samples using the alignment of the human and orangutan 12s rRNA genes analysed in Tables 1.3 and 1.4. We then estimate the sequence distance under the JC69 model and obtain 1,000 estimates. We sort them and find the 25th and 975th values, which form the 95% confidence interval for d: $(0.0801, 0.1223)$. This is similar to the approximate confidence interval based on $\hat{d} \pm 1.96 \times \text{SE}$, which gives $(0.0801, 0.1229)$ (see Table 1.4), or the likelihood interval $(0.0817, 0.1245)$. If we use the

K80 model to estimate the sequence distance, we obtain the bootstrap confidence interval (0.0818, 0.1276) for d, and (15.39, 107.30) for κ. The confidence interval (CI) for d is close to the CI given by $\hat{d} \pm 1.96 \times \text{SE}$, which is (0.0819, 0.1273). The CI for κ is very different. The normal approximation is expected to be unreliable for κ. □

5.4.1.2 *Bootstrap for phylogenies*

The bootstrap method is now the most commonly used method for assessing uncertainties in phylogenies estimated using parsimony, distance and likelihood methods. (In the Bayesian method, the posterior probability is the natural measure of reliability.) Bootstrap was introduced into phylogenetics by Felsenstein (1985a) as a straightforward application of the bootstrap technique in statistics.

In most models used in molecular phylogenetics, data at different sites are i.i.d., so that one site corresponds to one data point. Thus to generate bootstrap pseudo-datasets, one should resample, with replacement, sites from the original alignment (Figure 5.4). Each bootstrap sample has the same number of sites as the original data. Each site in the bootstrap sample is chosen at random from the original dataset, so that some sites in the original data may be sampled multiple times while others may not be sampled at all. Each bootstrap dataset is then analysed in the same way as the original dataset to reconstruct the phylogenetic tree (the plug-in principle). This process generates a set of trees from the bootstrap samples, which can then be summarized.

Different approaches can be used to summarize the bootstrap trees. One approach, rarely if ever used, is to use the bootstrap trees to construct a *confidence set* or *confidence region* of trees. Each tree has a *bootstrap support value* attached, which is the percentage of bootstrap datasets in which the tree is the estimated tree. Trees with highest support values are collected into a set until accumulatively they encompass > 95% of total support.

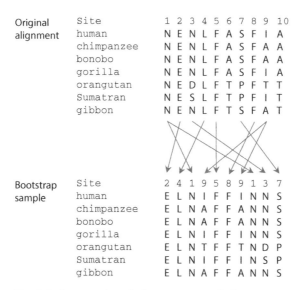

Fig. 5.4 Construction of a bootstrap sample, by sampling sites in the original sequence alignment with replacement.

174   5 COMPARISON OF PHYLOGENETIC METHODS AND TESTS ON TREES

The more common approach is to summarize splits shared among the bootstrap trees. There are two variations to this. The first is to attach a bootstrap support value for every split in the tree estimated from the original dataset, which is simply the proportion of bootstrap trees that include that split (Figure 5.5). This is called the bootstrap support or bootstrap proportion for the split. The second is to construct a majority-rule consensus tree using the bootstrap trees and attach support values for splits in the consensus tree. This second approach, although more commonly used, appears to be less sound if the consensus tree differs from the tree estimated from the original dataset. See Sitnikova et al. (1995) and Whelan et al. (2001) for similar comments.

As Felsenstein (1985a) emphasized, the tree reconstruction method must be consistent; otherwise high support values may be generated by bootstrap for incorrect trees or splits.

The RELL approximation. The bootstrap method involves intensive computation when used with the likelihood method of tree reconstruction, as in theory each bootstrap data sample has to be analysed in the same way as the original data, which in theory involves a full-scale tree search. To analyse 100 bootstrap samples, the computation will be ~100 times that needed to analyse the original dataset. For the likelihood method, an approximation suggested by Kishino and Hasegawa (1989) is to use the MLEs of branch

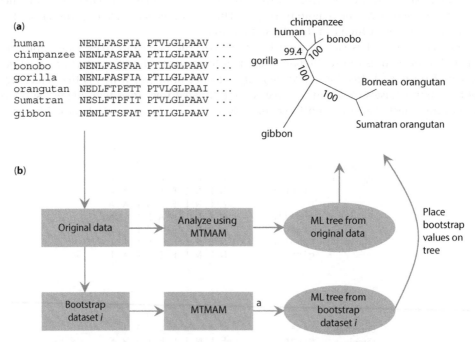

Fig. 5.5 The bootstrap method is used to calculate support values for splits on the ML tree. The mitochondrial proteins of seven ape species are analysed under the MTMAM model. From the original data, an exhaustive tree search is performed, identifying the ML tree, as shown. Bootstrap is then used to attach support values (in percentages) for splits on the ML tree. A number (say, 1,000) of bootstrap datasets are generated by resampling sites in the original alignment (Figure 5.4). Each bootstrap dataset is analysed in the same way as the original dataset, yielding an ML tree for each bootstrap sample. The proportion of bootstrap trees that contain a clade on the original ML tree is calculated and placed on the original ML tree. Constructed following Whelan et al. (2001).

lengths and substitution parameters from the original data to calculate the log likelihood values in each bootstrap sample; in theory one should re-estimate those parameters by likelihood iteration for every tree in every bootstrap sample. Resampling sites in the alignment is then equivalent to resampling the logarithms of the probabilities of data at the sites in the original dataset. The method is thus called the RELL bootstrap (for Resampling Estimated Log Likelihoods). Hasegawa and Kishino's (1994) computer simulation suggests that the RELL bootstrap provides a good approximation to Felsenstein's real bootstrap. The RELL method is convenient for evaluating a fixed set of trees; one simply evaluates all trees in the set using the original dataset, stores the logarithms of site probabilities for every tree, and then resamples them to calculate log likelihood values for trees in each bootstrap sample. The method is not applicable if heuristic tree search has to be performed for every bootstrap sample, since different trees may be visited during the tree search in different bootstrap samples.

5.4.1.3 *Interpretations of bootstrap support values*

Intuitively, if the dataset is large and the different sites have consistent phylogenetic signals, there will be little conflict among the bootstrap samples, resulting in high bootstrap support for the splits in the original tree. If the data lack information or different sites contain conflicting signals, there will be more variation among the bootstrap samples, resulting in lower bootstrap proportions for most splits. Thus higher bootstrap values indicate stronger support for the concerned split. However, the precise interpretation of bootstrap support values is not very clear. In the literature, at least three interpretations have been offered (see, e.g. Berry and Gascuel 1996).

The first is *confidence level* or *repeatability*. According to this interpretation, the bootstrap proportion P for a split is the confidence level for the split (Felsenstein 1985a). With a confidence interval for a conventional parameter, one imagines taking many samples from the same data-generating process that generated the original dataset. From each such dataset, one constructs a 95% confidence interval for the parameter, and then 95% of those confidence intervals will include the true value. Here 95% is called the *confidence level*. (See §6.2.3.1 and Figure 6.4 later in Chapter 6 for a discussion of confidence intervals.) Similarly we imagine here using the true evolutionary model and the true tree to simulate many datasets (sequence alignments), of which the original dataset is just one example, and estimating a tree from each dataset. Then a split with bootstrap support P in the original dataset will be found in those estimated trees with probability P. Note that there are some differences between the bootstrap support P and the confidence level. First in a conventional parameter estimation problem, the parameter is identified *a priori* and the method for constructing the confidence interval is given before the analysis of the data, while a split is often identified from the estimated tree, after the analysis of the data. Second, the confidence interval varies among the datasets with the confidence level fixed, while the split support P varies among datasets with the split fixed. Bootstrap support, interpreted as a confidence level, is called *repeatability* by Hillis and Bull (1993). Simulations of Hillis and Bull (1993), however, suggest that the bootstrap support for a split varies so much among replicate datasets that it has little value as a measure of repeatability.

A second interpretation is the *type-I error rate* or *false positive rate* in hypothesis testing; if a split has bootstrap proportion P, then $1 - P$ may be considered the *p*-value for the null hypothesis that the split is absent in the true tree (Felsenstein and Kishino 1993). Note that in conventional parameter estimation, there is no real difference between the confidence level and the *p*-value interpretations: the 95% confidence interval for a parameter θ

includes all values that are not rejected at the 5% level by hypothesis testing when compared with the MLE ($\hat{\theta}$). See, for example, Figure 1.7 for the construction of the confidence (likelihood) interval for the sequence distance under the JC69 model. However, the two interpretations are different in the context of phylogeny reconstruction. The p-value interpretation for bootstrap support was constructed by Felsenstein and Kishino (1993), using an analogy to model comparison involving the mean of a normal distribution, and was further discussed by Efron and Tibshirani (1998), again by analogy. It appears that those analogies do not adequately account for the complex nature of the parameter space in phylogeny reconstruction, since Susko (2009) has showed that $1 - P$ for a split with bootstrap support P is not a correct p-value to a first-order approximation. Here first-order correctness means that the interpretation is correct for large sequence lengths, a fairly weak requirement.

Susko (2010) has provided a correction so that the p-value interpretation is first-order correct for large sequence lengths. Even so there are at least two issues. First, there is ambiguity concerning the specification of the null hypothesis being tested (that a split is absent in the true tree). Both Felsenstein and Kishino (1993) and Susko (2010) represented the null hypothesis by fixing the branch length for the split to zero. It is not immediately obvious that this is the correct null hypothesis. It is true that in the test of the null hypothesis $H_0: \mu > 0$ against the alternative $H_1: \mu \leq 0$, the null H_0 is typically represented by $\mu = 0$ (for example, in the calculation of the likelihood). However, the space of different trees is more complex than different intervals of the same numerical parameter. For example, tree τ_2 in Figure 4.14 is not equivalent to tree τ_1 with negative branch length t_{10}. Nevertheless this null hypothesis appears to represent the *least favourable configuration* (Shimodaira and Hasegawa 1999): with this choice the type-I error will be controlled to be no larger than the allowed significance level, no matter what the null hypothesis is and what the parameter values in the null hypothesis are. A second issue is *selection bias* or *multiple testing*. One often does not have a split specified *a priori* for testing; one tests the significance of a split because it is in the estimated tree. Thus the null hypothesis is selected or identified from an analysis of the same data that is used to test the hypothesis. The problem is of the same nature as multiple comparisons or multiple testing. Susko's (2010) simulation suggests that correction for such selection bias can have a major impact.

Interpreted as a p-value for a pre-specified split, the bootstrap proportion tends to be conservative in general (Susko 2009). If one accepts splits with bootstrap support at least 95%, the false positive rate is in general lower than 5%. This conservative nature of the bootstrap (when viewed as a hypothesis testing procedure) appears to be due to the complex nature of the parameter space and to the choice of the null hypothesis as discussed above.

A third interpretation is *accuracy*; the bootstrap proportion is the probability that the tree or split is true given the data being analysed. This is a Bayesian interpretation and appears to be the one that most empirical phylogeneticists use or would like to use, perhaps in the same spirit that Frequentist confidence levels and p-values are often given Bayesian misinterpretations. Efron et al. (1996) have argued that bootstrap proportions could be interpreted as posterior probabilities under an uninformative prior. The concerned prior appears to be a uniform Dirichlet distribution on the probabilities for the 4^s site patterns (for s species). When this is translated into a prior on branch lengths, most of the prior probability mass is assigned to regions of the site pattern probabilities that correspond to trees with infinite branch lengths. Rather than being uninformative, this prior is biologically unreasonable. Yang and Rannala (2005) attempted to match the bootstrap proportions with posterior probabilities by adjusting the prior on branch lengths in the Bayesian analysis, and found that a match is in general unachievable. Based on simulation studies, many authors suggested that the bootstrap proportion, if interpreted as the probability that the split is correct, tends to be conservative. For example, Hillis and Bull (1993) found that in their simulations, bootstrap proportions of ≥70% usually correspond to a probability of ≥95% that the corresponding split is true. However, a number of authors have pointed out that this is not always the case; bootstrap proportions (interpreted as accuracy) can be either too conservative or too liberal (e.g. Efron et al. 1996; Yang and Rannala 2005; Susko 2010).

Several refinements to the bootstrap procedure of Felsenstein (1985a) have been suggested, including the complete-and-partial bootstrap of Zharkikh and Li (1995) and the modified method of Efron et al. (1996). Those methods involve more intensive computation and do not appear to have been used in real data analysis. Furthermore, they fail to correct for the first-order errors mentioned above (Susko 2009).

5.4.2 *Interior-branch test*

Another procedure for evaluating the significance of the estimated tree is to test whether the length of an interior branch is significantly greater than zero. This is known as the *interior-branch test*, and appears to have been first suggested by Felsenstein (1981) as a test of the reliability of the ML tree. An interior branch length not significantly greater than zero may be taken to mean that alternative branching patterns in that portion of the tree are not ruled out by the data. One can construct an LRT, calculating the log likelihood either with or without constraining the interior branch length to zero (Felsenstein 1988). In this case the null distribution of the test statistic (twice the log likelihood difference between the two models) is the 50:50 mixture of 0 and χ_1^2, as the branch length must be ≥0 and the value 0 is at the boundary of the parameter space (Self and Liang 1987).

The same test may be applied if a distance-based method is used for tree reconstruction. Nei et al. (1985) tested whether the estimated interior branch length in the UPGMA tree is significantly greater than zero by calculating the standard error of the estimated branch length and then applying a normal approximation. Li and Gouy (1991) and Rzhetsky and Nei (1992) discussed the use of this test when the tree is inferred without assuming the clock, for example, by using the NJ method.

The interior-branch test involves some difficulties. First, the hypothesis is not specified *a priori* and is instead derived from the data, since the ML or NJ tree is unknown until after the data are analysed (Sitnikova et al. 1995). Second, if one applies the same test to all interior branch lengths, multiple hypotheses are being tested using the same dataset, and a correction for multiple testing is called for (e.g. Li 1989). Third, the rationale of the test is not so straightforward; it is not obvious how testing an interior branch length on a tree that may be wrong should inform us about the reliability of the tree. Yang (1994c) considered the case of infinite data. With an infinite amount of data, one should expect all interior branch lengths in the true tree to be positive, but it is not true that interior branch lengths in a wrong tree are all zero. In this regard, there appear to be qualitative differences between ML and least-squares estimates of interior branch lengths in wrong trees. For the ML method, Yang (1994c) found the MLEs of interior branch lengths in wrong trees when $n \to \infty$ to be often strictly positive. Similarly in computer simulations, the interior branch length in each of the three trees for four species may be significantly positive (Tateno et al. 1994), casting doubt on the use of the positivity of the interior branch length as evidence for the significance of the ML tree. Correction for multiple testing is considered by Anisimova and Gascuel (2006) in their approximate likelihood ratio test (aLRT). For distance-based methods, Yang (1994c) found that in a few cases of small trees, all least-squares estimates of interior branch lengths on wrong trees are zero in infinite datasets (under the constraint that they are nonnegative). Sitnikova et al. (1995) allowed branch lengths to be negative and showed that the *expectations* of estimated interior branch lengths in a wrong tree in finite datasets can be positive, but that a tree with negative expected interior branch lengths must be wrong. Simulations of Sitnikova et al. (1995) suggest that the test of interior branch lengths provides a useful measure of accuracy for trees estimated using distance methods.

5.4.3 K-H test and related tests

Kishino and Hasegawa (1989) suggested an approximate test, known as the K-H test, for comparing two candidate phylogenetic trees in the likelihood framework. The log likelihood difference between the two trees is used as the test statistic and its approximate standard error is calculated by applying a normal approximation to the estimated log likelihoods. Suppose the two trees are 1 and 2. The test statistic is $\Delta = \ell_1 - \ell_2$. The log likelihood for tree 1 is

$$\ell_1 = \sum_{h=1}^{n} \log\{f_1(\mathbf{x}_h|\hat{\theta}_1)\}, \tag{5.13}$$

where $\hat{\theta}_1$ are the MLEs of branch lengths and other parameters for tree 1. The log likelihood ℓ_2 for tree 2 is defined similarly. Now let

$$d_h = \log(f_1(\mathbf{x}_h|\hat{\theta}_1)) - \log(f_2(\mathbf{x}_h|\hat{\theta}_2)) \tag{5.14}$$

be the difference between the two trees in the logarithm of the probability of data at site h, calculated at the MLEs. The mean is $\bar{d} = \Delta/n$. Kishino and Hasegawa (1989) suggest that the differences d_h are approximately i.i.d., so that the variance of $\bar{d}$ and thus of Δ can be estimated by the sample variance

$$\text{var}(\Delta) = \frac{n}{n-1} \sum_{h=1}^{n} (d_h - \bar{d})^2. \tag{5.15}$$

The two trees are significantly different if Δ is greater than twice (or 1.96 times) its standard error, $[\text{var}(\Delta)]^{1/2}$.

The K-H test is only valid if the two compared models or trees are specified beforehand. In molecular phylogenetics, the more common practice is to estimate the ML tree from the data and then test every other tree against the ML tree. As pointed out by Shimodaira and Hasegawa (1999) and Goldman et al. (2000), such use of the K-H test is not valid, tending to cause false rejections of non-ML trees. In other words, the test suffers from a *selection bias*, since the ML tree to be tested is identified from an analysis of the data, which are used to conduct the test. Shimodaira and Hasegawa (1999) developed a test, known as the S-H test, that corrects for selection bias. To ensure correct overall type-I error rates under all true models (true trees and parameter values), the test is constructed by considering the worst-case scenario or the least favourable condition, i.e. under the null hypothesis $E(\ell_1)/n = E(\ell_2)/n = \ldots$, with the likelihood defined for infinite data and the expectations taken over the true model. As a result, the S-H test is very conservative. Later, Shimodaira (2002) developed an approximately unbiased test or AU test, which is less conservative. The unbiasedness here simply means that the test is less conservative than the S-H test; the power (or the probability of rejecting the null when the null is false) of an unbiased test is higher than the significance level of the test. The AU test controls the overall type-I error rate in most but not all cases. The AU test is implemented in the CONSEL program (Shimodaira and Hasegawa 2001).

As pointed out by Shimodaira and Hasegawa (1999), the idea that underlies the K-H test was independently suggested by Linhart (1988) and Vuong (1989). The null hypothesis underlying the K-H and S-H tests is not entirely clear. In Vuong's (1989) version, the two models compared are both wrong and the null hypothesis is explicitly stated as $E(\ell_1)/n = E(\ell_2)/n$. Vuong also contrasted the test with Cox's (1961, 1962) test, in which the two compared models are treated in turn as the null hypothesis. In the context of tree comparison, one tree should be true so that the null hypothesis $E(\ell_1)/n = E(\ell_2)/n = \ldots$ appears unreasonable. Alternatively, one may consider one of the possible trees to be true in the null hypothesis, but the least favourable condition $E(\ell_1)/n = E(\ell_2)/n = \ldots$ is simply used to ensure that the test controls the overall type-I error rate under all possible model (tree) and parameter combinations.

5.4.4 Example: phylogeny of apes

We apply bootstrap to attach support values for clades on the ML tree inferred in subsection §4.2.4 for the mitochondrial protein sequences. The ML tree inferred under the MTMAM model, referred to as τ_1, is shown in Figure 5.5. The second best tree (τ_2) groups human and gorilla together, and the third tree (τ_3) groups the two chimpanzees with gorilla. These two trees are worse than the ML tree τ_1 by 35.0 and 38.2 log likelihood units, respectively (Table 5.1).

We generate 1,000 bootstrap datasets. Each dataset is analysed using an exhaustive tree search under the MTMAM model, and the resulting ML tree is recorded. Thus 1,000 bootstrap trees are generated from the 1,000 bootstrap datasets. We then find, for every split on τ_1, the percentage of the bootstrap trees that include that split, and present it as the bootstrap support for it. The support value for one split is 99.4%, while all others are 100%. Similarly, the bootstrap proportions for trees τ_1, τ_2, and τ_3 are 99.4%, 0.3% and 0.3%, respectively, and are ~0% for all other trees (see Figure 5.5).

The RELL approximation to the bootstrap uses the parameter estimates obtained from the original dataset to calculate the log likelihood values for the bootstrap samples. Application of this approach to evaluate all 945 possible trees leads to approximate bootstrap proportions 98.7%, 1.0%, and 0.3% for trees τ_1, τ_2, and τ_3, respectively, and 0% for all other trees. Thus the RELL bootstrap provides a good approximation to the bootstrap in this analysis.

Use of the K-H test leads to rejection of all trees at the 1% level except for τ_2, for which the p-value is 1.4%. As discussed above, the K-H test fails to correct for multiple

Table 5.1 Tests of trees for the mitochondrial protein dataset

Tree	ℓ_i	$\Delta\ell_i - \ell_{max}$	SE	K-H	S-H	Bootstrap	RELL
τ_1: ((human, chimps), gorilla)	−14,558.6	0	0	NA	NA	0.994	0.987
τ_2: ((human, gorilla), chimps)	−14,593.6	−35.0	16.0	0.014	0.781	0.003	0.010
τ_3: ((chimps, gorilla), human)	−14,596.8	−38.2	15.5	0.007	0.754	0.003	0.003

Note: p-values are shown for the K-H and S-H tests, while bootstrap proportions are shown for the bootstrap (Felsenstein 1985a) and RELL approximate bootstrap (Kishino and Hasegawa 1989).

comparisons and tends to be overconfident. The S-H test is far more conservative; at the 1% level, it fails to reject 27 trees when compared with the ML tree.

5.4.5 Indexes used in parsimony analysis

In parsimony analysis, several indexes have been suggested to measure split support levels or, more vaguely, 'phylogenetic signal'. They do not have a straightforward statistical interpretation, but are mentioned here as they are often reported in parsimony analyses.

5.4.5.1 Decay index

The *decay index*, also called *branch support* or *Bremer support*, is the difference between the tree length of the globally most parsimonious tree and the tree length achievable among trees that do not have a particular split (Bremer 1988). It is the 'cost', in terms of the extra number of changes, incurred by removing a particular split on the most parsimonious tree. In general, Bremer support does not have a clear-cut statistical interpretation; its statistical significance depends on other factors such as the size of the tree, the tree length or overall sequence divergence, and the evolutionary model (e.g. Cavender 1978; Felsenstein 1985c; Lee 2000; DeBry 2001; Wilkinson et al. 2003).

5.4.5.2 Winning-sites test

The winning-sites test is used to compare two trees. For every site, one can score which of the two compared trees has the shorter character length and thus 'wins'. One can represent the outcome by '+' or '−' with '0' representing a tie. A binomial distribution may be used to test whether the numbers of '+' and '−' deviate significantly from 1:1. A test of this sort was discussed by Templeton (1983) for restriction site data. Note its similarity to the K-H test for the likelihood method. In theory, the binomial is not the correct distribution, since the total number of '+' and '−' is not fixed. Instead the trinomial distribution $M_3(n, p_+, p_-, p_0)$ may be used to test the null hypothesis $p_+ = p_-$, with all sites in the sequence used, including constant and parsimony-noninformative sites. The results are nevertheless expected to be very similar to those from the binomial test. To extend the test to more than two trees, one has to account for the problem of multiple comparisons.

5.4.5.3 Consistency index and retention index

The consistency index for a character measures the 'fit' of the character to a tree. It is defined as m/s, the minimum possible number of changes on any tree (m) divided by the step length of the current tree (s) (Kluge and Farris 1969; Maddison and Maddison 1982). The consistency index for the whole dataset for a tree is defined as $\sum m_i / \sum s_i$, with the sum taken over characters (sites). If the characters are perfectly congruent with each other and with the tree, the consistency index will be one. If there is a lot of convergent evolution or homoplasy, the consistency index will be close to zero.

The retention index for a dataset is defined as $\sum (M_i - s_i) / \sum (M_i - m_i)$, where m_i and s_i are defined above while M_i is the maximum conceivable number of steps for character i on any tree (Farris 1989). Like the consistency index, the retention index also ranges from 0 to 1, with 0 meaning a lot of homoplasies and 1 meaning perfect congruence among characters and between characters and the tree.

For molecular data, homoplasy is not a good indicator of phylogenetic information content in the dataset even for parsimony, and the consistency index and retention index are not very useful despite their pleasant-sounding names. For example, Figure 5.6 shows a simulation study on a four-species tree. The probability that parsimony recovers the true

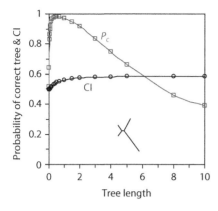

Fig. 5.6 The probability of recovering the correct tree by parsimony (P_c) and the consistency index of the most parsimonious tree (CI), plotted against the true tree length. Both P_c and CI range from 0 to 1, so the same y-axis is used for both. Data are simulated using the tree topology of Figure 5.1a: $((a:0.5, b:0.5):0.1, c:0.6, d:1.4)$, but all five branch lengths are multiplied by a constant so that the tree length (the expected number of changes per site on the tree) is as specified.

tree increases very rapidly with the increase of sequence divergence when the sequences are highly similar, peaks at an optimal divergence level, and then drops slowly with further increase in sequence divergence. The consistency index for the most parsimonious tree, however, increases very slowly over the whole range of the sequence divergence, and thus tells us nothing about how successful parsimony is in recovering the true tree.

5.5 Problems

5.1 Conduct a computer simulation to examine the efficiency of tree reconstruction methods. Use SEQ-GEN or EVOLVER to generate 1,000 datasets, each of 1,000 sites, under the JC69 model, and then construct the maximum parsimony and maximum likelihood trees from each to calculate the probability of recovering the correct tree. You can use PHYLIP or some other programs for tree reconstruction. Use three different shapes of four-species trees:

   i. $((a:0.1, b:0.1):0.05, c:0.1, d:0.1)$;
   ii. $((a:0.01, b:0.01):0.01, c:0.05, d:0.05)$;
   iii. $((a:0.01, b:0.05):0.01, c:0.01, d:0.05)$.

5.2 Conduct a computer simulation to examine the robustness of tree reconstruction methods to transition–transversion rate difference. Redo the simulation of Problem 5.1, but simulate the data under the K80 model with transition/transversion rate ratio $\kappa = 5$. Analyse the data using three methods: (i) parsimony, (ii) ML assuming JC69, and (iii) ML assuming K80 (with κ estimated).

5.3 Conduct a computer simulation to examine the robustness of tree reconstruction methods to rate variation among sites. Redo the simulation of Problem 5.1, but simulate the data under the JC69 + Γ_5 model with the gamma shape parameter $\alpha = 0.5$. Analyse the data using three methods: (i) parsimony, (ii) ML assuming JC69, and (iii) ML assuming JC69 + Γ_5 (with α fixed at 0.5).

5.4 Conduct a computer simulation to examine the relative performance of parsimony and likelihood methods of tree reconstruction when the sequences are highly divergent. Redo the simulation of Problem 5.1 using trees (ii) and (iii), with all branch lengths in the trees multiplied by 5.

CHAPTER 6

Bayesian theory

6.1 Overview

There are two principal philosophies in statistical data analysis: the *classical* or *Frequentist* and the *Bayesian*. They use different definitions of probability. The *Frequentist* defines the probability of an event as the expected frequency of occurrence of that event in repeated random draws from a real or imaginary population. When we say that the probability of heads for a coin toss is $1/2$, we mean that the frequency of heads will approach $1/2$ when the number of tosses is large. The performance of an inference procedure is judged by its properties in repeated sampling from the data-generating model (i.e. the likelihood model), with the parameters fixed. Important concepts include bias and variance of an estimator, confidence intervals, and p-values; these are all covered in a typical biostatistics course. Maximum likelihood (ML) estimation and likelihood ratio test (LRT) figure prominently in classical statistics; for example, t tests of population means, the χ^2 test of association in a contingency table, analysis of variance, linear correlation, and regression are either likelihood methods or their approximations.

In Bayesian statistics, probability is defined to represent one's *degree of belief* rather than a frequency. When we say that a hypothesis (e.g. that the extinction of the dinosaurs was caused by a meteorite hitting the earth) has probability 0.9, we mean that the hypothesis is very likely to be true, judged by currently available evidence. A key feature of Bayesian statistics is the use of probability distributions to describe uncertainties in the parameter. In classical statistics, only random variables have distributions while parameters are fixed although unknown constants and cannot have distributions. The distribution of the parameter before the data are collected and analysed is called the *prior distribution*, while the distribution of the parameter after the data are analysed is called the *posterior distribution*, which combines the information in the prior and information in the data.

While probability theory has been a subject of study for several hundred years, most conspicuously as it is related to games of chance (gambling), statistics is a much younger field. Concepts of regression and correlation were invented by Francis Galton and Karl Pearson around 1900 in studies of human inheritance. The field blossomed in the 1920s and 1930s when Ronald A. Fisher developed many of the techniques of classical statistics, such as analysis of variance, experimental design, and likelihood. The theories of hypothesis testing and confidence intervals, including such concepts as simple and composite hypotheses, type-I and type-II errors, etc., were developed by Jerzy Neyman and Egon Pearson (Karl Pearson's son) at about the same time. These contributions completed the foundation of classical statistics.

In contrast, Bayesian ideas are much older, dating to an essay published posthumously by Thomas Bayes in 1763, in which he calculated the posterior distribution of a binomial probability, using a $U(0, 1)$ prior. The idea was developed further by Pierre-Simon

Molecular Evolution: A Statistical Approach. Ziheng Yang. © Ziheng Yang 2014.
Published 2014 by Oxford University Press.

Laplace and others in the 19th century but was not popular among statisticians of the early 20th century. There are two major reasons for this lack of acceptance. The first is philosophical: the method's reliance and indeed insistence on a prior for unknown parameters was fiercely criticized by prominent statisticians, such as R.A. Fisher. The second is computational: except for simple toy examples that are analytically tractable, computation of posterior probabilities in practical applications almost always requires numerical calculation of integrals, often high-dimensional integrals. Nowadays, controversies concerning the prior persist, but Bayesian computation has been revolutionized by Markov chain Monte Carlo (MCMC) algorithms (Metropolis et al. 1953; Hastings 1970). Interestingly, the power of MCMC was not generally appreciated until around 1990; for example, Gelfand and Smith (1990) demonstrated the feasibility of such algorithms to solve nontrivial practical problems (see, e.g. Tanner and Wong 2000). MCMC has made it possible to implement sophisticated, parameter-rich models, for which the likelihood analysis would not be feasible. Since the 1990s, Bayesian inference has been applied to various scientific areas. Recent years have seen the overzealous excitement about the method subsiding to a more reasonable level; partly the method is now too popular to need any more promotion, and partly frustrated programmers have started to appreciate the complexities and difficulties of implementing and validating MCMC algorithms.

This chapter provides an overview of the theory and computation of Bayesian statistics. There are now many excellent textbooks on Bayesian theory (e.g. Leonard and Hsu 1999; O'Hagan and Forster 2004) and computation (e.g. Gilks et al. 1996; Robert and Casella 2004). In this chapter and the next I will use simple examples (toy examples in statistics and distance estimation under the JC69 model) to introduce the general principles. The biologist reader may wish to read §6.1–6.2 and §6.3.1, and then jump to Chapter 7 on MCMC. Chapter 8 discusses the application of Bayesian inference to phylogeny reconstruction, and Chapter 9 covers Bayesian computational methods in population genetics and phylogeography under the coalescent model. Another application, that of dating species divergences under the clock and relaxed clock models, is discussed in Chapter 10.

6.2 The Bayesian paradigm

6.2.1 *The Bayes theorem*

Suppose the occurrence of a certain event B may depend on whether another event A occurs. Then the probability that B occurs is given by the *law of total probability* as

$$P(B) = P(AB) + P(\bar{A}B) = P(A) \times P(B|A) + P(\bar{A}) \times P(B|\bar{A}). \tag{6.1}$$

Here $\bar{A}$ stands for 'non A' or 'A does not occur', and AB stands for the event that both A and B occur. Bayes' theorem, also known as the *inverse-probability theorem*, gives the conditional probability that A occurs given that B occurs:

$$P(A|B) = \frac{P(AB)}{P(B)} = \frac{P(A) \times P(B|A)}{P(B)} = \frac{P(A) \times P(B|A)}{P(A) \times P(B|A) + P(\bar{A}) \times P(B|\bar{A})}. \tag{6.2}$$

Equations (6.1) and (6.2) should be quite obvious when one examines the Venn diagram (Figure 6.1). We illustrate the theorem using an example.

Example 6.1. False positives of a clinical test. Suppose a new clinical test has been developed to screen for an infection in the population. If a person has the infection, the test accurately reports a positive 99% of the time, and if a person does not have the infection, the

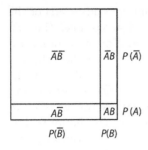

Fig. 6.1 A Venn diagram to illustrate the law of total probability (equation (6.1)) and Bayes' theorem (equation (6.2)). The square has a total area of 1 corresponding to the total probability. We have $P(B) = P(AB) + P(\bar{A}B)$, and the conditional probability $P(A|B)$ is the ratio of the area of AB to the area of B.

test falsely reports a positive only 2% of the time. The test appears to be quite reliable with low false positives and low false negatives. Suppose that 0.1% of the population have the infection. What is the probability that a person who has tested positive actually has the infection?

Let A be the event that a person has the infection, and $\bar{A}$ no infection. Let B stand for test positive. Then $P(A) = 0.001, P(\bar{A}) = 0.999, P(B|A) = 0.99, P(B|\bar{A}) = 0.02$. The probability that a random person from the population tests positive is, according to equation (6.1),

$$P(B) = 0.001 \times 0.99 + 0.999 \times 0.02 = 0.02097. \tag{6.3}$$

This is close to the proportion among the noninfected individuals of the population. Equation (6.2) then gives the probability that a person who has tested positive has the infection as

$$P(A|B) = \frac{0.001 \times 0.99}{0.02097} = 0.0472. \tag{6.4}$$

Thus among the positives, only 4.72% are true positives while 95.28% (= 1 − 0.0472) are false positives. Despite the apparent high accuracy of the test, the prevalence of the infection is so low (0.1%) that most people who test positive are healthy without the infection. □

6.2.2 The Bayes theorem in Bayesian statistics

When the Bayes theorem is used in Bayesian statistics, A and $\bar{A}$ correspond to different hypotheses H_1 and H_2, while B corresponds to the observed data (X). The Bayes theorem then specifies the conditional probability of hypothesis H_1 given the data as

$$P(H_1|X) = \frac{P(H_1) \times P(X|H_1)}{P(X)} = \frac{P(H_1) \times P(X|H_1)}{P(H_1) \times P(X|H_1) + P(H_2) \times P(X|H_2)}. \tag{6.5}$$

Here $P(H_1)$ and $P(H_2)$ are called *prior probabilities*. They are probabilities assigned to the hypotheses before the data are observed or analysed. The conditional probabilities $P(H_1|X)$ and $P(H_2|X) = 1 - P(H_1|X)$ are called *posterior probabilities*. $P(X|H_1)$ and $P(X|H_2)$ are the likelihoods under the two hypotheses. The extension to more than two hypotheses is obvious.

There are two philosophies of Bayesian statistics, i.e. the *objective* and *subjective* Bayesian, which differ in their interpretation of probability distributions. In objective Bayesian, probability represents the researcher's *rational* degree of belief, while in subjective Bayesian, probability represents the researcher's *personal* degree of belief. The interpretation applies to all probability distributions including the prior and posterior distributions. Even though the philosophical interpretations differ, the laws of probability are the same in classical statistics and in objective and subjective Bayesian statistics.

Much effort has been taken by objective Bayesians (Pierre-Simon Laplace, Harold Jeffreys, etc.) to represent ignorance, in the so-called noninformative priors. Nowadays it is generally accepted that this is an unachievable goal. Uniform priors are not noninformative and truly noninformative priors do not exist.

The difficulty of objective Bayesians to represent ignorance prompted the rise of *subjective Bayesian*, developed by Leonard Jimmie Savage, Bruno de Finetti, etc. This appears now to be the more popular version of Bayesian statistics. Nevertheless, most biologists are dismayed at the suggestion that a Bayesian analysis of their data calls for a psychoanalytical assessment of their personal beliefs, and that the conclusion drawn from their experiment may be influenced by subjective beliefs; to them the objective Bayesian idea of letting the data dominate the inference is appealing. In practice, it is important to evaluate the impact of the prior on the posterior in a Bayesian robustness analysis.

Note that in the infection-testing example discussed above, $P(A)$ and $P(\bar{A})$ are frequencies of infected and noninfected individuals in the population. There is no controversy concerning the use of Bayes' theorem in such situations. However, in Bayesian statistics, the prior probabilities $P(H_1)$ and $P(H_2)$ most often do not have such a frequency interpretation. The use of Bayes' theorem in such a context is controversial. We will return to this in the next subsection.

When the hypothesis concerns unknown continuous parameters, probability densities are used instead of probabilities. The Bayes theorem then takes the following form:

$$f(\theta|X) = \frac{f(\theta)f(X|\theta)}{f(X)} = \frac{f(\theta)f(X|\theta)}{\int f(\theta)f(X|\theta)\,d\theta}. \tag{6.6}$$

Here $f(\theta)$ is the *prior distribution*, $f(X|\theta)$ is the likelihood (the probability of data X given the parameter θ), and $f(\theta|X)$ is the *posterior distribution*. The *marginal probability* of the data, $f(X)$, is a normalizing constant, to make $f(\theta|X)$ integrate to one. Equation (6.6) thus says that the posterior is proportional to the prior times the likelihood, or equivalently, the posterior information is the sum of the prior information and the sample information. In the following, we focus on the continuous version, but take it for granted that the theory applies to the discrete version as well. Also, when the model involves more than one parameter, θ will be a vector of parameters.

The posterior distribution is the basis for all inference concerning θ. For example, the mean, median, or mode of the distribution can be used as the point estimate of θ. For interval estimation, one can use the 2.5% and 97.5% quantiles of the posterior density to construct the 95% *equal-tail credibility interval* (CI) (Figure 6.2a). When the posterior density is skewed or has multiple peaks, this interval has the drawback of including values of θ that are less supported than values outside the interval. One can then use the 95% *highest posterior density* (HPD) interval, which includes values of θ of the highest posterior density that encompass 95% of the density mass. In the example of Figure 6.2b, the HPD region consists of two disconnected intervals. Compared with the equal-tail CI, the HPD CI has two advantages:

i. every point inside the interval has higher density than every point outside the interval.
ii. For any given probability level (say, $1 - \alpha$), the HPD interval has the shortest length.

Nevertheless, if the data are informative and the posterior is nearly symmetrical, the two intervals should be very similar.

186   6 BAYESIAN THEORY

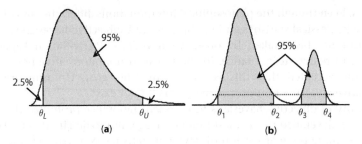

Fig. 6.2 (a) The 95% equal-tail credibility interval (θ_L, θ_U) is constructed by locating the 2.5% and 97.5% quantiles of the posterior probability distribution. (b) The 95% HPD interval includes values of θ that have the highest density and that cover 95% of the probability mass. With multiple peaks in the density, the HPD region may consist of disconnected intervals: (θ_1, θ_2) and (θ_3, θ_4).

An important strength of the Bayesian approach is that it provides a natural way of dealing with nuisance parameters, through *marginalization*. As discussed in §1.4.4, nuisance parameters are hard to deal with in the likelihood method. Let $\theta = (\lambda, \eta)$, with λ to be the parameters of interest and η the nuisance parameters. The joint posterior density of λ and η is

$$f(\lambda, \eta | X) = \frac{f(\lambda, \eta) f(X | \lambda, \eta)}{f(X)} = \frac{f(\lambda, \eta) f(X | \lambda, \eta)}{\int f(\lambda, \eta) f(X | \lambda, \eta) \, d\lambda \, d\eta}, \quad (6.7)$$

from which the (marginal) posterior density of λ can be obtained as

$$f(\lambda | X) = \int f(\lambda, \eta | X) \, d\eta. \quad (6.8)$$

This approach of dealing with nuisance parameters η by integrating them out is called marginalization. By averaging over the prior distribution on η, the approach accommodates uncertainties in the nuisance parameters.

Below we apply the Bayes theorem to a few simple examples.

Example 6.2. Estimation of the binomial probability.   Let x be the number of successes in a series of n trials with probability of success in each to be θ. Given θ, x has the binomial probability

$$f(x|\theta) = \binom{n}{x} \theta^x (1-\theta)^{n-x}, x = 0, 1, \ldots, n. \quad (6.9)$$

Now let θ have a beta prior distribution, $\theta \sim \text{beta}(a, b)$, with density

$$f(\theta) = \frac{1}{B(a,b)} \theta^{a-1} (1-\theta)^{b-1}, 0 < \theta < 1, \quad (6.10)$$

for some $a, b > 0$. The beta distribution has mean $a/(a+b)$ and variance $ab/[(a+b)^2(a+b+1)]$. Here

$$B(a,b) = \int_0^1 t^{a-1}(1-t)^{b-1} dt = \frac{\Gamma(a)\Gamma(b)}{\Gamma(a+b)}, a, b > 0, \quad (6.11)$$

is the beta function and $\Gamma(a)$ is the gamma function (see equation (1.34)). The marginal likelihood is thus

$$f(x) = \int_0^1 f(\theta)f(x|\theta)\,d\theta$$

$$= \binom{n}{x}\frac{B(x+a, n-x+b)}{B(a,b)} \times \int_0^1 \frac{1}{B(x+a, n-x+b)}\theta^{x+a-1}(1-\theta)^{n-x+b-1}d\theta \quad (6.12)$$

$$= \binom{n}{x}\frac{B(x+a, n-x+b)}{B(a,b)}.$$

The Bayes theorem then gives the posterior distribution as

$$f(\theta|x) = \frac{f(\theta)f(x|\theta)}{f(x)} = \frac{1}{B(x+a, n-x+b)}\theta^{x+a-1}(1-\theta)^{n-x+b-1}. \quad (6.13)$$

This is beta $(x+a, n-x+b)$ (cf.: equation (6.10)). A comparison of the prior and the posterior suggests that the information contained in the prior beta (a, b) is equivalent to observing a successes in $a+b$ trials. □

Example 6.3. Laplace's rule of succession. Suppose an event has occurred in x of the past n trials. What is the probability that it will occur in the next trial?

We use the uniform prior for the probability of the event: $\theta \sim U(0, 1)$, which is a special case of beta (a, b) with $a = b = 1$. From Example 6.2, the posterior is $\theta|x \sim$ beta $(x+1, n-x+1)$. The probability that the next trial will be a success is simply the posterior mean, $E(\theta|x) = (x+1)/(n+2)$. In other words, let $Y_i = 1$ if the ith trial is a success or 0 if it is a failure. Then we have

$$\Pr\{Y_{n+1} = 1 | Y_1 + Y_2 + \cdots + Y_n = x\} = \frac{x+1}{n+2}. \quad (6.14)$$

This is known as the *rule of succession*, due to Laplace.

In particular, if the event has occurred in each of the past n trials, the probability that it will occur in the next trial is $(n+1)/(n+2)$. Laplace used this formula to calculate the probability that the sun will rise the next day, given that it has risen every day for the past 5000 years: this is $1 - 1/(5000 \times 365.25 + 2) \approx 1 - 5.5 \times 10^{-7}$. This calculation then became a matter of ridicule as well as a focus of many religious, philosophical, and statistical discussions! Note that underlying the calculation is the prior for the probability that the sun will rise tomorrow: $\theta \sim U(0, 1)$. □

It may be fitting to discuss the justifications for the uniform prior on binomial probability θ here. The $U(0, 1)$ prior on θ was used in the original Essay of Thomas Bayes (1763), who intended his method to apply to any case where we 'absolutely know nothing' about θ. The uniform prior is often used to represent total ignorance based on the so-called *principle of insufficient reason* or *principle of indifference*, due to Laplace: if we know nothing about θ, it should have equal chance of being in any equi-width interval, hence the uniform distribution. This reasoning is flawed, however, as the prior is not invariant to nonlinear transforms. As pointed out by Edgeworth (1885), if θ is unknown, so is any monotone function of θ, say $\phi = \frac{1}{2}\cos^{-1}(1-2\theta)$. Yet treating ϕ as uniformly distributed yields a very different answer to treating θ as uniformly distributed. According to the introduction of Richard Price, who read Bayes' essay in the Royal Society on 23 December 1763, Bayes initially assumed the uniform distribution on θ as a postulate, but afterwards considered it 'might not perhaps be looked upon by all as reasonable'. He therefore wrote a separate section (a scholium), to justify the prior using an analogy with what is now described as a 'billiard table'. He reasoned that if we know nothing at all about θ, then if we take n trials we should not prefer one number of successes ($x = 0, 1, \ldots, n$) to any other; in other words, the $(n + 1)$ possible numbers of successes in n trials should each occur with probability $1/(n + 1)$. The 'prior of ignorance' $\pi(\theta)$ should thus satisfy

$$\int_0^1 \pi(\theta) \times \binom{n}{x} \theta^x (1-\theta)^{n-x} d\theta = \frac{1}{n+1}, \tag{6.15}$$

for all n. The prior that meets this requirement is the uniform $\theta \sim U(0, 1)$. This is in the same spirit as the principle of insufficient reason, which assigns uniform probabilities on the parameter. Edwards (1974) pointed out that Bayes' argument is equally arbitrary. He asks: 'Why not equal probability for each of the 2^n sequences instead? That would lead to the fixed prior value $\theta = \frac{1}{2}$ and all our inference problems would be over.' Stigler (1982) considered Bayes' argument, based on observable data instead of an unobservable parameter, to carry more force than the principle of insufficient reason. He appears to have missed Edwards' point that the data outcome may be represented in different ways, which would imply different priors on θ, in the same way that different choices of parameters imply different priors.

It is unclear whether Bayes was aware of this criticism. Some writers (e.g. Fisher) suggest that Bayes had misgivings about his use of the uniform prior, as he placed his justification in a separate section to the main argument of the paper. It seems that we will never know whether Bayes is truly Bayesian.

Example 6.4. Bayesian estimation of sequence distance under the JC69 model. Consider the use of the JC69 model to estimate the distance θ between the human and orangutan 12S rRNA genes from the mitochondrial genome (see Example 1.5 in §1.4.1). The data are summarized as $x = 90$ differences out of $n = 948$ sites. The MLE was found to be $\hat{\theta} = 0.1015$, with the 95% confidence (likelihood) interval to be $(0.0817, 0.1245)$. To apply the Bayesian approach, we have to specify a prior. We use an exponential prior

$$f(\theta) = \frac{1}{\mu} e^{-\theta/\mu}, \tag{6.16}$$

with mean $\mu = 0.2$. The posterior distribution of θ is

$$f(\theta|x) = \frac{f(\theta)f(x|\theta)}{f(x)} = \frac{f(\theta)f(x|\theta)}{\int_0^\infty f(\theta)f(x|\theta)\, d\theta}. \tag{6.17}$$

We consider the data to have a binomial distribution, with probability

$$p = \frac{3}{4} - \frac{3}{4} e^{-4\theta/3}, 0 \le p \le \frac{3}{4}, \tag{6.18}$$

for a difference and $1 - p$ for an identity. The likelihood is thus

$$f(x|\theta) = p^x (1-p)^{n-x} = \left(\frac{3}{4} - \frac{3}{4} e^{-4\theta/3}\right)^x \left(\frac{1}{4} + \frac{3}{4} e^{-4\theta/3}\right)^{n-x}. \tag{6.19}$$

This is the same as the likelihood of equation (1.47) except for a scale constant $4^n/3^x$, which cancels in calculation of the posterior (equation (6.17)). The integral in the denominator of equation (6.17) is calculated numerically, to be $f(x) = 5.16776 \times 10^{-131}$ (see §6.4). Figure 6.3 shows the posterior density, plotted together with the prior and likelihood. In this case the posterior is dominated by the likelihood, and the prior is nearly flat at

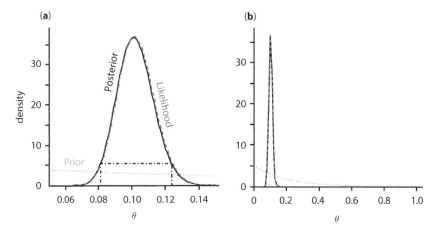

Fig. 6.3 Prior and posterior densities for sequence distance θ under the JC69 model. The likelihood is shown as well, rescaled to match up the posterior density. Plots (a) and (b) show the same three curves, with different scales for the θ axis. Note that the posterior density is the prior density times the likelihood, followed by a change of scale to make the area under the posterior density curve equal to one. The data are the human and orangutan mitochondrial 12S rRNA genes, with $x = 90$ differences at $n = 948$ sites. The 95% HPD interval, (0.08116, 0.12377), is indicated in (a).

the neighbourhood of the peak of the likelihood. By numerical integration the mean of the posterior distribution is found to be $E(\theta|x) = \int_0^\infty \theta f(\theta|x) \, d\theta = 0.10213$, with standard deviation 0.01091. The mode is at $\theta = 0.10092$. The 95% equal-tail CI can be constructed by calculating numerically the 2.5% and 97.5% percentiles of the posterior density, to be (0.08191, 0.12463). This is very similar to the confidence interval (0.0817, 0.1245) in the likelihood analysis, despite their different interpretations (see below). The 95% HPD interval, (0.08116, 0.12377), is found by lowering the posterior density function from the maximum (at 36.7712) until the resulting interval encompasses 95% of the density mass (Figure 6.3). As the posterior density has a single peak and is nearly symmetrical, the HPD and the equal-tail intervals are nearly identical. □

*6.2.3 Classical versus Bayesian statistics

There is a large body of statistics literature about the controversy between the Frequentist/likelihood methods and Bayesian methods. See, for example, Lindley and Phillips (1976) for a Bayesian introduction and Efron (1986) from the classical viewpoint. In this subsection, I will provide a summary of the major criticisms from both schools.

6.2.3.1 Criticisms of Frequentist statistics

A major Bayesian criticism of Frequentist statistics is that it does not answer the right question. Frequentist methods provide probability statements about the data or the method for analysing the data, but not about the parameter, even though the data have already been observed and our interest is in the parameter. We illustrate these criticisms using the concepts of confidence intervals and p-values. These concepts are known to be confusing; in molecular phylogenetics, published misinterpretations are embarrassingly common.

---
* indicates a more difficult or technical section.

Consider first the confidence interval. Suppose the data are a sample $(x_1, x_2, \ldots, x_n)$ from the normal distribution $N(\mu, \sigma^2)$, with unknown parameters μ and σ^2. Provided n is large (say, > 50), a 95% confidence interval for μ is

$$(\bar{x} - 1.96s/\sqrt{n}, \bar{x} + 1.96s/\sqrt{n}), \quad (6.20)$$

where $s = \left[\sum (x_i - \bar{x})^2/(n-1)\right]^{1/2}$ is the sample standard deviation. What does this mean? Many of us would want to say that a 95% confidence interval means that there is a 95% chance that the interval includes the true parameter value. But this interpretation is wrong.

To appreciate the correct interpretation, consider the following artificial example. Suppose we take two random draws x_1 and x_2 from a discrete distribution $f(x|\theta)$, in which the random variable x takes two values $\theta - 1$ and $\theta + 1$, each with probability $\frac{1}{2}$. The unknown parameter θ is in the range $-\infty < \theta < \infty$. Then

$$\hat{\theta} = \begin{cases} (x_1 + x_2)/2, & \text{if } x_1 \neq x_2, \\ x_1 - 1, & \text{if } x_1 = x_2 \end{cases} \quad (6.21)$$

defines a 75% confidence set for θ. Here we consider $\hat{\theta}$ as an interval or set even though it contains a single value. Let $a = \theta - 1$ and $b = \theta + 1$ be the two values that x_1 and x_2 can take. There are four possible data outcomes, each occurring with probability $\frac{1}{4}$: aa, bb, ab, and ba. Only when the observed data are aa does $\hat{\theta}$ differ from θ. Thus in 75% of the datasets, the confidence set $\hat{\theta}$ includes (equals) the true parameter value. In other words, $\hat{\theta}$ is a 75% confidence set for θ. The probability 75% is called the *confidence level* or *coverage probability*. If we sample repetitively from the data-generating model to generate many datasets and construct a confidence interval $\hat{\theta}$ for each dataset according to equation (6.21), then 75% of those confidence intervals will include the true parameter value.

Now suppose in the observed dataset, the two values x_1 and x_2 are distinct (say, $x_1 = 1.0$ and $x_2 = 3.0$). We then know for sure that $\hat{\theta} = \theta$ (or the true θ must be 2). It appears counterintuitive, if not absurd, for the confidence set to have only 75% coverage probability when we know that it includes the true parameter value. At any rate, the confidence level (which is 0.75) is not the probability that the interval we calculated includes the true parameter value (which is 1 for the observed data). One can also construct examples in which a 95% confidence interval clearly does not include the true parameter value.

To return to the normal distribution example, the confidence interval of equation (6.20) can be interpreted correctly in two ways. The first is based on repeated sampling (Figure 6.4). With μ and σ^2 fixed, imagine taking many samples (of the same size as the

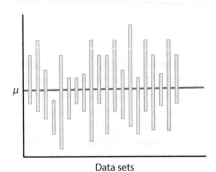

Fig. 6.4 Interpretation of the confidence interval. Many datasets are generated from the probability model $x_i \sim N(\mu, \sigma^2)$, with parameters μ and σ^2 fixed. A 95% confidence interval is constructed for μ in each dataset. Then 95% of those confidence intervals will include the true μ whatever the true value of μ is.

observed data) from the same population that the observed data are from, and constructing a confidence interval for each sample according to equation (6.20). Then 95% of those confidence intervals will contain the true parameter value μ. Note that the confidence intervals vary among datasets while parameters μ and σ^2 are fixed. The second interpretation is that confidence intervals are *pre-trial betting* (as opposed to *after-trial evaluations*) (Hacking 1965); before the data are collected, the confidence interval we will construct using equation (6.20) will include the true parameter value with probability 0.95. This statement is true whatever the true value of μ is. However, after the data were observed, the 95% confidence interval we did construct using the said procedure may or may not include the true value, and in general no probability (except 0 or 1) can be attached to the event that our interval includes the true parameter value. The criticism is that the confidence interval theory dodges the question of what information is available about the parameter and makes a roundabout probability statement about the data or the procedure instead; such a statement does not give much consolation to the biologist, who is interested not in the procedure of constructing confidence intervals but in the parameter and the particular confidence interval constructed from her data. In contrast, the Bayesian credibility interval gives a straightforward answer to this question. Given the data, a 95% Bayesian credibility interval includes the true parameter value with probability 95%. Bayesian methods are *post-trial* evaluations, making probabilistic statements about the parameters conditioned on the observed data.

Next we turn to the *p*-value in the Neyman–Pearson framework of hypothesis testing. We use the example of Problem 1.9 of Chapter 1, constructed by Lindley and Phillips (1976). Suppose $x = 9$ heads and $r = 3$ tails are observed in $n = 12$ independent tosses of a coin, and we wish to test the null hypothesis H_0: $\theta = 1/2$ against the alternative H_1: $\theta > 1/2$, where θ is the probability of heads. Suppose the number of tosses n is fixed, so that x has a binomial distribution, with probability

$$P_\theta(x) = \binom{n}{x} \theta^x (1-\theta)^{n-x}, x = 0, 1, \ldots, n. \tag{6.22}$$

The probability distribution under the null hypothesis is shown in Figure 6.5a. Note that the probability of the observed data $x = 9$ under H_0, which is 0.05371, is not the *p*-value. When there are many possible data outcomes, the probability of observing one particular dataset, even if it is the most likely data outcome, can be extremely small. Instead the *p*-value is a sum over all data outcomes that are at least as extreme as the observed data

$$p = P_{\theta=1/2}(x=9) + P_{\theta=1/2}(x=10) + P_{\theta=1/2}(x=11) + P_{\theta=1/2}(x=12) = 0.073, \tag{6.23}$$

where the subscript means that the probability is calculated under the null hypothesis $\theta = 1/2$. One is tempted to interpret the *p*-value as the probability that H_0 is true, but this interpretation is incorrect.

Now consider a different experimental design, in which the coin was tossed until we saw $r = 3$ tails, at which point $x = 9$ heads were observed. Then x has a negative binomial distribution, with probability (Figure 6.5b)

$$P_\theta(x) = \binom{r+x-1}{x} \theta^x (1-\theta)^r, x = 0, 1, 2, \ldots. \tag{6.24}$$

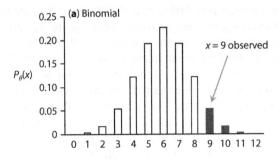
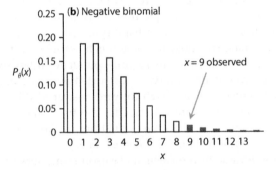

Fig. 6.5 The p-value for the test of a fair coin ($\theta = 1/2$) against the alternative $\theta > 1/2$ under **(a)** the binomial and **(b)** the negative binomial models. The data are $x = 9$ heads out of $n = 12$ tosses. **(a)** Under the binomial mode, the total number of tosses n is fixed. **(b)** Under the negative binomial model, the number of tails is fixed and the coin is tossed until $r = 3$ tails is observed. The p-value is the tail probability $\Pr\{x > 9\}$ under the null hypothesis $\theta = 1/2$, i.e. the probability of data that are at least as extreme as the observed value.

If we use this model, the p-value becomes

$$p = P_{\theta=1/2}(x=9) + P_{\theta=1/2}(x=10) + \cdots = 1 - \sum_{j=0}^{8} \binom{3+j-1}{j} \left(\frac{1}{2}\right)^j \left(\frac{1}{2}\right)^3 = 0.0327. \quad (6.25)$$

Thus, at the 5% significance level, we reject H_0 under the negative binomial model but not under the binomial model.

As equations (6.22) and (6.24) differ only by a proportionality constant that is independent of θ, and the likelihood is the same under the two models, the p-value thus violates the *likelihood principle*, which says that the likelihood function contains all information in the data about θ and the same inference should be made from two experiments that have the same likelihood (see §1.4.1).

The different p-values for the two models are due to the fact that the *sample space*, i.e. the space of all data outcomes, differs between the two models. Under the binomial model, the possible data outcomes are $x = 0, 1, \ldots, 9, \ldots, 12$ heads. Under the negative binomial model, the sample space consists of $x = 0, 1, 2, \ldots, 9, \ldots, 12, \ldots$, heads. The hypothesis testing approach allows unobserved data outcomes to affect our decision to reject H_0. The small probabilities of x values more extreme than the value actually observed ($x = 9$) are used as evidence against H_0, even though these values did not occur. Jeffreys (1961, p. 385) caricatured the approach as requiring 'that a hypothesis that may be true may be rejected because it has not predicted observable results that have not occurred'!

In the extreme case, Bayesian inference can proceed with no problem if the coin is tossed until the p-value of equation (6.23) is <5%. In classical statistics, this would be cheating, but such choices of the stopping rule do not affect Bayesian analysis. Bayesian estimation of θ depends on the results of the experiment but not how the experiment is monitored.

6.2.3.2 Criticisms of Bayesian methods

All criticisms of Bayesian methods are levied on the prior or the need for it. The objective Bayesians consider the prior to be a representation of prior objective information about the parameter. The approach runs into deep trouble when no prior information is available about the parameter and the prior is supposed to represent total ignorance. For a continuous parameter, the principle of insufficient reason assigns a uniform distribution over the range of the parameter. However, such so-called *noninformative priors* lead to contradictions. Earlier, we discussed the case of estimating the binomial probability. As another example, suppose we want to estimate the size of a square. We know its side a is between 1 and 2m, so we can assign the uniform prior $a \sim U(1, 2)$. On the other hand, the area A lies between 1 and $4m^2$, so we can assign the uniform prior $A \sim U(1, 4)$. Those two priors contradict each other. For example, by the uniform prior for a, we have $\Pr\{1 < a < 1.5\} = \frac{1}{2} = \Pr\{1 < A < 2.25\}$. However, by the uniform prior for A, $\Pr\{1 < A < 2.25\} = (2.25 - 1)/(4 - 1) = 0.417$, not $\frac{1}{2}$! This contradiction is due to the fact that the prior is not invariant to nonlinear transforms: if θ has a uniform distribution, θ^2 cannot have a uniform distribution. Similarly, a uniform prior for the probability of different sites p is very different from a uniform prior for sequence distance θ under the JC69 model (see Example A7 in Appendix A).

For a discrete parameter that takes m possible values, the principle of insufficient reason assigns probability $1/m$ for each value. This is not as simple as it may seem, as often it is unclear how the parameter values should be partitioned. For example, suppose we are interested in deciding whether a certain event occurred during weekdays or weekends. We can assign prior probabilities $\frac{1}{2}$ for weekdays and $\frac{1}{2}$ for weekends. An alternative is to consider each day of the week as having an equal probability and to assign 5/7 for weekdays and 2/7 for weekends. With no information about the parameter, it is unclear which prior is more reasonable. Such difficulties in representing total ignorance have caused the objective Bayesian approach to fall out of favour. Nowadays it is generally accepted that uniform priors are not noninformative and that indeed no prior can represent total ignorance.

The *subjective* Bayesians consider the prior to represent the researcher's *subjective belief* about the parameter before seeing or analysing the data. One cannot really argue against somebody else's subjective beliefs, but 'classical' statisticians reject the notion of subjective probabilities and of letting personal prejudices influence scientific inference. Even though the choice of the likelihood model involves some subjectivity as well, the model can nevertheless be checked against the data, but no such validation is possible for the prior.

If one accepts the prior, calculation of the posterior, dictated by the probability calculus, is automatic and self-consistent, free of the criticisms levied on classical statistics discussed above. When prior information is available about the parameter, the use of a prior in the Bayesian framework provides a natural way of incorporating such information. As mentioned already, Bayesian inference produces direct answers about the parameters that are easy to interpret. Bayesians argue that the concerns about the prior are technical issues in a fundamentally sound theory, while classical statistics is a fundamentally flawed theory, with *ad hoc* fixes often producing sensible results for the wrong reasons!

6.2.3.3 Does it matter?

Thus classical (Frequentist) and Bayesian statistics are based on different philosophies. To a biologist, an important question is whether the two approaches produce similar answers. This depends on the nature of the problem. Here we consider three kinds of problems.

In so-called *stable estimation problems* (Savage 1962), a well-formulated model $f(X|\theta)$ is available, and we want to estimate parameters θ from a large dataset. The prior will have little effect, and both likelihood and Bayesian estimates will be close to the true parameter value. Furthermore, classical confidence intervals in general match posterior credibility intervals under vague priors. A good example is distance estimation of Example 6.4 (Figure 6.3). To see more clearly how Bayesian parameter estimation may be insensitive to the prior, consider a uniform prior $f(\theta) = 1/(2c), -c < \theta < c$. The posterior is

$$f(\theta|x) = \frac{f(\theta)f(x|\theta)}{\int_{-c}^{c} f(\theta)f(x|\theta)\,d\theta} = \frac{\frac{1}{2c}f(x|\theta)}{\int_{-c}^{c} \frac{1}{2c}f(x|\theta)\,d\theta} = \frac{f(x|\theta)}{\int_{-c}^{c} f(x|\theta)\,d\theta}. \quad (6.26)$$

Note that the constant c in the prior density cancels, while in a large dataset, the integral in the denominator (i.e. the area under the curve $f(x|\theta)$ from $-c$ to c) is insensitive to the precise value of c as long as the likelihood $f(x|\theta)$ is well inside the prior interval (Figure 6.6). Thus the posterior will be insensitive to c or to the prior.

The second kind includes estimation problems in which both the prior and the likelihood exert substantial influence on the posterior. Bayesian inference will then be sensitive to the prior, and classical and Bayesian methods are also likely to produce different answers. The sensitivity of the posterior to the prior can be due to ill-formulated models that are barely identifiable, which lead to strongly correlated parameters, or to paucity of data containing little information about the parameters. Increasing the amount of data will help in this situation.

The third kind includes the most difficult problems, hypothesis testing or model selection when the models involve unknown parameters for which only vague prior information is available. In such cases, Bayesian analysis can be very sensitive to the prior, and

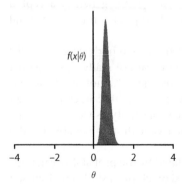

Fig. 6.6 The integral $\int_{-c}^{c} f(x|\theta)\,d\theta$ is the area under the likelihood curve $f(x|\theta)$ over the prior interval $(-c, c)$. When the dataset is large, the likelihood $f(x|\theta)$ is highly concentrated in a small interval of θ. Thus as long as c is large enough so that the peak of the likelihood curve is well inside the interval $(-c, c)$, the integral will be insensitive to the precise value of c. In the example, the likelihood peaks around $\theta = 0.2$ and is vanishingly small outside the interval $(0, 2)$. The area under the curve over the intervals $(-10, 10)$ or $(-100, 100)$ are virtually the same, and both are essentially identical to the area over the interval $(0, 2)$.

Bayesian analysis and classic hypothesis testing can produce opposite conclusions for the same dataset.

Suppose we take a sample, $x = \{x_1, x_2, \ldots, x_n\}$, from the normal distribution $N(\mu, 1)$ to compare models $H_0: \mu = 0$ against $H_1: \mu \neq 0$. The data are summarized by the sample mean $\bar{x}$ (this carries all information about μ in the dataset and is called the *sufficient statistic*), with $\bar{x} \sim N(0, 1/n)$ under H_0 and $\bar{x} \sim N(\mu, 1/n)$ under H_1. An observed $\bar{x}$ very different from 0 will be evidence against H_0. The p-value for the test of H_0 against H_1 is based on the fact that $(\bar{x} - 0)/\frac{1}{\sqrt{n}} = \sqrt{n}\bar{x} \sim N(0, 1)$ under H_0 and is $p = 2\Phi(-\sqrt{n}|\bar{x}|)$, where $\Phi(\cdot)$ is the cumulative distribution function (CDF) of the standard normal distribution. Alternatively we can conduct an LRT to compare H_0 and H_1. The likelihood under $H_0: \mu = 0$ is

$$L_0 = \frac{1}{\sqrt{2\pi/n}} \exp\left\{-\frac{n}{2}\bar{x}^2\right\}. \tag{6.27}$$

The (maximized) likelihood under $H_1: \mu \neq 0$ is

$$L_1 = L_1(\hat{\mu}) = \frac{1}{\sqrt{2\pi/n}} \exp\left\{-\frac{n}{2}(\bar{x} - \hat{\mu})^2\right\} = \frac{1}{\sqrt{2\pi/n}}, \tag{6.28}$$

evaluated at the MLE $\hat{\mu} = \bar{x}$. Thus the LRT statistic is

$$2\Delta\ell = -2\log\frac{L_0}{L_1} = n\bar{x}^2. \tag{6.29}$$

Note that if $\sqrt{n}\bar{x} \sim N(0, 1)$, then $n\bar{x}^2 \sim \chi_1^2$, so that the p-values based on the two statistics ($\bar{x}$ and $2\Delta\ell$) are the same.

In the Bayesian analysis, we calculate the posterior probabilities for the two models. We assign prior probabilities $\pi_0 = \pi_1 = 1/2$ for H_0 and H_1, and $\mu \sim N(\mu_0, \sigma_0^2)$ under H_1. As H_0 does not have any unknown parameters, the marginal likelihood under H_0 is $M_0 = f(x|H_0) = L_0$. The marginal likelihood M_1 under H_1 is an average of the likelihood $L_1(\mu)$ over the prior on μ:

$$M_1 = f(x|H_1) = E(L_1(\mu))$$
$$= \int_{-\infty}^{\infty} \frac{1}{\sqrt{2\pi/n}} \exp\left\{-\frac{n}{2}(\bar{x} - \mu)^2\right\} \times \frac{1}{\sqrt{2\pi\sigma_0^2}} \exp\left\{-\frac{1}{2\sigma_0^2}(\mu - \mu_0)^2\right\} d\mu. \tag{6.30}$$

As $n(\bar{x} - \mu)^2 + \frac{1}{\sigma_0^2}(\mu - \mu_0)^2 = (n + \frac{1}{\sigma_0^2})\left(\mu - \frac{n\sigma_0^2\bar{x}+\mu_0}{n\sigma_0^2+1}\right)^2 + \frac{1}{\frac{1}{n}+\sigma_0^2}(\bar{x} - \mu_0)^2$, we have

$$M_1 = \frac{1}{\sqrt{2\pi(1/n + \sigma_0^2)}} \exp\left\{-\frac{1}{2(1/n + \sigma_0^2)}(\bar{x} - \mu_0)^2\right\} = \phi(\bar{x}; \mu_0, \frac{1}{n} + \sigma_0^2), \tag{6.31}$$

where $\phi(x; \mu, \sigma^2)$ is the probability density function (PDF) of the normal distribution $N(\mu, \sigma^2)$. The posterior probability for model H_0 is given by the Bayes theorem as

$$P_0 = \frac{\pi_0 M_0}{\pi_0 M_0 + \pi_1 M_1} = \frac{1}{1 + M_1/M_0} = \frac{1}{1 + \frac{1}{\sqrt{1+n\sigma_0^2}} \exp\left\{\frac{n^2\sigma_0^2\bar{x}^2 + 2n\mu_0\bar{x} - n\mu_0^2}{2(1+n\sigma_0^2)}\right\}}. \tag{6.32}$$

Here the ratio of the marginal likelihoods $M_1/M_0 = B_{10}$ is called the Bayes factor, which we will discuss in more detail in §7.4.3. For the special case $\mu_0 = 0$ in the prior, we have

$$P_0 = \frac{1}{1 + \frac{1}{\sqrt{1+n\sigma_0^2}} \exp\left\{-\frac{n^2\sigma_0^2 \bar{x}^2}{2(1+n\sigma_0^2)}\right\}}. \tag{6.33}$$

Now suppose that in a particular dataset the observed $\bar{x}$ is fairly different from 0, so that we reject H_0 with a small p-value. For example, if $n = 10^6$ and $\bar{x} = 0.003$, we reject H_0 with $p = 2\Phi(-3) = 0.27\%$. However, if a diffuse prior is used, e.g., with $\mu_0 = 0$ and $\sigma_0^2 = 1000$, equation (6.33) gives $P_0 = 0.9972$. Thus while the test rejects H_0 with a small p-value, the Bayesian analysis of the same data strongly supports H_0. This contradiction between hypothesis testing and Bayesian model choice is known as *Lindley's paradox* (Lindley 1957; see also Jeffreys 1935). In this example, the observed effect, at $\bar{x} = 0.3\%$ of the standard deviation, may have little biological significance, but it is significant statistically because of the large data size. It is generally felt that hypothesis testing such as the LRT is too liberal and tends to reject the null too often, especially in large datasets. Note that if larger and larger datasets are simulated under H_0, a test at the 5% significance level will reject H_0 incorrectly in 5% of the datasets. In contrast, the Bayesian analysis will converge to the true model H_0, with $P_0 \to 1$ as $n \to \infty$.

We now turn to the sensitivity of the Bayesian model choice to the prior on model parameters. Note that P_0 in equations (6.32) and (6.33) can be made as close to 1 as one likes by letting $\sigma_0^2 \to \infty$ or by letting $\mu_0 \to \infty$. Both a very diffuse prior on μ (a very large σ_0^2) or an informative prior in conflict with the likelihood (with μ_0 far away from $\bar{x}$) make model H_1 look poor, leading to a reduced marginal likelihood M_1 and increased P_0. For the lower bound on P_0, note that the marginal likelihood M_1 of equation (6.31) is no larger than the maximized likelihood L_1 of equation (6.28):

$$M_1 = E(L_1(\mu)) \leq L_1(\hat{\mu}), \tag{6.34}$$

and thus

$$P_0 = \frac{M_0}{M_0 + M_1} \geq \frac{L_0}{L_0 + L_1}. \tag{6.35}$$

This lower bound is approached if $\mu_0 \approx \bar{x}$ and $\sigma_0^2 \approx 0$, i.e. if the prior on μ is highly concentrated around the MLE and is thus highly consistent with the data. Suppose in a particular dataset (which is not necessarily large), the LRT is significant with the p-value at exactly 5% (i.e. with $L_1/L_0 = e^{1.92}$). In such a dataset, $P_0 \geq L_0/(L_1 + L_1) = 0.128$. This value is quite high and may not be considered strong evidence against H_0. With changes to the prior on μ, P_0 can change from ~0.128 to ~1. If the prior information is weak, one may not be able to decide whether $\sigma_0^2 = 10$ or 100 is more appropriate, even though the posterior model probability may differ considerably between the two.

Why is Bayesian model selection sensitive to the prior while Bayesian parameter estimation is not? Again this is easy to see with the uniform prior. Suppose in the above example of testing we use the uniform prior $\mu \sim U(-c, c)$ under H_1 instead of the normal. Then instead of equation (6.32), we have

$$P_0 = \frac{M_0}{M_0 + M_1} = \frac{M_0}{M_0 + \int_{-c}^{c} \frac{1}{2c} f(x|\mu)\, d\mu} = \frac{M_0}{M_0 + \frac{1}{2c}\int_{-c}^{c} f(x|\mu)\, d\mu}. \tag{6.36}$$

As μ is in model H_1 but not in H_0, the constant c in the prior for μ does not cancel, in contrast to equation (6.26) for parameter estimation. P_0 may thus be sensitive to c.

Lindley's paradox, as a manifestation of the sensitivity of posterior model probabilities to the prior, arises whenever we wish to compare different models with unknown parameters that are in one model but not the other (e.g. O'Hagan and Forster 2004). For such difficulties to arise, the compared models can have one or more parameters, or one model can be sharp (with no parameters), and the prior can be proper and informative.

6.3 Prior

6.3.1 *Methods of prior specification*

Whether we adopt the objective or subjective views, Bayesian inference requires the specification of a prior. A prior should reflect our degree of belief about the parameter before collection and analysis of the data. Thus, we can construct the prior by making use of information gained in past experiments under similar conditions or from other independent evidence. The prior can also be specified by modelling the physical/biological process. For example, the Yule branching process (Edwards 1970) and the birth–death process (Rannala and Yang 1996) can be used to specify the probability distribution of phylogenies.

When we do not have prior information, the objective Bayesians idea of specifying *vague* or *diffuse* priors is attractive. Criteria proposed for specifying an objective prior include the principle of insufficient reason, which underlies the flat or uniform prior used by Bayes (1763) and Laplace (1812); invariance to reparametrization, which underlies the Jeffreys prior (Jeffreys 1961); and maximization of missing information, which underlies the reference prior of Bernardo (1979). In this section we illustrate those ideas using simple examples in statistics and the problem of estimation of sequence distance under the JC69 model. The parameter space of phylogenetic problems is complex, making it demanding to specify priors for each phylogenetic analysis. Automatic rules for specifying default priors in the computer program are thus appealing.

When the prior distribution involves unknown parameters, one can assign priors for them, called *hyper-priors*. Unknown parameters in the hyper-prior can have their own priors. This is known as the *hierarchical Bayesian* or *full Bayesian* approach. Typically, one does not go beyond two or three levels, as the effect will become unimportant. For example, the mean μ in the exponential prior in our example of distance calculation under JC69 (Example 6.4) can be assigned a hyper-prior. An alternative is to estimate the hyper-parameters from the marginal likelihood, and use them in posterior probability calculation for parameters of interest. This is known as the *empirical Bayes* approach. For example, μ can be estimated by maximizing $f(x|\mu) = \int f(\theta|\mu) f(x|\theta) d\theta$, and the estimate can be used to calculate $f(\theta|x)$ in equation (6.17). The empirical Bayes approach has been widely used in molecular phylogenetics, for example, to estimate evolutionary rates at sites (Yang and Wang 1995, see §4.3.1), to reconstruct ancestral DNA or protein sequences on a phylogeny (Yang et al. 1995a; Koshi and Goldstein 1996a, see §4.4.2), to identify amino acid residues under positive selection (Nielsen and Yang 1998, see §11.4.3), to infer secondary structure categories of a protein sequence (Goldman et al. 1998), and to construct sequence alignments under models of insertions and deletions (Thorne et al. 1991; Thorne and Kishino 1992).

An important question in real data analysis is whether the posterior is sensitive to the prior. It is always prudent to assess the influence of the prior through a Bayesian robustness analysis. If the posterior is dominated by the data, the choice of the prior is inconsequential. When this is not the case, the effect of the prior has to be assessed carefully and reported. Due to advances in computational algorithms, the Bayesian methodology has enabled researchers to fit sophisticated parameter-rich models. As a result, one may be tempted to add parameters that are barely identifiable (Rannala 2002), and the posterior may be unduly influenced by some aspects of the prior even without the knowledge of the researcher. In our example of distance estimation under the JC69 model, a problem of identifiability will arise if we attempt to estimate both the substitution rate r and the divergence time t instead of the distance $\theta = 2rt$. It is thus important to understand which aspects of the data provide information about the parameters, which parameters are knowable and which are not, to avoid overloading the model with too many parameters.

6.3.2 Conjugate priors

Under certain likelihood models, the prior and posterior can have the same distributional form, and the role of the data or likelihood is to update the parameters in the distribution. Such priors are called conjugate priors.

Beta prior for the binomial probability. From Example 6.2, we know that the beta distribution is a conjugate prior for the binomial probability θ. Here we use this case to illustrate how to identify a conjugate prior. With the data to be x successes in n trials, the likelihood is

$$f(x|\theta) \propto \theta^x (1-\theta)^{n-x}. \tag{6.37}$$

By looking at the likelihood as a function of θ, we see this is proportional to a beta density. Thus beta(a, b) should be a conjugate prior

$$f(\theta) \propto \theta^{a-1}(1-\theta)^{b-1}. \tag{6.38}$$

By multiplying the prior with the likelihood, the posterior becomes $f(\theta|x) \propto \theta^{a+x-1}(1-\theta)^{b+n-x-1}$, which is beta($x + a$, $n - x + b$). This has the mean

$$E(\theta|x) = \frac{a+x}{a+b+n}, \tag{6.39}$$

which always lies between the prior mean $a/(a+b)$ and the sample mean x/n.

We mentioned in §6.2.2 that beta(1, 1), or $U(0, 1)$, is suggested to be a 'noninformative prior' for the binomial probability θ. Two other conjugate priors have also been suggested to represent total ignorance: beta$\left(\frac{1}{2}, \frac{1}{2}\right)$ and beta(0, 0). The former, beta$\left(\frac{1}{2}, \frac{1}{2}\right)$, results from assigning a uniform prior for $z = \sin^{-1}\sqrt{\theta}$. This is also Jeffreys' prior (see §6.3.4). The latter, beta(0, 0) or $f(\theta) \propto \theta^{-1}(1-\theta)^{-1}$, suggested by Haldane (1931), is an improper prior but has the property that the posterior mean $E(\theta|x)$ equals the MLE, x/n. While in practical data analysis when n is not extremely small, the three priors may not make any noticeable difference to the posterior, the example does underline the difficulty of representing total ignorance.

Gamma prior for the Poisson rate. Suppose the number of events (such as mutations) x has a Poisson distribution with mean λ:

$$f(x|\lambda) = \frac{\lambda^x e^{-\lambda}}{x!}, \text{ for } x = 0, 1, \ldots. \tag{6.40}$$

For a vector $x = \{x_1, x_2, \ldots, x_n\}$ of i.i.d. observations, the likelihood is

$$f(x|\lambda) = \prod_{i=1}^{n} \frac{\lambda^{x_i} e^{-\lambda}}{x_i!} \propto \lambda^{n\bar{x}} e^{-n\lambda}, \tag{6.41}$$

where $\bar{x}$ is the sample mean. The form $\lambda^{n\bar{x}} e^{-n\lambda}$, as a function of λ, is proportional to a gamma density. Thus the conjugate prior on λ is the gamma distribution $\lambda \sim G(\alpha, \beta)$:

$$f(\lambda) = \frac{\beta^{\alpha}}{\Gamma(\alpha)} \lambda^{\alpha-1} e^{-\beta\lambda}. \tag{6.42}$$

The posterior is then proportional to $\lambda^{\alpha+n\bar{x}-1} e^{-(\beta+n)\lambda}$. In other words,

$$\lambda|x \sim G(\alpha + n\bar{x}, \beta + n). \tag{6.43}$$

Comparing the prior and posterior reveals that the information in the prior is equivalent to a total of α events in β observations.

Normal prior for normal mean μ. Suppose the data are an i.i.d. sample from the normal distribution: $x = \{x_i\}$, where $x_i \sim N(\mu, \sigma^2)$ with σ^2 known. The data are summarized by the sample mean $\bar{x} \sim N(\mu, \sigma^2/n)$, with the likelihood $L(\mu) \propto \exp\left\{-\frac{n}{2\sigma^2}(\bar{x}-\mu)^2\right\}$. Viewed as a function of μ, this is proportional to a normal density. Thus $N(\mu_0, \sigma_0^2)$ is a conjugate prior for μ. The posterior is then $f(\mu|x) \propto \exp\left\{-\frac{n}{2\sigma^2}(\bar{x}-\mu)^2 - \frac{1}{2\sigma_0^2}(\mu-\mu_0)^2\right\} \propto \exp\left\{-\frac{1}{2\sigma_n^2}(\mu-\mu_n)^2\right\}$, where

$$\mu_n = \frac{\frac{1}{\sigma_0^2}\mu_0 + \frac{n}{\sigma^2}\bar{x}}{\frac{1}{\sigma_0^2} + \frac{n}{\sigma^2}}, \text{ and } \frac{1}{\sigma_n^2} = \frac{1}{\sigma_0^2} + \frac{n}{\sigma^2}. \tag{6.44}$$

In other words, the posterior is $N(\mu_n, \sigma_n^2)$. The reciprocal of the variance is called the *precision*. Thus the posterior mean μ_n is a weighted average of the prior mean μ_0 and the sample mean $\bar{x}$, with the weights to be the precisions of the prior and sample distributions ($\frac{1}{\sigma_0^2}$ and $\frac{n}{\sigma^2}$). Furthermore, the posterior precision is the sum of the prior precision and sample precision.

Conjugate priors are possible for certain likelihood models only. They are convenient, as the integrals are tractable analytically, but they may not be realistic models for the problem at hand. Models in phylogenetics are in general too complex so that conjugate priors are nearly never used.

6.3.3 Flat or uniform priors

It is common to assign a flat or uniform prior on a parameter when little information is available about it. If the prior does not integrate to 1 it is called an improper prior. An improper prior is permissible if the posterior is proper. However, uniform priors and in particular, improper priors may not be reasonable biologically, in which case nonuniform priors are preferable.

For example, if the data are a sample from the normal distribution, $x_i \sim N(\mu, \sigma^2)$, with $-\infty < \mu < \infty, 0 < \sigma < \infty$, one can specify a flat prior on μ and a flat prior on $\log \sigma$, so that $\pi(\mu) = 1$, and $\pi(\sigma) = 1/\sigma$. Neither of those priors integrates to 1, but they lead to a proper posterior.

In contrast, we cannot use an improper prior for JC69 distance θ in Example 6.4. Note that the likelihood

$$f(x|\theta) = \left(\frac{3}{4} - \frac{3}{4}e^{-4\theta/3}\right)^x \left(\frac{1}{4} + \frac{3}{4}e^{-4\theta/3}\right)^{n-x} \tag{6.45}$$

approaches a nonzero constant $(\frac{3}{4})^x(\frac{1}{4})^{n-x}$ when $\theta \to \infty$. If we assign the improper flat prior $f(\theta) = 1, 0 \leq \theta < \infty$, the marginal likelihood $f(x) = \int_0^\infty f(\theta)f(x|\theta)\,d\theta = \infty$, and the posterior will be improper.

If we use a uniform prior $f(\theta) = 1/A, 0 \leq \theta < A$, with a large upper bound A, the posterior becomes

$$f(\theta|x) = \frac{f(x|\theta)}{\int_0^A f(x|\theta)\,d\theta}, \quad 0 \leq \theta < A. \tag{6.46}$$

For the human–orangutan mitochondrial 12S rRNA genes, we have $x = 90$ and $n = 948$. With $A = 10$, the integral in the denominator is $f(x) = 1.72481 \times 10^{-132}$. The posterior mean and 95% equal-tail CI are found numerically to be 0.1027 (0.0824, 0.1254). The results are the same if $A = 100$ is used. While here the results are similar to those from the exponential prior (Example 6.4), we stress that the uniform prior is generally a poor choice for sequence distance, and the exponential prior is preferable. Most sequence distances estimated from real data are small (say, < 1) and the uniform prior with mean at 5 or 50 (i.e. with $A = 10$ or 100) is not reasonable. The uniform prior may also cause convergence problems for the MCMC algorithm (see §7.3.1).

If we assign a uniform prior on p instead: $p \sim U(0, \frac{3}{4})$, we will have an exponential prior on θ with mean $\frac{3}{4}$ in Example 6.4:

$$f(\theta) = f_p(p) \times \left|\frac{dp}{d\theta}\right| = \frac{4}{3}e^{-4\theta/3}, \quad 0 \leq \theta < \infty. \tag{6.47}$$

This favours small distances and is more reasonable than the uniform prior on θ. For our example 12S rRNA data, the marginal likelihood is $f(x) = 2.0056 \times 10^{-131}$, and the posterior mean and 95% equal-tail CI are 0.1026 (0.0823, 0.1252).

*6.3.4 The Jeffreys priors

The Jeffreys prior is based on the Fisher (expected) information and is invariant to reparametrization. Let $I(\theta) = \{I_{ij}(\theta)\}$ be the Fisher information matrix, with

$$I_{ij}(\theta) = -E_\theta\left(\frac{\partial^2 \log f(x|\theta)}{\partial \theta_i \partial \theta_j}\right), \tag{6.48}$$

where the expectation is taken over all possible datasets. The Jeffreys prior is then given as

$$f(\theta) \propto [\det I(\theta)]^{\frac{1}{2}}, \tag{6.49}$$

where $\det[I(\theta)]$ is the determinant of $I(\theta)$.

Consider as an example the binomial probability θ of Example 6.2. From the likelihood $f(x|\theta) = \theta^x(1-\theta)^{n-x}, 0 \leq \theta < 1$, we have

$$\log f(x|\theta) = x\log\theta + (n-x)\log(1-\theta) \tag{6.50}$$

---

* indicates a more difficult or technical section.

and

$$\frac{d^2 \log f(x|\theta)}{d\theta^2} = -\frac{x}{\theta^2} - \frac{n-x}{(1-\theta)^2}. \tag{6.51}$$

Since $E(x) = n\theta$, we have

$$I(\theta) = \frac{n}{\theta(1-\theta)}. \tag{6.52}$$

Thus the Jeffreys prior on θ is $f(\theta) \propto \theta^{-\frac{1}{2}}(1-\theta)^{-\frac{1}{2}}, 0 \leq \theta \leq 1$, which is beta $\left(\frac{1}{2}, \frac{1}{2}\right)$. This is equivalent to assigning a uniform prior on the arcsine transform $y = \sin^{-1}(\sqrt{\theta}) \sim U(0, \pi/2)$.

For the normal likelihood, $x_i \sim N(\mu, \sigma^2)$, the Jeffreys prior is the improper priors $f(\mu) = 1$ and $f(\sigma) = 1/\sigma$.

For the JC69 distance θ (Example 6.4), we have

$$\log f(x|\theta) = x \log p + (n-x) \log(1-p) = x \log\left(\frac{3}{4} - \frac{3}{4}e^{-4\theta/3}\right) + (n-x) \log\left(\frac{1}{4} + \frac{3}{4}e^{-4\theta/3}\right), \tag{6.53}$$

where p is the probability of observing a difference at any site. Then

$$\frac{d^2 \log f(x|\theta)}{d\theta^2} = -x\left[\frac{e^{-\frac{8\theta}{3}}}{\left(\frac{3}{4} - \frac{3}{4}e^{-\frac{4\theta}{3}}\right)^2} + \frac{\frac{4}{3}e^{-\frac{4\theta}{3}}}{\frac{3}{4} - \frac{3}{4}e^{-\frac{4\theta}{3}}}\right] - (n-x)\left[\frac{e^{-\frac{8\theta}{3}}}{\left(\frac{1}{4} + \frac{3}{4}e^{-\frac{4\theta}{3}}\right)^2} - \frac{\frac{4}{3}e^{-\frac{4\theta}{3}}}{\frac{1}{4} + \frac{3}{4}e^{-\frac{4\theta}{3}}}\right]. \tag{6.54}$$

As $E(x) = np = n\left(\frac{3}{4} - \frac{3}{4}e^{-4\theta/3}\right)$, we have

$$I(\theta) = -E_\theta\left(\frac{d^2 \log f(x|\theta)}{d\theta^2}\right) = \frac{16n}{3(e^{\frac{8\theta}{3}} + 2e^{\frac{4\theta}{3}} - 3)}. \tag{6.55}$$

Thus the Jeffreys prior is

$$f(\theta) \propto \left(e^{\frac{8\theta}{3}} + 2e^{\frac{4\theta}{3}} - 3\right)^{-\frac{1}{2}}. \tag{6.56}$$

This is a proper density. This prior is derived by Ferreira and Suchard (2008) (their α should be 4/3 instead of 1/3).

The Jeffreys prior is designed to be invariant to reparametrization. Suppose we consider the probability p of difference as the parameter and assign the Jeffreys prior to it. From the likelihood $f(x|\theta) = p^x(1-p)^{n-x}, 0 \leq p < \frac{3}{4}$, we have (see equation (6.52))

$$I(p) = \frac{n}{p(1-p)}. \tag{6.57}$$

Thus the Jeffreys prior on p is

$$f(p) \propto p^{-\frac{1}{2}}(1-p)^{-\frac{1}{2}}, \ 0 \leq p < \frac{3}{4}. \tag{6.58}$$

This is beta$\left(\frac{1}{2}, \frac{1}{2}\right)$, truncated at $p = \frac{3}{4}$. We can see that the Jeffreys priors on θ and p (equations (6.56) and (6.58)) are equivalent in that one implies the other (Problem 6.2).

When the Jeffreys prior is applied to the JC69-distance data of $x = 90$ and $n = 948$, the posterior mean and 95% equal-tail CI for θ is 0.1021 (0.0818, 0.1247). This interval is nearly identical to the likelihood interval, (0.0817, 0.1245), which is also invariant to reparametrization (see Example 1.5 in Chapter 1).

The Jeffreys prior may not perform well when there are multiple parameters in the model, for example, when there are nuisance parameters (Datta and Ghosh 1996).

*6.3.5 The reference priors

The reference prior was first proposed by Bernardo (1979), and further developed by Berger et al. (1992). See Bernardo (2005) and Berger et al. (2009) for recent reviews. The reference prior maximizes the expected distance between the prior and the posterior. The distance between two distributions f and g are often measured by the Kullback–Leibler (K-L) divergence, defined as

$$D(f,g) = \int f(y) \log \frac{f(y)}{g(y)} dy. \quad (6.59)$$

Usually f is the true distribution and g is an estimate. Note that $D(f,g) \neq D(g,f)$ and $D(f,f) = 0$. It may be interesting to note that if f is the observed site pattern frequencies and g is the expected site pattern probabilities under the model, then ML estimation of parameters (such as branch lengths) will be equivalent to minimization of the K-L divergence between the observed and expected site pattern distributions.

Let the prior be $q(\theta)$, the likelihood be $f(x|\theta)$, and the posterior be $f(\theta|x) = q(\theta)f(x|\theta)/f(x)$. Then the K-L divergence between the prior and the posterior for the given dataset x is

$$D_x(f(\cdot|x), q(\cdot)) = \int f(\theta|x) \log \frac{f(\theta|x)}{q(\theta)} d\theta. \quad (6.60)$$

The expected K-L divergence is an average over the marginal distribution of the data $f(x)$:

$$D(q) = \int f(x) D_x(f(\cdot|x), q(\cdot)) dx = \int \int q(\theta) f(x|\theta) \log \frac{f(\theta|x)}{q(\theta)} d\theta dx. \quad (6.61)$$

The prior q that maximizes this expected K-L divergence is the reference prior. Intuitively D measures the difference between prior and the posterior and thus the reference prior maximizes the missing information in the prior.

A formal procedure for deriving reference priors is described by Berger et al. (2009). If there is only one parameter in the model, and if the model is regular (in the sense that the MLE has an asymptotic normal distribution when the data size approaches infinity), the reference prior is known to be the Jeffreys prior (Berger et al. 2009).

While the Jeffreys and reference priors are important in objective Bayesian statistics, they are rarely used in phylogenetics. The only exception is Ferreira and Suchard (2008), who discussed the use of reference priors for parameters in continuous-time Markov chain models of nucleotide substitution in pairwise sequence comparisons. It is unclear whether reference or Jeffreys priors may be useful in estimation of the d_N/d_S ratio (ω) discussed in Chapter 2, as there is much interest in ω. The MLE of ω (Goldman and Yang 1994) can be 0 or ∞ in some datasets, so that $\hat{\omega}$ does not have finite mean or variance. Use of a prior in Bayesian analysis will shrink the estimates away from such extreme values, so that the method will have better Frequentist properties than the MLE.

---

* indicates a more difficult or technical section.

6.4 Methods of integration

In most Bayesian inference problems, the prior and the likelihood are easy to calculate, but the marginal probability of the data $f(X)$, i.e. the normalizing constant in equation (6.6), is hard to calculate as it involves an integral. Except for trivial problems such as those involving conjugate priors, analytical results are unavailable. Models used in Bayesian phylogenetics may involve hundreds or thousands of parameters and high-dimensional integrals. For example, in the calculation of posterior probabilities for phylogenetic trees, the marginal likelihood $f(X)$ will be a sum over all possible tree topologies and, for each tree, an integration over branch lengths and substitution parameters.

The breakthrough is the development of MCMC algorithms, which provide a powerful framework for Bayesian computation. Even though MCMC algorithms were also suggested for calculating the likelihood function, to integrate over random variables in the model (Geyer 1991; Kuhner et al. 1995), they are not very successful in that application (Stephens and Donnelly 2000).

Before introducing MCMC algorithms in the next chapter, we will now discuss several classical methods for calculating integrals, such as Laplace's large-sample approximation, numerical integration (Gaussian quadrature), Monte Carlo (MC) integration, and importance sampling. We have two motivations here. First, Laplace's asymptotic expansion and Gaussian quadrature may be very useful in applications that involve low-dimensional integrals. Second, MC integration is closely related to MCMC, and it is important to appreciate their similarities and differences.

We illustrate the different methods by applying them to calculate the marginal likelihood for Bayesian estimation of the JC69 distance of Example 6.4 (see equation (6.17)). The data are $x = 90$ differences at $n = 948$ sites, and the prior is exponential with mean $\mu = 0.2$. The true value of the integral is

$$I = \int_0^\infty f(\theta) f(x|\theta) \, d\theta = \int_0^\infty \frac{1}{\mu} e^{-\theta/\mu} \left(\frac{3}{4} - \frac{3}{4} e^{-4\theta/3} \right)^x \left(\frac{1}{4} + \frac{3}{4} e^{-4\theta/3} \right)^{n-x} d\theta \qquad (6.62)$$

$$= 5.167762 \times 10^{-131},$$

or $\log I = -299.9962$ at the logarithmic scale.

*6.4.1 Laplace approximation

If the data are an i.i.d. sample of size n and if n is large, the likelihood in the neighbourhood of the MLE $\hat{\theta}$ is well approximated by a normal density. In other words, the log likelihood curve is approximately a parabola around $\hat{\theta}$. This property can be used to approximate the marginal likelihood in large datasets.

Rewrite the likelihood as $L(\theta) = f(x|\theta) = e^{nh(\theta)}$, and apply Taylor expansion to $h(\theta)$ around the MLE $\hat{\theta}$:

$$h(\theta) \approx h(\hat{\theta}) + \frac{dh}{d\theta}(\theta - \hat{\theta}) + \frac{1}{2}\frac{d^2h}{d\theta^2}(\theta - \hat{\theta})^2 = h(\hat{\theta}) + \frac{1}{2}\frac{d^2h}{d\theta^2}(\theta - \hat{\theta})^2, \qquad (6.63)$$

where $\frac{dh}{d\theta}$ and $\frac{d^2h}{d\theta^2}$ are the first and second derivatives of h with respect to θ, evaluated at the MLE $\hat{\theta}$. Note that $\frac{dh}{d\theta} = 0$ at $\hat{\theta}$, as we assume that $\hat{\theta}$ is inside the parameter space. Then

---

* indicates a more difficult or technical section.

$$I = \int f(\theta)e^{nh(\theta)}\,d\theta$$

$$\approx f(\hat{\theta})e^{nh(\hat{\theta})}\int \exp\left\{n\cdot\frac{1}{2}\cdot\frac{d^2h}{d\theta^2}(\theta-\hat{\theta})^2\right\}\,d\theta \qquad (6.64)$$

$$= f(\hat{\theta})L(\hat{\theta})\sqrt{2\pi V}\int \frac{1}{\sqrt{2\pi V}}\exp\left\{-\frac{1}{2V}(\theta-\hat{\theta})^2\right\}\,d\theta$$

$$= f(\hat{\theta})L(\hat{\theta})\sqrt{2\pi V}.$$

The first approximation replaces $f(\theta)$ by $f(\hat{\theta})$ because the likelihood is highly concentrated in a narrow interval around $\hat{\theta}$, within which the prior is nearly constant with $f(\theta) \approx f(\hat{\theta})$. The integrand in the third equation is the normal density $\theta \sim N(\hat{\theta}, V)$, where $V = -1\big/\left(n\cdot\frac{d^2h}{d\theta^2}\right)$ is the asymptotic variance of $\hat{\theta}$. Note that this is the same normal approximation to the likelihood as discussed in §1.4.1 (see equation (1.50)). Equation (6.64) is known as Laplace's approximation.

The case with multiple parameters, where θ is a vector, is very similar. Instead of equation (6.63), we have the multivariate Taylor expansion

$$h(\theta) \approx h(\hat{\theta}) + \frac{1}{2}(\theta-\hat{\theta})^{\mathrm{T}}H(\theta-\hat{\theta}), \qquad (6.65)$$

where $H = \left\{\frac{\partial^2 h}{\partial \theta_i \partial \theta_j}\right\}$ is the Hessian matrix of second derivatives, evaluated at the MLE $\hat{\theta}$. Then

$$I = \int f(\theta)e^{nh(\theta)}\,d\theta$$

$$\approx f(\hat{\theta})e^{nh(\hat{\theta})}\int \exp\left\{n\cdot\frac{1}{2}(\theta-\hat{\theta})^{\mathrm{T}}H(\theta-\hat{\theta})\right\}\,d\theta \qquad (6.66)$$

$$= f(\hat{\theta})L(\hat{\theta})\sqrt{2\pi|V|},$$

where $V = -H^{-1}$ is the variance–covariance matrix and $|V|$ its determinant.

To apply Laplace's method to calculate the integral of equation (6.62), note that the log likelihood is

$$h(\theta) = \log\{L(\theta)\}/n = \frac{x}{n}\log p + \left(1-\frac{x}{n}\right)\log(1-p), \qquad (6.67)$$

where $p = \frac{3}{4}(1-e^{-4\theta/3})$. The second derivative is $\frac{d^2h}{d\theta^2} = -\frac{1}{\hat{p}(1-\hat{p})}\times\left(1-\frac{4}{3}\hat{p}\right)^2 = -8.87832$, and the variance is $V_{\hat{\theta}} = -1\big/\left(n\cdot\frac{d^2h}{d\theta^2}\right) = 0.000118812$ (see equation (1.51)). Equation (6.64) then gives $I = 5.18427 \times 10^{-131}$, with a relative error of 0.32%. This is a fairly accurate approximation.

6.4.2 Mid-point and trapezoid methods

Mid-point method. Let the integrand be $f(x)$. The integral $\int_a^b f(x)\,dx$ is the area under the curve $f(x)$ between a and b (Figure 6.7). To approximate it, we break the interval (a, b) into n segments, each of width $h = (b-a)/n$. We approximate the area on each segment by the area of the rectangle, with the height given by the function value at the mid-point of the segment.

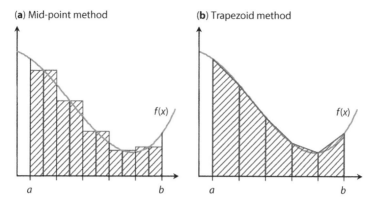

Fig. 6.7 The mid-point and trapezoid methods for numerical calculation of a definite integral (i.e. the area under the curve $f(x)$ between a and b). The interval (a, b) is divided into n (=5) segments, and the area above each segment is approximated by (**a**) the rectangle or (**b**) the trapezoid.

Label the n mid-points as $x_1 = a + h/2, x_2 = a + 3h/2, \ldots, x_n = a + (n - \frac{1}{2})h$, and let $f_i = f(x_i)$, $i = 1, 2, \ldots, n$. Then

$$\int_a^b f(x)\, dx \approx \frac{(b-a)}{n}(f_1 + f_2 + \cdots + f_n) = \frac{b-a}{n} \sum_{i=1}^n f\left(a + \left(i - \frac{1}{2}\right)\frac{(b-a)}{n}\right). \tag{6.68}$$

The trapezoid method. We break the interval (a, b) into n segments, each of width $h = (b-a)/n$. Label the $(n+1)$ break points as $x_0 = a, x_1 = a + h, x_2 = a + 2h, \ldots, x_n = b$, and let $f_i = f(x_i)$, $i = 0, 1, 2, \ldots, n$. Then we approximate the area on each segment by the area of the trapezoid, to get

$$\begin{aligned}\int_a^b f(x)\, dx &\approx \frac{h}{2}(f_0 + f_1) + \frac{h}{2}(f_1 + f_2) + \cdots + \frac{h}{2}(f_{n-1} + f_n) \\ &= \frac{h}{2}(f_0 + 2f_1 + 2f_2 + \cdots + 2f_{n-1} + f_n).\end{aligned} \tag{6.69}$$

6.4.3 *Gaussian quadrature*

Gaussian quadrature refers to a class of numerical integration methods that approximate the integrand using a polynomial. Note that a polynomial, $p(x) = a_0 + a_1 x + a_2 x^2 + \cdots + a_m x^m$, is integrable analytically. For example, the Gaussian–Legendre quadrature approximates the integral over $(-1, 1)$ as

$$\int_{-1}^1 f(x)\, dx \approx \sum_{i=1}^n w_i f(x_i), \tag{6.70}$$

where the points x_i and weights w_i are pre-determined (Abramowitz and Stegun 1972, p. 887). For instance, with $n = 8$, the points are $x_i = \pm 0.18343, \pm 0.52553, \pm 0.79667$, and ± 0.96029, with weights $w_i = 0.36268, 0.31370, 0.22238$, and 0.10122, respectively. Equation (6.70) will be exact if $f(x)$ is a polynomial of degree $2n - 1$ or less; otherwise it will be approximate. Other quadrature rules may be designed to work with integrands of the forms $f(x) = p(x)e^{-x}, f(x) = p(x)e^{-x^2}$, etc., where $p(x)$ is a polynomial.

6 BAYESIAN THEORY

An integral over (a, b) can easily be converted into an integral over $(-1, 1)$ through a transform (mapping or change of variables). For example, one may use the linear transform $y = \frac{b-a}{2}x + \frac{a+b}{2}$, with $-1 < x < 1$ and $a < y < b$. As $\frac{dy}{dx} = \frac{b-a}{2}$, we have

$$\int_a^b f(y)\,dy = \frac{b-a}{2} \int_{-1}^1 f\left(\frac{b-a}{2}x + \frac{a+b}{2}\right) dx \approx \frac{b-a}{2} \sum_{i=1}^n w_i f\left(\frac{b-a}{2}x_i + \frac{a+b}{2}\right), \quad (6.71)$$

where x_i and w_i are the points and weights given by the quadrature rule. There are many ways of mapping (a, b) to $(-1, 1)$, which may lead to different points in the original variable $y \in (a, b)$ being sampled, and to very different accuracies for the quadrature approximation. If a transform causes the integrand of the new integral to become a low-order polynomial, the quadrature approximation will become exact. In general one should aim to sample points more densely in regions in which the integrand changes rapidly. We discuss the effect of transforms using an example in §6.4.4.

Similarly, a two-dimensional integral $\int_a^b \int_c^d f(x, y)\,dy\,dx$ is the volume between the x–y plane and the surface $f(x, y)$ over the rectangle defined by $a < x < b$ and $c < y < d$. This can be approximated by applying the quadrature rule iteratively, first to the 1-D integral over x, where the integrand is an integral over y and is also approximated by the quadrature rule. This so-called product rule leads to

$$\int_a^b \int_c^d f(x, y)\,dy\,dx \approx \frac{b-a}{2}\frac{d-c}{2} \sum_{i=1}^n \sum_{j=1}^n w_i w_j \cdot f\left(\frac{b-a}{2}x_i + \frac{a+b}{2}, \frac{d-c}{2}x_j + \frac{c+d}{2}\right), \quad (6.72)$$

where x_i and x_j are the points and w_i and w_j are the weights, all pre-determined.

While the computation for a 1-D integral is proportional to the number of points n, the computation for integrals of d dimensions is proportional to n^d, which quickly becomes unmanageable as d increases. This so-called *curse of dimension* means that numerical integration is feasible for integrals of low (1–3) dimensions only.

6.4.4 Marginal likelihood calculation for JC69 distance estimation

In this subsection we apply the numerical integration methods discussed above to calculate the integral of equation (6.62) (see also Example 6.4). We need convert the infinite integral range $\theta \in (0, \infty)$ to a finite range (a, b) through a transform. Here we consider four transforms (two naïve and two sophisticated). The results are summarized in Table 6.1.

The first transform is

$$y = \frac{\theta - 1}{\theta + 1}, \quad \theta = \frac{1 + y}{1 - y}, \quad (6.73)$$

with $-1 < y < 1$. As $d\theta/dy = 2/(1-y)^2$, the integral is

$$I_1 = \int_0^\infty f(\theta) f(x|\theta)\,d\theta = \int_{-1}^1 \frac{1}{\mu} e^{-\theta/\mu} \left(\frac{3}{4} - \frac{3}{4}e^{-4\theta/3}\right)^x \left(\frac{1}{4} + \frac{3}{4}e^{-4\theta/3}\right)^{n-x} \times \frac{2}{(1-y)^2}\,dy, \quad (6.74)$$

with θ given in equation (6.73).

The second transform is based on the fact that p goes from 0 to $3/4$ when θ goes from 0 to ∞:

$$y = \frac{8}{3}p - 1 = 1 - 2e^{-4\theta/3}, \quad \theta = -\frac{3}{4}\log\frac{1-y}{2}, \quad (6.75)$$

Table 6.1 Numerical integration calculations of the marginal likelihood for the estimation of JC69 distance ($\times 10^{-131}$)

n	Transform 1	Transform 2 (p)	Transform 3 (t_2)	Transform 4 (Logistic)
Mid-point				
4	0.180	40.6	5.202	5.369
8	0.037	0.002	5.110	5.184
16	14.08	1.043	5.171	5.166
32	3.259	4.736	5.167732	5.167386
64	5.171	5.167828	5.167762	5.167736
128	5.167762	5.167762	5.167762	5.167761
Trapezoid				
4	0.000	0.000	5.234	4.945
8	0.090	20.3	5.218	5.157
16	0.064	10.2	5.164	5.170
32	7.071	5.599	5.167792	5.168198
64	5.165	5.167696	5.167762	5.167792
128	5.167762	5.167762	5.167762	5.1677164
Gaussian–Legendre quadrature				
4	1.032	0.000	4.918	5.104
8	14.5	2.185	5.156	5.1685
16	0.225	9.710	5.167403	5.167764
32	4.241	5.215	5.167762	5.167762
64	5.170	5.16782	5.167762	5.167762
128	5.167762	5.16776	5.167762	5.167762

Note: The correct value is $5.167762 \times 10^{-131}$.

with $-1 < y < 1$. As $d\theta/dy = 3/[4(1-y)]$, the integral becomes

$$I_2 = \int_{-1}^{1} \frac{1}{\mu} e^{-\theta/\mu} \left(\frac{3}{4} - \frac{3}{4} e^{-4\theta/3} \right)^x \left(\frac{1}{4} + \frac{3}{4} e^{-4\theta/3} \right)^{n-x} \times \frac{3}{4(1-y)} dy, \quad (6.76)$$

with θ given in equation (6.75).

The third transform uses a sigmoid function (an S-shaped curve) that rises steeply in the region of the likelihood spike (Figure 6.3b) to flatten the integrand. A good choice of such a function is the CDF for a density that resembles the likelihood (posterior density) curve. Indeed if we use the CDF of the posterior as the transform, the resulting integrand will be flat. Furthermore, we want the CDF to be invertible analytically. Here we use a heavy-tailed t_2 distribution (t distribution with two degrees of freedom) to fit to $\log \theta$ to construct a transform. From the earlier analysis (Example 1.1), we have roughly $\hat{\theta} \sim N = (0.1, 0.01^2)$, so that $\log \hat{\theta}$ should have mean $\mu \approx \log 0.1$ and variance $\sigma^2 \approx 0.01^2 \times (1/0.1)^2 = 0.1^2$ (see equation (13.24) for the variance of a function of a random variable in Appendix B). We thus let $x = (\log \theta - \mu)/\sigma$, with $-\infty < x < \infty$ to be approximated by t_2, with CDF

$$F(x) = \frac{1}{2}\left(1 + \frac{x}{\sqrt{2+x^2}}\right). \quad (6.77)$$

Then

$$y = 2F - 1 = \frac{x}{\sqrt{2 + x^2}}, \quad x = \frac{2y}{\sqrt{1-y^2}}, \qquad (6.78)$$

with $-\infty < x < \infty$ and $-1 < y < 1$ is the transform we seek. In terms of the original variable θ, we have

$$y = \frac{(\log \theta - \mu)/\sigma}{\sqrt{2 + [(\log \theta - \mu)/\sigma]^2}}, \quad \theta = \exp\left\{\mu + 2\sigma \frac{y}{\sqrt{1-y^2}}\right\}. \qquad (6.79)$$

This is plotted in Figure 6.8c. Note that y rises steeply in the region 0.1 ± 0.02 for θ, where the spike in the likelihood is located (Figure 6.8a). As $d\theta/dy = 2\theta\sigma/(1-y)^{3/2}$, the integral becomes

$$I_3 = \int_{-1}^{1} \frac{1}{\mu} e^{-\theta/\mu} \left(\frac{3}{4} - \frac{3}{4} e^{-4\theta/3}\right)^x \left(\frac{1}{4} + \frac{3}{4} e^{-4\theta/3}\right)^{n-x} \times \frac{2\theta\sigma}{(1-y^2)^{3/2}} dy, \qquad (6.80)$$

with θ given in equation (6.79). The new integrand is shown in Figure 6.8d, which is much flatter than the original integrand (Figure 6.8a): note that $\theta \in (0, \infty)$ while $y \in (-1, 1)$.

The fourth transform uses a logistic function to approximate $\log\theta$. Again let $\mu = \log 0.1$ and $\sigma^2 = 0.1^2$. We let $x = \log\theta$, with $-\infty < x < \infty$, to be a logistic variable, with the CDF to be the logistic function

$$F(x) = \frac{1}{1 + e^{-(x-\mu)/\sigma}}. \qquad (6.81)$$

We have the transform

$$y = 2F - 1 = \frac{2}{1 + e^{-(\log\theta - \mu)/\sigma}} - 1, \quad \theta = \exp\left\{\mu + \sigma \log\left(\frac{1+y}{1-y}\right)\right\}. \qquad (6.82)$$

As $d\theta/dy = \theta\sigma \times \frac{2}{1-y^2}$, the integral becomes

$$I_4 = \int_{-1}^{1} \frac{1}{\mu} e^{-\theta/\mu} \left(\frac{3}{4} - \frac{3}{4} e^{-4\theta/3}\right)^x \left(\frac{1}{4} + \frac{3}{4} e^{-4\theta/3}\right)^{n-x} \times \frac{2\theta\sigma}{1-y^2} dy, \qquad (6.83)$$

where θ in the integrand is given in equation (6.82).

The results of these calculations are summarized in Table 6.1. The mid-point and trapezoid methods have similar performance, while the quadrature methods perform better, especially with suitable transforms. Transform has a huge effect on the accuracy of all numerical integration methods. If we require a relative error of $\leq 1\%$, the estimates should be in the interval $(5.116, 5.219) \times 10^{-131}$. This level of accuracy is achieved by Gaussian–Legendre quadrature with 64, 32, 8 and 8 points for transforms I_1, I_2, I_3 (log-t_2), and I_4 (log-logistic), respectively. Even with four points, the log-t_2 or log-logistic transforms for the quadrature method give reasonable estimates. Tests in other contexts (e.g. Yang 2010; Zhu and Yang 2012) also suggest that Gaussian quadrature, if used with appropriate transforms, can achieve good accuracy in one dimension with 8 or 16 points. With two dimensions, 16 or 32 points should give good performance.

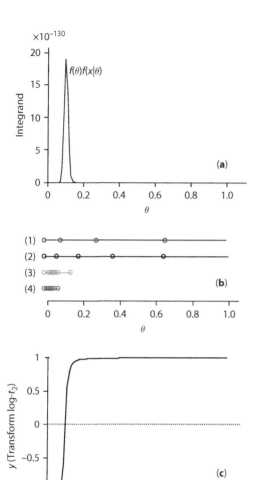
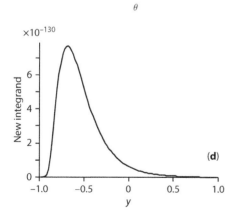

Fig. 6.8 Transforms are important before applying numerical integration algorithms. (**a**) The integrand of equation (6.62). Note that the integral is over $0 \leq \theta < \infty$ (**b**) The points of θ sampled by an eight-point Gaussian–Legendre quadrature when the four transforms discussed in the text are applied. For the first and second transforms, 4 and 3 points are for $\theta > 1$ and thus not shown. (**c**) A sigmoid function based on the log-t_2 distribution is used to transform $0 \leq \theta < \infty$ into $-1 \leq y < 1$ (equation (6.79)). This is transform 3 discussed in the text. (**d**) The resulting new integrand (equation (6.80)).

6.4.5 Monte Carlo integration

MC integration is a simulation method for calculating multidimensional integrals. Suppose we want to compute the expectation of $h(\theta)$ over the density $f(\theta)$, with θ possibly multidimensional:

$$I = E_f\{h(\theta)\} = \int h(\theta)f(\theta)d\theta. \tag{6.84}$$

We draw independent samples $\theta_1, \theta_2, \ldots, \theta_n$ from $f(\theta)$. Then

$$\hat{I} = \frac{1}{n}\sum_{i=1}^{n} h(\theta_i) \tag{6.85}$$

is the MLE of I. Since $h(\theta_i)$ are independent and identically distributed (i.i.d.), $\hat{I}$ has an asymptotic normal distribution with mean I and variance

$$\text{var}(\hat{I}) = \frac{1}{n(n-1)}\sum_{i=1}^{n}\left(h(\theta_i) - \hat{I}\right)^2. \tag{6.86}$$

The 95% confidence interval for I is thus $\hat{I} \pm 1.96\sqrt{\text{var}(\hat{I})}$.

One advantage of MC integration is that the variance of the estimate depends on the sample size n, but not on the dimension of the integral, unlike numerical integration, the performance of which deteriorates rapidly with the increase of dimension. For our problem, $f(\theta)$ is the prior and $h(\theta) = f(x|\theta)$ is the likelihood. This use of MC integration is typically very inefficient, as the prior $f(\theta)$ is often very different from the posterior or the integrand, which is dominated by the likelihood, and as a result, most values of θ sampled from the prior miss the spike in the integrand.

We apply MC integration to calculate the marginal likelihood of equation (6.62). The method is very inefficient and takes about 10^6 samples to reach the level of accuracy of the quadrature method (with log-t_2 and log-logistic transforms) with only 8 or 16 points. While the exponential prior spans the whole positive real line with a slow decay from $1/\mu$ at $\theta = 0$ to 0 at $\theta = \infty$, the likelihood has a spike around $\theta = 0.1$ and the likelihood (posterior) interval (0.082, 0.125) is quite narrow (Figure 6.3). Thus 87% $\left(= e^{-0.082/0.2} - e^{-0.125/0.2}\right)$ of θ values sampled from the prior will be outside this interval and correspond to very small values of $h(\theta)$ while the rest will correspond to high values of $h(\theta)$. The large variation among the $h(\theta_i)$ values means a large variance $\text{var}(\hat{I})$. The problem will be worse if the dataset is larger so that the likelihood is more spiked, or if we use uniform prior, such as $\theta \sim U(0, 100)$. In general, a huge number of samples are needed to obtain acceptable estimates by MC integration.

6.4.6 Importance sampling

Instead of $f(\theta)$, we may sample from a different distribution $g(\theta)$, which is defined on the same parameter space, and then the integral is given as

$$I = E_f\{h(\theta)\} = \int h(\theta)\frac{f(\theta)}{g(\theta)}g(\theta)\,d\theta = E_g\left(h(\theta)\frac{f(\theta)}{g(\theta)}\right), \tag{6.87}$$

where E_f and E_g are expectations over the distributions f and g, respectively. Thus we draw independent samples $\theta_1, \theta_2, \ldots, \theta_n$ from $g(\theta)$, and estimate I by

$$\hat{I}_{IS} = \frac{1}{n}\sum_{i=1}^{n} h(\theta_i)w(\theta_i) = \frac{1}{n}\sum_{i=1}^{n} h(\theta_i)\frac{f(\theta_i)}{g(\theta_i)}, \tag{6.88}$$

where $w(\theta_i) = f(\theta_i)/g(\theta_i)$ is the weight. Similarly to equation (6.86), the variance is given by noting that $h(\theta_i)f(\theta_i)/g(\theta_i)$ are i.i.d. variables:

$$\text{var}(\hat{I}_{IS}) = \frac{1}{n(n-1)} \sum_{i=1}^{n} \left[h(\theta_i) \frac{f(\theta_i)}{g(\theta_i)} - \hat{I}_{IS} \right]^2. \tag{6.89}$$

The density $g(\theta)$ is called the *importance sampling distribution* or *importance function*. By choosing $g(\theta)$ to approximately match the integrand, the importance sampling method can be much more efficient than the simple MC. The optimal sampling distribution is $g(\theta) \propto f(\theta)h(\theta)$, i.e. $g(\theta) = f(\theta)h(\theta)/I$; then each of the n random variables $h(\theta_i)f(\theta_i)/g(\theta_i)$ will equal I, with $\text{var}(\hat{I}_{IS}) = 0$. If $f(\theta)$ is the prior and $h(\theta)$ the likelihood, this $g(\theta)$ will be the posterior. Such a $g(\theta)$ is typically unavailable as it requires knowledge of I, but it highlights the fact that one should construct the importance function to be similar to the posterior.

While the sampling distribution g can be quite arbitrary, those with unbounded ratios f/g are not appropriate. If f/g may be ∞, the weights $f(\theta_i)/g(\theta_i)$ will vary widely, giving too much importance to a few values θ_i. The estimator $\hat{I}_{IS}$ may then have infinite variance and may be unstable. Thus one should use a g that is heavier-tailed than f. Of course we also want $g(\theta)$ in analytical form and want to be able to sample from g.

An alternative form of importance sampling works when we can sample from an unnormalized density g:

$$I'_{IS} = \frac{\sum_{i=1}^{n} h(\theta_i)f(\theta_i)/g(\theta_i)}{\sum_{i=1}^{n} f(\theta_i)/g(\theta_i)}. \tag{6.90}$$

Compared with equation (6.88), n is replaced by the sum of weights. When $n \to \infty$, equation (6.90) also converges to the integral I. This form is not sensitive to unbounded f/g ratios and may perform better than equation (6.88) in some settings (e.g. Robert and Casella 2004, p. 95).

We use importance sampling (equation (6.88)) with log-t_2 or log-logistic as sampling distributions to calculate the marginal likelihood of equation (6.62). In §6.4.4 we used the CDFs for those distributions to transform the parameter from $\theta \in (0, \infty)$ to $y \in (-1, 1)$ for numerical integration. Both are heavier-tailed than the normal distribution. First consider the log-t_2 distribution, with parameters $\mu \approx \log 0.1$ and $\sigma^2 = 0.1^2$. The density is

$$g(t; \mu, \sigma) = \frac{1}{\sigma t} \left[2 + \left(\frac{\log t - \mu}{\sigma} \right)^2 \right]^{-3/2}, \tag{6.91}$$

and the CDF is

$$G(t) = \frac{1}{2} + \frac{1}{2} \frac{\frac{\log t - \mu}{\sigma}}{\sqrt{\left(\frac{\log t - \mu}{\sigma}\right)^2 + 2}}. \tag{6.92}$$

To sample from log-t_2, we use the inversion method (see §12.4.1). Let $u \sim U(0, 1)$. Then

$$\theta = G^{-1}(u) = \exp\left\{ \mu + \sigma \frac{2u - 1}{\sqrt{2u(1-u)}} \right\} \tag{6.93}$$

is a log-t_2 variable. Thus we simulate n random numbers u_i, calculate θ_i according to equation (6.93), and average $f(x|\theta_i)f(\theta_i)/g(\theta_i)$ to calculate I (equation (6.88)).

Table 6.2 MC integration calculation of the marginal likelihood for the estimation of JC69 distance (mean ± 2SE, ×10^{-131})

		Importance sampling	
n	Monte Carlo	log-t_2	log-logistic
10^4	5.270 ± 0.287	5.211 ± 0.039	5.144 ± 0.051
10^5	5.112 ± 0.090	5.170 ± 0.013	5.176 ± 0.016
10^6	5.181 ± 0.029	5.168 ± 0.004	5.163 ± 0.005
10^7	5.1660 ± 0.0090	5.1678 ± 0.0012	5.1678 ± 0.0016
10^8	5.1658 ± 0.0029	5.1675 ± 0.0004	5.1679 ± 0.0005

Note: The correct value is 5.167762 × 10^{-131}.

Next we consider the log-logistic distribution, with density

$$g(t; \mu, \sigma) = \frac{1}{\sigma t} \cdot \frac{e^{-(\log t - \mu)/\sigma}}{\left[1 + e^{-(\log t - \mu)/\sigma}\right]^2}, \qquad (6.94)$$

and CDF

$$G(x) = \frac{1}{1 + e^{-(\log t - \mu)/\sigma}}. \qquad (6.95)$$

To sample from the log-logistic, we again use inversion. Let $u \sim U(0, 1)$. Then

$$\theta = G^{-1}(u) = \exp\left\{\mu + \sigma \log\left(\frac{u}{1-u}\right)\right\} \qquad (6.96)$$

is a log-logistic variable.

The numerical results are summarized in Table 6.2. Importance sampling using either the log-t_2 or log-logistic sampling distributions performed much better than MC integration. In particular, the log-t_2 sampling function is 56 times more efficient than simple MC (efficiency being measured by the ratio of the variances in the estimates). Note that for each method the SE decreases at the rate of $\sqrt{n}$: a 100-fold increase in n leads to a 10-fold reduction in SE.

6.5 Problems

6.1 In Example 6.1 of testing for infection, suppose that a person was tested twice and found to be positive both times. What is the probability that he has the infection?

6.2* The sequence distance θ under JC69 and the probability of difference (p) are related by $p = \frac{3}{4} - \frac{3}{4}e^{-4\theta/3}$, with $0 \leq \theta < \infty$ and $0 \leq p < \frac{3}{4}$. (a) Given $p \sim U(0, \frac{3}{4})$, derive the density for θ. (b) Given that p has the truncated beta distribution of equation (6.58), derive the density for θ. (Hint. Use Theorem 1 in Appendix A.)

6.3* Use the example of normal distributions to study the sensitivity of Bayesian model selection to the prior on parameters in the models, in contrast to the sensitivity of Bayesian parameter estimation to the prior. We use an i.i.d. sample from the normal distribution $N(\mu, 1)$ to estimate the population mean μ and to compare the null hypothesis H_0: $\mu = 0$ against the alternative H_1: $\mu \neq 0$. Let the sample size be $n = 100$

---
* indicates a more difficult or technical problem.

and the sample mean be $\bar{x} = 0.3$. Calculate the p-value for the LRT. Assign the prior probability $1/2$ for each of the two models, and $\mu \sim N(\mu_0, \sigma_0^2)$ under H_1 to calculate the posterior probability for H_0. Use different priors, for example, (i) $\mu_0 = 0$, $\sigma_0^2 = 0.3$; (ii) $\mu_0 = 0$, $\sigma_0^2 = 9$; and (iii) $\mu_0 = -2.95$, $\sigma_0^2 = 1$. Review the theory of §6.2.3.3 and use equation (6.32). Also calculate the posterior of μ under H_1 (equation (6.44)).

6.4 We use a prior in our analysis but take great effort to ensure that it does not have any undue influence on our results. Why don't we avoid the prior in the first place?

CHAPTER 7

Bayesian computation (MCMC)

7.1 Markov chain Monte Carlo

7.1.1 Metropolis algorithm

Markov chain Monte Carlo (MCMC) is a simulation algorithm that generates a dependent sample $\theta_1, \theta_2, \ldots, \theta_n$ from the target density $\pi(\theta)$. Indeed, $\theta_1, \theta_2, \ldots$ form a stationary (discrete-time) Markov chain, with the possible values of θ being the states of the chain. The expectation of a function $h(\theta)$ over π (i.e. an integral)

$$I = E_\pi\{h(\theta)\} = \int h(\theta)\pi(\theta)\,d\theta \qquad (7.1)$$

can then be estimated using the average of $h(\theta)$ over the sample

$$\tilde{I} = \frac{1}{n}\sum_{i=1}^{n} h(\theta_i). \qquad (7.2)$$

Here we use $\tilde{I}$ for the average over a dependent sample from the MCMC algorithm whereas $\hat{I}$ is an average over an independent sample (equation (6.85)). Just like Monte Carlo integration, which generates an independent sample, $\tilde{I}$ is an unbiased estimate of the integral I. However, the variance $\text{var}(\tilde{I})$ is not given by equation (6.86) any more, as we have to account for the fact that the sample is dependent. Let ρ_k be the coefficient of autocorrelation of $h(\theta_i)$ over the Markov chain at lag k: $\rho_k = \text{corr}(h(\theta_i), h(\theta_{i+k}))$. Then the large-sample variance from the MCMC sample is given as

$$\text{var}(\tilde{I}) = \text{var}(\hat{I}) \times [1 + 2(\rho_1 + \rho_2 + \rho_3 + \cdots)], \qquad (7.3)$$

where $\text{var}(\hat{I})$ is the variance for an independent sample (equation (6.86)) (e.g. Diggle 1990, pp. 87–92). Here $\tau = [1 + 2(\rho_1 + \rho_2 + \ldots)]$ is known as the *autocorrelation time*. The variance ratio

$$E = \frac{\text{var}(\hat{I})}{\text{var}(\tilde{I})} = \frac{1}{\tau} = \frac{1}{1 + 2(\rho_1 + \rho_2 + \rho_3 + \cdots)} \qquad (7.4)$$

measures the efficiency of the MCMC, and $nE = n/\tau$ is the *effective sample size* (ESS). A dependent sample of size n is as informative as an independent sample of size n/τ.

When MCMC is used in Bayesian computation, the target density is the posterior $\pi(\theta) = f(\theta|X)$. Note that various quantities of interest are all given in the form of the integral of equation (7.1). For example, if $h(\theta) = \theta$, then $I = E(\theta|X)$, the posterior mean; and if $h(\theta) = \theta^2$, then $I = E(\theta^2|X)$, from which the posterior variance can be calculated. While it is in general hard to generate *independent* samples from the posterior, an MCMC algorithm allows us to generate *dependent* samples.

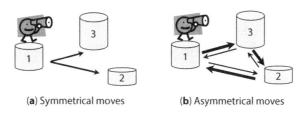

(a) Symmetrical moves   (b) Asymmetrical moves

Fig. 7.1 MCMC algorithm for a discrete parameter that can take three values: 1, 2, and 3. (a) From the current state (say, 1), a new state is proposed by choosing one out of the two alternative states at random, each with probability ½. In the Metropolis algorithm, the proposal density is symmetrical; i.e. the probability of going from box 1 to box 2 is the same as the probability of going from box 2 to box 1. (b) In the Metropolis-Hastings algorithm, the proposal density is asymmetrical. The robot has a 'left bias' and proposes the box on the left with probability 2/3 and the box on the right with probability 1/3. Then a proposal ratio is used in calculating the acceptance ratio to correct for the asymmetry of the proposal.

Here we illustrate the main features of the algorithm of Metropolis et al. (1953) by using a simple example of a robot jumping on three boxes (Figure 7.1a). Parameter θ can take three values: 1, 2, or 3, corresponding to the three boxes, while the heights of the boxes are proportional to the target (posterior) probabilities $\pi(\theta_i) = \pi_i$, for $i = 1, 2, 3$, which we wish to estimate. We give the algorithm before discussing its features:

1. Set initial state (say, box $\theta = 1$).
2. Propose a new state θ', by choosing one of the two alternative states, each with probability ½.
3. Accept or reject the proposal. If $\pi(\theta') > \pi(\theta)$ accept θ'. Otherwise accept θ' with probability $\alpha = \dfrac{\pi(\theta')}{\pi(\theta)}$. If the proposal is accepted, set $\theta = \theta'$. Otherwise set $\theta = \theta$.
4. Print out θ.
5. Go to step 2.

In step 3, one can decide whether to accept or reject the proposal θ' by generating a random number $u \sim U(0, 1)$. If $u < \alpha$, accept θ', or otherwise reject θ'. It is easy to see that in this way the proposal is accepted with probability α. Chapter 12 provides more detailed discussions of random numbers and simulation techniques.

The above is a version of the algorithm of Metropolis et al. (1953), which generates a Markov chain whose states are the possible values of the parameter θ, and whose steady-state distribution is the posterior distribution $\pi(\theta) = f(\theta|X)$. Here we note several important features of the algorithm. First, knowledge of the density ratio $\frac{\pi(\theta')}{\pi(\theta)}$, as opposed to the density $\pi(\theta)$ itself, is sufficient to implement the algorithm. This is the reason why MCMC algorithms can be used to generate (dependent) samples from the posterior. Note that with $\pi(\theta) = f(\theta|X) = f(\theta)f(X|\theta)/f(X)$, the new state θ' is accepted with probability

$$\alpha = \min\left(1, \frac{\pi(\theta')}{\pi(\theta)}\right) = \min\left(1, \frac{f(\theta')f(X|\theta')}{f(\theta)f(X|\theta)}\right). \tag{7.5}$$

Here α is called the *acceptance ratio*. Importantly, the normalizing constant $f(X)$, which is difficult to compute, cancels in calculation of α. In practical implementations, we calculate the logarithms of the prior ratio and likelihood ratio to avoid underflows or overflows, and then $\alpha < 1$ is equivalent to $\log \alpha < 0$.

Second, the sequence of visited states in the algorithm, say

$$1, 3, 3, 3, 2, 1, 1, 3, 2, 3, 3, \ldots$$

constitutes a Markov chain; given the current state, the next state which the chain will visit does not depend on past states.

Third, if we let the algorithm run for a long time, the robot will spend more time on a high box than on a low box. Indeed, the proportions of time the robot will spend on boxes 1, 2, 3 will be exactly π_1, π_2, π_3, so that π is the steady-state distribution of the Markov chain. Thus to estimate $\pi(\theta)$, one has only to run the algorithm for a long time and to calculate the frequencies at which the three states are visited.

It is easy to see that $\pi(\theta)$ is the steady-state distribution of the chain. The following proof is from Metropolis et al. (1953). Consider any two states i and j, and suppose $\pi_i \leq \pi_j$ so that all $i \to j$ moves are accepted with probability 1, while all $j \to i$ moves are accepted with probability π_i/π_j. Suppose the chain is currently visiting the two states with frequencies f_i and f_j. The amount of flow (transition) from i to j is thus $f_i \times 1$, and the amount of flow from j to i is $f_j \times \pi_i/\pi_j$. The net flow from i to j is thus $f_i - f_j \times \pi_i/\pi_j$. This is positive if and only if $f_i/f_j > \pi_i/\pi_j$. If the chain is currently visiting i too often relative to j (compared with the expectation from π_i/π_j), it will tend to move away from i to j. Thus f_i/f_j will change and approach π_i/π_j, and π is the steady-state distribution of the chain. Also the proof establishes that when the chain has reached the steady-state distribution, the amount of flow from any state i to any other state j will equal the amount of flow from j to i. In other words the detailed-balance condition of equation (1.68) holds and the Markov chain is reversible.

The algorithm for continuous parameters is essentially the same as for discrete parameters, except that the state space of the Markov chain is continuous. As an example, we apply the Metropolis algorithm to estimation of sequence distance under the JC69 model of Example 6.4. The reader is invited to write a small program to implement this algorithm (Problem 7.1).

Example 7.1. MCMC algorithm for Bayesian estimation of the JC69 distance between the human and orangutan 12S rRNA genes. The data are $x = 90$ differences out of $n = 948$ sites. The prior is $f(\theta) = \frac{1}{\mu}e^{-\theta/\mu}$, with $\mu = 0.2$. The likelihood is

$$f(x|\theta) = \left(\frac{3}{4} - \frac{3}{4}e^{-4\theta/3}\right)^x \left(\frac{1}{4} + \frac{3}{4}e^{-4\theta/3}\right)^{n-x}. \tag{7.6}$$

The proposal algorithm uses a sliding window of size w.

1. Initialize: $n = 948$, $x = 90$, $w = 0.01$. Set initial state: $\theta = 0.05$, say.
2. Propose a new state as $\theta' \sim U(\theta - w/2, \theta + w/2)$. That is, generate a $U(0, 1)$ random number u, and set $\theta' = \theta - w/2 + wu$. If $\theta' < 0$, set $\theta' = -\theta'$.
3. Calculate the acceptance ratio

$$\alpha = \min\left(1, \frac{\pi(\theta')}{\pi(\theta)}\right) = \min\left(1, \frac{f(\theta')f(x|\theta')}{f(\theta)f(x|\theta)}\right). \tag{7.7}$$

4. Accept or reject the proposal θ'. Draw $u \sim U(0, 1)$. If $u < \alpha$, set $\theta = \theta'$. Otherwise set $\theta = \theta$. Print out θ.
5. Go to step 2.

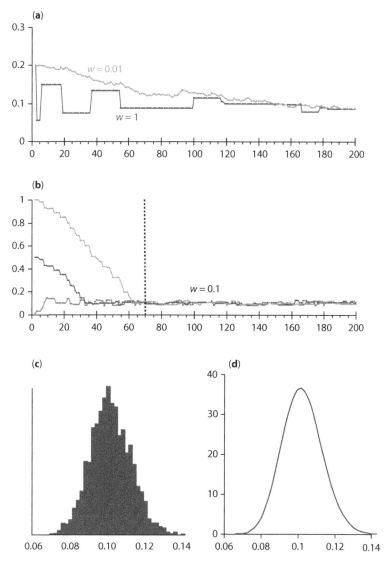

Fig. 7.2 MCMC for estimating sequence distance θ under the JC69 model. The data consist of $x = 90$ differences between two sequences out of $n = 948$ sites. (**a**) Two chains with the window size either too small ($w = 0.01$) or too large ($w = 1$). Both chains started at $\theta = 0.2$. The chain with $w = 0.01$ has an acceptance proportion $P_{\text{jump}} = 91\%$, so that almost every proposal is accepted. However, this chain takes tiny baby steps and mixes poorly. The chain with $w = 1$ has $P_{\text{jump}} = 7\%$, so that most proposals are rejected. The chain often stays at the same state for many iterations without a move. (**b**) Three chains started from $\theta = 0.01, 0.5,$ and 1, with window size $w = 0.1$ and $P_{\text{jump}} = 35\%$, which is nearly optimal. After about 70 iterations, the three chains become indistinguishable and have reached stationarity, so that a burn-in of 100 iterations appears sufficient for those chains. (**c**) Histogram constructed from a sample taken over 10^4 iterations. (**d**) Posterior density obtained from a long chain of 10^7 iterations, sampling every 10 iterations, estimated using a kernel density smoothing algorithm (Silverman 1986).

Figure 7.2a and b shows the first 200 iterations of five independent chains, started from different initial values and using different window sizes (w). Note that if w is too large most proposals will be from unreasonable regions of the parameter space, with $\pi(\theta') \ll \pi(\theta)$, and will be rejected. The chain will then stay in the same state for a long time, causing high correlation among the samples. On the other hand, if w is too small, the proposed states will be very close to the current state, and most proposals will be accepted. However, the chain baby-walks in the same region of the parameter space for a long time, leading again to high correlation. In both cases the variance of the estimate is large (equation (7.3)) and the chain does not mix well. Hence, the optimum should be somewhere between those two extremes. Later in §7.3.2 we will discuss how to determine the optimal window size to maximize the mixing efficiency of the MCMC algorithm. Here it suffices to note that one should adjust the window size so that the *acceptance proportion* or *jump probability* (P_{jump}), the proportion of proposals that are accepted, should be close to ~30%, or in the range 15–70%.

Samples taken before the chain has reached stationarity are usually discarded as *burn-in*. In the three runs of Figure 7.2b, the first 70 or 100 iterations may be so discarded. Figure 7.2c shows a histogram approximation to the posterior probability density estimated from a long chain, while Figure 7.2d is from a very long chain, which is indistinguishable from the distribution calculated using numerical integration (Figure 6.3). □

7.1.2 Asymmetrical moves and proposal ratio

In the Metropolis algorithm, the proposals are symmetrical; i.e. the probability of proposing θ' from θ is equal to the probability of proposing θ from θ'. Hastings (1970) extended the algorithm to allow asymmetrical proposals through a *proposal density* or *jump kernel* $q(\theta'|\theta)$. The resulting Metropolis–Hastings (MH) algorithm involves a simple correction in calculation of the acceptance ratio:

$$\begin{aligned}\alpha(\theta, \theta') &= \min\left(1, \frac{\pi(\theta')}{\pi(\theta)} \times \frac{q(\theta|\theta')}{q(\theta'|\theta)}\right) \\ &= \min\left(1, \frac{f(\theta')}{f(\theta)} \times \frac{f(X|\theta')}{f(X|\theta)} \times \frac{q(\theta|\theta')}{q(\theta'|\theta)}\right) \hspace{2cm} (7.8) \\ &= \min\left(1, \text{prior ratio} \times \text{likelihood ratio} \times \text{proposal ratio}\right).\end{aligned}$$

By using the *proposal ratio* or the *Hastings ratio*, $\frac{q(\theta|\theta')}{q(\theta'|\theta)}$, the correct target density is recovered, even if the proposal is biased.

Suppose that in the robot-on-box example, the robot chooses the left box with probability $\frac{2}{3}$ and the right box with probability $\frac{1}{3}$ (Figure 7.1b). Consider calculation of the proposal ratio for $\theta = 1$ and $\theta' = 2$. We have $q(\theta|\theta') = \frac{2}{3}$, $q(\theta'|\theta) = \frac{1}{3}$, so that the proposal ratio is $\frac{q(\theta|\theta')}{q(\theta'|\theta)} = 2$. Similarly, the proposal ratio for the move from boxes 2 to 1 is $\frac{1}{2}$. Thus by accepting right boxes more often than left boxes, the Markov chain recovers the correct target density even though the robot has a left 'bias' in proposing moves.

For the Markov chain to converge to $\pi(\theta)$, the proposal density $q(\cdot|\cdot)$ has to satisfy certain conditions; it has to specify an irreducible and aperiodic chain. In other words, $q(\cdot|\cdot)$ should allow the chain to reach any other state from any state, and the chain should not have a period. Those conditions are often easily satisfied and also easy to verify.

In most applications, the prior ratio $\frac{f(\theta')}{f(\theta)}$ is easy to calculate. The likelihood ratio $\frac{f(X|\theta')}{f(X|\theta)}$ is often easy to calculate as well, even though the computation may be expensive. The proposal ratio $\frac{q(\theta|\theta')}{q(\theta'|\theta)}$ greatly affects the efficiency of the MCMC algorithm. Therefore a lot of effort is spent on developing good proposal algorithms.

7.1.3 The transition kernel

A Markov chain on a continuous state space is characterized by the transition probability density or transition kernel $p(x, y)$. This is a probability density for y with the current state of the chain x given. If $p(x, y)$ is continuous it will have the usual interpretation that $p(x, y)dy$ is the probability that the chain will move into the interval $(y, y + dy)$ in the next step. Here and in the rest of this chapter, we may use x and y instead of θ to refer to the state of the chain. The stationary distribution $\pi(x)$ is the posterior, and is proportional to the product of the prior and likelihood.

The transition kernel for the Markov chain generated by the MH algorithm is

$$p(x, y) = \begin{cases} q(y|x) \cdot \alpha(x, y), & y \neq x, \\ 1 - \int q(y|x) \cdot \alpha(x, y) \, dy, & y = x, \end{cases} \quad (7.9)$$

where

$$\alpha(x, y) = \min\left(1, \frac{\pi(y)}{\pi(x)} \times \frac{q(x|y)}{q(y|x)}\right) \quad (7.10)$$

is the acceptance ratio of equation (7.8). In other words, given the current state of the chain x,

$$p(x, y)dy = q(y|x) \alpha(x, y) \, dy \quad (7.11)$$

is the probability that the chain will move to the interval $(y, y + dy)$ in the next step, for $y \neq x$. To see that this is the case, note that to move into the interval $(y, y + dy)$ in the next step, two things must happen: (i) a value y from the interval must be proposed, and (ii) this proposed value must be accepted. The first occurs with probability $q(y|x)dy$ and the second with probability $\alpha(x, y)$. The product of those two probabilities gives the transition probability $p(x, y)dy$. The transition kernel $p(x, y)$ generated by an MH algorithm is typically discontinuous at $y = x$, as there is a point mass at that point: $p(x, x)$ in equation (7.9) that is the probability that the proposed value y is rejected so that the chain remains at x.

The *acceptance proportion* or *jump probability*, i.e. the probability that proposals are accepted, is given as

$$P_{\text{jump}} = \iint \pi(x)q(y|x)\alpha(x, y)dx \, dy = \int \pi(x)(1 - p(x, x))dx. \quad (7.12)$$

An example density $p(x, y)$ is plotted later in Figure 7.11a, where the target (posterior) distribution $\pi(x)$ is the standard normal $N(0, 1)$ and the proposal is a normal distribution around the current value: $y|x \sim N(x, \sigma^2)$, with $\sigma = 2.5$. This proposal is described later in §7.2.2. The current value is fixed at $x = 1$ in the plot. Viewed as a density function for y with x given, $p(x, y)$ is not smooth at the point where $\pi(x)q(y|x) = \pi(y)q(x|y)$, i.e. when $y = -x = -1$. Furthermore it is not continuous at $y = x$. The point mass (or the rejection probability) at $x = 1$ can be calculated numerically using equation (7.9), to be $p(x, x) = 0.55$. Thus the distribution $p(1, y)$ is a mixture, with probability 0.55 for the point $y = 1$

and with probability 0.45 from the continuous distribution. The whole transition kernel $p(x, y)$ for different x and y is represented in a heat map in Figure 7.11b; the density of Figure 7.11a is a slice of the heat map across $x = 1$.

7.1.4 Single-component Metropolis–Hastings algorithm

There exist a number of special cases of the general MH algorithm. Below we discuss some commonly used ones, such as the single-component MH algorithm and the Gibbs sampler.

An advantage of Bayesian inference is the ease with which it can deal with sophisticated multi-parameter models. In particular, Bayesian 'marginalization' to deal with nuisance parameters (equation (6.8)) provides an attractive way of accommodating variation in the data that we are not interested in but cannot ignore. In MCMC algorithms for such multi-parameter models, it is often unfeasible or computationally too complicated to update all parameters simultaneously. Instead, it is more convenient to divide parameters into components or blocks, of possibly different dimensions, and then update those components one by one. Different proposals may be used to update different components. Many models have a structure of conditional independence, and this strategy often leads to computational efficiency. Also when one component is being updated, one can treat the other components as fixed constants to design efficient proposals.

Suppose the parameters are grouped into three blocks (x, y, z). One iteration of the MCMC algorithm changes (x, y, z) into (x', y', z'), in three steps.

In step 1, we generate x^* from $q(x^*|x, y, z)$, which is accepted with probability

$$\alpha = \min\left(1, \frac{\pi(x^*, y, z)}{\pi(x, y, z)} \times \frac{q(x|x^*, y, z)}{q(x^*|x, y, z)}\right). \tag{7.13}$$

Note here that $\pi(x, y, z)$ is the posterior of parameters x, y, z but is replaced by the product of the prior and likelihood, since the normalizing constant cancels in the calculation of α. Set $x' = x^*$ if the move is accepted or $x' = x$ otherwise. The state of the chain is now (x', y, z).

In step 2, we generate y^* from $q(y^*|x', y, z)$, and accept the proposal with probability

$$\alpha = \min\left(1, \frac{\pi(x', y^*, z)}{\pi(x', y, z)} \times \frac{q(y|x', y^*, z)}{q(y^*|x', y, z)}\right). \tag{7.14}$$

Set $y' = y^*$ if the move is accepted or $y' = y$ otherwise. The state of the chain is now (x', y', z).

In step 3, we generate z^* from $q(z^*|x', y', z)$, and accept the proposal with probability

$$\alpha = \min\left(1, \frac{\pi(x', y', z^*)}{\pi(x', y', z)} \times \frac{q(z|x', y', z^*)}{q(z^*|x', y', z)}\right). \tag{7.15}$$

Set $z' = z^*$ if the move is accepted or $z' = z$ otherwise. The state of the chain is now (x', y', z').

The iterations will produce a reversible Markov chain with steady-state distribution $\pi(x, y, z)$. In step 2, say, we have

$$\frac{\pi(x', y^*, z)}{\pi(x', y, z)} = \frac{\pi(x', y^*, z)/\pi(x', z)}{\pi(x', y, z)/\pi(x', z)} = \frac{\pi(y^*|x', z)}{\pi(y|x', z)}, \tag{7.16}$$

where $\pi(x, z)$ is the marginal posterior of x and z. The distribution of one component (y) given all other components is called the *full conditional distribution*. Equation (7.16) means that the ratio of the joint posteriors is also the ratio of the full conditionals, and that in each step we are simulating one component from the full conditional distribution. If we ran step 2 alone repetitively, we would be sampling y from $\pi(y|x', z)$.

One benefit of single-component MCMC is that one can use the current values of the other components (which are not changed during the step) to design efficient proposals. For example, when proposing y^* in step 2, we can treat x' and z as constants. Another benefit is that in real applications, the acceptance ratio α often simplifies, leading to computational savings. For example, some components may not be involved in the likelihood function, so there is no need for the expensive likelihood calculation when those components are updated.

A variety of strategies are possible concerning the order of updating the components. One can use a fixed order, or a random permutation of the components. One can also select components for updating with fixed probabilities. However, the probabilities should be fixed and not dependent on the current state of the Markov chain, as otherwise the stationary distribution may no longer be the target distribution $\pi(\cdot)$. It is advisable to group into one block those components that are highly correlated, and update them simultaneously using a proposal density that accounts for the correlation (see §7.2.4).

7.1.5 *Gibbs sampler*

The *Gibbs sampler* (Gelman and Gelman 1984) is a special case of the single-component MH algorithm discussed above. The proposal distribution for updating any component is the conditional distribution of that component given all other components (i.e. the full conditional). For example, to replace step 2 of the algorithm in §7.1.4 by a Gibbs step, we use the proposal density or jump kernel $q(y^*|x', y, z) = \pi(y^*|x', z)$. From equations (7.14) and (7.16), it is clear that this proposal leads to an acceptance ratio of $\alpha = 1$; i.e. all proposals are accepted.

The Gibbs sampler is widely used in inference under linear models in which the prior, the likelihood, and the posterior are all normal distributions. It is seldom used in molecular phylogenetics, as it is typically hard to derive the conditional distributions (but see Lartillot 2006).

7.2 Simple moves and their proposal ratios

The proposal ratio is separate from the prior or the likelihood and is solely dependent on the proposal algorithm. Thus, the same proposals can be used in a variety of Bayesian inference problems. As mentioned earlier, the proposal density has only to specify an aperiodic recurrent Markov chain to guarantee convergence of the MCMC algorithm to the correct target. It is typically easy to construct such a chain and to verify that it satisfies those conditions. For a discrete parameter that takes a set of values, calculation of the proposal ratio often amounts to counting the number of candidate values in the source and target. The case of continuous parameters requires more care. This section lists a few commonly used proposals and their proposal ratios.

Two general results in probability theory are very useful in deriving proposal ratios, and are presented as two theorems in Appendix A. Theorem 1 specifies the probability density of functions of random variables, which are themselves random variables. Theorem 2 gives the proposal ratio when the proposal is formulated as changes to certain functions of the variables (rather than the variables themselves) in the Markov chain. We will often refer to those two theorems.

7.2.1 Sliding window using the uniform proposal

This proposal chooses the new state x' as a random variable from a uniform distribution around the current state x (Figure 7.3a):

$$x' \sim U(x - w/2, x + w/2). \tag{7.17}$$

The window size w is a fixed constant, chosen to achieve a good acceptance rate (see §7.3.2 later). The proposal ratio is 1 since $q(x'|x) = q(x|x') = 1/w$.

Some parameters have constraints. For example, branch lengths must be positive, and the probability has to be in the interval $(0, 1)$. If x is constrained to be in the interval (a, b) but the proposed value is outside this range, the excess is reflected back into the interval. In other words, if $x' < a$, x' is reset to $a + (a - x') = 2a - x'$; and if $x' > b$, x' is reset to $b - (x' - b) = 2b - x'$. The process is repeated until x' is inside the interval. The proposal ratio is still 1, because if x can reach x' through a number of reflections, it must be possible for x' to reach x through the same number of reflections, so that $q(x'|x) = q(x|x')$ always holds. Note that it is incorrect to simply set the unfeasible proposed value to a or b. If the window is much larger than the range $b - a$, it is tedious to jump over the interval (a, b) many times. In such cases it is simpler to calculate the final value x' directly.

Let y be the initial proposed value. Define e to be the excess and note whether the excess is on the left or right. In other words, if $y < a$, the excess $e = a - y$ is on the left; and if $y > b$, the excess $e = y - b$ is on the right. Then $n = \lfloor e/(b-a) \rfloor$, where $\lfloor x \rfloor$ is the integer part of x, is the number of jumps over the interval, and the final excess is $e' = e - n(b - a)$. If n is odd, one changes sides. If the final excess e' is on the left, set $x' = a + e'$; and if it is on the right, set $x' = b - e'$.

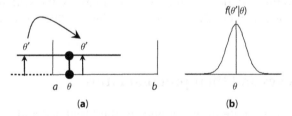

Fig. 7.3 (a) Sliding window using a uniform distribution. The current state is θ. A new value θ' is proposed by sampling uniformly from a sliding window of width w centred at the current value. The window width affects the acceptance proportion. If the proposed value is outside the feasible range (a, b), it is reflected back into the interval; for example, if $\theta' < a$, it is reset to $\theta' = a + (a - \theta') = 2a - \theta'$. (b) Sliding window using a normal distribution. The new value θ' is proposed by sampling from a normal distribution $N(\theta, \sigma^2)$. The standard deviation σ influences the size of the steps taken and plays the same role as the window size w in the proposal of (a).

A special case of the general constraint is that $x \geq 0$. If the proposed value $x' < 0$, one simply resets it to $-x'$. This has been used in Example 7.1.

7.2.2 Sliding window using the normal proposal

This algorithm uses a normal proposal density centred at the current state

$$x'|x \sim N(x, \sigma^2), \tag{7.18}$$

where σ controls the step size (Figure 7.3b). As $q(x'|x) = 1/\sqrt{2\pi\sigma^2} \times \exp\{-(x'-x)^2/(2\sigma^2)\} = q(x|x')$, the proposal ratio is 1. This proposal also works if x is constrained in an interval $x \in (a, b)$. If x' is outside the range, the excess is reflected back into the interval, and the proposal ratio is again 1. Even with reflection, the number of routes from x to x' is the same as from x' to x, and the densities are the same in the opposite directions for each route, even if not between the routes. Thus $q(x'|x) = q(x|x')$ always holds.

Note that sliding window algorithms using either uniform or normal jump kernels are Metropolis algorithms with symmetrical proposals.

7.2.3 Bactrian proposal

The Bactrian proposal density has the shape of a two-humped camel, and avoids states that are very close to the current state (Yang and Rodriguez 2013). Note that a $p:(1-p)$ mixture of two normal distributions $N(m_1, s_1^2)$ and $N(m_2, s_2^2)$ has mean $E(x) = pm_1 + (1-p)m_2$ and variance $V(x) = E(x^2) - [E(x)]^2 = [p(m_1^2 + s_1^2) + (1-p)(m_2^2 + s_2^2)] - [pm_1 + (1-p)m_2]^2 = ps_1^2 + (1-p)s_2^2 + p(1-p)(m_1 - m_2)^2$. In particular, the 1:1 mixture of $N(-m, 1-m^2)$ and $N(m, 1-m^2)$ has mean 0 and variance 1, with parameter m ($0 \leq m < 1$) controlling how spiky the two humps are (Figure 7.4). This may be called the standard Bactrian distribution. To simulate a variable y from it, generate $z \sim N(0, 1)$, and set

$$y = \begin{cases} m + z\sqrt{1-m^2}, & \text{with probability } \frac{1}{2}, \\ -m + z\sqrt{1-m^2}, & \text{with probability } \frac{1}{2}. \end{cases} \tag{7.19}$$

The Bactrian distribution can be used as a sliding window to propose new values around the current value x. We generate y as in equation (7.19), and propose the new value as $x' = x + y\sigma$, where the step size (standard deviation σ) is adjusted to achieve good mixing. The proposal density is

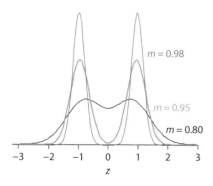

Fig. 7.4 The Bactrian proposal is a 1:1 mixture of two normal distributions (equation 7.20). When used as a proposal in an MCMC algorithm, it favours values that are different from the current value. From Yang and Rodriguez (2013).

$$q(x'|x; m, \sigma^2) = \frac{1}{2\sqrt{2\pi(1-m^2)\sigma^2}} \left[\exp\left\{ -\frac{(x'-x+m\sigma)^2}{2(1-m^2)\sigma^2} \right\} + \exp\left\{ -\frac{(x'-x-m\sigma)^2}{2(1-m^2)\sigma^2} \right\} \right]. \quad (7.20)$$

This is a symmetrical move, with $q(x'|x) = q(x|x')$, so the proposal ratio is 1. The proposal requires generation of an $N(0, 1)$ variate (z) and is as costly as the normal sliding window of equation (7.18). Tests suggest that $m = 0.95$ is a good choice. Numerical tests using different target densities suggest that the Bactrian proposal is more efficient than the uniform or normal sliding windows. We will discuss the mixing efficiency of different proposals in §7.3.2.

7.2.4 Sliding window using the multivariate normal proposal

To change k parameters all at once, one can use a k-dimensional proposal density. For example, one can use the multivariate normal density $q(x'|x) = N_k(x, I\sigma^2)$, where I is the $k \times k$ identity matrix. The proposal ratio is 1. For the case of multinormal target $N_k(0, I)$, Gelman et al. (1996) found through computer simulation the optimal scale factor σ to be 2.4, 1.7, 1.4, 1.2, 1, 0.9, 0.7 for $k = 1, 2, 3, 4, 6, 8, 10$, respectively, with an optimal acceptance proportion $P_{\text{jump}} \approx 0.44$ for $k = 1$, decreasing to about 0.26 for $k > 6$ and with a limit of 0.23 for very large k. It is interesting to note that at low dimensions, one should use over-dispersed proposals relative to the target and take big steps, while at high dimensions, one should use under-dispersed proposals and take small steps.

The proposal $q(x'|x) = N_k(x, I\sigma^2)$ may be inefficient. First, different variables may have different scales (variances), so that use of one scale factor σ may cause the proposal step to be too small for variables with a large variance and too large for those with a small variance. Second, the variables may be strongly correlated, so that ignoring such correlations in the proposal may cause most proposals to be rejected, leading to poor mixing (Figure 7.5). If the posterior density has the variance–covariance matrix S, one may reparametrize the model to use $y = S^{-1/2} x$ as parameters, where $S^{-1/2}$ is the square root of matrix inverse S^{-1}. Then y will have unit variance, so that the above proposal can be used. The second approach is to propose new states using the transformed variables y, i.e. $q(y'|y) = N_k(y, I\sigma^2)$, and then derive the proposal ratio in the original variables x. The proposal ratio is 1 according to Theorem 2 in Appendix A. A third approach is to use the proposal $x' \sim N_k(x, S\sigma^2)$, where σ^2 is chosen to achieve a good acceptance rate. The three approaches should be equivalent and all of them take care of possible differences in scales and possible correlations among the variables. In real data analysis, S is unknown. One may perform short runs of the Markov chain or use the burn-in to obtain an estimate of S, and use it in the proposal. If the normal distribution is a good approximation to the posterior density, those guidelines should be useful.

Fig. 7.5 When two parameters are strongly correlated, it is very inefficient to change one variable at a time in the MCMC, since such proposals will have great difficulty in moving along the ridge of the posterior density. Changing both variables but ignoring the correlation (**a**) is inefficient as well, while accommodating the correlation in the proposal by matching the proposal density to the posterior (**b**) leads to an efficient algorithm. The correlation structure among the parameters may be estimated by running a short chain before MCMC sampling takes place.

7.2.5 Proportional scaling

This proposal, also called multiplier, modifies the current parameter value by multiplying it with a random variable that is around 1. It is useful when the parameter is always positive or always negative. The proposed value is

$$x' = x \cdot c = x \cdot e^{\varepsilon(u-1/2)}, \qquad (7.21)$$

where $c = e^{\varepsilon(u-1/2)}$ and $u \sim U(0, 1)$. Here $\varepsilon > 0$ is a fine-tuning parameter, similar to the window size w in the sliding window proposal. Note that x is shrunk or expanded depending on whether u is $<$ or $> 1/2$. Also x and x' always have the same sign.

The proposal ratio is $|x'/x| = c$. To see this, derive the proposal density $q(x'|x)$ through variable transform, considering the random variable x' to be a function of the random variable u, while treating ε and x as fixed. Note that $f_u(u) = 1, 0 \le u \le 1$, and $dx'/du = x \cdot e^{\varepsilon(u-1/2)} \cdot \varepsilon = \varepsilon x'$. We have, from Theorem 1 in Appendix A,

$$q(x'|x) = f_u(u(x')) \times \left|\frac{du}{dx'}\right| = \frac{1}{\varepsilon |x'|}, \quad xe^{-\varepsilon/2} \le x' \le xe^{\varepsilon/2}. \qquad (7.22)$$

Similarly $q(x|x') = 1/(\varepsilon|x|)$, so the proposal ratio is $q(x|x')/q(x'|x) = |x'/x| = c$.

An alternative approach to deriving this proposal ratio is to view the proposal as the sliding window on the transformed variable $y = \log(x)$. Here we assume x is positive; otherwise use $y = \log(-x)$. In other words, equation (7.21) is equivalent to $y' = y + \varepsilon(u - 1/2)$ or $y' \sim U(y - 1/2\varepsilon, y + 1/2\varepsilon)$; i.e. y' is generated from a sliding window of width ε around y. The proposal ratio on the transformed variable y is 1. Since $\frac{dx}{dy} = e^y$, by Theorem 2 in Appendix A, the proposal ratio for the original variable x is $\left|\frac{dx'}{dy'}\right|/\left|\frac{dx}{dy}\right| = \frac{e^{y'}}{e^y} = x'/x = c$. This argument suggests that the proposal ratio is also $x'/x = c$ if y' is generated from the normal or Bactrian proposals instead of the uniform.

This multiplier proposal is convenient for modifying a variable that is always positive (or always negative). However, if the current value x is 0, it will never move away from 0 by this proposal. If the maximum likelihood estimate (MLE) of x or the posterior mode of x is 0, then almost all moves that reduce x (with $c < 1$) will be accepted, and x will become smaller and smaller during the chain, becoming 0 because of underflow. One way of dealing with this problem is to apply bounds on the variable. If $0 < a < x < b$, the transformed variable $y = \log(x)$ will have bounds $\log\{a\} < y < \log\{b\}$. If the proposed value y' is outside the bounds, the excess is reflected back into the interval; i.e. if $y' < \log\{a\}$, set $y' = 2\log\{a\} - y'$. This is equivalent to setting $x' = a^2/x'$ if $x' < a$, and setting $x' = b^2/x'$ if $x' > b$. As the proposal ratio for the transformed variable y (with reflections) is 1, the proposal ratio is given by the Jacobian term as x'/x (which now differs from c because of the reflections).

One use of the multiplier proposal is to shrink or expand many variables by the same factor c: $x'_i = cx_i$, $i = 1, 2, \ldots, m$, where c is from equation (7.21). If the variables have a fixed order, as in the case of the ages of nodes in a phylogenetic tree (Thorne et al. 1998), the order of the variables will remain unchanged. The proposal is also effective in bringing all variables, such as branch lengths on a phylogeny, to the right scale if all of them are too large or too small. Although m variables are altered, the proposal is in one dimension (along a curve in the m-dimensional space). We can derive the proposal ratio by using the transform: $y_1 = x_1, y_i = x_i/x_1, i = 2, 3, \ldots, m$. The proposal changes y_1, but $y_2, \ldots, y_m$ are unaffected. The proposal ratio in the transformed variables is c. The Jacobian of the transform is $|J(\mathbf{y})| = \left|\frac{\partial \mathbf{x}}{\partial \mathbf{y}}\right| = y_1^{m-1}$. The proposal ratio in the original variables is thus $c \times (y_1'/y_1)^{m-1} = c^m$, according to Theorem 2 in Appendix A.

A related proposal generates c as in equation (7.21), and then multiplies m variables by c and divides n other variables by c (Rannala and Yang 2003). Again, this is a one-dimensional move even though it alters $m + n$ variables. By using a variable transform similar to the one discussed above, one can derive the proposal ratio to be c^{m-n}. This move may be used to deal with strong negative correlations between variables in the posterior. For example, in Bayesian divergence time estimation, the times and rates are strongly correlated. We can generate c, and then multiply all m times by c and divide all n rates by c. As the branch lengths, which are products or rates and times, remain unchanged there is no need to calculate the likelihood for this move.

7.2.6 Proportional scaling with bounds

Suppose a set of m variables has bounds: $a < x_i < b, i = 1, 2, \ldots, m$. Then the transformed variables

$$y_i = (b - x_i) / (x_i - a) \tag{7.23}$$

will all be positive: $0 < y_i < \infty$. We can apply the proportional scaling on the y_is. Let

$$y_i' = y_i \cdot c = y_i \cdot e^{\varepsilon(u-1/2)}, \tag{7.24}$$

where $c = e^{\varepsilon(u-1/2)}$ and $u \sim U(0, 1)$, with $\varepsilon > 0$ to be the step length. Equivalently

$$\frac{b - x_i'}{x_i' - a} = \frac{b - x_i}{x_i - a} \cdot c. \tag{7.25}$$

Then $\frac{dy_i}{dx_i} = -\frac{b-a}{(x_i-a)^2}$. According to Theorem 2 in Appendix A, the proposal ratio is

$$\frac{q(\mathbf{x}|\mathbf{x}')}{q(\mathbf{x}'|\mathbf{x})} = \frac{q(\mathbf{y}|\mathbf{y}')}{q(\mathbf{y}'|\mathbf{y})} \times \frac{|J(\mathbf{y}')|}{|J(\mathbf{y})|} = c^m \times \prod_{i=1}^{m} \left(\frac{x_i' - a}{x_i - a}\right)^2. \tag{7.26}$$

This proposal may be effective for bringing the variables x_is into the right range if they are all too small or too large. It should also be useful when all the x variables are positively correlated.

7.3 Convergence, mixing, and summary of MCMC

7.3.1 Convergence and tail behaviour

7.3.1.1 Rejection rate and light and heavy tails

In Figure 7.2b, the three chains that started from widely different initial values become indistinguishable after ~70 iterations. They have all converged to the stationary distribution of the Markov chain. One may wonder whether this is the behaviour of all Markov chains or what determines the convergence rate of a chain. Here we discuss these issues (Rannala et al. 2012).

We consider first a smooth 1-D unimodal posterior $\pi(x)$, and focus on the right tail where $x \gg 1$. The discussion for the left tail, if the parameter can take negative values, should be similar. Given that an MCMC is a stochastic hill-climbing algorithm, one might think that the behaviour at the tail should depend on how flat the posterior curve is, and that a steeper curve at the tail means the chain will move out of the tail more quickly. We usually measure the flatness of a curve by its gradient (slope), $\nabla \pi = \frac{d\pi}{dx}$. For any proper posterior distribution, the gradient $\nabla \pi \to 0$ as $x \to \infty$, as otherwise the density will not integrate to 1. Thus every posterior surface is nearly flat in the tail and $\nabla \pi$ cannot possibly determine the behaviour of the chain. Instead the tail behaviour of the MCMC is determined by the gradient of the logarithm of the posterior, $\nabla \log(\pi) = \frac{d\log(\pi)}{dx} = \frac{1}{\pi} \cdot \frac{d\pi}{dx}$. This is due to the fact that the acceptance or rejection of a proposal depends on whether the ratio of the posteriors $\frac{\pi(x+\Delta x)}{\pi(x)}$ is close to 1 rather than whether their difference $\pi(x + \Delta x) - \pi(x)$ is close to 0. Here $x' = x + \Delta x$ is the new proposed state. As the chain is far out in the right tail, a move to the left ($\Delta x < 0$) is always uphill, with $\frac{\pi(x+\Delta x)}{\pi(x)} > 1$, and is always accepted. A move to the right ($\Delta x > 0$) is downhill and will be accepted with probability $\frac{\pi(x+\Delta x)}{\pi(x)}$. The chain will behave differently depending on whether the acceptance ratio $\frac{\pi(x+\Delta x)}{\pi(x)}$ for the right move is less than 1 or nearly equal to 1. In the former case, the chain will move out of the tail quickly (the chain is said to have a *geometric convergence*). If the latter case, both left and right moves are accepted with near certainty; the chain then behaves like a random walk, taking 'long excursions' away from the centre of the distribution, leading to an algorithm with serious convergence problems. Note that a random walk along a line is guaranteed to return to the origin but the expected time for it to do so is infinite.

More formally, Mengersen and Tweedie (1996) defined *geometric convergence* as follows. The Markov chain is said to have a geometric convergence if the distance between the stationary distribution $\pi(x)$ and the distribution that the MCMC reaches in n steps decreases to 0 at least as fast as r^n, with a certain $r < 1$. They showed that in one dimension, the MCMC algorithm (such as a symmetrical sliding window) achieves geometric convergence if and only if

$$\lim_{x \to \infty} \nabla \log \pi = \lim_{x \to \infty} \frac{d \log \pi(x)}{dx} < 0. \tag{7.27}$$

In contrast, if $\nabla \log \pi \to 0$ as $x \to \infty$, geometric convergence will not be possible. Note that heuristically, $\frac{\pi(x+\Delta x)}{\pi(x)} < 1$ (or ≈ 1) means that $\frac{\log \pi(x+\Delta x) - \log \pi(x)}{\Delta x} < 0$ (or ≈ 0). Criterion (7.27) is met if and only if the posterior is light-tailed. The exponential, normal, and gamma are examples of light-tailed distributions while the Cauchy, the t, and inverse gamma are heavy-tailed.

As an example of light-tailed posteriors, consider the standard normal distribution

$$\pi(x) \propto e^{-\frac{1}{2}x^2}. \tag{7.28}$$

The gradient is $\nabla \pi = -x\pi(x)$, and $\nabla \log \pi = -x$. Suppose the current state is $x = 10$, far away from the posterior mode 0, and consider a proposal that changes the current x by $\Delta x = 1$. We have $R_{\text{left}} = \frac{\pi(9)}{\pi(10)} = 1.3 \times 10^4 > 1$ so the left move is always accepted. As $R_{\text{right}} = \frac{\pi(11)}{\pi(10)} = 2.8 \times 10^{-5}$, the right move is nearly always rejected. This means that half the time the chain moves to the left (towards the mode), and half the time it stays still. The chain will thus quickly move out of the tail. Further out in the tail, at $x = 100$, the ratios are $R_{\text{left}} = \frac{\pi(99)}{\pi(100)} = 1.6 \times 10^{43}$ and $R_{\text{right}} = \frac{\pi(101)}{\pi(100)} = 2.3 \times 10^{-44}$, which are even more extreme than for $x = 10$. The farther away from the mode, the flatter the tail is (i.e. the gradient $\nabla \pi$ is closer to 0) but the faster the MCMC moves out of the tail. This behaviour may appear paradoxical and is referred to as the 'tail paradox' by Rannala et al. (2012).

The behaviour is similar with the estimation of the JC69 distance θ using an exponential prior (Example 7.1). Far out in the tail (say, with $\theta = 100$ or 10), the likelihood is flat so that the likelihood ratio is ~1. The acceptance ratio is then determined by the prior ratio. Since the exponential is light-tailed, the chain will converge quickly (See Problem 7.3).

As an example of heavy-tailed posteriors, consider the inverse gamma distribution invG(3, 1). The density is

$$\pi(x) \propto e^{-\beta/x} x^{-\alpha-1}, \quad x > 0, \tag{7.29}$$

with $\alpha = 3$ and $\beta = 1$. This distribution has mode at $\beta/(\alpha+1) = 0.25$ and mean $\beta/(\alpha-1) = 0.5$. Suppose the current state is $x = 100$, and a sliding window changes x by $\Delta x = 0.1$. We have $R_{\text{left}} = \pi(99.9)/\pi(100) = 1.004$ and $R_{\text{right}} = \pi(100.1)/\pi(100) = 0.996$. Both left and right moves are accepted with near certainty, and the chain behaves like a random walk. It may take a very long time for the chain to move to the centre of the distribution.

The same situation occurs with the estimation of the JC69 distance θ of Example 7.1 if a uniform prior, say $U(0, 200)$, is used instead of the exponential. Far out in the tail, the likelihood is flat so that the likelihood ratio is ~1, and with a flat prior, the posterior will be flat as well. Convergence will be very slow if the chain is started far out in the tail, say, with $\theta = 100$ (see Problem 7.3).

In multiple dimensions, the posterior $\pi(x)$ must be light-tailed and in addition must be sufficiently 'smooth' in the tail to achieve geometric convergence (Roberts and Tweedie 1996). We will not go into the technical definition of 'smoothness' here, but will study a counter-example provided by the authors. This has the density

$$\pi(x, y) \propto e^{-(x^2 + x^2 y^2 + y^2)}, \quad -\infty < x, y < \infty. \tag{7.30}$$

A contour plot is shown in Figure 7.6a. Even though the posterior surface is smooth in the sense that all derivatives exist, there are sharp ridges in the tails along the x- and y-axes, which can cause serious convergence problems. Suppose the current state is $(x, y) = (100, 0)$ and we use a 2-D sliding window of side ε to propose moves: $x' \sim U(x - \frac{\varepsilon}{2}, x + \frac{\varepsilon}{2})$ and $y' \sim U(y - \frac{\varepsilon}{2}, y + \frac{\varepsilon}{2})$. As long as ε is fixed and finite, a random point in the window

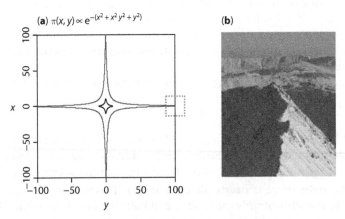

Fig. 7.6 (a) The contour for the 2-D posterior density $\pi(x, y) \propto \exp\{-(x^2 + x^2 y^2 + y^2)\}$, which causes convergence problems for the 2-D sliding window move. Roberts and Tweedie (1996) show that far out in the tail, at $(x, 0)$, essentially every point (proposal) in the sliding window is a fall off the cliff and will be rejected. (b) On Striding Edge (Lake District), a random step is very likely to lead to a fall off the cliff.

has probability ~1 of falling off the cliff and being rejected. At such a point, the expected acceptance proportion is $P_{\text{jump}} \approx 0$, and the chain is stuck in the tail. In fact, Roberts and Tweedie (1996) used an infinite size for the sliding window at the point $(k, 0)$, with the side of the window being $\sqrt{k}$, to prove that $P_{\text{jump}} \to 0$ as $k \to \infty$. Note that in this example, two 1-D moves along the x- and y-axes, respectively, will quickly bring the chain out of the tail.

The situation appears to be similar if one estimates both time t and rate r from a pair of sequences in Example 7.1. When the sequences are long, the likelihood is highly informative, so that the posterior is concentrated along the ridge where $2 \cdot t \cdot r = \hat{\theta}$, the MLE of the sequence distance (Figure 7.7). It is unclear whether the smoothness conditions of Roberts and Tweedie (1996, p. 105) are satisfied in this case. At any rate, if we use a 2-D sliding window far out in the tail, say at $t = 5$ and $r = 0.01$, the chain will be in serious trouble, as nearly all proposed points lead to a fall off the cliff and are rejected. However, if we change t and r simultaneously, with their product fixed, the chain will move out of the tail quickly. See Problem 7.5.

7.3.1.2 Multiple modes in the posterior

Multiple modes in the posterior can cause serious convergence and mixing problems. One such case is a serious conflict between the prior and the likelihood. For example, in the estimation of sequence distance between two sequences, if the prior for θ has a mode that is distant from the mode of the likelihood, it is possible for the posterior to have two modes, one near the mode of the likelihood and another near the mode of the prior. Although the likelihood-induced mode may be many orders of magnitude higher than the prior-induced mode, it is possible for the MCMC to be trapped at the lower mode. MCMC algorithms may then have serious convergence problems.

Example 7.2. We use a highly informative but unreasonable prior to estimate the distance between two sequences under the JC69 model, so that the prior and likelihood are in serious conflict, leading to two modes in the posterior density (Rannala et al. 2012). The

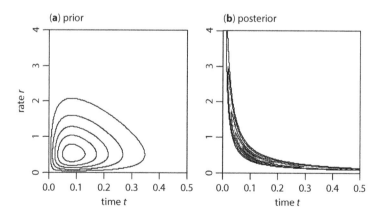

Fig. 7.7 The (a) prior and (b) posterior densities for time t and rate r when the two sequences of Example 7.1 are analysed, with gamma priors $t \sim G(2, 12)$ and $r \sim G(2, 2)$. There is a ridge on the posterior along the line $2 \cdot t \cdot r = \hat{\theta}$, where $\hat{\theta} = 0.1015$ is the MLE of the sequence distance. A 2-D sliding window will have a very low acceptance probability in the tail (Problem 7.5).

data are the human and chimpanzee mitochondrial 12S rRNA genes, with $x = 11$ differences at $n = 946$ sites. This is the same segment of the mitochondrial genome as was used in Example 7.1, but the species are different. The MLE of sequence distance θ is 0.01172. For Bayesian estimation of θ, we assign a gamma prior

$$f(\theta) = \frac{\beta}{\Gamma(\alpha)} e^{-\beta\theta} \theta^{\alpha-1}, \qquad (7.31)$$

and let $\alpha = 100$ and $\beta = 10$, with the prior mean 10 and variance 1, in place of the exponential prior of Example 7.1. This gamma prior is informative and unreasonable, in serious conflict with the likelihood. The posterior is thus

$$\pi(\theta) = f(\theta|x) = \frac{1}{C} f(\theta) f(x|\theta), \qquad (7.32)$$

where the likelihood $f(x|\theta)$ is given in equation (7.6). The normalizing constant is calculated numerically to be $C = 3.83834 \times 10^{-208}$. The likelihood and posterior are plotted in Figure 7.8. The posterior mean is 0.1217 with the 95% equal-tail credibility interval (CI) to be (0.0997, 0.1460). The posterior estimates are 10 times too large compared with the MLE, due to the impact of the prior. Furthermore, the conflicting prior creates two modes in the posterior, at $\theta = 0.120468$ with $\pi(\theta) = 33.9$ and at $\theta = 9.89311$ with $\pi(\theta) = 5.26 \times 10^{-358}$. While the lower peak is so much lower than the higher peak that its contribution to the posterior density is negligible, it can nevertheless cause convergence problems if the chain starts far out in the tail, say, with $x > 7$ (see Problem 7.6). □

Another example involving a large real dataset is discussed later in Chapter 8 (Problem 8.2), where the default prior on branch lengths in MrBayes (version 3.2 or earlier) induces a strong unreasonable prior on tree length (sum of branch lengths), leading to conflict with the likelihood and creating a second peak. Again the lower peak is much lower than the high peak. Later in §7.4.1, we will discuss the parallel tempering algorithm, which is designed to help the MCMC to navigate over a multimodal posterior.

7.3.2 Mixing efficiency, jump probability, and step length

Following Example 7.1, we briefly discussed the impact of the window size in the sliding window proposal on the mixing efficiency and acceptance proportion of the MCMC algorithm. We pointed out that both very large and very small windows lead to strong positive correlations in the sample and to a poorly mixing Markov chain. In this subsection we discuss the mixing efficiency of MCMC algorithms in detail. As mentioned earlier, the efficiency of an MCMC algorithm may be measured by the ratio of variances calculated using dependent versus independent samples (equation (7.4)). As a Markov chain is characterized by its transition matrix (or the transition kernel for continuous states, equation (7.11)), we study the efficiency of an MCMC algorithm by examining the transition matrix for the Markov chain that the algorithm generates. Discrete state Markov chains are more tractable than continuous state chains, so we consider discrete state chains first and later use the theory to study the performance of continuous state chains.

7.3.2.1 Discrete state chains

The setup. Let $\pi = (\pi_1, \pi_2, \ldots, \pi_K)$ be a probability distribution on the discrete states $\{1, 2, \ldots, K\}$. We assume that $\pi_i > 0$ for all i. We wish to estimate the expectation

$$I = E_\pi\{h(X)\} = \sum_{i=1}^{K} \pi_i h(i). \qquad (7.33)$$

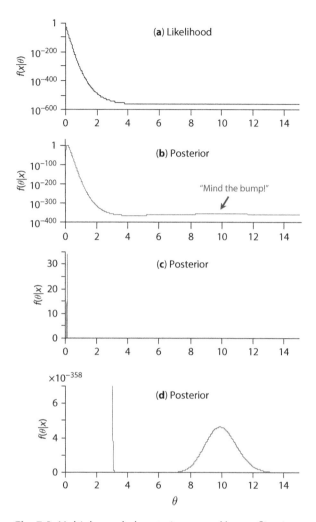

Fig. 7.8 Multiple-moded posterior created by conflicts in the prior and data. The human and chimpanzee mitochondrial 12S rRNA genes, with $x = 11$ differences at $n = 946$ sites, are used to estimate the sequence distance θ under the JC69 model, with the gamma prior $\theta \sim G(100, 10)$. The likelihood is shown in (**a**). The posterior is shown in **b–d**, using different scales for the y-axis. There are two modes in the posterior: at $\theta = 0.120468$ with $\pi(\theta) = 33.9$ and $\theta = 9.89311$ with $\pi(\theta) = 5.26 \times 10^{-358}$. Between the two modes, there is a minimum at $\theta = 4.204258$ with $\pi(\theta) = 1.17 \times 10^{-365}$. Redrawn following Rannala et al. (2012).

Suppose an irreducible reversible Markov chain has a $K \times K$ transition matrix $P = \{p_{ij}\}$, which has π as the steady-state distribution. This Markov chain may be generated by an MCMC algorithm; see Problem 7.7 for the P matrix generated for the robot-on-box example in §7.1. We use a sample $\{X_1, X_2, \ldots, X_n\}$ from the Markov chain to estimate I:

$$\tilde{I} = \frac{1}{n} \sum_{i=1}^{n} h(X_i). \tag{7.34}$$

As in the case of continuous states (equations (7.1) and (7.2)), this setup is general. For example, if the K states represent different models, I will be the posterior probability of model k if we let $h(X) = 1$ for $X = k$ and 0 otherwise. One benefit of this setup is that the asymptotic variance var($\tilde{I}$) of equation (7.3) can be calculated analytically using the transition matrix P.

Let $h = (h_1, h_2, \ldots, h_K)^T$ be the column vector of function values, with $h_i = h(i)$. Define $A = \{a_{ij}\}$, with $a_{ij} = \pi_j$, to be the 'limiting matrix', $Z = [I - (P - A)]^{-1}$ to be the 'fundamental matrix' (Kemeny and Snell 1960, p. 74; Peskun 1973), and $B = \text{diag}\{\pi_1, \pi_2, \ldots, \pi_K\}$. Then var($\tilde{I}$) of equation (7.3) is equal to v/n, with

$$v = h^T \cdot B \, (2Z - I - A) \cdot h, \tag{7.35}$$

where h^T is the transpose of h (Kemeny and Snell 1960, p. 84; Peskun 1973).

This can also be calculated using the eigenvalues and eigenvectors of P. Let $1 = \lambda_1 > \lambda_2 \geq \lambda_3 \geq \ldots \lambda_K \geq -1$ be the eigenvalues of P and columns of E be the corresponding (right) eigenvectors. As the chain is irreducible, only one eigenvalue is equal to 1, with $\lambda_i < 1$ for all $i \geq 2$. Also the value -1 is possible only if the chain is periodic. The eigenvalues and eigenvectors of P can be calculated as follows (this is the same theory of §2.6 for the diagonalization of the rate matrix for a reversible Markov chain). From the detailed-balance condition, $T = B^{1/2} P B^{-1/2}$ is a symmetrical matrix so that its eigenvalues are all real and can be calculated using standard algorithms. Let $T = R \Lambda R^T$, where $\Lambda = \text{diag}\{\lambda_1, \lambda_2, \ldots, \lambda_K\}$, and $R^T = R^{-1}$. Then

$$P = B^{-1/2} T B^{1/2} = \left(B^{-1/2} R\right) \Lambda \left(R^T B^{1/2}\right) = E \Lambda E^T B, \tag{7.36}$$

with $E = B^{-1/2} R$. Note that the eigenvectors are normalized such that $E^T B E = I$ and $P^n = E \Lambda^n E^{-T} = E \Lambda^n E^T B$ for any integer n. Then equation (7.35) can also be written as

$$v = \sum_{k \geq 2}^{K} \frac{1 + \lambda_k}{1 - \lambda_k} \left(E^T B h\right)_k^2, \tag{7.37}$$

where $(a)_k$ is the kth element of vector a (Sokal 1989; Frigessi et al. 1992; Green and Han 1992). This formula underlies the notion that a small λ_2 (the second largest eigenvalue of P) is associated with efficient mixing.

The acceptance proportion (the proportion of proposals that are accepted) for the MCMC algorithm is given as

$$P_{\text{jump}} = \sum_{i=1}^{K} \pi_i (1 - p_{ii}), \tag{7.38}$$

which is an average over the states (cf.: equation (7.12)). Note that p_{ii} is the probability that the proposals are rejected when the chain is in state i.

Peskun's theorem. Consider two irreducible reversible Markov transition matrices $P^{(1)}$ and $P^{(2)}$, both having the steady-state distribution π. If $p^{(1)}_{ij} \geq p^{(2)}_{ij}$ for all $i \neq j$, i.e. if $P^{(1)}$ has off-diagonal elements that are larger than or equal to the corresponding off-diagonal elements in $P^{(2)}$, then $P^{(1)}$ is more efficient than $P^{(2)}$, irrespective of the function h. In other words, var($\tilde{I}$) based on the sample from $P^{(1)}$ will be smaller than that based on $P^{(2)}$. This is known as Peskun's (1973) theorem. While in the long run the two chains visit the states with the same frequencies (as they have the same steady-state distribution π), the efficient chain moves between states very often while the inefficient chain stays in the same state for a long time before it moves (and then it stays in the new state for a long time). In general a more mobile chain tends to be more efficient.

The case of two states. We now consider the case of two states in more detail. The transition matrix

$$P = \begin{bmatrix} 1-p_{12} & p_{12} \\ p_{21} & 1-p_{21} \end{bmatrix} \qquad (7.39)$$

has the steady-state distribution $\pi_1 = p_{21}/(p_{12}+p_{21})$ and $\pi_2 = p_{12}/(p_{12}+p_{21})$. Here (π_1, π_2) may represent the posterior probabilities for two models H_1 and H_2. Assume $\pi_1 \geq \pi_2 > 0$. From $0 \leq p_{12} \leq p_{21} \leq 1$ and $p_{12} = p_{21}\frac{\pi_2}{\pi_1}$, we have the maximum off-diagonal elements to be $p_{21} = 1$ and $p_{12} = \pi_2/\pi_1$. In other words, to achieve the highest P_{jump}, the chain should move to H_1 with certainty if it is in H_2, and move to H_2 with probability π_2/π_1 if it is in H_1. Such a chain has $P_{\text{jump}} = \pi_1 \times \pi_2/\pi_1 + \pi_2 \times 1 = 2\pi_2$.

As there are only two values for function h, the precise function is unimportant when we compare the efficiency of different chains. Suppose we are interested in estimating π_1; in other words, $h = 1$ if $X = 1$ and $h = 0$ if $X = 2$. The variance of the estimate from an independent sample of size n is $\pi_1\pi_2/n$. Equation (7.35) then gives the efficiency for a dependent sample as v/n, with $v = \pi_1\pi_2(2\pi_2 - p_{12})/p_{12}$. The efficiency of the dependent sample is then

$$E = \frac{\pi_1 \pi_2}{v} = \frac{p_{12}}{2\pi_2 - p_{12}}. \qquad (7.40)$$

The larger the p_{12}, the more efficient the Markov chain. The optimum is achieved when $p_{12} = \pi_2/\pi_1$ and $p_{21} = 1$, which gives the highest $P_{\text{jump}} = 2(1-\pi_1)$ and maximum efficiency $E = 1/(2\pi_1 - 1)$. In the special case $\pi_1 = \pi_2 = 1/2$, we can achieve $P_{\text{jump}} = 1$ (i.e. $p_{12} = p_{21} = 1$), with every proposal being accepted, as well as maximal efficiency $E = \infty$ (and $v = 0$). The chain is then periodic.

The case of more than two states. The case of more than two states is more complex. For convenience suppose the states are ordered so that $\pi_1 \geq \pi_2 \geq \ldots \geq \pi_K$. First, we note that the highest jump probability is $P_{\text{jump}} = 2(1-\pi_1)$. This can be proved as follows. Since π is the steady-state distribution, we have

$$\begin{aligned} & \sum_i \pi_i p_{i1} = \pi_1, \\ & \pi_1 p_{11} + \sum_{i \geq 2} \pi_i p_{i1} = \pi_1, \qquad (7.41) \\ & \pi_1(1 - p_{11}) = \sum_{i \geq 2} \pi_i p_{i1} \leq \sum_{i \geq 2} \pi_i = 1 - \pi_1. \end{aligned}$$

Then $P_{\text{jump}} = \pi_1(1-p_{11}) + \sum_{i \geq 2} \pi_i(1-p_{ii}) \leq (1-\pi_1) + \sum_{i \geq 2} \pi_i = 2(1-\pi_1)$. In other words, the highest state probability π_1 places a limit on the acceptance proportion; if the chain jumped too often it would not be spending enough time in state 1 to achieve the correct π_1. If $\pi_1 \geq 1/2$, the highest jump probability, $2(1-\pi_1)$, is achieved by having $p_{i1} = 1$ and $p_{1i} = \pi_i/\pi_1$ for all $i \geq 2$, so that

$$P = \begin{bmatrix} 1-\frac{1-\pi_1}{\pi_1} & \frac{\pi_2}{\pi_1} & \frac{\pi_3}{\pi_1} & \cdots & \frac{\pi_K}{\pi_1} \\ 1 & 0 & 0 & \cdots & 0 \\ 1 & 0 & 0 & \cdots & 0 \\ \vdots & \vdots & \vdots & \ddots & 0 \\ 1 & 0 & 0 & \cdots & 0 \end{bmatrix}. \qquad (7.42)$$

If $\pi_1 < 1/2$, there are many ways of achieving $P_{\text{jump}} = 1$.

In contrast to P_{jump}, the efficiency E is less understood (Mira 2001). While Peskun's theorem implies that a high P_{jump} is in general preferable, efficiency may depend on the function h and may not be entirely determined by P_{jump}. For example, the most efficient chains for estimating the smallest and the largest of π_is may well be different. Here we mention two constructions.

Frigessi et al. (1992) have shown that the second largest eigenvalue for the transition matrix P (which is reversible with steady-state distribution π) has the limit

$$\lambda_2 \geq -\frac{\pi_K}{1 - \pi_K}. \tag{7.43}$$

Furthermore, the matrix P that achieves this lower bound, $\lambda_2 = -\frac{\pi_K}{(1-\pi_K)}$, has its last column (and last row) uniquely determined, as follows:

$$P = \begin{bmatrix} & & & -\lambda_2 \\ & P_2 & & \vdots \\ & & & -\lambda_2 \\ -\frac{\pi_1}{\pi_K}\lambda_2 & \cdots & -\frac{\pi_{K-1}}{\pi_K}\lambda_2 & 0 \end{bmatrix}. \tag{7.44}$$

The diagonal has $p_{KK} = 0$, all other elements on the last column are $-\lambda_2$, and the last row is given by the detailed-balance condition: $p_{Kj} = p_{jK}\frac{\pi_j}{\pi_K}, j = 1, \ldots, K-1$.

The same argument is applied successively to submatrices of smaller sizes ($P_2, P_3, \ldots$), so that the eigenvalues, $\lambda_2 \geq \lambda_3 \geq \ldots \geq \lambda_K$, are uniquely determined:

$$\lambda_{j+1} = -\frac{\pi_{K-j+1}}{\pi_{K-j} + \pi_{K-j-1} + \cdots + \pi_1} \cdot \prod_{i=1}^{j-1}\left(1 - \frac{\pi_{K-i+1}}{\pi_{K-i} + \pi_{K-i-1} + \cdots + \pi_1}\right). \tag{7.45}$$

Also $\lambda_{j+1} = \lambda_j$ if and only if $\pi_{K-j+1} = \pi_{K-j+2}$. The procedure leads to the following P matrix:

$$P = \begin{bmatrix} 1 + \lambda_2 + \cdots \lambda_K & -\lambda_K & \cdots & -\lambda_3 & -\lambda_2 \\ -\frac{\pi_1}{\pi_2}\lambda_K & 0 & \cdots & -\lambda_3 & -\lambda_2 \\ \vdots & \vdots & \ddots & \vdots & \vdots \\ -\frac{\pi_1}{\pi_{K-1}}\lambda_3 & -\frac{\pi_2}{\pi_{K-1}}\lambda_3 & \cdots & 0 & -\lambda_2 \\ -\frac{\pi_1}{\pi_K}\lambda_2 & -\frac{\pi_2}{\pi_K}\lambda_2 & \cdots & -\frac{\pi_{K-1}}{\pi_K}\lambda_2 & 0 \end{bmatrix}. \tag{7.46}$$

All diagonal elements of P are 0 except p_{11}, which is zero if and only if $\pi_1 = \pi_2$. Each column has the same entry above the diagonal, which is $-\lambda_{K-j+2}$. Note that $\lambda_j < 0$, for all $j \geq 2$. The average jump probability is $P_{\text{jump}} = 1 - \pi_1 \sum_{j \geq 1} \lambda_j$.

By minimizing the eigenvalues, the matrix achieves good efficiency without knowledge of the function h. However, it is possible to construct more efficient chains if knowledge of the function h is available (Frigessi et al. 1992).

Another construction is the P matrix defined in equation (7.42), under the assumption that $\pi_1 \geq \frac{1}{2}$. This seems to be the most efficient chain if we want to estimate π_1. It can be show that the eigenvalues of this P matrix are $\lambda_1 = 1, \lambda_2 = \lambda_3 = \ldots = \lambda_{k-1} = 0$, and $\lambda_k = 1 - \frac{1}{\pi_1}$ (Problem 7.8). The asymptotic variance for estimating π_1 is $v = \pi_1(1 - \pi_1)(2\pi_1 - 1)$, so that efficiency is $E = \pi_1(1 - \pi_1)/v = 1/(2\pi_1 - 1)$ (Problem 7.9). This chain is more efficient than the independent sampler, with $E > 1$, and appears to have the maximum efficiency possible, although a proof is yet to be found. When $\pi_1 = \frac{1}{2}$, this chain achieves $P_{\text{jump}} = 1$ and $v = 0$ and is the most efficient possible.

7.3 CONVERGENCE, MIXING, AND SUMMARY OF MCMC

Here we characterize a few example proposals for the case of $K = 4$ in Table 7.1. Note that, given the stationary distribution π, different proposals correspond to different MCMC algorithms and different transition matrices (P), with different mixing efficiencies (E). Again label the states in the order of decreasing posterior probabilities.

a. Algorithm P_0 is the independent sampler, $q_{ij} = \pi_j$, with acceptance rate 1. However, we count only changes to different states so that $P_{\text{jump}} = 1 - \sum_i \pi_i^2$.

b. Algorithm P_1 proposes states different from the current state i with equal probability, with $q_{ij} = \frac{1}{K-1}, j \neq i$. The proposal from $i \to j$ is accepted with probability $\alpha_{ij} = \min\{1, \frac{\pi_j}{\pi_i}\}$. Thus $p_{ij} = q_{ij}\alpha_{ij} = \frac{1}{K-1}\frac{\pi_j}{\pi_i}$ if $i < j$ or $\frac{1}{K-1}$ if $i > j$. The average jump probability, $P_{\text{jump}} = \frac{1}{K-1}\sum_{i>j} 2\pi_i = \frac{2}{K-1}[\pi_2 + 2\pi_3 + \cdots + (K-1)\pi_K]$.

c. Algorithm P_2 is the construction of Frigessi et al. (1992) (equation (7.46)).

d. Algorithm P_3 is the one with maximum P_{jump} (equation (7.42)).

The transition matrices and calculations for the posterior target $\pi = (0.6, 0.2, 0.1, 0.1)$ are shown in Table 7.1. In this example, P_2 and P_3 are 1.5 and 5 times as efficient as the independent sampler, respectively. The highest P_{jump} is $2(1 - 0.6) = 0.8$.

Table 7.1 Characterizations of a few Markov chains with stationary distribution $\pi = \{0.6, 0.2, 0.1, 0.1\}$ for estimating π_1

MCMC algorithm (transition matrix P)	P_{jump}	v	E	Eigenvalues
$P_0 = \begin{bmatrix} \pi_1 & \pi_2 & \pi_3 & \pi_4 \\ \pi_1 & \pi_2 & \pi_3 & \pi_4 \\ \pi_1 & \pi_2 & \pi_3 & \pi_4 \\ \pi_1 & \pi_2 & \pi_3 & \pi_4 \end{bmatrix}$	0.580	0.240	1	$(1, 0, 0, 0)$
$P_1 = \begin{bmatrix} 1-\frac{1-\pi_1}{3\pi_1} & \frac{\pi_2}{3\pi_1} & \frac{\pi_3}{3\pi_1} & \frac{\pi_4}{3\pi_1} \\ \frac{1}{3} & \frac{2}{3}-\frac{\pi_3+\pi_4}{3\pi_2} & \frac{\pi_3}{3\pi_2} & \frac{\pi_4}{3\pi_2} \\ \frac{1}{3} & \frac{1}{3} & \frac{1}{3}-\frac{\pi_4}{3\pi_3} & \frac{\pi_4}{3\pi_3} \\ \frac{1}{3} & \frac{1}{3} & \frac{1}{3} & 0 \end{bmatrix}$	0.467	0.624	0.385	$(1, 0.444, 0, -0.333)$
$P_2 = \begin{bmatrix} 1+\lambda_2+\lambda_3+\lambda_4 & -\lambda_4 & -\lambda_3 & -\lambda_2 \\ -\frac{\pi_1}{\pi_2}\lambda_4 & 0 & -\lambda_3 & -\lambda_2 \\ -\frac{\pi_1}{\pi_3}\lambda_3 & -\frac{\pi_2}{\pi_3}\lambda_3 & 0 & -\lambda_2 \\ -\frac{\pi_1}{\pi_4}\lambda_2 & -\frac{\pi_2}{\pi_4}\lambda_2 & -\frac{\pi_3}{\pi_4}\lambda_2 & 0 \end{bmatrix}$	0.689	0.160	1.498	$(1, -0.111, -0.111, -0.259)$
$P_3 = \begin{bmatrix} 1-\frac{1-\pi_1}{\pi_1} & \frac{\pi_2}{\pi_1} & \frac{\pi_3}{\pi_1} & \frac{\pi_4}{\pi_1} \\ 1 & 0 & 0 & 0 \\ 1 & 0 & 0 & 0 \\ 1 & 0 & 0 & 0 \end{bmatrix}$	0.8	0.048	5	$(1, 0, 0, -0.667)$

Note: In P_2, the eigenvalues are $\lambda_1 = 1$, $\lambda_2 = -\frac{\pi_4}{\pi_3+\pi_2+\pi_1}$, $\lambda_3 = -\frac{\pi_3}{\pi_2+\pi_1}\left(1-\frac{\pi_4}{\pi_3+\pi_2+\pi_1}\right)$, $\lambda_4 = -\frac{\pi_2}{\pi_1} \times \left(1-\frac{\pi_4}{\pi_3+\pi_2+\pi_1}\right)\left(1-\frac{\pi_3}{\pi_2+\pi_1}\right)$.

7.3.2.2 Efficiency of continuous state chains

To study the mixing efficiency of MCMC on a continuous state space, Gelman et al. (1996) discretized the state space, and then used asymptotic variance $\text{var}(\tilde{I})$ to measure efficiency (equation (7.35)). Consider the normal target $N(0, 1)$, with density $\phi(x) \propto \exp\{-\frac{1}{2}x^2\}$, and normal proposal $x'|x \sim N(x, \sigma^2)$. The space $x \in (-\infty, \infty)$ can be represented by $K = 500$ bins over the range $(x_L, x_U) = (-5, 5)$, each of width $\delta = (x_U - x_L)/K$. Bin i is represented by the mid-value $x_i = x_L + (i - 1/2)\delta$. The steady-state distribution on the discrete space is then $\pi_i \propto \phi(x_i)\delta, i = 1, 2, \ldots, K$, renormalized to sum to 1. The $K \times K$ transition matrix $P = \{p_{ij}\}$ is constructed to mimic the MCMC algorithm:

$$p_{ij} = q(x_j|x_i) \cdot \min\left(1, \frac{\pi(x_j)}{\pi(x_i)}\right) \cdot \delta, 1 \leq i, j \leq K. \tag{7.47}$$

Note that this matches equation (7.11). Efficiency is then calculated using P (equation (7.35)). The calculation can be confirmed by using a different range and a larger K, and by using MCMC simulation (equation (7.4)).

By minimizing the asymptotic variance for estimating the mean of the $N(0, 1)$ target, Gelman et al. (1996) found the optimal σ to be ~ 2.5, with the optimal efficiency to be 0.23. Thus if the target is $N(\theta, \tau^2)$, the optimal normal proposal should be $x'|x \sim N(x, \tau^2\sigma^2)$ with $\sigma = 2.5$; the optimal proposal should have a standard deviation 2.5 times as large as that of the target. At this optimal step length, the dependent sample of size n is roughly equivalent to an independent sample of size $n/4$. It is also noted that at the optimal step length, $P_{\text{jump}} \approx 43\%$.

Using the same procedure, Yang and Rodríguez (2013) examined a number of target-proposal combinations. The results for three target densities and five proposal kernels are summarized in Table 7.2 and Figure 7.10. The three target densities with different shapes are as follows (Figure 7.9):

a. Normal distribution $N(0, 1)$, with mean 0 and variance 1.
b. A mixture of two normal distributions, $\frac{1}{4}N(-1, \frac{1}{4}) + \frac{3}{4}N(1, \frac{1}{4})$, with mean $\frac{1}{2}$ and variance 1.

Table 7.2 Approximate optimal step length σ^* (standard deviation), optimal efficiency E^*, and optimal jump probability under different combinations of target density and proposal kernel

Proposal kernel	(a) Normal				(b) 2 Normals				(c) 2 t_4			
	σ^*	E^*	P_{jump}	λ_2	σ^*	E^*	P_{jump}	λ_2	σ^*	E^*	P_{jump}	λ_2
(1) Uniform	2.2	0.28	0.41	0.67	1.9	0.23	0.39	0.75	2.2	0.23	0.37	0.76
(2) Normal	2.5	0.23	0.43	0.66	2.2	0.17	0.39	0.75	2.6	0.20	0.38	0.75
(3) t_4	3.2	0.21	0.42	0.68	3.0	0.15	0.37	0.77	3.2	0.19	0.39	0.76
(4) t_1 (Cauchy)	2.0	0.16	0.37	0.74	1.8	0.12	0.33	0.81	2.0	0.15	0.34	0.80
(5) Bactrian												
$m = 0.80$	2.3	0.27	0.41	0.67	2.1	0.21	0.36	0.77	2.4	0.22	0.35	0.76
$m = 0.90$	2.3	0.33	0.35	0.77	2.2	0.26	0.31	0.83	2.4	0.26	0.30	0.82
$m = 0.95$	2.3	0.38	0.30	0.83	2.3	0.30	0.26	0.88	2.3	0.30	0.27	0.88
$m = 0.98$	2.2	0.41	0.29	0.88	2.4	0.33	0.23	0.90	2.2	0.31	0.26	0.90
$m = 0.99$	2.2	0.41	0.28	0.92	2.5	0.32	0.22	0.94	2.1	0.31	0.27	0.93

Note: The three target distributions are shown in Figure 7.9. The Cauchy kernel does not have a finite variance, so σ is the scale parameter.

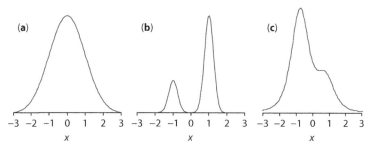

Fig. 7.9 Three target distributions used to evaluate the efficiency of different MCMC proposals: (**a**) The normal $N(0, 1)$; (**b**) A mixture of two normals: $\frac{1}{4}N(-1, \frac{1}{4}) + \frac{3}{4}N(1, \frac{1}{4})$; and (**c**) A mixture of two t_4 distributions: $\frac{3}{4}t_4(-\frac{3}{4}, s) + \frac{1}{4}t_4(\frac{3}{4}, s)$, with $s = 0.760345$ (see equation (7.48)). All three targets have variance 1.

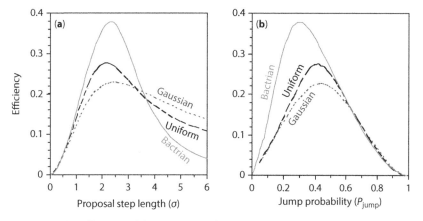

Fig. 7.10 The efficiency of the normal, uniform and Bactrian proposals plotted (**a**) against the proposal step length (σ) and (**b**) against the jump probability. The target is the normal distribution $N(0, 1)$. From Yang and Rodriguez (2013).

c. A mixture of t_4 distributions (t distributions with df = 4): $\frac{3}{4}t_4(-\frac{3}{4}, s) + \frac{1}{4}t_4(\frac{3}{4}, s)$, with $s = \sqrt{37}/8$. This has mean $-\frac{3}{8}$ and variance 1. The t_4 distribution with mean μ and variance σ^2 has density

$$f(t; \mu, \sigma) = \frac{3}{4\sqrt{2}\sigma}\left[1 + \frac{(t-\mu)^2}{2\sigma^2}\right]^{-\frac{5}{2}}. \tag{7.48}$$

This is heavy-tailed, so $K = 1000$ bins are used over the range $(-10, 10)$.

The five different proposals are as follows.

1. Uniform sliding window with window size $\sqrt{12}\sigma$ and variance σ^2.
2. Normal sliding window with variance σ^2.
3. A t_4 sliding window: $x'|x \sim t_4(x, \sigma)$ with variance σ^2 (see equation (7.48)). This has a similar bell shape to the normal sliding window but is heavier-tailed.

4. A Cauchy sliding window $x'|x \sim C(x, \sigma)$. The Cauchy distribution $C(\mu, \sigma)$ with location parameter μ and scale parameter σ has density

$$f(x; \mu, \sigma) = \frac{1}{\pi \sigma \left[1 + \left(\frac{x-\mu}{\sigma}\right)^2\right]}. \qquad (7.49)$$

This does not have finite mean or variance, and σ^2 is not the variance.

5. The Bactrian proposal with parameters m and σ^2 (equation (7.20) and Figure 7.4).

The impact of step length (σ) on efficiency is illustrated in Figure 7.10a. The efficiency at the optimal step length for the kernel is shown in Table 7.2, together with the corresponding σ and P_{jump}. The relative performance of the jumping kernels does not depend on the target. For all three targets, the uniform kernel is more efficient than the normal, while the Bactrian is the best. The Bactrian proposal is better if the two modes are far apart (e.g. when m is large). However, when m is close to 1, efficiency is noted to be more sensitive to the step length and drops off quickly with the increase of σ (Figure 7.10a). It appears that $m = 0.95$ strikes the right balance, where the optimal jump probability is $P_{\text{jump}} \approx 0.3$. With this m, the Bactrian move has better performance than the uniform and normal proposals for almost the whole range of jump probability for the normal target (Figure 7.10b). For the normal-mixture target, the Bactrian kernel (with $m \geq 0.95$) is nearly twice as efficient as the normal. For all targets evaluated, the Bactrian proposal is at least 50% more efficient than the normal proposal.

The Bactrian kernel is superior presumably because it proposes values different from the current value, reducing autocorrelation in the sample. *To propose something new, propose something different.* The Markov chain transition kernels (equations (7.9) and (7.47)) for the optimal normal proposal (at $\sigma = 2.5$) and for the optimal Bactrian proposal with $m = 0.95$ (at $\sigma = 2.3$) are illustrated in the heat maps of Figure 7.11b and b', for the normal target $N(0, 1)$. Figure 7.11a and a' shows the transition density when the current state is $x = 1$. The two proposals generate very different Markov chains. With the normal proposal, the next step x' tends to be around the current value x, leaning towards the mode at 0. With the Bactrian the next step x' has a good chance of being in the interval $(-1, 1)$ if x is outside it, but if x is around 1 (or -1), x' tends to be around -1 (or 1). The Bactrian proposal has the flavour of so-called antithetic variables, commonly used in Monte Carlo simulation for variance reduction.

7.3.2.3 Convergence rate and step length

While our focus is on the mixing efficiency, measured by the asymptotic variance, the relationship between the convergence rate and the step length may also be important. The convergence rate may be measured by the distance between $\pi^{(n)}$, the distribution reached after n steps, and the steady-state distribution π, when the chain is started from an arbitrary distribution $\pi^{(0)} \neq \pi$. Example measures may include $\sum_{i,j} \left|p_{ij}^n - \pi_j\right|$ and $\max_i \sum_j \left|p_{ij}^n - \pi_j\right|$, where p_{ij}^n is the ijth element of P^n (e.g. Green and Han 1992; Besag and Green 1993). As $P^n = E\Lambda^n E^{-T}$, it is clear that the rate of convergence is dominated by the second largest eigenvalue in absolute value: $R = \max_{k \geq 2} |\lambda_k|$. If R is small, the chain converges to its stationary distribution quickly. Often a chain with good mixing efficiency will also have a high convergence rate, although discrepancies are possible.

Our tests (Yang and Rodríguez 2013) suggest that for the uniform and normal proposals, the optimal step length for fast convergence is slightly greater than for efficient mixing, and one should take larger steps during the burn-in than after the burn-in. For the Bactrian

7.3 CONVERGENCE, MIXING, AND SUMMARY OF MCMC

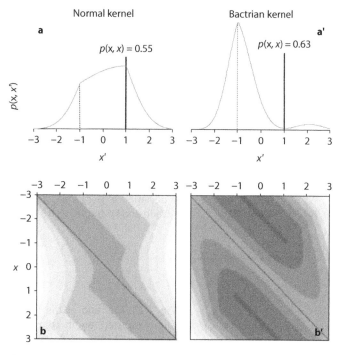

Fig. 7.11 A slice of the transition kernel $p(x, x')$, with $x = 1$, for (**a**) the normal kernel with $\sigma = 2.5$ and (**a'**) the Bactrian kernel with $m = 0.95$ and $\sigma = 2.3$, when the target is the standard normal $N(0, 1)$. The density is unsmooth at $x' = -x$, where $\pi(x') = \pi(x)$, and is discontinuous at $x' = x$, where there is a point mass due to rejected proposals. (**b** and **b'**) Heat map representations of the transition kernel $p(x, x')$ for the two proposals. $K = 200$ bins are used to discretize the state space $(-3, 3)$. The $K \times K$ transition probability matrix is constructed according to equation (7.47) and plotted here, with red (dark) for high values and white (light) for low values. In the normal kernel, the accepted values are close to the current value, while in the Bactrian kernel, the accepted values tend to be far away from the current value. The contours have sharp corners along the line $x' = -x$, while the high values along the main diagonal are due to the discontinuity at $x' = x$.

kernels, the optimal step length for fast convergence is about the same as that for mixing. As the burn-in typically constitutes a small portion of the computational effort, it appears to be adequate to optimize the step length for fast mixing only.

7.3.2.4 Automatic adjustment of step length

As the jump probability P_{jump} typically has a monotonic relationship with the step length σ, one can use the observed P_{jump} of an MCMC proposal to adjust σ to achieve maximum mixing efficiency. When the normal jump kernel $x'|x \sim N(x, \sigma^2)$ is applied to the normal target $N(0, 1)$, Gelman et al. (1996) noted that P_{jump} has a simple relationship with σ:

$$P_{\text{jump}} = \frac{2}{\pi}\tan^{-1}\left(\frac{2}{\sigma}\right) \qquad (7.50)$$

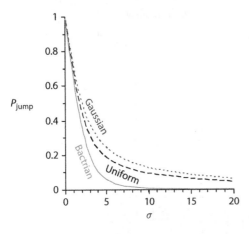

Fig. 7.12 The acceptance probability (P_{jump}) as a function of the step length σ for three proposal kernels when the target is $N(0, 1)$. The results for the normal kernel are calculated using equation (7.50) and confirmed by simulation, while those for the uniform and Bactrian kernels are based on simulation. If the target does not have unit variance, σ will be the ratio of the proposal standard deviation to the target standard deviation. From Yang and Rodriguez (2013).

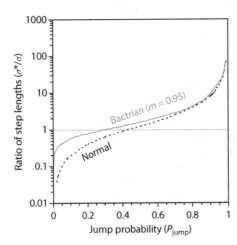

Fig. 7.13 The ratio of the step lengths (σ^*/σ) for automatic step length adjustment for the normal (equation (7.51)) and Bactrian proposals. The target distribution is the normal. The x-axis shows the observed acceptance proportion (P_{jump}), and the y-axis shows the factor by which one should multiply the step length (σ) to obtain the optimal step length (σ^*). For example, if the observed acceptance proportion is 0.2, one should multiply the step length by 0.39 for the normal proposal (to achieve the optimal P_{jump} of 0.44) or by 0.77 for the Bactrian proposal (to achieve the optimal P_{jump} of 0.30).

(Problem 7.11). This is plotted in Figure 7.12. Suppose that σ and P_{jump} are the current values, which can be estimated during the burn-in. Then

$$\sigma^* = \sigma \times \frac{\tan\left(\frac{\pi}{2} P_{\text{jump}}\right)}{\tan\left(\frac{\pi}{2} P^*_{\text{jump}}\right)}, \tag{7.51}$$

with the optimal jump probability $P^*_{\text{jump}} \approx 0.44$ should give the optimal step length σ^* (Figure 7.13).

For the Bactrian proposal (equation (7.20)) applied to the normal target $N(0, 1)$, P_{jump} is given by the following 1-D integral:

$$P_{\text{jump}} = \frac{2}{\pi} \int_0^a \frac{1}{1+t^2} \exp\left\{-\frac{b^2(1+t^2)}{2(1+at)^2}\right\} dt, \tag{7.52}$$

where $a = \frac{2}{\sigma\sqrt{1-m^2}}, b = \frac{m}{\sqrt{1-m^2}}$. This is plotted in Figure 7.12. One can use equation (7.52) for automatic scale adjustment, using a linear search or a look-up table to calculate the step length σ for a given P_{jump}. It is nevertheless noted that the P_{jump} versus σ curves for the Bactrian and normal kernels have very similar shapes over a large range of σ

(Figure 7.12), so that equation (7.51) can be used quite reliably to calculate the optimal σ for the Bactrian move as well, using $P^*_{\text{jump}} \approx 0.3$.

It is convenient to use the burn-in to estimate P_{jump} and to adjust the step length. We have focused on the normal target because in large datasets the posterior should be approximately normal around the mode. Far away from the mode (i.e. before the Markov chain has converged), the posterior may differ a lot from the normal. Thus a few rounds of adjustment may be necessary. In my tests four rounds of adjustment during the burn-in (which should be longer than 400 iterations to allow a trustable estimate of P_{jump}) have been effective.

7.3.3 Validating and diagnosing MCMC algorithms

Developing a correct and efficient MCMC algorithm for a practical application is a challenging task. A great benefit of MCMC is that it enables sophisticated parameter-rich models to be applied to real data analysis, liberating the researcher from the limitations of mathematically tractable but biologically unrealistic models. However, parameter-rich models often cause problems for both inference and computation. There is often a lack of information for estimating the multiple parameters, resulting in nearly flat or ridged likelihood (and thus posterior) surfaces or strong correlations between parameters. It is usually impossible to independently calculate the posterior probability distribution, making it hard to validate an implementation, i.e. to confirm the correctness of the computer program. A Bayesian MCMC program tends to be harder to debug than an ML program implementing the same model. In likelihood iteration, the convergence is to a point, and in most optimization algorithms, the log likelihood should always go up, with the gradient approaching zero when the algorithm approaches the MLE. In contrast, a Bayesian MCMC algorithm converges to a statistical distribution, with no statistics having a fixed direction of change.

An MCMC algorithm, even if correctly implemented, can suffer from two problems: slow convergence and poor mixing. Slow convergence means that it takes a very long time for the chain to reach stationarity. Early in this section, we discussed the idea that convergence problems have to do with the lightness of the tail of the posterior and with the existence of ridges on the posterior surface. Poor mixing means that the sampled states are highly correlated over iterations, and the chain is inefficient in exploring the parameter space. Step lengths that are either too large or too small may cause mixing problems but they can be fixed easily. Ridges and multiple local peaks can cause serious problems to both convergence and mixing.

While it is often obvious that the proposal density $q(\cdot|\cdot)$ satisfies the required conditions so that the MCMC is in theory guaranteed to converge to the target distribution, it is much harder to determine in real data problems whether the chain has reached stationarity. A number of heuristic methods have been suggested to diagnose an MCMC run. Some of them are described below. However, those diagnostics are able to reveal certain problems but are unable to prove the correctness of the algorithm or implementation. Often when the algorithm converges slowly or mixes poorly, it is difficult to decide whether this is due to faulty theory, buggy programs, or inefficient but correct algorithms. In short, diagnostic tools are very useful, but one should bear in mind that they may fail.

In the following we discuss a few strategies for validating and diagnosing MCMC programs. Free software tools are available that implement many more diagnostic tests.

1. *Time-series plots* or *trace plots* are a very useful tool for detecting lack of convergence and poor mixing (see, e.g. Figure 7.2a and b). One can plot parameters of interest or their

functions against the iterations. Note that the chain may appear to have converged with respect to some parameters but not to others, so it is important to monitor many or all parameters.
2. For most proposals, the acceptance proportion should be neither too high nor too low.
3. Multiple chains run from different starting points should all converge to the same distribution. Gelman and Rubin's (1992) statistic can be used to analyse multiple chains; see below.
4. Another strategy is to run the chain with no data, i.e. with the likelihood $f(X|\theta) = 1$ fixed. The posterior should then be the same as the prior, which may be analytically available for comparison. Also, theoretical expectations are often available for infinite data. Therefore one can simulate larger and larger datasets under a fixed set of parameters and analyse the simulated data under the correct model, to confirm that the Bayesian point estimate becomes closer and closer to the true value. This test relies on the fact that Bayesian estimates are consistent.

One can also conduct so-called *Bayesian simulation*, to confirm theoretical expectations. To generate each replicate dataset, one samples parameter values from the prior, and then use them to generate one dataset under the likelihood model. For a continuous parameter, one can confirm that the $(1-\alpha)100\%$ posterior CI contains the true parameter value with probability $(1-\alpha)$. We construct the CI for each dataset, and examine whether the true value is included in the interval; the proportion of replicates in which the CI includes the true parameter value should equal $(1-\alpha)$. This is called the *hit probability* (Wilson et al. 2003). A similar test can be applied to a discrete parameter, such as the tree topology. In this case, each sequence alignment is generated by sampling the tree topology and branch lengths from the prior and by then evolving sequences on the tree. The Bayesian posterior probabilities of trees or clades should then be the probability that the tree or clade is true. One can bin the posterior probabilities and confirm that among trees with posterior probabilities in the bin 94–96%, say, about 95% of them are the true tree (Huelsenbeck and Rannala 2004; Yang and Rannala 2005).

For a continuous parameter, a more sensitive test than the hit probability is the so-called *coverage probability*. Suppose a fixed interval (θ_L, θ_U) covers $(1-\alpha)100\%$ of the prior distribution. Then the posterior coverage of the same fixed interval should on average be $(1-\alpha)100\%$ as well (Rubin and Schenker 1986; Wilson et al. 2003). Thus we use Bayesian simulation to generate many datasets, and in each dataset, calculate the posterior coverage probability of the fixed interval, i.e. the mass of posterior density in the interval (θ_L, θ_U). The average of the posterior coverage probabilities over the simulated datasets should equal $(1-\alpha)$.

7.3.4 Potential scale reduction statistic

Gelman and Rubin (1992) suggested a diagnostic statistic called 'estimated potential scale reduction', based on variance-component analysis of samples taken from several chains run using 'over-dispersed' starting points. The rationale is that after convergence, the within-chain variation should be indistinguishable from the between-chain variation, while before convergence, the within-chain variation should be smaller than the between-chain variation. The statistic can be used to monitor any or every parameter of interest or any function of the parameters. Let x be the parameter being monitored, and τ^2 be its posterior variance. Suppose there are m chains, each run for n iterations, after the burn-in is discarded. Let x_{ij} be the parameter sampled at the jth iteration from the ith chain. Gelman and Rubin (1992) defined the between-chain variance to be

$$B = \frac{n}{m-1} \sum_{i=1}^{m} (\bar{x}_{i\cdot} - \bar{x}_{\cdot\cdot})^2, \tag{7.53}$$

and the within-chain variance to be

$$W = \frac{1}{m(n-1)} \sum_{i=1}^{m} \sum_{j=1}^{n} (x_{ij} - \bar{x}_{i\cdot})^2, \tag{7.54}$$

where $\bar{x}_{i\cdot} = \frac{1}{n} \sum_{j=1}^{n} x_{ij}$ is the mean within the ith chain, and $\bar{x}_{\cdot\cdot} = \frac{1}{m} \sum_{i=1}^{m} \bar{x}_{i\cdot}$ is the overall mean. If all m chains have reached stationarity and x_{ij} are samples from the same target density, both B and W are unbiased estimates of τ^2, and so is their weighted mean:

$$\hat{\tau}^2 = \frac{n-1}{n} W + \frac{1}{n} B. \tag{7.55}$$

If the m chains have not reached stationarity, W will be an underestimate of τ^2, while B will be an overestimate. Gelman and Rubin (1992) showed that in this case $\hat{\tau}^2$ is also an overestimate of τ^2. The *estimated potential scale reduction* is defined as

$$\hat{R} = \frac{\hat{\tau}^2}{W}. \tag{7.56}$$

This should get smaller and approach one when the parallel chains reach the same target distribution. In real data problems, values of $\hat{R} < 1.1$ or 1.2 indicate convergence.

7.3.5 Summary of MCMC output

Before we process the output, the beginning part of the chain before it has converged to the stationary distribution is often discarded as burn-in. Often we do not sample every iteration but take a sample only for every certain number of iterations. This is known as *thinning* the chain, as the thinned samples have reduced autocorrelations across iterations. In theory, it is always more efficient (producing estimates with smaller variances) to use all samples even if they are correlated. However, MCMC algorithms typically create huge output files, and thinning reduces disk usage and makes the output small enough for processing.

Samples taken after the burn-in can be summarized in a straightforward manner. The marginal distribution for a parameter, say, θ, is generated by simply ignoring the other parameters. Suppose θ is the parameter of interest, and λ are the other parameters. If $\{(\theta_i, \lambda_i), i = 1, 2, \ldots, n\}$ is a sample from the joint posterior for (θ, λ), then $\{\theta_i, i = 1, 2, \ldots, n\}$ will be a sample from the marginal posterior for θ. One can generate a histogram, which can be smoothed to generate an estimate of the (marginal) posterior distribution, using, e.g. the R function hist(). Two-dimensional joint densities can be estimated as well. The sample mean or median can be used as a point estimate of the parameter. The sample median is the mid-point after the sample is sorted.

The 95% equal-tail CI is given by the 2.5% and 97.5% percentiles of the sample. We sort the sample in the increasing order, and remove 2.5% from the left tail and 2.5% from the right tail and what is left will be the 95% CI (Figure 7.14). The highest probability density (HPD) interval is more complicated to calculate, but Chen and Shao (1999) developed a simple algorithm based on the fact that the HPD CI is the shortest. Let $\theta_{(j)}$ be the jth smallest value in the sample. Then $(\theta_{(j)}, \theta_{(j+0.95n)})$ brackets 95% of the sample and is a

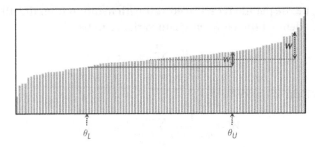

Fig. 7.14 Construction of the 50% equal-tail CI and the 50% HPD interval. The sampled values from the MCMC are sorted, and the 25% point (θ_L) and the 75% point (θ_U) constitute the 50% equal-tail CI: (θ_L, θ_U). This has the width $w = \theta_U - \theta_L$. To construct the 50% HPD interval, we slide the same window to the left and to the right until the induced interval width (w) reaches the minimum.

95% interval. By varying j so that the interval width $\theta_{(j+0.95n)} - \theta_{(j)}$ is minimized, we find the HPD interval (Figure 7.14). This method can be used to generate the HPD interval for any function of the parameters, $h(\theta)$. From the MCMC sample $\{\theta_i, i = 1, \ldots, n\}$, we can calculate a sample for h: $\{h_i, i = 1, \ldots, n\}$ and sort the sample to construct the HPD interval. Note that while the equal-tail interval is invariant to nonlinear transform, the HPD interval is not: if (θ_L, θ_U) is the HPD interval for θ, the HPD interval for h is in general not $(h(\theta_L), h(\theta_U))$.

This algorithm assumes that the HPD interval consists of one interval rather than several disconnected sub-intervals. The latter may result from multiple modes in the posterior and should be obvious from an inspection of the histogram (see Figure 6.2b). If there are two modes, one may search for two non-overlapping sub-intervals that have the smallest total length (Chen and Shao 1999).

The autocorrelation time (τ) or the ESS may be calculated from an MCMC sample based on their definitions (equation (7.4)). Geyer (1992) points out that for reversible Markov chains, the sum of two consecutive autocorrelation coefficients is always positive, $\rho_k + \rho_{k+1} > 0$. One can thus obtain a consistent estimator of τ by truncating the sum of equation (7.4) until $\rho_k + \rho_{k+1}$ becomes negative. This is the so-called *initial positive sequence* method (Geyer 1992). The ESS is given as n/τ.

7.4 Advanced Monte Carlo methods

In this section, we discuss a few advanced MCMC algorithms. Existence of multiple local peaks on the posterior can cause a lot of problems for MH algorithms (Figure 7.15). As they are stochastic hill-climbing algorithms, MH algorithms are prone to getting trapped in local optima. A number of methods have been developed to overcome this problem. See the books by Liu (2001), Robert and Casella (2004), and Liang et al. (2010). The best known such method in molecular phylogenetics is the Metropolis-coupled MCMC or MC^3 algorithm, also known as parallel tempering (Geyer 1991; Marinari and Parisi 1992), due to its implementation in MrBayes (Altekar et al. 2004). We will discuss this method in §7.4.1. We will also discuss reversible-jump MCMC (rjMCMC), which allows moves between models of different dimensions, so that in one MCMC run one can calculate the

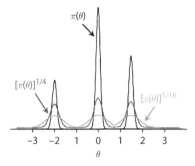

Fig. 7.15 A density $\pi(\theta)$ with three peaks. This is a mixture of three normal densities: $0.2N(-2, 0.1^2) + 0.5N(0, 0.1^2) + 0.3N(1.5, 0.1^2)$. This has mean 0.05 and variance 1.254^2. Two 'flattened' densities, proportional to $[\pi(\theta)]^{1/4}$ and $[\pi(\theta)]^{1/16}$, are also shown to illustrate parallel tempering (MC^3).

posterior model probabilities as well as the posterior of parameters within each model (Green 1995) (§7.4.2). Alternatively when only a few models are under consideration, one may calculate the marginal likelihood or the Bayes factor. This is discussed in §7.4.3.

7.4.1 Parallel tempering (MC^3)

If the target distribution has multiple peaks, separated by low valleys, the Markov chain may have difficulty moving from one peak to another, even though it can move around each peak with ease. As a result, the chain may get stuck on one peak and the resulting samples will not approximate the posterior density correctly. For example, the density $\pi(\theta)$ in Figure 7.15 is a mixture of three normal densities with modes at $-2, 0$, and 1.5, each with variance 0.01, resulting in three sharp peaks in the posterior surface. If one uses a sliding window to sample from this target, the algorithm will tend to stay in the neighbourhood of one peak but will have trouble moving between peaks (Problem 7.12). The transition matrix P for a normal proposal $\theta'|\theta \sim N(\theta, 0.25^2)$ is shown in Figure 7.16a. If the current

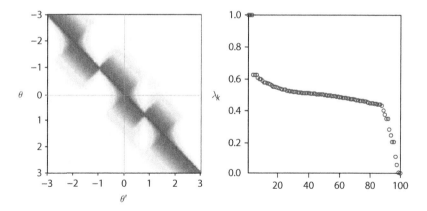

Fig. 7.16 (a) The heat map representation of the transition matrix P when a normal proposal kernel $\theta'|\theta \sim N(\theta, \sigma^2)$, with $\sigma = 0.25$, is applied to sample from the target of Figure 7.15. The state space is discretized using $K = 100$ bins, as in Figure 7.11. Moves within each of the three regions around the peaks: $(-\infty, -1)$, $(-1, 1)$, and $(1, \infty)$, are frequent, but moves between the regions are rare. (b) The eigenvalues of the transition matrix P. The first three eigenvalues are all close to 1, while all others are much less than 1, indicating the presence of the three peaks.

state is around the peak at −2, the next state that the chain will visit will almost certainly be around the same peak as well. For example, if we define the neighbourhood of −2 as the interval (−2.5, −1.5), we have $\Pr\{-2.5 < \theta' < -1.5 \mid -2.5 < \theta < -1.5\} \approx 1$ (Figure 7.16a). Similarly if the chain is currently around 0 (or 1.5), it will almost certainly remain in the same neighbourhood in the next step. In such cases, P will have more than one eigenvalue very close to 1 (recall the largest eigenvalue is always $\lambda_1 = 1$), and the number of such eigenvalues equals the number of such trapping peaks, while all other eigenvalues are much less than 1 (Figure 7.16b). The term *metastasis* is used to describe the transitions between the peaks.

Whether multiple local peaks exist on the posterior surface depends on the nature of the inference problem. For example, in molecular dynamic simulations of protein folding, many configurations of the atoms may represent locally optimal energy states, creating numerous local peaks on the posterior surface. In molecular phylogenetics, most within-model parameter estimation problems appear to be well behaved, with a single peak on the posterior surface. However, local peaks are known to exist in the space of trees, and the problem is more serious for tree-perturbation algorithms such as nearest neighbour interchange that do not induce many neighbours (see §3.2.4).

Developing effective simulation algorithms for a rugged posterior with multiple local peaks is a major research focus in Bayesian computation. Numerous methods have been developed, such as simulated tempering (Marinari and Parisi 1992), parallel tempering (Geyer 1991), evolutionary Monte Carlo (Liang and Wong 2001), dynamic weighting (Wong and Liang 1997), multicanonical sampling (Berg and Neuhaus 1991), the Wang–Landau algorithm (Wang and Landau 2001; Liang 2005), equi-energy sampler (Kou et al. 2006), stochastic approximation Monte Carlo (Liang et al. 2007), etc. Bayesian phylogenetics has also become a rich application area for testing new computational algorithms (e.g. Cheon and Liang 2009; Bouchard-Coté et al. 2012). It is as yet too early to say which of these methods will stand the test of time. Unfortunately, most of these algorithms are designed to help the chain to move from one peak to another on the same posterior surface in problems of parameter estimation under a well-specified likelihood model, but are not useful for moving from one model to another in model selection problems. In phylogenetics, trees are different models and the major computational challenge for a Bayesian MCMC algorithm is the difficulty of moving from one tree to another.

Below we discuss parallel tempering (Geyer 1991), which involves running multiple Markov chains in parallel, which have steady-state distributions that are flattened versions of the posterior. This is similar to the simulated annealing algorithm (see §3.2.5). The method is also known as Metropolis-Coupled MCMC or MCMCMC (MC^3).

In MC^3, m chains are run in parallel. The jth chain has the stationary distribution

$$\pi_j(\theta) \propto \pi(\theta)^{1/T_j} = \pi(\theta)^{1/[1+\lambda(j-1)]}, \tag{7.57}$$

where $T_j = 1 + \lambda(j-1)$, with $\lambda > 0$, is the temperature for chain j. The first chain has $\pi_1(\cdot) = \pi(\cdot)$, so it samples from the target posterior density and is called the *cold chain*. The other chains are designed to improve mixing and are called *hot chains*. Note that raising the density $\pi(\cdot)$ to the power $1/T$ with $T > 1$ has the effect of flattening out the surface, making it easier for the Markov chain to cross valleys (see Figure 7.15).

Within each chain j, the ordinary MH algorithm is used to update the parameters. For example, a symmetrical move that changes θ to θ' is accepted with probability

$$\alpha = \min\left\{1, \frac{\pi_j(\theta')}{\pi_j(\theta)}\right\} = \min\left\{1, \left[\frac{\pi(\theta')}{\pi(\theta)}\right]^{1/T_j}\right\}. \tag{7.58}$$

Furthermore, with a certain probability, we attempt to swap the states of two chains chosen at random through an MH step. Let the two chosen chains be i and j and let their current states be θ_i and θ_j. The proposal for the chain swap is accepted with probability

$$\alpha = \min\left\{1, \frac{\pi_i(\theta_j)}{\pi_i(\theta_i)} \times \frac{\pi_j(\theta_i)}{\pi_j(\theta_j)}\right\} = \min\left\{1, \left[\frac{\pi(\theta_j)}{\pi(\theta_i)}\right]^{\frac{1}{T_i}} \times \left[\frac{\pi(\theta_i)}{\pi(\theta_j)}\right]^{\frac{1}{T_j}}\right\} = \min\left\{1, \left[\frac{\pi(\theta_j)}{\pi(\theta_i)}\right]^{\frac{1}{T_i}-\frac{1}{T_j}}\right\}. \quad (7.59)$$

Intuitively $\frac{\pi_i(\theta_j)}{\pi_i(\theta_i)}$ is the acceptance ratio for chain i and $\frac{\pi_j(\theta_i)}{\pi_j(\theta_j)}$ is the ratio for chain j; the swap is accepted only if it is accepted in both chains.

The hot chains will visit the local peaks easily, and swapping states between chains will let the cold chain occasionally jump across valleys, leading to better mixing. To increase the acceptance rate, one may build a temperature ladder with multiple hot chains and attempt to swap only adjacent chains in the ladder. Atchadé et al. (2011) has shown that under certain conditions it is optimal to space the temperatures in the different chains so that 23.4% of chain swaps are accepted.

At the end of the run, output from only the cold chain is used, while outputs from the hot chains are discarded. An obvious disadvantage of the algorithm is that m chains are run but only one chain is used for inference. MC^3 is well suited to implementation on multi-processor multi-core servers, since each chain will require about the same amount of computation per iteration, and there is very little communication between chains.

7.4.2 Trans-model and trans-dimensional MCMC

7.4.2.1 General framework

The MCMC algorithms we have described up to now are useful for simulating the posterior of parameters under one well-specified model. Sometimes we may want to compare several different models (data-generating mechanisms), which may have different dimensions (different numbers of parameters). Ordinary MCMC (the MH algorithm) will then not work. Green (1995, 2003) developed the reversible-jump MCMC algorithm, which allows moves between models of different dimensions. The resulting Markov chain visits the models with the frequencies equal to the posterior probabilities of the models. The name reversible-jump is curious, since all MCMC algorithms generate reversible Markov chains. The fundamental idea is to match up the dimensions between the models using dummy random variables.

The cross-model inference falls naturally into the framework of hierarchical Bayesian analysis. Let the data be X. We assign prior probabilities π_k to the models, with $k = 1, 2, \ldots, K$. Given each model H_k, we have the prior $f(\theta_k|H_k)$ for its parameters θ_k and the likelihood $f(X|H_k, \theta_k)$. The joint posterior of the model and its parameters is then

$$f(H_k, \theta_k|X) = \frac{1}{Z}\pi_k f(\theta_k|H_k) f(X|H_k, \theta_k). \quad (7.60)$$

The normalizing constant

$$Z = \sum_{j=1}^{K} \pi_j M_j = \sum_{j=1}^{K} \pi_j \times \left[\int f(\theta_j|H_j) f(X|H_j, \theta_j) \, d\theta_j\right] \quad (7.61)$$

is a sum over the models and, within each model, an integral over its parameters. Here the integral M_j is the marginal likelihood for model H_j. Equation (7.60) can also be written as

$$f(H_k, \theta_k|X) = f(H_k|X) f(\theta_k|H_k, X). \quad (7.62)$$

The joint posterior of the model and its parameters is the product of posterior model probability and the within-model parameter posterior.

We now consider constructing a Markov chain to sample from the joint posterior (7.60). The chain will move between models as well as between parameter values within each model. The state of the chain is represented by (k, θ_k), for model H_k and parameters θ_k, so a move will be from (k, θ_k) to $(k', \theta'_{k'})$. If $k = k'$, the move will be a within-model MH move. Here we focus on between-model moves, with $k \neq k'$. Suppose model k has d parameters and k' has d' parameters. We use random variables u and u' to match up the dimensions between the two models: $(\theta_k, u) \leftrightarrow (\theta'_{k'}, u')$. Let the dimension of $\theta_k, \theta'_{k'}, u$, and u' be d, d', r, r', respectively, with $d + r = d' + r'$. This matching is flexible. For example, if H_k has one parameter and $H_{k'}$ has three, we may have $r = 2$ and $r' = 0$ (in which case the single parameter in H_k matches one parameter in $H_{k'}$) or $r = 3$ and $r' = 1$ (in which case each parameter in one model matches a random variable in the other). To propose a move from H_k to $H_{k'}$, generate $u \sim g(u)$, and apply a deterministic transform to calculate the new parameter values: $(\theta'_{k'}, u') = T(\theta_k, u)$. When we move back from $H_{k'}$ to H_k, generate random variables $u' \sim g'(u')$, and apply the inverse transform: $(\theta_k, u) = T^{-1}(\theta'_{k'}, u')$. Note that $g(\cdot)$ and $g'(\cdot)$ are r-variate and r'-variate probability densities, respectively.

The proposed move from (k, θ_k) to $(k', \theta'_{k'})$ is accepted with probability

$$\alpha_{k,k'} = \min\left\{1, \frac{f(H_{k'}, \theta_{k'}|X)}{f(H_k, \theta_k|X)} \times \frac{r_{k'k}}{r_{kk'}} \cdot \frac{g'(u')}{g(u)} \cdot \left|\frac{\partial(\theta'_{k'}, u')}{\partial(\theta_k, u)}\right|\right\}. \qquad (7.63)$$

Here the posterior ratio is replaced by the product of the prior ratio and the likelihood ratio, because $f(H_k, \theta_k|X) \propto \pi_k f(\theta_k|H_k)f(X|H_k, \theta_k)$, and the normalizing constant of equation (7.61) cancels. The proposal ratio comprises three parts: the model-proposal probability ratio $r_{k'k}/r_{kk'}$, the random variable density ratio $g'(u')/g(u)$, and the absolute value of the Jacobian determinant for the transform T.

The model-proposal probability $r_{kk'}$ is the probability of proposing the move to $H_{k'}$ when the chain is in H_k. This may be used to divide the computational effort into within-model moves and cross-model moves. Nevertheless, one should note that if the interest is in estimating the posterior model probability, one should spend most of the effort on cross-model moves. It is easy to see that if θ_k is from the posterior $f(\theta_k|H_k, X)$, then the accepted values of $\theta'_{k'}$ in the cross-model move will be from the posterior $f(\theta_{k'}|H_{k'}, X)$ as well. In other words, given that the parameter values for the current model are from its posterior, the parameter values accepted by the algorithm for the new model will be automatically from the stationary (posterior) distribution for the new model as well, and there is no need to run many iterations of within-model moves under the new model. Indeed the algorithm will still be correct if within-model moves are attempted when the chain is in one model only, and only cross-model moves are attempted when the chain is in other models.

Green's (2003) formulation is very general and includes the ordinary MH algorithms as special cases. Here we use this formulation to derive the proposal ratio for the familiar proportional scaling move (equation (7.21)), in addition to the two derivations given right below equation (7.21). Recall that in this move we generate a random number u to convert the current value x into the new value x', and in the reverse move we generate u' to convert x' into x. The matching may thus be $(x, u) \leftrightarrow (x', u')$.

$$\begin{cases} x' = x \cdot c = x \cdot e^{\varepsilon(u-1/2)}, \\ x = x' \cdot c' = x' \cdot e^{\varepsilon(u'-1/2)}, \end{cases} \qquad (7.64)$$

where $\varepsilon > 0$ is a fine-tuning step length. The transform $(x, u) \leftrightarrow (x', u')$ is thus

$$\begin{cases} x' = x \cdot e^{\varepsilon(u-1/2)}, \\ u' = \frac{1}{2} + \frac{1}{\varepsilon} \log \frac{x}{x'} = \frac{1}{2} - (u - \frac{1}{2}) = 1 - u, \end{cases} \quad (7.65)$$

which gives the Jacobi determinant as

$$\frac{\partial(x', u')}{\partial(x, u)} = \begin{vmatrix} \frac{\partial x'}{\partial x} & \frac{\partial x'}{\partial u} \\ \frac{\partial u'}{\partial x} & \frac{\partial u'}{\partial u} \end{vmatrix} = \begin{vmatrix} e^{\varepsilon(u-\frac{1}{2})} & x\varepsilon e^{\varepsilon(u-\frac{1}{2})} \\ 0 & -1 \end{vmatrix} = -e^{\varepsilon(u-\frac{1}{2})} = -\frac{x'}{x}. \quad (7.66)$$

The proposal ratio is thus $1 \cdot 1 \cdot \left|\frac{\partial(x',u')}{\partial(x,u)}\right| = \frac{x'}{x} = c$, as before (see §7.2.5). Note that the model-jump probability ratio $r_{k'k}/r_{kk'} = 1$ and the random variable density ratio $g'(u')/g(u) = 1$ since both u and u' are $U(0, 1)$ random numbers.

Alternatively one may treat the multipliers c and c' in equation (7.64) as the random variables and use the matching $(x, c) \leftrightarrow (x', c')$. This makes the transform and the resulting Jacobi determinant look simpler, but the densities of the random variables look more complex. The transform is

$$\begin{cases} x' = x \cdot c, \\ c' = x/x' = 1/c, \end{cases} \quad (7.67)$$

and the Jacobi determinant is

$$\left|\frac{\partial(x', c')}{\partial(x, c)}\right| = \begin{vmatrix} c & x \\ 0 & -\frac{1}{c^2} \end{vmatrix} = -\frac{1}{c}. \quad (7.68)$$

The random variable densities are $g(c) = g_u(u) \cdot \left|\frac{du}{dc}\right| = 1/(c\varepsilon)$, for $e^{-\varepsilon/2} \leq c \leq e^{\varepsilon/2}$, and $g'(c') = g(c') = 1/(c'\varepsilon)$. Thus the random variable density ratio is $g'(c')/g(c) = c/c' = c^2$. The proposal ratio is thus $1 \cdot c^2 \cdot (1/c) = c$, as before.

7.4.2.2 Trans-model MCMC

If the different models have the same number of parameters, one can use the ordinary MCMC (the MH algorithm) to move between them if we match up their parameters. Consider for instance two models H_1 and H_2, each with two parameters: α_1 and α_2 for H_1, and β_1 and β_2 for H_2. We match up the parameters, with $\alpha_1 \leftrightarrow \beta_1$ and $\alpha_2 \leftrightarrow \beta_2$, say. When the chain moves between the models, the model identifier k in the state (k, θ_k) determines how the two parameters are interpreted in the calculation of the prior $f(\theta_k|H_k)$ and likelihood $f(X|H_k, \theta_k)$: they are α_1 and α_2 if $k = 1$, and β_1 and β_2 if $k = 2$.

The strategy is best illustrated using an example. Suppose we observe $n = 5$ data points $X = \{5, 6, 8, 9, 12\}$, and want to fit two models. Model H_1 assumes that the data are from the normal distribution $x_i \sim N(\mu, \sigma^2)$ with parameters $\theta_1 = \{\mu, \sigma\}$ and likelihood

$$f(X|H_1, \theta_1) = \prod_{i=1}^{n} f(x_i|\mu, \sigma) = \prod_{i=1}^{n} \frac{1}{\sqrt{2\pi\sigma^2}} \exp\left\{-\frac{1}{2\sigma^2}(x_i - \mu)^2\right\}. \quad (7.69)$$

Model H_2 assumes the gamma distribution $x_i \sim G(\alpha, \beta)$ with parameters $\theta_2 = \{\alpha, \beta\}$ and likelihood

$$f(X|H_2, \theta_2) = \prod_{i=1}^{n} f(x_i|\alpha, \beta) = \prod_{i=1}^{n} \frac{\beta^\alpha e^{-\beta x_i} x_i^{\alpha-1}}{\Gamma(\alpha)}. \quad (7.70)$$

Note that in such trans-model analysis, model-specific constants such as $\sqrt{2\pi}$ in the likelihood function cannot be ignored. We assign prior probabilities $\pi_1 = \pi_2 = \frac{1}{2}$ for the two

Table 7.3 Prior and posterior for two models used in Bayesian model selection via trans-model MCMC

Model (H_k)	Prior	Posterior mean (SD)	Marginal likelihood (M_k)
H_1: Normal $N(\mu, \sigma^2)$	$\mu \sim G(2, 0.5)$	7.3684 (1.3543)	4.0018×10^{-7}
	$\sigma \sim G(2, 1)$	2.8750 (0.9180)	
H_2: Gamma $G(\alpha, \beta)$	$\alpha \sim G(2, 0.4)$	8.6543 (3.5170)	2.7218×10^{-7}
	$\beta \sim G(2, 0.5)$	1.1178 (0.4648)	

Note: In this example, we assume that all parameters including μ are >0. M_k is calculated using 2-D numerical integration. Posterior means and SDs are calculated by both 2-D numerical integration and by trans-model MCMC.

Table 7.4 Trans-model moves and proposal ratios for Bayesian comparison of the normal and gamma models

Algorithm	$N(\mu, \sigma) \to G(\alpha, \beta)$	R_{12}	$G(\alpha, \beta) \to N(\mu, \sigma)$	R_{21}
1	$\alpha = \mu \times c$ $\beta = \sigma$	c	$\mu = \alpha/c$ $\sigma = \beta$	$1/c$
2	$\alpha = (\mu/\sigma)^2$ $\beta = \mu/\sigma^2$	$2\beta^3/\sqrt{\alpha}$	$\mu = \alpha/\beta$ $\sigma = \sqrt{\alpha}/\beta$	$\sqrt{\alpha}/(2\beta^3)$
3	$m \sim G(10, 10/\mu)$ $s \sim G(10, 10/\sigma)$ $\alpha = (m/s)^2$ $\beta = m/s^2$	Equation (7.80)	$\mu \sim G(10, 10/(\alpha/\beta))$ $\sigma \sim G(10, 10/(\sqrt{\alpha}/\beta))$	Equation (7.81)

Note: The acceptance ratio includes the prior ratio times the likelihood ratio in every algorithm.

models. All parameters are assumed to be positive and assigned gamma priors, listed in Table 7.3. The posterior means and standard deviations (SDs) of the parameters as well as the marginal likelihoods (M_1 and M_2), calculated numerically, are listed in the same table. The posterior model probability is $f(H_1|X) = M_1/(M_1 + M_2) = 0.5953$ for H_1: normal.

Below, we describe three trans-model MCMC algorithms to sample from the posterior, summarized in Table 7.4. In the first, we simply match μ with α and σ with β. The algorithm is as follows.

1. Set initial model and parameters (say, $k = 1, \mu = 5, \sigma = 3$).
2. Within-model move. Use MH moves to update θ_k. Change the two parameters by two 1-D proportional moves (equation (7.21)).
3. With a certain probability (0.8, say), perform a trans-model move from (k, θ_k) to $(k', \theta'_{k'})$. If $k = 1$, set $k' = 2$, and $\theta'_{k'} = (\alpha', \beta') = (\mu, \sigma)$. Otherwise if $k = 2$, set $k' = 1$ and $\theta'_{k'} = (\mu', \sigma') = (\alpha, \beta)$. Accept the new model with its parameters, $(k', \theta'_{k'})$, with probability

$$\alpha = \min\left\{1, \frac{f(k', \theta'_{k'}|X)}{f(k, \theta_k|X)}\right\}. \tag{7.71}$$

Note that the proposal ratio is 1 since no random variables are created and no transforms are applied. If the move is accepted, set $(k, \theta_k) = (k', \theta'_{k'})$. Otherwise keep the current state.

4. Print out model identifier and model parameters (k, θ_k).
5. Go to step 2.

At the end of the run, the frequency at which H_1 is visited will be an estimate of the posterior probability of H_1, while the subset of samples in which $k = 1$ (or 2) can be summarized to generate the posterior of the parameters under model H_1 (or H_2). If H_k is rarely visited during the MCMC (because H_k has a very small posterior probability), this parameter posterior may be poorly estimated. Nevertheless, it is the same posterior as can be generated from a within-model MCMC with model H_k fixed.

The example illustrates the arbitrariness of the mapping, since parameter μ (or σ) in the normal model has nothing to do with parameter α (or β) in the gamma model. In practice, such arbitrary mapping will most likely cause mixing problems since the matched parameters may have very different posterior distributions, and as a result such cross-model moves will most likely be rejected because the proposed parameter values for the new model are poor. In this example, it will be beneficial to reparametrize the gamma model using the mean and variance as parameters and match them with the mean and variance of the normal.

Indeed we can replace step 3 of the algorithm above by a move that is equivalent to such a reparametrization. To move from H_1: normal $N(\mu, \sigma^2)$ to H_2: gamma $G(\alpha, \beta)$, we set

$$\alpha = (\mu/\sigma)^2, \beta = \mu/\sigma^2, \tag{7.72}$$

so that the gamma distribution has mean $\alpha/\beta = \mu$ and variance $\alpha/\beta^2 = \sigma^2$. The mean and variance of a distribution are the first two moments, so this move is known as a moment-matching move. The absolute value of the Jacobi determinant for the transform is

$$\left|\frac{\partial(\alpha, \beta)}{\partial(\mu, \sigma)}\right| = \left\|\begin{array}{cc} \frac{2\mu}{\sigma^2} & -\frac{2\mu^2}{\sigma^3} \\ \frac{1}{\sigma^2} & -\frac{2\mu}{\sigma^3} \end{array}\right\| = \frac{2\mu^2}{\sigma^5} = \frac{2\beta^3}{\sqrt{\alpha}}. \tag{7.73}$$

Since we do not generate any random variables, we have $g = g' = 1$. The move from H_1 to H_2 is thus accepted with probability $\alpha_{12} = \min\{1, R_{12}\}$, where

$$R_{12} = \frac{f(H_2, \alpha, \beta|X)}{f(H_1, \mu, \sigma|X)} \times \frac{2\beta^3}{\sqrt{\alpha}}, \tag{7.74}$$

(see equation (7.63)). Similarly to move from H_2: gamma $G(\alpha, \beta)$ to H_1: normal $N(\mu, \sigma^2)$, we set $\mu = \alpha/\beta, \sigma = \sqrt{\alpha}/\beta$. This is the reverse transform of equation (7.72). The acceptance ratio for the move is thus

$$R_{21} = \frac{f(H_1, \mu, \sigma|X)}{f(H_2, \alpha, \beta|X)} \times \frac{\sqrt{\alpha}}{2\beta^3} = \frac{1}{R_{12}}. \tag{7.75}$$

As a third algorithm, we generate the new mean and standard deviation for the new model as random variables centred around the current values. Suppose we use the gamma distribution to propose new values. Step 3 of the algorithm above is then replaced by the following.

To move from H_1 normal to H_2 gamma, we generate two random variables u_1 and u_2:

$$u_1 \sim G(10, 10/\mu), u_2 \sim G(10, 10/\sigma), \tag{7.76}$$

and set $\alpha = (u_1/u_2)^2, \beta = u_1/u_2^2$ as the parameters for the new model H_2 gamma. Here u_1 is the new mean and u_2 is the new standard deviation for the gamma model. The shape parameters (10 in the example) control how highly concentrated the proposed values will be around the current values (μ and σ).

To come back from H_2 gamma to H_1 normal, note that the mean and standard deviation under H_2 are α/β and $\sqrt{\alpha}/\beta$, respectively. We generate two random variables v_1 and v_2 as the new mean and standard deviation:

$$v_1 \sim G(10, 10/(\alpha/\beta)), \quad v_2 \sim G\left(10, \frac{10}{\sqrt{\alpha}/\beta}\right), \tag{7.77}$$

and set $\mu = v_1, \sigma = v_2$ for H_1 normal.

The dimension matching between the two models is $(\mu, \sigma, u_1, u_2) \leftrightarrow (v_1, v_2, \alpha, \beta)$, with

$$\begin{aligned} v_1 &= \mu, \\ v_2 &= \sigma, \\ \alpha &= (u_1/u_2)^2, \\ \beta &= u_1/u_2^2. \end{aligned} \tag{7.78}$$

Thus the Jacobian determinant of the transform is

$$\left|\frac{\partial(v_1, v_2, \alpha, \beta)}{\partial(\mu, \sigma, u_1, u_2)}\right| = \begin{Vmatrix} 1 & 0 & 0 & 0 \\ 0 & 1 & 0 & 0 \\ 0 & 0 & \frac{2u_1}{u_2^2} & -\frac{2u_1^2}{u_2^3} \\ 0 & 0 & \frac{1}{u_2^2} & -\frac{2u_1}{u_2^3} \end{Vmatrix} = \frac{2u_1^2}{u_2^5} = \frac{2\beta^3}{\sqrt{\alpha}}. \tag{7.79}$$

The acceptance ratio for the move from H_1 normal to H_2 gamma is thus (see equation (7.63))

$$R_{12} = \frac{f(H_2, \alpha, \beta|X)}{f(H_1, \mu, \sigma|X)} \times \frac{g\left(v_1; 10, \frac{10}{\alpha/\beta}\right) \cdot g\left(v_2; 10, \frac{10}{\sqrt{\alpha}/\beta}\right)}{g\left(u_1; 10, \frac{10}{\mu}\right) \cdot g\left(u_2; 10, \frac{10}{\sigma}\right)} \times \frac{2\beta^3}{\sqrt{\alpha}}, \tag{7.80}$$

where the posterior ratio is the product of the prior ratio and likelihood ratio and where $g(x; a, b)$ is the density function for gamma $G(a, b)$. Similarly the acceptance ratio for the move from H_2 gamma to H_1 normal is

$$R_{21} = \frac{f(H_1, \mu, \sigma|X)}{f(H_2, \alpha, \beta|X)} \times \frac{g\left(u_1; 10, \frac{10}{\mu}\right) \cdot g\left(u_2; 10, \frac{10}{\sigma}\right)}{g\left(v_1; 10, \frac{10}{\alpha/\beta}\right) \cdot g\left(v_2; 10, \frac{10}{\sqrt{\alpha}/\beta}\right)} \times \frac{\sqrt{\alpha}}{2\beta^3} = \frac{1}{R_{12}}. \tag{7.81}$$

The three trans-model algorithms are summarized in Table 7.4. All three algorithms generate the correct posterior. However, the acceptance probability is 1.1% for Algorithm 1, while it is 70.3% for Algorithm 2 and 31.4% for Algorithm 3. As the posterior probability for model 1 normal is 0.5953, the highest model-jump probability is 80.9%, so Algorithm 2 is very efficient. In comparison, the most efficient chain (with the highest P_{jump}) samples the parameter values from the posterior under the new model, and will be 5.2 times as efficient as the independent sampler (see discussion below equation (7.40)).

7.4.2.3 Trans-dimensional MCMC (rjMCMC)

Here we consider reversible-jump moves between two models H_1 and H_2, with parameters θ_1 and θ_2, respectively. Assume $d_1 < d_2$, so that H_1 has fewer parameters than H_2. We use a vector of random variables u, of size $d_2 - d_1$, for dimension matching: $(\theta_1, u) \leftrightarrow (\theta_2)$. The first d_1 elements of θ_2 match θ_1 even though they may have different biological interpretations. This is a special case of the general framework considered earlier but is the most common type of reversible-jump moves. To move from H_1 to H_2, generate

$u \sim g(u)$. Usually g is the product of $d_2 - d_1$ independent densities. Use a deterministic transform to generate

$$\theta_2 = T(\theta_1, u). \tag{7.82}$$

When we move back from H_2 to H_1, use the inverse transform $(\theta_1, u) = T^{-1}(\theta_2)$ and then drop u. The move from H_1 to H_2 is accepted with probability $\alpha_{12} = \min\{1, R_{12}\}$, where

$$R_{12} = \frac{f(H_2) f(\theta_2|H_2) f(X|H_2, \theta_2)}{f(H_1) f(\theta_1|H_1) f(X|H_1, \theta_1)} \cdot \frac{r_{21}}{r_{12}} \cdot \frac{1}{g(u)} \cdot \left\| \frac{\partial \theta_2}{\partial (\theta_1, u)} \right\|, \tag{7.83}$$

in which r_{12} is the probability of attempting to jump to H_2 when the chain is in H_1 and r_{21} is defined similarly. For the reverse move from H_2 to H_1, we have

$$R_{21} = \frac{f(H_1) f(\theta_1|H_1) f(X|H_1, \theta_1)}{f(H_2) f(\theta_2|H_2) f(X|H_2, \theta_2)} \cdot \frac{r_{12}}{r_{21}} \cdot g(u) \cdot \left\| \frac{\partial (\theta_1, u)}{\partial \theta_2} \right\| = \frac{1}{R_{12}}. \tag{7.84}$$

As an example, we calculate the posterior probabilities for the JC69 and K80 models fitted to data of a pair of sequences. The fictitious data are $n = 100$ sites, $n_S = 5$ transitional differences, and $n_V = 8$ transversional differences. (We do not use the human and orangutan 12S rRNA genes of Example 7.1 here since that dataset is so informative that JC69 has posterior probability ~ 0.) Model H_1 (JC69) has one parameter: $\theta_1 = (d)$, while model H_2 (K80) has two: $\theta_2 = (d, \kappa)$, where d is the sequence distance and κ is the transition/transversion rate ratio. We assign prior probabilities $\pi_1 = \pi_2 = 1/2$ for the two models. We assign the prior $d \sim \exp(10)$ with mean 0.1 under both models and $\kappa \sim G(2, 1)$ under H_2 (K80). The likelihood function is

$$f(x|d) = \left(\frac{1}{16} - \frac{1}{16} e^{-4d/3} \right)^{n_S + n_V} \left(\frac{1}{16} + \frac{3}{16} e^{-4d/3} \right)^{n - (n_S + n_V)}, \tag{7.85}$$

for JC69. For K80, it is

$$f(x|d, \kappa) = \left(\frac{p_0}{4} \right)^{n - (n_S + n_V)} \left(\frac{p_1}{4} \right)^{n_S} \left(\frac{p_2}{4} \right)^{n_V}, \tag{7.86}$$

where p_0, p_1, and p_2 are given in equation (1.11). The posterior probability for JC69 is calculated to be $f(H_1|x) = 0.6090$.

To construct a rjMCMC algorithm, we use a random variable u for dimension matching, with $(d_{\text{JC69}}, u) \leftrightarrow (d_{\text{K80}}, \kappa)$. An iteration of the algorithm may involve the following steps.

1. Within-model move. Propose an MH move to change parameters in the model. If the chain is in model H_1 (JC69), use a sliding window to update d_{JC69}. If it is in H_2 (K80), use two 1-D sliding windows to update d_{K80} and κ.
2. Trans-model move.

We describe three rjMCMC algorithms for step 2, which are summarized in Table 7.5. In Algorithm 1, one generates $\kappa \sim G(\alpha, \beta)$ to move from JC69 to K80 and simply drops κ to move from K80 to JC69. The proposal ratio is simply given by the gamma density $g(\kappa; \alpha, \beta)$. Algorithm 2 is similar except that the distance d is multiplied by a factor (1.2) in the move from JC69 to K80 and divided by 1.2 in the opposite move. The Jacobi term for the transform is then 1.2, and the acceptance ratios are

$$R_{12} = \frac{f(H_2, d, \kappa|x)}{f(H_1, d|x)} \times \frac{1.2}{g(\kappa; \alpha, \beta)}, \tag{7.87}$$

from JC69 to K80, and $R_{21} = 1/R_{12}$ from K80 to JC69. In Algorithm 3, the distance d is not changed between models, but in the move from JC69 to K80 the new κ for K80 is

Table 7.5 Reversible-jump moves and proposal ratios for Bayesian comparison of the JC69 and K80 models using two sequences

Algorithm	JC69 → K80	R_{12}	K80 → JC69	R_{21}	P_{jump}
1	$d = d$, $\kappa \sim G(5,5)$	$1/g(\kappa; 5, 5)$	drop κ	$g(\kappa; 5, 5)$	0.57
2	$d_{K80} = 1.2 d_{JC69}$, $\kappa \sim G(5,5)$	$1/g(\kappa; 5, 5) \times 1.2$	$d_{JC69} = d_{K80}/1.2$	$g(\kappa; 5, 5)/1.2$	0.52
3	$d = d$, $\kappa = d e^{5(u-0.5)}$	5κ	Move (drop κ) only if $d/e^{5/2} < \kappa < d e^{5/2}$	$1/(5\kappa)$	0.27

Note: $g(x; \alpha, \beta)$ is the gamma density function. P_{jump} is the acceptance proportion for cross-model moves.

generated from around the current value of d (i.e. d_{JC69}). The κ value generated this way is in the range $(d/e^{\varepsilon/2}, d e^{\varepsilon/2})$. In the reverse move, κ is simply dropped if it is in the range $(d/e^{\varepsilon/2}, d e^{\varepsilon/2})$, but if it is not, the move is disallowed. In general, a move from model 2 to model 1 is allowed only if the current parameter values in model 2 are reachable in the reverse move from model 1.

All three algorithms produce the same correct results. Judged by the cross-model acceptance proportion, Algorithm 1 is the most efficient among the three, while Algorithm 3 is the poorest. The highest P_{jump} is ~78%, reachable if parameter values are proposed from the posterior under the new model. The different algorithms illustrate the great flexibility of rjMCMC proposals. For example, the density $g(u)$ for generating random variables can be specified using the current parameter values in the model. The deterministic transform T is also arbitrary (such as the use of the factor 1.2).

7.4.2.4 Mixing problems of rjMCMC

While rjMCMC is very flexible in terms of permissible proposals, it often has mixing problems in real applications, especially when the data are informative so that the within-model parameter posteriors are highly concentrated. The cross-model moves tend to be rejected and the chain is stuck in the current model, not because the new model has low posterior probabilities but because the parameter values proposed for the new model are poor. Efficient proposals for rjMCMC are distinctively harder to construct than for within-model MCMC. For the latter, the metric structure of the state space guides the construction of the proposal. For instance with the sliding window proposal on a continuous target density, very small windows will lead to small jumps with $P_{\text{jump}} \approx 1$, whereas large windows will lead to large jumps with a low P_{jump}. By adjusting the window size to obtain a near-optimal P_{jump}, the MH algorithm will work well whether the posterior is highly concentrated or quite diffuse. In rjMCMC, there is no direct analogue to such scale adjustment, since there is no natural notion of a 'local' move when the chain jumps from one model into another.

Here we note a few differences between within-model MCMC and cross-model rjMCMC. First, for within-model MCMC, there is an optimal acceptance proportion (say ~30–40% for 1-D moves; see §7.3.2). For cross-model rjMCMC, the higher P_{jump} is, the more efficient the chain tends to be. This is an implication of Peskun's theorem (§7.3.2). Second, with within-model MCMC, P_{jump} can be made arbitrarily close to 1 by taking small steps. With rjMCMC, the posterior model probabilities place an upper bound on P_{jump}. Thus $P_{\text{jump}} \approx 0$ does not necessarily mean poor mixing. If a model has posterior probability 99.9%, the chain has to stay in the model 99.9% of the time and it cannot accept more than ~0.1% of the proposals to move away from it. Third, the highest P_{jump} is achieved by proposing parameters for the new model from the posterior. With such a proposal, the acceptance ratio of equation (7.63) becomes

$$\alpha_{k,k'} = \frac{f(H_{k'})f(X|H_{k'})}{f(H_k)f(X|H_k)} = \frac{f(H_{k'}|X)}{f(H_k|X)}. \tag{7.88}$$

This is the acceptance ratio when the models do not have parameters.

Fourth, with within-model MCMC, taking a baby step nearly guarantees acceptance (although the resulting Markov chain mixes slowly). One might think that the same would apply to rjMCMC, and that to move from a small model (with fewer parameters) to a large one (with more parameters), proposing parameter values that correspond to the small model would guarantee acceptance. Indeed Brooks et al. (2003) discussed this idea as a way of ensuring an acceptance ratio of ~1, and called it the 'weak non-identifiability centring' method. However, this is a misconception. If one model dominates the posterior, the chain must stay in that model most of the time and the jump probability must not be high. To see why such a proposal leads to near certain rejection, consider Algorithm 1 of Table 7.5, which moves from H_1 (JC69) to H_2 (K80), by keeping the distance d unchanged and sampling κ from the gamma $\kappa \sim G(\alpha, \beta)$. Note that JC69 is a special case of K80 with $k = 1$. We use $\alpha = \beta = 1000$ so that the gamma is approximately normal with mean 1 and variance 0.001. With such a highly concentrated proposal, most values are close to 1: indeed 95% of κ values will be in the interval (0.94, 1.06). However, the gamma density $g(\kappa; \alpha, \beta)$, which appears in the denominator of the acceptance ratio α, is large ($g = 12.6$ for $\kappa = 1$), leading to a small α and likely rejection of the proposal. The prior, if it is diffuse, also works against the parameter-rich model. Even though the likelihood stays essentially the same, the proposal is very likely to be rejected. Bayesian inference favours the small model if both the small and large models fit the data equally well.

The mixing problems discussed here apply to both trans-model MCMC algorithms and trans-dimensional rjMCMC algorithms. Nevertheless, trans-model MCMC algorithms may be less problematic if the parameters matched across models have similar biological interpretations and similar posterior distributions.

While one may combine heating (MC^3) with rjMCMC to move between models, it does not appear to be effective. Heating is useful to flatten one surface for within-model MCMC moves, but is not so effective when we are using rjMCMC to move between different surfaces corresponding to different models.

One alternative to rjMCMC is the so-called product-space method, suggested by Carlin and Chib (1995; see also Godsill 2001). This collects the parameters in all models in one super vector, and updates all of them in every iteration of the MCMC, even though only parameters in the current model are relevant to the likelihood calculation. The advantage of the algorithm is that the chain retains a memory of the parameters in the more complex model when the chain last left it. However, Bayesian phylogenetic and phylogeographic inference involves huge dimensions in the space of species trees or gene genealogies and the idea does not appear easily applicable.

7.4.2.5 Model averaging

If we are interested in a quantity that can be calculated under each of the models under consideration, and there is uncertainty concerning model choice, one approach is to average over the models. To be specific, let the quantity of interest be parameter θ, present in all models. We run a cross-model Markov chain (a rjMCMC, say), and sample θ throughout the MCMC irrespective of the model the chain is in. This sample will provide an estimate of the posterior of θ, which is an average over the models, with the weight to be the posterior model probability:

$$f(\theta|X) = \sum_{k=1}^{K} f(H_k|X) f(\theta|H_k, X). \tag{7.89}$$

This approach is known as *model averaging*. Its great appeal is that it accounts for uncertainties in the model. For instance, if we are interested in distance d but are uncertain about whether JC69 or K80 is a better-fitting model to the sequence data, we can run a rjMCMC, moving between the two models, and then use a sample of d to generate our posterior estimate. This has the following mixture distribution:

$$f(d|x) = f(H_1|x) f(d|H_1, x) + f(H_2|x) f(d|H_2, x). \tag{7.90}$$

In other words, the posterior of d is a weighted average of the posteriors under JC69 and K80, with the weights to be the posterior model probabilities. In real data analysis, neither JC69 nor K80 may be an adequate fit even if K80 may fit the data much better than JC69. Furthermore, those two models typically produce very similar estimates of sequence distance d.

Model averaging requires sophisticated algorithms such as cross-model rjMCMC, and thus suffers from similar computational difficulties. The benefits of model averaging appear to be overstated in the literature; something technically challenging often has its own appeal even if it may not be useful. If all models under consideration fit the data poorly, model averaging is unlikely to help much. If one model fits the data much better than the others, that model will have a posterior probability of nearly 100%, so that model averaging will produce essentially the same result as use of the single best-fitting model (Dornburg et al. 2008). Determining the best-fitting model using criteria such as Akaike information criterion (AIC) or Bayesian information criterion (BIC) and then running a within-model MCMC will be less costly in terms of code development and computational demand. If several models fit the data nearly equally well and they make essentially the same inference concerning the quantity of interest, there is not much point in averaging; use of any of those models will give us similar results. The above cases may cover most situations in practical phylogenetic analysis. The only case where model averaging is useful should be that in which several models fit the data nearly equally well but they make different inferences concerning the quantity of interest. In such a case, the model-averaged posterior may even be multi-moded, highlighting the uncertainty of model choice. Such a case may presumably be detected by a Bayesian sensitivity analysis. The case implies that the data are not informative about the quantity of interest: one may have to consider collecting different types of data or changing the research problem.

7.4.3 Bayes factor and marginal likelihood

Consider two models H_0 and H_1, which specify the probability of the data X as $f(X|H_0, \theta_0)$ and $f(X|H_1, \theta_1)$, respectively. The prior probabilities are $f(H_0)$ and $f(H_1) = 1 - f(H_0)$. The Bayesian approach to hypothesis testing or model selection naturally uses the posterior model probabilities. Earlier literature often uses odds to discuss probabilities (e.g. Bayes 1763; Jeffreys 1935, 1961), which may explain the current focus on the odds. (If p is the probability, $p/(1-p)$ is called the odds.) The *Bayes factor* is defined as the ratio of the posterior odds to the prior odds and is equal to the ratio of the marginal likelihoods (Kass and Raftery 1995):

$$B_{01} = \frac{M_0}{M_1} = \frac{f(X|H_0)}{f(X|H_1)} = \frac{\int f(X|\theta_0, H_0) f(\theta_0|H_0) \, d\theta_0}{\int f(X|\theta_1, H_1) f(\theta_1|H_1) \, d\theta_1}. \tag{7.91}$$

The posterior probabilities of the models are then

$$\frac{f(H_0|X)}{f(H_1|X)} = \frac{f(H_0)}{f(H_1)} \times \frac{f(X|H_0)}{f(X|H_1)}. \tag{7.92}$$

In words,

$$\text{Posterior odds} = \text{Prior odds} \times \text{Bayes factor},$$

or

$$f(H_0|X) = \frac{f(H_0)f(X|H_0)}{f(H_0)f(X|H_0) + f(H_1)f(X|H_1)} = \frac{1}{1 + \frac{f(H_1)}{f(H_0)}/B_{01}}. \tag{7.93}$$

If the prior probabilities are uniform, $f(H_0) = f(H_1) = 1/2$, then $f(H_0|X) = 1/(1 + 1/B_{01})$. Interpretation of Bayes factor is largely through the posterior model probabilities (Kass and Raftery 1995) (Table 7.6).

The Bayes factor is a likelihood ratio. Nevertheless it has a few important differences from the likelihood ratio in hypothesis testing. First, in the likelihood ratio test (LRT), the likelihood is *optimized* over the model parameters, while in calculation of the Bayes factor or in Bayesian model comparison, the marginal likelihood is *averaged* over the model parameters. As a result, the prior may have considerable influence on the marginal likelihood, besides the data-generating model. Second, hypothesis testing and posterior model probabilities may lead to different numerical results when applied to the same data. In §6.2.3.3, we discussed the case of comparing a sharp null hypothesis H_0 with no free parameter against an alternative hypothesis H_1 with one free parameter. It is seen there that a p-value of 5% means a posterior probability $f(H_0|x) \geq 0.128$. Relative to Bayesian model selection, hypothesis testing is too liberal and tends to reject H_0 too often, especially in large datasets. Third, Bayes factors can be applied easily to non-nested models and to comparison of more than two models, while hypothesis testing under such situations is difficult.

In phylogenetic problems, the marginal likelihood is defined only if proper priors are assigned on the branch lengths. This is because the likelihood approaches a non-zero constant (rather than zero) when the branch lengths approach infinity, so that an improper prior on branch lengths would mean an infinite marginal likelihood (see §6.3.3). The same applies to certain parameters in the substitution model, such as the α parameter in the gamma model for rates among sites or among site partitions: when $\alpha \to \infty$, the

Table 7.6 Interpretation of the Bayes factor

B_{01}	Evidence against H_0	Posterior model probability $\Pr\{H_0|X\} = 1/(1 + 1/B_{01})$
1 to 3	Not worth more than a bare mention	0.5 to 0.25
3 to 20	Positive	0.25 to 0.048
20 to 150	Strong	0.048 to 0.0066
>150	Very strong	< 1/151 = 0.0066

likelihood approaches a nonzero constant. Some Bayesian phylogenetic programs use improper priors by default, which may cause problems to marginal likelihood calculation (Baele et al. 2012; Li and Drummond 2012). They can also cause convergence and mixing problems to MCMC algorithms. Furthermore, improper priors are often unreasonable biologically. They should be avoided.

Here we describe four methods for calculating the marginal likelihood under a model. The first two methods are the arithmetic mean and harmonic mean methods. These are well known but do not provide usable results. The last two methods, called path sampling (or thermodynamic integration) and stepping stone methods, involve much more computation but can provide accurate estimates.

The first method is known as the (*prior*) *arithmetic mean* and is based on the definition of the marginal likelihood:

$$z = f(x) = \int f(x|\theta) f(\theta) \, d\theta. \tag{7.94}$$

Here and below we suppress the model in the notation as only one model is considered. One takes a sample $\{\theta_1, \theta_2, \ldots, \theta_n\}$ from the prior $f(\theta)$, and calculates the arithmetic mean of the likelihood:

$$\hat{z} = \frac{1}{n} \sum_{i=1}^{n} f(x|\theta_i). \tag{7.95}$$

This is the Monte Carlo integration discussed in §6.4.5. It is inefficient, as the prior $f(\theta)$ typically differs from the likelihood or posterior considerably, so that most samples taken from the prior may fall outside the region where the likelihood is large.

The second method calculates the integral using importance sampling, with the sampling distribution to be the posterior $f(\theta|x) \propto f(x|\theta) f(\theta)$. In equation (6.90), replace $h(\theta)$ by $f(x|\theta)$ and $g(\theta)$ by $f(x|\theta) f(\theta)$. Then we get the (*posterior*) *harmonic mean* method (Newton and Raftery 1994)

$$\hat{z} = \frac{\sum_{i=1}^{n} f(x|\theta_i) f(\theta_i) / f(\theta_i|x)}{\sum_{i=1}^{n} f(\theta_i) / f(\theta_i|x)} = \frac{n}{\sum_{i=1}^{n} \frac{1}{f(\theta_i|x)}}, \tag{7.96}$$

where $\theta_i, i = 1, \ldots, n$, are a sample from the posterior $f(\theta|x)$. The advantage of this method is that we already have a sample from the posterior generated by the MCMC algorithm, so we get z for free. Its disadvantage is that it is dominated by the smallest likelihood values and is thus very unstable. For many applications it has an infinite variance. Furthermore, it has a positive bias and tends to overestimate the marginal likelihood (Xie et al. 2011). Tests suggest that it does not usually produce usable results (Lartillot and Philippe 2006; Calderhead and Girolami 2009), despite the initial optimism (Kass and Raftery 1995).

More recent methods include path sampling (Gelman and Meng 1998) or thermodynamic integration (Lartillot and Philippe 2006; Friel and Pettitt 2008) and stepping stone sampling (Xie et al. 2011). Those methods generate MCMC samples from a series of densities that lie between the prior and the posterior. Here we describe thermodynamic integration and stepping stone sampling.

Consider the unnormalized density

$$q_\beta(\theta) = f(x|\theta)^\beta f(\theta), \quad 0 \leq \beta \leq 1, \tag{7.97}$$

with a normalizing constant

$$z_\beta = f_\beta(x) = \int q_\beta(\theta) \, d\theta = \int f(x|\theta)^\beta f(\theta) \, d\theta. \tag{7.98}$$

We write the normalized density as $p_\beta(\theta) = \frac{1}{z_\beta} q_\beta(\theta)$. This is called the power posterior (Friel and Pettitt 2008), and is equivalent to the prior if $\beta = 0$ or to the posterior if $\beta = 1$. Raising the likelihood to a power $\beta < 1$ flattens the likelihood surface, making it easy for the Markov chain to move around. This is analogous to a thermodynamic system, with a smaller β corresponding to a high temperature. The marginal likelihood of the model we seek to calculate is then $z = z_1 = z_1/z_0$, with $z_0 = 1$. We take a series of β values between 0 and 1 to form a sample path between the prior and the posterior (or they are like stepping stones); for example, one may set $\beta_k = k/K, k = 1, 2, \ldots, K$.

The thermodynamic integration method calculates the logarithm of the marginal likelihood as

$$\log\{f(x)\} = \log \frac{z_1}{z_0} = \log z_\beta \Big|_0^1 = \int_0^1 \frac{\partial \log z_\beta}{\partial \beta} \, d\beta, \tag{7.99}$$

where the integrand

$$\frac{\partial \log z_\beta}{\partial \beta} = \frac{1}{z_\beta} \cdot \frac{\partial z_\beta}{\partial \beta} = \frac{1}{z_\beta} \int \frac{\partial q_\beta(\theta)}{\partial \beta} \, d\theta = \int \frac{1}{q_\beta(\theta)} \frac{\partial q_\beta(\theta)}{\partial \beta} \frac{q_\beta(\theta)}{z_\beta} \, d\theta$$

$$= \int \frac{\partial \log q_\beta(\theta)}{\partial \beta} p_\beta(\theta) \, d\theta = E_\beta \left[\frac{\partial \log q_\beta(\theta)}{\partial \beta} \right]. \tag{7.100}$$

Here $E_\beta [\cdot]$ stands for the expectation with respect to $p_\beta(\theta)$. Equation (7.99) then becomes

$$\log\{f(x)\} = \int_0^1 E_\beta \left[\frac{\partial \log q_\beta(\theta)}{\partial \beta} \right] d\beta = \int_0^1 E_\beta [\log f(x|\theta)] \, d\beta. \tag{7.101}$$

For any value of β, one runs an MCMC to generate a sample from $p_\beta(\theta)$. The expectation $E_\beta[\cdot]$ in equation (7.101) is then estimated as an average over this sample. This computation is done for a series of values of β, and then the integral of equation (7.101) is approximated using numerical integration. Lartillot and Philippe (2006) used Simpson's method. The method requires K MCMC runs. If $K = 10$, the method will require 10 times as much computation as the original MCMC for parameter estimation under the model.

The last method we discuss is the stepping stone sampling method of Xie et al. (2011). Given the power posteriors at different β values, the marginal likelihood $z_1 = f(x)$ can be expressed as a product of K ratios:

$$z = \frac{z_1}{z_0} = \prod_{k=1}^{K} \frac{z_{\beta_k}}{z_{\beta_{k-1}}}. \tag{7.102}$$

Each ratio in the product,

$$r_k = \frac{z_{\beta_k}}{z_{\beta_{k-1}}} = \frac{f_{\beta_k}(x)}{f_{\beta_{k-1}}(x)}, \tag{7.103}$$

is a ratio of two integrals (two normalizing constants). We use importance sampling to approximate both integrals, using $g(\theta) = p_{\beta_{k-1}}(\theta)$ as the sampling distribution (see equation (6.90)). One gets

$$\hat{r}_k = \frac{1}{n} \sum_{i=1}^{n} \frac{f(x|\theta_i)^{\beta_k}}{f(x|\theta_i)^{\beta_{k-1}}} = \frac{1}{n} \sum_{i=1}^{n} f(x|\theta_i)^{\beta_k - \beta_{k-1}}, \tag{7.104}$$

where $\{\theta_i\}$ are an MCMC sample from $p_{\beta_{k-1}}(\theta)$. As $\beta_{k-1} < \beta_k$, we expect $p_{\beta_{k-1}}(\theta)$ to be slightly more dispersed than $p_{\beta_k}(\theta)$, so it should be a good importance function for the integral $f_{\beta_k}(x)$ (see discussion in §6.4.6).

The marginal likelihood for the model, $z = f(x)$, is then given by the product of the K ratios (equation (7.102)). This method requires $K - 1$ MCMC runs if one can sample from the prior directly. The method reduces to the arithmetic mean method if $K = 1$.

In both thermodynamic integration and stepping stone sampling, one does not have to use uniformly spaced β values. Choices of β correspond to different temperature schedules. Placing most points on small β values (close to the prior) is noted to increase accuracy (Lepage et al. 2007; Friel and Pettitt 2008; Calderhead and Girolami 2009; Xie et al. 2011).

7.5 Problems

7.1 Write a program to implement the MCMC algorithm of Example 7.1 to estimate the distance between the human and orangutan 12S rRNA genes under the JC69 model. Use any programming language of your choice, such as C/C++, Java, or R. Investigate how the acceptance proportion changes with the window size w. (Note: Calculate the logarithms of the prior, the likelihood and the acceptance ratio to avoid overflows and underflows. If the logarithm of the acceptance ratio is ≥ 0, the proposal is accepted. Otherwise take the exponential to calculate the acceptance ratio to decide whether the proposal is accepted.)

7.2 Modify the program of Problem 7.1 to estimate the sequence distance θ under the K80 model. Use the exponential prior $f(\theta) = \frac{1}{\mu} e^{-\theta/\mu}$ with mean $\mu = 0.2$ for distance θ and exponential prior with mean 5 for the transition/transversion rate ratio κ. Implement two proposal steps, one for updating θ and another for updating κ. Compare the posterior estimates with the MLEs of §1.4.2.

7.3 Study the tail behaviour of the Markov chain for estimation of the sequence distance under JC69 of Example 7.1. Suppose the current state is $\theta = 10$. Calculate the acceptance ratios $\alpha_{\text{left}} = \pi(9.9)/\pi(10)$ and $\alpha_{\text{right}} = \pi(10.1)/\pi(10)$ for moves of size $\Delta\theta = 0.1$. Use different starting values such as $\theta = 10$ and 100 to run the MCMC program of Problem 7.1 to examine convergence of the chain. Then do the same calculation using the uniform prior $\theta \sim U(0, 200)$, and modify the MCMC program to do the same test. Use a sliding window with window size $w = 0.2$.

7.4 Write an MCMC program to sample from the 2-D posterior density of equation (7.30). Use a 2-D sliding window to propose moves: $x' \sim U(x - \frac{\varepsilon}{2}, x + \frac{\varepsilon}{2})$ and $y' \sim U(y - \frac{\varepsilon}{2}, y + \frac{\varepsilon}{2})$. Start the chain at $(10000, 0)$. Observe the fraction of the proposals that are accepted. Then use two 1-D sliding windows to propose moves, changing x and y, respectively.

7.5 Modify the program of Problem 7.1 to estimate two parameters under the JC69 model: the time of species divergence t and the substation rate $r = 3\lambda$, instead of the distance $\theta = 2t \times r$. Consider one time unit to be 100 million years. Assign the gamma prior $t \sim G(2, 12)$ with mean 0.167 (meaning 16.7 million years for the human–orangutan divergence) and another gamma prior for the rate $r \sim G(2, 2)$ with mean 1 (meaning a prior mean rate of 10^{-8} substitutions/site/year). Implement one or more of the following proposals:

   i. A 2-D sliding window updating t and r in one step.
   ii. Two separate 1-D sliding windows updating t and r, respectively.
   iii. In both the above cases add an extra step of multiplying t by a random variable c that is close to 1 and dividing r by the same c (see §7.2.5).

Study the convergence of the chain at the tail by using different initial states, such as (100, 100), (5, 0.01), etc., and by comparing different proposals. Study the sensitivity of the posterior to the prior by changing the parameters in the priors.

7.6 Modify the program of Problem 7.1 to estimate the JC69 distance with a gamma prior and confirm the two modes in the posterior of Example 7.2 by using different initial values (0.01, 1, 10, 100) and window sizes (0.01, 0.1, or 1). Note that the constant in the gamma prior cancels so that you need not calculate $\Gamma(100) = 99!$. As in Problem 7.1, calculate the logarithms of the prior and likelihood to avoid numerical problems.

7.7 Write down the transition matrix P for the Markov chain generated in the MCMC algorithm for the robot-on-box example of §7.1. Consider both the symmetrical move of §7.1.1 and the asymmetrical move of §7.1.2. Assume $\pi_1 \geq \pi_2 \geq \pi_3$.

7.8* Show or confirm that the eigenvalues and eigenvectors of the P matrix of equation (7.42), as defined in equation (7.36), are given as

$$\Lambda = \begin{bmatrix} 1 & & & & \\ & 0 & & & \\ & & 0 & & \\ & & & \ddots & \\ & & & & 0 \\ & & & & & 1 - \frac{1}{\pi_1} \end{bmatrix},$$

$$E = \begin{bmatrix} 1 & 0 & 0 & \cdots & 0 & \left(1 - \frac{1}{\pi_1}\right) a_K \\ 1 & a_2 & a_3 & \cdots & a_{K-1} & a_K \\ 1 & -\frac{\pi_2}{\pi_3} a_2 & 0 & \cdots & 0 & a_K \\ 1 & 0 & -\frac{\pi_2}{\pi_4} a_3 & \cdots & 0 & a_K \\ \vdots & \vdots & \vdots & \ddots & \vdots & \vdots \\ 1 & 0 & 0 & \cdots & -\frac{\pi_2}{\pi_K} a_{K-1} & a_K \end{bmatrix}, \quad (7.105)$$

where the factors $a_k = \left[\frac{\pi_{k+1}}{\pi_2(\pi_2 + \pi_{k+1})}\right]^{\frac{1}{2}}$, $k = 2, \ldots, K-1$, and $a_K = \left[\frac{\pi_1}{1-\pi_1}\right]^{\frac{1}{2}}$ are for normalizing the eigenvectors so that $E^T B E = I$ or $\sum_{i=1}^{K} \pi_i e_{ik}^2 = 1$ for each k. Note that the kth column in E is the eigenvector corresponding to λ_k.

[Hint. (**a**) To confirm the results, simply check that Λ and E above satisfy $Px = \lambda x$, with x to be a column in E. (**b**) To derive the eigenvalues, one way is to solve the characteristic equation $|P - \lambda I| = 0$. By Laplace's formula, the determinant $|P - \lambda I| = p_{11} \cdot |P_{11}| - p_{12} \cdot |P_{12}| + p_{13} \cdot |P_{13}| - \ldots$, where P_{1k} is the $(K-1) \times (K-1)$ matrix that results from removing the 1st row and kth column of $P - \lambda I$. Thus show that $|P - \lambda I| = \frac{1}{\pi_1}(-\lambda)^{K-2}(\lambda - 1)(\pi_1 \lambda + 1 - \pi_1)$. Note that the determinant of a triangular matrix is the product of the diagonal elements and that interchanging two rows of a matrix multiplies its determinant by -1.]

7.9* Show that the asymptotic variance when a Markov chain sample from P of equation (7.42) is used to estimate π_1 is $v = \pi_1(1 - \pi_1)(2\pi_1 - 1)$. [Hint: One way is to use equation (7.37) with the eigenvalues and eigenvectors given in Problem 7.8.]

---

* indicates a more difficult or technical problem.

7.10* Calculate the efficiency of the following Markov chain sampler for estimating $\pi_1 = 1/K$:

$$P = \begin{bmatrix} 0 & \frac{1}{K-1} & \cdots & \frac{1}{K-1} \\ \frac{1}{K-1} & 0 & \cdots & \frac{1}{K-1} \\ \vdots & \vdots & \ddots & \vdots \\ \frac{1}{K-1} & \frac{1}{K-1} & \cdots & 0 \end{bmatrix}. \quad (7.106)$$

Note that efficiency is defined as $\pi_1(1-\pi_1)/v$, where v is the asymptotic variance of equation (7.35).

7.11* Show that if the target density is $N(0,1)$ and the MCMC proposal density is $x'|x \sim N(x, \sigma^2)$, the acceptance proportion is given by equation (7.50).

7.12* Write an MCMC program to sample from the posterior target $\pi(\theta)$ of Figure 7.15, which is a mixture of three normal distributions. Use a uniform sliding window of size 1. Record the frequencies at which the chain moves between the three regions of the parameter space corresponding to the three peaks: $(-\infty, -1)$, $(-1, 1)$, and $(1, \infty)$. Also pay attention to whether the chain visits those three regions with the correct probabilities (i.e. 0.2, 0.5, and 0.3). Then implement the parallel tempering algorithm of §7.4.1 with a second hot chain at temperature T, and see whether it helps the cold chain to mix.

* indicates a more difficult or technical problem.

CHAPTER 8

Bayesian phylogenetics

8.1 Overview

8.1.1 *Historical background*

In the early 1960s, Anthony Edwards and Luca Cavalli-Sforza made an effort to apply R.A. Fisher's likelihood method to estimate genealogical trees of human populations using gene frequency data (Edwards and Cavalli-Sforza 1964; Cavalli-Sforza and Edwards 1966). They used the Yule process (pure birth process) to describe the probabilities of rooted trees (or labelled histories) (see §3.1.1.5) and the Brownian motion process to describe the drift of transformed gene frequencies over time. As mentioned in Chapter 3, this effort led them to the development of the distance (additive-tree) method and the parsimony method (minimum-evolution principle), both as approximations to maximum likelihood (ML). Nevertheless they experienced considerable difficulties in applying ML itself, including 'singularities' in what they thought was the likelihood function (Cavalli-Sforza and Edwards 1966). The nature of the estimation problem was clarified by Edwards (1970), who pointed out that the tree or labelled history, which has a distribution specified by the Yule process, should not be estimated by maximizing the 'likelihood' but should instead be estimated from the conditional distribution of the trees given the genetic data. In modern terminology, the Yule process specifies a prior on the trees, and the tree should be estimated from the posterior distribution (i.e. the conditional distribution given the data). Thus Edwards (1970) may be considered the first attempt to apply Bayesian statistics to phylogeny reconstruction.

Sequence data with discrete nucleotides are easier to deal with mathematically than continuous gene frequencies. In the 1990s, three groups worked independently to introduce Bayesian inference to molecular phylogenetics: Rannala and Yang (1996; Yang and Rannala 1997), Mau and Newton (1997), and Li et al. (2000). The early studies assumed a constant rate of evolution (the molecular clock) as well as an equal probability prior for rooted trees, either with or without the node ages ordered (i.e. either labelled histories or rooted trees; see §3.1.1.5). Later, more efficient MCMC algorithms were implemented in the computer programs BAMBE (Larget and Simon 1999) and MrBayes (Huelsenbeck and Ronquist 2001). The clock constraint was relaxed, enabling phylogenetic inference to be conducted under more realistic evolutionary models. A number of innovations have been introduced in these programs, adapting tree perturbation algorithms such as nearest neighbour interchange (NNI), subtree pruning and regrafting (SPR), and tree bisection and reconnection (TBR) (see §3.2.3) for use as MCMC proposals to move between trees. Later versions of MrBayes (3.0 and 3.2) (Ronquist and Huelsenbeck 2003; Ronquist et al. 2012b) incorporated many evolutionary models developed for likelihood inference, and can accommodate heterogeneous datasets from multiple gene loci in a combined analysis. The more recent program BEAST allows the estimation of rooted trees under clock and relaxed-clock models (Drummond et al. 2006; Drummond and Rambaut 2007), while PhyloBayes implements sophisticated models to deal with substitution heterogeneity among lineages and along the sequence that may be important for Bayesian inference of deep phylogenies using protein sequences (Lartillot et al. 2007, 2009; Lartillot and Philippe 2008). Nowadays those Bayesian programs are standard tools in molecular phylogenetics, together with fast likelihood programs such as RAxML (Stamatakis 2006) and PHYML (Guindon and Gascuel 2003).

Several reviews of Bayesian phylogenetics have been published, including Huelsenbeck et al. (2001), Holder and Lewis (2003), Zwickl and Holder (2004), Yang (2005), and Yang and Rannala (2012). A book on Bayesian phylogenetics is being edited by Chen et al. (2014).

8.1.2 A sketch MCMC algorithm

The problem of phylogeny reconstruction fits in a straightforward manner into the general framework of hierarchical Bayesian inference. Let X be the sequence data. Let τ_i be the ith tree topology, $i = 1, 2, \ldots, T_s$, where T_s is the total number of tree topologies for s species (see §3.1.1.5). We leave it open whether the tree is rooted or unrooted. Usually a uniform prior $f(\tau_i) = 1/T_s$ is assumed. Let t_i be the vector of branch lengths on tree τ_i, with prior $f(t_i)$. While branch lengths are defined only in the context of a specific tree topology, the common practice has been to treat all branch lengths in the same way irrespectively of the tree topology, so $f(t_i|\tau_i) = f(t_i)$. Let θ represent the parameters in the substitution model, with the prior $f(\theta)$. The joint posterior distribution of the tree τ_i, branch lengths t_i, and parameters θ is

$$f(\tau_i, t_i, \theta | X) \propto f(\theta) f(\tau_i) f(t_i|\tau_i) f(X|\theta, \tau_i, t_i). \tag{8.1}$$

The normalizing constant, or the marginal probability of the data $f(X)$, is a sum over all possible tree topologies and, for each tree topology τ_i, an integral over all branch lengths t_i and substitution parameters θ. This is impossible to calculate numerically except for very small trees. The MCMC algorithm avoids direct calculation of $f(X)$, and achieves the sum and integration through MCMC.

A sketch of an MCMC algorithm may look like the following:

1. Start with a random tree τ, with random branch lengths t, and random substitution parameters θ.
2. Iterate using the following steps:
   a. Propose a change to the tree, by using tree rearrangement algorithms (such as NNI, SPR, or TBR). This step may change branch lengths t as well.
   b. Propose changes to branch lengths t.
   c. Propose changes to parameters θ.
   d. Every k iterations, sample the chain: save τ, t, and θ to disk.
3. At the end of the run, summarize the results.

8.1.3 The statistical nature of phylogeny estimation

A phylogenetic tree is a statistical model rather than a parameter, and phylogeny reconstruction is a model selection problem rather than a within-model parameter estimation problem. While some writers consider the distinction between a model and a parameter as merely a semantic one, it is important in two situations. First, in discussions of ML estimation of the phylogeny, one should note that the well-known asymptotic efficiency of maximum likelihood estimates (MLEs) applies to parameter estimation within a well-specified model, such as estimation of branch lengths and substitution parameters when the tree topology is given, but not to ML phylogeny reconstruction (see §5.2.3). Second, in designing MCMC algorithms for Bayesian phylogenetic inference, one should note that proposals that change the tree topology are cross-model moves, and they should be designed and optimized differently from within-model moves (see §7.4.2.4 and below).

We note here a few differences between phylogeny reconstruction and 'typical' statistical model selection or hypothesis testing. In typical model selection, one is not really interested in the model itself although one may be concerned about the impact of assuming an inadequate model. In hypothesis testing, the model (hypothesis) is the focus of interest but one does not often entertain more than a few models. Indeed one is often interested in only two hypotheses, with one (the null) being a special case of the other (the alternative). In phylogenetics, the tree is the focus of interest, but there are a huge number of trees, with intricate neighbourhood relationships among them. Strictly speaking, one needs both the phylogeny and the model of sequence evolution to specify the likelihood for the sequence data, so that the likelihood model is really a tree-process combined model. However it is convenient to distinguish the two components. A common situation is that the violation of the process model causes all the combined models to be wrong to some extent, but one would hope that the combined model fits the data better if the tree is correct than if the tree is wrong.

As discussed earlier, MCMC algorithms for within-model parameter estimation (§7.3.2) and for trans-model calculation of posterior model probabilities (§7.4.2) have important differences. A within-tree move, i.e. a proposal that modifies the branch lengths or substitution parameters without changing the tree topology, is a within-model move. Its objective is to traverse the parameter space for the tree. By reducing the step length of the proposals to nearly 0, one can accept nearly all proposals, with $P_{\text{jump}} \approx 100\%$. However, to traverse the parameter space efficiently, one should aim to achieve an intermediate P_{jump} of about 30–40%. A cross-tree move, i.e. a proposal that changes the tree topology, is a move between different models. For such a move, it may be impossible to achieve a high acceptance rate (P_{jump}). For example, if one tree has posterior probability 99%, the chain has to stay in that tree 99% of the time and should not move too often; it is then impossible for P_{jump} to be higher than 2% (= $2 \times (1 - 0.99)$). For efficient mixing, one should in general attempt to maximize P_{jump}. Branch lengths should be proposed in the new tree to maximize the acceptance rate rather than to traverse the parameter space for the new tree.

As pointed out by Lakner et al. (2008), there is a general perception that intermediate jump probabilities are optimal. However, this is true for within-tree moves only but not for cross-tree moves. Many proposals used in current phylogenetic programs are mixtures of within-tree moves and cross-tree moves, making it awkward to adjust the step length to achieve optimal performance. For example, Lakner et al. (2008) found in their extensive tests that the overall acceptance rate was not a good indicator of efficiency of the algorithm, whereas the acceptance rate of topology changes was. This is apparently because the acceptance rate that Lakner et al. calculated averages over within-tree moves (for which intermediate acceptance rates are optimal) and cross-tree moves (for which high acceptance rates are good), so that the overall acceptance rate is not a very useful indicator of performance.

As all binary trees have the same number of branch lengths and substitution parameters, the different trees have the same dimension. We can thus match up the parameters to use trans-model MCMC to move between trees, with no need for trans-dimensional rjMCMC (see §7.4.2.1–3). To maximize the acceptance rate of cross-tree moves, one should match up the parameters with similar posterior distributions across trees. Substitution parameters such as the transition/transversion rate ratio κ and the gamma shape parameter α for rate variation among sites have the same biological interpretations irrespective of the tree topology and it is natural to match them up across trees. It is also natural to match up the external branch lengths. The lengths of the internal branches that define the same splits (bipartitions) may be matched up across tree topologies as well. Such matching is used in Bayesian programs such as MrBayes.

All models developed for ML phylogenetic analysis can be implemented in the Bayesian framework. Likelihood calculation under those models has been described in Chapters 1, 2, and 4. The Bayesian framework also allows us to implement sophisticated parameter-rich models that are impractical for ML implementation. Bayesian marginalization through the prior provides an attractive way of dealing with nuisance parameters, while implementation of such models by ML is hard or unfeasible, as the likelihood function involves high-dimensional integrals or may suffer from the problem of infinitely many parameters (see §1.4.4).

To implement a model in the Bayesian framework, one has (i) to assign priors on the model parameters, and (ii) to design proposal algorithms to alter the parameters during the MCMC. In §8.2 we discuss the priors assigned on parameters in the substitution models. While ML estimation is invariant to reparametrization, priors on parameters are not, and assigning uniform priors on poorly chosen parametrizations may bias Bayesian inference. Thus care is needed for prior specification. Because of the high dimension and complex nature of the parameter space in Bayesian phylogenetics, diffuse or minimally informative priors that do not have an undue influence on the posterior are desirable.

In §8.3 we discuss MCMC proposals in Bayesian phylogenetics. We first discuss within-tree moves. These are straightforward Metropolis–Hastings (MH) algorithms discussed in §7.2. The rest of §8.3 discusses cross-tree proposals, which are far more complex.

8.2 Models and priors in Bayesian phylogenetics

8.2.1 *Priors on branch lengths*

A binary unrooted tree for s species has $2s-3$ branches. Let the branch lengths be $\mathbf{t} = \{t_1, t_2, \ldots, t_{2s-3}\}$. Bayesian phylogenetic programs such as MrBayes have used independent and identically distributed (i.i.d.) priors on the branch lengths, with each branch length having either a uniform or exponential distribution. The uniform prior $t_i \sim U(0, A)$ has an upper bound A specified by the user (the default in MrBayes is 100). The exponential prior $t_i \sim \exp(\beta)$ has density

$$f(t|\beta) = \beta e^{-\beta t}, t > 0. \tag{8.2}$$

In MrBayes the default rate parameter $\beta = 10$, with mean $1/\beta = 0.1$.

Those i.i.d. priors on branch lengths are problematic. While each one may appear innocent, collectively they may make a strong and unreasonable statement concerning the average branch length or equivalently the sum of branch lengths (called tree length) (Rannala et al. 2012). It has been observed that MrBayes with its default i.i.d. prior generates exceptionally long trees in some empirical datasets (Brown et al. 2010;

Marshall 2010). In some cases, the posterior branch lengths are an order of magnitude too large and the 95% credibility interval (CI) of the tree length excludes the MLE (the sum of the MLEs of the branch lengths) under the same model. This is particularly a problem in the analysis of data from closely related species with multiple sequences from the same species, in which the sequence divergence is very low (e.g. Leaché and Mulcahy 2007; Schmidt-Lebuhn et al. 2012).

The distribution of the tree length $T = \sum_{i=1}^{2s-3} t_i$ induced by the branch length prior can be derived by considering T as a function of the branch lengths (Theorem 1 in Appendix A). Note that the sum of two gamma variables $x_1 \sim G(\alpha_1, \beta)$ and $x_2 \sim G(\alpha_2, \beta)$ is itself a gamma variable, $x_1 + x_2 \sim G(\alpha_1 + \alpha_2, \beta)$. Also the exponential distribution $\exp(\beta)$ is $G(1, \beta)$. Thus the i.i.d. exponential prior on t_i means that T has the gamma distribution $G(2s - 3, \beta)$, with density

$$f(T|\beta, s) = \frac{\beta^{2s-3}}{(2s-4)!} T^{2s-4} e^{-\beta T}. \tag{8.3}$$

With $2s - 3 \gg 1$, the gamma distribution is approximately normal, with a very large mean $E(T) = (2s - 3)/\beta$ and a moderate standard deviation $\sqrt{2s - 3}/\beta$. For example, if $s = 100$ and $\beta = 10$, the prior mean of tree length is 19.7 and the 99% interval is (16.3, 22.9). For datasets of similar sequences (say, with $T < 1$), this will be an extremely informative prior, favouring unreasonably large tree lengths.

The situation is worse for the i.i.d. uniform prior $t_i \sim U(0, A)$. While most branch lengths in phylogenetic trees are small (< 1 or even < 0.1), large values (such as 5 or 10) occasionally occur. To avoid precluding long branches, a large bound A is used. Since t_i has mean $A/2$ and variance $A^2/12$, the induced prior on T is approximately normal, with $E(T) = (2s - 3)A/2$, and $V(T) = (2s - 3)A^2/12$. With $s = 100, A = 100$, the 99% prior interval will be $9850 \pm 2.576 \times 405.2$ or $(8806, 10894)$. Such long trees, with $\sim 10,000$ substitutions at an average site, are not expected in real datasets and are unreasonable. While uniform priors are sometimes advertised as being noninformative, here they make strong and unreasonable statements about the tree length. Besides generating long and unreasonable trees in the posterior, the i.i.d. priors may potentially create a local peak in the posterior (because the likelihood is flat at very large branch lengths), and cause convergence problems for MCMC algorithms (see Problem 8.2).

One modification to the exponential prior is to assign a hyper prior on the exponential parameter (Suchard et al. 2001). While $t_i \sim \exp(\beta)$, parameter β can be assigned a gamma hyper prior (or equivalently the prior mean $\mu = 1/\beta$ is assigned an inverse gamma hyperprior). This is helpful by increasing the variance in the induced prior on T (Rannala et al., 2012).

A better prior is the compound Dirichlet distribution, suggested by Rannala et al. (2012). A prior such as the gamma is assigned on the tree length T, with density

$$f(T; \alpha_T, \beta_T) = \frac{\beta_T^{\alpha_T}}{\Gamma(\alpha_T)} e^{-\beta_T T} T^{\alpha_T - 1}, \quad \alpha_T > 0, \beta_T > 0, \tag{8.4}$$

with mean α_T/β_T and variance α_T/β_T^2. Then T is broken into component branch lengths by using a symmetric Dirichlet distribution on $x_1 = t_1/T, x_2 = t_2/T$, etc.

The Dirichlet is a multivariate distribution for a set of proportions $\mathbf{x} = \{x_1, x_2, \ldots, x_K\}$, with $x_i > 0$ and $\sum_i x_i = 1$. The density is

$$f(\mathbf{x}; \alpha_1, \alpha_2, \ldots, \alpha_K) = \frac{\Gamma(\alpha_0)}{\prod_{i=1}^{K} \Gamma(\alpha_i)} \prod_{i=1}^{K} x_i^{\alpha_i - 1}, \quad \alpha_i > 0, \tag{8.5}$$

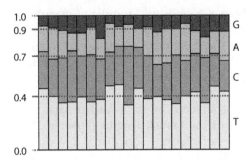

Fig. 8.1 Fifteen samples from the Dirichlet distribution Dir(40, 30, 20, 10). Here the variables represent the frequencies of the four nucleotides T, C, A, and G, with averages 0.4, 0.3, 0.2, and 0.1. In this breaking-stick representation of the Dirichlet distribution, a stick of fixed length (1 in the example) is broken into different segments, with given average lengths, allowing variation in the relative sizes of the segments.

where $\alpha_0 = \sum_i \alpha_i$. This is a $(K-1)$-dimensional generalization of the beta distribution, with mean $E(x_i) = \alpha_i/\alpha_0$ and variance $V(x_i) = \alpha_i(\alpha_0 - \alpha_i)/[\alpha_0^2(\alpha_0 + 1)]$. See Figure 8.1 for a so-called broken-stick construction of the distribution. If $K = 2$, this is the familiar beta distribution beta(α_1, α_2). If all the α_i parameters are equal, with $\alpha_i = \alpha$, the distribution is called symmetric Dirichlet, with a larger α meaning smaller variances. If $\alpha_i = 1$, the density reduces to $f(\mathbf{x}) = (K-1)!$, which is a multivariate generalization of the uniform $U(0, 1)$.

In our case, the number of proportions is $K = 2s - 3$, and with the sum being one, $\mathbf{x}$ is $(2s-4)$-dimensional. The $(2s-3)$-dimensional joint density of tree length T and the proportions $\mathbf{x}$ is:

$$f(T, \mathbf{x}) = f(T; \alpha_T, \beta_T) f(\mathbf{x}). \tag{8.6}$$

Suppose the proportions have the symmetric Dirichlet distribution with $\alpha_i = 1$, with density $f(\mathbf{x}) = (2s-4)!$. The joint density of the branch lengths $\mathbf{t}$ can be derived by applying the variable transform $(T, \mathbf{x}) \leftrightarrow (\mathbf{t})$, with

$$\begin{aligned} x_1 &= t_1/T, \\ &\cdots \\ x_{2s-4} &= t_{2s-4}/T, \\ T &= t_1 + t_2 + \cdots + t_{2s-3}, \end{aligned} \tag{8.7}$$

The Jacobian determinant of the transform is $\left|\frac{\partial(T,\mathbf{x})}{\partial(\mathbf{t})}\right| = T^{-(2s-4)}$. Thus the prior on branch lengths is (see Theorem 1 in Appendix A)

$$f(\mathbf{t}|\alpha_T, \beta_T) = \frac{\beta_T^{\alpha_T}}{\Gamma(\alpha_T)} e^{-\beta_T \sum t_i} \times \left(\sum t_i\right)^{\alpha_T - 1 - (2s-4)} (2s-4)! \tag{8.8}$$

(Rannala et al. 2012). Here given T, the branch lengths have a Dirichlet distribution, but T itself is compounded with a gamma distribution. Thus the branch lengths are said to have a compound Dirichlet distribution. One may use $\alpha_T = 1$ for a diffuse prior on T, while β_T should be chosen so that the prior mean of T is reasonable for the dataset being analysed.

Rannala et al. (2012) presented a more general prior, with different means for the internal and external branch lengths (Yang and Rannala 2005). Tests by Zhang et al. (2012) on six problematic datasets identified by Brown et al. (2010) suggest that the posterior was quite stable when β_T varied by a few orders of magnitude, so a rough estimate of β_T should be adequate. The compound Dirichlet prior was found to be effective in correcting the branch length overestimation problem in empirical data analysis (Zhang et al. 2012).

Rannala et al. (2012; Zhang et al. 2012) also implemented an alternative compound Dirichlet prior, which assigns, instead of the gamma, an inverse gamma on the tree length $T \sim \text{invG}(\alpha_T, \beta_T)$:

8.2 MODELS AND PRIORS IN BAYESIAN PHYLOGENETICS

$$f(T; \alpha_T, \beta_T) = \frac{\beta_T^{\alpha_T}}{\Gamma(\alpha_T)} e^{-\beta_T/T} T^{-\alpha_T - 1}, \quad \alpha_T > 2, \beta_T > 0. \tag{8.9}$$

This has mean $\beta_T/(\alpha_T - 1)$ and variance $\beta_T^2/[(\alpha_T - 1)^2(\alpha_T - 2)]$, for $\alpha_T > 2$. With the same symmetric Dirichlet distribution used to partition T into branch lengths, the joint prior of branch lengths becomes

$$f(\mathbf{t}|\alpha_T, \beta_T) = \frac{\beta_T^{\alpha_T}}{\Gamma(\alpha_T)} e^{-\beta_T/\sum t_i} \times \left(\sum t_i\right)^{-\alpha_T - 1 - (2s-4)} (2s-4)!, \tag{8.10}$$

in place of equation (8.8). For a diffuse prior, one can use $\alpha_T = 3$, while β_T should be chosen to have a reasonable prior mean. While the gamma is light-tailed, the inverse gamma has a heavy tail.

Example 8.1. Leaché and Mulcahy (2007) used mitochondrial and nuclear gene sequences from spiny lizards of the *Sceloporus magister* species group (squamata: phrynosomatidae) to test phylogeographic hypotheses concerning the timing and origins of diversity across mainland deserts and the Baja California Peninsula. The mtDNA data consist of two fragments, for 12S rRNA and NADH4, with 1,606 nucleotide sites in a 123-sequence alignment. This is one of the six datasets analysed by Brown et al. (2010) to demonstrate the problem of extremely long trees. We present some results for this dataset in Table 8.1 (Zhang et al. 2012). The sequences are from six closely related species in the *Sceloporus magister* species group, with multiple individuals for most species/subspecies. The MLE of tree length is 2.48 under GTR + I + Γ_4 and 2.11 under GTR + Γ_4. Table 8.1 summarizes results for MrBayes analysis of the dataset using three priors:

i. The i.i.d. exponential prior, with $\beta = (2s - 3)/\bar{T}$;
ii. The gamma-Dirichlet prior, with $\alpha_T = 1$ and $\beta_T = 100, 10, 1, 0.1, 0.01$; and
iii. The inverse gamma-Dirichlet prior, with $\alpha_T = 3$ and $\beta_T = 0.02, 0.2, 2, 20, 200$.

Here parameter β is adjusted to match the prior mean $\bar{T}$ of tree length. For the i.i.d. exponential prior, the default value $\beta = 10$ (with mean 0.1) in MrBayes gives $\bar{T} = 24.3$. At such a prior tree length, the program generates very long trees. Overall, the posterior tree length is quite sensitive to the prior when the i.i.d. exponential prior is used. It is much more robust under the compound Dirichlet priors, when the prior mean changed over 3–4 orders of magnitude (Table 8.1 and Figure 8.2). For this dataset, the posterior means of tree length are smaller than the MLEs, apparently because the prior has the effect of shrinking the branch lengths in a Bayesian analysis: note that strictly speaking, the MLEs of branch lengths do not have finite means or variances. □

Inference of the tree topology appears to be much less affected by the branch length prior than are branch lengths or the tree length. For example, Marshall (2010, Figure 1) presented an example in which two runs of the same analysis using MrBayes 'converged' to (or get stuck at) apparently two modes in the posterior, at which the tree lengths are very different (1.50 and 7.70, respectively), but the consensus tree topologies are identical, with the branch lengths being nearly proportional.

8.2.2 Priors on parameters in substitution models

8.2.2.1 Nucleotide substitution models

Nucleotide substitutions over time are described by a continuous-time Markov chain with instantaneous rate matrix $Q = \{q_{ij}\}$, where $q_{ij}, i \neq j$, is the rate of change from nucleotides

Table 8.1 Posterior means and 95% CIs of the tree length for the lizard dataset

Prior mean $\bar{T}$	Exp(β)	GammaDir (1, β_T, 1, 1)	invGamDir (3, β_T, 1, 1)
The GTR + I + Γ_4 model			
0.01	0.034 (0.032, 0.037)	1.361 (1.243, 1.484)	1.939 (1.755, 2.147)
0.1	0.303 (0.281, 0.325)	1.858 (1.677, 2.044)	1.938 (1.747, 2.155)
1	1.457 (1.351, 1.565)	1.948 (1.758, 2.147)	1.935 (1.744, 2.140)
10	3.379 (3.006, 3.800)	1.947 (1.751, 2.164)	1.986 (1.789, 2.189)
100	57.38 (19.82, 74.43)	1.956 (1.748, 2.176)	2.448 (2.221, 2.699)
The GTR + Γ_4 model			
0.01	0.066 (0.062, 0.070)	1.498 (1.404, 1.596)	1.817 (1.690, 1.958)
0.1	0.479 (0.457, 0.501)	1.785 (1.661, 1.920)	1.821 (1.692, 1.963)
1	1.535 (1.452, 1.623)	1.825 (1.693, 1.970)	1.822 (1.701, 1.963)
10	2.549 (2.336, 2.786)	1.828 (1.697, 1.968)	1.847 (1.717, 1.989)
100	68.03 (57.65, 80.30)	1.831 (1.701, 1.972)	2.108 (1.957, 2.274)

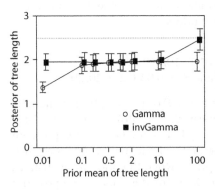

Fig. 8.2 The posterior mean and 95% CIs for the tree length (sum of branch lengths) for the lizard dataset (Leaché and Mulcahy 2007) calculated using the two compound Dirichlet priors for branch lengths. The GTR + I + Γ_4 model is assumed, following Brown et al. (2010). For the gamma-Dirichlet prior, the shape parameter is fixed at $\alpha_T = 1$ while for the inverse gamma-Dirichlet prior, $\alpha_T = 3$. The scale parameter β_T in both priors is chosen to achieve the designed prior mean for tree length. The concentration parameter of the Dirichlet distribution is fixed at $\alpha = 1$. The plots for the two priors are shifted slightly for clarity. The MLE of tree length (dotted line) is 2.48, while the posterior mean under GammaDir(1, 1, 1, 1) is 1.95. Under GTR + Γ_4, the MLE becomes 2.11 while the posterior mean is 1.83. Redrawn after Zhang et al. (2012). See also Table 8.1.

i to j. All the models of nucleotide substitution developed for likelihood calculation and discussed in Chapters 1 and 4 can be used in a Bayesian analysis, ranging from the simplest Jukes–Cantor model (Jukes and Cantor 1969) to the general time-reversible model (GTR or REV) (Tavaré 1986; Yang 1994b; Zharkikh 1994). The JC69 model assumes the same rate between any two nucleotides as well as equal equilibrium base frequencies, and involves no parameter. The K80 (Kimura 1980), HKY85 (Hasegawa et al. 1984, 1985), and TN93 (Tamura and Nei 1993) models involve the transition/transversion rate ratio κ. This can be assigned a gamma prior $\kappa \sim G(\alpha, \beta)$, say, $G(2, 1)$ with mean 2. Alternatively $\frac{\kappa}{1+\kappa}$ can be assigned a beta prior such as beta(a, b) or even beta(1, 1). In other words, $y = \frac{\kappa}{1+\kappa}$ has density

$$f(y) = \frac{1}{B(a,b)} y^{a-1}(1-y)^{b-1}, \ 0 < y < 1. \tag{8.11}$$

Then as $\frac{dy}{d\kappa} = \frac{1}{(1+\kappa)^2}$, the prior density for κ is (see Theorem 1 in Appendix A)

$$f(\kappa) = f(y) \times \left|\frac{dy}{d\kappa}\right| = \frac{1}{B(a,b)}\kappa^{a-1}(1+\kappa)^{-a-b}, \quad 0 < \kappa < \infty. \tag{8.12}$$

The equilibrium base frequencies ($\pi_T, \pi_C, \pi_A, \pi_G$) in models such as F81 (Felsenstein 1981), HKY85, TN93, and GTR are typically assigned a Dirichlet prior, Dir($\alpha_T, \alpha_C, \alpha_A, \alpha_G$), or simply Dir(1, 1, 1, 1). However, those frequency parameters can be reliably estimated using the frequencies in the observed data, so it should be adequate just to fix them at the observed frequencies.

The GTR model is specified using substitution rate parameters and equilibrium base frequency parameters, with the matrix of relative rates given as

$$Q = \begin{bmatrix} \cdot & a\pi_C & b\pi_A & c\pi_G \\ a\pi_T & \cdot & d\pi_A & e\pi_G \\ b\pi_T & d\pi_C & \cdot & f\pi_G \\ c\pi_T & e\pi_C & f\pi_A & \cdot \end{bmatrix}, \tag{8.13}$$

where the nucleotides are ordered T, C, A, and G. Note that only the relative rates are important. In ML estimation, the parametrization does not matter. One can fix $f = 1$ and estimate a, b, c, d, and e (as in Yang 1994b), or estimate the relative rates a–f under the constraint that their sum or average is 1. Both lead to the same inference. In a Bayesian analysis, the parametrization may matter and one should avoid assigning uniform priors on inappropriately chosen parameters. One may assign a Dirichlet prior on the relative rates a–f under the constraint that they sum to 1. For example, Dir(1, 1, 1, 1, 1, 1) ignores the transition/transversion rate ratio, while Dir(2, 1, 1, 1, 1, 2) assigns a prior mean 2 for κ. An alternative is to fix one rate at 1 (say, $f = 1$), and assign gamma priors on the other rates (which are then rate ratios). Assigning uniform priors such as $U(0, 100)$ on rate ratios a–e, with $f = 1$ fixed, is ill-advised and may bias the estimates (Suchard et al. 2001; Zwickl and Holder 2004).

8.2.2.2 Amino acid and codon models

Most models of amino acid replacement are empirical models generated from large databases (e.g. Jones et al. 1992; Whelan and Goldman 2001), with no free parameters. The '+F' models (§2.2.1) treat the stationary amino acid frequencies as free parameters. These can be fixed at the observed frequencies in the sequence data or estimated in the MCMC by assigning a Dirichlet prior, as in the treatment of nucleotide frequencies. The general time-reversible model for amino acid replacement (Yang et al. 1998) can be implemented in the same way as the GTR model for nucleotides, with the Dirichlet prior used to partition the relative rates. There are $19 \times 20/2 - 1 = 189$ relative rate (exchangeability) parameters and 19 amino acid frequency parameters, with a total of 208 free parameters in the Q matrix (§2.2.1).

Parameters in the codon models of Goldman and Yang (1994) and Muse and Gaut (1994), such as the nucleotide/codon frequencies, the transition/transversion rate ratio κ, and the nonsynonymous/synonymous rate ratio ω, can be assigned similar priors. It is natural to assign a Dirichlet prior on the frequency parameters although these can be estimated using the observed frequencies. Parameter ω can be assigned a gamma prior, e.g., $G(2, 4)$ with mean 0.5; or $\omega/(1 + \omega)$ can be assigned a beta prior. ML implementations of more sophisticated codon substitution models are described in Chapter 11; these can be implemented similarly in the Bayesian framework.

8.2.2.3 Models of variable substitution rates among sites

Models of variable substitution rates across sites are described in §1.3 and §4.3.1. The commonly used gamma model assumes that the branch lengths at site i are multiplied by a random variable r_i, which has a gamma distribution $r_i \sim G(\alpha, \beta)$ (Yang 1993, 1994a). As the rates are relative, we fix $\beta = \alpha$ so that the mean of the gamma distribution is 1. In a Bayesian implementation, the gamma shape parameter α can be assigned a gamma prior, e.g., $G(2, 4)$ with mean 0.5. An exponential prior may also be reasonable. The uniform prior, say $U(0, 200)$ used in MrBayes, seems to be a poor choice since estimates of α from real datasets are small (<1) (see Table 2.2), and furthermore the likelihood is nearly flat for large values of α (say, $\alpha > 10$), so the use of the uniform prior may potentially cause convergence problems for the MCMC.

The 'I + Γ' models, such as HKY85 + I + Γ and GTR + I + Γ for nucleotides and WAG + I + Γ for amino acids, assume a proportion of invariable sites (p_0) with rate zero, in addition to the gamma distributed rates for sites. Those models are pathological as both the gamma and the proportion of invariable sites are used to deal with the variable rates, leading to unstable parameter estimates that are hard to interpret (see §4.3.1.4). These models are however widely used in empirical data analysis due to mechanical application of model-test procedures, which favour the most complex models due to the huge size of sequence datasets. It is common to assign a uniform prior on the proportion of invariable sites: $p_0 \sim U(0, 1)$. Ideally the prior should accommodate the correlation between p_0 and the gamma shape parameter α.

Finite-mixture models. Instead of the parametric gamma distribution, one can also use several site classes with different rates (Yang 1995a, see §4.3.1.1). Both the probabilities and the rates for the site classes are parameters. With the constraints that the probabilities sum to 1 and that the mean rate is 1, the model involves $2K - 2$ parameters for K classes. This is known as a finite-mixture model. To implement it, one may assign a uniform Dirichlet prior to the site class probabilities: $(p_1, p_2, \ldots, p_K) \sim \text{Dir}(1, 1, \ldots, 1)$. Another Dirichlet prior may be used to generate the relative rates for the site classes, $(r_1, r_2, \ldots, r_K) \sim \text{Dir}(1, 1, \ldots, 1)$; the relative rates are then rescaled so that their mean is one, $\sum_{i=1}^{K} p_i r_i = 1$, before used in the likelihood calculation. The number of site classes K can be fixed, and a small number (2–4, say) should be sufficient for phylogenetic analysis. If one assigns a discrete distribution on K, one can in theory estimate K by generating a posterior distribution for K. However models of different K have different numbers of probability and rate parameters, so that rjMCMC algorithms become necessary to move between models of different dimensions. However, the precise value of K is unlikely to matter to the tree topology or branch lengths as long as more than a few classes are included in the model, and such a computational effort to estimate K appears to be ill-spent.

8.2.2.4 Dirichlet process models of among site heterogeneity

The Dirichlet process is an important stochastic process that is widely used in Bayesian clustering and Bayesian nonparametric statistics (Ferguson 1973). Suppose we seek to partition n data points into K clusters, such that data points in the same cluster are generated from the same parametric model while those in different clusters are generated from different models. The Dirichlet process specifies a prior on the partitions, i.e. on the assignments of the n data points to different clusters, which induces a prior on the number of clusters (K). As a result, the model allows Bayesian inference of K, as well as the probabilities for the clusters and the assignments of the n data points to the clusters. Rate variation among sites can be accommodated within such a Bayesian clustering framework, in which the n sites in the sequence alignment are partitioned into K site classes evolving at different rates. The model is appealing in that the user does not have to know or specify the number of site classes K; the model allows estimation of K from the data. In practice, this appeal is often illusory.

The Dirichlet process is often characterized by using a Chinese restaurant analogy. Imagine a fanciful Chinese restaurant with an infinite number of round tables, each with infinite capacity. The first customer sits at the first table. The second customer sits at the first table with probability $1/(\alpha + 1)$ or at the next empty table with probability $\alpha/(\alpha + 1)$. In general, the $(n+1)$th customer sits at an occupied table i with probability $n_i/(n + \alpha)$, where n_i is the number of customers already sitting at table i, or at the next empty table with probability $\alpha/(n + \alpha)$. Here α is called the concentration parameter. A smaller α means that customers tend to sit together, requiring fewer tables. Note that the first two customers sit at the same table with probability $1/(\alpha + 1)$, and this is true for any two customers. Also a table with many customers tend to attract more customers, so the process shows a 'rich-gets-richer' phenomenon. In Bayesian clustering, customers are data points and tables are clusters. Customers at the same table are served the same menu, in the same way that data points in the same cluster are generated from the same model. The Chinese restaurant process thus defines a probabilistic model for assigning data points to clusters.

An assignment of all n data points to clusters (which also determines the number of clusters) is called a *partition*. This can be represented by an *assignment* or *allocation vector*, $\mathbf{z} = \{z_1, z_2, \ldots, z_n\}$, where z_i takes values from among the consecutive numbers $1, 2, \ldots, K$, with $z_i = k$ meaning that data point i is in cluster k. Thus if $n = 5$, $\mathbf{z} = \{1, 2, 1, 1, 2\}$ means $K = 2$ clusters, with data points 1, 3, and 4 in cluster 1 and data points 2 and 5 in cluster 2. Obviously $\mathbf{z}_1 = \{1, 2, 1, 1, 2\}$ and $\mathbf{z}_2 = \{2, 1, 2, 2, 1\}$ represent the same partition although the labels for the two clusters are switched. Actually this label-switching problem causes a lot of trouble in Bayesian MCMC algorithms, because if one partition of K clusters is optimal, there will be $K!$ peaks in the posterior with exactly the same height (e.g. Stephens 2000; Jasra et al. 2005).

The number of possible partitions for a given K (i.e. the number of ways of partitioning n data points to K clusters) is given by the Stirling number of the second kind:

$$S_2(n, K) = \frac{1}{K!} \sum_{i=0}^{K-1} (-1)^i \binom{K}{i} (K-i)^n. \tag{8.14}$$

Note that there is only one partition for $K = 1$ or $K = n$, so that $S_2(n, 1) = S_2(n, n) = 1$. With $n = 5$ and $K = 3$, there are $S_2(5, 3) = 25$ partitions. The total number of partitions for a given n is given by summing over K and is called the Bell number:

$$B_n = \sum_{k=1}^{N} S_2(n, k). \tag{8.15}$$

For example, $n = 5$ data points can be partitioned into 1, 2, 3, 4, or 5 clusters, with the total number of partitions to be $B_5 = \sum_{i=1}^{5} S_2(5, i) = 1 + 15 + 25 + 10 + 1 = 52$.

The Dirichlet process or the Chinese restaurant process specifies the probability of $\mathbf{z}$ (which is also the joint probability of K and $\mathbf{z}$, since $\mathbf{z}$ fully determines K) as

$$f(K, \mathbf{z}|\alpha, n) = \alpha^K \frac{\prod_{i=1}^{K}(n_i - 1)!}{\prod_{i=1}^{n}(\alpha + i - 1)}, \tag{8.16}$$

where n_i is the size of cluster i (Ferguson 1973; Antoniak 1974). The probability for the number of clusters K is obtained by summing equation (8.16) over all partitions that have K clusters:

$$f(K|\alpha, n) = \frac{\alpha^K S_1(n, K)}{\prod_{i=1}^{n} (\alpha + i - 1)}, \qquad (8.17)$$

where $S_1(n, K)$ is the absolute value of the Stirling number of the first kind. This prior on K is nonparametric; as a result Bayesian clustering using the Dirichlet process is considered a nonparametric method. The expectation of K is

$$E(K|\alpha, n) \approx \alpha \log(1 + \frac{n}{\alpha}), \qquad (8.18)$$

which grows with n roughly at the rate of log n.

Neal (2001) describes a Gibbs sampling scheme that updates the assignment vector **z** in an MCMC algorithm. It picks a data point i at random and places it back into a cluster according to the full conditional distribution, i.e. the distribution of z_i given the rest of the vector **z**. If the data point is alone in a cluster, the move reduces K by one. If the item is placed into a new cluster, the move increases K by one. Thus the Gibbs sampler updates K as well. By summarizing the assignment vectors visited by the MCMC algorithm, one can generate a posterior distribution of K. Similarly the posterior of the mixing probabilities for the clusters, the model parameters for the clusters, as well as the assignments of data points to the clusters are generated by the MCMC as well.

When the Dirichlet process is used to model variable rates among sites, the data points are the sites, and the clusters are the site classes with different rates. The rate for each class can be generated from a parametric distribution (called base distribution), such as the gamma with mean 1. This model of variable rates for sites is implemented by Huelsenbeck and Suchard (2007).

The Dirichlet process is also known as an infinite mixture model. It is a discrete mixture model because it allows two sites to be in the same site class with the same rate, whereas in a continuous model (such as the gamma) of variable rates for sites, the probability that two sites have the same rate is zero. It is an infinite mixture because, as K is variable and approaches infinity, the model can generate infinitely many different rates, whereas a finite-mixture model with K classes allows only K different rates.

An attractive feature of the Dirichlet process is that it allows one to estimate the number of clusters K from the data. In some applications, the precise value of K is unimportant. For example, in phylogenetic analysis, we are often interested in the tree topology and branch lengths, but not the number of site classes (cf. Huelsenbeck and Suchard 2007). In other applications, knowledge of K may be very important. For example, in species delimitation, one clusters sequence samples into different populations that represent distinct species, and then knowledge of the number of populations or species is biologically very important. One should then bear in mind that the Dirichlet process may or may not be the most suitable model for estimating K. The Dirichlet process predicts more clusters for larger n. Furthermore its rich-gets-richer property means that it favours partitions with very large or very small clusters over partitions with intermediate-sized clusters (Green and Richardson 2001). Those properties may not always be desirable. Indeed one of the surprising findings from the use of the Dirichlet prior to deal with variable substitution rates at sites is that the Bayesian analysis favoured many site classes (30 or 50) (Lartillot and Philippe 2004). In contrast, with ML under finite-mixture models, one cannot typically fit more than a few (3 or 4, say): the use of more site classes simply leads to the collapse of the model into one of fewer classes (see §4.3.1.1 and Table 4.2).

The conflicting results are apparently due to the impact of the Dirichlet process prior, which favours more site classes for long sequences. At any rate, it is essential to examine the robustness of the posterior of K to the concentration parameter α (Huelsenbeck and Suchard 2007).

In addition to variable rates for sites, other aspects of among-site heterogeneity can be accommodated through using the Dirichlet process in a straightforward manner. Lartillot and Philippe's (2004) CAT model accommodates variable nucleotide or amino acid frequencies among site classes. Similar models are developed to account for the 'pattern heterogeneity' across sites, in which sites in different classes evolve according to different rate matrices (Pagel and Meade 2004, 2008).

8.2.2.5 Nonhomogeneous models

For distantly related species, different sequences may have different nucleotide or amino acid compositions, indicating that the substitution process is not stationary. In the ML framework, Yang and Roberts (1995) and Galtier and Gouy (1998) implemented models that allow the base frequencies to drift in different directions on the tree, by using different base frequency parameters in the rate matrix for different branches. However, the number of sets of base frequency parameters and the assignment of branches to those sets have to be pre-specified. The general version of the model uses one set of frequency parameters for every branch as well as another set for the root in the rooted tree, and involves too many parameters. Foster (2004) implemented a Bayesian version of this model, assuming a fixed number of sets of base frequencies and with the assignment of the branches to the sets varied in an MCMC algorithm. Blanquart and Lartillot (2006) used a compound Poisson process to describe the occurrences of so-called breakpoints along branches of the tree. At a breakpoint, the substitution process shifts and a new set of base frequency parameters is used in the rate matrix, generated from a uniform Dirichlet prior. Unlike Foster's model, the breakpoints can occur within a branch. The compound Poisson process was used earlier by Huelsenbeck et al. (2000a) to model the 'evolution' of substitution rate over time. It allows both the number of breakpoints and their positions on the tree to be generated at random. One difficulty is that the model dimension changes with the number of breakpoints so that rjMCMC becomes necessary, with its computational difficulties. As Blanquart and Lartillot (2006) commented, if our interest is in tree reconstruction instead of the history of compositional shifts, it may be simpler to allow compositional shift to occur only at a node of the tree so that every branch is associated with one set of base frequencies. Such a simplified version of the model may produce similar results concerning tree reconstruction and branch length estimation.

Blanquart and Lartillot (2008) combined the CAT mixture model of Lartillot and Philippe (2004) with the nonstationary breakpoint model of Blanquart and Lartillot (2006) into a new model, called CAT-BP. This accounts for heterogeneity in the evolutionary process both along the sequence and across lineages. As in CAT, the model implements a mixture of Markov substitution models with different base (or amino acid) frequencies among site classes, thus accommodating site-specific selective constraints induced by protein structure and function. Furthermore, as in the breakpoint model, those processes are nonstationary, and the equilibrium frequencies are allowed to change along lineages. The model is constructed such that the breakpoint events affect the protein sequence globally: for example, at a breakpoint all sequence categories may shift towards amino acids coded by high-AT codons. The CAT-BP model was implemented using MCMC. Tests on real data suggest that joint modelling of heterogeneities across sites and among lineages is useful for reconstructing deep phylogenies using protein sequences, where amino acid frequencies may be drifting in different directions along different lineages and where there may be strong compositional heterogeneity in different domains of the protein (Blanquart and Lartillot 2008).

8.2.2.6 Partition and mixture models for analysis of large genomic datasets

Modern phylogenetic analysis typically uses many genes or proteins, and partitioned analyses are now routine. Nylander et al. (2004) and Brown and Lemmon (2007) have emphasized the importance of data partitioning in Bayesian phylogenetic analysis and discussed the use of Bayes factors to compare different partition models (see also Fan et al. 2011, and §7.4.3). Also Bayesian programs such as MrBayes and BEAST have implemented models for partitioned analysis.

Determining the best partitioning strategy is part science and part art. Partitioning by gene (protein) or by codon position are two commonly used strategies. Suppose there are K partitions, and the overall substitution rate differs among partitions, with rate r_k for partition k. In an ML implementation (see §4.3.3), one can let $r_1 = 1$ so that branch lengths are defined with reference to the first partition and the rates for all other partitions are relative to the rate for the first. Or one can let the average rate over all K partitions be 1. Such different parametrizations lead to the same inference, as the likelihood ratio test (LRT) comparing different partition models remains the same. In Bayesian inference, the parametrization may matter when one specifies the prior. For example, fixing $r_1 = 1$ and assigning gamma priors on $r_2, \ldots, r_K$ may not be ideal as the results may depend on the order of the partitions, and seriously skewed inference may be drawn concerning the rate ratio r_2/r_1, say. Here we mention two better choices. The first is the Dirichlet distribution for all K relative rates. Another is the i.i.d. prior $r_i \sim G(\alpha, \alpha)$, with mean one, for $i = 1, \ldots, K$. With a large number of partitions, those two priors should give very similar results.

Differences among partitions in other aspects of the substitution process, such as the base or amino acid frequencies, may be dealt with in similar ways.

In a partitioned analysis, one may in addition use a random-effects (mixture) model to deal with the heterogeneity of the substitution process among sites *within* the same partition. Examples of such models include the parametric gamma or the nonparametric Dirichlet process models of rates among sites discussed above. In such a partition-mixture model, the overall rates for partitions accommodate large-scale rate variations *among* partitions, while the random-effect model such as the gamma rates for sites deals with the fine-scale variation within a partition.

Some authors (e.g. Pagel and Mead 2004) suggest treating all sites as one partition and using a mixture model to deal with all heterogeneity among sites in the dataset. For the present it is unclear whether partition or mixture models should be preferred, or how the hundreds or even thousands of genes should be partitioned in phylogenetic analysis. Such decisions may depend on the particular datasets being analysed (Yang and Rannala 2012).

8.2.3 Priors on tree topology

8.2.3.1 Prior on rooted trees

If the trees are rooted and the nodes are associated with times (either absolute geological times or relative times measured by sequence distance), the probabilistic distribution of trees can be generated from a stochastic model of cladogenesis, i.e. a model of speciation, extinction, and sampling of species. Such a process often generates a distribution of divergence times (node ages) as well. For example, Edwards (1970) used the Yule process to generate a distribution of the trees (labelled histories; see §3.1.1.5) as well as node ages. Yang and Rannala (1997) used a birth-death-sampling process as a prior for phylogenetic trees, specified by a per-lineage birth rate λ, a per-lineage death rate μ, and a sampling fraction ρ. All those processes generate a prior with equal probabilities for all possible labelled histories. The birth–death process model has been extended to incorporate sequential sampling by Stadler (2010). Those processes specify a distribution of ages of nodes on the rooted tree, which serves as a natural prior in Bayesian estimation of divergence times under the clock and relaxed-clock models. We will revisit such priors in Chapter 10.

8.2.3.2 *Prior on unrooted trees*

Most phylogenetic analyses do not assume the molecular clock and infer unrooted trees. A commonly used prior on unrooted tree topologies is the uniform prior. The total number of unrooted trees U_s is given in equation (3.1), so the prior probability for each tree is $1/U_s$.

Pickett and Randle (2005) pointed out that equal prior probabilities for all trees induce unequal prior probabilities on splits or clades (calculated by summing over prior probabilities for trees containing the split), with very small or very large splits having higher probabilities than intermediate-sized splits. This is due to the fact that combinatorially there are more splits of small or large sizes than of intermediate sizes. Indeed no matter what reasonable priors are assigned to trees, it is effectively impossible to achieve equal split probabilities (Steel and Pickett 2006). Of course if the dataset is very large, the likelihood will eventually dominate the posterior, and the effect of unequal split probabilities in the prior will disappear (Brandley et al. 2006). However, for finite data there is the concern that if the prior split probabilities vary greatly depending on the split size, the posterior split probabilities may not be comparable.

Goloboff and Pol (2005) constructed a scenario in which the uniform prior on trees, which induces a non-uniform prior on splits, leads to spuriously high posterior probabilities for incorrect splits. This is illustrated in Figure 8.3a. Suppose in the tree for s species, all branches have intermediate lengths except branch a, which is very long so that sequence a is nearly random relative to the other sequences. Thus an informative dataset will allow the backbone tree for $s-1$ species excluding a to be correctly recovered, but the placement of a is nearly random and can be on any of the $(2s-5)$ branches. Thus the resulting $(2s-5)$ trees will receive nearly equal support. By summing up probabilities across those trees to calculate split probabilities, the posterior probability for the incorrect split (bc) will be about $(2s-7)/(2s-5)$, because only two out of the $(2s-5)$ trees do not include that split; these two trees have species a joining either b or c. Similarly splits (bcd), $(bcde)$, $(bcdef)$, etc. will receive support values around $(2s-9)/(2s-5)$, $(2s-11)/(2s-5)$, and so on. Those splits are incorrect, but their posterior probabilities can be close to 1 if s is large.

Example 8.2: The impact of rogue species on phylogenetic support. A dataset of 20 sequences is simulated under the JC69 model using EVOLVER, based on the tree of Figure 8.3a, with all branch lengths to be 0.1 except branch a, which has length 10. The sequence length is 10,000 sites. The data are very informative about the relationships among sequences b–t:

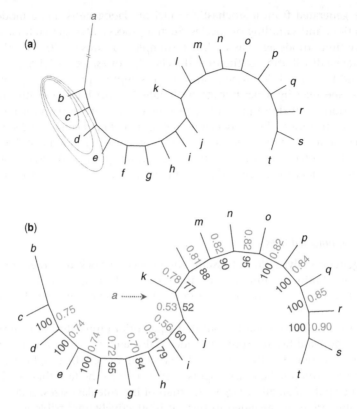

Fig. 8.3 (a) A model tree that can lead to high posterior probabilities for incorrect splits because a uniform prior on trees induces a non-uniform prior on splits, constructed following Goloboff and Pol (2005). All branch lengths on the tree of $s = 20$ species are 0.1 changes per site except branch a, which has 10 changes per site. (b) ML bootstrap support values (shown as percentages) calculated using PHYML/BASEML and posterior split probabilities calculated using MrBayes 3.2.1 from a dataset of 10,000 sites simulated using the tree of panel (a). The data are very informative about the relationships among species b-t, but the location of species a is nearly random. The JC69 model is used in both simulation and analysis.

analysis of those 19 sequences recovers the correct tree with full support by both ML and Bayesian methods.

The 20 species dataset is then analysed under JC69 using ML and Bayesian methods. First BASEML is used to evaluate all the 35 trees that differ in the placement of sequence a. The ML tree joins branch a with the tip branch k. The maximum difference in log likelihood among those 35 trees is 0.541, and the true tree is worse than the ML tree by 0.469. Next PHYML is used to conduct a bootstrap analysis, and MrBayes is used to calculate the consensus tree and the split probabilities. Tree search for this small dataset has proved to be hard for both programs, and different runs produced different results. Thus the precise support values (bootstrap proportions or posterior probabilities) of Figure 8.3b should be treated with caution. Nevertheless, in every run the results showed the same pattern: the support values are much higher for the small splits such as $(b,c),(b,c,d),(t,s),(t,s,r)$, etc. than for intermediate-sized splits. One should be aware of rogue sequences. □

8.3 MCMC proposals in Bayesian phylogenetics

8.3.1 Within-tree moves

8.3.1.1 Updating branch lengths

Without changing the tree topology, standard MH algorithms can be used to update parameters, such as branch lengths, the transition/transversion rate ratio κ, the shape parameter α for gamma rates at sites, etc. For example, branch lengths can be updated one by one, using a sliding window around the current branch length based on the uniform, normal or Bactrian distributions, or using a multiplier (§7.2.1–5). One can also choose one branch length at random for updating, but updating all of them in a fixed order may have a computational advantage as it is easy to save on the conditional probability calculation in the pruning algorithm. The step length in the proposal can be adjusted using the burn-in period of the chain to achieve good mixing (§7.3.2.3). While larger branch lengths tend to have larger variances and should in theory use larger steps, it is impractical to use different step lengths for different branch lengths given the huge number of branches in the different trees. Instead the same step length is typically used to update all branch lengths. In this sense, the multiplier, which is a sliding window on the logarithmic scale and tends to take larger steps for longer branches, may be more suitable for changing branch lengths than a sliding window on the branch lengths themselves.

One can also scale the whole tree by applying a multiplier to all branch lengths (§7.2.5). Such a tree length scaling move is useful to bring branch lengths into the right range if all of them are either too small or too large. It may also be useful in dealing with the correlation between the branch lengths and the gamma shape parameter for rate variation among sites (α). Similarly one can select a subtree and apply a multiplier to all branch lengths within the subtree. On a rooted tree, the ages of all nodes in the subtree can be scaled so that the relative ages of nodes inside the subtree do not change (Figure 8.4).

Branch lengths cannot be negative. There are two problems when the branch lengths are too close to zero. First, if two sequences have differences between them but are separated by a distance on the tree (sum of branch lengths on the path connecting the two sequences) that is too close to zero, the probability of data may become zero or negative due to rounding errors, making it impossible to calculate the log likelihood. Second, as soon as a branch length becomes zero, some moves such as the multiplier will never move away from it. This is a problem especially if the likelihood or posterior has a mode at zero for a branch length, and then the multiplier move will make the branch length smaller and smaller, eventually becoming zero due to underflow. The common practice is to apply minimum bounds on branch lengths. External branch lengths may use a bound such as $t > 10^{-6}$. Internal branch lengths can have a weaker bound, such as $t > 10^{-8}$, because the distance between any two sequences will not be zero as long as the external branches are bounded away from zero. Nevertheless underflow can occur for internal branch lengths if one uses the multiplier. Application of bounds means that the prior for branch lengths is truncated. The effect of the truncation on the prior may be ignored as the normalizing constant will be very close to one and cancel in the MCMC algorithm. However the

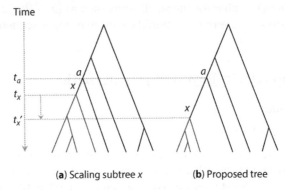

Fig. 8.4 Scaling a whole clade in a rooted tree. A non-root internal node x is chosen at random, with mother node a. A sliding window is used to propose a new age for node x, reflected into the range $(0, t_a)$. Then all node ages inside the clade are scaled proportionally, with $t'_j = t_j \times \frac{t'_x}{t_x}$, so that the subtree retains its shape. The proposal ratio is $\left[\frac{t'_x}{t_x}\right]^m$, where m is the number of internal nodes inside the clade ($m = 2$ in the example). This move changes the branch lengths (node ages) but not the tree topology.

bounds have to be respected in the MCMC proposals by using appropriate reflection (see §7.2.1–5). For example, it is incorrect to simply set the proposed negative branch length to the lower bound.

8.3.1.2 Updating substitution parameters

Parameters in the substitution model, such as the transition/transversion rate ratio (κ) and the gamma shape parameter for variable rates among sites (α), can be updated using standard MH proposals.

The nucleotide frequency parameters $\pi = (\pi_1, \pi_2, \pi_3, \pi_4)$ have the constraint that they sum to 1. One way of updating them is to sample new frequencies from a Dirichlet distribution centred around the current values: $\pi' = (\pi'_1, \pi'_2, \pi'_3, \pi'_4) \sim \text{Dir}(\alpha_0\pi_1, \alpha_0\pi_2, \alpha_0\pi_3, \alpha_0\pi_4)$. See equation (8.5) for the density function. Here the *concentration parameter* $\alpha_0 > 1$ plays the role of the step length, with a large α_0 meaning small steps; note that $E(\pi'_j) = \pi_j$ and $V(\pi'_j) = \pi_j(1-\pi_j)/(\alpha_0+1)$. Figure 8.1 shows 15 samples of π' for $\alpha_0 = 100$ and $\pi = (0.4, 0.3, 0.2, 0.1)$. The proposal ratio for the move is simply the ratio of the Dirichlet densities

$$\frac{q(\pi|\pi')}{q(\pi'|\pi)} = \prod_i \frac{\Gamma(\alpha_0\pi_i) \cdot \pi_i^{\alpha_0\pi'_i-1}}{\Gamma(\alpha_0\pi'_i) \cdot (\pi'_i)^{\alpha_0\pi_i-1}}. \tag{8.19}$$

Sequence data are typically very informative about the frequency parameters, so the optimal α_0 should be large, in the order of the sequence length. The move is three-dimensional, so α_0 should be adjusted to achieve an acceptance proportion of 20–30% (§7.3.2.3).

This proposal may be expensive as it involves sampling Dirichlet random variables (see §12.4.5). A simpler proposal may be to pick up two frequencies at random, say, π_i and π_j, and change their values with their sum ($s = \pi_i + \pi_j$) kept fixed. Let π'_i be generated using a sliding window around π_i based on the uniform or Bactrian distributions, reflected into the range $0 < \pi'_i < s$, if necessary. We set $\pi'_j = s - \pi'_i$. The proposal ratio is one (see Problem 8.1). This is a 1-D move, so the step length should be adjusted to achieve an acceptance proportion of 30–40%.

Amino acid frequencies in amino acid replacement models or codon frequencies in codon substitution models can be updated using the same methods.

8.3.2 Cross-tree moves

8.3.2.1 Proposals and proposal ratios

Proposals that change the tree topology are more complex than the simple moves discussed in §7.2. In deriving the proposal ratio, it often simplifies the procedure to break the proposal into several component steps, and use the product rule of probabilities across the steps. Suppose the move $x \to x'$ can be broken into three component steps: $x \to y \to z \to x'$, with $q(x'|x) = q(y|x)q(z|y)q(x'|z)$. The proposal ratio of the move is then

$$\frac{q(x|x')}{q(x'|x)} = \frac{q(z|x')q(y|z)q(x|y)}{q(y|x)q(z|y)q(x'|z)} = \frac{q(x|y)}{q(y|x)} \times \frac{q(y|z)}{q(z|y)} \times \frac{q(z|x')}{q(x'|z)}. \tag{8.20}$$

If any one of the component steps is symmetrical, the corresponding 'proposal ratio' will be 1, thus leading to simplification. Here the state x of the chain may consist of the tree τ and branch lengths $\boldsymbol{t}$: $x = (\tau, \boldsymbol{t})$. For instance, a tree rearrangement move such as NNI, $(\tau, \boldsymbol{t}) \to (\tau', \boldsymbol{t}')$, may be achieved in two steps. The first step $(\tau, \boldsymbol{t}) \to (\tau', \boldsymbol{t})$ changes the current tree topology τ to a new tree topology τ' through NNI, and it is clear that the move is symmetrical with 'proposal ratio' 1. The second step $(\tau', \boldsymbol{t}) \to (\tau', \boldsymbol{t}')$ modifies the branch lengths with the tree topology fixed at τ', and the proposal ratio may be easy to derive as well. We will use this strategy to derive the proposal ratios for the moves in this section.

Cross-tree moves are instances of cross-model proposals of §7.4.2.2. From equation (7.63), the acceptance rate for the move $(\tau, \boldsymbol{t}) \to (\tau', \boldsymbol{t}')$ is

$$\alpha = \frac{f(\tau', \boldsymbol{t}'|X)}{f(\tau, \boldsymbol{t}|X)} \times \frac{r_{\tau'\tau}}{r_{\tau\tau'}} \times \frac{q(\tau, \boldsymbol{t}|\tau', \boldsymbol{t}')}{q(\tau', \boldsymbol{t}'|\tau, \boldsymbol{t})}. \tag{8.21}$$

Here the posterior ratio equals the prior ratio times the likelihood ratio, as before. The tree proposal probability $r_{\tau\tau'}$ is the probability of proposing tree τ' given that the chain is in the current tree τ, with the possibility that $\tau = \tau'$. This characterizes the division of computational effort into within-tree versus cross-tree moves, as well as our choice of tree perturbation algorithms (such as NNI, SPR, and TBR), which incur different probabilities for candidate trees. The proposal ratio $\frac{q(\tau, \boldsymbol{t}|\tau', \boldsymbol{t}')}{q(\tau', \boldsymbol{t}'|\tau, \boldsymbol{t})}$ implies a mapping of branch lengths between the current tree and the new tree.

The separation of a cross-tree proposal into several component steps discussed above can be carried one step further, in that the performance of the different component steps can be assessed more or less independently. In other words the following questions can be answered largely separately.

i. How should one divide the computational effort between the cross-tree and within-tree moves? In general, a highly mobile chain that moves frequently between trees should be more efficient (producing estimates of posterior tree or split probabilities with smaller variances). One might think that for each cross-tree move, many rounds of within-tree moves are necessary for the chain to reach stationarity in the new tree. However, this is not the case. If the branch lengths (and other parameters) are from the posterior for the current tree, the branch lengths in the new tree (when the algorithm moves to the new tree through a cross-tree move) will automatically be from the stationary (posterior) distribution for the new tree. There is no need to run many iterations of within-tree moves for the new tree.

ii. Given that we have decided to change the tree topology, which tree perturbation algorithms are efficient? And given our choice of the tree rearrangement algorithm, which defines a set of neighbours for the current tree, how should we select a tree from the set? The answers to these questions are not obvious, but appear to depend on the particular datasets, in particular, on the presence or absence of local optima in the space of trees. For example, NNI moves are local and generate close neighbours, while SPR and TBR moves are global and can introduce large changes to the tree. Close neighbours are more likely to be accepted, because the trees may have similar (optimized) likelihood scores, and because it is easier to propose good branch lengths for a highly similar target tree. However, if we limit our proposals to close neighbours the chain may get stuck in a local optimum (tree island).

iii. Given the proposed tree topology, how do we generate branch lengths for the new tree? Note that the objective is to increase the chance of getting the new proposed tree accepted, rather than traversing the parameter space of the new tree. As the branch lengths in the current tree are nearly optimized to fit the sequence data, we should aim to maintain features of the tree, such as the pairwise sequence distances implied by the tree. In this regard, experience with MLEs of branch lengths in different trees may be useful. It is noted that in general branch lengths inside the same subtree tend to be similar across different trees all of which contain that subtree. Thus branch lengths within a subtree that are not changed by the tree rearrangement algorithm should be retained.

Note that both (i) and (ii) concern the model proposal probability $r_{\tau\tau'}$ of equation (8.21). Optimal choices of $r_{\tau\tau'}$ appear to depend on the landscape of the tree space. If there are no tree islands or local optima, a local move such as NNI may be very efficient. Otherwise global moves such as SPR and TBR may be necessary to avoid the algorithm getting trapped in a local optimum. The discussion of tree rearrangement proposals in the rest of the section focuses on (iii), the generation of branch lengths in the new tree.

8.3.2.2 Criteria for evaluating cross-tree moves

To evaluate the mixing efficiency of the MCMC algorithm across trees, one can use the posterior probabilities for trees. For example, one can evaluate the variance for the MCMC estimate of the posterior probability for the MAP tree, in comparison with the estimate based on an independent sample. However there are too many trees and most of them may have extremely small probabilities, which are of no practical significance. For some analyses, all trees may have very small probabilities. Alternative criteria, used by Lakner et al. (2008) and Höhna et al. (2008), are instead based on probabilities for splits. An extremely long 'reference chain' can be run to collect the splits and their 'true' probabilities. Splits not visited during the MCMC with very small posterior probabilities are unimportant and can be ignored. One can then run the MCMC algorithm for a fixed number of iterations (n), and calculate the distance between the 'true split probabilities' and the estimates from the chain. For instance, the distance may be calculated as

$$\delta_n = \max_i |\hat{p}_i - p_i|, \tag{8.22}$$

where the maximization is over split i, and where p_i is the true split probability from the reference chain and $\hat{p}_i$ the estimate from the test chain. One can run the test algorithm many times to calculate the average $E(\delta_n)$ over replicates runs. Note that this criterion is not very different from the variance measure of §7.3.2.1: if we set $h(i) = 1$ if a split exists in tree i and 0 otherwise, then I in equation (7.33) will be the posterior probability for that split.

Lakner et al. (2008) reviewed cross-tree proposals for unrooted trees and Höhna et al. (2008) reviewed proposals for rooted trees under the clock or local-clock models. For the present, good cross-tree proposals are still a matter of research. Below we consider proposals for unrooted trees first and then those for rooted trees. Some proposals described below are not yet tested. Note that while tree perturbation algorithms (such as NNI and SPR) are used both in parsimony and likelihood-based tree search (see §3.2.3) and as MCMC proposals, important differences exist in the two applications. In optimality-based tree search, the algorithm is used to generate candidate tree topologies only, and branch lengths in the new tree are recalculated. In a Bayesian MCMC algorithm, one has to specify a mechanism for generating branch lengths for the new tree based on the branch lengths of the current tree and to calculate the proposal ratio correctly. Only then will the MCMC algorithm visit the different trees in proportion to their posterior probabilities.

8.3.2.3 *Empirical observations on branch lengths in different trees*

Here we provide a brief discussion of empirical observations on MLEs of branch lengths in different tree topologies. The dataset consists of 985bp of mitochondrial DNA for five species: human (H), chimpanzee (C), gorilla (G), orangutan (O), and gibbon (B) (Brown et al. 1982). This dataset was used in Table 3.3 (see also Problem 3.1). The MLEs of branch lengths in the 15 unrooted trees estimated under JC69 are shown in Figure 8.5. The three trees which have the OB split (i.e. which group human, chimpanzee, and gorilla into one split) have the highest likelihood. The other trees are much worse. The ML tree length ranges from 0.429 to 0.477, and the maximum parsimony (MP) tree length from 355 to 389. One obtains the same ranking of the trees by the ML tree lengths as by the MP tree length except that MP produces a few ties. ML (using the log likelihood values) and MP rank the trees similarly, especially the good trees. Such high similarities between ML and MP are common, especially when sequence divergence is low (see Figure 4.6 for another example).

Under JC69, all branch lengths in all trees are strictly positive for this dataset. The external branch lengths are quite similar across trees. Good trees tend to have slightly longer internal branches and shorter external branches than poor trees. For example, the human branch length ranges from 0.036 to 0.046 in the 15 trees, with the SE 0.007–0.008, with the long human branches found in the poor trees in which the human joins orangutan or gibbon. The gorilla branch ranges from 0.058 to 0.067, with SEs 0.009–0.010. The gibbon branch ranges from 0.123 to 0.167, with SEs over 0.013–0.016.

τ_1: ((C: 0.049, G: 0.064): 0.015, H: 0.036, (O: 0.091, B: 0.124): 0.051) 355
τ_2: ((H: 0.040, C: 0.052): 0.016, G: 0.059, (O: 0.090, B: 0.125): 0.047) 357
τ_3: ((H: 0.041, G: 0.068): 0.009, C: 0.050, (O: 0.091, B: 0.123): 0.052) 358
τ_4: ((H: 0.039, C: 0.054): 0.018, O: 0.126, (G: 0.061, B: 0.155): 0.008) 377
τ_5: ((H: 0.039, C: 0.054): 0.018, B: 0.159, (G: 0.063, O: 0.123): 0.006) 378
τ_6: ((H: 0.044, B: 0.161): 0.003, O: 0.131, (C: 0.053, G: 0.060): 0.015) 382
τ_7: ((H: 0.046, O: 0.131): 0.000, B: 0.163, (C: 0.053, G: 0.060): 0.016) 384
τ_8: ((H: 0.043, G: 0.066): 0.010, O: 0.130, (C: 0.057, B: 0.161): 0.004) 385
τ_9: ((C: 0.052, B: 0.166): 0.009, H: 0.041, (G: 0.060, O: 0.127): 0.015) 385
τ_{10}: ((H: 0.038, B: 0.168): 0.009, C: 0.055, (G: 0.059, O: 0.128): 0.015) 386
τ_{11}: ((C: 0.057, O: 0.136): 0.004, H: 0.043, (G: 0.059, B: 0.158): 0.016) 386
τ_{12}: ((H: 0.042, O: 0.136): 0.005, C: 0.056, (G: 0.058, B: 0.159): 0.016) 387
τ_{13}: ((H: 0.043, G: 0.066): 0.011, B: 0.163, (C: 0.061, O: 0.131): 0.000) 387
τ_{14}: ((H: 0.045, O: 0.133): 0.005, G: 0.067, (C: 0.055, B: 0.163): 0.008) 389
τ_{15}: ((H: 0.042, B: 0.163): 0.008, G: 0.067, (C: 0.059, O: 0.133): 0.004) 389

Fig. 8.5 MLEs of branch lengths in the 15 unrooted trees for five species: human (H), chimpanzee (C), gorilla (G), orangutan (O), and gibbon (B), estimated under JC69 from the mitochondrial dataset of Brown et al. (1982). The number next to the tree is the parsimony tree length (the minimum number of changes).

With five species, each internal branch, which splits the species into two on one side and three on the other, is found in three trees. The MLEs of branch length are very similar among those three trees. For example, for the HC split, the range is 0.016–0.018, with SE 0.055–0.057; and that for the OB split is 0.047–0.052, with SE at 0.009. When a split is not supported by the data (when trees with the split have low likelihood), the branch tends to be short, while the branches inside the subtree defined by the split tend to be long.

The patterns described above may be representative of datasets of similar sequences. When the sequences are divergent, the branch lengths may vary more across trees. The nucleotide substitution model appears also to exert some influence. For example, with the gamma model of rates for sites, the log likelihood values tend to become more similar among trees and zero length internal branches become common. Zero-length internal branches are also common in poor trees for large datasets.

A cross-tree move is likely to be accepted if the branch lengths for the new tree are close to the MLEs. Nevertheless, the easiest way of generating branch lengths for the new tree may be to manipulate the branch lengths in the current tree. By examining how the MLEs of branch lengths change when one alters the tree topology through local tree rearrangements, one may hopefully design better algorithms for branch length generation.

8.3.3 NNI for unrooted trees

The basic NNI algorithm. Each internal branch defines the relationships among four subtrees, say, a, b, c, and d (Figure 8.6). The simplest adaptation of the NNI algorithm for use as an MCMC proposal is to select an internal branch at random (called *focal branch*), choose one of the two alternative trees at random, and transfer the branch lengths from the current tree to the new tree without altering them (for example, the old internal branch length will become the internal branch length in the new tree). The proposal ratio is 1.

8.3 MCMC PROPOSALS IN BAYESIAN PHYLOGENETICS 285

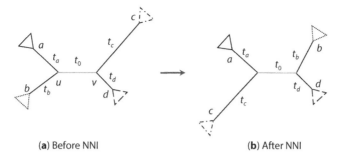

(a) Before NNI (b) After NNI

Fig. 8.6 Nearest neighbour interchange (NNI) as an MCMC proposal for unrooted trees. An internal branch in the current tree is selected at random to be the *focal branch* (u-v). This defines three possible trees relating subtrees a, b, c, and d. The current tree is $((a, b), c, d)$, while the other two trees are $((a, c), b, d)$ and $((a, d), b, c)$. Choose one of those two alternative trees at random as the new tree. In the LOCAL move of Larget and Simon (1999), the focal branch is extended on both sides to form a three-branch backbone b-u-v-c. One end of the focal branch is selected at random (node u, say) and the connecting subtree (subtree a, say) is moved to a random location along the backbone.

It is possible to modify the branch lengths at the time of changing the tree topology, but it is beneficial to do so only if it improves the chance of the new tree being accepted. For example, one can apply a multiplier (§7.2.5) to the internal branch length (see equation (7.21)). Generate a random number $u \sim U(0, 1)$ and multiply the internal branch length t_0 of Figure 8.6 by

$$c = e^{\varepsilon(u-1/2)}, \tag{8.23}$$

where $\varepsilon > 0$ is a step length, to be adjusted to maximize the acceptance rate of the proposal. The proposal ratio will then be c (see §7.2.5). However, observations such as discussed in the last subsection suggest that such random modifications are likely to reduce rather than increase the acceptance rate.

The LOCAL move of Larget and Simon. The LOCAL move for unrooted tree of Larget and Simon (1999) is a variant of the NNI algorithm. It modifies three branch lengths by a multiplier c (equation (8.23)) and induces a proposal ratio of c^3. The move can be achieved by the following two component steps.

i. Choose an internal branch on the tree at random as the focal branch, say, u-v (Figure 8.6a). Select at random one further branch from each end of the focal branch, say b-u and v-c, to form a three-branch backbone b-u-v-c. Multiply the three branch lengths t_b, t_0, t_c by c, generated from equation (8.23). Note that the choice of the focal branch, choice of the branches at the ends of the focal branch to form the backbone, etc. are all symmetrical moves. This step thus contributes a factor c^3 to the proposal ratio, due to the multiplier applied to three branch lengths.

ii. Select either u or v at random. Suppose u is selected. Move u together with the subtree a to a random location along the backbone b-u-v-c. (If v is selected, move node v and subtree *d* instead.) This move is symmetrical and contributes a factor 1 to the proposal ratio. The tree topology is changed if the new position for *u* is on the *v-c* branch. Otherwise this will be a within-tree move.

The proposal ratio for the whole move is given by multiplying the factors for the two component steps, and is c^3. An incorrect ratio (c^2) was published by Larget and Simon (1999) and corrected by Holder et al. (2005).

The LOCAL move is a mixture of within-tree and cross-tree moves, which makes it awkward to adjust the step length to improve mixing efficiency. Also, random modification of three branch lengths by the multiplier appears to be a bad idea as it is expected to reduce the acceptance rate for cross-tree moves. If the branch lengths are not modified, the proposal ratio will be 1. If the move changes the tree topology, it will break one branch (*v-c* in Figure 8.6a) and merge two other branches (*b-u* and *u-v*). This way of generating branch lengths for the new tree may not be as good as direct transfer of branch lengths without any alteration.

The NNI variant of Lakner et al. The NNI algorithm that Lakner et al. (2008) implemented chooses each of the three tree topologies around the focal branch with probability 1/3. Then all five branch lengths around the focal branch are modified by using five independent multipliers (with the proposal ratio to be the product of the multipliers). Modification of all five branch lengths in cross-tree moves appears to be a poor idea, because random modifications are likely to generate poor branch lengths, leading to reduced acceptance rate. In the test of Lakner et al. (2008), modifying branch lengths at the same time of changing the topology in general led to reduced acceptance probability. In any case, one should record the acceptance probabilities for within-tree and between-tree moves separately. The step length for the move should be adjusted to achieve intermediate acceptance rate (30–40%) for the within-tree moves and maximum acceptance rate for cross-tree moves. Given the ease with which one can design and optimize within-tree moves (see §7.3.2), it may be advisable to disable the within-tree moves in the NNI algorithm.

Selection of the target tree. In the NNI algorithm described above, each internal branch is chosen as the focal branch with equal probability and the target tree is chosen at random from the neighbour trees. Those choices concern the tree proposal probability ratio $r_{\tau'\tau}/r_{\tau\tau'}$ of equation (8.21), and the random choices result in the ratio $r_{\tau'\tau}/r_{\tau\tau'} = 1$. Those choices may not be optimal. Theory for guiding our choice of models in cross-model moves (that is in our choice of $r_{\tau'\tau}/r_{\tau\tau'}$) is not well developed. However, efficient cross-model moves tend to be associated with high acceptance rates (Peskun 1973; Mira 2001). As suggested by Höhna and Drummond (2012), it may be advisable to use features of the trees to guide our proposals. For example, one may give higher weights to target trees if they have much better parsimony scores than the current tree. It is yet unclear how parsimony tree lengths should be effectively used in such proposals. To be concrete about how such ideas might work, let s_τ and $s_{\tau'}$ be the parsimony scores for current tree τ and target tree τ'. One may define the weight to be $w_{\tau\tau'} = e^{-c(s_{\tau'}-s_\tau)}$, so that the tree proposal probability is

$$r_{\tau,\tau'} = \frac{w_{\tau,\tau'}}{\sum_{j'} w_{\tau,j'}}, \qquad (8.24)$$

where the summation is over all neighbour trees j' of the current tree τ. Note that trees τ and τ' have different neighbours. Such weights can be used to calculate the probabilities with which different neighbour trees are proposed. The idea works with other tree-arrangement algorithms as well.

8.3.4 *SPR for unrooted trees*

The SPR algorithm prunes a subtree and reattaches it to a branch on the remaining part of the tree. This is illustrated in Figure 8.7a and b. A branch (say, a) in the current tree (either internal or external) is chosen at random to be the focal branch. One end of the focal branch is chosen at random to be subtree A; together with branch a, subtree A will be pruned off and reattached to a random branch (r, say) in the rest of the tree (subtree B) (Figure 8.7a). If branch a is terminal, subtree A will be the tip node. Branches on the path from a to r form the backbone. Connected to branch a are branches x on the backbone and p not on the backbone. The path then passes zero or more branches (b_1, b_2, etc.) before reaching r. Branches p and x are excluded when one selects the regrafting branch r,

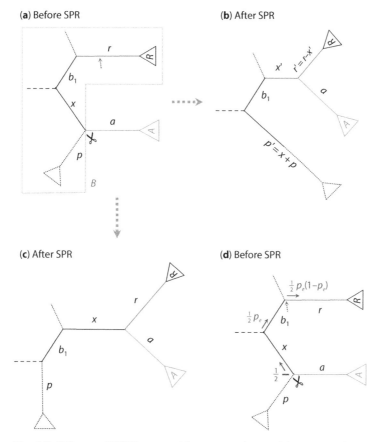

Fig. 8.7 SPR as an MCMC proposal for unrooted trees. (**a**) An internal branch (branch a) in the current tree is chosen at random to be the focal branch. This defines subtrees A and B. Branch a and subtree A are pruned and reattached to a random branch r on subtree B, resulting in a new tree (**b**). Panel (**c**) shows the transfer of branch lengths from the source tree to the target tree in the algorithm of Lakner et al. (2008). Note the transfer of branch x along the backbone (which connects the focal branch a to the regrafting branch r). Panel (**d**) shows a weighting scheme (extension mechanism) used by Lakner et al. (2008) to make the algorithm favour local moves.

as otherwise one will get back the original tree. When subtree A is pruned off, branches p and x merge into one branch. Also re-attaching branch a (and subtree A) back into branch r breaks branch r into two branches: x' and r' (Figure 8.7b). We generate a random variable $u \sim g(u)$ and let $x' = ru$ and $r' = r(1-u)$. Here and below we abuse the notation and use the same symbol to refer both to the branch and to the branch length.

To derive the proposal ratio, note that the steps for changing the tree topology are symmetrical and incur a 'proposal ratio' of 1. The reverse move involves selecting branch a of Figure 8.7b as the focal branch and branch p' as the regrafting branch. The probabilities for those choices are identical in the forward and reverse moves. Thus we have only to consider the changes of branch lengths in the current tree to those in the new tree to derive the proposal ratio. Three branch lengths (p, x, r) are affected and a random variable (u) is generated. The other branch lengths are unchanged and need not be considered. The mapping is thus $(p, x, r, u) \leftrightarrow (p', x', r', u')$ (see Figure 8.7b)

$$\begin{aligned} p' &= x + p, \\ x' &= ru, \\ r' &= r(1-u), \\ u' &= x/(x+p). \end{aligned} \qquad (8.25)$$

The last equation is because the reverse move involves selecting branch a as the focal branch and inserting branch a (and subtree A) into branch p', with random variable u' used to break p' into x and p. The Jacobi determinant of the transform is

$$\frac{\partial(p', x', r', u')}{\partial(p, x, r, u)} = \begin{vmatrix} 1 & 1 & 0 & 0 \\ 0 & 0 & u & r \\ 0 & 0 & 1-u & -r \\ -\frac{x}{(x+p)^2} & \frac{p}{(x+p)^2} & 0 & 0 \end{vmatrix} = -\frac{r}{x+p}. \qquad (8.26)$$

Thus the proposal ratio is $r/(x+p)$. In words, the SPR move breaks one branch into two and merges two branches into one, and the proposal ratio is the ratio of the length of the broken branch to the length of the merged branch. If u is not a $U(0, 1)$ random variable, the proposal ratio will be multiplied by the random variable density ratio $g(u')/g(u)$.

The SPR algorithms of Lakner et al. (2008). We now describe the variant of SPR implemented by Lakner et al. (2008, Figure 6). The authors implemented two versions. The first, called rSPR (for random SPR), chooses the pruning and regrafting branches at random with equal probabilities. The second, called eSPR (for extending SPR), uses a scheme to give different probabilities to different neighbour trees. This is discussed later. We note that Lakner et al.'s algorithms (rSPR and eSPR) select internal branches (but not external branches) as focal branches. Also they are mixtures of within-tree and cross-tree moves as they allow branches p and x of Figure 8.7a to be chosen as the regrafting branch, in which case the tree topology does not change.

Here we focus on the generation of branch lengths when the new tree topology is already chosen. Lakner et al.'s procedure is illustrated in Figure 8.7c, and moves branch x together with the pruned subtree A along the backbone, passes branches b_1, b_2, etc., and inserts it right before the regrafting branch r. Transfer of branch lengths without alteration (Figure 8.7a and c) incurs a proposal ratio of 1. If there is only one branch (x) between branches a and r, the SPR move will be equivalent to an NNI move, and this transfer of branch lengths appears to be a good idea (see discussions of NNI and variants). For example, the top three trees in Figure 8.5 are different through an NNI move, and the external branch lengths for H, C, and G are very similar among those trees. Even the internal branch lengths are similar in the three trees (0.009–0.016). In contrast, the algorithm of Figure 8.7b is similar to the LOCAL move of Largett and Simon (1999) in that one branch is broken into two and two branches are merged into one. Applied to moves between the top three trees of Figure 8.5, this move of breaking and merging branches is seen to be a poor idea.

When the SPR move is not an NNI move (in other words, when there are other branches b_1, b_2, etc. besides branch x between a and r), the transfer of branch x along the backbone of the tree as in Figure 8.7c appears to be a poor idea, since it disturbs the current tree too much. For example, the move from $\tau_2 \to \tau_{10}$ of Figure 8.5 is an SPR move in which B is the pruning branch and H is the regrafting branch. The mapping of the algorithm of Figure 8.7c gives $0.04 \leftrightarrow 0.038, 0.047 \leftrightarrow 0.009$, and $0.090 \leftrightarrow 0.128$, while that of Figure 8.7b gives $0.090 + 0.047 \leftrightarrow 0.128$, and $0.04 \leftrightarrow 0.009 + 0.038$, which seems better. The move of Figure 8.7b retains the relationships inside subtree B, as all pairwise sequence distances inside subtree B are retained. Another difference is that in the SPR algorithm of Lakner et al. (2008), random multipliers are applied to branch lengths a and x. This is likely to lead to reduced acceptance rate and is not desirable. In view of the discussions above, it may be advisable to use different strategies to propose branch lengths in the SPR move. If the move is an NNI move, use direct transfer of branch lengths, and otherwise use the strategy of Figure 8.7b.

Proposing target trees using the extension mechanism. In the algorithm of Figure 8.7b, the focal branch and the regrafting branch are selected at random, and neighbour trees are sampled with equal probability, resulting in the tree proposal probability ratio $r_{\tau'\tau}/r_{\tau\tau'} = 1$. This uniform choice of the target tree may not be optimal, since a random proposal may well generate very different trees and have low acceptance rates. First, a distant tree may be poor even if good branch lengths are generated. Second with large changes it is difficult to suggest good branch lengths. Nevertheless, it may be necessary to propose big moves occasionally to avoid getting stuck at a tree island.

Similarly to the NNI move, one can use parsimony weights to weight different neighbour trees generated by the SPR move. An alternative strategy is the so-called extension mechanism of Lakner et al. (2008, Figure 6), which assigns higher weights to close neighbours and favours local moves. This is illustrated in Figure 8.7d. After the focal branch is selected, one chooses a random direction to extend the backbone, with probability p_E of extending to the next branch. This way, the regrafting branch has a high probability of being close to the focal branch. Lakner et al. allowed the regrafting branch to coincide with the focal branch, in which case the target tree has the same topology as the current tree.

8.3.5 *TBR for unrooted trees*

With the TBR algorithm, we cut a branch so that the tree falls into two subtrees. Then we pick a reconnection point from each subtree and connect the two subtrees back into a new tree. This is illustrated in Figure 8.8. We choose a random internal branch in the tree (let this be a) to be the bisection branch. Cutting branch a results in two subtrees: X on the left and Y on the right. A random branch (either internal or external) is chosen in subtree X (let this be x) and another random branch y is chosen in subtree Y as the two reconnecting branches. Random reconnection points are chosen on x and y (indicated by arrows in Figure 8.8a) and reconnected to form the new tree (Figure 8.8b and c).

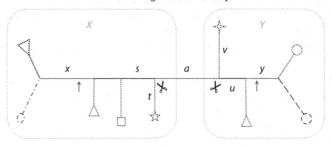
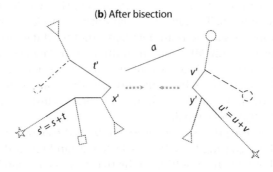
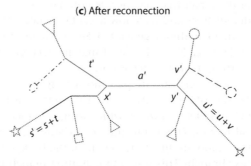

Fig. 8.8 TBR as an MCMC proposal for unrooted trees. (**a**) A focal branch (*a*) is chosen at random for cutting the tree into subtrees *X* and *Y*. The reconnecting branches (*x* and *y*) are chosen at random in subtrees *X* and *Y*, respectively. Branches on the path *x-a-y* (drawn as solid lines) form the backbone for the move.
(**b**) Subtrees *X* and *Y* are reconnected at the reconnecting branches to form the new tree of (**c**). Note the transfer of branches from the current tree to the new proposed tree.

To derive the proposal ratio for the move, we need label a few more branches. From branch *a* extend left towards *x* and right towards *y*. The path *x-a-y* is called the backbone (Figure 8.8a). The left end of branch *a* connects two other branches: *s* on the backbone and *t* not on the backbone. Further left, one may pass zero or more branches before reaching branch *x*. Similarly the right end of branch *a* connects two other branches: *u* on the backbone and *v* not on the backbone. Further right, one may pass zero or more branches before reaching y. Transfer or generation of branch lengths for the new tree is summarized in Figure 8.8a–c. Branches s and t are merged into one branch s', and branches u and v are merged into u'. We generate a random variable $0 < r_1 < 1$ to break the reconnecting branch x into two pieces: $x' = xr_1$ and $t' = x(1-r_1)$. Similarly we generate $0 < r_2 < 1$ to break y into $y' = yr_2$ and $v' = y(1-r_2)$. (Here again we use the same symbol to refer to both the branch and the branch length, and use prime to refer to the branch lengths in the new tree.) Also we let $a' = a$.

The choices of the bisection branch and reconnecting branches are symmetrical with the same probabilities in the forward and reverse moves, and incur a factor of 1 in the proposal ratio. We have only to consider changes in the branch lengths. The move affects six branch lengths: x, t, s, y, v, u, and involves the generation of two random variables r_1 and r_2. The transform is

$$\begin{cases} x' = xr_1, \quad t' = x(1-r_1), \quad s' = s+t \text{ for subtree } X; \\ y' = yr_2, \quad v' = y(1-r_2), \quad u' = u+v \text{ for subtree } Y; \\ r_1' = s/(s+t), \quad r_2' = u/(u+v) \text{ for the reverse move}. \end{cases} \quad (8.27)$$

The last two equations, for r_1' and r_2', are because the reverse move involves selecting branch a' in the tree of Figure 8.8c as the bisection branch, selecting branches s' and u' as the reconnecting branches, and generating random variables r_1' and r_2' to move back to the current tree of Figure 8.8a, with $s'r_1' = s$ and $u'r_2' = u$. The Jacobi determinant of the transform is

$$\left| \frac{\partial(x',t',s',y',v',u',r_1',r_2')}{\partial(x,t,s,y,v,u,r_1,r_2)} \right| = \frac{xy}{(s+t)(u+v)}. \quad (8.28)$$

Suppose $r_1, r_2 \sim U(0, 1)$. Then the proposal ratio for the move is $xy/[(s+t)(u+v)]$.

The TBR move is a substantial one, affecting many branch lengths. However, it is noticeable that the relationships inside subtrees X and Y are not disturbed. Inside each of the two subtrees, the topology and branch lengths are not changed and neither are the pairwise distances (sum of branch lengths).

If both reconnecting branches x and y are adjacent to branch a (i.e. if they share a node with branch a), the TBR move returns the same tree topology. Such a proposal may be disabled. In the case of four species, TBR does not lead to any topological change and is thus not useful.

The random TBR (rTBR) and extending TBR (eTBR) algorithms of Lakner et al. (2008) are designed similarly to the corresponding rSPR and eSPR moves. Both are mixtures of within-tree and cross-tree moves, which makes it difficult to optimize the moves, and both also introduce large changes to the tree by moving branches along the backbone.

8.3.6 Subtree swapping

This move picks two branches (x and y) and their subtrees (X and Y) at random and swaps them. This is illustrated in Figure 8.9. Branches x and y may be external branches but they should not be adjacent on the tree (i.e. they should not share the same node) as otherwise swapping does not change the tree topology. As the branch lengths are transferred to the new tree without any alteration, the proposal ratio is one. Note that NNI is a special case of subtree swapping.

The algorithm implemented by Lakner et al. (2008), called eSTS for extending subtree swapping, uses an extension mechanism to favour local changes that is similar to that used for the eSPR move. Adjacent branches are allowed when branches x and y are chosen

(a) Before subtree swapping   **(b)** After subtree swapping

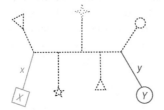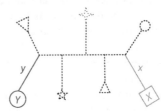

Fig. 8.9 Subtree swapping for proposing new unrooted trees. Two subtrees X and Y are chosen at random and, together with their branches (x and y), are swapped to generate the new tree. Branch lengths are transferred without change.

so that it is possible for the topology to stay the same, and the proposal is a mixture of within-tree and cross-tree moves. Given the proposed tree topology, eSTS applies independent multipliers to all branches on the backbone (i.e. on the path from branches x to y, Figure 8.9). The random modification of many branch lengths may be undesirable and lead to reduced acceptance rates; indeed in the test of Lakner et al., eSTS did not work well.

8.3.7 NNI for rooted trees

Rooted trees are used in phylogenetic analysis under the clock and local-clock models (§4.2.3 and Chapter 10). The parameters on the tree are the ages of the internal nodes rather than the branch lengths. The node ages may represent absolute times, as in estimation of divergence times using fossil calibrations or serially sampled sequences (to be discussed in Chapter 10) or may be relative, measured by sequence distance. In either case, a node on a rooted tree must not be older than its mother node or ancestral nodes. Proposals for topological changes in rooted trees have to respect this constraint.

Variants of the NNI algorithm have been implemented as an MCMC proposal for rooted trees by a number of authors, including Kuhner et al. (1995), Larg101 and Simon (1999), Li et al. (2000), and Drummond et al. (2002). On a rooted tree, each internal branch (u-v, say) defines the relationships among three subtrees: a, b, and c (Figure 8.10). Suppose the current tree is $\tau_1 : ((ab)c)$, and the two alternative trees are $\tau_2 : ((ca)b)$ and $\tau_3 : ((bc)a)$, which are generated by swapping c with b or with a in τ_1. Without altering the ages of any nodes on the tree, such swapping is possible only if node c is younger than node v: $t_c < t_v$. (Here t_c is the age of node c, and so on.) This algorithm is implemented by Drummond et al. (2002), called Narrow Exchange. More specifically, one selects a random internal branch u-v. Let the daughter nodes of u be v and c, and the daughter nodes of v be a and b. Then choose a or b at random to swap with c if $t_c > t_v$. As there is no change to any of the node ages, the proposal ratio is one. If $t_c < t_v$, the move is impossible and abandoned.

The algorithms implemented by Kuhner et al. (1995; see also Li et al. 2000) and Largel and Simon (1999) all modify t_u and t_v while other node ages (such as t_a, t_b, t_c, t_w) (see Figure 8.10) are kept fixed. Kuhner et al.'s algorithm moves between gene genealogies for a DNA sample from a population to estimate parameter θ (population size parameter; see Chapter 9). It generates t_u and t_v by first erasing nodes u and v in the tree and then

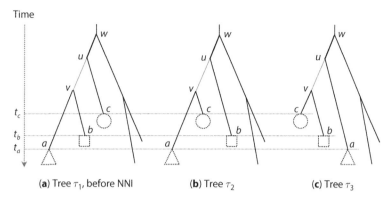

Fig. 8.10 Nearest neighbour interchange (NNI) on a rooted tree. Each internal branch (say, u-v) defines three possible trees relating the subtrees a, b, and c. The current tree is τ_1: $((a, b), c)$, while the other two trees are τ_2: $((c, a), b)$ and τ_3: $((b, c), a)$. The NNI move chooses one of those two alternative trees at random as the new tree if the node ages do not violate the parent-child age constraint. Node ages are retained during transfer from the source tree to the target tree.

simulating the coalescent process using the current value of θ, to join subtrees a, b, and c to form nodes u and v up to node w. The algorithm should generate moves with good acceptance rates but is complex, as it uses the coalescent prior to simulate the erased part of the gene tree. The algorithm of Larg64 and Simon, called LOCAL with molecular clock, generates t_u and t_v as uniform random variables between t_w and the two older ages of nodes a, b, and c. This is a mixture of within-tree and cross-tree moves. The random values for t_u and t_v are unlikely to fit the new tree well. For a within-tree move, it should be better to propose new values for t_u and t_v around their current values. For a cross-tree move, it should be better to keep t_u unchanged: the height of a clade tends to be stable when the relationships inside the clade change (for example, the MLEs of t_u in trees τ_1, τ_2, and τ_3 of Figure 8.10 tend to be similar).

8.3.8 SPR on rooted trees

Variants of the SPR algorithm have been implemented for rooted trees by Wilson and Balding (1998), Drummond et al. (2002), and Rannala and Yang (2003). The algorithm of Wilson and Balding is illustrated in Figure 8.11, which prunes off subtree x and reattaches it above node y. Choose a random node x in the tree that is not the root. (Wilson and Balding restricted x to be an internal node but this restriction is unnecessary.) Let a be the mother node of x. Subtree x is pruned off the tree. Then choose another node y at random from the remaining part of the tree for regrafting. For this to be feasible, the mother node (b) of y should be older than node x. Also y should not be the sister node of x as otherwise one recovers the original tree at reattachment. In the example of Figure 8.11a, 10 nodes in the remaining part of the tree satisfy those requirements, and y is chosen at random from among them. If the root is chosen, node a will become the new root. Reattach subtree x above node y (i.e. insert inside branch b-y). The new age of node a is generated as follows. If y is not the root (Figure 8.11b), sample $t'_a \sim U(\max(t_x, t_y), t_b)$. If y is the root (Figure 8.11c), generate t'_a from an exponential distribution above the current

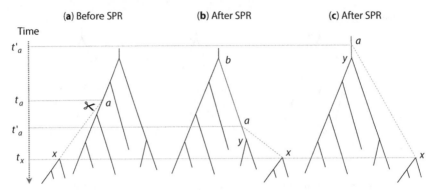

Fig. 8.11 SPR on a rooted tree. Subtree x is pruned off by cutting a random branch a-x (**a**), and reattached above a random node y selected in the remaining part of the tree (**b**). Node y may be the current root (**c**), in which case node a becomes the new root. The proposal changes the age of node a. In the example, the tips are not contemporary and the ages of the tip nodes are not all zero.

root age. If the move does not change the root (i.e. if neither a nor y is the root of the current tree), the proposal ratio will be one. Otherwise the proposal ratio will be the ratio of the uniform and exponential densities for the age of node a:

$$\frac{g(t_a)}{g(t'_a)} = \begin{cases} 1, & \text{if neither } a \text{ nor } y \text{ is the root in the current tree,} \\ \dfrac{g(t_a)}{1/(t_b - \max\{t_x, t_y\})}, & \text{if } a \text{ is root in the current tree,} \\ \dfrac{1/(t_b - \max\{t_x, t_y\})}{g(t'_a)}, & \text{if } y \text{ is root in the current tree,} \end{cases} \quad (8.29)$$

where $g(t_a)$ is the exponential density for t_a, and so on.

The algorithm of Rannala and Yang (2003) selects node x and prunes off subtree x as above, and then generates a new t'_a using a sliding window around the current t_a, reflected appropriately so that $t'_a > t_x$. The rest of the tree is then scanned to determine the number of branches feasible for reattachment. If t'_a is greater than the age of the root, node a will become the new root. In the example of Figure 8.11b, the newly generated t'_a will be compatible with four branches. One of them is chosen at random for reattachment. The proposal ratio is n/m, where m is the number of feasible branches for attachment in the forward move (with age t'_a) and n is the number of feasible branches in the reverse move (with age t_a). For example, $m = 4, n = 3$ for the proposal of Figure 8.11b, and $m = 1, n = 3$ for Figure 8.11c.

8.3.9 Node slider

This proposal slides an internal node along the tree, and may or may not change the tree topology. First, select at random an internal node x that is not the root (Figure 8.12). Let its mother node be a, with age t_a. Second, generate a new age t^* for node a using a sliding window around the current age t_a based on the uniform or Bactrian distributions, reflected so that $t^* > 0$. This move is symmetrical and incurs a factor 1 in the proposal ratio. Third, slide the node (together with the subtree x) up or down the tree according to t^*. If $t^* > t_a$, move up the tree towards the root. When we meet a node, choose the

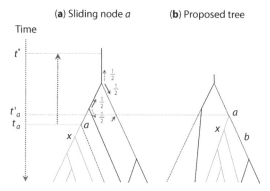

Fig. 8.12 Node slider as an MCMC proposal on rooted trees. A subtree x is chosen at random, and the age (t_a) of its mother node a is changed using a sliding window around the current age. Node a is then slid up or down the tree according to t'_a. When the final resting place of node a is determined, all node ages in subtree x are rescaled by t'_a/t_a.

upwards route (along the mother branch) and the downwards route (along the sibling branch) with equal probability. If we meet the root, move up to build a new root with probability $\frac{1}{2}$ and move down the sibling branch with probability $\frac{1}{2}$. When we move down and meet a node, take one of the two daughter branches with equal probability. If we reach a tip, reflect the remaining branch length back and move up the tree. The process is continued until the branch length is exhausted and then we know the final branch for re-attaching node a. Let the final age of node a be t'_a. Fourth, scale all nodes inside subtree x by the factor $\frac{t'_a}{t_a}$; i.e. $t'_j = t_j \times \frac{t'_a}{t_a}$ for every node j inside subtree x. Suppose there are m nodes inside subtree x, the proportional scaling incurs a factor $\left(\frac{t'_a}{t_a}\right)^m$ in the proposal ratio (see §7.2.5). This is also the proposal ratio for the entire move.

A drawback of this move is that it is a mixture of within-tree and cross-tree moves, so optimizing it may be complicated. Nevertheless, the move can introduce either small or large changes to both the tree topology and the branch lengths inside subtree x, and is thus a mixture of local moves and global moves. In practice, the move is likely to be accepted if the changes are small.

The node slider works with unrooted trees as well. One chooses a node and subtree and slide the node along branches on the tree, reflecting when hitting a tip. There is no need to rescale the branch lengths inside the subtree. The move is then very similar to the LOCAL move of Larget and Simon (1999), which slides a node on a three-branch backbone, and to the SPR move, which moves a subtree to a random location in the rest of the tree.

8.4 Summarizing MCMC output

Several procedures have been suggested to summarize the posterior probability distribution of phylogenetic trees. One can take the tree topology with the maximum posterior probability as a point estimate of the true tree. This is called the maximum *a posteriori* tree or *MAP tree* (Rannala and Yang 1996), and should be similar to the ML tree under the same model if the data are informative. One can also collect the trees with the highest posterior probabilities into a set until the total probability exceeds a preset threshold, such as 95%. This is known as the 95% credibility set of trees (Rannala and Yang 1996; Mau et al. 1999).

Posterior probabilities of single (whole) trees can be very low and the 95% credibility set may include many trees, especially when there are a large number of sequences and the data are not very informative. Thus alternative procedures summarize shared splits among trees visited during the MCMC. For example, one can construct a majority-rule consensus tree and, for each split on the consensus tree, report the proportion of sampled trees that include the split (Larget and Simon 1999). Such a proportion is known as the *posterior split* or *clade probability*. The posterior probability for a tree is equivalent to the posterior probability for a model, which is the standard measure of confidence in Bayesian model selection. The posterior probability for a split may be given a model-averaging interpretation, as a summary of common features among the well-supported models.

A few practical concerns may be raised concerning the use of split probabilities. First, as discussed in §3.1.1.7, a majority-rule consensus tree may not recognize certain similarities among them. Second, instead of attaching probabilities on the splits on the consensus tree, one may attach probabilities to splits on the MAP tree. This appears to be easier to justify when the consensus tree and MAP tree differ. The situation is similar to attaching bootstrap support values to either the consensus tree or the best estimated tree (the ML or MP tree). See §5.4.1. Third, Pickett and Randle (2005) pointed out that the uniform prior on trees induces a non-uniform prior on splits, with splits of very small and very large sizes having higher probabilities than splits of intermediate sizes. Such prior influence may persist in the posterior (see Example 8.2 earlier).

The posterior means and CIs for branch lengths on the consensus tree had better be ignored. The posterior mean of a branch is calculated by averaging over the sampled trees that share the concerned branch (split). If branch lengths are of interest, a more proper way is to run another MCMC, sampling branch lengths with the tree topology fixed.

8.5 High posterior probabilities for trees

To many, Bayesian inference of molecular phylogenies enjoys a theoretical advantage over ML with bootstrapping. The posterior probability for a tree or split has an easy interpretation: it is the probability that the tree or split is correct given the data, model, and prior. In contrast, the interpretation of the bootstrap in phylogenetics has been marred with difficulties (§5.4.1.3). As a result, posterior probabilities for trees can be used in a straightforward manner in a variety of phylogeny-based evolutionary analyses to accommodate phylogenetic uncertainty (Huelsenbeck et al. 2000b, 2001). Methods that rely on knowledge of the phylogeny can now proceed as usual, using the best trees sampled during the MCMC run, averaging over those trees with their posterior probabilities as weights.

However, it has been observed that posterior probabilities for trees or splits calculated from real datasets are often extremely high. In this section, we discuss the issue of high posterior probabilities for trees or splits.

8.5.1 *High posterior probabilities for trees or splits*

In the very first Bayesian inference of molecular phylogenies, the posterior probability for the best tree was calculated to be 0.9999 (Rannala and Yang 1996). The dataset consisted of 11 mitochondrial tRNA genes (739 bp) from five ape species, and is very

8.5 HIGH POSTERIOR PROBABILITIES FOR TREES

small by today's standards. While the tree was reasonable, the support seemed very high, especially as the human-chimpanzee-gorilla relationship was considered a difficult phylogenetic problem. Similarly, numerous analyses of modern large datasets using the program MrBayes often produced high posterior probabilities. Simulation studies and empirical data analyses have repeatedly found posterior split probabilities to be much higher than bootstrap support values (e.g. Cummings et al. 2003; Douady et al. 2003; Erixon et al. 2003; Simmons et al. 2004). It may be observed that while bootstrap proportions are published only if they are >50% (as otherwise the relationships are untrustworthy) posterior split probabilities are reported only if they are <100% (as most of them are 100%). The difference between the two measures of support may not suggest anything inappropriate about the Bayesian probabilities, especially given the difficulties in the interpretation of the bootstrap. However, in some cases the trees are decidedly wrong and the high posterior probabilities must be spurious. Sometimes different models applied to the same data produced different trees, each with high posterior probabilities (e.g. Yang 2008). Sometimes changes to taxon sampling led to conflicting trees, each with high support (Bourlat et al. 2006). Yet sometimes different genes or proteins produced different trees, all with high support (Rokas et al. 2005). Certain biological processes, such as horizontal gene transfer, gene duplications followed by gene losses, incomplete lineage sorting due to ancestral polymorphism, etc., can cause the gene trees to differ from the species tree. However, this explanation may be untenable, as in the analysis of mitochondrial genes that share the same history, or when phylogenies inferred using the three codon positions or using the DNA and protein data differ (Yang 2008; Butler et al. 2009). The problem is not so much that the trees differ in different analyses as that very high posterior probabilities are attached to them.

As the posterior probability for a tree is the probability that the tree is correct given the prior and data, there can be only three possible reasons for spuriously high posterior tree probabilities: (i) errors, which may be due to theory fault, program bugs, or convergence or mixing problems in the MCMC algorithm, (ii) violation of the substitution (likelihood) model, and (iii) the impact of the prior and the asymptotic behaviour of Bayesian model selection.

First, numerically incorrect posterior probabilities may be caused by errors in the theoretical formulation or in the computer program, or by computational problems in the MCMC algorithm, such as lack of convergence or poor mixing. If the MCMC fails to explore the parameter space properly and only visits a small portion of the space, the posterior probabilities for the visited trees will be too high. This problem may be a concern in Bayesian analysis of large datasets, but in principle may be resolved by running longer chains and designing more efficient algorithms. While errors are possible in isolated cases, they do not appear to be the fundamental reason for the problem. Second, model misspecification is always a concern in real data analysis. Computer simulation suggests that use of a simplistic and unrealistic model may lead to inflated posterior probabilities for trees (e.g. Lemmon and Moriarty 2004; Huelsenbeck and Rannala 2004). In theory, the problem can be resolved by implementing more realistic substitution models or taking a model-averaging approach (Huelsenbeck et al. 2004). Nevertheless, high posterior probabilities for trees were observed in simulations even when the correct model was assumed (Cummings et al. 2003; Lewis et al. 2005; Yang and Rannala 2005). Lewis et al. (2005) and Yang and Rananla (2005; Yang 2007a) suggest that the phenomenon may have to do with the asymptotic behaviour of Bayesian model selection and its sensitivity to the prior.

It is well known that Bayesian model selection is statistically consistent (e.g. Dawid 1992). When the data size $n \to \infty$, the model closest to the true model, as measured by the Kullback–Leibler divergence, dominates, with its posterior probability approaching 1. If several models have the same K-L divergence from the true model, the one with the smallest size (i.e. with the smallest number of parameters) dominates. The method thus automatically penalizes parameter-rich models, and this is widely recognized as an attractive property. Nevertheless, the case in which several models of the same size have the same distance from the true model (so that they are equally correct or equally wrong) is not well understood. This is the case we explore here. Note that binary trees for a fixed number of species are equivalent to models of the same size as they have the same number of parameters.

Below we discuss a simple case of phylogeny reconstruction known as the star tree paradox (§8.5.2), and a few simple examples of Bayesian model selection (§8.5.3) to examine the behaviour of posterior model probabilities in large datasets. The main message from this discussion is that statistical methods, Bayesian in particular, tend to give answers with extreme confidence when the dataset is large, whether or not the answer is correct. This may provide an explanation for the high posterior tree or split probabilities since molecular phylogenetics is a field of huge datasets. It is less clear how to resolve the problem.

8.5.2 Star tree paradox

In a simulation study, Suzuki et al. (2002) generated sequence datasets under the star tree for four species and analysed them using MrBayes, which compares unrooted binary trees (Figure 8.13a). The posterior probability for the inferred binary tree was often found to be very high. The study used a wrong and simplistic model so that the problem was partly due to model violation. However, the pattern is similar even when the true model is used to analyse the simulated data (Cummings et al. 2003; Lewis et al. 2005; Yang and Rannala 2005). If datasets are simulated under the star tree and if the data size (the

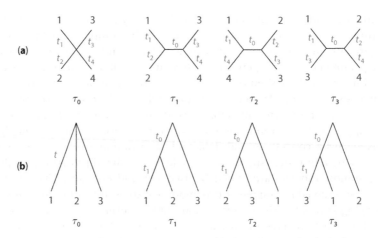

Fig. 8.13 (a) The three unrooted trees for four species and (b) the three rooted trees for three species, for illustrating the star tree paradox. Note that in each of the two cases, the star tree is a special case of the three binary trees. In the case of rooted trees, each binary tree has two branch length parameters (t_0 and t_1), while the star tree has one parameter (t).

number of sites) $n \to \infty$, a common intuition is that the posterior probabilities for the three binary trees should converge to $\left(\frac{1}{3}, \frac{1}{3}, \frac{1}{3}\right)$. However this is not how the Bayesian method behaves, and the counterintuitive result is called the *star tree paradox* (Lewis et al. 2005). The concern is not so much that the posterior tree probabilities differ from $\left(\frac{1}{3}, \frac{1}{3}, \frac{1}{3}\right)$ as that they are sometimes either very small or very large when in fact the data should contain no information beyond random noise either for or against any binary tree.

Here we consider the case of rooted trees for three species (Figure 8.13b), which is simpler than the case of unrooted trees for four species (Yang and Rannala 2005; Yang 2007a). The data are three sequences of binary characters, which are assumed to be evolving at a constant rate (i.e. under the molecular clock) (Yang 2000a). The data may be summarized as counts $\boldsymbol{n} = \{n_0, n_1, n_2, n_3\}$ of site patterns *xxx*, *xxy*, *yxx*, and *xyx*, where x and y are any two distinct characters, with $n_0 + n_1 + n_2 + n_3 = n$ to be the total number of sites. Each binary tree has two parameters t_0 and t_1. Intuitively, the three variable patterns *xxy*, *yxx*, and *xyx* 'support' trees τ_1, τ_2, and τ_3, respectively; indeed τ_i ($i = 1, 2, 3$) is the ML tree if n_i is the largest among n_1, n_2, n_3. The likelihood functions for the three trees, $f(\boldsymbol{n}|\tau_i, t_0, t_1)$, are given in equation (4.36), and the parameter space for this problem is illustrated in Figure 4.15.

In a Bayesian analysis, we assign equal probabilities $\left(\frac{1}{3}\right)$ to the three binary trees, and exponential priors with means μ_0 and μ_1 on the two branch lengths t_0 and t_1 in each binary tree: $f(t_0) = \frac{1}{\mu_0} e^{-t_0/\mu_0}$ and $f(t_1) = \frac{1}{\mu_1} e^{-t_1/\mu_1}$. The marginal likelihood under tree τ_i is

$$M_i = f(\boldsymbol{n}|\tau_i) = \int_0^\infty \int_0^\infty f(t_0) f(t_1) f(\boldsymbol{n}|\tau_i, t_0, t_1) \, dt_0 \, dt_1, \quad i = 1, 2, 3. \tag{8.30}$$

The posterior tree probability is then

$$P_i = \Pr(\tau_i|\boldsymbol{n}) = \frac{M_i}{M_1 + M_2 + M_3}, \quad i = 1, 2, 3. \tag{8.31}$$

If we simulate many datasets of a fixed size n under the star tree with a fixed branch length t, the posterior tree probabilities (P_1, P_2, P_3) will vary among datasets, according to a distribution. Yang (2007a) generated this distribution by computer simulation, with the integrals of equation (8.30) calculated numerically using Gaussian quadrature. For large n (>3,000, say), the distribution is independent of the branch length t in the star tree and of the means μ_0 and μ_1 in the priors (Figure 8.14). Note that (P_1, P_2, P_3) does not converge to $(\frac{1}{3}, \frac{1}{3}, \frac{1}{3})$ or to any other point. Susko (2008) has shown that the limiting distribution (P_1, P_2, P_3) when $n \to \infty$ can be simulated efficiently as follows. Generate (z_1, z_2, z_3) as $N(0, 1)$ random variables, with $\text{cov}(z_i, z_j) = -\frac{1}{2}$ for any $i \neq j$. Then calculate (P_1, P_2, P_3) as

$$P_i \propto e^{z_i^2/2} \Phi(z_i), \quad i = 1, 2, 3, \tag{8.32}$$

where $\Phi(\cdot)$ is the cumulative distribution function (CDF) of the standard normal distribution and the proportionality constant is to ensure that $P_1 + P_2 + P_3 = 1$. There appear to be four modes in the density of Figure 8.14, one near the centre and three at the corners. Thus in most datasets, either the three posterior probabilities are all close to $\frac{1}{3}$, or one of them is close to 1 while the other two are close to 0. Suppose we consider very high and very low posterior probabilities for binary trees as 'errors'. In 4.2% (or 0.8%) of datasets, at least one of the three posterior probabilities is >0.95 (or >0.99%), and in 17.3% (or 2.6%) of datasets, at least one of the three posterior probabilities is <0.05 (or <0.01). Those 'error' rates may be considered too high since we cannot have more than infinite data to reduce the errors further.

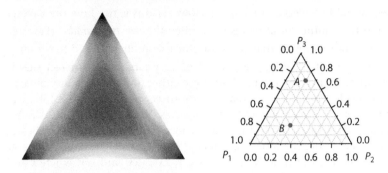

Fig. 8.14 The star tree distribution, i.e. the distribution of the posterior probabilities for the binary trees τ_1, τ_2, and τ_3 of Figure 8.13b, when large datasets are generated using the star tree τ_0. The distribution is independent of the branch length t in the generating tree τ_0 or the prior for t_0 and t_1 in the binary trees. The two points shown in the key have the coordinates $A(0.1, 0.2, 0.7)$ and $B(0.5, 0.3, 0.2)$, while the centre point is $(\frac{1}{3}, \frac{1}{3}, \frac{1}{3})$.

*8.5.3 Fair coin paradox, fair balance paradox, and Bayesian model selection

To gain insights into the behaviour of Bayesian model selection in large datasets, we now consider a few simple cases that are analytically tractable. The dynamics depends on whether there are free parameters in the compared models. We consider first the case of no parameters and then the case with unknown parameters. A summary of those different scenarios is in Figure 8.15.

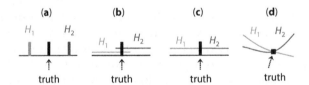

Fig. 8.15 Different scenarios of comparing two equally correct or equally wrong models using Bayesian model selection. (**a**) When the two compared models (H_1 and H_2) do not have any free parameters, $\Pr(H_1|x)$ approaches 0 or 1, each with probability $\frac{1}{2}$ (e.g. Figure 8.16). (**b**) When the two models have parameters and overlap in the region of the true model, $\Pr(H_1|x) \to \frac{1}{2}$. (**c**) When the two models partition a parameter space into two parts and the truth is at the boundary, $\Pr(H_1|x)$ converges to a distribution, such as $U(0, 1)$. (**d**) When the truth is at the cross point of two models and is inside the parameter space of each model, $\Pr(H_1|x)$ converges to a distribution, such as the one shown in Figure 8.18.

---

* indicates a more difficult or technical section.

8.5.3.1 Simple models with no free parameters

We first consider the case of comparing two models which are equally wrong and which have no unknown parameters (Figure 8.15a). In this case, Bayesian model selection has an extreme and unpleasant behaviour: it favours one model in some datasets and the other in other datasets, always with posterior probability ≈ 1.

Example 8.3. Fair balance paradox with simple models. The data are an i.i.d. sample of size n from $N(\mu_0, 1)$ with $\mu_0 = 0$. We compare two models H_1: $\mu = \mu_1$ and H_2: $\mu = \mu_2$, with $\mu_1 < \mu_2$ given. We assign the prior $\pi_1 = \pi_2 = \frac{1}{2}$ to the two models. Given the data or the sample mean $\bar{x}$, the (marginal) likelihoods are

$$L_1 = f(\bar{x}|H_1) = \frac{1}{\sqrt{2\pi/n}} \times \exp\left\{-\frac{n}{2}(\bar{x}-\mu_1)^2\right\},$$
$$L_2 = f(\bar{x}|H_2) = \frac{1}{\sqrt{2\pi/n}} \times \exp\left\{-\frac{n}{2}(\bar{x}-\mu_2)^2\right\}.$$
(8.33)

Thus the posterior probability for model H_1 is

$$P_1 = f(H_1|\bar{x}) = \frac{L_1}{L_1+L_2} = \frac{1}{1+L_2/L_1} = \frac{1}{1+e^{-z}},$$
(8.34)

where $z = n(\mu_1 - \mu_2)(\bar{x} - \frac{\mu_1+\mu_2}{2})$. Note that $z = \log\{P_1/(1-P_1)\}$ is the log odds since $P_1/(1-P_1)$ is known as the posterior odds. As $\bar{x} \sim N(0, 1/n)$, we have $z \sim N\left(-\frac{n}{2}(\mu_1^2 - \mu_2^2), n(\mu_1 - \mu_2)^2\right)$, with mean $m = -\frac{n}{2}(\mu_1^2 - \mu_2^2)$ and variance $s^2 = n(\mu_1 - \mu_2)^2$. By viewing P_1 as a function of z, where $z \sim N(m, s^2)$, one can derive the density of P_1 through a variable transform (Theorem 1 in Appendix A).

As $\frac{dz}{dP_1} = \frac{1}{P_1(1-P_1)}$, the density of P_1 is

$$f(P_1) = \frac{1}{\sqrt{2\pi s^2}} \times \exp\left\{-\frac{1}{2s^2}(z-m)^2\right\} \times \frac{1}{P_1(1-P_1)}$$
$$= \frac{1}{\sqrt{2\pi n(\mu_1-\mu_2)^2} \cdot P_1(1-P_1)} \times \exp\left\{-\frac{1}{2n(\mu_1-\mu_2)^2}\left[\log\frac{P_1}{1-P_1} + \frac{1}{2}n(\mu_1^2-\mu_2^2)\right]^2\right\}.$$
(8.35)

This is a density with two parameters m and s^2.

In the case $\mu_1 = -\mu$ and $\mu_2 = \mu > 0$, the two compared models are equally wrong. We have

$$f(P_1) = \frac{1}{\sqrt{8\pi n\mu^2} \cdot P_1(1-P_1)} \times \exp\left\{-\frac{1}{8n\mu^2}\left[\log\frac{P_1}{1-P_1}\right]^2\right\}.$$
(8.36)

This has one parameter $s^2 = 4n\mu^2$. The problem is analogous to using n measurement errors observed on a fair balance to test two hypotheses that the balance has either negative or positive bias.

Figure 8.16a plots the density (8.36) when $\mu_1 = -0.1$, $\mu_2 = 0.1$, so that the two models are equally wrong. With a large sample size n, $P_1 = f(H_1|\bar{x})$ converges to a two-point distribution with probability $\frac{1}{2}$ each for 0 and 1. The result may be paraphrased as follows. One collects some data from the world that is grey, and asks a sage whether the world is black or white. The sage answers black half of the time and white the other half, always with absolute certainty.

Figure 8.16b plots the density (8.35) for $\mu_1 = -0.1$ and $\mu_2 = 0.2$ so that model H_1 is less wrong than H_2. When $n \to \infty$, P_1 will eventually converge to 1, selecting the better (less

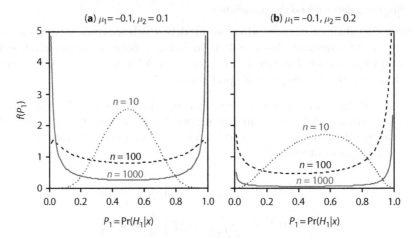

Fig. 8.16 *Fair balance paradox for simple models with no free parameters*. The distribution of posterior model probability $P_1 = \Pr(H_1|x)$ when data of i.i.d. samples of size n from $N(0, 1)$ are used to compare two models H_1: $\mu = \mu_1$ and H_2: $\mu = \mu_2$. In (a), $\mu_1 = -0.1$ and $\mu_2 = 0.1$ so that the two models are equally wrong. When the sample size $n \to \infty$, P_1 is 0 in half of the datasets and 1 in the other half. In (b), $\mu_1 = -0.1$ and $\mu_2 = 0.2$ so that model H_1 is less wrong than H_2; then when $n \to \infty$, P_1 converges to 1.

wrong) model with certainty. However, in some large and finite datasets, P_1 may be close to 0. For example, with $n = 100$, $\Pr(P_1 < 1\%) = 2.1\%$, so that in 2.1% of datasets, the more-wrong model H_2 is favoured with probability higher than 99%. With the increase of n, Bayesian model selection becomes confident before it becomes reliable. □

The general case of comparing K equally wrong simple models is similar. When the amount of data approaches infinity, Bayesian analysis tends to favour one model attaching posterior probability 1 to it. However, the favoured model varies among datasets, so that each model has probability $1/K$ of being favoured. With no data, the posterior model probabilities are given by the prior, which is $1/K$ for each model, which may be considered the ideal answer. However when faced with a large dataset, Bayesian model selection becomes polarized and holds extreme views with full conviction, even if the data contain no information for or against any of the K models.

8.5.3.2 Composite models with free parameters

When the compared models involve free parameters, the asymptotics of posterior model probabilities is more complex. Below we focus on the case of two competing models.

If the two models overlap and the truth is in the area of overlap, the posterior model probability will approach $\frac{1}{2}$ when $n \to \infty$. This is illustrated in Figure 8.15b. For example, in a coin-tossing experiment suppose the true probability of heads is $\theta_0 = 0.5$, and the two compared models are H_1: $0 \leq \theta \leq 0.6$ and H_2: $0.4 \leq \theta \leq 1.0$. Then as $n \to \infty$, $\Pr(H_1|X) \to \frac{1}{2}$. This performance is desirable. However, the construction is unusual with the two models being the same when the true $\theta \in (0.4, 0.6)$.

Another case is illustrated in Figure 8.15c. Here the two models border each other, and the true model lies at the boundary of both models. The star tree paradox discussed in §8.5.2 (Figure 8.13) is one such case. We describe another case here, in which the distribution of the posterior model probability can be derived analytically.

Example 8.4: Fair balance paradox. Suppose that the data are an i.i.d. sample of size n from $N(\mu, 1)$, with the true mean $\mu_0 = 0$. We compare two models $H_1: \mu < 0$ and $H_2: \mu > 0$, with the prior $\pi_1 = \pi_2 = \frac{1}{2}$ and $\mu \sim N(0, \xi^2)$ truncated to the appropriate range under each model. Note that the data can be summarized as the sample mean $\bar{x}$. As the prior precision is $1/\xi^2$, and the data precision is n, we have $\mu|\bar{x} \sim N\left(\frac{n\bar{x}}{n+1/\xi^2}, \frac{1}{n+1/\xi^2}\right)$. The posterior model probability is thus

$$P_1 = \Pr(H_1|\bar{x}) = \Pr(\mu < 0|\bar{x}) = \Phi(z) = \Phi\left(-\frac{\sqrt{n}\bar{x}}{\sqrt{1 + \frac{1}{n\xi^2}}}\right). \tag{8.37}$$

Note that $\bar{x}$ has the density $f(\bar{x}) = \frac{1}{\sqrt{2\pi/n}} e^{-\frac{n}{2}\bar{x}^2}$. Consider P_1 as a function of $\bar{x}$ related by equation (8.37), with

$$\left|\frac{dP_1}{d\bar{x}}\right| = \phi\left(-\frac{\sqrt{n}\bar{x}}{\sqrt{1 + \frac{1}{n\xi^2}}}\right) \times \sqrt{\frac{n}{1 + \frac{1}{n\xi^2}}}, \tag{8.38}$$

where $\phi(\cdot)$ is the probability density function (PDF) of $N(0, 1)$. The density of P_1 can be derived through variable transform (Theorem 1 in Appendix A) as

$$f(P_1) = f(\bar{x}) \bigg/ \left|\frac{dP_1}{d\bar{x}}\right| = \sqrt{1 + \frac{1}{n\xi^2}} \exp\left\{-\frac{1}{2n\xi^2}\left[\Phi^{-1}(P_1)\right]^2\right\}, \tag{8.39}$$

where $\Phi^{-1}(\cdot)$ is the inverse CDF of the standard normal distribution (Yang and Rannala 2005, Equation 12).

Note that in equation (8.37), when $n \to \infty$, $z \to -\sqrt{n}\bar{x} \sim N(0, 1)$ so that $P_1 = \Phi(z) \sim U(0, 1)$, because the CDF of any random variable has a uniform distribution. Figure 8.17 shows the distribution of P_1 (equation (8.39)) for sample sizes $n = 2, 10$, and 100, when $\xi^2 = 1$ in the prior. Even with a small sample size $n = 10$, the distribution is very close to uniform. □

The final case of Bayesian model selection we consider here is when the truth lies inside the parameter space of the two compared models and is at the cross point of the two

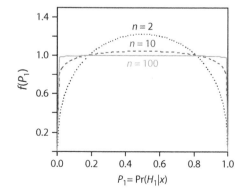

Fig. 8.17 *Fair balance paradox with parameters.* The distribution of posterior model probability $P_1 = \Pr(H_1|x)$ when data of i.i.d. samples of size n from $N(0, 1)$ are used to compare two models $H_1: \mu < 0$ and $H_2: \mu > 0$. The two models are assigned equal prior probability (1/2), and under each model, the prior is $\mu \sim N(0, \xi^2)$ truncated to the right range. In the plots, $\xi^2 = 1$. When the data size $n \to \infty$, P_1 behaves like a random number, with $P_1 \sim U(0, 1)$.

models. In this case the two models are equally correct (Figure 8.15d). We provide a numerical example below.

Example 8.5: Posterior model probabilities when the two models are crossing. The data are an i.i.d. sample of size n from $N(0, 1)$, and we compare two models H_1: $x_i \sim N(\mu, 1)$ with unknown mean and H_2: $x \sim N(0, 1/\tau)$ with unknown precision τ. Each model has one parameter and the two models cross at the point of truth. We assign equal prior probabilities $\left(\frac{1}{2}\right)$ to the two models. Under H_1 we assign the conjugate prior $\mu \sim N(0, \xi^2)$ with ξ^2 given, while under H_2 we assign the conjugate prior $\tau \sim \exp(m_0)$ with mean m_0 given. The dataset $x = (x_1, x_2, \ldots, x_n)$ can be summarized as the sample mean $\bar{x} = \frac{1}{n}\sum_{i=1}^{n} x_i$ and sample variance $s^2 = \frac{1}{n}\sum_i (x_i - \bar{x})^2 = \frac{1}{n}(\sum_i x_i^2 - n\bar{x}^2)$. The marginal likelihood under H_1 is an average of the likelihood over the prior on μ:

$$M_1 = \int_{-\infty}^{\infty} \frac{1}{\sqrt{2\pi\xi^2}} \exp\left\{-\frac{1}{2\xi^2}\mu^2\right\} \times \frac{1}{(2\pi)^{n/2}} \exp\left\{-\frac{1}{2}\sum_i (x_i - \mu)^2\right\} d\mu$$

$$= \frac{1}{\sqrt{2\pi\xi^2} \times (2\pi)^{n/2}} \int_{-\infty}^{\infty} \exp\left\{-\frac{1}{2}\left(n + \frac{1}{\xi^2}\right)\left(\mu - \frac{n\bar{x}}{n + \frac{1}{\xi^2}}\right)^2\right\} \times \exp\left\{-\frac{1}{2}\sum_i x_i^2 + \frac{1}{2}\frac{(n\bar{x})^2}{n + \frac{1}{\xi^2}}\right\} d\mu$$

$$= \frac{1}{(2\pi)^{n/2} \times \sqrt{\xi^2} \times \sqrt{n + \frac{1}{\xi^2}}} \exp\left\{-\frac{1}{2}(ns^2 + n\bar{x}^2) + \frac{1}{2}\frac{(n\bar{x})^2}{n + \frac{1}{\xi^2}}\right\}$$

$$= \frac{1}{(2\pi)^{n/2}\sqrt{n\xi^2 + 1}} \exp\left\{-\frac{n\bar{x}^2}{2(n\xi^2 + 1)} - \frac{ns^2}{2}\right\}. \tag{8.40}$$

$\square$

The marginal likelihood under H_2 is

$$M_2 = \int_0^{\infty} \frac{1}{m_0} e^{-\tau/m_0} \times \frac{1}{(2\pi/\tau)^{n/2}} \exp\left\{-\frac{\tau}{2}\sum_i x_i^2\right\} d\tau$$

$$= \frac{1}{(2\pi)^{n/2} m_0} \times \int_0^{\infty} \tau^{n/2} \exp\left\{-\tau\left(\frac{1}{m_0} + \frac{1}{2}n\bar{x}^2 + \frac{1}{2}ns^2\right)\right\} d\tau \tag{8.41}$$

$$= \frac{1}{(2\pi)^{n/2} m_0} \times \frac{\Gamma\left(\frac{n}{2} + 1\right)}{\left(\frac{1}{m_0} + \frac{1}{2}n\bar{x}^2 + \frac{1}{2}ns^2\right)^{\frac{n}{2}+1}}.$$

Then the posterior model probability

$$P_1 = f(H_1|x) = \frac{M_1}{M_1 + M_2} = \frac{1}{1 + M_2/M_1}, \tag{8.42}$$

where

$$\frac{M_2}{M_1} = \frac{\Gamma\left(\frac{n}{2} + 1\right)\sqrt{n\xi^2 + 1}}{m_0\left(\frac{1}{m_0} + \frac{1}{2}n\bar{x}^2 + \frac{1}{2}ns^2\right)^{\frac{n}{2}+1}} \times \exp\left\{\frac{n\bar{x}^2}{2(n\xi^2 + 1)} + \frac{ns^2}{2}\right\}. \tag{8.43}$$

The distribution of P_1 depends on n, ξ^2, and m_0.

While the distribution of P_1 is not tractable, one can easily generate a sample of P_1 by simulation. One samples n times from $N(0, 1)$ to form one dataset $x = (x_1, x_2, \ldots, x_n)$,

8.5 HIGH POSTERIOR PROBABILITIES FOR TREES

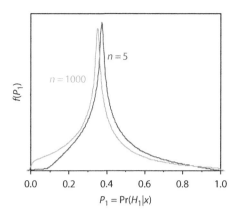

Fig. 8.18 The distribution of posterior model probability $P_1 = \Pr(H_1|x)$ when data of i.i.d. samples of size n from $N(0, 1)$ are used to compare two models H_1: $N(\mu, 1)$ and H_2: $x \sim N(0, 1/\tau)$, each with one unknown parameter. The two models are assigned equal prior probability (1/2). Under $H_1, \mu \sim N(0, \xi^2)$, while under $H_2, \tau \sim \exp(m_0)$ with mean m_0. Each dataset consists of the sample mean and variance, generated by using equation (8.44). These are then used to calculate P_1 by equations (8.42) and (8.43). The density of P_1 is estimated by simulating 10^6 datasets. Parameters $\xi^2 = 2$ and $m_0 = 1$ are fixed, while the sample size $n = 5$ or 1000.

calculate $\bar{x}$ and s^2 and then P_1 using equations (8.42) and (8.43). Or equivalently one can, instead of the sample from $N(0, 1)$, generate $\bar{x}$ and s^2 directly as independent variables

$$\begin{cases} \bar{x} \sim N(0, \ 1/n), \\ ns^2 \sim \chi^2_{n-1}. \end{cases} \quad (8.44)$$

Then calculate P_1 using equations (8.42) and (8.43).

Figure 8.18 shows the estimated densities of P_1 for two sample sizes $n = 5$ and 1000 when $\xi^2 = 2$ and $m_0 = 1$ are fixed. The densities are estimated by generating 10^6 datasets or P_1 values. The densities are indistinguishable for sample sizes of $n \geq 100$. □

8.5.4 Conservative Bayesian phylogenetics

From the last subsection, Bayesian model selection appears to have undesirable behaviour when the compared models have the same size and are nearly equally wrong. It is an open question whether anything should be done about this behaviour. I suggest that the asymptotic behaviour of Bayesian model selection may partly explain the high posterior probabilities for trees and splits observed in Bayesian phylogenetic analysis.

In molecular phylogenetics, two priors have been suggested to alleviate the problem of very high posterior probabilities for trees or splits, by what Alfaro and Holder (2006) called 'conservative Bayesian inference'. Yang and Rannala (2005) assigned exponential priors with means μ_0 and μ_1 for internal and external branch lengths, respectively, and suggested the use of very small μ_0. Only binary trees are considered in the Bayesian analysis. Based on analogous results for the normal distribution (the fair balance problem of Example 8.4), Yang (2007a) suggests that when the data size $n \to \infty$, the prior mean μ_0 should approach zero at a rate faster than $1/\sqrt{n}$ but more slowly than $1/n$. In particular, Yang (2008) used $\mu_0 = 0.1 \, n^{-2/3}$. This prior has been implemented in combination with the compound Dirichlet prior for branch lengths, assuming different α_i parameters for internal and external branch lengths in equation (8.5). We note that using the data size to specify the prior is against the Bayesian philosophy, and furthermore the posterior tree probabilities are sensitive to μ_0.

Another prior is the polytomy prior suggested by Lewis et al. (2005). This includes the multifurcating trees (polytomies) in the Bayesian tree search by assigning non-zero prior probabilities to them. Reversible jump MCMC is then used to deal with the different numbers of branch length parameters in the bifurcating and multifurcating trees. Here multifurcating trees are used as a means for reducing posterior probabilities for the binary

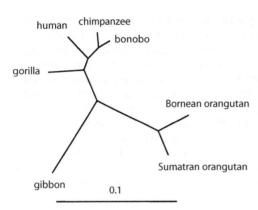

Fig. 8.19 The maximum posterior probability (MAP) tree for the mitochondrial protein sequences from seven ape species. Posterior split probabilities are calculated under the MTMAM model using MrBayes 3.2.1. Independent exponential priors are assumed for branch lengths, with the prior mean for external branch lengths fixed at $\mu_1 = 0.1$, while three values are used for the prior mean μ_0 for internal branch lengths: $10^{-1}, 10^{-3}$, and 10^{-5}. The posterior split probabilities are 1.0 for all splits under all three priors. The branches are drawn in proportion to the posterior means of the branch lengths calculated using the prior mean $\mu_0 = 0.1$.

trees and not because a model of simultaneous speciations is believed to be biologically reasonable. The polytomy prior will resolve the star tree paradox: if large datasets are simulated under the star tree, the star tree will dominate the posterior. Conceptually the polytomy prior is equivalent to using a mixture-distribution prior for internal branch lengths in the comparison of binary trees only, with a component of zero and another component from a continuous distribution. Similarly, the results tend to be sensitive to the prior probabilities assigned to the multifurcating versus binary trees. The approach of Yang and Rannala is computationally simpler as it avoids the rjMCMC.

Example 8.6: Phylogeny of apes. We apply the Bayesian approach with different priors on branch lengths to the dataset of the 12 mitochondrial proteins from seven ape species, analysed in §4.2.4. MrBayes version 3.2.1 (Ronquist et al. 2012b) is used for the analysis, modified to implement the gamma-Dirichlet prior with different prior means for internal and external branch lengths (Yang and Rannala 2005; Rannala et al. 2012; Zhang et al. 2012). The first prior is the default i.i.d. prior, which assumes that all branch lengths have the exponential distribution with mean 0.1. The second prior is independent exponential priors with mean $\mu_0 = 0.1 \, n^{-2/3} = 0.1 \times 3331^{-2/3} = 0.00045$ and $\mu_1 = 0.1$ for the internal and external branch lengths, respectively. The third prior uses the gamma-Dirichlet with parameters $\alpha_T = 1$, $\beta_T = 1$, and $c = 0.00215$, where c is the ratio of the prior means of internal and external branch lengths: this gives a prior mean of internal branch lengths of ~0.00045. All analyses used the MTMAM model for mitochondrial proteins (Yang et al. 1998). The MAP tree is shown in Figure 8.19, the same as the ML tree of Figure 4.5 in Chapter 4. The posterior probability is 100% for every split in every analysis □

8.6 Problems

8.1 A proposal modifies two variables x and y with their sum ($s = x + y$) fixed. Generate $x' \sim U(x - \varepsilon/2, x + \varepsilon/2)$, reflected into the range $0 < x' < s$, if necessary. Set $y' = s - x'$. Derive the proposal ratio for this move. Also explain why the proposal ratio remains the same if x' is generated from the normal or Bactrian proposals around x (§7.2.2–3). [Hint: One method is to use Theorem 2 in Appendix A. Another is to use Green's formulation and the mapping $(x, y) \leftrightarrow (x', y')$.]

8.2 Multiple peaks can be caused by conflicting prior and likelihood. Download the lizard dataset of Leaché and Mulcahy (2007) from the book's website (file name

Sceloporus.nex). This is the same data as were analysed in Example 8.1, with 123 sequences and 1,606 alignment columns. Run MrBayes (Ronquist et al. 2012b) under JC69+Γ_5 with the tree topology fixed at the ML tree, and using different initial branch lengths. The default prior for branch lengths is i.i.d. exponential with mean 0.1, so that the prior mean of tree length (sum of branch lengths) is $0.1 \times (2 \times 123 - 3) = 24.3$. The ML estimate of tree length under the model is 2.2. The prior and the likelihood are thus in conflict. Try to confirm the existence of two peaks in the posterior: one higher peak with tree length $T \approx 2.2$ and log likelihood $\ell \approx -1310$ and another much lower peak at $T \approx 18.3$ with $\ell \approx -1350$. This dataset is problematic for MrBayes 3.1, which by default use 0.1 for initial branch lengths and almost always gets stuck at the lower peak. MrBayes 3.2.1, with the tree length multiplier, almost always converges to the higher peak, but can get stuck at the lower peak if the starting tree is sampled around the lower leak generated from MrBayes 3.1 (see notes in the data file).

8.3* *Fair coin paradox with no parameter*. Confirm the following either analytically or using computer simulation. Suppose a coin is fair with the probability of heads to be $\theta_0 = \frac{1}{2}$. Flip the coin n times and observe x heads. Calculate the posterior probabilities for two hypotheses: H_1: $\theta = 0.4$ and H_2: $\theta = 0.6$, by assigning equal prior probabilities for the two models. Confirm that when n is large, the posterior model probability $P_1 = \Pr(H_1|x)$ is 0 or 1, each half of the times. [Hint. There is no need to write an MCMC program, but you will need to write down the likelihood functions for the two models. To simulate in R, use runif to generate $U(0, 1)$ random numbers, and hist to plot the histogram.]

8.4* *Fair coin paradox* (Lewis et al. 2005; Yang and Rannala 2005). Redo Problem 8.3 except that the two hypotheses now have unknown parameters: H_1: $\theta < \frac{1}{2}$ and H_2: $\theta > \frac{1}{2}$. Assign uniform priors for θ in each model. Confirm that when n is large, the posterior model probability $\Pr(H_1|x)$ behaves like a $U(0, 1)$ random number. [Hint. Given the prior $\theta \sim U(0, 1)$, the posterior $\theta|x \sim \text{beta}(x + 1, n - x + 1)$. Then $P_1 = \Pr(H_1|x) = \Pr(\theta < \frac{1}{2}|x)$. Also, to simulate in R, use `runif` to generate $U(0, 1)$ random numbers, pbeta to calculate $\Pr(H_1|x)$, and hist to plot the histogram.]

8.5* *Two equally wrong exponential models*. Conduct a computer simulation to explore the posterior model probability in large datasets, when two equally wrong exponential models are compared. Suppose we take a sample of size $n = 10$ from the exponential distribution with mean $\mu_0 = \log 4$, and use the data to compare two models H_1: $\mu_1 = 1$, and H_2: $\mu_2 = 2$. Those two models are equally wrong. [Note that the two models are equally wrong if $\mu_0 = \log \left\{ \frac{\mu_2}{\mu_1} \right\} / \left(\frac{1}{\mu_1} - \frac{1}{\mu_2} \right)$.] We assign the prior $\pi_1 = \pi_2 = \frac{1}{2}$ for the two models. Calculate the posterior model probability $P_1 = \Pr(H_1|x)$. Simulate 10,000 datasets and plot the histogram of the 10,000 P_1 values. Repeat the analysis using $n = 100$ and 1,000 to see the impact of the sample size. [Hint. In R, use rexp to generate exponential random variables.]

8.6 *Phylogenetic reconstruction using Bayesian inference*. Use the same data of Problem 4.6 to infer the phylogeny using the Bayesian method under the same substitution models, in comparison with the likelihood analysis. You can use MRBAYES to run tree search under those models.

* indicates a more difficult or technical problem.

CHAPTER 9

Coalescent theory and species trees

9.1 Overview

In this chapter we review several major computational methods for analysing genetic and genomic sequence data under the coalescent model. The data may be sequence samples from one single species or from several closely related species. The chapter begins with an introduction to the basic coalescent model for a single population (§9.2) and extends it to multiple closely related species (§9.3). The multispecies coalescent model provides a framework for estimating the species tree despite conflicting gene trees and for Bayesian species delimitation. These topics are discussed in §9.4 and §9.5. My focus will be on likelihood-based inference methods, which are closely related to the likelihood and Bayesian algorithms in phylogenetics (Chapters 4 and 8). Indeed statistical analysis of sequence data from a few closely related species (called statistical phylogeography in the broad sense of Knowles 2009) lies at the interface between population genetics and phylogenetics. Bayesian Markov chain Monte Carlo (MCMC) algorithms in those disciplines have a lot of similarities; for example, they all involve sampling in the space of trees (genealogies or phylogenies) and calculating the likelihood function for a sequence alignment on a given tree. Recombination and selection are not covered in this chapter. We focus on sequence data and ignore other types of population genetics data such as restriction fragment length polymorphisms (RFLPs), microsatellites and single nucleotide polymorphisms (SNPs).

For general background in population genetics, the reader should consult one of the many excellent textbooks available (e.g. Gillespie 1998; Hartl and Clark 1997). Hein et al. (2005) and Wakeley (2009) provide good introductions to coalescent theory, as well as the reviews of Hudson (1990), Donnelly and Tavaré (1997, 2005), Rosenberg and Nordborg (2002), and Nordborg (2007).

It may be fitting to note here a few trends in theoretical population genetics. First, in the time of R.A. Fisher, J.B.S. Haldane, and S. Wright (1920–1930s), or of G. Malecot and M. Kimura (1950–1970s), there was much theory in population genetics but little data, and the work was mostly concerned with probabilistic predictions of the model behaviour, i.e. how allele frequencies change over generations when the parameters in the model take certain values. Nowadays there is more data than we can analyse, and the focus of the field has shifted to statistical inference, i.e. parameter estimation and hypothesis testing using genomic sequence data. To a large extent, methodological developments in the field are now driven by data, in particular, genomic sequence data. Second, the coalescent approach has taken the centre stage in statistical methods of comparative data analysis.

Molecular Evolution: A Statistical Approach. Ziheng Yang. © Ziheng Yang 2014.
Published 2014 by Oxford University Press.

Third, most modern inference methods, developed in the coalescent framework, involve intensive computation.

9.2 The coalescent model for a single species

9.2.1 *The backward time machine*

To 'coalesce' means to 'merge' or to 'join'. When one traces the genealogical relationships of a sample of chromosomes (or genes from a particular genomic region) backwards in time over the generations, the lineages join or coalesce when they meet their common ancestors. The coalescent is the genealogical process of joining lineages when one traces the genealogy of the sample backwards in time. The coalescent theory, also known as Kingman's coalescent, was developed in the early 1980s (Kingman 1982a, 1982b; Hudson 1983b; Tajima 1983), and has since played a central role in theoretical population genetics. It underlies modern computational methods for the comparative analysis of genetic data, in particular, genomic sequence data, including both polymorphism data from the same species and divergence data from multiple closely related species. A classical population genetics model typically focuses on the population and takes a forward approach, making predictions about allele frequency changes over generations in a population under the influence of various factors such as mutation, genetic drift, population subdivision, and natural selection (Fisher 1930a; Wright 1931; Haldane 1932; Kimura and Ohta 1971a). In contrast, the coalescent approach focuses on the data sample and 'runs the time machine backwards', tracing the genealogical relationships of the sample until the most recent common ancestor (MRCA) is reached.

The coalescent theory is the result of several fundamental insights, realized around 1980 (Ewens 1990; Kingman 2000). Under many population genetic models, the genealogy for a sample of genes is usually easier to model backward than forward in time. Under the assumption of random mating and neutral mutations, each individual 'picks' its parent at random from the previous generation when one traces the genealogy of the sample backwards in time. Thus we can ignore the rest of the population and simply focus on the lineages that are ancestral to the sample. Second, we can focus on the genealogical tree structure and branch lengths (coalescent times) while ignoring neutral mutations. After the genealogical tree has been generated, we 'drop' neutral mutations onto branches of the tree. This strategy allows us to derive the probability distribution of the data sample under many population genetics models. The probability of the data, also known as the likelihood (see §1.4), is the basis for statistical inference from the genetic data.

Our interest is in the biological process that affects the genealogical relationships of the sample, such as population demographic changes or population subdivision. The genetic data or mutations act as markers, providing information about the unobserved genealogy. The genealogy is of interest only because it contains information about the evolutionary process and the population parameters. This is a major difference from molecular phylogenetics, where the species phylogeny is often the focus of study.

9.2.2 *Fisher–Wright model and the neutral coalescent*

Fisher–Wright model. The Fisher–Wright model (Fisher 1930a; Wright 1931) is an idealized model in population genetics, characterized by a constant population size, non-overlapping generations, random mating (panmixia), and neutral evolution (Figure 9.1). The population size N is finite and constant over time. All N individuals in the population

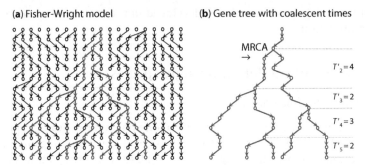

Fig. 9.1 (a) The Fisher–Wright model for a population of $2N = 20$ gene sequences. Each row corresponds to a generation. The genealogical relationship for a sample of $n = 5$ genes from the present generation is highlighted. (b) The genealogical tree for the sample, with the rest of the population ignored. The coalescent times are waiting times until coalescent events, measured in generations; i.e. T'_j is the number of generations when there are j lineages ancestral to sequences in the sample.

are assumed to die each generation and are replaced by offspring in the next generation through random mating and reproduction. This is true for annual plants, which have exactly one generation per year, but it is only approximate for species with overlapping generations. There is no population subdivision, and mating is random. The model applies to both haploid and diploid populations. The dynamics depend on the total number of genes and is essentially the same for a diploid population of size N and a haploid population of size $2N$. In this book, we will focus on the diploid system and sequence data, so that there are $2N$ genes for a population of size N. The term 'gene' or allele refers to any locus or a segment of the DNA sequence in the genome, often noncoding. We assume there is no recombination within a locus and free recombination between any two loci.

The coalescent model is known to be quite robust, and certain departures from the Fisher–Wright model can be dealt with using Sewall Wright's concept of *effective population size* (N_e) (Hartl and Clark 1997, Chapter 7). This is defined as the size of an idealized (Fisher–Wright) population that would show the same magnitude of genetic drift as the real population under study. Note that the smaller a population is, the more effective genetic drift is in removing genetic variation. There are different measures of the magnitude of genetic drift and thus different definitions of N_e. One such measure is the coalescent waiting time between two sequences taken at random from the population, with the waiting time being shorter the smaller the population is. For example, biased sex ratio in a diploid organism, overlapping generations, etc. often leads to a change in the coalescent time scale but not the dynamics of the system. If the sex ratio is biased in a diploid organism, with N_m and N_f to be the numbers of males and females, then the effective population size is $N_e = 4N_m N_f/(N_m + N_f)$. This is twice the harmonic mean of N_m and N_f and is smaller than twice the arithmetic mean or the census size $N_m + N_f$. The interpretation is that the expected coalescent waiting time between two sequences taken at random from a population with N_m males and N_f females is approximately equal to the coalescent time between two sequences taken at random from an ideal population with $N_e/2$ males and $N_e/2$ females. Similarly if the population size is changing over generations, N_e is given by the harmonic mean, which is dominated by small values (population bottlenecks) and is in general much smaller than the arithmetic mean. In this chapter, we will use population size N but it should usually be interpreted as the effective population size N_e. Of course some departures from the Fisher–Wright model will change the dynamics of the system and may have to be dealt with explicitly. For example, to use genetic data to infer demographic changes in population size, one should consider N explicitly in the model instead of N_e, which is a long-term average.

Coalescent of two genes. Viewed backwards in time, the process of random mating and reproduction becomes the genealogical process of coalescent or lineage joining. Consider two genes sampled at random from the population. How many generations do we have to go back before we find their common ancestor? The probability that the two lineages 'pick' the same parent and coalesce in the previous generation is $1/(2N)$. To see this, let the first gene pick its parent and then note that the second gene will pick the same parent with probability $1/(2N)$. The probability that the two genes pick different parents and remain distinct lineages in the previous generation is $1 - 1/(2N)$. Continue going backwards in time. The probability that two genes do not find the common ancestor in the first i generations is

$$\Pr\{T_2' > i\} = \left(1 - \frac{1}{2N}\right)^i, \tag{9.1}$$

and the probability that they find the common ancestor exactly i generations back is

$$\Pr\{T_2' = i\} = \left(1 - \frac{1}{2N}\right)^{i-1} \times \frac{1}{2N}. \tag{9.2}$$

Equations (9.1) and (9.2) mean that the coalescent waiting time T_2' has a geometric distribution, with mean $2N$. On average, it takes $2N$ generations for two lineages to coalesce.

Strictly speaking, the above description applies to an asexual haploid species of size $2N$ but not to a sexual diploid species of size N. For example, the two alleles carried by the same individual from a diploid species are from the father and mother respectively, and cannot possibly coalesce in the previous generation. However the difference between the two systems is negligibly small and is ignored.

As $2N$ is large, it is common to change the time scale so that one time unit is $2N$ generations. Let the rescaled time be $T_2 = T_2'/(2N)$. Since

$$\Pr\left\{T_2 > \frac{i}{2N}\right\} = \Pr\{T_2' > i\} = \left(1 - \frac{1}{2N}\right)^i \approx e^{-\frac{i}{2N}}, \tag{9.3}$$

T_2 is an exponential variable with mean 1 (in $2N$ generations) and variance 1, with density

$$f(T_2) = e^{-T_2}. \tag{9.4}$$

Recall that if $\Pr\{X > x\} = e^{-\lambda x}$, random variable X has an exponential distribution with mean $1/\lambda$.

In comparison of DNA sequences, it is convenient to multiply time by the mutation rate per site per generation (μ) and define one time unit as the expected time to accumulate one mutation per site. Note that in classic population genetics, the mutation rate is typically defined as the rate for the whole locus, but for DNA sequence data, the per site rate is far more convenient. For example, it is the per site rate, not the per locus rate, that is nearly constant throughout the noncoding genome. At this time scale, the coalescent time $t_2 = T_2'\mu$ has expectation $2N\mu = \frac{\theta}{2}$, where $\theta = 4N\mu$, with probability density

$$f(t_2) = \frac{2}{\theta} e^{-\frac{2}{\theta} t_2}. \tag{9.5}$$

In other words, coalescent events occur at the rate $\frac{2}{\theta}$ when time is measured as the number of mutations.

Thus three time scales are commonly used in the literature, and they are distinguished here using different symbols: T'_2 is measured in generations with mean $2N$, $T_2 = T'_2/(2N)$ is measured in $2N$ generations with mean 1, and $t_2 = T'_2 \mu = T_2 \times 2N\mu = T_2 \times \frac{\theta}{2}$ is measured in the number of mutations per site with mean $\frac{\theta}{2}$.

Population size parameter θ. Parameter $\theta = 4N\mu$, where N is the (effective) population size and μ is the mutation rate per site per generation, is known as the population size parameter. Suppose we take two genes (two DNA sequences from a particular genomic region) at random from a species/population. How different should we expect them to be? As it takes on average $2N$ generations for the two genes to coalesce, the expected number of mutations per site between the two genes will be $E(2T'_2 \times \mu) = 4N\mu = \theta$. The factor 2 is needed because there are two lineages (branches) going from the present time to the common ancestor in the tree of two sequences. Parameter θ is thus a measure of the genetic diversity in the population. It reflects the balance between genetic drift causing the loss of polymorphism, and mutations introducing new variation. For the human, estimates of θ are about 0.0006 or 0.6 per kilobase. In other words, two homologous sequences sampled at random from the human population will have a difference about every 1,700 base pairs. If the generation time is $g \approx 20$ years, and the mutation rate is $\mu/g \approx 1.2 \times 10^{-9}$ mutations per site per year (e.g. Kumar and Subramanian 2002) or $\mu \approx 2.4 \times 10^{-8}$ per site per generation, we have an estimate of the effective population size $N = \theta/(4\mu) \approx 6{,}250$. This estimate is extremely small compared with the current size of the species ($\approx 7 \times 10^9$ at the time of writing). It is generally believed that the human species went through bottlenecks in the early history and dramatic expansion since the advent of agriculture.

9.2.3 A sample of n genes

We now consider a sample of n genes. The probability that no coalescent event occurs between any two of the n lineages in the previous generation is

$$\left(1 - \frac{1}{2N}\right)\left(1 - \frac{2}{2N}\right) \cdots \left(1 - \frac{n-1}{2N}\right) \approx 1 - \frac{1 + 2 + \cdots + (n-1)}{2N} = 1 - \binom{n}{2}\frac{1}{2N}, \tag{9.6}$$

where $\binom{n}{2} = \frac{n(n-1)}{2}$, read '$n$ choose 2', is the number of possible pairs. Note that $\left(1 - \frac{1}{2N}\right)$ is the probability that the second gene picks a different parent than the first gene, $\left(1 - \frac{2}{2N}\right)$ is the probability that the third gene picks a different parent than the first two genes, and so on, and finally $\left(1 - \frac{n-1}{2N}\right)$ is the probability that the nth gene picks a different parent than the first $(n-1)$ genes. The approximation in equation (9.6) is because we assume that $n \ll N$, so that the probability of more than two genes coalescing in the same generation is negligible and terms involving $1/(2N)^2$, $1/(2N)^3$, etc. are ignored.

Using the same argument as above for two genes, we have the probability that the first coalescent event (which reduces n genes to $n-1$ genes) occurs exactly i generations ago to be

$$\Pr\{T'_n = i\} = \left[1 - \binom{n}{2}\frac{1}{2N}\right]^{i-1} \times \binom{n}{2}\frac{1}{2N}. \tag{9.7}$$

Thus T'_n, the waiting time until the next coalescent event when there are n genes in the sample, has a geometric distribution with mean $2N / \binom{n}{2}$. In general the waiting time until the next coalescent when there are j genes in the sample has the geometric distribution with mean $2N / \binom{j}{2}$. In other words, each pair of genes coalesces at the rate $\frac{1}{2N}$ per generation, and j genes coalesce at the rate of $\binom{j}{2} \cdot \frac{1}{2N}$ since there are $\binom{j}{2}$ pairs.

Again, let $T_j = T'_j/(2N)$ be the scaled waiting time (in $2N$ generations) until the next coalescent event, given that there are currently j lineages in the sample. By the same argument as for the case of two genes, T_j has the exponential distribution with mean $\frac{2}{j(j-1)}$ and variance $\left[\frac{2}{j(j-1)}\right]^2$, with density

$$f(T_j) = \frac{j(j-1)}{2} \exp\left\{-\frac{j(j-1)}{2} T_j\right\}. \tag{9.8}$$

In summary, the genealogy for a sample of n genes is a random bifurcating tree, generated by random joining of the lineages. The genealogical tree differs from a rooted tree topology in that the ranking of nodes by age is considered in addition to the rooted tree topology; different rankings mean different sequences of coalescent events. In other words, the genealogical tree is a *labelled history* (see §3.1.1.5). The number of possible labelled histories for n sequences is

$$H_n = \binom{n}{2} \cdot \binom{n-1}{2} \cdots \binom{2}{2} = \frac{n!(n-1)!}{2^{n-1}}, \tag{9.9}$$

and each labelled history has the same probability $1/H_n$ under the model. On any labelled history G, the $n-1$ coalescent times $T_n, T_{n-1}, \ldots, T_2$ are independent exponential variables, with density

$$f(T_n, T_{n-1}, \ldots, T_2 | G) = \prod_{j=2}^{n} \frac{j(j-1)}{2} \exp\left\{-\frac{j(j-1)}{2} T_j\right\}. \tag{9.10}$$

The joint distribution of the genealogical tree G and the coalescent times is

$$f(G, T_n, T_{n-1}, \ldots, T_2) = f(G) f(T_n, T_{n-1}, \ldots, T_2 | G) = \prod_{j=2}^{n} \exp\left\{-\frac{j(j-1)}{2} T_j\right\}. \tag{9.11}$$

We now discuss a few characterizations of the coalescent process: the tree height (T_{MRCA}), the tree length (T_{total}), and the probability that the sample contains the population root. The time to the MRCA of the sample, also called the tree height, is

$$T_{\text{MRCA}} = T_n + T_{n-1} + \cdots + T_2. \tag{9.12}$$

This is a sum of $(n-1)$ independent exponential waiting times (Figure 9.2). Thus its expectation and variance are

$$E(T_{\text{MRCA}}) = E(T_n + T_{n-1} + \cdots + T_2) = \sum_{j=2}^{n} \frac{2}{j(j-1)} = 2 \sum_{j=2}^{n} \left(\frac{1}{j-1} - \frac{1}{j}\right) = 2\left(1 - \frac{1}{n}\right),$$

$$V(T_{\text{MRCA}}) = \sum_{j=2}^{n} V(T_j) = \sum_{j=2}^{n} \left(\frac{2}{j(j-1)}\right)^2 = 8 \sum_{j=1}^{n-1} \frac{1}{j^2} - 4\left(3 - \frac{2}{n} - \frac{1}{n^2}\right). \tag{9.13}$$

Note that $E(T_{\text{MRCA}}) \approx 2$ (in $2N$ generations) for large n. Since $E(T_2) = 1$, it takes on average only twice as long for the whole sample to coalesce as for the last two lineages to coalesce. Also $V(T_{\text{MRCA}}) \approx \frac{8\pi^2}{6} - 12 \approx 1.16$ for large n. Since $V(T_2) = 1$, the variation in

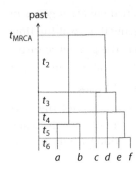

Fig. 9.2 A coalescent tree for six sequences with coalescent times. Different time scales are commonly used for the coalescent time, with T'_j in generations, $T_j = T'_j/(2N)$ in $2N$ generations, and $t_j = T'_j \mu$ in the expected number of mutations per site. All of them are exponential variables, with means $E(T'_j) = 2N/\binom{j}{2}$, $E(T_j) = 1/\binom{j}{2}$, and $E(t_j) = \frac{\theta}{2}/\binom{j}{2}$.

T_{MRCA} for the whole sample is mostly due to the variation in T_2, the waiting time for the last two lineages.

The tree length is defined as the sum of branch lengths in the genealogical tree: $T_{\text{total}} = \sum_{j=2}^{n} jT_j$. This has the mean and variance

$$E(T_{\text{total}}) = E\left(\sum_{j=2}^{n} jT_j\right) = \sum_{j=2}^{n} \frac{2}{j-1} = 2\sum_{j=1}^{n-1} \frac{1}{j},$$

$$V(T_{\text{total}}) = \sum_{j=2}^{n} j^2 V(T_j) = \sum_{j=2}^{n} j^2 \left(\frac{2}{j(j-1)}\right)^2 = 4\sum_{j=1}^{n-1} \frac{1}{j^2}. \qquad (9.14)$$

For large n, the mean is $E(T_{\text{total}}) \approx 2(\gamma + \log n)$, with $\gamma = \lim_{n \to \infty} \sum_{j=1}^{n} \frac{1}{j} - \log(n) \approx 0.577$ to be Euler's constant, while the variance is $V(T_{\text{total}}) \approx 2\pi^2/3 \approx 6.579$. Thus the expected tree length grows very slowly with n since increasing the sample size only adds short twigs to the tree.

Now consider the time to the MRCA of the whole population. The probability that the MRCA for a sample of size n is also the MRCA for the whole population is $\frac{n-1}{n+1}$ (Saunders et al. 1984). This is $\frac{1}{3}$, $\frac{1}{2}$, and 0.8 for $n = 2, 3$, and 9, respectively. Thus even a small sample is likely to contain the root of the population tree.

Figure 9.3 shows three coalescent trees for $n = 20$ lineages. The trees are extremely variable, both in topology and branch lengths. They have very short branches at the tips and long branches near the root. Note that the coalescent rate, $\binom{n}{2} \frac{2}{\theta} = \frac{n(n-1)}{\theta}$, is roughly in the order of n^2, so that the number of distinct lineages decreases rapidly as one goes back in time.

A consequence of this is that increasing the size n in the data sample is very ineffective for many inference problems, such as estimation of θ or inference of ancient demographic changes. As a rough calculation, the human effective population size is estimated to be $N \approx 10^4$. Suppose the generation gap is $g = 20$ years. Then it will be very unlikely

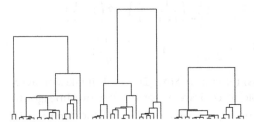

Fig. 9.3 Three realizations of the coalescent process for $n = 20$, drawn on the same time scale.

for T_{MRCA} to exceed $E(T_{MRCA}) + 2\sqrt{V(T_{MRCA})} \approx 2 + 2\sqrt{1.16} \approx 4.15$ (i.e. $4.15 \times 2N$ generations), and DNA samples from the human population will be uninformative about early human history $4.15 \times 2Ng \approx 1.7$ million years ago.

9.2.4 Simulating the coalescent

The differences between classical population genetic modelling and the coalescent approach may be appreciated by considering simulations in the two frameworks. Consider simulating a DNA sample of size n drawn at random from a population of size N under the Fisher–Wright model. In the classical (or so-called 'forward') simulation, one keeps track of all gene copies in the whole population, uses the binomial or multinomial sampling of gametes from the current generation to form the genes of the next generation. After simulating the process over a large number of generations, one takes a sample of size n from the population. One then has to trace back the genealogical history of the sample. Forward simulation is thus severely limited by computational resources, as the population is typically large ($N = 10^4$–10^8, say), even though the sample is small ($n = 2$–30, say).

In the coalescent simulation, time runs backward. One generates the genealogical tree (both the tree topology and the coalescent times) for the sample by randomly joining the genes, until one reaches the MRCA for the sample. The rest of the population is ignored. Then one runs the time machine forward and 'drops' mutations onto the branches of the genealogical tree. One generates a sequence at the root of the gene tree (at the MRCA), and 'evolves' it forward in time along branches of the tree until all n sequences at the tips are generated. Compared with forward simulation, coalescent simulation requires little computational resource, as most of the genealogical history of the population is irrelevant and hence ignored.

We outline here two coalescent simulation algorithms for generating a sample of n DNA sequences, each of l sites long, from a population with population size parameter $\theta = 4N\mu$. The first algorithm simulates mutations and coalescent events as independent competing Poisson processes. With one time unit being $2N$ generations, the coalescent rate is $k(k-1)/2$ and the mutation rate is $kl\theta/2$ if there are k genes in the sample. Note that the mutation rate or the expected number of mutations per site per unit time ($2N$ generations) is $\theta/2$. There are k sequences each of l sites, so the total mutation rate is $kl\theta/2$. Two competing Poisson processes with rates λ_1 and λ_2 can be simulated as follows. We generate an exponential variable with rate $\lambda_1 + \lambda_2$ as the waiting time until the occurrence of any of the two events. Given the occurrence of an event, it is event 1 with probability $\lambda_1/(\lambda_1 + \lambda_2)$ and event 2 with probability $\lambda_2/(\lambda_1 + \lambda_2)$. The correctness of the algorithm can be seen as follows. Let $X_1 \sim \exp(\lambda_1)$ and $X_2 \sim \exp(\lambda_2)$ be the waiting times for the two events. Let $X = \min(X_1, X_2)$. Then $\Pr\{X > x\} = \Pr\{X_1 > x \text{ and } X_2 > x\} = \Pr\{X_1 > x\} \times \Pr\{X_2 > x\} = e^{-\lambda_1 x} \times e^{-\lambda_2 x} = e^{-(\lambda_1 + \lambda_2)x}$, which means $X \sim \exp(\lambda_1 + \lambda_2)$. In other words, the waiting time until any of the two events occurs is an exponential variable with rate $\lambda_1 + \lambda_2$. It is also easy to see that, given the occurrence of an event, it is of event 1 with probability $\lambda_1/(\lambda_1 + \lambda_2)$.

Algorithm 9.1. Simulation of a DNA sample of size n under the coalescent model.

1. Set the number of lineages $k \leftarrow n$.
2. Loop through the following steps until $k = 1$.
    - Generate an exponential waiting time (in $2N$ generations) with rate $k(k-1)/2 + kl\theta/2$.
    - The event is a coalescent with probability $(k-1)/(k-1+l\theta)$ or a mutation with probability $l\theta/(k-1+l\theta)$. If the event is a coalescent, choose a pair from the k genes at

random to join, and set $k \leftarrow k-1$. If the event is a mutation, choose a lineage (a sequence) and a site at random to mutate, and leave k unchanged. □

One can keep a record of mutations that have occurred along branches of the genealogical tree. After the tree is generated, one can generate a sequence at the root and drop mutations onto the branches of the tree. This algorithm can be used to simulate sequence data under simple mutation models such as JC69 or K80 in which all nucleotides have the same mutation rate. The algorithm does not work for more complex mutation models such as HKY85 or GTR, as under those models the mutation rate varies with the nucleotide and the mutation process is not Poisson.

The second algorithm generates the genealogical tree first, ignoring mutations, and then simulates the mutation process along branches of the tree.

Algorithm 9.2. Simulation of a DNA sample of size n under the coalescent model.

1. Set $k \leftarrow n$.
2. Loop through the following steps until $k = 1$.
   - Generate an exponential waiting time (in $2N$ generations) with mean $2/[k(k-1)]$.
   - Choose a random pair from the k lineages to join. Decrease k by one.
3. Generate a random DNA sequence for the root of the gene tree. Evolve the sequence along branches of the tree until all tip sequences are generated. From step 2, each branch on the tree will have a length (let it be T, in $2N$ generations). The expected number of mutations per site on the branch is then $T \times \theta/2$. □

Step 2 generates the genealogical tree and branch lengths (coalescent times in $2N$ generations). Step 3 simulates sequences given the gene tree topology and branch lengths (see §12.6.1 in Chapter 12 later). Algorithm 9.2 is more flexible as it works for all mutation models. This is the algorithm commonly used in simulation programs.

9.2.5 Estimation of θ from a sample of DNA sequences

Estimation of θ from a sample of DNA sequences may be the simplest inference problem under the coalescent model. We review simple estimates under the infinite-site model, as well as maximum likelihood (ML) and Bayesian methods under a finite-site model.

9.2.5.1 Estimation under the infinite-site model

One locus, two sequences. The infinite-site model assumes that every new mutation occurs at a different site. Let the sequence length be l sites so that $l\mu$ is the mutation rate per generation for the locus. Given the coalescent time T (in $2N$ generations), the number of mutations or differences between two genes sampled at random in the population, S_2 or x, follows a Poisson distribution with both mean and variance being $T \times 2Nl\mu \times 2 = l\theta T$. By averaging over the coalescent time T, which is an exponential variable with $E(T) = V(T) = 1$, we find the mean and variance of x to be

$$E(x) = E(E(x|T)) = E(l\theta T) = l\theta \cdot E(T) = l\theta,$$
$$V(x) = E(V(x|T)) + V(E(x|T)) = E(l\theta T) + V(l\theta T) = l\theta \cdot E(T) + (l\theta)^2 \cdot V(T) = l\theta + (l\theta)^2. \quad (9.15)$$

Here $E(x|T)$ is the conditional expectation of x given T, etc. In fact the probability for x can be obtained by averaging over T as

$$p_x = \int_0^\infty e^{-T} \times \frac{e^{-l\theta T}(l\theta T)^x}{x!} dT$$

$$= \frac{(l\theta)^x}{(1+l\theta)^{x+1}} \times \int_0^\infty \frac{(1+l\theta)^{x+1}}{\Gamma(x+1)} e^{-(1+l\theta)T} T^{x+1-1} dT \qquad (9.16)$$

$$= \frac{(l\theta)^x}{(1+l\theta)^{x+1}} = \frac{1}{1+l\theta}\left(\frac{l\theta}{1+l\theta}\right)^x.$$

Note that the integrand in the second integral is the density function for the gamma distribution $G(x+1, 1+l\theta)$ and thus integrates to 1. Equation (9.16) suggests that the number of mutations between two sequences has a geometric distribution, with mean $l\theta$ and variance $l\theta(1+l\theta)$ (Watterson 1975).

One locus, many sequences. For a sample of n sequences, the number of segregating sites or the number of variable sites in the alignment, S_n, is the number of mutations under the infinite-site model. Given the tree length T_{total} (in $2N$ generations; see equation (9.14)), S_n is a Poisson variable with mean $T_{\text{total}} \times 2N l\mu = T_{\text{total}} l\theta/2$. By averaging over T_{total}, one obtains the mean and variance of S_n as

$$E(S_n) = E(E(S_n|T_{\text{total}})) = E\left(\frac{1}{2}l\theta T_{\text{total}}\right) = l\theta \sum_{j=1}^{n-1} \frac{1}{j},$$

$$V(S_n) = E(V(S_n|T_{\text{total}})) + V(E(S_n|T_{\text{total}})) = E\left(\frac{1}{2}l\theta T_{\text{total}}\right) + V\left(\frac{1}{2}l\theta T_{\text{total}}\right) \qquad (9.17)$$

$$= l\theta \sum_{j=1}^{n-1} \frac{1}{j} + (l\theta)^2 \sum_{j=1}^{n-1} \frac{1}{j^2}.$$

Thus an estimate of θ based on S_n is

$$\hat{\theta}_S = \frac{S_n}{l \sum_{j=1}^{n-1} \frac{1}{j}}. \qquad (9.18)$$

This is known as Watterson's estimator.

Another estimate is the average pairwise proportion of differences: $\hat{\theta}_\pi = \pi$ (Tajima 1983). This is an average over all $\binom{n}{2}$ pairs. Both $\hat{\theta}_S$ and $\hat{\theta}_\pi$ are based on the infinite-site model.

Many loci, two sequences. If only two genes are sampled at every locus, θ can be estimated from multiple loci by ML under the infinite-site model (Takahata et al. 1995). Suppose there are L loci. Let there be l_i sites and x_i differences at the ith locus, so that the data can be represented as $X = \{x_1, x_2, \ldots, x_L\}$. We assume that the mutation rate per site (μ) is constant among loci, so that the total mutation rate for locus i is $l_i\mu$ while the same θ applies to all loci. Under the assumption of free recombination between loci, the sequence data are independent among loci and the log likelihood is simply given by using equation (9.16) as

$$\ell(\theta) = \sum_{i=1}^L \log(p_{x_i}) = \sum_{i=1}^L \left\{ x_i \log\left(\frac{l_i\theta}{1+l_i\theta}\right) - \log(1+l_i\theta) \right\}. \qquad (9.19)$$

The maximum likelihood estimate (MLE) of θ can be found numerically.

9.2.5.2 Estimation under a finite-site model

We now turn to ML and Bayesian methods for estimating θ under a finite-site model, which make use of the sequence alignment rather than the number of segregating sites. The finite-site model allows multiple mutations to occur at the same site and can be any of the Markov chain models discussed in Chapter 1, such as JC69 (Jukes and Cantor 1969), K80 (Kimura 1980), or HKY85 + Γ_5 (Hasegawa et al. 1985; Yang 1994a). Calculation of the likelihood function on a sequence alignment under those models is discussed in §4.2 and §4.3. Since the main role of the mutation model is to correct for multiple hits at the same site, we do not expect different finite-site models to produce substantially different results in such highly similar sequences. As a result, the simple JC69 model appears to be adequate for such data, although there is no conceptual difficulty in applying more complex models like HKY85 + G_5. In many analyses, even the infinite-site model produces results that are very similar to those obtained under a finite-site model (e.g. Satta et al. 2004). However, when the sequences are moderately divergent, for example, if three or more nucleotides are observed at the same site in the alignment, the data will contradict the infinite-site model. To apply the infinite-site model to such data one has to remove sequences or variable sites to avoid the contradiction, which is arbitrary and produces biased estimates. Use of a finite-site model in such data is then highly desirable.

Here we use the JC69 model. First we consider only two sequences at each locus. As before, we assume that the mutation rate per site (μ) is constant among loci so that the same θ applies to all of them. Let the time unit be the expected time to accumulate one mutation per site, so that the coalescent time t between the two sequences at any locus is an exponential variable with mean $\theta/2$, with density

$$f(t|\theta) = \frac{2}{\theta} e^{-\frac{2}{\theta}t}. \tag{9.20}$$

This is equation (9.5). Given the coalescent time t, the expected number of mutations per site between the two sequences is $d = 2t$, as there are two branches in the tree of two sequences. The probability that the nucleotides at any site are different between the two sequences given the coalescent time t is

$$p = \frac{3}{4} - \frac{3}{4}e^{-4d/3} = \frac{3}{4} - \frac{3}{4}e^{-8t/3} \tag{9.21}$$

(see equation (1.45)). The probability of observing x_i differences at l_i sites at the locus is then given by averaging the binomial probability $f(x_i|t) = p^{x_i}(1-p)^{l_i-x_i}$ over the distribution of t:

$$f(x_i|\theta) = \int_0^\infty f(t|\theta)f(x_i|t)dt = \int_0^\infty \frac{2}{\theta}e^{-\frac{2}{\theta}t} \times \left(\frac{3}{4} - \frac{3}{4}e^{-8t/3}\right)^{x_i} \left(\frac{1}{4} + \frac{3}{4}e^{-8t/3}\right)^{l_i-x_i} dt, \tag{9.22}$$

in place of equation (9.16). The 1-D integral may be calculated using numerical integration (§6.4). Note that under the assumption of no recombination within a locus, all sites in the gene sequence at the locus share the same coalescent time t.

Under the assumption of free recombination across loci, the log likelihood for the whole dataset is a sum over the L loci:

$$\ell(\theta) = \sum_{i=1}^{L} \log f(x_i|\theta). \tag{9.23}$$

If there are $n_i > 2$ sequences at each locus, the data X_i will be the sequence alignment at the locus. With coalescent time measured by the number of mutations per site

(Figure 9.2), the joint distribution of the genealogical tree G_i and coalescent times t_i is given by equation (9.11), with a change of time scale $t_j = T_j \times 2N\mu = T_j \times \theta/2$, as

$$f(G_i, \boldsymbol{t}_i | \theta) = \prod_{j=2}^{n_i} \frac{2}{\theta} \exp\left\{-\frac{j(j-1)}{2} \times \frac{2}{\theta} t_j\right\}. \tag{9.24}$$

The probability of data at the locus is then an average over all genealogical trees and coalescent times. The log likelihood for the multi-locus sequence data $X = \{X_i\}$ is then

$$\ell(\theta) = \sum_{i=1}^{L} \log f(X_i | \theta) = \sum_{i=1}^{L} \log \left\{ \sum_{G_i} \int_{\boldsymbol{t}_i} f(G_i, \boldsymbol{t}_i | \theta) f(X_i | G_i, \boldsymbol{t}_i) \right\}. \tag{9.25}$$

where $f(X_i | G_i, \boldsymbol{t}_i)$, the probability of the sequence alignment at locus i given the gene tree and branch lengths at the locus, is Felsenstein's (1981) phylogenetic likelihood (see §4.2).

While all the terms in equation (9.25) are straightforward to calculate, the sums and integrals pose an insurmountable hurdle to ML estimation of θ unless there are only two or three sequences at every locus. Kuhner et al. (1995) used MCMC to generate the log likelihood curve, using a rough estimate θ_0 as the 'driver' parameter to run an MCMC, and then calculated the log likelihood values for other values of θ through an *importance-sampling adjustment* using the same MCMC sample. Unfortunately the strategy does not work well, and, for example, produces confidence intervals that are too narrow (Stephens and Donnelly 2000).

As an alternative, it is simpler to estimate θ in the Bayesian framework, using an MCMC algorithm to generate a sample from the posterior of θ. With a prior $f(\theta)$, the joint posterior is

$$f(\theta, G, \boldsymbol{t} | X) \propto f(\theta) \prod_{i=1}^{L} f(G_i, \boldsymbol{t}_i | \theta) f(X_i | G_i, \boldsymbol{t}_i), \tag{9.26}$$

where $G = \{G_i\}$ and $\boldsymbol{t} = \{\boldsymbol{t}_i\}$ are the collections of gene genealogies and coalescent times for all loci. The normalizing constant is a sum over all possible genealogies at every locus i and an $(n_i - 1)$-dimensional integral over coalescent times $\boldsymbol{t}_i$ in each gene genealogy G_i. The MCMC algorithm achieves the integration and summation through the Markov chain. The algorithm samples θ in proportion to its posterior distribution. For definiteness, we outline the algorithm of Rannala and Yang (2003; see also Yang and Rannala 2010), implemented in the BPP program, which includes the following steps in each iteration.

Algorithm 9.3. MCMC algorithm for estimating θ from a DNA sample.

1. Propose changes to coalescent times in each gene tree one by one, under the constraint that each node must be older than its descendent nodes and younger than its mother node.
2. Propose changes to the genealogy, by using tree-rearrangement algorithms (such as subtree pruning and regrafting (SPR); see §8.3.6–7). This step may change coalescent times $\boldsymbol{t}$ as well.
3. Propose a change to parameter θ.
4. Propose a change to all coalescent times, using a multiplier (proportional scaling). □

The single parameter θ is easy to estimate and often simple estimates such as $\hat{\theta}_S$ and $\hat{\theta}_\pi$ (equation (9.18) and below) are reasonably good. Nevertheless, Felsenstein (1992) has demonstrated that gene genealogy-based estimates of θ can be far more efficient than simple estimates such as $\hat{\theta}_S$.

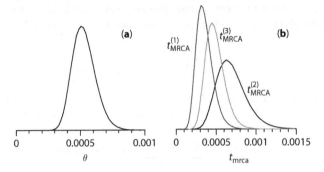

Fig. 9.4 (a) Posterior distribution of $\theta = 4N\mu$ for modern humans estimated from three neutral loci. (b) Posterior distributions of t_{MRCA} for the three loci, measured by the expected number of mutations per site.

Example 9.1. Estimation of θ for the human population using three neutral loci (Rannala and Yang 2003). The samples are from both African and non-African populations. The three loci are from 1q24 (about 10 kb, 61 sequences) (Yu et al. 2001), 16q24.3 (~6.6 kb, 54 sequences) (Makova et al. 2001), and 22q11.2 (10 kb, 63 sequences) (Zhao et al. 2000). Since all three loci are noncoding, we assume the same mutation rate to estimate a common θ for all loci. The prior used is $\theta \sim G(2, 2000)$. Here the shape parameter ($\alpha = 2$) means a fairly diffuse prior, while the prior mean ($2/2000 = 0.001$, which means one mutation per kilobase between two random sequences) appears reasonable for the human population. We also estimate t_{MRCA} for each locus, but note that t_{MRCA} is not a parameter and does not need a prior. The likelihood is calculated under the JC69 model. The BPP/MCMCCOAL program (Rannala and Yang 2003) is used to run the MCMC. The posterior distribution of θ is shown in Figure 9.4a. The posterior mean and 95% credibility interval (CI) are 0.00053 (0.00037, 0.00072). If $\mu = 10^{-9}$ mutations/site/year and the generation time is $g = 20$ years, the effective population size is estimated to be $N = 0.00053/(4g\mu) \approx 6600$, with the 95% CI (4600, 9000). Those values are much smaller than the current census population size. The estimated posterior densities of t_{MRCA} for the three loci are shown in Figure 9.4b. The posterior means and CIs for the three loci are 0.00036 (0.00020, 0.00060); 0.00069 (0.00040, 0.00110); and 0.00049 (0.00030, 0.00076), corresponding to the posterior mean root ages for the three gene trees to be 0.36Ma, 0.69Ma, and 0.49Ma. □

9.3 Population demographic process

In this section we discuss the use of multi-locus genetic sequence data to estimate parameters in a demographic model of population size change. When we trace the genealogy of a DNA sample backwards in time, every pair of lineages coalesces at the rate $\frac{1}{2N}$. If the population size N has been changing in the past, lineages will coalesce faster when N is small and more slowly when N is large. Thus the shape of the genealogical tree and the distribution of node ages on the tree will reflect the demographic history of the population. For example, a sample taken from a population that has been expanding rapidly will tend to have a star-shaped genealogy, with coalescent events pushed towards the root.

Multi-locus genetic sequence data, by providing information about the underlying genealogical trees, can be used to infer past demographic history of the species. For example, it is generally believed that the human population went through bottlenecks before expanding after the advent of agriculture. The number and timings of the bottlenecks, as well as the timing and rate of the expansion, are of great interest for our understanding of the origin and migration patterns of our species.

Slatkin and Hudson (1991) considered the simple model of exponential population growth, while the general theory is described by Griffiths and Tavaré (1994). The coalescent with a deterministically variable population size is a straightforward application of the theory of variable-rate Poisson processes. Thus we describe this theory first. We will discuss a simple population demographic model, that of exponential growth, as well as two nonparametric models. The latter are useful for analysing real data without making strong assumptions about the past demography.

9.3.1 *Homogeneous and nonhomogeneous Poisson processes*

Suppose that a particular event (for example, accidents, mutations, or, in the present case, coalescent events) occurs at random at a constant rate λ. The number of events over time $(0, t)$ constitutes a Poisson process if:

i. The numbers of events occurring in disjoint time intervals are independent;
ii. The probability that an event occurs in a small time interval Δt is $\approx \lambda \Delta t$, the probability that no event occurs in the interval is $\approx 1 - \lambda \Delta t$, and the probability that two or more events occur in the interval is negligibly small.

Here λ is called the *rate or intensity parameter* of the Poisson process. Two important results about the Poisson process are (i) that the number of events over the time interval $(0, T)$ has the Poisson distribution with mean λT:

$$p_k = \frac{e^{-\lambda T}(\lambda T)^k}{k!}, \quad (9.27)$$

and (ii) that the waiting times between successive events are independent exponential variables with mean $1/\lambda$.

If the rate λ varies deterministically over time, the process is called a *nonhomogeneous, time-dependent,* or *variable-rate Poisson process*, and $\lambda(t)$ is called the *rate or intensity function*. In a similar manner to the homogeneous Poisson process, events in disjoint subintervals are independent, and in a small time interval $(t, t + \Delta t)$, we expect either one event, with probability $\lambda(t)\Delta t$, or no event, with probability $1 - \lambda(t)\Delta t$. Then the number of events over the time interval $(0, T)$ has the Poisson distribution:

$$p_k = \frac{(\bar{\lambda}T)^k}{k!} e^{-\bar{\lambda}T}, \quad (9.28)$$

where

$$\bar{\lambda} = \frac{1}{T}\int_0^T \lambda(t)\,dt \quad (9.29)$$

is the average rate over $(0, T)$.

Subsection §12.5.2 in Chapter 12 discusses the simulation of the variable-rate Poisson process. Here we consider the probability density for k events that occur at time points $0 \leq y_1, y_2, \ldots, y_k \leq T$. We break the interval $(0, T)$ into a large number m of subintervals, each of length $\Delta t = T/m$. Let the subintervals be $[t_j, t_j + \Delta t), j = 1, \ldots, m$. Then k of them will each contain an event while the other $m - k$ are empty and do not contain any event. Note that the probability that an interval $[t_j, t_j + \Delta t]$ contains an event (at time point y_i, say) is $\lambda(t_j) \Delta t \approx \lambda(y_i) \Delta t$, and the probability that it does not contain any event is $1 - \lambda(t_j) \Delta t$. Multiplying the probabilities over all the disjoint subintervals gives the probability that k events occur at time points $y_1, y_2, \ldots, y_k$ as

$$\prod_{i=1}^{k} [\lambda(y_i)\Delta t] \times \prod_{j}^{*} [(1 - \lambda(t_j)\Delta t)]. \quad (9.30)$$

The notation here is intuitive. The first product is over the k nonempty subintervals while the second product is over the $m - k$ empty subintervals. As $\Delta t \to 0$, the second product approaches $1 - \sum_{j=1}^{m} \lambda(t_j)\Delta t \approx \exp\left\{-\int_0^T \lambda(s)\,ds\right\}$. Thus we obtain the probability density function (PDF) for k events at time points $y_1, y_2, \ldots, y_k$ as

$$\left(\prod_{i=1}^{k} \lambda(y_i)\right) \times \exp\left\{-\int_0^T \lambda(s)\,ds\right\} \quad (9.31)$$

(see, e.g. Cox and Hinkley 1974, p. 15). Recall that a PDF $f(x, y)$ has the interpretation that $\Pr\{x < X < x + \Delta x, y < Y < y + \Delta y\} = f(x, y)\Delta x \Delta y$. Note that the density of Equation (9.31) is given as the rates of the k events that have occurred, multiplied by the probability of no events over the whole time interval $(0, T)$. If $\lambda(t) = \lambda$, the density becomes $\lambda^k e^{-\lambda T}$.

The coalescent process, even when population size N is constant, is a variable-rate Poisson process, with the rate per generation being $\binom{j}{2} \times \frac{1}{2N}$ when there are j lineages in the sample. The rate is constant over each coalescent interval in which the number of lineages is constant.

9.3.2 Deterministic population size change

We now use the theory of the variable-rate Poisson process to derive the joint probability distribution of the genealogical tree and coalescent times under a demographic model of population size change. Let $N(t)$ be the population size time t ago, with $N(0) = N_0$ to be the present size. Here we take the coalescent worldview, running time backwards. Similarly let $\theta(t) = 4N(t)\mu$ be the population size parameter time t ago, with $\theta(0) = \theta_0$ for the present time. The mutation rate μ is constant over time and may be assumed as known, so that the temporal fluctuation in N is reflected in the fluctuation in θ. The coalescent process can then be described by a variable-rate Poisson process, with rate function

$$\lambda(t) = \binom{j}{2} \times \frac{2}{\theta(t)}, \quad (9.32)$$

if there are j lineages in the sample at time t. Here one unit of time is the expected time to accumulate one mutation per site. We note that in §9.3.1, the rate function $\lambda(t)$ is a deterministic function of t, while here $\lambda(t)$ depends on j as well and is a deterministic function of t only within each coalescent time interval when j is constant. This difference is nevertheless of no consequence.

Let $t_n, t_{n-1}, \ldots, t_2$ be the time points at which coalescent events occur, with $t_2 = t_{\text{MRCA}}$ to be the root of the gene tree. In other words, t_j is the time point at which the number of lineages ancestral to the sample changes from j to $j - 1$. We define $t_{n+1} = 0$ to be the present time. As in the case of a constant population size, the gene genealogies (labelled histories) have equal probabilities. The joint density of the gene tree and coalescent times is then given by a direct application of equation (9.31):

$$f(G, t_n, t_{n-1}, \ldots, t_2) = \left(\prod_{j=2}^{n} \lambda(t_j)\right) \times \exp\left\{-\int_0^{t_2} \lambda(s)\,ds\right\} = \prod_{j=2}^{n} \left[\frac{2}{\theta(t_j)} \times \exp\left\{-\int_{t_{j+1}}^{t_j} \frac{j(j-1)}{\theta(s)}\,ds\right\}\right]. \tag{9.33}$$

Note that this becomes equation (9.24) if $\theta(t) = \theta$ is constant over time.

The model is straightforward to implement if the integral is analytically tractable. This is the case with the exponential growth model. Suppose the exponential growth rate is r, so that the population size parameter time t ago is $\theta(t) = \theta_0 e^{-rt}$. Equation (9.33) then becomes

$$f(G, t_n, t_{n-1}, \ldots, t_2) = \prod_{j=2}^{n} \left[\frac{2}{\theta_0 e^{-rt_j}} \times \exp\left\{-\frac{j(j-1)}{\theta_0} \times \frac{1}{r}\left(e^{rt_j} - e^{rt_{j+1}}\right)\right\}\right]. \tag{9.34}$$

This can be used in equations (9.25) and (9.26). MCMC implementation of the model would allow the estimation of the exponential growth rate r and the initial population size θ_0. The exponential growth model was implemented by Beaumont (1999) and Drummond et al. (2002).

9.3.3 Nonparametric population demographic models

Used for describing population size changes, simple mathematical functions such as the exponential have an advantage in that they involve only a few parameters. However, they make strong assumptions about the population demographic process that may be highly unrealistic. Testing many candidate functional forms will be computationally expensive and impractical. As alternatives, nonparametric models of population size change are attractive. Here we describe two such models: a piecewise constant model and a piecewise linear model.

The first model is piecewise constant and is also known as a change-point model. N changes at several time points but otherwise stays constant, so that the function $N(t)$ or $\theta(t)$ is piecewise constant (Figure 9.5a) (Drummond et al. 2005). The positions of the change points and the population size during each time segment are shared across all loci. These are the parameters to be estimated from the data. In the example of Figure 9.5a, there are $K = 2$ change points (s_1 and s_2) and $K + 1 = 3$ population sizes (N_0, N_1, N_2, or equivalently $\theta_0, \theta_1, \theta_2$), with $2K + 1$ parameters to be estimated. The number of change points K may be fixed or assigned a prior and estimated from the data, although in the latter case reversible-jump MCMC (rjMCMC) may be necessary to move between models of different dimensions (Opgen Rhein et al. 2005). The precise value of K may be arbitrary and hard to estimate (for example, it may be sensitive to the prior). It should be adequate to use a relatively large fixed K (5 or 10) to avoid over-smoothing.

The distribution of the gene tree and coalescent times (branch lengths) at any locus, $f(G_i, t_i)$, is straightforward following the general framework described above. Suppose there are n_i sequences in the sample for locus i. The $n_i - 1$ coalescent events on the gene tree and the K change points in the demographic model break the interval from $(0, t_{MRCA})$ into at most $K + n_i - 1$ subintervals, within which the coalescent rate is constant. For the example of Figure 9.5a, $n_i = 6$ and $K + n_i - 1 = 7$. Note that there is no need to trace the genealogy beyond the MRCA, and if this occurs before time point s_K, there will be fewer subintervals to consider.

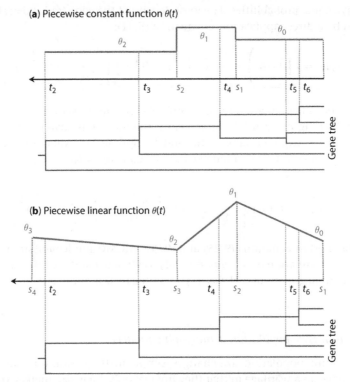

Fig. 9.5 Two nonparametric models of population size change: (a) the piecewise constant model or change-point model and (b) the piecewise linear model. In the analysis of the sequence data, the mutation rate is assumed to be constant, so the model defines the change of the population size parameter $\theta = 4N\mu$.

A second nonparametric demographic model is the piecewise *linear* model, in which the rate function $\theta(t)$ is continuous and piecewise linear (Opgen-Rhein et al. 2005; Heled and Drummond 2008) (Figure 9.5b). Let $0 = s_1 < s_2 < \ldots < s_K = T$ be the change points, at which $\theta(t)$ changes from one linear function to another. Here T should be larger than the root age t_{MRCA} at every locus. Parameters in this model include the locations of the K change points (s_1, s_2, s_3, and s_4 in Figure 9.5b), and the population sizes at those points, with $2K$ parameters. To calculate the probability density of the gene tree and coalescent times (branch lengths) for any locus, we use the K change points as well as the $n_i - 1$ coalescent times to break the interval $(0, t_{MRCA})$ into at most $K + n_i - 3$ subintervals, within which $N(t)$ is linear. For the example of Figure 9.5b, $K = 4$ and $n_i = 6$ so that $K + n_i - 3 = 7$. Within each subinterval, $\theta(t)$ is a linear function, $\theta(t) = at + b$, and the integral over $1/\theta(t)$ of equation (9.33) can be calculated analytically:

$$\int_{t_0}^{t_1} \frac{1}{\theta(s)} ds = \int_{t_0}^{t_1} \frac{1}{as + b} ds = \frac{1}{a} \log \frac{at_1 + b}{at_0 + b}. \tag{9.35}$$

Then the probability density of the gene tree and coalescent times for any locus, $f(G_i, t_i | \theta)$ of equation (9.33), is calculated in a straightforward manner.

Note that, given the sampled values of model parameters (the time points and θ values of Figure 9.5), one can calculate the value of θ at any time point in the interval $(0, T)$. Thus at the end of the MCMC run, the MCMC sample can be processed to obtain a point estimate (posterior mean or median) and associated CI for any time point in the interval $(0, T)$ (Drummond et al. 2005). By doing this at many points, perhaps uniformly spaced, one can generate a credibility band for the past demographic history of the population.

We now note the differences in the implementations in the literature. Drummond et al. (2005) implemented the piecewise constant model, in the so-called Bayesian skyline plot in the program BEAST. This is a Bayesian adaptation of early heuristic methods called a 'skyline plot' (Pybus et al. 2000) or a 'generalized skyline plot' (Strimmer and Pybus 2001), which make use of estimated gene trees with branch lengths, with errors in the estimates ignored. The errors in branch length estimates are expected to be large for population data due to the high sequence similarity and low information content. The Bayesian skyline plot of Drummond et al. (2005) calculates the likelihood of the sequence data and thus accommodates phylogenetic estimation errors. The method produces CIs for the estimated effective population size at any point in time, back to the MRCA of the gene sequences. It works for one locus, and has been extended to multiple loci by Heled and Drummond (2008), who used the piecewise linear model. Another extension along this line of skyline methods is the Bayesian skyride algorithm of Minin et al. (2008), which uses a 'time-aware' Gaussian random field model to generate a prior for the population size over time. This is the same prior as was used for smoothing substitution rates in relaxed-clock models (Thorne et al. 1998, see §10.4.3), with the logarithm of θ assumed to drift over time according to a Brownian motion process.

The piecewise linear model implemented by Opgen-Rhein et al. (2005) works for one locus, and uses gene trees with branch lengths estimated by the program BEAST. Heled and Drummond's (2008) implementation works for multiple loci.

The number of change points (K) is fixed in the implementation of Drummond et al. (2005). Minin et al. (2008) considered this to be a major issue but there is little evidence that K is important (as long as it is not too small). Opgen-Rhein et al. (2005) use rjMCMC (see §7.4.2) to allow the number of change points to be estimated from the data.

One concern about the Bayesian skyline implementations in the program BEAST (Drummond et al. 2005; Heled and Drummond 2008; Minin et al. 2008) is that population size changes are assumed to always coincide with coalescent events on the gene trees. This assumption is biologically unreasonable and mathematically unnecessary. The impact of the assumption is not well understood. Most of the coalescent events occur near the tips, so that the implemented model will tend to use more change points close to time 0. This may be less of a problem for the Bayesian skyride method (Minin et al. 2008), which uses a geometric Brownian motion process to describe population size drift over time, so smaller drifts are expected over shorter time intervals. At any rate, simulations by those authors demonstrate that the nonparametric methods can reliably reconstruct demographic histories and are clearly more useful than arbitrary mathematical functions such as the exponential, which may not describe complex demographic processes well.

9.4 Multispecies coalescent, species trees and gene trees

9.4.1 *Multispecies coalescent*

The genealogical relationships for a sample of DNA sequences taken from several species are described by the *interspecific coalescent* (Takahata 1989) or *censored coalescent* (Rannala

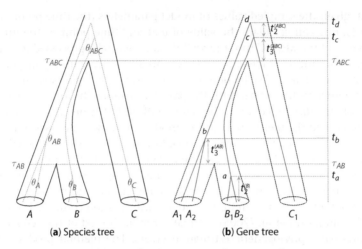

Fig. 9.6 (a) A species tree for three species illustrating the parameters in the multispecies coalescent model: two species divergence times τ_{AB} and τ_{ABC}, and five population size parameters θ_A, θ_B, θ_C, θ_{AB}, and θ_{ABC}. (b) A gene tree for a locus with 2, 2, and 1 sequences from species A, B, and C, respectively. Both the speciation times (the τs) and the coalescent times (the t's) are measured by the expected number of mutations per site. The multispecies coalescent model specifies the probability distribution for the gene tree (including the tree topology and coalescent times) (equation (9.40)).

and Yang 2003), now widely known as the *multispecies coalescent* (e.g. Liu et al. 2009b). In this subsection we discuss the probability density of the gene trees under the multispecies coalescent model and its use for parameter estimation using multi-locus sequence data. The species phylogeny is assumed to be known. Each branch on the species tree represents one species, either extant or extinct. For example, in the species tree of Figure 9.6a, A, B, and C are the extant species, while AB and ABC are the extinct ancestral species. We use the terms species and population interchangeably, and use AB to refer to the population that is ancestral to A and B, and so on. Gene trees are embedded inside the species tree, generated by the stochastic coalescent process running backwards in time within each branch (Maddison 1997). We assume complete isolation after species divergence, with no migration, hybridization, or introgression. As a result, sequences from two different species cannot coalesce until they are in a population that is a common ancestor of the two species. For example, in Figure 9.6b, genes A_1 and C_1 can coalesce in population ABC, but not in any other population. Parameters in the model include the species divergence times (τs) and the population sizes (θs) for the species on the species tree. The former are reflected in the length of the branch, while the latter may be represented by the width of the branch. Both the τs and θs are measured by the expected number of mutations per site. For the species tree of Figure 9.6, the parameters are $\Theta = \{\theta_A, \theta_B, \theta_C, \theta_{AB}, \theta_{ABC}, \tau_{AB}, \tau_{ABC}\}$. The population size parameter for a modern species (θ_A, θ_B, or θ_C) can be estimated only if two or more sequences are sampled at some loci from that species.

Suppose n_i sequences are sampled from a modern species (a tip of the species tree) at a neutral locus i. We assume no recombination, so that all sites within the locus share the same gene tree (topology and coalescent times). We are interested in $f(G_i, \boldsymbol{t}_i|\Theta)$, the probability of the gene tree topology G_i and coalescent times $\boldsymbol{t}_i$, given the species tree and parameters Θ. This is specified by the multispecies coalescent model. Different aspects of this density have been studied by numerous authors (Hudson 1983a; Tajima 1983; Pamilo and Nei 1988; Takahata and Nei 1985; Takahata 1989; Degnan and Salter 2005), and a general form is given by Rannala and Yang (2003).

For the case of three species and one sequence for each species, Yang (2002b; see also Takahata et al. 1995) derived the (marginal) probability of the gene tree topology (rooted tree), $f(G_i|\Theta)$, and the conditional density of the coalescent times given the gene tree topology, $f(\boldsymbol{t}_i|G_i, \Theta)$. Then

$$f(G_i, \boldsymbol{t}_i|\Theta) = f(G_i|\Theta) f(\boldsymbol{t}_i|G_i, \Theta). \tag{9.36}$$

For an arbitrary species tree, Degnan and Salter (2005) developed an algorithm to calculate $f(G_i|\Theta)$, with one sequence sampled from each species. The algorithm is fairly complex and its extension to arbitrary gene trees with arbitrary numbers of sequences from the species is not obvious. Thus the strategy of equation (9.36) is not feasible in general. It is far simpler to derive the joint distribution $f(G_i, \boldsymbol{t}_i|\Theta)$ directly by considering the independent coalescent processes running on different branches of the species tree (Rannala and Yang 2003).

In each population, we trace the genealogy backwards in time, until the end of the population (or, in the real world, until its birth). Let the duration of the population (branch length on the species tree) be τ, and its effective size be N, with $\theta = 4N\mu$. Let the number of lineages entering the population be m and the number of lineages leaving it be $n \leq m$. For example, population AB in Figure 9.6 begins at time τ_{AB} and ends at time τ_{ABC}, with the duration (branch length) $\tau = \tau_{ABC} - \tau_{AB}$ and population size $\theta = \theta_{AB}$. On the gene tree of Figure 9.6b, $m = 3$ lineages enter population AB and $n = 2$ lineages leave it, with one coalescent event reducing the number of lineages from 3 to 2. With one time unit defined as the time taken to accumulate one mutation per site, any two lineages coalesce at the rate $2/\theta$ in the population. The waiting time t_j until the next coalescent event, which reduces the number of lineages from j to $j-1$, has the exponential density

$$f(t_j) = \frac{j(j-1)}{2} \times \frac{2}{\theta} \exp\left\{-\frac{j(j-1)}{2} \times \frac{2}{\theta} t_j\right\}, \quad j = m, m-1, \ldots, n+1. \tag{9.37}$$

If $n > 1$, we have to consider the probability that no coalescent event occurs between the last one and the end of the population at time τ; i.e. during the time interval $\tau - (t_m + t_{m-1} + \cdots + t_{n+1})$. This probability is $\exp\left\{-\frac{n(n-1)}{\theta}[\tau - (t_m + t_{m-1} + \cdots + t_{n+1})]\right\}$ and is 1 if $n = 1$. (Recall that the probability of no events over time interval t for a Poisson process with rate λ is $e^{-\lambda t}$ and here the coalescent rate when there are n lineages is $n(n-1)/\theta$.)

Furthermore, the probability of a particular gene tree topology in the population is given by random joining of lineages; given the occurrence of a coalescent event, the probability that it joins a particular pair of lineages is $1 \big/ \binom{j}{2} = \frac{2}{j(j-1)}$, as there are $\binom{j}{2}$ possible pairs when there are j lineages in the population ancestral to the sample. Multiplying those probabilities together, we obtain the joint probability of the gene tree topology and coalescent times in the population as

$$\left[\prod_{j=n+1}^{m} \frac{2}{\theta} \exp\left\{-\frac{j(j-1)}{\theta} t_j\right\}\right] \times \left[\exp\left\{-\frac{n(n-1)}{\theta}(\tau - (t_m + t_{m-1} + \cdots + t_{n+1}))\right\}\right]. \tag{9.38}$$

As an example, population AB in Figure 9.6 has the duration $\tau_{ABC} - \tau_{AB}$ and population size θ_{AB}, and there is one coalescent event in that population on the gene tree of Figure 9.6b, with $m = 3$ and $n = 2$. The contribution from population AB to the multispecies coalescent density for the locus, given by equation (9.38), is thus

$$\left[\frac{2}{\theta_{AB}} e^{-6t_3^{(AB)}/\theta_{AB}}\right] \times \left[e^{-2\left(\tau_{ABC}-\tau_{AB}-t_3^{(AB)}\right)/\theta_{AB}}\right]. \tag{9.39}$$

In the above description, we have somewhat abused the terminology. The 'gene tree topology in the population' refers to the part of the gene tree (labelled history) at the locus that resides in that population. If $n > 1$, this will consist of n disconnected subtrees or branches. For example, the gene tree in population AB of Figure 9.6b consists of two disconnected parts (a subtree and a single branch). Furthermore, the order of the coalescent events matters, so the gene tree refers to a labelled history rather than a rooted tree (see §3.1.1.5 for the distinction).

The density for the whole gene tree at the locus, $f(G_i, t_i|\Theta)$, is simply the product of such probability densities across all populations. For the gene tree of Figure 9.6b, this is

$$\begin{aligned}
f(G_i, t_i|\Theta) &= e^{-2\tau_{AB}/\theta_A} &&\text{(population } A: 2 \to 2)\\
&\times \frac{2}{\theta_B} e^{-2t_2^{(B)}/\theta_B} &&\text{(population } B: 2 \to 1)\\
&\times \frac{2}{\theta_{AB}} e^{-6t_3^{(AB)}/\theta_{AB}} \times e^{-2\left(\tau_{ABC}-\tau_{AB}-t_3^{(AB)}\right)/\theta_{AB}} &&\text{(population } AB: 3 \to 2)\\
&\times \frac{2}{\theta_{ABC}} e^{-6t_3^{(ABC)}/\theta_{ABC}} \times \frac{2}{\theta_{ABC}} e^{-2t_2^{(ABC)}/\theta_{ABC}} &&\text{(population } ABC: 3 \to 1).
\end{aligned} \tag{9.40}$$

The multispecies coalescent is very similar to the single-species process discussed in §9.2. One difference is that it is now possible for two or more lineages to leave a population (if it is not the root node on the species tree) to enter the ancestral population ($n \geq 1$). The process is called a *censored* coalescent by Rannala and Yang (2003) because the coalescent process for one population may be terminated before all lineages that entered the population have coalesced.

The multispecies coalescent may be viewed as a variable-rate Poisson process, in which the Poisson rate is the total coalescent rate (summed over all contemporary populations) and changes both at a coalescent event (when reduces the number of lineages by one) and at a speciation event (which reduces the number of populations by one and resets the population sizes). From equation (9.31), the density $f(G_i, t_i|\Theta)$ is given by multiplying two terms: (i) the product of rates for the $n_i - 1$ coalescent events and (ii) the probability of no events over the whole time period until the root of the gene tree (except for the coalescent events that do occur). For the first term, recall the rate for each coalescent event is $2/\theta$, where θ is the size of the population in which the coalescent event occurs. For the second term, note that the rate for the Poisson process is constant over the time segments marked by nodes on both the gene tree and the species tree, e.g. $0 < t_a < \tau_{AB} < t_b < \tau_{ABC} < t_c < t_d$ for the gene tree of Figure 9.6b. For example, in the time segment from τ_{AB} to t_b, the total rate is $\frac{6}{\theta_{AB}}$, so that the probability of no events in this segment is $e^{-6(t_b-\tau_{AB})/\theta_{AB}} = e^{-6t_3^{(AB)}/\theta_{AB}}$. We can add up the time segments which have the same rate when calculating the probability of no events. The density is then

$$f(G_i, \mathbf{t}_i|\Theta) = \frac{2}{\theta_B} \cdot \frac{2}{\theta_{AB}} \cdot \frac{2}{\theta_{ABC}} \cdot \frac{2}{\theta_{ABC}}$$

$$\times \exp\left\{-\left[\frac{2\tau_{AB}}{\theta_A} + \frac{2t_2^{(B)}}{\theta_B} + \frac{6t_3^{(AB)}}{\theta_{AB}} + \frac{2\left(\tau_{ABC} - \tau_{AB} - t_3^{(AB)}\right)}{\theta_{AB}} + \frac{6t_3^{(BC)} + 2t_2^{(ABC)}}{\theta_{ABC}}\right]\right\}$$

$$= \frac{2}{\theta_B} \cdot \frac{2}{\theta_{AB}} \cdot \left(\frac{2}{\theta_{ABC}}\right)^2$$

$$\times \exp\left\{-\left[\frac{2\tau_{AB}}{\theta_A} + \frac{2t_a}{\theta_B} + \frac{6(t_b - \tau_{AB})}{\theta_{AB}} + \frac{2(\tau_{ABC} - t_b)}{\theta_{AB}} + \frac{6(t_c - \tau_{ABC}) + 2(t_d - t_c)}{\theta_{ABC}}\right]\right\}. \tag{9.41}$$

This is equal to equation (9.40).

We now consider estimation of parameters Θ using sequence data sampled from multiple species and multiple loci. We assume free recombination between loci, so that the gene trees (and sequence alignments) are independent across loci given the species tree and parameters Θ. The log likelihood function for the multi-locus sequence data $X = \{X_i\}$ is then:

$$\ell(\Theta) = \sum_{i=1}^{L} \log f(X_i|\Theta) = \sum_{i=1}^{L} \log \left\{ \sum_{G_i} \int_{\mathbf{t}_i} f(G_i, \mathbf{t}_i|\Theta) f(X_i|G_i, \mathbf{t}_i) \, d\mathbf{t}_i \right\}. \tag{9.42}$$

Note that this has the same form as equation (9.25) for one species, except that now we have more parameters, with Θ replacing the single θ for one population, and that sequences at the locus may now be from the multiple species rather than one single species. Again $f(X_i|G_i, \mathbf{t}_i)$ is the probability of the sequence alignment given the gene tree and branch lengths (coalescent times) or Felsenstein's phylogenetic likelihood.

Conceptually, equation (9.42) is the log likelihood function for ML estimation of Θ using multi-locus sequence data. However, the integration over the coalescent times and the summation over all possible gene trees is not feasible computationally except for simple cases where there are only two or three sequences at each locus (Yang 2002b). A more practical approach is to assign priors on the parameters and estimate them using a Bayesian approach, using MCMC to achieve the integration and summation. This is implemented by Rannala and Yang (2003) in the BPP program (see also Yang and Rannala 2010). The joint posterior is

$$f(\Theta, G, \mathbf{t}|X) \propto f(\Theta) \prod_{i=1}^{L} f(G_i, \mathbf{t}_i|\Theta) f(X_i|G_i, \mathbf{t}_i), \tag{9.43}$$

where $G = \{G_i\}$ and $\mathbf{t} = \{\mathbf{t}_i\}$ are, respectively, gene tree topologies and coalescent times across all loci. The state of the MCMC will include the species divergence times (τs) and population sizes (θs) on the species tree, as well as the gene tree topologies and coalescent times (branch lengths) at all loci. For example, the algorithm implemented in BPP includes the following steps:

Algorithm 9.4. Algorithm 9.4. MCMC estimation of τ s and θs under the multispecies coalescent model using multi-locus sequence data.

1. Update the coalescent times (node ages) at interior nodes in the gene tree one by one, under the constraint that each node must be older than its descendent nodes and younger than its mother node, and that genes split earlier than species.
2. SPR to change the gene tree topology (see §8.3.7–8). This step changes coalescent times $\mathbf{t}$ as well.

3. Update population size parameters (θs) one by one.
4. Update species divergence times in the species tree (τs) one by one. As the node ages in the gene trees may severely constrain the species divergence time, the new species divergence time (τ) is proposed without considering the gene trees, and a 'rubber-band' algorithm is then used to modify the node ages in the gene trees to remove conflicts in the new proposed state (i.e. to ensure that genes split before species).
5. Use a multiplier to change all coalescent times (see §7.2.5). □

Example 9.2. We use BPP to estimate parameters in the multispecies coalescent model from the hominoid genomic sequences compiled by Burgess and Yang (2008). Sequence alignments from 14,663 autosomal loci and from 783 X-linked loci for four ape species: human (H), chimpanzee (C), gorilla (G), and orangutan (O), are analysed as two datasets. The average length of the nuclear loci is 508 bp. Alignment gaps are treated as missing data (ambiguity nucleotides). The JC69 mutation model is used and the clock is assumed to hold. While Burgess and Yang (2008) modelled sequence errors and slight violations of the molecular clock, those factors were found to have only minor effects on the analysis and are here ignored. The species tree is shown in Figure 9.7, illustrating the parameters in the model: the species divergence times τ_{HC}, τ_{HCG}, and τ_{HCGO} and the population sizes θ_{HC}, θ_{HCG}, and θ_{HCGO}. As only one sequence is used from each species at every locus, θ parameters for modern species cannot be estimated from the data. We assign the priors $\theta \sim G(2, 2000)$ with mean 0.001 (1 mutation per kb) for all three θ parameters. The divergence time for the root of the species tree is assigned the prior $\tau_{HCGO} \sim G(2, 120)$ with mean 0.0167. With a mutation rate of 10^{-9} changes/site/year, this implies a prior mean of the H-O divergence time of 16.7 Ma. The other divergence times (τ_{HCG} and τ_{HC}) are assigned a uniform Dirichlet prior given the root age τ_{HCGO} (see equation (8.5)) (Yang and Rannala 2010). These priors are slightly different from those in Burgess and Yang (2008), but they have very little influence.

The estimates are summarized in Table 9.1. The differences between the estimates from the HCGO and HCGOM (including macaque) data are due to species sampling and not to other differences such as prior specifications. Compared with estimates of θ_H for modern humans (~0.6 per kb), the ancestors had much larger population sizes. It may be noteworthy that population sizes for old ancestors are not necessarily harder to estimate than for recent ones. Here θ_{HC} is estimated with lower precision than θ_{HCG}. This is because population HC has a shorter branch (small $\tau_{HCG} - \tau_{HC}$) and thus there are not many coalescent events in that population, which provide information about θ_{HC}. Use of the mutation rate of 10^{-9} changes/site/year would translate the τ_{HC} and τ_{HCG} estimates into ~4.0–4.2 Ma and 6.6–6.7 Ma for the H-C and H-G divergences, respectively. These

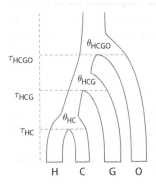

Fig. 9.7 The species phylogeny for four ape species: human (H), chimpanzee (C), gorilla (G), and orangutan (O), illustrating the parameters. Their estimates from the genomic sequence data are shown in Table 9.1.

Table 9.1 Posterior estimates (means and 95% CIs) of parameters in the multispecies coalescent model for the hominoid genomic data

Data	θ_{HC}	θ_{HCG}	θ_{HCGO}	τ_{HC}	τ_{HCG}	τ_{HCGO}
Nuclear loci						
HCGO	6.0 (5.6, 6.4)	3.5 (3.4, 3.6)	8.1 (7.8, 8.4)	4.1 (4.0, 4.2)	6.6 (6.6, 6.7)	13.8 (13.6, 13.9)
HCGOM[a]	6.1 (5.7, 6.6)	3.3 (3.2, 3.4)	4.9 (4.7, 5.2)	3.9 (3.8, 4.0)	6.3 (6.2, 6.4)	14.3 (14.2, 14.5)
X-linked loci						
HCGO	2.4 (1.7, 3.0)	2.4 (1.9, 2.9)	5.2 (4.1, 6.3)	3.3 (3.1, 3.6)	5.6 (5.4, 5.9)	11.6 (11.0, 12.1)
HCGOM[a]	2.6 (2.0, 3.3)	2.0 (1.6, 2.5)	3.2 (1.9, 4.4)	3.1 (2.8, 3.9)	5.4 (5.1, 5.6)	11.7 (11.1, 12.3)

Note: Parameters θs and τs are measured by the number of mutations per kilobase.
[a] Estimates from the five species data (HCGOM) including macaque (M) are from Burgess and Yang (2008, Table 7).

dates are recent, and are in conflict with certain interpretations of the fossil record, but are consistent with other molecular estimates (e.g. Scally et al. 2012). Recent estimates of the human mutation rate based on mutation counts over generations are much lower, at 0.5–0.6×10^{-9} mutations/site/year (Lynch 2010; Roach et al. 2010). Such rates would give much older species divergence time estimates. Despite much genetic data, it remains difficult to resolve the differences in mutation rate estimates or the conflicts between molecular and fossil date estimates. □

9.4.2 *Species tree–gene tree conflict*

Earlier in §3.1.2, we introduced the concepts of species trees and gene trees. A species tree or species phylogeny represents the relationships among a set of species, generated by the speciation process, while a gene tree represents the genealogical relationships among the gene sequences sampled from the species. It is generally accepted that the primary objective of molecular phylogenetics is to reconstruct the species tree rather than the gene tree (e.g. Nei 1977; Edwards 2009b). When one species diverges into two, the gene in the ancestral species will split into two genes in the descendent species, which will then evolve independently. As a result, the gene tree should usually match the species tree. However, this is not always the case. Several biological processes can cause the gene tree to differ from the species tree, including (i) polymorphism in the ancestral species and the resulting incomplete lineage sorting, (ii) gene duplications followed by gene losses and misidentification of orthologues, and (iii) introgression or horizontal gene transfer (Nichols 2001). In this section we focus on ancestral polymorphism and incomplete lineage sorting as the source of species tree–gene tree conflicts (see §3.1.2 and Figure 3.12b). Such conflicts are widely observed in empirical data analyses, for example, in hominoids (Chen and Li 2001), birds (Jennings and Edwards 2005), and rice (Cranston et al. 2009).

Polymorphism in ancestral species can cause the gene genealogy to differ from the species phylogeny even when there is no migration, hybridization, or introgression (Maddison and Knowles 2006). Tajima (1983) calculated the probabilities for different gene tree topologies relating four sequences sampled from two species (with two sequences from each species) and noted that when the two species diverged only recently, an incorrect gene tree topology (one that does not match the species tree) may have a higher probability than the correct one. Takahata and Nei (1985) studied samples of arbitrary sizes from two species. The case of multiple sequences sampled from three species was considered by Takahata (1989).

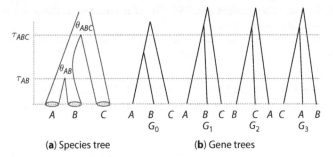

Fig. 9.8 (a) The species tree for three species and (b) the four possible gene trees for a locus, with one sequence sampled from each species. The probability that the gene tree topology differs from the species tree (i.e. the probability for gene trees G_2 and G_3) is equal to the probability that sequences A and B do not coalesce in the ancestral population AB times $\frac{2}{3}$.

The case of three species, with one sequence from each, is of particular interest due to its use in estimating population sizes in the common ancestors of the human, chimpanzee, and gorilla (Figure 9.8). The species tree-gene tree mismatch probability, i.e. the probability that the species tree (S) and the gene tree (G) differ from each other, is

$$P_{SG} = \frac{2}{3} e^{-\frac{(T_{ABC}-T_{AB})}{2N_{AB}}} = \frac{2}{3} e^{-\frac{2(\tau_{ABC}-\tau_{AB})}{\theta_{AB}}}, \qquad (9.44)$$

where the species divergence times T_{AB} and T_{ABC} are in generations while τ_{AB} and τ_{ABC} are in the number of mutations, and N_{AB} is the effective population size of species AB, with $\theta_{AB} = 4N_{AB}\mu$. Note that P_{SG} is a product of two terms: $e^{-(T_{ABC}-T_{AB})/2N_{AB}}$ is the probability that sequences A and B do not coalesce in population AB while $\frac{2}{3}$ is the probability that the random joining of sequences A, B, and C in population ABC results in gene trees G_2 or G_3 but not G_1 (Figure 9.8b). Equation (9.44) was derived by Hudson (1983a) and later by Nei (1987, p. 402; see also Pamilo and Nei 1988). For the human, chimpanzee, and gorilla, this probability has been estimated using different methods from genomic sequence data to be ~30% (Burgess and Yang 2008; Scally et al. 2012).

The mismatch probability P_{SG} has been used to estimate the population sizes in hominoid ancestors (Chen and Li 2001; Satta et al. 2004). P_{SG} of equation (9.44) is equated to the estimated proportion of loci at which the gene tree differs from the species tree, to obtain estimates of N_{AB}, with the species divergence times and mutation rate assumed known. This is known as the 'mismatch' or 'trichotomy' method. A serious drawback of the method is that it ignores errors in gene tree reconstruction, which lead to overestimation of N_{AB}; Yang (2002b) shows that phylogeny reconstruction errors always lead to inflated species tree–gene tree conflict, with $P_{SE} > P_{SG}$, where P_{SE} is the probability that the species tree (S) and the estimated gene tree (E) differ (Figure 9.9).

More complex cases involving four or five species were studied by Pamilo and Nei (1988). The recent work by Degnan and Salter (2005) and Degnan and Rosenberg (2006) has highlighted the implications of species tree–gene tree conflicts for modern phylogenetic analysis using genome-scale datasets. When multiple gene loci are available for phylogenetic analysis to estimate the species tree, two commonly used approaches are (i) 'majority vote' of the gene trees inferred at the individual loci and (ii) concatenation (also called 'supermatrix' method). If short internal branches in the species tree cause an

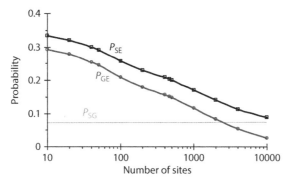

Fig. 9.9 Phylogeny reconstruction errors tend to inflate the species tree–gene tree conflict. Plotted are the mismatch probabilities between any two of the species tree (S), the (true) gene tree (G), and the estimated gene tree (E). P_{SG} is the probability that the species tree and the (true) gene tree are different, while P_{SE} and P_{GE} are defined similarly. P_{GE} is the phylogenetic error, and approaches zero when the sequence length approaches infinity. The results are calculated using computer simulation, using parameter estimates from a small dataset for the human, chimpanzee, and gorilla: $\theta_{HC} = 0.003057$, $\theta_{HCG} = 0.00099$, $\tau_{HC} = 0.005194$, and $\tau_{HCG} = 0.006283$ (see Figure 9.8). For those parameter values, the probability that the species tree and gene tree differ is $P_{SG} = 0.0739$, while the probability (P_{SE}) is much higher if estimated gene trees are used. Note that estimates of the mismatch probability P_{SG} using the HCG genomic datasets are much higher, at ≈ 0.3 (Burgess and Yang 2008; Scally et al. 2012). Redrawn following Yang (2002b, Figure 3).

incorrect gene tree topology (one that does not match the species tree) to have a higher probability than the correct tree topology, then the method of choosing the most commonly supported gene tree as the estimate of the species tree will be inconsistent, and will converge to an incorrect species tree when the number of loci increases. Species trees (with parameters τs and θs) that can generate such *anomalous gene trees* are said to be in the *anomaly zone*. The method of concatenation may also be inconsistent in the anomaly zone (Kubatko and Degnan 2007).

We will now take a tour of the anomaly zone. Note that a species tree is represented by the rooted tree topology with the rank order of node ages ignored. Thus $((A, B), (C, D))$ is considered one species tree whether or not the common ancestor of A and B is younger than the common ancestor of C and D. The asymmetrical species tree for four species of Figure 9.10a is in the anomaly zone. To see that, consider a locus with one sequence sampled from each species. Because the three speciation events in the species tree are very close to each other, there is little time for coalescent events to occur in populations AB or ABC, so that all three coalescent events in the gene tree are most likely to occur in the common ancestor of all four species, $ABCD$. The resulting gene tree will be generated in effect by the random coalescent process of joining lineages in that population, so

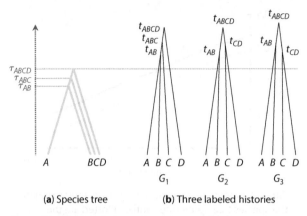

Fig. 9.10 (a) An unbalanced species tree for four species with the three speciation events very close in time. This species tree generates so-called anomalous gene trees and is said to be in the anomaly zone. (b) Three labelled histories (rooted gene trees with interior nodes ordered by age) for a locus with four sequences, one from each of the four species. Because τ_{AB}, τ_{ABC}, and τ_{ABCD} on the species tree are nearly equal, all three coalescent events on the gene tree very likely occur in the common ancestor $ABCD$. Then the 18 possible labelled histories have nearly equal probability. Labelled history G_1, with probability ~1/18, has the same tree topology as the species tree. Labelled histories G_2 and G_3, with a total probability ~2/18, have the same rooted tree topology $((A, B), (C, D))$, which differs from the species tree.

that the 18 possible labelled histories will have nearly equal probabilities (1/18 each). A labelled history (Edwards 1970), also called a ranked gene tree by Degnan et al. (2012), distinguishes the rank order of nodes by age, while the rooted tree topology does not (§3.1.1.5). For example, G_2 (with $t_{AB} < t_{CD}$) and G_3 (with $t_{AB} > t_{CD}$) in Figure 9.10b are different labelled histories but they share the same rooted tree topology: $((A, B), (C, D))$. The coalescent model places equal probabilities on the 18 labelled histories but not on the 15 rooted gene tree topologies. As a result, the 12 asymmetrical rooted gene tree topologies each receive probability 1/18 while the three symmetrical rooted gene tree topologies each receive probability 2/18. If one uses the most common rooted gene tree as an estimate of the species tree, one will obtain the incorrect species tree $((A, B), (C, D))$, instead of the correct species tree $(((A, B), C), D)$ (Figure 9.10). For four species, anomalous gene trees can occur on the asymmetrical species tree only. For five or more species, they may occur for any species tree; in other words, there exist branch lengths (τs) that create anomalous gene trees on any species tree of five or more species (Degnan and Rosenberg 2006).

Note that the discussion above assumes that the true gene trees are known without error. Estimated gene trees may involve substantial phylogenetic errors and uncertainties, due to high sequence similarity and low information content in sequences from closely related species. Indeed, simulations by Huang et al. (2009) suggest that in the anomaly zone, phylogenetic analysis of typical datasets tends to return unresolved gene trees rather than resolved incorrect trees, and the difficulty posed by anomalous gene trees to practical phylogenetic analysis is rather limited. Note that the theory of anomalous gene trees concerns the asymptotic property of the simple estimation methods when the data size is large. If typical datasets are not informative and the estimates are dominated by sampling errors, the large sample theory will not be very relevant.

Anomalous gene trees exist because the coalescent model assigns equal probabilities to labelled histories whereas we focus on rooted gene tree topologies as estimates of the species tree. In this setting the species tree (the species tree topology as well as parameters τs and θs) represent the parameter in the model while the gene trees (the labelled histories and coalescent times) are random variables, with distributions specified by the species tree. The appropriate statistical procedure is then to estimate the parameters using the likelihood function rather than 'estimating' the random variables and then attempting to summarize such estimates to generate parameter estimates. A likelihood-based approach (ML or Bayesian) to species tree estimation will average over the distribution of the gene trees under the multispecies coalescent model in the likelihood calculation, and will be consistent whatever the shape of the species tree. Furthermore such an approach will accommodate uncertainties in the gene trees.

While the importance of ancestral polymorphism and incomplete lineage sorting to inference of shallow phylogenies (for closely related species) is well recognized (e.g. Takahata et al. 1995; Edwards and Beerli 2000), its relevance to deep phylogenies (for distantly related species) is less obvious. Indeed many of us have the misconception that species tree–gene tree conflicts due to ancestral polymorphism are not relevant to deep phylogenies. Edwards et al. (2005; see also Degnan and Rosenberg 2009; Oliver 2013) have pointed out that this is not the case; the problem lies with short internal branches in the species tree, rather than how deep in the tree the short branches are. For example, in the case of three species (Figure 9.8), the species tree-gene tree mismatch probability (equation (9.44)) depends on the internal branch length for species AB ($\tau_{ABC} - \tau_{AB}$) but not the external branch lengths (τ_{AB} or τ_{ABC}). If the two speciation events are separated by only N_{AB} generations, 40% ($= \frac{2}{3}e^{-1/2}$) of the genome will have gene trees different from the species tree. This is the case even if the three species diverged more than 300 million years ago. Thus incomplete lineage sorting may leave permanent marks in our genomes (in the form of conflicting gene trees), rather than eventually 'sorting things out' as lineages continue to diverge. Such ancient radiations and the resulting incomplete lineage sorting may cause major problems for phylogenetic inference. Sampling multiple individuals in the same species will not help, as the multiple sequences from the same species will have long coalesced before reaching the deep short internal branch. Sampling multiple loci may be the only approach to resolving the species tree despite gene tree conflict.

9.4.3 Estimation of species trees

9.4.3.1 Methods using gene tree topologies

A number of methods have been developed to estimate the species tree despite gene tree conflict, using multi-locus genetic data. They differ in the information used and in the extent by which the uncertainties involved are accounted for. The simplest class of methods includes the tree-mismatch method mentioned earlier, as well as gene tree parsimony (e.g. Page and Charleston 1997) and the minimizing-deep-coalescent (MDC) method (Maddison 1997). The maximum pseudo-likelihood (MPEST) method of Liu et al. (2010b), as well as species tree estimation using average ranks of coalescences (STAR) or using average coalescent times (STEAC) (Liu et al. 2009a), all fall into this category. These methods use the gene tree topologies only, and ignore information in the branch lengths and uncertainties in the estimated gene trees. Failure to use the branch length information means that not all parameters in the multispecies coalescent model are identifiable. For example, the species tree of three species involves at least four parameters (θ_{AB}, θ_{ABC}, τ_{AB}, τ_{ABC}) (Figure 9.8a), but only the ratio ($\tau_{ABC} - \tau_{AB}$)/θ_{AB} is estimable by the tree-mismatch method using one sequence from each species (equation (9.44)). In contrast, methods that use the branch lengths as well as gene tree topologies can identify all parameters in the model. Furthermore, for closely related species, the sequences may contain little phylogenetic information and the gene tree topologies may be unresolved or highly uncertain (e.g. Chen and Li 2001). Overall, methods based on gene tree topologies alone are expected to be far less efficient than methods using sequence alignments, which incorporate the branch length information. The latter are discussed in §9.4.3.4.

9.4.3.2 Methods using gene trees with branch lengths

Liu et al. (2010a) have developed an ML method for estimating the species tree using gene trees with branch lengths (rooted trees with node ages) as given data. The likelihood function is the probability of observing the gene trees with branch lengths given the species tree and parameters, and is simply the multispecies coalescent density of Rannala and Yang (2003). Liu et al. assumed that all species have the same population size (the same θ). Under this assumption, the ML estimate of the species tree can be obtained analytically. Indeed the ML species tree is the one that achieves the largest species divergence times (τs), under the constraints of the gene trees from all loci. The rest of this subsection demonstrates this result, and describes the Maximum Tree algorithm of Liu et al. (2010a), which produces the ML species tree.

The simple case of three species, with one sequence from each species, is dealt with in Problem 9.3. Here we deal with the general case. Initially we consider the maximization of the likelihood function on a fixed species tree S and then compare the maximized likelihood values across different species tree. Let $G = \{G_i\}$ be the gene tree topologies and $t = \{t_i\}$ the node ages at multiple loci, and let n_i be the number of sequences in the alignment at locus i. The parameters (Θ) in the species tree (S) includes the species divergence times (τs) and the common θ. Using the formulation of the variable-rate Poisson process, the likelihood can be written as

$$f(G, t|S, \Theta) = \left(\frac{2}{\theta}\right)^C \times e^{-\frac{2}{\theta}T}, \tag{9.45}$$

where $C = \sum_i (n_i - 1)$ is the total number of coalescent events on the gene trees, $e^{-\frac{2}{\theta}T}$ is the probability of no coalescent events in all populations and at all loci, and T may be called the 'total per-lineage-pair coalescent time', summed over all populations and all gene trees. To calculate T, a time interval of length t during which there are k lineages in a population contributes $\binom{k}{2} \times t$, with the convention $\binom{0}{2} = \binom{1}{2} = 0$.

For example, if we use the single gene tree of Figure 9.6b with node ages t_a, t_b, t_c, and t_d to estimate the parameters $\Theta = \{\tau_{AB}, \tau_{ABC}, \theta\}$ on the species tree of Figure 9.6a, assuming one θ, the likelihood is, from equation (9.41),

$$f(G_i, t_i|\tau_{AB}, \tau_{ABC}, \theta) = \left(\frac{2}{\theta}\right)^C \times e^{-\frac{2}{\theta}T}, \tag{9.46}$$

with $C = 4$ and $T = \tau_{AB} + t_a + 3(t_b - \tau_{AB}) + (\tau_{ABC} - t_b) + 3(t_c - \tau_{ABC}) + (t_d - t_c)$.

We now maximize the likelihood of equation (9.45) as a function of parameters on the species tree, i.e. the divergence times τs and the common θ. Viewed as a function of θ,

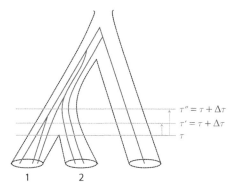

Fig. 9.11 A gene tree embedded in a species tree to illustrate the Maximum Tree algorithm. At a speciation time τ, the two daughter populations have $k_1 = 3$ and $k_2 = 2$ lineages, which merge into $k = k_1 + k_2 = 5$ lineages in the ancestor population. Increasing τ, either with or without passing a node on the gene tree (to τ'' or τ') reduces the total per-lineage-pair waiting time and improves the likelihood.

the likelihood of equation (9.45) is maximized by setting $\frac{\partial f}{\partial \theta} = 0$, which gives $\hat{\theta} = 2T/C$, with the likelihood $\left(\frac{C}{T}\right)^C e^{-C}$. Thus the MLE of θ is simply given by the fact $\frac{1}{2}\hat{\theta} = \frac{T}{C}$ is the average waiting time for a pair of lineages to coalesce.

As C is fixed, the likelihood is larger the smaller T is. Now we show that T becomes smaller whenever we increase any species divergence time τ (within the constraints of the gene trees). We will use the following result, which the reader may wish to confirm:

$$\binom{k_1+k_2}{2} > \binom{k_1}{2} + \binom{k_2}{2}, \text{ for } k_1, k_2 \geq 1. \quad (9.47)$$

When we increase a species' divergence time τ, we may or may not pass any node on the gene trees (Figure 9.11). Suppose at this speciation event, a gene tree has k_1 and k_2 lineages in the two daughter species, and these merge into $k = k_1 + k_2$ lineages in the ancestral species. If we increase τ by $\Delta\tau$ without passing any node in the gene tree, the per-lineage-pair coalescent waiting time in species 1 and 2 will increase, by $\binom{k_1}{2}\Delta\tau$ and $\binom{k_2}{2}\Delta\tau$ respectively, while that in the ancestral species will decrease by $\binom{k_1+k_2}{2}\Delta\tau$, with a net decrease in T of $\left[\binom{k_1+k_2}{2} - \binom{k_1}{2} - \binom{k_2}{2}\right]\Delta\tau > 0$. If we pass a node on the gene tree, T will decrease by an even greater amount.

In summary, given the species tree S, the likelihood is maximized by having the species divergence times (τs) as large as possible, given the constraints of the gene trees. Now the likelihood of equation (9.45) applies to all possible species trees. If one can alter the species tree to achieve a smaller T, the likelihood will increase as well. It thus follows that the ML estimate of the species tree is the one that achieves the smallest T, which implies the largest species divergence times (τs).

This reasoning immediately leads to the Maximum Tree algorithm of Liu et al., which produces the ML species tree given the gene trees and branch lengths. Note that for any pair of species A and B, the gene trees at multiple loci place a maximum bound for their species divergence time τ_{AB}: this is the smallest sequence divergence time between two sequences from the two species among all gene trees. As there are $\binom{s}{2}$ species pairs, there are $\binom{s}{2}$ such maximum bounds. For example, given the single gene tree of Figure 9.6b, we have $\tau_{AB} < t_d$, $\tau_{AC} < t_d$, and $\tau_{BC} < t_c$, irrespective of the species tree. The Maximum Tree algorithm works by first finding all the $\binom{s}{2}$ bounds (constraints), reducing them into $s-1$ equivalent bounds and then producing a rooted species tree that satisfies those bounds.

Algorithm 9.5. Maximum Tree algorithm to estimate species tree using gene tree constraints.

1. Find the minimum of the $\binom{s}{2}$ constraints. Let this be c_1 and the two species involved be s_1 and s_2. Join s_1 and s_2 at a new node, with age c_1.
2. Find the minimum of the $\binom{s}{2}$ constraints involving one of species s_1 or s_2 and another new species s_3. Let the minimum be c_2. Join the two species at a new node with age c_2.
3. Find the minimum of the $\binom{s}{2}$ constraints involving one of species $s_1, s_2,$ and s_3, and another new species s_4. Let the constraint be c_3. Join the two species at a new node with age c_3.
4. Repeat until all species are in the constraints and in the tree. □

The algorithm produces $s-1$ constraints, which are equivalent to the $\binom{s}{2}$ constraints: in other words, if the $s-1$ constraints are satisfied, all the $\binom{s}{2}$ constraints are satisfied as well. For example, it is easy to check that after step 1, the three constraints involving species $s_1, s_2,$ and s_3, are all satisfied. The algorithm also produces an ultrametric species tree with divergence times set at the $s-1$ maximum bounds. The tree is thus called the Maximum Tree. It is the same tree as inferred by the program GLASS (for Global LAteSt Split) (Mossel and Roch 2010). The Maximum Tree is the largest species tree under the constraints of the gene trees, while the GLASS tree is the minimum 'gene tree' that is consistent with all the given gene trees. Under the assumption of the same θ for all species, the two are exactly the same.

Note that in the case of one locus, with one sequence from each species, the Maximum Tree is exactly the gene tree.

The Maximum Tree algorithm is implemented in the program STEM (for Species Tree Estimation using Maximum likelihood) by Kubatko et al. (2009). This uses PAUP* to generate gene trees with branch lengths under the clock, and then use them as input data to estimate the species tree and branch lengths (τs) by ML.

A major criticism of the method (Maximum Tree or GLASS tree) is that it fails to accommodate phylogenetic uncertainties in the estimated gene trees. If individual loci are informative and gene trees are reconstructed with high confidence, the method is expected to have good performance. However, if the species are closely related, the gene trees may involve considerable uncertainties, and the performance of the method may suffer. For example, simulation by Leaché and Rannala (2011) suggests that the STEM program (Kubatko et al. 2009), which implements the ML method (Liu et al. 2010a), is more efficient than concatenation and simple methods based on summary statistics such as the MDC method (Maddison 1997), but is less efficient than a Bayesian method that fully incorporates the uncertainties from different sources (BEST) (Liu and Pearl 2007; Liu 2008). The Bayesian method is discussed in §9.4.3.4.

9.4.3.3 Singularities on the likelihood surface

The Maximum Tree algorithm discussed above assumes the same population size parameter for all populations. This assumption is of course not essential; one can use ML to estimate a different θ for each population. In this case, Liu et al. (2010b) have noted that singularities may exist on the multispecies coalescent likelihood of Rannala and Yang (2003); in other words, the likelihood $f(G, \boldsymbol{t}|\Theta)$ may become infinite for certain values of parameters Θ, even though those parameter values may be unreasonable and are not good estimates.

Consider the use of the gene tree of Figure 9.6b to estimate parameters in the species tree of Figure 9.6a. The gene tree topology and node ages (t_a, t_b, t_c, t_d) are fixed data, and the parameters are $\Theta = \{\theta_A, \theta_B, \theta_{AB}, \theta_{ABC}, \tau_{AB}, \tau_{ABC}\}$. The likelihood is given in equation (9.41). The term involving θ_{AB} is $\frac{2}{\theta_{AB}} e^{-\frac{2}{\theta_{AB}}[2t_b - 3\tau_{AB} + \tau_{ABC}]}$, with $\tau_{AB} < t_b < \tau_{ABC}$. Now let $\tau_{AB} \to t_b$, $\tau_{ABC} \to t_b$, and $\theta_{AB} = 2(t_b - 3\tau_{AB} + \tau_{ABC}) \to 0$. Then this term approaches ∞ while other terms in the likelihood are positive so that the likelihood $\to \infty$. In other words, when we collapse population AB onto a single coalescent event (at node b) by shrinking the branch length ($\tau_{ABC} - \tau_{AB}$) to zero and increasing the coalescent rate ($2/\theta_{AB}$) to infinity, the likelihood becomes infinite. Similar singularities can be constructed for arbitrary species trees and multiple gene trees as long as one can collapse an internal branch on the species tree onto one single coalescent event. If multiple coalescent events occur in the population, we collapse the population onto the most recent one. See Problem 9.4 for an example involving only three species.

The situation is very similar to the famous singularity in the likelihood function under a normal mixture model. Suppose we use a data sample $y = \{y_1, y_2, \ldots, y_n\}$ to fit a 1:1 mixture of two component normal distributions: $N(\mu_1, \sigma_1^2)$ and $N(\mu_2, \sigma_2^2)$, with μ_1, μ_2, σ_1^2, and σ_2^2 to be unknown parameters. The likelihood is

$$f(y|\mu_1, \sigma_1^2, \mu_2, \sigma_2^2) = \prod_{i=1}^{n} \left[\frac{1}{2} \cdot \frac{1}{\sqrt{2\pi\sigma_1^2}} \exp\left\{-\frac{1}{2\sigma_1^2}(y_i - \mu_1)^2\right\} + \frac{1}{2} \cdot \frac{1}{\sqrt{2\pi\sigma_2^2}} \exp\left\{-\frac{1}{2\sigma_2^2}(y_i - \mu_2)^2\right\} \right]. \tag{9.48}$$

Now fix $\mu_1 = y_1$ (or any of the n data points) and let $\sigma_1^2 \to 0$. In other words, let the component $N(\mu_1, \sigma_1^2)$ become $N(y_1, 0)$ and collapse onto the data point y_1. We see that the likelihood becomes infinite because the term $\frac{1}{\sigma_1} \to \infty$. The problem does not arise if one fits a single normal distribution $N(\mu, \sigma^2)$. If this collapses onto a data point, with $\mu = y_1$ and $\sigma^2 \to 0$, the probabilities for other data points will approach 0 much faster than $1/\sigma$ approaches ∞, causing the likelihood to approach 0 rather than ∞. However, in the mixture model, one component can have a finite variance and therefore assigns finite probabilities to all data points while the other components can collapse onto particular data points, creating singularity points.

In the case of species tree estimation, singularities arise because an ancestral population collapses onto a coalescent event, which is given a zero-length time interval but infinite rate to occur. The problem is nevertheless not as serious as in the case of the normal mixture. Singularities cannot occur if all populations are assumed to have the same size (the same θ), as implemented in STEM. They cannot occur if one uses ML to analyse sequence alignments rather than treating the gene trees with branch lengths as observed data: the likelihood is then an average over coalescent times (equation (9.42)) and cannot collapse onto a single time point. They cannot occur if one uses the Bayesian approach, whether the data are gene trees with branch lengths or sequence alignments, because the prior constrains θs away from zero and forces τs to be different. Approaches using sequences will also have the major advantage of accommodating uncertainties in the gene trees.

9.4.3.4 Methods using sequence alignments

Methods that analyse sequence alignments under a mutation model account for the uncertainties in the gene tree due to limited sequence data. In the Bayesian framework, one assigns a prior distribution for the species tree and parameters and calculates their posterior. The posterior of species tree S and parameters Θ given data X is then

$$f(S, \Theta|X) \propto f(S, \Theta) \times \prod_{i=1}^{L} \sum_{G_i} \int_{t_i} f(G_i, t_i|S, \Theta) f(X_i|G_i, t_i) \, dt_i, \tag{9.49}$$

where $f(G_i, \boldsymbol{t}_i|S, \Theta)$ is the multispecies coalescent density for the gene tree at locus i. The summation is over all possible gene tree topologies for each locus and the $(n_i - 1)$-dimensional integral is over the $(n_i - 1)$ coalescent times in the gene tree if locus i has n_i sequences. As before, we assume free recombination among loci so that the gene trees are independent across loci given the species tree. The prior on the species tree S (rooted tree topology and branch lengths) may be specified using a Yule process or birth–death process (Yang and Rannala 1997; Stadler 2010). Parameters θs may be assigned gamma priors as in section §9.4.1.

The MCMC algorithm is more complex than those in Bayesian phylogenetics, in that the gene trees are embedded in the species tree and may place strong constraints on the species tree when one proposes changes to the species divergence times. Such constraints may cause mixing problems for the MCMC.

At the time of writing, there are two implementations of Bayesian estimation of species tree under the multispecies coalescent model. The first is the program BEST (for Bayesian Estimation of Species Trees), developed by Liu and Pearl (2007; Liu 2008). This estimates the species tree topology, divergence times (τs), and ancestral population sizes (θs), using the posterior estimates of gene trees (topologies and branch lengths) generated by MrBayes. As the posterior for gene trees produced by MrBayes are not generated under the multispecies coalescent prior, an importance-sampling strategy is used to make corrections. All populations are assumed to have the same size (same θ), and the prior on θ is noted to affect the posterior probabilities for species trees (Leaché and Rannala 2011). The algorithm is complex and does not mix well (e.g. Cranston et al. 2009; Leaché 2009). MrBayes 3.2 (Ronquist et al. 2012b) attempts an improved implementation, explicitly incorporating the multispecies coalescent model in the prior for species tree estimation. The second implementation is in the program *BEAST (Heled and Drummond 2010). This estimates the species tree as well as gene trees simultaneously in one Bayesian MCMC analysis, using the MCMC machinery in the program BEAST.

Another Bayesian MCMC method is implemented in the program BUCKy (for Bayesian untangling of Concordance Knots) (Ané et al. 2007), which conducts the so-called Bayesian concordance analysis (Baum 2007). The *concordance factor* of a clade is defined as the proportion of the genome for which the clade is true. For example, in comparisons of the human, chimpanzee, and gorilla genomes, 70% of the genome is estimated to have the gene tree ((H, C), G), while ~15% of the genome has one of the two other gene trees. Thus the concordance factor for the clade (H, C) is 70%. The concordance tree is composed of clades that are true for more than half of the genome, and is considered an estimate of the species tree. The concordance method of Ané et al. (2007) uses the Dirichlet process to partition different gene loci into clusters, within each of which the genes share the same tree topology. The inference follows the general form of equation (9.49), where S represents the concordance tree and where the prior for gene trees given the concordance tree $f(G, \boldsymbol{t}|S, \Theta)$ is specified using the Dirichlet process rather than the multispecies coalescent process. In the BUCKy implementation, the individual genes are analysed separately using MrBayes to estimate posterior probabilities for gene trees. These are then used as input for a second-stage MCMC to provide revised posterior probabilities for each gene, taking account of concordance. The Dirichlet process model has no clear biological interpretations and ignores the fact that the gene trees should be similar as they are embedded in the same species tree and share many common features such as species divergence times. This should be the case whether gene tree discordance is caused by ancestral lineage sorting or by horizontal gene transfer. As the concordance analysis does not use the multispecies coalescent model to specify the distribution of gene trees, it is prone to the problem of anomalous gene trees.

Computer simulations have been conducted to compare different species tree inference methods. The Bayesian method based on the multispecies coalescent model (BEST) was found to perform better than BUCKy and STEM, with concatenation being the worst (Leaché and Rannala 2011). However, if the data are generated under a model of horizontal gene transfer, BUCKy may be more robust than BEST (Chung and Ané 2011). *BEAST has better accuracy in estimating parameters on the species tree than BEST (Heled and Drummond 2010).

With the advancement of new sequencing technologies, SNP data are becoming common. Bryant et al. (2012) developed a Bayesian MCMC sampling algorithm to analyse such data, in a program called SNAPP. The method computes the likelihood of a species tree directly from the markers under a finite-sites mutation model, effectively integrating out all possible gene trees. The method applies to independent (unlinked) biallelic markers such as well-spaced SNPs. The MCMC sampler infers the species tree, the divergence times, and population sizes.

Example 9.3. Inference of the species tree despite gene tree conflicts. The *Sceloporus undulatus* species group is a group of small insectivorous lizards widely distributed in the United States and Mexico, which originated through a recent radiation. Leaché (2009) conducted phylogenetic and phylogeographic analyses of a mitochondrial gene (NADH1) as well as 29 nuclear loci (>23.6 kb) from 21 specimens, inferring the species tree for nine species (Figure 9.12a). The mtDNA has a mutation rate nearly ten times as high as that for the nuclear genes, and its analysis by Leaché and Reeder (2002) identified four species within the previously recognized single species *S. 'undulatus'*: *S. consobrinus*, *S. cowlesi*, *S. tristichus*, and *S. undulatus* – the status of these species was also confirmed in a species delimitation analysis by Yang and Rannala (2010). The 21 specimens include four from each of those four mtDNA-based species, as well as one from each of other five species in the group, with nine species in total. Leaché's phylogenetic analysis identified four mtDNA introgression events at the contact zones between adjacent phylogeographic groups. As inaccurate species assignments may affect species tree inference, Leaché recommends reassigning specimens with introgressed mtDNA or excluding them from the analysis. Thus a subset of the data is used here to illustrate species tree inference under the multispecies coalescent model in presence of gene tree conflicts.

We use eight loci (>5.5 kb) at which sampling is complete and 18 specimens, with three specimens involving mtDNA introgression excluded. The ML trees for the individual genes (see Leaché 2009, Figure S1) highlight two typical features of phylogeographic data: (a) lack of resolution of individual gene trees due to the high sequence similarity and thus low information content in the data at each locus and (b) considerable conflicts among gene trees. The data are analysed to infer the species tree using two coalescent-based Bayesian methods: BEST and *BEAST. The results (Figure 9.12b) suggest that many of the clades on the species tree receive high posterior support despite the weak or conflicting signals in individual genes. The tree topologies inferred by the two programs have differences, although the associated posterior probabilities are low (≤ 0.91) and may reflect the impact of priors.

We then use BPP to estimate parameters in the multispecies coalescent model (τs and θs) on the species tree inferred by *BEAST. The priors are $G(2, 1000)$ for all the 12 θ parameters (for four extant species and eight ancestral species), and $\tau \sim G(2, 2000)$ for the root of the species tree with a uniform Dirichlet prior for the other τ parameters (Rannala and Yang 2003; Yang and Rannala 2010). The estimates of θs are around 2–4 mutations per kb, while the age of the root of the species tree is about 4.6 mutations per kb, with the 95% CI (3.5, 5.5) (Figure 9.12c). The estimates of τs are found to be stable but the estimates

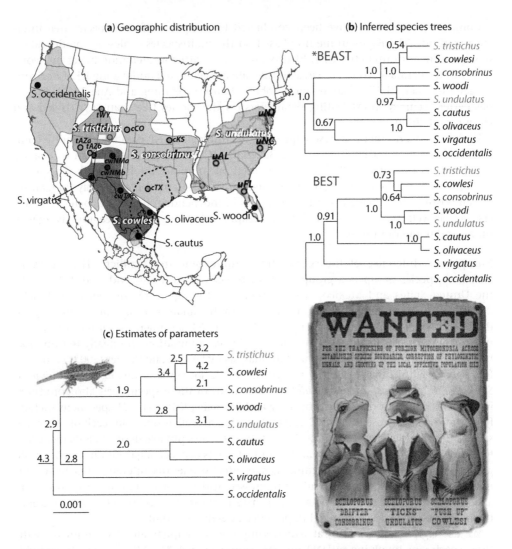

Fig. 9.12 (a) The geographic distribution of North American fence lizards in the *Sceloporus undulatus* species group, illustrating the sampling localities of the specimens (Leaché et al. 2009). The four mtDNA-based species have distributions shaded in colour, with samples indicated by ◯, while the single samples for five other species are indicated by ●. Three specimens in the mtDNA-based species showing mitochondrial introgression (*tCO*, *cwAZ*, and *cMS*) are excluded in the analysis (the fourth introgression involves *S. cautus* and is not removed). (b) Species trees inferred using *BEAST and BEST, with posterior probabilities shown on the branches. The results for BEST are from figure 5a of Leaché (2009), while those for *BEAST are calculated by Adam Leaché. (c) The species divergence time (τs) and population size parameters (θs) are estimated using BPP on the fixed species tree from *BEAST. The τ estimates are used to draw the tree while the θ estimates (in units of mutations per kb) are shown on the branches. The inset picture is that of a *S. tristichus* × *S. cowlesi* hybrid. Thanks to Der-shing Helmer for the lizard artwork.

of θs are sensitive to the priors. The *BEAST estimates of the τs appear different and may reflect the fact that they are averages across different species trees. Ideally those parameters should be estimated with the species tree fixed. □

9.4.4 Migration

Migration, or gene flow (the transfer of genes from one population into another), is an important biological process. For example, different models of speciation make different predictions about whether gene flow is possible at the time of species formation (Coyne and Orr 2004; Wu and Ting 2004). Genomic sequences can be used to estimate migration rates and infer patterns of migration between populations, offering exciting opportunities for testing theories of speciation and understanding the forces and mechanisms of species divergence (Feder et al. 2012). Furthermore, migration may influence many aspects of population genetic and phylogeographic inference, such as species tree estimation (Leaché et al. 2014) and species delimitation (to be discussed later in section §9.5) (Zhang et al. 2011). There may thus be a need to accommodate migration in the model even if our interest is not on migration per se.

Theoretical studies of migration or population subdivision have a long history in population genetics (Wright 1931, 1943). A number of models have been developed to accommodate population structure. For example, in the stepping-stones model, only neighbouring populations are allowed to exchange migrants. In the finite islands model any population can exchange migrants with any other. Those models are studied in the framework of *structured coalescent* (e.g. Li 1976; Strobeck 1987; Takahata 1988; Notohara 1990; Nath and Griffiths 1993; Wilkinson-Herbots 1998). Implementations of such models in GENETREE (Bahlo and Griffiths 2000) and MIGRATE (Beerli and Felsenstein 1999, 2001; Beerli 2006) have made it possible to estimate the population size parameters (θs) and migration rates (Ms) jointly from genetic data. However, those models fail to accommodate phylogenetic relationships among populations. They assume that population subdivision promoting population divergence and migration homogenizing the populations have reached the equilibrium, or that in effect the populations diverged an infinitely long time ago. In the real world, the history and timings of divergences are important factors that affect the genetic variation we observe today. Such considerations have motivated the development of the so-called isolation-with-migration (IM) models, which incorporate the population/species phylogeny in a model of migration (Nielsen and Wakeley 2001; Hey and Nielsen 2004; Wilkinson-Herbots 2008, 2012). Also the IM model allows one to evaluate the role of gene flow during speciation.

Here we focus on calculation of the probability density of the gene tree under the IM model, and its use to estimate parameters from multi-locus genetic sequence data when the species phylogeny is fixed. In theory, when the species phylogeny is allowed to vary in the MCMC algorithm, the same framework will allow species tree estimation under a model of migration. Similarly, if species assignment and species delimitation are allowed to vary in the MCMC algorithm, one will have a method of species delimitation under a migration model. However, currently computation even on a fixed species tree is challenging.

9.4.4.1 *Definitions of migration rates*

The important parameter in a model of migration is the migration rate. Unfortunately, the migration rate is defined in different ways in the literature, which can be confusing. Suppose populations 1 and 2 have sizes N_1 and N_2, and on average M_{12} individuals migrate from population 1 to population 2 per generation. Then M_{12}/N_1 is the proportion of emigrants in population 1 and is sometimes called the 'forward migration rate', while M_{12}/N_2 is the proportion of immigrants in population 2 from population 1 and is sometimes called the 'backwards migration rate' (e.g. Notohara 1990). We do not use those definitions here. In most population genetics models, the proportion of individuals that emigrate out of a population is either irrelevant or hard to deal with. Emigrants may get drowned trying to cross the ocean or they may move to a location unsampled, and furthermore losing a few emigrants will have virtually no impact on the source population. It is often easier to deal with the proportion of immigrants in the receiving population and furthermore, even a few immigrants per generation may have a profound impact on the genetic composition of the receiving population.

We define the migration rate from population i to population j to be the proportion of individuals in population j that are immigrants from population i, or $m_{ij} = M_{ij}/N_j$. This is the 'backwards migration rate' of Notohara (1990) but we simply call it the migration rate. This definition is used in the programs MIGRATE (Beerli and Felsenstein 2001; Beerli 2006) and BPP or 3S (Zhang et al. 2011; Zhu and Yang 2012). The scaled migration rate $M_{ij} = N_j m_{ij}$ is the expected number of immigrants in population j (from population i) per generation. In a general migration model, migration rates are specified in a matrix $M = \{M_{ij}\}$, of size $k \times k$ if there are k populations.

Some authors (e.g. Nath and Griffiths 1993; Bahlo and Griffiths 2000; Wang and Hey 2010) use the coalescent worldview migration rate so that an $i \to j$ migration is in the real world a migration from population j to population i. This is also known as the 'backwards migration rate': it is backwards because the coalescent process runs the 'time machine' backwards. Programs GENETREE (Nath and Griffiths 1993; Bahlo and Griffiths 2000) and IMa2 (Hey 2010) use this definition. Thus the 'migration rate' $M_{i \to j}$ in IMa2 corresponds to m_{ji} here. The migration rate M_{ij} in MIGRATE corresponds to $m_{ij}/\mu = 4M_{ij}/\theta_j$ here.

9.4.4.2 Probability density of gene tree with migration trajectory

Two species, with two sequences per locus. The IM model in the case of only two populations/species is illustrated in Figure 9.13a (Nielsen and Wakeley 2001). The model involves six parameters: three population size parameters ($\theta_1 = 4N_1\mu$ and $\theta_2 = 4N_2\mu$ for the two modern populations and $\theta_a = 4N_a\mu$ for the common ancestor), the species divergence time (τ), and two scaled migration rates in the two directions ($M_{12} = N_2 m_{12}$ and $M_{21} = N_1 m_{21}$). As before, both τ and the θs are measured by the expected number of mutations per site. Let $\Theta = \{\theta_1, \theta_2, \theta_a, \tau, M_{12}, M_{21}\}$.

We first consider the case of only two sequences at each locus. When we trace the genealogy of the sample backwards in time, the process of coalescent and migration can be described by a Markov chain with five states: $S_{11}, S_{12}, S_{22}, S_1$, and S_2, where S_{11} means that both sequences are in population 1, S_{12} means that the two sequences are in the two populations, and so on. State S_1 means that the two sequences have coalesced and the single ancestral sequence is in population 1, while state S_2 means that the coalesced sequence is in population 2. The initial state of the chain is one of S_{11}, S_{12}, and S_{22}, meaning that the two sampled sequences are from the same population (S_{11} or S_{22}) or from the two different populations (S_{12}). With only two sequences sampled at the locus, there is no need to distinguish between states S_{12} and S_{21}.

Let one time unit be the expected time to accumulate one mutation per site. Consider population i. The coalescent rate is $\frac{2}{\theta_i}$ if the two sequences are in population i and is 0 otherwise. The (real-world) migration rate from population j into population i when

9.4 MULTISPECIES COALESCENT, SPECIES TREES AND GENE TREES

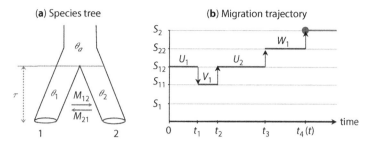

Fig. 9.13 (a) There are six parameters in the IM model for two species: $\theta_1, \theta_2, \theta_a, \tau, M_{12} = N_2 m_{12}$, and $M_{21} = N_1 m_{21}$, where the migration rate m_{ij} is defined as the proportion of individuals in population j that are immigrants from population i, per generation, and M_{ij} is the expected number of migrants from population i to population j per generation. (b) A migration trajectory for a locus with two sequences, with one sequence from each species (i.e. the initial state is S_{12}). The total amount of time spent in the states S_{12}, S_{11}, and S_{22} are $U = U_1 + U_2, V = V_1$, and $W = W_1$, respectively. The divergence time between the two sequences at the locus (i.e. t_{MRCA} for the sample) is $t = t_4 = U + V + W$. The likelihood of the sequence data depends on t only and not on the migration trajectory (such as U, V, and W).

there are n_i lineages in population i is $n_i m_{ji}/\mu = n_i \times 4M_{ji}/\theta_i$. The rate matrix Q for the Markov chain is thus

	S_{11}	S_{12}	S_{22}	S_1	S_2
S_{11}	$-2\left(m_1 + \frac{1}{\theta_1}\right)$	$2m_1$	0	$\frac{2}{\theta_1}$	0
S_{12}	m_2	$-(m_1 + m_2)$	m_1	0	0
S_{22}	0	$2m_2$	$-2\left(m_2 + \frac{1}{\theta_2}\right)$	0	$\frac{2}{\theta_2}$
S_1	0	0	0	$-m_1$	m_1
S_2	0	0	0	m_2	$-m_2$

(9.50)

where $m_1 = \frac{4M_{21}}{\theta_1}$ and $m_2 = \frac{4M_{12}}{\theta_2}$ are the migration rates per time unit (per mutation) (Wang and Hey 2010; Hobolth et al. 2011). For example, the transition from S_{11} to S_{12} in the (backward) coalescent process means an immigration from populations 2 to 1 in the real world, and since either of the two sequences in population 1 can be the immigrant, the rate is $2m_1$. Note that states S_1 and S_2 constitute a closed set of absorbing states.

A migration trajectory, including the number, times and directions of all migration events, is shown in Figure 9.13b. Given the trajectory, the density function for the gene tree (which includes the migration trajectory as well as the coalescent time) can be calculated easily. Let the total time the chain spends in the states S_{12}, S_{11}, and S_{22} be U, V, and W, respectively. Let x be the number of $1\to 2$ (real-world) migration events (each with rate m_2), and y the number of $2\to 1$ migration events (each with rate m_1). Using the idea of variable-rate Poisson process, one has the density for the gene tree as

$$f(G|\Theta) = \left(\frac{2}{\theta_2} \cdot m_2^x \cdot m_1^y\right) \times e^{-(m_1 + m_2)U - 2\left(m_1 + \frac{1}{\theta_1}\right)V - 2\left(m_2 + \frac{1}{\theta_2}\right)W}. \quad (9.51)$$

As in equations (9.31) and (9.41), this density consists of two terms: the first term is the product of the rates for all the coalescent and migration events, while the second term (the exponential term) is the probability of no event over the whole gene tree. Gene tree G here contains all information about the coalescent and migration history at the locus, such as x, U, V, W, t_1–t_4 (note that y is determined by x and the initial state). Note that the sequence likelihood or the probability of observing the aligned pair of sequences given the gene tree depends on the time of divergence between the two sequences, $t = t_4 = U + V + W$, but not on the detailed migration history (Figure 9.13b). For example, the likelihood remains unchanged if the migration time t_1 changes arbitrarily in the interval $(0, t_2)$ in the migration trajectory.

The general case of multiple species and multiple sequences per locus. In general, the coalescent process with migration is a simple modification of the multispecies coalescent process of §9.4.1, and can be described as a variable-rate Poisson process. The gene tree G_h for locus h with n_h sequences includes the $n_h - 1$ coalescent times and the full migration trajectory. The density of the gene tree is easily calculated using the same strategy; it is given by the rates of coalescent and migration events that have occurred, multiplied by the probability of no event over the whole gene tree (Beerli and Felsenstein 2001). To calculate the probability of no event, one can use the species phylogeny to define different time epochs within which the populations are fixed so that the coalescent rate per lineage pair and the migration rate per lineage remain constant (Figure 9.14). In population i with n_i sequences in the sample, the coalescent rate is $\binom{n_i}{2}\frac{2}{\theta_i}$, and the migration rate from population j (into population i) is $n_i m_{ji}/\mu = n_i \times 4M_{ji}/\theta_i$.

Estimation of parameters on the species tree (Θ) can in principle proceed as in the multispecies coalescent model without migration (§9.4.1). For example, the posterior for the parameters is given by equation (9.43), with the modifications that Θ now includes the migration rates as well as the species divergence times (τs) and population size parameters (θs) and that the gene tree G_h at locus h now includes the migration history at the locus as well as the coalescent times. While there is no conceptual difficulty, the computation is much more demanding. The MCMC algorithm has to average over the gene tree topologies, the coalescent times, and the migration trajectory including the number, directions, and times of migration events at every locus. While there are $n_h - 1$ coalescent times to integrate over in a gene tree for n_h sequences, there is no upper limit to the number of migration events, so that the space the MCMC algorithm has to average over can be huge, especially at high migration rates. Furthermore MCMC proposals altering migration histories without changing the gene tree topology or branch lengths will not affect the likelihood (i.e. the probability of the sequence data), so that the MCMC may be

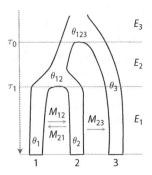

Fig. 9.14 Parameters in an IM model for three species, used to illustrate the calculation of the probability density for the gene tree as well as simulation of gene trees or sequence alignments. The two speciation events on the species tree define three time epochs within which the species or populations are fixed: E_1: $(0, \tau_1)$, E_2: (τ_1, τ_0), and E_3: (τ_0, ∞).

averaging over a huge, nearly flat surface. As a result, the MCMC algorithm under the IM model tends to be very inefficient. The major Bayesian implementation of the IM model through MCMC is the IM program (Nielsen and Wakeley 2001; Hey and Nielsen 2004; Hey 2010), and it is currently infeasible to analyse more than a few dozen loci. Gronau et al. (2011) modified the Bayesian MCMC program of Rannala and Yang (2003) for multispecies coalescent to include migration. This can deal with many loci in genomic-scale datasets. However, the implementation, through so-called migration bands, is heuristic.

9.4.4.3 Inference using data from only a few genomes

When only a few genomes are compared, it may be feasible computationally to apply the ML method. The recent appearance of whole-genome sequence data for a few individuals has prompted the development of methods that can analyse tens of thousands of loci from just a few genomes (Wang and Hey 2010; Lohse and Barton 2011; Zhu and Yang 2012; Andersen et al. 2014).

For example, with only two sequences at a locus, the likelihood or the probability of the sequence alignment at the locus depends on the sequence divergence time t only, measured by the expected number of mutations per site, independent of the migration trajectory such as the number and times of the migration events. For the model of equation (9.50), one can calculate the density for t directly by using the transition probability matrix: $P(t) = \{P_{ij}(t)\} = e^{Qt}$ (Hobolth et al. 2011). Let the initial state be a, which is one of S_{11}, S_{12}, and S_{22}. The probability density for the divergence time t between the two sequences is then

$$f(t|\Theta) = \begin{cases} P_{aS_{11}}(t) \times \dfrac{2}{\theta_1} + P_{aS_{22}}(t) \times \dfrac{2}{\theta_2}, & t < \tau, \\ \left(P_{aS_{11}}(\tau) + P_{aS_{12}}(\tau) + P_{aS_{22}}(\tau)\right) \times \dfrac{2}{\theta_a} e^{-\frac{2}{\theta_a}(t-\tau)}, & t \geq \tau. \end{cases} \quad (9.52)$$

First, we consider the case $t < \tau$, which means that the two sequences coalesce before reaching τ, either in population 1 or in population 2. For the two sequences to coalesce in population 1 in the small time interval $(t, t + \Delta t)$, with $t < \tau$, two things must happen: (i) the chain must be in the state S_{11} at time t, which occurs with probability $P_{aS_{11}}(t)$; and (ii) the two sequences must coalesce in the time interval, which occurs with probability $\frac{2}{\theta_1}\Delta t$. Similarly, the probability that the two sequences coalesce in population 2 in the interval $(t, t + \Delta t)$ is $P_{aS_{22}}(t) \times \frac{2}{\theta_2}\Delta t$. Adding those two probabilities together, we get $f(t|\Theta)\Delta t = \left(P_{aS_{11}}(t) \times \frac{2}{\theta_1} + P_{aS_{22}}(t) \times \frac{2}{\theta_2}\right)\Delta t$, for $t < \tau$, as in equation (9.52).

Second, we consider the case $t > \tau$. The probability density is then given by the probability that there is no coalescent event before τ, which is $P_{aS_{11}}(\tau) + P_{aS_{12}}(\tau) + P_{aS_{22}}(\tau)$, multiplied by the exponential density for the coalescent time $(t-\tau)$ in the ancestral population, as in equation (9.52).

Under the symmetrical IM model for two species (SIM2s), with $\theta_1 = \theta_2 = \theta$ and $M_{12} = M_{21} = M$, the Q matrix of equation (9.50) can be diagonalized analytically (Zhu and Yang 2012), leading to analytical calculation of $P(t)$ and density $f(t|\Theta)$. Some examples of the density $f(t|\Theta)$ are shown in Figure 9.15a and b for two sequences taken from the same population or from the two different populations. For the general Q matrix of equation (9.50), $P(t)$ can be calculated using standard algorithms such as scaling and squaring (see §2.6).

Note that in equation (9.51), the gene tree G is the full history of Figure 9.13b, while in equation (9.52), t is the divergence time between the two sequences, $t = U + V + W$

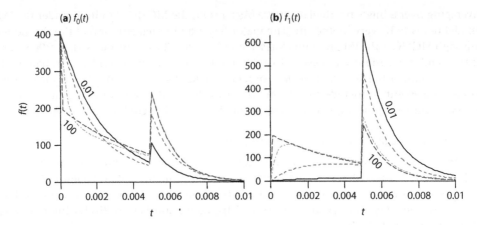

Fig. 9.15 Probability density of divergence time t at different migration rates under the symmetrical IM model with two species (SIM2s), which assumes $\theta_1 = \theta_2 = \theta$, and $M_{12} = M_{21} = M$ (see Figure 9.13a). The values of parameters used in the calculation are $\theta_1 = \theta_2 = 0.005$, $\theta_a = 0.003$, and $\tau = 0.005$. (a) The density $f_0(t)$ is for two sequences sampled from the same population (with the initial state S_{11} or S_{22}), while (b) $f_1(t)$ is for two sequences sampled from the two different populations (with the initial state S_{12}). The migration rates for the four curves in each plot are $M = 0.01, 0.1, 1$, and 10, but only curves for 0.01 and 10 are labelled.

(Figure 9.13b). In effect, the approach of equation (9.52) integrates out the migration history analytically in the calculation of $P(t)$.

The log likelihood function is then given as

$$\ell(\Theta) = \sum_{h=1}^{L} \log f(X_h|\Theta) = \sum_{h=1}^{L} \log \left\{ \int_0^\infty f(t|\Theta) f(X_h|t) \, dt \right\}, \quad (9.53)$$

where X_h is the sequence alignment at locus h and, under JC69, can be summarized as the number of differences between the two sequences at the locus, $f(t|\Theta)$ is the prior on t (equation (9.52)), while the probability of data X_h at the locus, $f(X_h|t)$, is given by equation (1.46), say. The 1-D integral over t can be calculated easily using numerical integration (§6.4.3).

This approach of integrating out the migration history in the calculation of $P(t)$ can be generalized to more than two sequences at each locus (Andersen et al. 2013). An ML implementation in the 3s program (Zhu and Yang 2012) deals with three species on the species tree, with one sequence sampled from each at every locus. The model assumed is the same SIM2s model, but a third outgroup species is included in the data to provide additional information in the form of gene tree topologies and branch lengths. The likelihood for three sequences given the gene tree (rooted tree topology and node ages), $f(X|G)$, is calculated efficiently under JC69 (Yang 2002b). The advantage of this approach is that the computation is feasible for genome-scale datasets consisting of many thousands of loci. Its drawback is that the number of states explodes and the dimension of the numerical integration becomes unmanageable when there are more than three sequences at each locus.

In summary, inference under a migration model using genomic sequence data remains challenging. The approach of averaging over migration trajectories in a Bayesian MCMC algorithm works for multiple species and sequences but is computationally very expensive. The approach of integrating out the migration trajectories analytically in the likelihood calculation can deal with many loci but works with only a few sequences per locus and is suitable for comparisons of only two or three genomes.

9.4.4.4 Simulation under the IM model

As discussed above, the multispecies coalescent with migration is a variable-rate Poisson process. Gene trees and sequence alignment can thus be generated by simulating migration and coalescent events as competing Poisson processes (Zhang et al. 2011). We use the species/population tree to define time epochs, so that within each time epoch the populations are fixed, as is the per lineage migration rate (Figure 9.14). In each time epoch, the waiting time until the next event is sampled from an exponential distribution with the intensity parameter (total rate) to be the sum of the coalescent rates and the migration rates. Suppose population i has the size parameter θ_i and contains n_i lineages ancestral to the sample. With one time unit being the expected time to accumulate one mutation per site, the total coalescent rate is $\binom{n_i}{2} \frac{2}{\theta_i}$, while the migration rate from population j (into population i) is $n_i m_{ji}/\mu = n_i \times 4M_{ji}/\theta_i$. The coalescent and migration rates are summed over all populations for the time epoch, and the total rate is used to sample the waiting time until the next event. Given the occurrence of the event, the event type (coalescent or migration) is sampled in proportion to their rates. This process is repeated until the time epoch is exhausted or until the MRCA for the whole sample is reached. With the migration trajectory ignored, this process generates a gene tree with node ages, which can be used to simulate a sequence alignment.

Heled et al. (2013) discussed simulation under the IM model with migration when the population size and the migration rate change deterministically over time. The process is simulated as two competing variable-rate Poisson processes, one for the coalescent process and another for migration (see §9.3.1 and §12.5.2). We simulate the next waiting time for coalescent and the next waiting time for migration, and let the first event occur, at which point the process is reset.

9.5 Species delimitation

9.5.1 *Species concept and species delimitation*

Species delimitation is the process of inferring the number of species, determining species boundaries, and discovering new species (Sites and Marshall 2004; De Queiroz 2007). Wiens (2007) describes species delimitation and phylogeny reconstruction as the two major goals of systematics. Species delimitation differs from species identification; the latter refers to the assignment of a specimen to one of several known and well-characterized species.

In discussing species delimitation, one finds it hard to avoid the tricky business of defining species. There are well over 20 definitions of 'species' (De Queiroz 2007; Mallet 2013). Here we mention just two. The biological species concept (BSC), due to Mayr (1942), defines a species as a group of individuals that actually or potentially interbreed in nature, while distinct species are kept apart by reproductive isolation (Coyne and Orr 2004). A common scenario for species divergence is that a species splits into two populations due to a certain geological event, such as the rise of a mountain range (Zhang et al. 2013), river reversal or river capture due to glaciation or tectonic movement (Burridge et al. 2006), or even the building of a railway. The geographical barrier causes cessation or reduction of interbreeding between the two populations. Over time the two groups develop morphological, behavioural, or genomic incompatibilities so that they will not be able to interbreed even if they are later brought together; they have become 'good' biological species. This scenario is probably the most common mode of species formation, and is called *allopatric speciation*. In contrast, with *sympatric speciation*, species divergence takes place even though all individuals live in the same area. While mathematical theories and computer simulations predict that sympatric speciation can occur (Dieckmann and Doebeli 1999; Kondrashov and Kondrashov 1999), how often it actually does in nature is still a matter of debate.

BSC thus defines a clear criterion for species status and is consistent with the most common mode of speciation. Nevertheless, BSC cannot be applied to asexual species. In plants, hybridization or introgression between distinct species is quite common. Thus BSC is often applied in a pragmatic manner, with distinct species characterized by *substantial but not necessarily complete reproductive isolation* (Coyne and Orr 2004, p. 30).

The second species concept we mention here is the phylogenetic, genealogical, or evolutionary species concept, according to which species are units in phylogenetic analysis (Baum 1992). Shaw (1998) defines a genealogical species as 'a basal, exclusive group of organisms, whose members are all more closely related to each other than they are to any organisms outside the group, and that contains no exclusive group within it'. In practice, two groups are recognized as distinct genealogical species if the gene trees show reciprocal monophyly (Figure 9.16) at all or most gene loci.

While there is no consensus on what a species is, most biologists acknowledge that species are real. There are continuities within species (say, between the males and females of the same species) and discontinuities between species that are consistent when one examines morphological, genetic, ecological, and reproductive characters (Dobzhansky 1937; Coyne and Orr 2004). Thus species delimitation is very much a meaningful and important exercise despite the controversy concerning species definition.

Species have traditionally been described and identified using morphological characters such as plumage, mouthparts, mating behaviour, etc. In recent years, use of genetic data for species delimitation has become more common, with proposals such as DNA barcoding (e.g. Hebert et al. 2004) and DNA taxonomy (Tautz et al. 2003). To a large part this is due to the ease with which genetic data can now be gathered. Taxomonic descriptions are time consuming and require much experience, while assessing reproductive isolation is sometimes nearly impossible. Nevertheless, ease of data acquisition is not the only reason for the popularity of genetic data. Genetic data are numerous, and track the population divergence and migration history more faithfully than morphology. They can be used

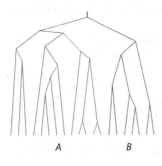

Fig. 9.16 The gene tree is said to be reciprocally monophyletic if all sequences from each of the two species form a monophyletic clade.

to delimit cryptic species, which are indistinguishable morphologically. Morphological characters often undergo convergent evolution if the species live in similar habitats.

It may be debatable whether genetic data alone are sufficient for delimiting species, but there is no doubt that genetic data can provide important information on the divergence and migration patterns of species and populations, and genetic information should be integrated with other evidence to decide on the species status of particular groups of organisms. For a recent debate concerning the relative merits of morphological characters versus genetic sequences for species delimitation, see Leaché and Fujita (2010), Bauer et al. (2011), Fujita and Leaché (2011), and Fujita et al. (2012).

9.5.2 Simple methods for analysing genetic data

A number of simple methods for analysing genetic data to delimit species have been proposed (Wiens 2007). Some use cutoffs on certain indicators of species status such as sequence divergence and migration rate. For example, the '10× rule' requires the between-species divergence at a mitochondrial locus to be at least ten times as large as the within species polymorphism (Hebert et al. 2004). This criterion is arbitrary and often fails to identify new species (Hickerson et al. 2006). Another criterion is reciprocal monophyly of the mitochondrial gene tree (Figure 9.16), which may be used to identify species based on the genealogical species concept (Baum and Shaw 1995). This simple criterion fails to accommodate errors in phylogeny reconstruction, and ignores the fact that conflicts between the species tree and the gene tree are expected even for good species, due to ancestral polymorphism and incomplete lineage sorting (Hudson and Coyne 2002) (§9.4.2). In simulation studies, reciprocal monophyly has very lower power and often fails to identify species (Hickerson et al. 2006; Zhang et al. 2011). It is also easy to construct situations in which reciprocal monophyly will tend to over-split.

Knowles and Carstens (2007) and O'Meara (2010) used inferred gene tree topologies as data to construct LRTs to compare the one-species and two-species models. The methods accommodate species tree–gene tree conflicts due to incomplete lineage sorting, but ignore phylogenetic errors in gene tree reconstruction and branch length estimation. Also Knowles and Carstens's use of the chi square with one degree of freedom as the null distribution for the LRT is incorrect. Carstens and Dewry (2010) discussed another method based on reconstructed gene trees. Given the gene trees with branch lengths (node ages) and the assignment of individuals to populations (which are potential species), a species tree is fitted to the gene trees by ML using the program STEM (Kubatko et al. 2009) (see §9.4.3.2), and the different species delimitation models are compared using the AIC. Similarly phylogenetic errors in gene tree reconstruction are ignored. This method is implemented in the program SpedeSTEM (Ence and Carstens 2011).

Pons et al. (2006; see also Fujisawa and Barraclough 2013) argue that between-species divergences should be described by a Yule pure-birth process while within species lineage joining should be described by the coalescent model, and suggested an LRT to identify the time point at which a switch from coalescent to Yule processes occurs on the estimated gene tree. The method, called GMYC (for Generalized Mixed Yule Coalescent), works on a single gene tree with estimated node ages but ignores sampling errors in node age estimation. It also fails to accommodate ancestral polymorphism and assumes that sequences from every species form a monophyletic clade, with all coalescent events being more recent than all species divergences. The method does not require assignments and may be useful for initial discovery analysis when only single-locus data are available.

A proper analysis of the multi-locus genetic data should take into account several sources of information as well as the uncertainties involved. First, ancestral polymorphism and incomplete lineage sorting can cause the gene tree to differ from the species tree. Second, sequence data for recently diverged species, for which species delimitation is likely to be an issue, tend to be highly similar and produce poorly resolved gene trees. As a result, estimated gene tree topologies and branch lengths involve considerable errors and uncertainties. Third, branch lengths on gene trees are informative about population divergence times and ancestral lineage sorting and should be used as well as gene tree topologies.

All these can be accommodated in a straightforward manner by using the multispecies coalescent model (see §9.4.1) (Rannala and Yang 2003) in a Bayesian modelling framework. Such a method has been implemented by Yang and Rannala (2010). This is the same implementation as discussed in §9.4.1 (in particular, Algorithm 9.4), except that the species delimitation and species phylogeny are allowed to change in the MCMC (only to some extent, see below). rjMCMC (see §7.4.2) is used to estimate the posterior probabilities of the different species delimitation and species phylogeny models, which have different numbers of parameters. As the multispecies coalescent model of Rannala and Yang (2003) assumes complete isolation following species divergence, with no migration, hybridization, or introgression, this Bayesian method of species delimitation effectively adopts the strict BSC. Calculation of the likelihood on the sequence alignments naturally accounts for uncertainty in gene trees and branch lengths, while the multispecies coalescent accommodates ancestral polymorphism and incomplete lineage sorting.

9.5.3 Bayesian species delimitation

A full Bayesian solution to the problem of species delimitation should involve both assignment of individuals to different species/populations and inference of the species phylogeny. The former is a problem of species delimitation and the latter one of phylogenetic inference. The number of possible species delimitations given the number of individuals (n) is the number of possible ways that n items can be partitioned into nonempty groups (clusters). This is the Bell number (see equation (8.15)). For three individuals (a, b, c), this number is 5, and the five species delimitations are {abc}, {$ab|c$}, {$bc|a$}, {$ca|b$}, and {$a|b|c$}. If a species delimitation involves three or more species, there will be different phylogenies. For example, three distinct phylogenies are possible for the delimitation with three species: {$a|b|c$}. In total there are seven models of species delimitation and species phylogeny for three individuals. The parameters involved in these models are illustrated in Figure 9.17.

With more individuals, the possible species delimitations and phylogenies constitute a huge parameter space. Comparing so many models in a Bayesian rjMCMC algorithm is very demanding computationally. Furthermore, it may be unnecessary to evaluate all possible models, since strong evidence may indicate that certain individuals belong to the same species. To reduce the number of models to be evaluated, Yang and Rannala (2010) implemented a simplified approach that requires the user to assign the individuals to populations and to supply a phylogeny for the populations, called the 'guide tree'. The populations on the guide tree can be subspecies, distinct geographical populations, morphotypes, etc. The rjMCMC algorithm then attempts to merge the populations at internal nodes on the guide tree into one species. The current implementation evaluates only those species delimitation models that can be generated by collapsing nodes on the guide tree; it never alters the relationships among the populations on the guide tree or

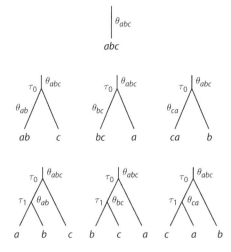

Fig. 9.17 In the general model of species delimitation and phylogenetic inference, there are five species delimitations for three individuals, with one (top), two (middle) or three (bottom) species. For the delimitation of three species, there are three distinct phylogenies. In total there are seven possible phylogenetic models. From Yang and Rannala (2010).

splits one population into two species. An example is shown in Figure 9.18. Note that the models evaluated by the algorithm differ concerning both species delimitation and species phylogeny.

Let Λ be a species delimitation model. Within the model, the parameters are those in the multispecies coalescent, including the species divergence times (τs) and population sizes (θs), collectively denoted Θ. The data $X = \{X_i\}, i = 1, \ldots, L$, consists of sequence alignments at the L loci. The Bayesian rjMCMC algorithm generates the posterior distribution:

$$f(\Lambda, \Theta|X) \propto f(\Lambda)f(\Theta|\Lambda) \times \prod_{i=1}^{L} \sum_{G_i} \int_{t_i} f(G_i, t_i|\Lambda, \Theta) f(X_i|G_i, t_i) \, dt_i, \quad (9.54)$$

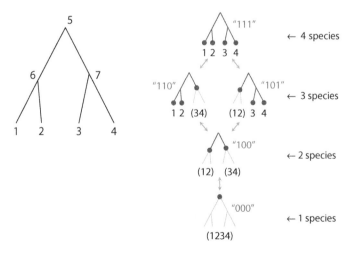

Fig. 9.18 (a) An example guide tree used for species delimitation. (b) The five species delimitation models are obtained by collapsing none or several nodes (5, 6, 7) on the guide tree, as indicated by the flags 000, 100, etc. A pair of rjMCMC moves ('split' and 'join') are used to move between models.

where $f(\Lambda)$ and $f(\Theta|\Lambda)$ are the priors, $f(G_i, \boldsymbol{t}_i|\Lambda, \Theta)$ is the prior density of G_i and $\boldsymbol{t}_i$ under the multispecies coalescent model (e.g. equation (9.40)), and $f(X_i|G_i, \boldsymbol{t}_i)$ is Felsenstein's phylogenetic likelihood on the sequence alignment at locus i given the gene tree G_i and node ages $\boldsymbol{t}_i$. Given the species delimitation model (Λ) and its parameters (Θ), the gene trees and sequence alignments are independent among loci. Note that Λ, Θ, and the gene trees at the loci $\{G_i, \boldsymbol{t}_i\}$ are all states of the MCMC, but we focus on the marginal probability $f(\Lambda, \Theta|X)$, integrating over the gene trees in the MCMC.

The population size parameters θs are positive and can be assigned gamma priors. The species divergence time τ_0 for the root of the species tree can be assigned a gamma prior as well, while the other divergence times may be assigned a Dirichlet prior conditional on τ_0. To specify the prior $f(\Lambda)$, one may assign equal probabilities for all possible species delimitation models implied by the guide tree (e.g. 1/5 for each of the five models of Figure 9.18). Another way is to assign a prior probability that each interior node on the guide tree exists and represents a speciation event (Rannala and Yang 2013).

The MCMC algorithm involves the usual Metropolis-Hastings moves that change the parameters on the species tree (Θ) and moves that change the gene tree topologies and coalescent times (see Algorithm 9.4). In addition, a pair of rjMCMC proposals allows the chain to move between different species delimitation models. A 'split' move splits one species into two and a 'join' move joins two species into one (Figure 9.19). Consider the split move, which chooses one of the n_{split} feasible nodes in the guide tree for splitting. A joined node on the guide tree is feasible for splitting if its mother node is already split or otherwise if it is the root. Let the chosen node for splitting be node i, and its two daughter nodes in the guide tree be j and k. Let the current and new models be Λ_0 and Λ_1, with parameters Θ_0 and Θ_1, respectively. The split move thus splits the species represented by node i in Λ_0 into two species j and k in Λ_1. The new model Λ_1 will have up to three new parameters: τ_i, θ_j, and θ_k. Parameter θ_j is present in the model only if j is an internal node on the guide tree or otherwise if there are at least two sequences sampled from population j at some loci. If population j is a tip on the guide tree and if there is at most one sequence from population j at every locus, the data will contain no information about θ_j so that θ_j will not be a parameter in the model. The same applies to θ_k. We

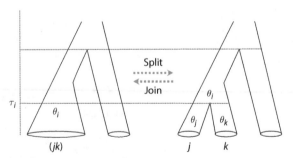

Fig. 9.19 The reversible-jump algorithm for Bayesian species delimitation on a guide tree (Yang and Rannala 2010) involves splitting and joining nodes on the guide tree. Node i on the guide tree, which is currently joined, has daughter nodes j and k. When node i is split, up to three new parameters are created for the new model: τ_i, θ_j, and θ_k.

generate the values of the new parameters in model Λ_1 from different proposal densities: $\tau_i \sim g(\tau_i), \theta_j \sim g(\theta_j)$, and $\theta_k \sim g(\theta_k)$ (see Yang and Rannala 2010 for examples of proposal densities). These densities may depend on unchanged parameters such as θ_i. In the reverse join move, we choose a feasible node for joining at random. Let n_{join} be previously split nodes in the guide tree that are feasible for joining. A split node is feasible for joining if all its descendent nodes are already joined. When one joins node i, one simply drops the extra parameters in the current model ($\tau_i, \theta_j, \theta_k$). The rjMCMC proposals are accepted with probability min$\{1, \alpha\}$, with

$$\alpha_{\text{split}} = \frac{n_{\text{split}}}{n_{\text{join}}} \cdot \frac{\pi(\Lambda_1, \Theta_1)}{\pi(\Lambda_0, \Theta_0)} \cdot \frac{1}{g(\tau_i)g(\theta_j)g(\theta_k)}, \quad \text{for the split move,}$$

$$\alpha_{\text{join}} = \frac{1}{\alpha_{\text{split}}}, \quad \text{for the join move.}$$
(9.55)

where $\frac{\pi(\Lambda_1, \Theta_1)}{\pi(\Lambda_0, \Theta_0)}$ is the posterior ratio between the two models, and is equal to the prior ratio times the likelihood ratio. This is a straightforward application of the rjMCMC methodology of section §7.4.2.3 (see, in particular, equation (7.83)). Note that if θ_j is not a parameter in the model, the $g(\theta_j)$ term will not appear in equation (9.55). The same applies to θ_k.

As discussed in section §7.4.2.4, rjMCMC algorithms often have mixing problems. In this case, the poor mixing appears to be caused by the strong constraint placed by the gene trees on the new parameter τ_i during the split move. According to the multi-species coalescent model, genes split before species. If a node on the guide tree is split without modifying the gene trees, the minimum coalescent time between sequences from the two descendent populations across all loci will constitute a maximum bound to the new species divergence time (τ_i). With multiple loci, this bound can be too tight (too close to zero). Rannala and Yang (2013) modified the algorithm to propose τ_i without considering the gene trees and then modify the node ages in the gene trees to avoid conflicts. Tests on simulated and real data suggest that the modification improves the mixing efficiency of the algorithm.

9.5.4 The impact of guide tree, prior, and migration

The guide tree may be generated based on morphological characters or geographic distributions of the populations. In vertebrates, the mitochondrial genome has a much higher mutation rate than nuclear loci and only $1/4$ of the population size for a nuclear locus, so that a mitochondrial gene harbours disproportionally more information about the population history than a nuclear locus (Avise 2000). Thus the mtDNA phylogeny can be used to generate a guide tree for species delimitation using nuclear loci. Similarly phylogenetic analysis of concatenated nuclear genes (the so-called supermatrix method) and even population-clustering analysis (Pritchard et al. 2000; Huelsenbeck and Andolfatto 2007) may be used to produce guide trees.

In a computer simulation study, Leaché and Fujita (2010) found that use of an incorrect guide tree may cause Bayesian species delimitation by BPP (Yang and Rannala 2010) to over-split. If two populations that belong to the same species are incorrectly separated on the guide tree and placed together with other more divergent species, BPP may misinterpret all of them as distinct species. The incorrect guide trees in this case were generated by randomizing species at the tips of the guide tree, which may result in guide trees that are too wrong. In practical data analysis, incorrect guide trees may reflect misplacement of populations separated by short branches, and their impact may not be quite as large (Zang et al. 2011). At any rate, it is prudent to evaluate the impact of the guide tree.

Similarly, the prior on the species divergence times (τs) and population sizes (θs) may affect the posterior probabilities for species delimitation models and should be assessed in the analysis. There is no simple trend concerning the change of posterior model probabilities when one changes the prior (Zhang et al. 2011). Instead the posterior for any model tends to increase if the prior becomes more and more highly concentrated around the MLEs of parameters under that model. See §6.2.3.3 for a discussion of the sensitivity of Bayesian model comparison to the prior.

The multispecies coalescent model (Rannala and Yang 2003), which underlies the Bayesian implementation in the program BPP by Yang and Rannala (2010), assumes no gene flow following species divergence. The impact of gene flow on species delimitation by BPP has been examined in a simulation study (Zhang et al. 2011). Figure 9.20 shows a set of typical results, where BPP is used to calculate the posterior probability for the one-species and two-species models when sequences are sampled from two species undergoing different levels of gene flow. Migration is measured by $M = Nm$, the expected number of immigrants per generation. Note that if $M = 0$, the two-species model is true, while if $M \to \infty$, the one-species model is true. In both cases the true model will dominate in the posterior if the dataset is not too small. When $0 < M < \infty$, however, neither the one-species model nor the two-species model is true. The 'phase change' occurs around $M = 1$ or in the range $0.1 < M < 10$. When $M < 0.1$, the method behaves as if there is no migration, while at $M \geq 10$, the method infers one species. Thus low levels of migration have little impact and the method infers distinct species with good power despite small amounts of gene flow. This appears to be consistent with biologists' common practice of identifying distinct species despite occasional hybridizations. At high levels of hybridization (e.g. with > 10 immigrants per generation), there is effectively one species, and the method indeed infers one species.

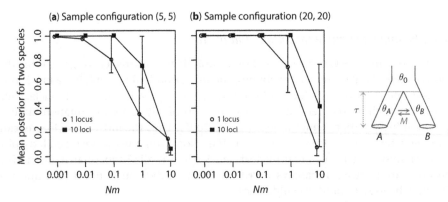

Fig. 9.20 Impact of migration on species delimitation. Mean posterior probability for the two-species model plotted against the migration rate M when the data are simulated under the two-species model with migration, with parameters $\theta = \tau = 0.01$. The scaled migration rate $M = Nm$ is the expected number of migrants per generation. The data consist of either one locus or ten loci, with (a) 5 or (b) 20 sequences from each population at each locus. The priors used in the analysis are $\theta \sim G(1, 100)$ and $\tau \sim G(1, 100)$. Redrawn following Zhang et al. (2011, Figure 4).

It is well known in population genetics theory that the impact of migration on population differentiation measured by F_{ST} depends on whether $M = Nm$ is much larger or much smaller than 1 (e.g. Wright 1931; Takahata 1983). If $M = Nm \ll 1$, the populations will be strongly differentiated, whereas if $M \gg 1$ (e.g. if there are more than ten or so migrants), migration will swamp the population and the population will behave as a panmictic unit. It is interesting that the phase change occurs around $M = 1$, in both Bayesian species delimitation and in the impact of migration on population differentiation, although the two perspectives are very different.

Figure 9.21 shows the behaviour of BPP when data are taken from a species that has a broad geographical distribution and is undergoing isolation by distance. Data are generated under a stepping stones model, with four populations (A, B, C, D) arranged on a line and with migration occurring between adjacent populations. The data were simulated under the equilibrium migration model, but samples were taken for analysis from two populations only, to calculate the posterior probabilities for the one-species and two-species models. The mean posterior probability for the two-species model (P_2) was very similar when the two sampled populations are next to each other (AB data) or far apart (AD data). In all cases, BPP tends to infer one species when $M \leq 0.1$ and two species when $M \gg 1$, where $M = Nm$ is the scaled migration rate between two adjacent populations. Many of us have the misconception that if the migration rate (proportion of immigrants) between two adjacent populations is $m \ll 1$, the effective migration rate between A and D which are three steps apart should be $\sim m^3$. However, the correct migration rate between A and D is $\sim m/3$ (Strobeck 1987; Slatkin 1991; Zhang et al. 2011). The simulation suggests that BPP is robust to complex population structures or presence of unsampled ghost populations (Wakeley and Aliacar 2001; Beerli 2004). Obviously the behaviour of the method will depend on the number of intermediate steps in the stepping stones model and the

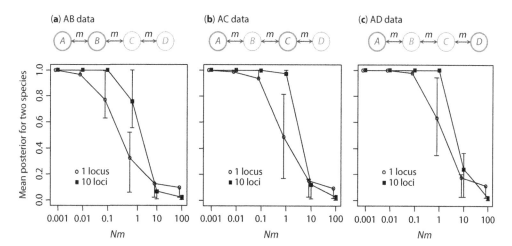

Fig. 9.21 Impact of isolation by distance on Bayesian species delimitation. Mean posterior probability for the two-species model is plotted against the migration rate M in the stepping stones model. The population size parameter for each of the four populations is $\theta = 0.01$, while $M = Nm$ is the scaled migration rate between two adjacent populations. The sample configuration is $(5, 5)$, with samples taken (**a**) from A and B, (**b**) from A and C, and (**c**) from A and D. The data were then analysed to compare the one-species and two-species models. The priors used in the analysis are $\theta \sim G(1, 100)$ and $\tau \sim G(1, 100)$. Redrawn following Zhang et al. (2011, Figure 7).

migration rates between adjacent populations. Note that it is often arbitrary to decide whether allopatric populations should be considered distinct species, geographical races, or subspecies.

In summary, for the purpose of delimiting species, there does not appear to be a need to explicitly account for migration in the model. However a model of migration (e.g. Hey 2010) is useful for estimating parameters such as the migration rates.

9.5.5 *Pros and cons of Bayesian species delimitation*

The Bayesian method of species delimitation has a few clear advantages over simple methods that use genetic data to delimit species, such as arbitrary cutoffs based on genetic divergence or estimated migration rates, reciprocal monophyly of gene trees, and coalescent-based methods that use estimated gene trees (either with or without using branch length information) without accommodating phylogeny reconstruction errors. The Bayesian method can analyse multiple loci, unlike DNA barcoding or mtDNA taxonomy which typically rely on one mitochondrial locus. The multispecies coalescent model accounts for ancestral polymorphism and incomplete lineage sorting and the resulting species tree–gene tree conflict. There is then no need to rely on reciprocal monophyly of gene trees to infer species status. The likelihood calculation on sequence alignments allows the method to make a full use of the information in the data while accounting for the uncertainties in the gene tree topologies and branch lengths.

Compared with traditional taxonomic practice for species delimitation, the Bayesian method may be considered more objective (Fujita and Leaché 2011). All its model assumptions are explicit and can be tested. In contrast, morphology-based taxonomic practices vary widely and are incomparable across different taxonomic groups. Issac et al. (2004), for example, observed that ant taxonomists tend to be 'splitters' while butterfly taxonomists are 'lumpers', and species counts are not comparable in those species groups. The Bayesian method infers species status from a genealogical and population genetic perspective. The method also allows biologists to incorporate information on plausible species memberships based on morphology, behaviour, etc., through the use of the prior. The prior information is then combined with the genetic evidence to generate posterior probabilities for species delimitation models. Simulation also suggests that Bayesian species delimitation using BPP has good power in identifying distinct species in presence of small amounts of gene flow, and is not easily misled to infer geographical populations as distinct species.

The Bayesian method of species delimitation has nevertheless a number of limitations. First, the multispecies coalescent model which underlies the method assumes neutral evolution and does not account for selection, in particular species-specific selection. The method is designed to analyse the neutral genome to infer species status based on information concerning the patterns and timings of population divergences. Protein-coding genes under similar selective constraints in different species may still be used in the analysis, as the effect of such selection is to reduce the neutral mutation rate. However, the method may not be useful for identifying speciation genes or genes that are involved in the establishment of reproductive barriers or that are responsible for species adaptation to different ecological habitats. These genes may show drastically different patterns of divergence from the neutral genome and may show up as outliers in the analysis. Methods for detecting adaptive evolution, to be discussed in Chapter 11, may be more suitable for identifying such adaptively evolving genes (Swanson and Vacquier 2002b). Second the different species delimitation models compared in the Bayesian method make different assumptions about isolation or random mating between populations but are reticent about the causes of isolation. Isolation due to geographical barriers that remove the chance of interbreeding and isolation due to intrinsic reproductive barriers (inability to mate, hybrid inviability or sterility, etc.) are drastically different to defining species by the BSC, but may have the same effect on the Bayesian analysis under the multispecies coalescent model. The Bayesian method may infer two allopatric populations that have accumulated neutral mutations and are thus genetically divergent but that have not established reproductive barriers as distinct species if the divergence time is long enough or if the dataset is large enough. Whether allopatric populations should be considered distinct species or merely subspecies or geographical races is often arbitrary. We note that this ambiguity of interpretation does not exist if sympatric populations are analysed using the Bayesian method.

9.6 Problems

9.1 Write a small simulation program to generate a sample of n (= 10, say) DNA sequences from a population. Use the JC69 mutation model. Simulate $L = 5$ loci, with $l = 1{,}000$ sites at each locus. For example, implement Algorithms 9.1 or 9.2.

9.2 Two species A and B, each of population size N, separated τ generations ago. We sample two sequences a_1 and a_2 from species A and one sequence b from species B. Derive the probability that the gene tree has the topology $((a_1, a_2), b)$.

9.3* ML estimation of the species tree for three species given the gene tree at one locus, with one sequence from each species. Use the gene tree $G = ((A, B), C)$, with node ages t_0 and t_1, of Figure 9.22a as given data to evaluate the likelihood for the species

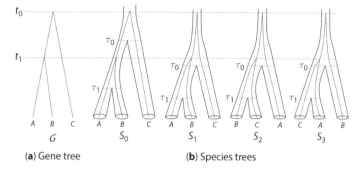

(a) Gene tree   (b) Species trees

Fig. 9.22 Estimation of the species tree for three species using a gene tree for one locus, with one sequence from each species. (a) The gene tree, with topology $G = ((A, B), C)$ and node ages t_0 and t_1, is the given data. (b) The species trees with their parameters. Species tree S_0 involves parameters τ_0 and τ_1, under the constraints $\tau_1 \leq t_1 \leq \tau_0 \leq t_0$, while each of species trees S_1, S_2, and S_3 involves parameters τ_0 and τ_1, with $\tau_1 \leq \tau_0 \leq t_1 \leq t_0$. Each of the species tree also involve two population size parameters θ_0 and θ_1, which are not shown. The Maximum Tree algorithm assumes $\theta_0 = \theta_1 = \theta$. Note that species trees S_0 and S_1 have the same topology.

* indicates a more difficult or technical problem.

trees of Figure 9.22b, under the assumption that all populations have the same θ. Treat species trees S_0 and S_1 separately even though they have the same tree topology. Show that the ML estimate of the species tree is S_0, with $\hat{\tau}_0 = t_0$ and $\hat{\tau}_1 = t_1$. [Hint: Write down the likelihood function for species tree S_0, which is the multispecies coalescent density for the gene tree, $f(G, t_0, t_1 | S_0, \tau_0, \tau_1, \theta)$, and maximize it by adjusting τ_0, τ_1, and θ under the constraints $\tau_1 \leq t_1 \leq \tau_0 \leq t_0$. Then repeat the analysis for species trees S_1, S_2, and S_3.]

9.4* *Singularity in the likelihood function in estimation of the species tree given the gene tree.* As in Problem 9.3, use the gene tree G with node ages t_0 and t_1 of Figure 9.22a to evaluate the likelihood for the species trees of Figure 9.22b, but allowing different population size parameters θ_0 and θ_1. Show that the likelihood for species tree S_0 can become infinite when $\tau_1 \to t_1, \theta_1 \to 0$, with $\tau_0 > \tau_1$ and $\theta_0 > 0$. In this particular case, species trees S_1–S_3 do not show singularity in the likelihood function, but in the general case of more than three species and sequences, multiple species trees can have infinite likelihood. Also convince yourself that singularity cannot occur if the likelihood is calculated using the sequence alignment (i.e. Yang 2002b, Equation 8) rather than the coalescent times on a gene tree.

9.5 Examine the probability densities $f_0(t)$ and $f_1(t)$ of Figure 9.15 for the divergence time t between two sequences at a locus under the symmetrical IM model for two species (SIM2s). Derive the densities $f_0(t)$ and $f_1(t)$ for the limiting cases of $M = 0$ and ∞.

9.6 Use the four nuclear autosomal loci from the two species/populations of butterflies (*Heliconius demeter* and *H. eratosignis*) to delimit species (Dasmahapatra et al. 2010; Zhang et al. 2011). The four loci are Mpi (9 sequences, 496 bp), Tektin (9 sequences, 733 bp), Rp15 (15 sequences, 713 bp), and Ef1a (18 sequences, 766 bp). Examine the effect of the number of loci by analysing 1, 2, 3, or 4 loci. Use different priors on τs and θs to run BPP to evaluate the impact of the prior.

* indicates a more difficult or technical problem.

CHAPTER 10

Molecular clock and estimation of species divergence times

10.1 Overview

The hypothesis of the *molecular clock* asserts that the rate of DNA or protein sequence evolution is constant over time or among evolutionary lineages. In the early 1960s, when protein sequences became available, it was observed that the number of differences between proteins from different species, such as haemoglobin (Zuckerkandl and Pauling 1962), cytochrome *c* (Margoliash 1963), and fibrinopeptides (Doolittle and Blomback 1964), was roughly proportional to the divergence time between the species. This observation led to the proposal of the *molecular evolutionary clock* hypothesis by Zuckerkandl and Pauling (1965).

A few clarifications are in order. First, the clock was envisaged as a stochastic clock due to the random nature of the amino acid or nucleotide substitution process. It 'ticks' at random time points rather than regularly as a clock does. Under the Markov models of nucleotide or amino acid substitution, substitution events arrive ('tick') at time intervals that are exponentially distributed. Second, it was noted from the beginning that different proteins or regions of a protein evolve at very different rates, and the clock hypothesis allows for such rate differences among proteins; the different proteins are said to have their own clocks, which tick at different rates. Third, rate constancy may not hold globally for all species and usually applies to a group of species. For example, the clock may hold roughly within primates but may be seriously violated when different vertebrate species are compared.

As soon as it was proposed, the molecular clock hypothesis had an immediate and tremendous impact on the burgeoning field of molecular evolution and has been a focus of controversy throughout its history. First, the utility of the molecular clock was immediately recognized. If proteins evolve at roughly constant rates, they can be used to reconstruct phylogenetic trees and to estimate divergence times among species (Zuckerkandl and Pauling 1965). Second, the reliability of the clock and its implications for the mechanism of molecular evolution were a focus of immediate controversy, entwined in the neutralist–selectionist debate. See §11.2 later for a brief review of the latter debate. At the time, the neo-Darwinian theory of evolution was generally accepted by evolutionary biologists, according to which the evolutionary process is dominated by natural selection. A constant rate of evolution among species as different as elephants and mice was incompatible with that theory, as species living in different habits with different life histories must be under very different regimes of selection. When the *neutral theory of molecular evolution* was proposed (Kimura 1968; King and Jukes 1969), the observed clock-like behaviour of molecular evolution became 'perhaps the strongest evidence for the theory' (Kimura and Ohta 1971b). This theory emphasizes random fixation of neutral or nearly neutral mutations. Suppose the mutation rate is μ, and among all new mutations a proportion f_0 of them are neutral while the others are strongly deleterious. The rate of molecular evolution is then equal to the neutral mutation rate $\mu_0 = \mu f_0$, independent of factors such as the environment or the population size. If the mutation rate (μ) is similar and the function of the protein remains the same so that the proportion of neutral mutations (f_0) is the same among species, a constant evolutionary rate will be expected by the theory (Ohta and Kimura 1971). Rate differences among proteins are explained by the presupposition that different proteins are under different functional constraints, with different fractions of neutral mutations. Nevertheless, the neutral theory is not the only one compatible with clock-like evolution, and neither does the neutral theory always predict a molecular clock. Controversies also exist concerning whether the neutral theory predicts rate constancy over generations or over calendar time, or whether the clock should apply to silent DNA changes or to protein evolution (Kimura 1983; Li and Tanimura 1987; Gillespie 1991; Li 1997).

Since the 1980s, DNA sequences have accumulated rapidly and have been used to conduct extensive tests of the clock and to estimate evolutionary rates in different groups of organisms. Wu and Li (1985) and Britten (1986) noted that primates have lower rates than rodents, and humans have lower rates than apes and monkeys, observations known as *primate slowdown* and *hominoid slowdown* (Li and Tanimura 1987). Two major factors that are proposed to account for between-species rate differences are generation time, with a shorter generation time associated with more germ-line cell divisions per calendar year and thus a higher substitution rate (Laird et al. 1969; Wilson et al. 1977; Wu and Li 1985; Li et al. 1987) and DNA repair, with a less reliable repair mechanism associated with a higher mutation or substitution rate (Britten 1986). Martin and Palumbi (1993) found that substitution rates in both nuclear and mitochondrial genes are negatively related to body size, with high rates in rodents, intermediate rates in primates, and slow rates in whales. Body size is not expected to affect substitution rate directly, but is highly correlated with a number of physiological and life history variables, notably generation time and metabolic rate. Part of the difficulty in this debate is that the multiple predictors are all correlated, and correlation analysis is ill-posed for detecting cause-effect relationships. For example, in mammals, many traits reflect different aspects of the mammalian life history. Small mammals tend to have short generation times, higher metabolic rates, short life spans, and large population sizes, and they tend to have higher molecular evolutionary rates (Bromham 2011).

Application of the clock to estimate divergence times began with Zuckerkandl and Pauling (1962), who used an approximate clock to date duplication events among α, β, λ, and δ globins of the haemoglobin family. The molecular clock has since been widely used to date species divergences, and has produced a steady stream of controversies, mostly because molecular dates are often at odds with the fossil records. The conflict is most evident concerning several major events in evolution (Cooper and Fortey 1998). The first is the origin of the major animal forms. Fossil forms of metazoan phyla appear as an 'explosion' about 500–600 million years ago (MYA) in the early Cambrian (Knoll and Carroll 1999), but most molecular estimates have been much older, sometimes twice as old (e.g. Wray et al. 1996). It may be the case that the currently available molecular data and fossil calibrations are simply not informative enough to provide meaningful date estimates for such deep divergences. Another major evolutionary event is the origins and divergences of modern mammals and birds following the demise of the dinosaurs about 66 MA at the Cretaceous–Paleogene boundary (the K–Pg boundary). Early molecular studies again produced much older dates than were expected from fossils (e.g. Hedges et al. 1996). The discrepancy is partly due to the incompleteness of fossil data; fossils represent the time when species developed diagnostic morphological characters and were fossilized, while molecules represent the time when the species became genetically isolated, so fossil dates have to be younger than molecular dates (Foote et al. 1999; Tavaré et al. 2002). Another part of the discrepancy is due to the inaccuracies and deficiencies of early molecular dating studies. More recent analyses integrating fossil and molecular data have produced date estimates that are far more in line with the fossil record (Meredith et al. 2011; dos Reis et al. 2012). It is noteworthy that interactions between molecules and fossils have prompted reinterpretations of the fossil record as well as critical evaluations of molecular dating techniques and developments of more advanced analytical methods.

A number of reviews have been published about the molecular clock and its use in the estimation of divergence times. See, for example, Morgan (1998) and Kumar (2005) for the history of the molecular clock, and Bromham and Penny (2003) for a discussion of the clock in relation to theories of molecular evolution. My focus in this chapter is on statistical methods for testing the clock hypothesis, and on likelihood and Bayesian methods for dating species divergences under global clock (strict clock) and local clock (relaxed clock) models. In such analysis, fossils are used to calibrate the clock, i.e. to translate sequence distances into absolute geological times and substitution rates. A similar situation concerns the dating of viral divergences, as viral genes evolve so fast that changes are observed over years. In these studies one can use the dates at which the viruses were isolated to calibrate the clock and to estimate divergence times, using very similar techniques to those for dating species divergences using fossil calibrations.

10.2 Tests of the molecular clock

10.2.1 *Relative-rate tests*

The simplest test of the clock hypothesis examines whether two species A and B evolve at the same rate by using a third outgroup species C (Figure 10.1). This has become known as the *relative-rate test*, even though almost all tests of the clock compare relative rather than absolute rates. The relative-rate test has been proposed by a number of authors, starting with Sarich and Wilson (1973; see also Sarich and Wilson 1967). If the clock hypothesis is true, the distances from ancestral node O to species A and B should be equal: $d_{OA} = d_{OB}$ or $a = b$ (Figure 10.1b). Equivalently one can formulate the clock hypothesis as $d_{AC} = d_{BC}$. Sarich and Wilson (1973) did not describe how to decide on the significance of the difference. Fitch (1976) used the number of differences to measure the distance between species, and calculated $a = (d_{AB} + d_{AC} - d_{BC})/2$ and $b = (d_{AB} + d_{BC} - d_{AC})/2$ as the numbers of changes along branches OA and OB. The clock was then tested by comparing $X^2 = (a-b)^2/(a+b)$ against χ_1^2. This is equivalent to using the binomial distribution $\text{bin}(a+b, \frac{1}{2})$ to test whether the observed proportion $a/(a+b)$ deviates significantly from $\frac{1}{2}$, with $\left(\frac{a}{a+b} - \frac{1}{2}\right) / \sqrt{\frac{1}{2} \times \frac{1}{2}/(a+b)} = (a-b)/\sqrt{a+b}$ compared against the standard normal distribution. This approach fails to correct for multiple hits, and furthermore, the χ^2 and binomial approximations may not be reliable if $a+b$ is small. Wu and Li (1985) corrected for multiple hits under the K80 model (Kimura 1980), and calculated $d = d_{AC} - d_{BC}$ and its standard error. Then $d/\text{SE}(d)$ was compared against the standard normal distribution.

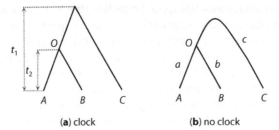

Fig. 10.1 Rooted and unrooted trees used to illustrate the relative-rate test. (**a**) Under the clock, the parameters are the ages of two ancestral nodes t_1 and t_2, measured by the expected number of substitutions per site. (**b**) Without the clock, the parameters are the three branch lengths a, b, and c, also measured by the expected number of substitutions per site. The clock model is a special case of the no-clock model with the constraint $a = b$; that is, the no-clock model reduces to the clock model when $a = b = t_2$ and $(a + c)/2 = t_1$.

Of course, calculation of the distances and their standard errors relies on a substitution model.

Another version of the relative-rate test that is not sensitive to the substitution model was proposed by Tajima (1993). This compares the counts (say, m_1 and m_2) of two site patterns xyy and xyx, where x and y are any two distinct nucleotides. Tajima suggested the comparison of $(m_1 - m_2)^2/(m_1 + m_2)$ against χ_1^2, like Fitch's test above.

The relative-rate test can also be conducted in a likelihood framework (Muse and Weir 1992). This is then a special case of the likelihood ratio test (LRT) of the clock discussed by Felsenstein (1981) (see below). One calculates ℓ_0 and ℓ_1, the log likelihood values with and without constraining $a = b$, respectively (Figure 10.1b). Then $2\Delta\ell = 2(\ell_1 - \ell_0)$ is compared against χ_1^2, with one degree of freedom.

10.2.2 Likelihood ratio test

The LRT of the clock (Felsenstein 1981) applies to a tree of any size. Under the clock (H_0), there are $s - 1$ parameters corresponding to the ages of the $s - 1$ internal nodes on the rooted tree with s species, measured by the expected number of changes per site (see §4.2.3). The more general no-clock model (H_1) allows every branch to have its own rate. Because time and rate are confounded, this model involves $2s - 3$ free parameters, corresponding to the branch lengths in the unrooted tree. In the example of Figure 10.2 with $s = 6$ species, there are $s - 1 = 5$ parameters under the clock (t_1–t_5), and $2s - 3 = 9$ parameters without the clock (b_1–b_9). The clock model is nested within the no-clock model: by applying $s - 2$ equality constraints, the no-clock model reduces to the clock model. In the example, the four constraints may be $b_1 = b_2, b_3 = b_7 + b_1, b_4 = b_8 + b_3$, and $b_5 = b_9 + b_4$. The inequality constraint $b_6 > b_5$ does not reduce the number of parameters. Let ℓ_0 and ℓ_1 be the log likelihood values under the clock and no-clock models, respectively. Then $2\Delta\ell = 2(\ell_1 - \ell_0)$ is compared with the χ^2 distribution with df = $s - 2$ to

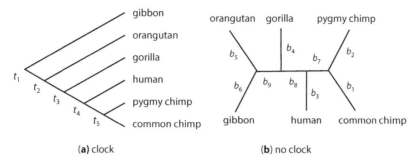

Fig. 10.2 Rooted and unrooted trees for six primate species for the data of Horai et al. (1995), used to illustrate the LRT of the molecular clock. (**a**) Under the clock model, the parameters are the ages of the five ancestral nodes t_1–t_5 in the rooted tree. (**b**) Without the clock, the parameters are the nine branch lengths b_1–b_9 in the unrooted tree. The two models can be compared using an LTR to decide whether the clock holds, with 4 (= 9 − 5) degrees of freedom. Note that both the node ages of (**a**) and the branch lengths of (**b**) are measured by the expected number of substitutions per site.

decide whether the clock is rejected. Note that to test the clock, there is no need to know or to estimate absolute divergence times.

Example 10.1: A test of the molecular clock in the mitochondrial 12s rRNA genes of human, common chimpanzee, pygmy chimpanzee, gorilla, orangutan, and gibbon. The rooted and unrooted trees are shown in Figure 10.2. See Horai et al. (1995) for the GenBank accession numbers of the sequences. There are 957 sites in the alignment, with the average base compositions to be 0.216 (T), 0.263 (C), 0.330 (A), and 0.192 (G). We use the K80 model to conduct the test. The log likelihood values are $\ell_0 = -2345.10$ under the clock and $\ell_1 = -2335.80$ without the clock. Comparison of $2\Delta\ell = 2(\ell_1 - \ell_0) = 18.60$ with the χ_4^2 distribution indicates significant difference, with $P < 0.001$. The clock is thus rejected for those data. □

10.2.3 Limitations of molecular clock tests

A few limitations of the molecular clock tests may be noted here. First, the null hypothesis examined by the tests discussed above is not exactly rate constancy over time. Instead it is a weaker hypothesis that all tips of the tree are equidistant from the root, with distance measured by the number of substitutions. If the evolutionary rate has been accelerating (or decelerating) over time in all contemporary lineages, the tree will look clock-like and the null hypothesis examined by those tests will be true, even though the rate is not constant. Similarly the tests examine rate differences between ingroup species and do not detect different rates between the ingroup and outgroup species. For example, in the relative-rate test using three species, the tests may detect a rate difference between species *A* and *B*, but not between species *C* and the ingroup species (Figure 10.1). Second, the tests of molecular clock cannot distinguish a constant rate from an average variable rate *within* a lineage, although the latter may be a more sensible explanation than the former when the clock is rejected and the rate is variable across lineages. Finally, as with any statistical test, failure to reject the clock may be due to lack of information in the data or lack of power of the test. For example, the relative-rate test applied to only three species is noted to have only limited power in simulations (Bromham et al. 2000). However, when applied to multiple species, the LRT of the clock appears very powerful, rejecting the clock before the violation of the clock is serious enough to have any major impact on phylogenetic analysis.

10.2.4 *Index of dispersion*

Many early studies tested the molecular clock using a so-called *index of dispersion* (R), defined as the variance to mean ratio of the number of substitutions among lineages (Ohta and Kimura 1971; Langley and Fitch 1974; Kimura 1983; Gillespie 1984, 1986b, 1991). When the rate is constant and the species are related to each other through a star tree, the number of substitutions on each lineage should follow a Poisson distribution, according to which the mean should equal the variance (Ohta and Kimura 1971). An R significantly greater than one, a phenomenon known as the *over-dispersed clock*, means that the expectation of the clock hypothesis is violated. The dispersion index is used more as a diagnosis of the relative importance of mutation and selection and as a test of the neutral theory than as a test of rate constancy over time. A dispersion index greater than one can be taken as evidence for selection, but it can also indicate non-selective factors, such as generation times, mutation rates, efficiency of DNA repair, and metabolic rate effect. Kimura (1983, 1987) and Gillespie (1986a, 1991) used gene sequences from different orders of mammals and found that R is often greater than one. However, the authors' use of the star phylogeny for mammals appears to have led to overestimates of R (Goldman 1994). Ohta (1995) used three lineages only (primates, artiodactyls, and rodents) to avoid the problem of phylogeny, but the analysis may be prone to sampling errors as the variance is calculated using only three lineages. The dispersion index appears somewhat obsolete (cf. Cutler 2000), as the hypotheses can often be tested more rigorously using LRTs, with more realistic models being used to estimate evolutionary rates (e.g. Yang and Nielsen 1998).

10.3 Likelihood estimation of divergence times

10.3.1 *Global clock model*

The molecular clock assumption provides a simple yet powerful approach to dating evolutionary events. Under the clock, the expected distance between sequences increases linearly with their time of divergence. When external information about the geological ages of one or more nodes on the phylogeny is available (typically based on the fossil record), sequence distances or branch lengths can be converted into absolute geological times. This method of estimating species divergence times is known as *molecular clock dating*. Both distance and likelihood methods can be used to estimate the distances from the internal nodes to the present time. The assumed substitution model may be important, as a simplistic model may not correct for multiple hits properly and may underestimate distances. Often the underestimation is more serious for large distances than for small ones, and the non-proportional underestimation may generate systematic biases in divergence time estimation. Nevertheless, there is no difficulty in using any of the substitution models discussed in Chapters 1, 2, and 4, so the effect of the substitution model can easily be assessed. My focus in this chapter is on assumptions about the rates, incorporation of fossil uncertainties, assessment of errors in time estimates, etc.

Similarly the rooted tree topology is assumed to be known in molecular clock dating. Uncertainties in the tree may have an impact on estimation of divergence times, but

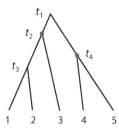

Fig. 10.3 A phylogenetic tree for five species, used to illustrate likelihood and Bayesian methods for divergence time estimation. Fossil calibrations are available for nodes 2 and 3.

the effects are expected to be complex, depending on whether the conflicts are around short internal branches, or on the relative locations of the nodes to be dated, the calibration nodes and the unresolved nodes. When the tree topology involves uncertainties, it appears inappropriate to use a consensus tree with poorly supported nodes collapsed into polytomies. A polytomy in a consensus tree indicates unresolved relationships rather than the best estimate of the true relationship. It is better to use a binary tree that is likely to be correct, such as the maximum likelihood (ML) tree. The use of several binary trees may provide an indication of the robustness of time estimation to uncertainties in the tree topology.

There are $s-1$ ancestral nodes in a rooted tree of s species. Suppose that the ages of c ancestral nodes are known without error, determined from fossil data. The model then involves $s-c$ parameters: the substitution rate μ and the ages of $s-1-c$ nodes that are not calibration points. For example, the tree of Figure 10.3 has $s = 5$ species, with four interior node ages: $t_1, t_2, t_3,$ and t_4. Suppose node ages t_2 and t_4 are fixed according to the fossil record. Then three parameters are estimated under the model: $\mu, t_1,$ and t_3. Given rate μ and the times, each branch length is just the product of the rate and the time duration of the branch, so that the likelihood can be calculated using standard algorithms (see §4.2). Times and rates are then estimated by maximizing the likelihood. The time parameters have to satisfy the constraints that any node should not be older than its mother node; in the tree of Figure 10.3, $t_1 > \max(t_2, t_4)$ and $0 < t_3 < t_2$. Numerical optimization of the likelihood function has to be performed under such constraints, which can be achieved by using constrained optimization or through variable transformations (see §4.5.2.4).

10.3.2 Local clock model

The molecular clock often holds in closely related species, for example, within the hominoids or even primates, but is clearly violated in distant comparisons, for example, among different orders of mammals (Yoder and Yang 2000; Hasegawa et al. 2003; Springer et al. 2003). Given that sequences provide information about distance but not time and rate separately, one may expect divergence time estimation to be sensitive to assumptions about the rate. This is indeed found to be the case in numerous studies.

One approach to dealing with violation of the clock is to remove some species so that the clock approximately holds for the remaining species (Li and Tanimura 1987; Takezaki et al. 1995). This may be useful if one or two lineages with grossly different rates can be identified and removed, but is difficult to use if the rate variation is more complex. Another approach is to take explicit account of among-lineage rate variation when estimating divergence times. This has been the focus of much research, with both likelihood and Bayesian methodologies employed. In this section we describe

Fig. 10.4 The tree of four species used in the quartet-dating method of Rambaut and Bromham (1998). Two substitution rates are assumed for branches on the left and on the right of the tree, respectively.

the likelihood approach, including heuristic rate-smoothing algorithms. The Bayesian approach is discussed in §10.4.

In the likelihood method, one may assign different rates to branches on the tree and then estimate, by ML, both the divergence times and the branch rates. The first application of such local clock models appears to be Kishino and Hasegawa (1990), who estimated divergence times within hominoids under models with different transition and transversion rates among lineages. The likelihood is calculated by using a multivariate normal approximation to the observed numbers of transitional and transversional differences between sequences. Rambaut and Bromham (1998) discussed a likelihood quartet-dating approach, which applies to a particular tree of four species: $((a, b), (c, d))$, shown in Figure 10.4, with a rate for the left part of the tree and another rate for the right part of the tree. This was extended by Yoder and Yang (2000) to trees of any size, with an arbitrary number of rates assigned to the branches. The implementation is very similar to that of the global clock model discussed above. The only difference is that under a local clock model with k branch rates, one estimates $k-1$ extra rate parameters.

A serious drawback of such local clock models is their arbitrariness in the number of rates to assume and in the assignment of rates to branches. Note that some assignments are not feasible as they cause the model to become unidentifiable. The approach may be straightforward to use if external biological reasons can be used to assign branches to rates; for example, two groups of species may be expected to be evolving at different rates and thus can be assigned different rates. However, in general too much arbitrariness may be involved. Yang (2004) tested a rate-smoothing algorithm to assist automatic assignment of branches to rates, which has been improved by Aris-Brosou (2007).

10.3.3 Heuristic rate-smoothing methods

Sanderson (1997) described a heuristic rate-smoothing approach to estimating divergence times under local clock models without *a priori* assignment of branches to rates. The approach follows Gillespie's (1991) idea that the rate of evolution may itself evolve, so that the rate is auto-correlated across lineages on the phylogeny, with closely related lineages sharing similar rates. The method minimizes changes in rate across branches, thus smoothing rates and allowing joint estimation of rates and times. Let the branch leading to node k be referred to as branch k, which has rate r_k. Let $\boldsymbol{t}$ be the times and $\boldsymbol{r}$ the rates. Sanderson (1997) used parsimony or likelihood to estimate branch lengths on the tree (b_k) and then fitted times and rates to them, by minimizing

$$W(\boldsymbol{t}, \boldsymbol{r}) = \sum_k \left(r_k - r_{\mathrm{anc}(k)}\right)^2, \qquad (10.1)$$

subject to the constraints

$$r_k T_k = b_k. \qquad (10.2)$$

Here anc(k) is the node ancestral to node k, so that $r_k - r_{\text{anc}(k)}$ is the change in rate between branch k and its mother branch, and T_k is the time duration of branch k, which is $t_{\text{anc}(k)}$ if node k is a tip and $t_{\text{anc}(k)} - t_k$ if node k is an interior node. The summation in equation (10.1) is over all the nodes except for the root. For the two daughter nodes of the root (let them be 1 and 2), which do not have an ancestral branch, the sum of squared rate differences may be replaced by $(r_1 - \bar{r})^2 + (r_2 - \bar{r})^2$, where $\bar{r} = (r_1 + r_2)/2$. The approach insists on a perfect fit to the estimated branch lengths (equation (10.2)) while minimizing changes in rate between the ancestral and descendent branches. Sampling errors in the estimated branch lengths are ignored.

An improved version was suggested by Sanderson (2002), called the penalized likelihood method, which maximizes the following 'log likelihood function':

$$\ell(\mathbf{t}, \mathbf{r}, \lambda; X) = \log\{f(X|\mathbf{t}, \mathbf{r})\} - \lambda \sum_k (r_k - r_{\text{anc}(k)})^2. \tag{10.3}$$

The second term is the same as in equation (10.1) and penalizes rate changes across branches. The first term is the log likelihood of the data, given $\mathbf{t}$ and $\mathbf{r}$. This could be calculated by applying Felsenstein's (1981) pruning algorithm to the sequence alignment. Sanderson (2002) used a Poisson approximation to the number of changes for the whole sequence along each branch estimated by parsimony or likelihood. The probability of the data is then given by multiplying Poisson probabilities across branches. The approximation ignores correlations between estimated branch lengths. Furthermore, the assumption of Poisson distribution for the number of substitution is valid only under simple models like JC69 and K80. Under more complex models such as HKY85 or GTR, different nucleotides have different rates and the process is not Poisson.

The smoothing parameter λ determines the relative importance of the two terms in equation (10.3), or how much rate variation should be tolerated. If $\lambda \to 0$, the rates are entirely free to vary, leading to a perfect fit to the branch lengths. If $\lambda \to \infty$, no rate change is possible and the model should reduce to the clock. There is no rigorous criterion for choosing λ. Sanderson (2002) used a cross-validation approach to estimate it from the data. For any value of λ, one sequentially removes small subsets of the data, estimates parameters $\mathbf{r}$ and $\mathbf{t}$ from the remaining data, and then uses the estimates to predict the removed data. The value of λ that gives the overall best prediction is used as the estimate. Sanderson used the so-called leave-one-out cross-validation and removed in turn the branch length (the number of changes) for every terminal branch.

A few modifications were introduced by Yang (2004), who suggested maximizing the following 'log likelihood function':

$$\ell(\mathbf{t}, \mathbf{r}, v; X) = \log\{f(X|\mathbf{t}, \mathbf{r})\} + \log\{f(\mathbf{r}|\mathbf{t}, v)\} + \log\{f(v)\}. \tag{10.4}$$

The first term is the log likelihood of the sequence data. Yang (2004) used a normal approximation to the maximum likelihood estimates (MLEs) of branch lengths estimated without the clock, following Thorne et al. (1998) (see below). The second term, $f(\mathbf{r}|\mathbf{t})$, is a prior density for rates. This is specified using the geometric Brownian motion model of rate drift of Thorne et al. (1998). The model is illustrated in Figure 10.5 and will be discussed in more detail in §10.4.3. Here it is sufficient to note that given the ancestral rate $r_{\text{anc}(k)}$, the current rate r has a log-normal distribution with density

$$f(r_k|r_{\text{anc}(k)}) = \frac{\exp\left\{-\frac{1}{2tv}\left(\log(r_k/r_{\text{anc}(k)}) + \frac{1}{2}tv\right)^2\right\}}{r_k\sqrt{2\pi tv}}, 0 < r_k < \infty, \tag{10.5}$$

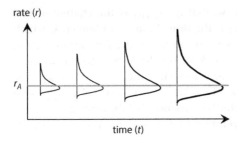

Fig. 10.5 The geometric Brownian motion model of rate drift. The mean rate stays the same as the ancestral rate r_A, but the variance parameter tv increases linearly with time t.

where parameter v specifies how clock-like the tree is, with a large v meaning highly variable rates and serious violation of the clock. The density $f(\boldsymbol{r}|\boldsymbol{t})$ of all rates is given by multiplying the log-normal densities across branches. Because the density of equation (10.5) is high when r_k is close to $r_{\text{anc}(k)}$ and low when the two rates are far apart, maximizing $f(\boldsymbol{r}|\boldsymbol{t})$ has the effect of minimizing rate changes over branches, similar to minimizing the sum of squared rate differences in equation (10.3). Compared with Sanderson's approach, this has the advantage of automatically taking account of different scales of rates and branch lengths, so that there is no need for the smoothing parameter λ or for cross-validation. If the data contain sequences from multiple genes, one simply multiplies the likelihood $f(X|\boldsymbol{t},\boldsymbol{r})$ and the rate prior $f(\boldsymbol{r}|\boldsymbol{t})$ across loci. Lastly, the third term in equation (10.4) applies an exponential prior with a small mean (0.001) to penalize large values of v.

A modification to equation (10.4) is to include an additional prior $f(\boldsymbol{t})$. The estimates may then have an approximate Bayesian interpretation. Akerborg et al. (2008) discussed faster algorithms for optimizing the resulting objective function.

The rate-smoothing algorithms (Sanderson 1997, 2002; Yang 2004) are *ad hoc*, and suffer from several problems. First, the 'log likelihood functions' of equations (10.3) and (10.4) are not log likelihood in the usual sense of the word since the rates $\boldsymbol{r}$ are unobservable random variables in the model and should ideally be integrated out in the likelihood calculation. While such 'penalized likelihood' is used by statisticians, for example, in smoothing empirical frequency densities (Silverman 1986, pp. 110–119), their statistical properties are uncertain. Second, the reliability of the Poisson or normal approximations to the likelihood is unknown, although in large datasets the normal approximation has been found to work well (dos Reis and Yang 2011). Exact calculation of sequence alignment is computationally expensive because of the large number of rates and the high dimension of the optimization problem.

In the rate-smoothing algorithm of Yang (2004), rate smoothing is used only to assign branches into rate groups for the local clock model; divergence times and branch rates are then estimated by ML from the original sequence alignment. Recently Heath and Huelsenbeck (2012) implemented a Bayesian version of the local clock models, using a Dirichlet process to partition the branches into groups with different rates. The number of rate groups, assignment of branches to rate groups, and the rates for the groups are estimated from the data. The authors' simulation shows that the model provides robust estimates of divergence times and branch rates without sacrificing much power.

10.3.4 Uncertainties in calibrations

10.3.4.1 Difficulty of dating with uncertain calibrations

How should we summarize information in the fossil record for use in molecular clock dating? The answer appears to be a statistical distribution. The distribution for the age of a node characterizes our best assessment of the fossil record in relation to the age of the node. However, the use of fossils to specify calibration information is a complicated process (Hedges and Kumar 2004; Yang and Rannala 2006; Benton et al. 2009). First, determining the date of a fossil is prone to errors, such as experimental errors in radiometric dating and mis-assignment of the fossil to the wrong stratum. Second, placing the fossil correctly on the phylogeny may be challenging. A fossil may be clearly ancestral to a clade but by how much the fossil species predates the common ancestor of the clade may be hard to determine. Misinterpretations of character state changes may also cause a fossil to be assigned to a wrong lineage. A fossil presumed to be ancestral may in fact represent an extinct side branch. Given the many sources of uncertainties involved in the fossil data, the right way forward appears to be probabilistic modelling of fossil discovery (considering factors such as fossil preservation potential, sampling intensities, etc.), as well as statistical analysis of the fossil data, such as fossil occurrences and morphological measurements on fossils. Such an analysis may generate posterior density curves for node ages, which can be used as calibrations in molecular clock dating (Tavaré et al. 2002; Wilkinson et al. 2011; Ronquist et al. 2012a).

Besides the fossil record, geological events such as continental breakups and other tectonic events may also be used to calibrate molecular trees. For example, the rise of the Isthmus of Panama at 2.7–3.5 MYA has been used to calibrate the molecular clock for dating divergences among marine species. Such a palaeogeographic approach is also thwarted with difficulties (Goswami and Upchurch 2010; Papadopoulou et al. 2010). First, transoceanic barriers can form over long periods, and modern and past ocean current regimes may be very different, so that continental breakup times may be even harder to determine than fossil occurrences. Second, the dispersal abilities of the species concerned are in general unknown, so that lineage divergence and barrier formation may not coincide (Heads 2005). For example, ancestral species may have diverged well before the closure of the Isthmus (Marko 2002).

For likelihood-based dating methods, it is unclear how to use uncertain calibrations represented by statistical distributions, even if they are available. The penalized likelihood method of Sanderson (2002) does not deal with uncertainties in fossil calibrations, despite claims to the contrary (see below). The estimation problem is unconventional, as the model is not fully identifiable and the errors in the estimates do not decrease to zero with the increase of (sequence) data. Yang (2006, pp. 236–245) discussed simple cases for which the answers may be judged intuitively. Although the idea appears to be sensible, computationally feasible algorithms are yet to be developed.

10.3.4.2 *Problems with naïve likelihood implementations*

The likelihood method under both the global clock and the local clock models as well as the rate-smoothing algorithms can use one or more fossil calibrations, but assume that the ages of calibration nodes are known constants, without error. The problem of ignoring uncertainties in calibrations is well appreciated (see, e.g. Graur and Martin 2004; Hedges and Kumar 2004). If a single fossil calibration is used and the fossil age represents the minimum age of the clade, the estimated ages for all other nodes will also be interpreted as minimum ages. However, with this interpretation, it will not be appropriate to use multiple calibrations simultaneously, because the minimum calibrations are treated by those methods as known node ages (rather than as constraints), and may not be compatible with each other or with the sequence data. Suppose that the true ages of two nodes are 100 and 200 MY, and fossils place their minimum ages at 50 and 130 MY. Both minimum constraints are correct as they do not violate the true ages. However, used as calibrations, those minimum ages may be interpreted by current dating methods as true node ages, and will imply, under the clock, a ratio of 1:2.6 for the distances from the two nodes to the present time, instead of the correct ratio 1:2. Time estimates will be systematically biased as a result and will not be correct even if interpreted as minimum age estimates.

Sanderson (1997) argued for the use of minimum and maximum bounds on node ages as a way of incorporating fossil date uncertainties, with constrained optimization algorithms used to obtain estimates of times and rates in his penalized likelihood method (equation (10.3)). This causes the model to become unidentifiable and is thus not a valid approach. Consider a single fossil calibration which constrains the node age t_C to lie in the interval (t_L, t_U). Suppose for a certain t_C inside the interval, the estimates of other node ages and of rates are $\hat{t}$ and $\hat{r}$, with the smoothing parameter $\hat{\lambda}$. If $t'_C = 2t_C$ is also in the interval, then $\hat{t}' = 2\hat{t}$, $\hat{r}' = \hat{r}/2$, and $\hat{\lambda}' = 4\hat{\lambda}$ will fit the likelihood (equation (10.3)) or the cross-validation criterion equally well; in other words, by making all divergence times twice as old and all rates half as fast, one achieves exactly the same fit and there is no way to distinguish between the two or many other sets of parameter values. If multiple fossil calibrations are used simultaneously and if they are in conflict with each other or with the molecules, the optimization algorithm may get stuck at the boundaries of the intervals, but the uncertainties in the fossil node ages are not properly accommodated. To avoid unidentifiability, at least one point calibration has to be used; that is, one node age must be known without error, in which case uncertainties in fossil calibrations are unaccounted for.

Thorne and Kishino (2005) also pointed out a problem with the use of the nonparametric bootstrap method to construct confidence intervals for estimated divergence times in Sanderson's approach. If one re-samples sites in the alignment to generate bootstrap datasets, and analyse each bootstrap sample using the same fossil calibrations, the method will fail to account for uncertainties in the fossils and lead to misleadingly narrow confidence intervals. This problem is obvious when one considers datasets of extremely long sequences. When the data size approaches infinity there will be no difference among time estimates from the bootstrap samples, and the constructed confidence intervals will have zero width. The correct intervals, however, should have positive width to reflect calibration uncertainties (see §10.4.5).

10.3.5 Dating viral divergences

Viruses, especially RNA viruses with high mutation rates, evolve fast enough to accumulate noticeable mutations or substitutions over a few years. When viral gene sequences are sampled at different time points, the sampling times may be used to calibrate the molecular clock. Suppose that the rate is constant and the clock holds. Then sequences sampled earlier will have shorter distances from the root of the tree because they have had a shorter time to accumulate changes, while sequences sampled more recently will have longer distances from the root. Such distance information, combined with the sampling times, may be used to estimate the mutation rate and thus the ages of nodes representing major viral divergence events. As a simple example, suppose sequences a and b are isolated at different time points t_a and t_b, and suppose sequence c is an outgroup (Figure 10.6). The rate of evolution can then be estimated as $(d_{ac} - d_{bc})/(t_b - t_a)$, and this rate can be used to estimate the ages of the two nodes on the tree. The sampling times of the sequences at the tips play the role of fossil calibrations. For a general tree, Rambaut (2000) developed an ML method, called TipDate, for estimating viral substitution rate and divergence times under the molecular clock.

10.3 LIKELIHOOD ESTIMATION OF DIVERGENCE TIMES 373

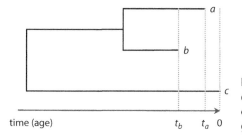

Fig. 10.6 A tree for three sequences with sampling dates. Under the molecular clock, the sample dates can be used to estimate the mutation rate and divergence times.

If the molecular clock is violated, one can in theory use the local clock models to assign rates to branches as before. However time estimation without the clock becomes much harder because of the arbitrariness of rate assignment. Also estimation using dated tips is clearly harder if the events to be dated are much older that the time period covered by the samples; for example, one should be cautious about estimating events that occurred more than 500 years ago when the sample times cover only 50 years.

10.3.6 Dating primate divergences

We apply the likelihood global and local clock models to analyse the dataset of Steiper et al. (2004) to estimate the times of divergence between hominoids (apes and

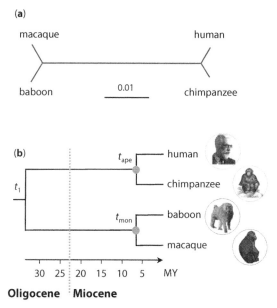

Fig. 10.7 (a) The unrooted phylogenetic tree for four primate species for the data of Steiper et al. (2004), with branch lengths estimated under the HKY85 + Γ_5 model without assuming the clock. (b) The rooted tree, with branch lengths representing the posterior means of divergence times estimated in a Bayesian analysis under the correlated-rate model (Table 10.1).

humans) and cercopithecoids (Old World monkeys) (Figure 10.7). The data consist of five genomic contigs (each of 12–64 k base pairs) from four species: human (*Homo sapiens*), chimpanzee (*Pan troglodytes*), baboon (*Papio anubis*), and rhesus macaque (*Macaca mulatta*). See Steiper et al. (2004) for the GenBank accession numbers. The five contigs are concatenated into one dataset; in a previous analysis, accommodating the differences among the contigs, such as differences in substitution rate, in base compositions, and in the extent of rate variation among sites, produced very similar results (Yang and Rannala 2006). The JC69 and HKY85+Γ_5 models are used for comparison. The branch lengths estimated under HKY85+Γ_5 on the unrooted tree without assuming the clock are shown in the tree of Figure 10.7a. The MLEs of parameters are $\hat{\kappa} = 4.4$ and $\hat{\alpha} = 0.79$, while the base compositions are estimated using their observed values: 0.327 (T), 0.177 (C), 0.312 (A), and 0.184 (G). HKY85+Γ_5 has four more parameters than JC69. The log likelihood difference between the two models is $>7,700$, so that JC69 is rejected by a huge margin. Test of the molecular clock under either JC69 or HKY85+Γ_5 fails to reject the clock.

Two fossil calibrations are used. The human–chimpanzee divergence was assumed to be between 6 and 8 MY (Brunet et al. 2002) and the divergence of baboon and macaque is assumed to be between 5 and 7 MY (Delson et al. 2000). See Steiper et al. (2004) and Raaum et al. (2005) for reviews of relevant fossil data. In the likelihood analysis, we fix those two calibration points at $t_{ape} = 7$MY and $t_{mon} = 6$MY. Under the clock, two parameters are involved in the model: the age of the root t_1 and the rate r. The estimates are $\hat{t}_1 = 33$–34 MY and $\hat{r} = 6.6 \times 10^{-10}$ substitutions/site/year under both JC69 and HKY85+Γ_5 (Table 10.1). A local clock model assumes that the apes and monkeys have different rates. This is the quartet-dating approach of Rambaut and Bromham (1998). Estimates of t_1 are largely the same as under the clock, while the rate is lower in the apes than

Table 10.1 Likelihood and Bayesian estimates of divergence times from the data of Steiper et al. (2004)

	ML		Bayesian			
	Global clock	Local clock	Prior	Global clock	Independent rates	Correlated rates
JC69						
t_1 (root)	32.8 (31.5, 34.1)	32.3 (31.0, 33.5)	34 (8, 60)	32 (28, 37)	33 (24, 45)	32 (25, 40)
t_{ape}	7	7	6.5 (5, 8)	5.7 (5.0, 6.6)	5.9 (5.0, 7.6)	6.0 (5.0, 7.7)
t_{monkey}	6	6	6.5 (5, 8)	7.1 (6.1, 8.0)	7.0 (5.3, 8.0)	6.9 (5.2, 8.0)
r_{ape}	6.6 (6.3, 6.9)	5.4 (5.1, 5.7)	99 (12, 278)	6.7 (5.9, 7.8)	6.9 (5.4, 8.9)	6.9 (5.2, 9.2)
r_{monkey}		8.0 (7.6, 8.4)				
HKY85+Γ_5						
t_1 (root)	33.8 (32.4, 35.2)	33.3 (31.9, 34.7)	34 (8, 60)	33 (29, 38)	34 (24, 46)	33 (26, 41)
t_{ape}	7	7	6.5 (5, 8)	5.7 (4.9, 6.6)	5.9 (5.0, 7.7)	6.0 (5.0, 7.8)
t_{monkey}	6	6	6.5 (5, 8)	7.1 (6.1, 8.1)	7.0 (5.3, 8.0)	6.9 (5.2, 8.0)
r_{ape}	6.6 (6.4, 6.9)	r_{ape}: 5.4 (5.1, 5.8)	99 (12, 278)	6.8 (5.9, 7.9)	6.9 (5.4, 9.1)	6.9 (5.2, 9.2)
r_{monkey}		r_{mon}: 8.0 (7.6, 8.5)				

Note: Divergence times (in million years) are defined in Figure 10.7. Rate is $\times 10^{-10}$ substitutions per site per year. For ML, the MLEs and the 95% confidence intervals are shown. For Bayesian inference, the posterior mean and 95% CIs are shown. Note that the ML analysis ignores uncertainties in the fossil dates and the CIs are too narrow. The likelihood and Bayesian results are obtained by using programs BASEML and MCMCTREE in the PAML package (Yang 2007b), respectively.

in the monkeys: $\hat{r}_{ape} = 5.4$ and $\hat{r}_{monkey} = 8.0$. Note that this likelihood analysis ignores uncertainties in the fossil calibrations and grossly overestimates the confidence in the point estimates. Steiper et al. (2004) used four combinations of the fossil calibrations, with $t_{ape} = 6$ or 8MY and $t_{monkey} = 5$ or 7MY, and used the range of estimates of t_1 as an assessment of the effect of fossil uncertainties. At any rate, the estimates of divergence time t_1 here between hominoids and cercopithecoids are close to those of Steiper et al.

10.4 Bayesian estimation of divergence times

10.4.1 *General framework*

A Bayesian Markov chain Monte Carlo (MCMC) algorithm was developed by Thorne and colleagues (Thorne et al. 1998; Kishino et al. 2001), implemented in the MULTIDIVTIME program to estimate species divergence times. Fossil calibrations are specified as minimum- or maximum-age bounds on nodes on the given tree, thus accounting for uncertainties in the fossil record. The substitution rate is assumed to drift according to a geometric Brownian motion model. A slightly different model using a compound-Poisson process to model rate drift was described by Huelsenbeck et al. (2000a). Yang and Rannala (2006; Rannala and Yang 2007) developed an algorithm to allow for 'soft bounds' or arbitrary distributions to be used as fossil calibrations, implemented in the MCMCTREE program. Similarly flexible calibration densities are implemented in programs such as BEAST (Drummond et al. 2006; Drummond and Rambaut 2007), MrBayes 3.2 (Ronquist et al. 2012b), and PhyloBayes (Lepage et al. 2007; Lartillot et al. 2009). Here we describe the general framework of Bayesian dating of species divergences and comment on some published differences in the different implementations.

Let X be the sequence data, $\boldsymbol{t}$ the $s-1$ divergence times on the rooted tree for s species, and $\boldsymbol{r}$ the rates. The rates can be either for the $(2s-2)$ branches, as in Thorne et al. (1998) and Yang and Rannala (2006), or for the $(2s-1)$ nodes, as in Kishino et al. (2001). In the former case, the rate at the midpoint of a branch is used as the approximate average rate for the whole branch. In the latter case, the average of the rates at the two ends of the branch is used to approximate the average rate for the branch, although this average rate can be calculated more accurately from the geometric Brownian motion model (Guindon 2013). Let $f(\boldsymbol{t})$ and $f(\boldsymbol{r}|\boldsymbol{t})$ be the priors for times and rates, respectively. Let θ be parameters in the substitution model and in the priors for $\boldsymbol{t}$ and $\boldsymbol{r}$, with prior $f(\theta)$. The joint posterior distribution is then given as

$$f(\boldsymbol{t}, \boldsymbol{r}, \theta|X) \propto f(\theta)f(\boldsymbol{t}|\theta)f(\boldsymbol{r}|\boldsymbol{t}, \theta)f(X|\boldsymbol{t}, \boldsymbol{r}, \theta). \qquad (10.6)$$

Here as in equation (10.4), $f(X|\boldsymbol{t}, \boldsymbol{r}, \theta)$ is the sequence likelihood, $f(\boldsymbol{r}|\boldsymbol{t}, \theta)$ is the prior on rates, and $f(\boldsymbol{t}|\theta)$ is the prior on divergence times. The normalizing constant or marginal probability of the data, $f(X)$, is a high-dimensional integral over $\boldsymbol{t}$, $\boldsymbol{r}$, and θ, and is avoided by the MCMC. The MCMC algorithm generates samples from the joint posterior distribution. The marginal posterior of $\boldsymbol{t}$,

$$f(\boldsymbol{t}|X) = \iint f(\boldsymbol{t}, r, \theta|X)\, d\boldsymbol{r}\, d\theta, \qquad (10.7)$$

can be constructed from the samples of $(\boldsymbol{t}, \boldsymbol{r}, \theta)$ taken during the MCMC.

The following is a sketch of the MCMC algorithm implemented in the MCMCTREE program in the PAML package (Yang 2007b; see also Thorne et al. 1998). See Chapter 7 for discussions of MCMC algorithms. In the next few subsections we discuss the individual terms involved in equation (10.6).

- Start with a random set of divergence times *t*, substitution rates *r*, and parameters *θ*.
- In each iteration do the following:
  - Propose changes to the divergence times *t*, under the constraint that any node must be older than its child nodes and younger than its mother node.
  - Propose changes to the substitution rates for different loci.
  - Propose changes to substitution parameters *θ*.
  - Propose a change to all times and rates, by multiplying all times by a random variable *c* close to one and dividing all rates by *c* (see §7.2.5).
  - For every *k* iterations, sample the chain: save *t*, *r*, and *θ* to disk.
- At the end of the run, summarize the results.

10.4.2 Approximate calculation of likelihood

The likelihood $f(X|t, r, θ)$ can be calculated under any substitution model on the sequence alignment, as discussed in Chapter 4. This is straightforward but expensive, although efficient implementations on multicore multiprocessor computers are easing the computational burden. To achieve computational efficiency, Thorne et al. (1998) and Kishino et al. (2001) used a normal approximation to the MLEs of branch lengths in the rooted tree for the ingroup species, estimated without the clock, with the variance–covariance matrix calculated using the local curvature of the log likelihood surface. The approximation means that the sequence alignment is not needed during the MCMC, after the MLEs of branch lengths and their variance–covariance matrix have been calculated. Thorne et al. used outgroup species to break the branch around the root into two parts, such as branch lengths b_7 and b_8 in the tree of Figure 10.8a. If no outgroups are available or if the outgroups are too far away from the ingroup species to be useful, an alternative may be to estimate the $2s - 3$ branch lengths in the unrooted tree for the *s* ingroup species only, without outgroups (Figure 10.8b). In the likelihood calculation, the predicted branch length around the ingroup root will be the sum of the two parts (i.e. b_7 in Figure 10.8b instead of b_7 and b_8 in Figure 10.8a).

In either case, the log likelihood is a function of the vector of branch lengths in the ingroup tree and can be approximated using the first three terms of the Taylor expansion of the log likelihood around the MLE $\hat{b}$:

$$\ell(b) = \ell(\hat{b}) + g(\hat{b})^T(b - \hat{b}) + \frac{1}{2}(b - \hat{b})^T H(\hat{b})(b - \hat{b}), \tag{10.8}$$

where both the gradient vector *g* and the Hessian matrix *H* are evaluated at the MLEs $\hat{b}$, prior to the MCMC run. If other parameters exist in the model, such as *κ*, they are fixed at the MLEs since we are interested in the impact of branch lengths only. The log likelihood at the MLEs, $\ell(\hat{b})$, is a constant and can be ignored. If the MLEs of all branch lengths are strictly positive, we have $g(\hat{b}) = 0$ and the second term disappears. In that case, equation (10.8) is the well-known normal approximation to the likelihood, with the variance–covariance matrix $V = -H^{-1}$. However when the MLEs of some branch lengths are 0, inclusion of the gradient term is important (dos Reis and Yang 2011). See Seo et al. (2004) and dos Reis and Yang (2011) for discussions of the calculation of *g* and *H*.

Better approximation may be achieved by applying the Taylor expansion, not on the branch lengths, but on their transformations (dos Reis and Yang 2011). For example, the

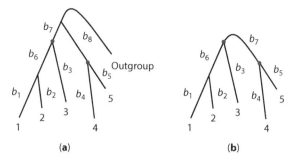

Fig. 10.8 Calculation of the likelihood function for data of s sequences (species) by the normal approximation to the branch lengths estimated without assuming the molecular clock. (a) Thorne et al. (1998) used outgroup species to locate the root on the ingroup tree, estimating $2s - 2$ branch lengths for the rooted tree of s ingroup species. Here the ingroup tree is that of Figure 10.3. The likelihood is then approximated by the normal density around the MLEs of those $2s - 2$ branch lengths ($b_1, b_2, \ldots, b_8$ in the example). (b) An alternative approach implemented in MCMCTREE (dos Reis and Yang 2011) is to estimate $2s - 3$ branch lengths in the unrooted tree of s ingroup species, without the need for an outgroup. The likelihood will then be approximated by the normal density around the MLEs of the $2s - 3$ branch lengths ($b_1, b_2, \ldots, b_7$ in the example).

log likelihood surface tends to be more symmetrical around the MLE and close to a parabola if it is plotted against the logarithm or square root of the branch lengths than against the branch lengths themselves. In particular an arcsine transform is found to have the best performance in the test of dos Reis and Yang (2011). The approximation makes it possible to analyse large genome-scale datasets (Meredith et al. 2011; dos Reis et al. 2012). The approximation is not easy to use if the tree topology varies, as in the implementation of the BEAST program.

10.4.3 Prior on evolutionary rates

Thorne et al. (1998) and Kishino et al. (2001) used a recursive procedure to specify the prior for rates, proceeding from the root of the tree towards the tips. The rate at the root is assumed to have a gamma prior. Then the rate at each node is specified by conditioning on the rate at the ancestral node. Specifically, given the logarithm of the rate for the ancestral node, $\log(r_A)$, the logarithm of the rate for the current node, $\log(r)$, follows a normal distribution with mean $\log(r_A) - \frac{1}{2}vt$ and variance vt, where t is the time duration separating the two nodes. Note that if the log rate $y \sim N(\mu, vt)$, then $E(e^y) = \exp(\mu + \frac{1}{2}vt)$, so the correction term $-\frac{1}{2}vt$ is used to remove the positive drift in rate, to have $E(r) = r_A$ (Kishino et al. 2001). In other words, given the rate r_A at the ancestral node, the rate r at the current node has a log-normal distribution with density

$$f(r|r_A) = \frac{1}{r\sqrt{2\pi vt}} \exp\left\{-\frac{1}{2vt}\left[\log\left(\frac{r}{r_A}\right) + \frac{1}{2}vt\right]^2\right\}, \quad 0 < r < \infty. \tag{10.9}$$

Parameter v controls how rapidly the rate drifts or how clock-like the tree is *a priori*. A large v means that the rates vary a lot over time or among branches and the clock is seriously violated, while a small v means that the clock roughly holds. The prior for rates on the tree $f(r)$ is calculated by multiplying the prior densities across the nodes.

This model is called the correlated-rate model or *geometric Brownian motion* model, as the logarithm of the rate drifts over time according to a *Brownian motion* process (Figure 10.5). Geometric Brownian motion is the simplest model for random drift of a positive variable, and is widely used in financial modelling and resource management, for example, to describe the fluctuations of the market value of a stock. In Rannala and Yang (2007), the model is used to specify the distribution of rates at the midpoints of the branches on the tree. The model assumes that closely related species tend to have similar rates.

As an alternative, Drummond et al. (2006) and Rannala and Yang (2007) implemented an independent-rate model, in which the rate for any branch is a random draw from the same distribution, such as the log-normal with density

$$f(r|\mu, \sigma^2) = \frac{1}{r\sqrt{2\pi\sigma^2}} \exp\left\{-\frac{1}{2\sigma^2}\left(\log(r/\mu) + \frac{1}{2}\sigma^2\right)^2\right\}, 0 < r < \infty. \qquad (10.10)$$

Here μ is the average rate for the locus, and σ^2 measures departure from the clock model. The prior density for all rates $f(r)$ is simply the product of the densities like equation (10.10) across all branches on the tree. Note that equation (10.10) does not involve t. Parameter σ^2 in this model may be used to measure how seriously violated the molecular clock hypothesis is. Estimates from real data suggest that $\sigma^2 = 0.2$ indicates fairly strong deviation from the model.

The formulation here assumes that the rates follow different and independent trajectories at different loci. With relaxed clock models, it is thus important to use data of multiple loci to accumulate information about the divergence times, which are shared across all loci. In real data analysis, the multiple loci may well have correlated rates, as the rates for all genes may be affected by life history traits of the species (such as generation time, population size, etc.).

Some studies used the Bayes factor to compare different rate-drift models, reaching conflicting conclusions as to whether the independent-rate or correlated-rate models are preferable (e.g. Lepage et al. 2007; Baele et al. 2012; Li and Drummond 2012). There are several issues with such comparisons. First, Bayes factors or posterior model probabilities are often very sensitive to the prior on parameters within each model (see §6.2.3). Second, in this case the sequence likelihood is the same, and the different rate-drift models are priors in the Bayesian analysis. While it is common to use the Bayes factor to compare different likelihood models, it is uncommon to use it to compare different priors. Instead, a more important question may be the robustness of the posterior time estimates to different prior assumptions.

10.4.4 Prior on divergence times and fossil calibrations

Bayesian estimation of divergence times also requires specification of a prior on the divergence times (node ages in the tree). Typically this time prior also incorporates fossil calibration information. It should be noted that as long as the uncertainties in fossil calibrations are incorporated in the time prior $f(t)$, they will be automatically incorporated in the posterior. There is then no need to use bootstrap resampling to incorporate fossil uncertainties in Bayesian time estimation, as did Kumar et al. (2005b). Effort should instead be spent on developing 'objective priors' that best summarize the fossil record to represent our state of knowledge concerning the ages of calibration nodes. Studies of fossil preservation, sampling intensities, and fossil discovery, and of errors in fossil dating techniques may all contribute to this goal (Tavaré et al. 2002; Wilkinson et al. 2011). See Ho and Phillips (2009) and Parham et al. (2012) for useful discussions about generating fossil calibration densities.

Kishino et al. (2001) introduced a recursive procedure for specifying the prior for divergence times, proceeding from the root towards the tips. A gamma density is used for the age of the root (t_1 in the example tree of Figure 10.3) and a uniform Dirichlet distribution is used to break the path from an ancestral node to the tip into time segments, corresponding to branches on that path. For example, along the path from the root to tip 1 (Figure 10.3), the proportions of the three time segments, $(t_1 - t_2)/t_1$, $(t_2 - t_4)/t_1$, and t_4/t_1, follow a Dirichlet distribution with equal means. Next, the two proportions $(t_1 - t_3)/t_1$ and t_3/t_1 follow a Dirichlet distribution with equal means.

To incorporate uncertain fossil calibrations in the prior $f(\boldsymbol{t}|\theta)$, Thorne et al. (1998) used lower (minimum) and upper (maximum) bounds on node ages, implemented in the MCMC algorithm by avoiding proposals of times that contradict such bounds. This strategy in effect specifies a uniform prior on the fossil calibration age: $t \sim U(t_L, t_U)$.

One could introduce a parameter into the Dirichlet distribution (see equation (8.5)) to generate trees of different shapes, e.g. trees with long old branches and short young branches versus star-like trees. However, then the prior density $f(\boldsymbol{t})$ becomes intractable analytically. Another drawback is that it is difficult to incorporate flexible statistical distributions (other than the minimum and maximum bounds) for fossil calibrations.

The minimum and maximum bounds specified by the uniform prior may be called 'hard' bounds, as they assign zero probability for any age outside the interval. Such priors represent strong conviction on the part of the researcher and may not always be appropriate. In particular, fossils often provide good minimum bounds and rarely good maximum bounds. As a result, the researcher may be forced to use an unrealistically high maximum bound to avoid precluding an unlikely (but not impossible) ancient age for the node. Such a 'conservative' approach may be problematic as the bounds imposed by the prior may influence posterior time estimation.

Yang and Rannala (2006) developed a strategy to incorporate arbitrary distributions for describing fossil calibration uncertainties. These are called 'soft' bounds and assign nonzero probabilities over the whole positive line ($t > 0$). A few examples are shown in Figure 10.9. The basic model used is the birth–death process (Kendall 1948), generalized to account for species sampling (Rannala and Yang 1996; Yang and Rannala 1997). Here we describe the main features of this model, as it is also useful for generating random trees with branch lengths. Let λ be the per-lineage birth (speciation) rate, μ the per-lineage death (extinction) rate, and ρ the sampling fraction. Then conditional on the root age t_1, the $(s-2)$ node ages of the tree are order statistics from the kernel density

$$g(t) = \frac{\lambda p_1(t)}{v_{t_1}}, \qquad (10.11)$$

where

$$p_1(t) = \frac{1}{\rho} P(0, t)^2 e^{(\mu - \lambda)t} \qquad (10.12)$$

is the probability that a lineage arising at time t in the past leaves exactly one descendent in the sample, and

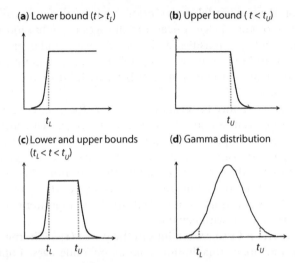

Fig. 10.9 Probability densities for node ages implemented by Yang and Rannala (2006) to describe uncertainties in fossils. The bounds are 'soft', and the node age has nonzero probability density over the whole positive line: that is, $g(t) > 0$ for $0 < t < \infty$. (a) A lower bound, specified by $t > t_L$, implemented with $P(t < t_L) = 2.5\%$. This is an improper prior density and should not be used alone. This is replaced by a different density based on the Cauchy distribution (Inoue et al. 2010). (b) An upper (maximum) bound, specified by $t < t_U$, with $P(t > t_U) = 2.5\%$. (c) Both lower (minimum) and upper (maximum) bounds, specified as $t_L < t < t_U$, with $P(t < t_L) = P(t > t_U) = 2.5\%$. (d) A gamma distribution, with (t_L, t_U) to be the 95% equal-tail prior interval.

$$v_{t_1} = 1 - \frac{1}{\rho} P(0, t_1) e^{(\mu-\lambda)t_1}, \tag{10.13}$$

with $P(0, t)$ to be the probability that a lineage arising at time t in the past leaves one or more descendents in a present-day sample

$$P(0, t) = \frac{\rho(\lambda - \mu)}{\rho\lambda + [\lambda(1-\rho) - \mu]e^{(\mu-\lambda)t}}. \tag{10.14}$$

When $\lambda = \mu$, equation (10.11) becomes

$$g(t) = \frac{1 + \rho\lambda t_1}{t_1(1 + \rho\lambda t)^2}. \tag{10.15}$$

The joint density of the $s - 2$ node ages $t_2, t_3, \ldots, t_{s-1}$ is thus

$$f(t_2, t_3, \ldots, t_{s-1}) = (s-2)! \prod_{j=2}^{s-1} g(t_j). \tag{10.16}$$

To generate a set of $s - 2$ random node ages when the root age t_1 is given, one generates $s - 2$ independent random variables from the kernel density (equation (10.11)) and

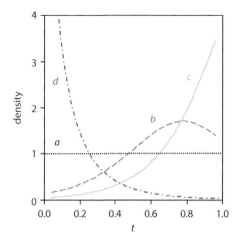

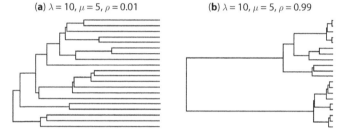

Fig. 10.10 Kernel densities of the birth–death process with species sampling, with per-lineage birth rate λ, death rate μ, and sampling fraction ρ (equation (10.11)). The parameters used are (**a**) $\lambda = 1, \mu = 1, \rho = 0$ (the uniform kernel); (**b**) $\lambda = 10, \mu = 5, \rho = 0.01$; (**c**) $\lambda = 10, \mu = 5, \rho = 0.001$; (**d**) $\lambda = 10, \mu = 5, \rho = 0.99$.

Fig. 10.11 Two trees generated from the birth–death process with species sampling, with parameters (**a**) $\lambda = 10, \mu = 5, \rho = 0.01$, and (**b**) $\lambda = 10, \mu = 5, \rho = 0.99$.

orders them. Sampling from the kernel density can be achieved by using the cumulative distribution function (CDF) given in Yang and Rannala (1997).

The prior density of times t can thus be calculated analytically under the model based on the theory of order statistics. It may be too optimistic to expect this model to provide an accurate description of the biological process of speciation, extinction, and sampling of species by biologists. However, by changing the parameters λ, μ, and ρ in the model, one can easily change the shape of the tree, and examine the sensitivity of posterior time estimates to prior assumptions about the tree shape. A few example densities are shown in Figure 10.10, with the age of the root fixed at $t_1 = 1$. In (a), the density when $\lambda = \mu = 1$ and $\rho = 0$ is uniform between 0 and 1 (equation (10.15)). This is close to the uniform Dirichlet prior of Thorne et al. (1998). In (b) and (c), the densities are skewed towards 1, so that the tree will be star-like, with short internal branches and long terminal branches. In (c), the density is highly skewed towards 0, so that the tree will tend to have long internal and short external branches. This is the shape of trees generated by the standard coalescent prior or the Yule process without species sampling. Figure 10.11 shows two trees with 20 tips generated from the birth–death sampling process with different parameter values.

Fossil calibration information is incorporated in the prior for times, $f(t)$. Let t_C be ages of nodes for which fossil calibration information is available, and t_{-C} be the ages of the other nodes, with $t = (t_C, t_{-C})$. The prior $f(t)$ is constructed as

$$f(\boldsymbol{t}) = f(\boldsymbol{t}_C, \boldsymbol{t}_{-C}) = f_{BD}(\boldsymbol{t}_{-C}|\boldsymbol{t}_C)f(\boldsymbol{t}_C), \tag{10.17}$$

where $f(\boldsymbol{t}_C)$ is the density of the ages of the fossil calibration nodes, specified by summarizing fossil information, and

$$f_{BD}(\boldsymbol{t}_{-C}|\boldsymbol{t}_C) = f_{BD}(\boldsymbol{t}_C, \boldsymbol{t}_{-C})/f_{BD}(\boldsymbol{t}_C) \tag{10.18}$$

is the conditional distribution of $\boldsymbol{t}_{-C}$ given $\boldsymbol{t}_C$, specified according to the birth–death process with species sampling.

In the BEAST program (Drummond et al. 2006), the time prior is constructed by multiplying the density $f(\boldsymbol{t}_C)$ for calibration nodes with the Yule process density $f(\boldsymbol{t}_C, \boldsymbol{t}_{-C})$. This is not a correct construction as it does not follow the rules of probability theory (Heled and Drummond 2012). Heled and Drummond (2012) referred to this as the 'multiplicative construction', as opposed to equation (10.17), the 'conditional construction' of Yang and Rannala (2006).

In all current Bayesian implementations, the user-specified calibration densities are not the real prior densities for the node ages used by the computer programs. Any node on the phylogeny must be older than its descendent nodes and younger than its ancestral nodes. The *user-specified prior*, in the form of densities for individual nodes, may not respect this biological constraint. Automatic application of this constraint by the program during the MCMC is equivalent to a truncation of the unfeasible part of the parameter space. The prior after this truncation, which is called *the effective prior* and is used by the computer program, may be very different from the user-specified prior (Inoue et al. 2010). Suppose two node ages on the tree are t_1 and t_2, with $t_1 > t_2$, but independent gamma priors are specified for them. After truncating the region where $t_1 \leq t_2$, the resulting marginal distributions of t_1 and t_2 are no longer gamma. The user should run the program without data to generate this effective prior and to confirm its reasonableness.

10.4.5 Uncertainties in time estimates

The importance of fossil calibrations to molecular dating analysis, especially when relaxed clocks are used, has been well appreciated. Many empirical dating studies have found that posterior divergence time estimates are often sensitive to changes to the fossil calibrations (e.g. Hurley et al. 2007; Magallon et al. 2013), that posterior intervals for divergence times often have comparable widths as the prior intervals despite the use of huge amounts of sequence data, and that posterior intervals obtained by analysing many loci are not much narrower than those from analysing very few loci (Mulcahy et al. 2012). Perhaps the most dramatic effect of this persisting uncertainty is the fact that even if an infinite amount of sequence data is available, the posterior will not converge to a point estimate (Britton 2005). As a statistical estimation problem, the model is not fully identifiable; the sequence data provide information about the distances, which are the products of times and rates, but not about times and rates separately. As a result, even if the amount of sequence data approaches infinity, there will still be uncertainties in the posterior and the posterior will still be sensitive to the prior.

Yang and Rannala (2006) and Rannala and Yang (2007) have derived the limiting distribution when the amount of sequence data approaches infinity: the joint posterior distribution of times does not converge to a point (i.e. the true divergence times) as occurs in a conventional estimation problem, but to a one-dimensional distribution. In other words, the root age t_1 has a posterior distribution, while the posterior distribution of any other time t_i in the tree is a simple linear transform of the distribution of t_1.

10.4 BAYESIAN ESTIMATION OF DIVERGENCE TIMES

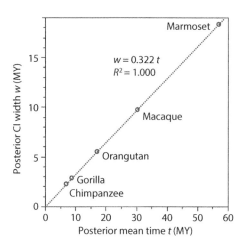

Fig. 10.12 In the infinite-site plot, the width of the posterior CI for each divergence time is plotted against the posterior mean. The slope measures the precision of fossil calibrations, while the correlation measures how close the amount of sequence data is to the infinite limit. If the amount of sequence data approaches infinity, the points will fall on a straight line. There are six primate species in the tree, so that the five divergence times are those between the human and five other primate species. The data are an alignment of 8,708,584 sites (the first and second codon positions of 14,631 genes), analysed using the Bayesian program MCMCTREE. The analysis produces a posterior mean and the 95% CI for each of the five interior nodes on the tree. The CI width (the difference of the 2.5% and 97.5% limits) is plotted against the posterior mean in the scatter plot. A line with intercept zero is fitted to the scatter plot. Redrawn following dos Reis and Yang (2013a, Figure 1).

The theory is called the 'infinite-site theory'. (Note that this has no connection with the infinite-sites mutation model in population genetics.) As a consequence, a plot of the posterior credibility interval (CI) widths against the posterior means of divergence times will approach a straight line when the amount of sequence data approaches infinity. This plot is known as the *infinite-site plot*. An example is shown in Figure 10.12, using posterior estimates of divergence times on a phylogeny of six primate species: human, chimpanzee, gorilla, orangutan, macaque, and marmoset. Despite the huge alignment, the posterior time estimates involve considerable uncertainties: for every million years of divergence, 0.322 million years are added to the CI width. The perfect linear relationship ($R^2 = 1$) indicates that essentially all uncertainties in the posterior are due to the uncertainty in the fossil calibrations, and adding more sequence data is unlikely to produce more precise estimates. The uncertainty of fossil information therefore imposes a theoretical limit on the precision that can be achieved in divergence time estimation when the amount of sequence data increases.

Consider the case of time estimation under the molecular clock (Yang and Rannala 2006). Suppose the distances from the interior nodes to the present time, measured by the expected number of substitutions per site, are $d_1, d_2, \ldots, d_{s-1}$. These are estimated without errors when the sequence length $n \to \infty$ and are thus fixed constants. Let the prior on the rate r be $g(r)$ so that the joint prior for the rate and times is $f(r, t_1, t_2, \ldots, t_{s-1}) = g(r) f_t(t_1, t_2, \ldots, t_{s-1})$. By treating the distances as fixed constants, one obtains the joint posterior of the rate and times as the joint prior of rate and times conditional on the fixed distances. Change variables from $(r, t_1, t_2, \ldots, t_{s-1})$ to $(r, d_1, d_2, \ldots, d_{s-1})$, with $d_j = t_j r$, for $j = 1, \ldots, s-1$. The Jacobi determinant for the transform is $\left| \frac{\partial(r, t_1, t_2, \ldots, t_{s-1})}{\partial(r, d_1, d_2, \ldots, d_{s-1})} \right| = r^{1-s}$. Then we have

$$f(r, d_1, d_2, \ldots, d_{s-1}) = g(r) f_t\left(\frac{d_1}{r}, \frac{d_2}{r}, \ldots, \frac{d_{s-1}}{r}\right) \cdot r^{1-s} \qquad (10.19)$$

(see Theorem 1 in Appendix A). The posterior of rate r is thus

$$f(r|d_1, d_2, \ldots, d_{s-1}) = \frac{f(r, d_1, d_2, \ldots, d_{s-1})}{\int f(r, d_1, d_2, \ldots, d_{s-1}) dr} = \frac{g(r) f_t\left(\frac{d_1}{r}, \frac{d_2}{r}, \ldots, \frac{d_{s-1}}{r}\right) r^{1-s}}{\int g(r) f_t\left(\frac{d_1}{r}, \frac{d_2}{r}, \ldots, \frac{d_{s-1}}{r}\right) r^{1-s} dr}. \quad (10.20)$$

The denominator is a normalizing constant and can be calculated using numerical integration, or the posterior density can be easily approximated using MCMC. The posterior for time t_j can be derived by using the transformation $t_j = d_j/r, j = 1, 2, \ldots, s-1$,

$$f(t_j|d_1, d_2, \ldots, d_{s-1}) \propto g\left(\frac{d_j}{t_j}\right) \cdot f_t\left(\frac{d_1}{d_j} t_j, \frac{d_2}{d_j} t_j, \ldots, \frac{d_{s-1}}{d_j} t_j\right) \left(\frac{d_j}{t_j}\right)^{2-s} \cdot \frac{1}{t_j}. \quad (10.21)$$

A few remarks are in order concerning this limiting distribution. First, with infinite sequence data, the branch lengths are estimated without error, and the enforcement of the molecular clock means that given the rate, all divergence times are fully determined. Second, the posterior means for all node ages t_j will lie on a straight line when plotted against the true ages, as will the percentiles and CIs. Thus in real data analysis, one can plot the posterior CI widths against the posterior means of divergence times to assess how much of the posterior uncertainty is due to the limited amount of sequence data (Figure 10.12). Third, if there is only one fossil calibration on the tree and if the prior on rate is not extremely informative, one expects the posterior for the age of the calibration node to be nearly identical to the prior. In that case, the infinite amount of sequence data has not helped to reduce the uncertainty in the age of the calibration node. When calibrations are placed on multiple nodes on the tree, the posterior for the age of each calibration node will be more informative (more highly concentrated) than the prior on that node because information is pooled across nodes.

The case of large but finite sequence datasets was considered by dos Reis and Yang (2013a). A finite-site theory was developed through an analogy with a simple case involving normal distributions and then confirmed by computer simulation. The theory predicts that the posterior variance for any node age obtained from analysing a finite amount of sequence data consists of two parts, due to uncertain fossil calibrations and due to finite sequence data, respectively. When the number of sites in the sequence alignment $n \to \infty$, the variance approaches the infinite-data limit at the rate of $1/n$. In other words, the posterior CI width for a node age in finite data (w_n) approaches the infinite-data limit (w_∞) at the rate $\frac{1}{\sqrt{n}}$.

The case of relaxed clock is more complex. Clearly more uncertainties exist when the rates vary over time. Rannala and Yang (2007) predicted that posterior time estimates will converge to the infinite-data limit when both the number of loci and the number of sites approach infinity. The number of loci appears to be far more important than the number of sites in affecting the precision of posterior time estimates. Thus partitioned analysis may be extremely important in relaxed clock dating analysis (dos Reis et al. 2012).

10.4.6 Dating viral divergences

The Bayesian method can also be used to estimate divergence times using viral gene sequences, using sampling times to calibrate the molecular evolutionary rate. Important to this inference is the prior on divergence times (node ages). Rodrigo and Felsenstein (1999) derived the distribution of coalescent times when the tips are sampled at different time points. The process is a straightforward extension of the single-population standard coalescent discussed in §9.2, and fits well into the framework of a variable-rate Poisson process (see §9.3.1). While in the standard coalescent for contemporary sequences, the total Poisson rate changes at every coalescent event, with dated tips the rate changes at each sampling time point as well as at each coalescent event. As the neutral coalescent may not be a good description of viral genealogical trees, the recent developments extend the birth–death process to include sampling through time (Stadler 2010). Priors based on such birth–death-sequential-sampling (BDSS) models are implemented in BEAST (Stadler et al. 2012) and MCMCTREE (Stadler and Yang 2013). Those models also allow the estimation of the basic reproductive number, a fundamental parameter in epidemiological models of viral transmission, using genetic sequence data (Stadler et al. 2012, 2013).

If the molecular clock is violated, one can in theory use the relaxed clock models to accommodate the variable rates among lineages. While such relaxed clock models are widely used to date important viral transmission events, the accuracies of such inference are not well established. In general, one should be wary of dating events that occurred hundreds or thousands of years ago when all samples are taken in the last few decades.

10.4.7 Application to primate and mammalian divergences

We apply the Bayesian methods to two datasets. First, we use the four nuclear loci of Steiper et al. (2004) to estimate divergence times of primates using the Bayesian method for comparison with the likelihood analysis discussed in §10.3.6. Second, we use the mitochondrial protein-coding genes of dos Reis and Yang (2011) to estimate divergence times among 36 mammalian species.

10.4.7.1 Analysis of the primate data of Steiper et al.

The MCMCTREE program in the PAML package is used to analyse the data of Steiper et al. under the JC69 and HKY85+Γ_5 models. Three rate-drift models are assumed: the global clock assuming one rate throughout the tree, the independent-rate model, and the autocorrelated-rate model. The overall rate is assigned the gamma prior $G(2, 2)$ with mean 1 and variance $1/2$. Here one time unit is 100 MY so a rate of 1 means 10^{-8} substitutions per site per year. Parameter v in equation (10.9) and σ^2 in equation (10.10) are assigned the gamma prior $G(1, 10)$. Under HKY85+Γ_5, the transition/transversion rate ratio κ is assigned the gamma prior $G(2, 0.5)$, with mean 4, while the shape parameter α for rate variation among sites is assigned the prior $G(2, 2)$, with mean 1. These two parameters are reliably estimated in the data, so that the prior has little significance.

The prior for divergence times is specified using the birth–death process with species sampling, with the birth and death rates $\lambda = \mu = 1$, and sampling fraction $\rho = 0$. Those parameters produce the uniform density $U(0, 1)$ as the kernel (equation (10.15)). To calibrate the tree, we use a pair of bounds 5–8 MY for both t_{ape} and t_{monkey}, as both divergences appear to have occurred at about the same time (Steiper et al. 2004; Raaum et al. 2005). Bounds are soft, with 2.5% probability on each tail (Figure 10.9c). In addition, the age of the root is constrained to be less than 60 MY: $t_1 < 0.6$. This is expected to be a weak constraint on the root age.

The posterior means and 95% CIs in the different analyses are listed in Table 10.1. First, posterior estimates of the root age t_1 under the three rate models are similar to each other and also to the ML estimates. In these data, the molecular clock does not appear to be violated, so that the different models are expected to produce similar results. Second, the Bayesian analysis has the advantage of providing CIs that take into account fossil uncertainties. Indeed the Bayesian CIs are all much wider than the likelihood CIs. For the two fossil calibration nodes (t_ape and t_monkey), the posterior CIs are not much narrower than the prior intervals. Third, the two substitution models provided almost identical estimates, with the posterior means of $t_1 \approx 33$ MY under both models. In all analyses, the posterior age of the root is older than 25 MY, consistent with the analysis of Steiper et al. (2004).

10.4.7.2 Divergence times of mammals

In the second example, we estimate the divergence times of 36 mammalian species using the first and second codon positions from the 11 protein-coding genes (>90 codons) on the H strand of the mitochondrial genome (dos Reis and Yang 2013a). The concatenated alignment has 3630 codons (or 7,260 nucleotide sites for positions 1 and 2). The tree is shown in Figure 10.13. The fossil calibrations include 24 minimum and 14 maximum constraints based on the fossil record, which we implement using soft bounds (see dos Reis and Yang 2013a for details). We use this example to illustrate the approximate method for likelihood calculation (dos Reis and Yang 2011) and the infinite-site plot. See Meredith et al. (2011) and dos Reis et al. (2012) for recent discussions of divergence times among mammals.

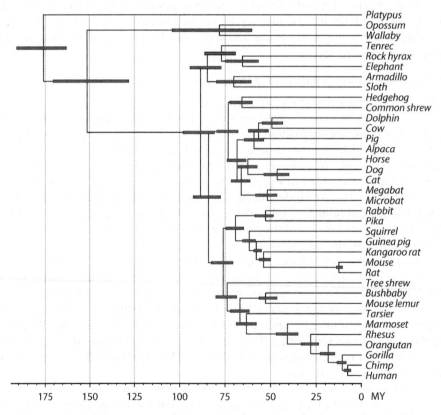

Fig. 10.13 The phylogenetic tree for 36 mammalian species for the mitochondrial data, with the branches drawn using the posterior means of divergence times estimated under the correlated-rate model.

We use $\lambda = \mu = 1$ and $\rho = 0$ for parameters in the birth–death process model to specify the prior for the ages of non-calibration nodes; these parameters specify the uniform kernel. We use a gamma prior $\mu \sim G(1, 1)$ for the mean substitution rate μ, with the mean at one change per site per 100 MY. We use the correlated-rate model to accommodate the violation of the clock, with a diffuse gamma prior $v \sim G(1, 1)$ for the rate-drift parameter v. The nucleotide substitution model used is HKY85 + Γ_5. The MLEs of branch lengths under the no-clock model and the LRT suggest that the strict clock is seriously violated for this dataset.

Figure 10.14a plots the log likelihood values calculated using the exact and approximate methods, for branch lengths visited during the MCMC under the correlated-rate model. The approximation applies the Taylor expansion to the arcsine transforms of branch lengths around the MLEs under the no-clock model (dos Reis and Yang 2011). The approximation is very good, especially at high log likelihood values (large $\Delta\ell$). The posterior means and 95% CIs obtained using the two methods are plotted in Figure 10.14b, which show that the estimates are very similar.

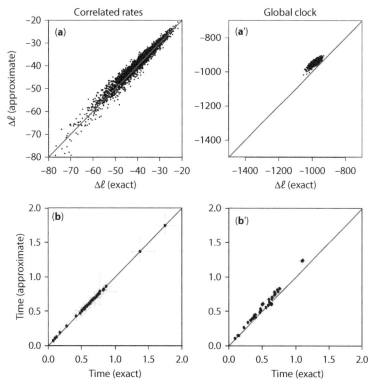

Fig. 10.14 (**a** and **a'**) The log likelihood (or the difference from the highest log likelihood at the MLEs) and (**b** and **b'**) posterior estimates (means and 95% CIs) of divergence times calculated using the exact and approximate likelihood methods under the relaxed clock and strict clock models. The mitochondrial data for 36 mammalian species are analysed. Redrawn from the plots for the arcsine transform in Figures 3 and 5 of dos Reis and Yang (2011).

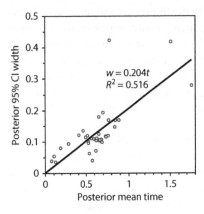

Fig. 10.15 Infinite-site plot for the Bayesian dating analysis of the mammalian data (Figure 10.13). There are 36 species in the tree, so that there are 35 points in the plot. See legend to Figure 10.12.

To test the approximation method, we also used the strict molecular clock in another analysis, and the corresponding results are shown in Figure 10.14a' and b'. The approximate likelihood calculation is not quite as accurate as under the relaxed clock models. This is because the approximation is expected to be good when the branch lengths are close to the MLEs under the no-clock model, but poor when they are far away. In this dataset, the molecular clock is seriously wrong, so that the MLEs of branch lengths under the no-clock model are unachievable under the clock. At any rate, the strict clock should not be used in a dating analysis if it is seriously violated as it is known that such use may lead to spurious time estimates.

The posterior mean and the 95% CI for the rate μ are 0.16 (0.091, 0.28) per site per 10^8 years and those for the drift parameter v are 0.61 (0.37, 0.97). There is substantial rate variation among lineages, consistent with the rejection of the clock by the LRT.

Figure 10.15 shows the infinite-site plot, with the posterior 95% CI widths (w) plotted against the posterior means of divergence times (t). The regression line $w = 0.204t$ is an indication of the precision of fossil calibrations: for every 100 MY of divergence, 20 MY of uncertainty will be added to the 95% CI. Unlike Figure 10.12, here the correlation is quite low, with $R^2 = 0.516$, indicating that the sequence data are far from saturation, and adding more sequence data is very likely to lead to more precise estimates with narrower CIs.

10.5 Perspectives

The greatest obstacle to the reliable estimation of divergence times is perhaps the confounding effect of time and rate; sequences provide information about the distance but not about time and rate separately. A fixed amount of evolution is compatible with infinitely many assumptions about the rates, as they can fit the data equally well with an appropriate adjustment of the times (Thorne and Kishino 2005). Relaxation of the clock thus makes time estimation a tricky business. This problem may be alleviated to some extent by analysing multiple gene loci simultaneously and by using multiple calibration points. If different genes have different patterns of evolutionary rate change but share the same divergence times, analysing multiple gene loci simultaneously may allow one locus to 'borrow information' from the other loci, making inference about times possible. A long branch at one locus may be due to either a long time duration or a high rate, but a high rate becomes a more likely explanation if the same branch is short at other loci. Similarly, simultaneous application of multiple fossil calibrations may be critical in characterizing local evolutionary rates on the phylogeny.

Recent methodological developments in Bayesian divergence time estimation make it possible to estimate divergence times without relying on a global molecular clock. The method allows integrated analysis of heterogeneous datasets from multiple loci and simultaneous incorporation of multiple fossil calibrations. The Bayesian method can naturally incorporate uncertainties in node ages provided by fossils, but developing objective approaches to describing fossil uncertainties using statistical distributions is challenging. The most promising approaches appear to be probabilistic modelling and statistical analysis of fossil data (either fossil preservation and occurrence data or morphological measurements from both extant and extinct species) (Wilkinson et al. 2011; Ronquist et al. 2012a).

10.6 Problems

10.1 Use the mitochondrial 12S rRNA genes from the apes (Horai et al. 1995) of Example 10.1 to test the molecular clock hypothesis under different substitution models to examine the sensitivity of the test to the assumed model. Use the following models to conduct the LRT: JC69, HKY85, GTR, JC69 + Γ_5, HKY85 + Γ_5, and GTR + Γ_5. Note that the phylogeny is shown in Figure 10.2.

10.2 Run a Bayesian MCMC program (such as MCMCTREE) to estimate the species divergence times using the data of Steiper et al. (2004) (see §10.4.7.1). Treat the five loci as independent loci with different rate trajectories. Compare the results with those of Table 10.1, which were obtained by concatenating the five loci as one partition. Are the CIs any narrower? Note that the tree is shown in Figure 10.7.

10.3 Use the mitochondrial data (codon positions 1 and 2) of §10.4.7.2 to examine the impact of priors on posterior divergence time estimation. Change the priors for the substitution rate (μ) and for the rate-drift parameter (ν) to examine the robustness of posterior time estimates.

10.4 Use the mitochondrial data (codon positions 1 and 2) of §10.4.7.2 to compare the independent-rates and correlated-rate models for relaxing the molecular clock.

10.5 Use the protein sequence data (36 species, 3,630 amino acid sites) of §10.4.7.2 to estimate the species divergence times under the independent and correlated-rate models, and compare results with those of Problems 10.3 and 10.4. You can use the MCMCTREE program to run the MCMC, with the approximation method (dos Reis and Yang 2011) for likelihood calculation under the MTMAM model (Yang et al. 1998).

CHAPTER 11

Neutral and adaptive protein evolution

11.1 Introduction

The adaptive evolution of genes and genomes is ultimately responsible for adaptation in morphology, behaviour, and physiology, and for species divergences and evolutionary innovations. Molecular adaptation is thus an exciting topic in molecular evolutionary studies. While natural selection appears to be ubiquitous in shaping morphological and behavioural evolution, its role in the evolution of genes and genomes is more controversial. Indeed the neutral theory of molecular evolution claims that most of the observed variation within and between species is not due to natural selection, but rather to random fixation of mutations with little fitness significance (Kimura 1968; King and Jukes 1969). A number of tests have been developed in population genetics to test the neutral model. In this chapter, we will introduce the basic concepts of negative and positive selection and the major theories of molecular evolution, and review such neutrality tests (§11.2). The rest and bulk of the chapter follows up on Chapter 2, in which we described Markov models of codon substitution and their use to estimate d_S and d_N, the synonymous and nonsynonymous distances between two protein-coding sequences. Here we discuss the use of codon models in phylogenetic analysis of multiple sequences to detect positive selection driving the fixation of advantageous replacement mutations.

To aid in understanding the roles of natural selection, protein-coding genes offer a great advantage over introns or noncoding DNA in that they allow us to distinguish between synonymous and nonsynonymous substitutions (Miyata and Yasunaga 1980; Gojobori 1983; Li et al. 1985; Nei and Gojobori 1986). If the synonymous rate is used as a benchmark, one can infer whether the fixation of nonsynonymous mutations is aided or hindered by natural selection that acts on the protein. The nonsynonymous/synonymous rate ratio, $\omega = d_N/d_S$, measures selective pressure at the protein level. If selection has no effect on fitness, nonsynonymous mutations will be fixed at the same rate as synonymous mutations, so that $d_N = d_S$ or $\omega = 1$. If nonsynonymous mutations are deleterious, purifying selection will reduce their fixation rate, so that $d_N < d_S$ or $\omega < 1$. If nonsynonymous mutations are favoured by Darwinian selection, they will be fixed at a higher rate than synonymous mutations, resulting in $d_N > d_S$ or $\omega > 1$. A significantly higher nonsynonymous rate than the synonymous rate is thus evidence for adaptive protein evolution.

In comparison, attempts to detect the adaptive evolution of noncoding DNA (e.g. Wong and Nielsen 2004; Andolfatto 2005) have been much less successful, due to the lack of a natural benchmark. One can identify rapidly evolving noncoding regions, by comparison

Molecular Evolution: A Statistical Approach. Ziheng Yang. © Ziheng Yang 2014.
Published 2014 by Oxford University Press.

with the synonymous rate in coding exons in the genomic neighbourhood, but it is difficult to rule out local mutation rate variation as the reason for the high variability in the noncoding DNA. As a result, the evidence is much less convincing.

Early studies using the d_N/d_S criterion took the approach of pairwise sequence comparison, averaging d_S and d_N over all codons in the gene sequence and over the whole time period separating the two sequences. However, one may expect most sites in a functional protein to be constrained during most of evolutionary time. Positive selection, if it occurs, should affect only a few sites and occur in an episodic fashion (Gillespie 1991). Thus, the pairwise averaging approach rarely detects positive selection (e.g. Sharp 1997). Later efforts have focused on detecting positive selection affecting particular lineages on the phylogeny or individual sites in the protein. These methods are discussed in §11.3–11.5. In §11.6, we discuss assumptions and limitations of methods based on the ω ratio, in comparison with neutrality tests. Section §11.7 provides a brief review of empirical findings concerning genes undergoing adaptive evolution.

A number of reviews have been published on codon models and their use in detecting molecular adaptation, including Yang (2000, 2002a), Jensen et al. (2007), Anisimova and Liberles (2007), Anisimova and Kasiol (2009, 2012), and Anisimova (2012). See also the recent book edited by Cannarozzi and Schneider (2012). This chapter focuses on maximum likelihood estimation (MLE) and likelihood ratio tests (LRTs) based on codon models. We note, however, that Bayesian methods implementing similar codon models through MCMC are becoming increasingly feasible computationally (Rodrigue et al. 2008, 2010).

11.2 The neutral theory and tests of neutrality

11.2.1 *The neutral and nearly neutral theories*

Here we introduce the basic concepts of positive and negative selection as well as the major theories of molecular evolution. We also briefly describe a few commonly used tests of neutrality developed in population genetics. The reader should consult a population genetics textbook for more details (e.g. Gillespie 1991, 1998; Hartl and Clark 1997; Li 1997). Ohta and Gillespie (1996) provide a historical perspective of developments in the neutral theory.

In population genetics, the relative fitness of a new mutant allele a relative to the predominant wild-type allele A is measured by the selective coefficient (Malthusian parameter) s. Let the relative fitness of genotypes AA, Aa, and aa be 1, $1 + s$, and $1 + 2s$, respectively. Then $s < 0$, $= 0$, > 0 correspond to negative (purifying) selection, neutral evolution, and positive selection, respectively. The frequency of the new mutant allele goes up or down over generations, affected by natural selection as well as random genetic drift. Whether random drift or selection is more important to the fate of the mutation depends on Ns, where N is the effective population size. If $|Ns| \gg 1$, natural selection dominates the fate of the allele, but if $|Ns|$ is close to 0, the mutation is neutral or nearly neutral and its fate is dominated by random drift. Note also that an allele at very low frequency has a high chance of getting lost by drift even if it is highly advantageous.

Much of the theoretical work studying the dynamics of allele frequency changes was completed in the early 1930s by Fisher (1930a), Haldane (1932), and Wright (1931). The view generally accepted up to the 1960s was that natural selection is the driving force of evolution. Natural populations were believed to be in nearly optimal states, with little genetic variation. Most new mutations were deleterious and quickly removed from the population. Occasionally an advantageous mutation occurred and spread over the entire population. However, in 1966, high levels of genetic variation were detected in allozymes using electrophoresis in *Drosophila* (Harris 1966; Lewontin and Hubby 1966). The *neutral theory*, or *the neutral mutation random drift hypothesis*, was proposed by Kimura (1968) and King and Jukes (1969) mainly to accommodate this surprising finding.

According to the neutral theory, the genetic variation we observe today – both the polymorphism within a species and the divergence between species – is not due to fixation of advantageous mutations driven by natural selection, but to random fixation of mutations with effectively no fitness effect, i.e. neutral mutations. Below are some of the claims and predictions of the theory (Kimura 1983):

- Most mutations are deleterious and are removed by purifying selection.
- The nucleotide substitution rate is equal to the neutral mutation rate, i.e. the total mutation rate times the proportion of neutral mutations. If the neutral mutation rate is constant among species (either in calendar time or in generation time), the substitution rate will be constant. This prediction was seen to provide an explanation for the molecular clock hypothesis.
- Functionally more important genes or gene regions evolve more slowly. A gene with a more important role, or under stronger functional constraint, will undergo a smaller proportion of neutral mutations so that the nucleotide substitution rate will be lower. The negative correlation between functional significance and substitution rate is now a general observation in molecular evolution. For example, replacement substitution rates are almost always lower than silent substitution rates; third codon positions evolve more quickly than first and second codon positions; and amino acids with similar chemical properties tend to replace each other more often than dissimilar amino acids. In contrast, if natural selection drives the evolutionary process at the molecular level, we would expect functionally important genes to have higher evolutionary rates than functionally unimportant genes.
- Within-species polymorphism and between-species divergence are two phases of the same process of neutral evolution.
- The evolution of morphological traits (including physiology, behaviour, etc.) is indeed driven by natural selection. The neutral theory concerns evolution at the molecular level.

The controversy surrounding the neutral theory has generated a rich body of population genetics theory and analytical tools. The neutral theory makes simple and testable predictions. This fact, together with the rapid accumulation of DNA sequence data in the last few decades, has prompted the development of a number of tests of the theory. Some of them are discussed below. For reviews, see Kreitman and Akashi (1995), Kreitman (2000), Nielsen (2001), Fay and Wu (2001, 2003), Ohta (2002), Hein et al. (2005), and Wakeley (2009).

The strict neutral model assumes only two kinds of mutations: the strictly neutral mutations with $s = 0$ and highly deleterious mutations that are wiped out by purifying selection as soon as they occur ($Ns \ll -1$). The strict neutral model is often rejected when tested against real data. Ohta (1973) proposed the *slightly deleterious mutation hypothesis*, which allows for mutations with small negative selective coefficients so that their fate is influenced by both random drift and selection. The fixation probability of such mutations is positive but smaller than for neutral mutations. This model later evolved into the *nearly neutral hypothesis* (Ohta and Tachida 1990; Ohta 1992), which allows both slightly advantageous and slightly deleterious mutations. In contrast to the strict neutral model,

11.2 THE NEUTRAL THEORY AND TESTS OF NEUTRALITY

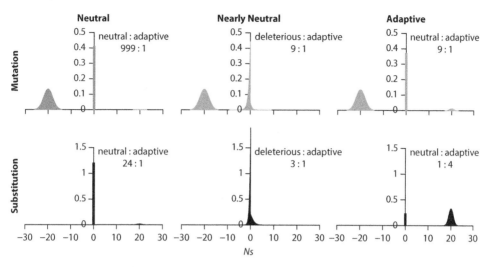

Fig. 11.1 Distributions of scaled selective coefficient (Ns) among mutations and substitutions under neutral, nearly neutral, and adaptive models of molecular evolution. One possible distribution is shown for each model, with the area under the curve being one. The bars at $Ns = 0$ for neutral mutations are broadened for visualization. The sampling formulae under the Poisson random field model of Sawyer and Hartl (1992) were used to transform the density of fitness effects of mutations to that of fixations (substitutions). Redrawn after Akashi (1999b); data courtesy of Hiroshi Akashi.

in which the dynamics depend on the neutral mutation rate alone and not on other factors such as population size and selective pressure, the dynamics of slightly deleterious or nearly neutral mutations depend on all these parameters. As a result, the modified theories are very difficult to test or refute. This is also the case for various selection models (Gillespie 1991). See Figure 11.1 for an illustration of those different theories.

11.2.2 Tajima's D statistic

The amount of genetic variation at a neutral locus maintained in a random mating population is determined by $\theta = 4N\mu$, where N is the (effective) population size and μ is the mutation rate per site per generation (see §9.2.2). Defined on a per site basis, θ is also the expected site heterozygosity between two sequences drawn at random from the population. Population data typically involve little variation, so that the *infinite-site model* is commonly used, which assumes that every mutation occurs at a different site in the DNA sequence and there is no need for correction for multiple hits. Two simple approaches can be used to estimate θ using a random sample of DNA sequences from the population. First the number of segregating (polymorphic) sites S in a sample of n sequences is known to have the expectation $E(S) = l\theta a_n$, where l is the number of sites in the sequence and $a_n = \sum_{i=1}^{n-1} \frac{1}{i}$, so that θ can be estimated by $\hat{\theta}_S = S/(la_n)$ (see equation (9.18)) (Watterson 1975). Second, the average proportion of nucleotide differences over all pairwise comparisons of the n sequences has the expectation θ and can thus be used as an estimate; let this be $\hat{\theta}_\pi$ (Tajima 1983). Both estimates of θ are unbiased under the neutral mutation model, assuming no selection, no recombination, no population subdivision or population size change, and equilibrium between mutation and drift. However, when the model assumptions are violated, different factors have different effects on $\hat{\theta}_S$ and $\hat{\theta}_\pi$. For example, slightly deleterious mutants are maintained in a population at low frequencies and can greatly inflate S and $\hat{\theta}_S$, while having little effect on $\hat{\theta}_\pi$. The direction and magnitude of the difference between the two estimates of θ can provide insights into factors and mechanisms that caused departure from the strict neutral model. Thus Tajima (1989) constructed the following test statistic:

$$D = \frac{\hat{\theta}_\pi - \hat{\theta}_S}{\text{SE}(\hat{\theta}_\pi - \hat{\theta}_S)}, \tag{11.1}$$

where SE is the standard error. Under the null neutral model, D has mean 0 and variance 1. Tajima suggested the use of the standard normal and beta distributions to determine whether D is significantly different from 0. Alternatively simulation can be used to generate the null distribution efficiently.

Statistical significance of Tajima's D test may be compatible with several different explanations, and it may be difficult to distinguish among them. As discussed above, a negative D is indicative of purifying selection or the presence of slightly deleterious mutations segregating in the population. However, a negative D can also be caused by population expansion. In an expanding population, many new mutations may be segregating and will be observed in the data as *singletons*, sites at which only one sequence has a different nucleotide while all other sequences are identical. Singletons inflate the number of segregating sites and cause D to be negative. Similarly, a positive D is compatible with balancing selection, which maintains mutations at intermediate frequencies. However, a shrinking population can also generate a positive D.

11.2.3 Fu and Li's D, and Fay and Wu's H statistics

Consider sites carrying mutations in a sample of n sequences taken at random from a population. Under the infinite-site model, the frequency of the mutant nucleotide can be $r = 1, 2, \ldots,$ or n. Let s_j be the number of sites at which exactly j sequences carry a mutation. The vector $(s_1, s_2, \ldots, s_n)$ is called the *site frequency spectrum* of the sample. This clearly carries more information than the number of segregating sites (S) in the sample. Often, a closely related outgroup species is used to infer the ancestral and derived nucleotide state. For example, if the observed nucleotides at a variable site in a sample of $n = 5$ are AACCC and if the outgroup has A, which is assumed to be the ancestral type, then the mutant nucleotide has frequency $r = 3$. If the observed nucleotides in the sample are AAAAA but the outgroup has G, the frequency is $r = 5$. Fu (1994) referred to r as the *size of the mutation*. If the ancestral state is unknown, it will be impossible to distinguish mutations of sizes r and $n-r$, so that those mutations are grouped into the same class. The site frequency spectrum is then said to be *folded* (Akashi 1999a). Of course, folded configurations are much less informative than unfolded ones. Thus use of the outgroup to infer ancestral states should improve the power of the analysis. Nevertheless, the test may be affected by errors in ancestral reconstruction (Baudry and Depaulis 2003; Akashi et al. 2007; Hernandez et al. 2007). Different types of natural selection will cause the site frequency spectrum to deviate from the neutral expectation in distinct ways. Thus, the site-spectrum distributions or their summary statistics can be used to construct statistical tests of neutrality. For example, Tajima's D is such a test; it contrasts singletons (which mostly contribute to S) with mutations of intermediate frequencies (which mostly contribute to π). Two further popular tests explore similar ideas.

Fu and Li (1993) distinguished *internal* and *external* mutations, which occur along internal or external branches of the genealogical tree, respectively. Let the numbers of such mutations be η_I and η_E. Note that η_E is the number of singletons. Fu and Li constructed the following statistic:

$$D = \frac{\eta_I - (a_n - 1)\eta_E}{\text{SE}(\eta_I - (a_n - 1)\eta_E)}, \quad (11.2)$$

where $a_n = \sum_{i=1}^{n-1} \frac{1}{i}$ and SE is the standard error. Similar to Tajima's D, this statistic is also constructed as the difference of two estimates of θ under the neutral model, divided by the standard error of the difference. Fu and Li (1993) argue that deleterious mutations segregating in the population tend to be recent and reside on external branches of the tree, and contribute to η_E, while mutations in the internal branches are most likely neutral and contribute to η_I (Williamson and Orive 2002). Besides D in equation (11.2), Fu and Li constructed several other tests of this kind. The power of such tests depends on how different the two estimates of θ used in the test are when there is selection. Braverman et al. (1995), Simonsen et al. (1995), Fu (1997), Akashi (1999a, 1999b), and McVean and Charlesworth (2000) conducted simulations to examine the power of the tests.

Fay and Wu (2000) explored a similar idea and constructed an estimate of θ as

$$\hat{\theta}_H = \sum_{i=1}^{n} \frac{2s_i i^2}{n(n-1)}, \quad (11.3)$$

where s_i is the number of mutations of size i. They then defined a statistic $H = \hat{\theta}_\pi - \hat{\theta}_H$, which has expectation 0 under the strict neutral model. The null distribution is generated by computer simulation. Note that in this notation, $\hat{\theta}_\pi = \sum_{i=1}^{n} \frac{2s_i i(n-i)}{n(n-1)}$, so that mutations of intermediate frequencies (with i close to $n/2$) make the greatest contribution to $\hat{\theta}_\pi$, while mutations of high frequencies (with i close to n) make the greatest contribution to $\hat{\theta}_H$. Thus Fay and Wu's H statistic compares mutations at intermediate and high frequencies. Under selective neutrality, the site frequency spectrum is L-shaped, with low-frequency mutants being common and high-frequency mutants rare. When a neutral mutation is tightly linked to a locus under positive selection, the mutant may rise to high frequencies if selection drives the advantageous allele at the selected locus to fixation. The neutral mutation is said to be under *genetic hitchhiking* (Maynard Smith and Haigh 1974; Braverman et al. 1995). Fay and Wu (2000) point out that an excess of high-frequency mutants, indicated by a significantly negative H, is a distinct feature of hitchhiking. The test requires the sequence from an outgroup species to be used to infer the ancestral and derived states at segregating sites.

11.2.4 McDonald–Kreitman test and estimation of selective strength

The neutral theory claims that both the diversity (polymorphism) within a species and the divergence between species are two phases of the same evolutionary process, i.e. both are due to random drift of selectively neutral mutations. Thus if both synonymous and nonsynonymous mutations are neutral, the proportions of synonymous and nonsynonymous polymorphisms within a species should be the same as the proportions of synonymous and nonsynonymous differences between species. The McDonald–Kreitman (1991) test examines this prediction.

Table 11.1 Numbers of silent and replacement divergences and polymorphisms in the *Adh* locus in *Drosophila* (from McDonald and Kreitman 1991)

Type of change	Fixed	Polymorphic
Replacement (nonsynonymous)	7	2
Silent (synonymous)	17	42

Variable sites in protein-coding genes from closely related species are classified into four categories in a 2 × 2 contingency table, depending on whether the site has a polymorphism or a fixed difference, and whether the difference is synonymous or nonsynonymous (Table 11.1). Suppose we sample five sequences from species 1 and four sequences from species 2. A site with data AAAAA in species 1 and GGGG in species 2 is called a fixed difference. A site with AGAGA in species 1 and AAAA in species 2 is called a polymorphic site. Note that under the infinite-site model, there is no need to correct for hidden changes. The neutral null hypothesis is equivalent to independence between the row and column in the contingency table and can be tested using the χ^2 distribution or Fisher's exact test if the counts are small. McDonald and Kreitman (1991) sequenced the alcohol dehydrogenase (*Adh*) gene from three species in the *Drosophila melanogaster* subgroup and obtained the counts of Table 11.1. The *p*-value is < 0.006, suggesting significant departure from the neutral expectation. There are far more replacement differences between species than within species. McDonald and Kreitman interpreted the pattern as evidence that positive selection drives species divergence.

To see the reasoning behind this interpretation, consider the effects of selection on nonsynonymous mutations that occur after the divergence of the species, assuming that synonymous mutations are neutral. Advantageous replacement mutations are expected to go to fixation quickly and become fixed differences between species. Thus an excess of replacement fixed differences, as observed at the *Adh* locus, is indicative of positive selection. As pointed out by McDonald and Kreitman, an alternative explanation for the pattern is a smaller population size in the past than in the present, combined with presence of slightly deleterious replacement mutations. Such mutations might have gone to fixation in the past due to random drift and became fixed differences between species, while they are removed by purifying selection in the large present-day populations. McDonald and Kreitman argued that such a scenario is unlikely for the *Adh* gene they analysed.

In mammalian mitochondrial genes, an excess of replacement polymorphism is observed (e.g. Rand et al. 1994; Nachman et al. 1996). This is indicative of slightly deleterious replacement mutations under purifying selection. Deleterious mutations are removed by purifying selection and will not be seen in between-species comparisons but might still be segregating within species.

The infinite-site model assumed in the McDonald and Kreitman test can break down if the different species are not very closely related. An LRT for testing the equality of the within-species ω_W and the between-species ω_B was implemented by Hasegawa et al. (1998), which uses the codon model to correct for multiple hits. However, the test uses a genealogical tree, and it is unclear how robust the test is to errors in the estimated tree topology.

An idea very similar to that behind the McDonald and Kreitman test is used by Akashi (1999b) to test for natural selection on silent sites driving synonymous codon usage in

Drosophila. Rather than the synonymous and nonsynonymous categories, Akashi classified variable silent sites according to whether the changes are to *preferred* (commonly used) or *unpreferred* (rarely used) codons. If both types of mutations are neutral, their proportions should be the same within species and between species. Akashi detected significant departure from this neutral expectation, providing evidence that evolution at silent sites is driven by natural selection, possibly to enhance the efficiency and accuracy of translation.

The idea underlying the McDonald and Kreitman test has been extended to estimate parameters measuring the strength of natural selection, using the so-called *Poisson random field* theory (Sawyer and Hartl 1992; Hartl et al. 1994; Akashi 1999a). The model assumes free recombination among sites within the same gene. Then the counts in the 2×2 contingency table (see Table 11.1) are independent Poisson random variables with means which are functions of parameters in the model, including the population sizes of the current species, the divergence time between the species, and the selection coefficient of new replacement mutations. The test of selection using this model becomes more powerful if multiple loci are analysed simultaneously, as the species divergence time and population sizes are shared among loci. Power is also improved by using the full site frequency spectrum at polymorphic sites instead of the counts in the 2×2 table. Likelihood and Bayesian methods can be used to test for adaptive amino acid changes and to estimate the strengths of selection (Bustamante et al. 2002, 2003). Bustamante et al. (2002) demonstrated evidence for beneficial substitutions in *Drosophila* and deleterious substitutions in the mustard weed *Arabidopsis*. They attributed the difference to partial self-mating in *Arabidopsis*, which makes it difficult for the species to weed out deleterious mutations.

The Poisson random field model is currently the only tractable framework that explicitly incorporates selection and is applicable to molecular sequence data; in contrast, the neutrality tests only detect violation of the null neutral model but do not consider selection explicitly in the model. While the theory provides a powerful framework for estimating mutation and selection parameters in various population genetics settings, the assumption of free recombination within a locus is highly unrealistic, and can considerably influence the test, especially if the full site frequency spectrum is analysed (Bustamante et al. 2001). The theory has been extended to improve the robustness of the inference to recombination (Zhu and Bustamante 2005) and to demographic processes (Williamson et al. 2007).

11.2.5 *Hudson–Kreitman–Aquade test*

The Hudson–Kreitman–Aquade test or HKA test (Hudson et al. 1987) examines the neutral prediction that polymorphism within species and divergence between species are two facets of the same process. The test uses sequence data at multiple unlinked loci (typically noncoding) from at least two closely related species, and tests whether the polymorphisms and divergences at those loci are compatible. The rationale is that at a locus with a high mutation rate, both polymorphism and divergence should be high, while at a locus with a low mutation rate, both polymorphism and divergence should be low.

Suppose there are L loci. Let the numbers of segregating sites in the two species A and B be S_i^A and S_i^B at locus i, and the number of differences between the two species at the locus be D_i. S_i^A and S_i^B measure within-species polymorphism, while D_i measures between-species divergence. Hudson et al. (1987) assumed that S_i^A, S_i^B, and D_i are independent normal variables and derived their expectations and variances under the neutral model to construct a goodness-of-fit test statistic:

$$X^2 = \sum_{i=1}^{L} (S_i^A - E(S_i^A))^2 \Big/ V(S_i^A) + \sum_{i=1}^{L} (S_i^B - E(S_i^B))^2 \Big/ V(S_i^B) + \sum_{i=1}^{L} (D_i - E(D_i))^2 \Big/ V(D_i). \quad (11.4)$$

The null neutral model involves $L + 2$ parameters: θ_i for each locus defined for species A, the ratio of the two population sizes, and the species divergence time T. There are $3L$ observations: S_i^A, S_i^B, and D_i for $i = 1, \ldots, L$. Thus the test statistic is compared with a χ^2 distribution with $3L - (L + 2) = 2L - 2$ degrees of freedom to test whether the data fit the neutral expectation.

11.3 Lineages undergoing adaptive evolution

11.3.1 *Heuristic methods*

In this section we discuss phylogenetic methods for detecting positive selection along pre-specified lineages on the phylogenetic tree, indicated by $d_N > d_S$. The classic example is the analysis of lysozyme evolution in primates, by Messier and Stewart (1997). The authors inferred the gene sequences in extinct ancestral species and used them to calculate d_N and d_S for every branch on the tree. (See §3.4 and §4.4 for ancestral sequence reconstruction under the parsimony and likelihood criteria.) Positive selection along a branch is identified by testing whether d_N is significantly greater than d_S for that branch, using a normal approximation to the statistic $d_N - d_S$. The authors were able to detect positive selection along the branch ancestral to the leaf-eating colobine monkeys, supporting the hypothesis that acquisition of a new function in the colobine monkeys (i.e. digestion of bacteria in the foreguts of these animals) caused accelerated amino acid substitutions in the enzyme. (Another branch, ancestral to the hominoids, was also found to have a very high d_N/d_S ratio, but no biological explanation has yet been offered for the result, which appears to be a chance effect due to multiple testing.) By focusing on a single branch rather than averaging over the whole time period separating the sequences, the approach of Messier and Stewart has improved power in detecting episodic adaptive evolution.

Zhang et al. (1997) were concerned about the reliability of the normal approximation in small datasets, as the lysozyme gene has only 130 codons. Instead, they suggested the use of Fisher's exact test applied to counts of synonymous and nonsynonymous sites and synonymous and nonsynonymous differences along each branch. A drawback of this approach is that it fails to correct for multiple hits, as the inferred number of differences may underestimate the number of substitutions. Both the approaches of Messier and Stewart (1997) and Zhang et al. (1997) use reconstructed ancestral sequences without accommodating their errors (see §4.4.4).

Another simple approach, suggested by Zhang et al. (1998), is to calculate d_N and d_S in all pairwise comparisons of current species, and to fit branch lengths using least squares for synonymous and nonsynonymous rates separately. The synonymous and nonsynonymous branch lengths, called b_S and b_N, can then be compared to test whether $b_N > b_S$ for the branch of interest, using a normal approximation to the statistic $b_N - b_S$. This approach has the benefit of avoiding the use of reconstructed ancestral sequences. It may be important to use a realistic model to estimate d_N and d_S in pairwise comparisons.

11.3 LINEAGES UNDERGOING ADAPTIVE EVOLUTION

11.3.2 Likelihood method

The simple approaches discussed above have an intuitive appeal and are useful for exploratory analysis. They can be made more rigorous by taking a likelihood approach under a model of codon substitution, analysing all sequences jointly on a phylogenetic tree (Yang 1998a; Yang and Nielsen 1998). The likelihood calculation averages over all possible states at ancestral nodes, weighting them according to their probabilities of occurrence. Thus random and systematic errors in ancestral sequence reconstruction are avoided if the assumed substitution model is adequate. It is also straightforward in a likelihood model to accommodate different transition and transversion rates and unequal codon usage; as a result, likelihood analysis can be conducted under more realistic models.

When a model of codon substitution (see §2.4) is applied to a tree, one can let the ω ratio differ among lineages. Yang (1998a) implemented several models that allow for different levels of heterogeneity in the ω ratio among lineages. The simplest model ('one-ratio') assumes the same ω ratio for all branches. The most general model ('free-ratio') assumes an independent ω ratio for each branch in the phylogeny. This model involves too many parameters except on small trees with few lineages. Intermediate models with two or three ω ratios are implemented as well. These models can be compared using the LRT to examine interesting hypotheses. For example, the one-ratio and free-ratio models can be compared to test whether the ω ratios are different among lineages. We can fit a two-ratio model in which the branches of interest (called foreground branches) have a different ω ratio from all other branches (called the background branches). This two-ratio model can be compared with the one-ratio model to test the null hypothesis that the two ratios are the same. Similarly, one can test the null hypothesis that the ω ratio for the lineages of interest is 1. This test directly examines the possibility of positive selection along specific lineages.

Likelihood calculation under such *branch models* is very similar to that under the standard codon model (one-ratio), discussed in §4.2; the only difference is that the transition probabilities for different branches need to be calculated from different rate matrices (Qs) generated using different ωs. As an illustration, consider the two models of Figure 11.2, used in Zhang's (2003) analysis of the *ASPM* gene, a major determinant of human brain size. In model 0, the same ω ratio is assumed for all branches in the tree, while in model 1, three different ω ratios ($\omega_H, \omega_C, \omega_O$) are assumed for the three branches. Let $x_H, x_C,$ and x_O

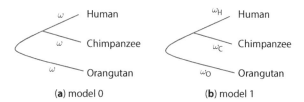

Fig. 11.2 The unrooted phylogenetic tree of human (H), chimpanzee (C), and orangutan (O), showing two branch models. (**a**) In model 0, the same ω ratio is assumed for all branches in the tree. (**b**) In model 1, three different ω ratios ($\omega_H, \omega_C, \omega_O$) are assumed for the three branches. Twice the log likelihood difference between the two models can be compared with the χ^2 distribution with df = 2. After Zhang (2003).

be codons observed at a particular site in the three sequences, and $t_H, t_C,$ and t_O be the three branch lengths in the tree. The probability of observing this site is

$$f(x_H, x_C, x_O) = \sum_i \pi_i p_{ix_H}(t_H; \omega) p_{ix_C}(t_C; \omega) p_{ix_O}(t_O; \omega) \qquad (11.5)$$

under model 0 and

$$f(x_H, x_C, x_O) = \sum_i \pi_i p_{ix_H}(t_H; \omega_H) p_{ix_C}(t_C; \omega_C) p_{ix_O}(t_O; \omega_O) \qquad (11.6)$$

under model 1. Here we use $p_{ij}(t; \omega)$ to emphasize that the transition probability from codon i to j over time t is calculated using the rate ratio ω. The likelihood, i.e. the probability of observing the whole sequence alignment, is the product of the probabilities across all sites in the sequence.

The estimates obtained by Zhang (2003) under model 1 for the *ASPM* gene are $\hat{\omega}_H = 1.03$, $\hat{\omega}_C = 0.66$, and $\hat{\omega}_O = 0.43$. Zhang provided evidence that the high ω_H estimate is due to positive selection driving enlargement of the human brain, and not due to relaxed selective constraint along the human lineage. Another application of the branch-based test is discussed later in §11.5, where the evolution of the angiosperm phytochrome gene family is analysed.

Two remarks are in order concerning the branch-based tests. First, the lineages of interest should be identified *a priori*. It is inappropriate, for example, to analyse the data to identify lineages of high ω ratios and then to apply the branch test to examine whether these ω ratios are significantly higher than one. Because of the problem of multiple testing, the null distribution will not be correct if the hypothesis is derived from the data. In this regard, it may be noted that Kosakovsky Pond and Frost (2005a) described a genetic algorithm to assign ω ratios to lineages on the tree, effectively 'evolving' the model to be tested out of the genetic algorithm by maximizing the model's fit to data. The approach does not require *a priori* specification of the lineages to be tested, but does not appear to be a valid statistical test as it fails to account for multiple testing.

Second, variation in the ω ratio among lineages is a violation of the strictly neutral model, but is itself not sufficient evidence for adaptive evolution. Similarly, if the ω ratio on the foreground branch is higher than that on the background branches but not higher than 1, the result may not be taken as convincing evidence for positive selection. Besides positive selection, two other compatible explanations are relaxation of purifying selection due to loss or diminishment of the protein function, and reduced efficacy of purifying selection removing deleterious mutations due to reduction in population size (Ohta 1973). The criterion that the ω ratio, averaged over all sites in the sequence, should be greater than 1 is very stringent; as a result, the branch test often has little power to detect positive selection.

11.4 Amino acid sites undergoing adaptive evolution

11.4.1 *Three strategies*

The assumption that all amino acid sites in a protein are under the same selective pressure, with the same underlying nonsynonymous/synonymous rate ratio (ω), is grossly unrealistic. Most proteins have highly conserved amino acid positions at which the underlying ω ratio is close to zero. The requirement that the ω ratio, averaged over all sites in the protein, is >1 is thus a very stringent criterion for detecting adaptive evolution. It seems most sensible to allow the ω ratio to vary among sites in the model. Positive selection is then indicated by presence of sites at which $\omega > 1$ rather than the ω ratio averaged over all sites being >1.

Three different strategies appear possible to detect positive selection affecting individual amino acid sites. The first is to focus on amino acid sites that are likely to be under positive selection, as indicated by external information such as the crystal structure of the protein. This is workable only if such information is available. The classic example is the analysis of the human major histocompatibility complex (MHC) loci by Hughes and Nei (1988) and Hughes et al. (1990). The d_N/d_S ratio for the entire gene, although higher than in most other protein-coding genes, is <1, providing no evidence for positive selection. However, studies of the tertiary structure of the molecule (Bjorkman et al. 1987a, 1987b) identified 57 amino acid residues that make up the antigen recognition site (ARS), a groove in the structure involved in binding foreign antigens. Hughes and Nei (1988) thus focused on these 57 codons only, and found that the d_N/d_S ratio for them was significantly >1. Hughes and Nei (1988) used an approach of pairwise sequence comparison. A likelihood approach for joint analysis of multiple sequences on a phylogeny was implemented by Yang and Swanson (2002), in which codons in different partitions are assigned different ω ratios, estimated from the data. In the case of the MHC, two independent ω ratios can be assigned and estimated for codons in the ARS region and those outside, and an LRT can be used to test whether ω_{ARS} is significantly greater than 1. Such models are referred to as *fixed-site* models by Yang and Swanson (2002). They are similar to the models of different substitution rates for different codon positions or different gene loci, discussed in §4.3.3. The likelihood calculation under the models is similar to that under the model of one ω ratio for all sites; the only difference is that the correct ω ratio is used to calculate the transition probabilities for data in each partition. Such models can also be used to test for differences in selective constraints, indicated by the ω ratio, among multiple genes. Muse and Gaut (1997) termed such tests *relative-ratio tests*.

The second strategy is to estimate one ω ratio for every site. There is then no need for *a priori* partitioning of sites. Fitch et al. (1997) and Suzuki and Gojobori (1999) reconstructed ancestral sequences on the phylogeny by parsimony, and used the reconstructed sequences to count synonymous and nonsynonymous changes at each site along branches of the tree. Fitch et al. (1997) analysed the human influenza virus type A haemagglutinin (HA) genes and considered a site to be under positive selection if the calculated ω ratio for the site is higher than the average across the whole sequence. Suzuki and Gojobori (1999) used a more stringent criterion and considered a site to be under positive selection only if the estimated ω ratio for the site is significantly greater than 1. The test is conducted by applying the method of Nei and Gojobori (1986) to analyse counts of sites and of differences at each site. Computer simulations (Suzuki and Gojobori 1999; Wong et al. 2004) demonstrate that a large number of sequences are needed for the test to have any power. The simple approach has intuitive appeal and is useful in the exploratory analysis of large datasets.

The use of reconstructed ancestral sequences in the approaches of Fitch et al. (1997) and Suzuki and Gojobori (1999) may be a source of concern, since these sequences are not real observed data. In particular, positively selected sites are often the most variable in the alignment, and are those at which ancestral reconstruction is the least reliable. This problem can be avoided by taking a likelihood approach, averaging over all possible ancestral states. Indeed Suzuki (2004), Massingham and Goldman (2005), and Kosakovsky Pond and Frost (2005b) implemented methods to estimate one ω parameter for each site using ML. Typically, other parameters in the model such as branch lengths are estimated for all sites and fixed when the ω ratio is estimated for every site. Use of a model of codon substitution enables the method to incorporate the transition–transversion rate difference and unequal codon frequencies. At each site, the null hypothesis $\omega = 1$ is tested using either the χ_1^2 distribution or a null distribution generated by computer simulation. Massingham and Goldman (2005) called this the *site-wise likelihood ratio* (SLR) test. While the model allows the ω ratio to vary freely among sites, the number of parameters increases without bound with the increase of the sequence length. It thus does not have the well-known asymptotic properties of MLEs (Stein 1956, 1964; Kalbfleisch 1985, pp. 92–95). The standard procedure for pulling ourselves out of this trap of infinitely many parameters is to assign a prior on ω, and to derive the conditional (posterior) distribution of ω given the data. This is the Bayesian (Lindley 1962) or empirical Bayes (EB) approach (Maritz and Lwin 1989; Carlin and Louis 2000), the third strategy to be discussed below. Despite these theoretical criticisms, computer simulations of Massingham and Goldman (2005) suggested that the SLR test achieved good false positive rates as well as reasonably high power, even though the ω ratios for sites were not estimated reliably.

Note that all methods discussed in this category (Fitch et al. 1997; Suzuki and Gojobori 1999; Suzuki 2004; Massingham and Goldman 2005; Kosakovsky Pond and Frost 2005b) are designed to test for positive selection on a single site. To test whether the sequence contains any sites under positive selection (i.e. whether the protein is under positive selection), a correction for multiple testing should be applied. For example, Wong et al. (2004) applied the modified Bonferroni procedure of Simes (1986) to the test of Suzuki and Gojobori (1999) to detect positive selection on the protein.

The third strategy, as indicated above, is to use a statistical distribution (prior) to describe the random variation of ω over sites (Nielsen and Yang 1998; Yang et al. 2000). The model assumes that different sites have different ω ratios but we do not know which sites have high ωs and which low ωs. The null hypothesis of no positive selection can be tested using an LRT comparing two statistical distributions, one of which assumes no sites with $\omega > 1$ while the other assumes the presence of such sites. When the LRT suggests the presence of sites with $\omega > 1$, an EB approach is used to calculate the conditional (posterior) probability distribution of ω for each site given the data at the site. Such models are referred to as the *random-site* models by Yang and Swanson (2002). They are discussed in the next two subsections.

11.4.2 Likelihood ratio test of positive selection under random-site models

We now discuss likelihood calculation under the random-site models. The ω ratio at any site is a random variable from a distribution $f(\omega)$. Thus inference concerning ω is based on the conditional (posterior) distribution $f(\omega|X)$ given data X. In simple statistical problems, the prior $f(\omega)$ can be estimated from the data without assuming any distributional form, leading to the so-called nonparametric EB method (Robbins 1955, 1983; Maritz and Lwin 1989, pp. 71–78; Carlin and Louis 2000, pp. 57–88). However, the approach appears intractable in the present case. Instead, Nielsen and Yang (1998) and Yang et al. (2000) implement parametric models $f(\omega)$ and estimate the parameters involved in those densities. The synonymous rate is assumed to be constant among sites, and only the nonsynonymous rates are variable. The branch length t is defined as the expected number of nucleotide substitutions per codon, averaged over all sites.

The model has the same structure as the gamma and discrete-gamma models of variable rates among sites (see §4.3.1). The likelihood is a function of the parameters in the ω distribution (but not of the ωs themselves), as well as other parameters in the model such as the branch lengths and the transition/transversion rate ratio κ. The probability of observing data at a site, say data x_h at site h, is an average over the ω distribution:

$$f(x_h) = \int_0^\infty f(\omega) f(x_h|\omega) \, d\omega \simeq \sum_{k=1}^K p_k f(x_h|\omega_k). \tag{11.7}$$

If $f(\omega)$ is a discrete distribution, the integral becomes a sum. The integral for a continuous $f(\omega)$ is analytically intractable. Thus we apply a discrete approximation, with $K = 10$ equal-probability categories used to approximate the continuous density, so that p_k and ω_k in equation (11.7) are all calculated as functions of the parameters in the continuous density $f(\omega)$.

Note that in Markov models of nucleotide, amino acid or codon substitution, one usually scales the rate matrix Q so that the expected number of substitutions per site per time unit is one, $-\sum_i \pi_i q_{ii} = 1$. Time t is then measured by distance, the expected number of changes per site. The codon-based site models assume that the synonymous rate is constant among site classes. This means that one should not scale the Q matrices for the different site classes separately. The correct procedure is as follows. Constructs the Q matrices for the site classes using equation (2.7) with the correct ωs. Calculate a single scale factor for all Q matrices. This may be calculated using an average Q matrix with ω replaced by an average over the site classes: $\bar{\omega} = \sum_k^K p_k \omega_k$. Alternatively one may choose not to scale the Q matrices at all: one will then obtain the correct estimates of parameters such as κ and ω and the correct log likelihood, although the branch lengths will be off by a scale factor.

To test for positive selection or the presence of sites with $\omega > 1$, we use an LRT to compare a null model that does not allow $\omega > 1$ with an alternative model that does. Two pairs of models are found to be particularly effective in computer simulations (Anisimova et al. 2001; Anisimova et al. 2002; Wong et al. 2004). They are summarized in Table 11.2. The first pair involves the null model M1a (neutral), which assumes two site classes in proportions p_0 and $p_1 = 1 - p_0$ with $0 < \omega_0 < 1$ and $\omega_1 = 1$, and the alternative model M2a (selection), which adds a proportion p_2 of sites with $\omega_2 > 1$ estimated from the data. M1a and M2a are slight modifications of models M1 and M2 of Nielsen and Yang (1998), which had $\omega_0 = 0$ fixed. As M2a has two more parameters than M1a, the χ_2^2 distribution may be used for the test. However, the regularity conditions for the asymptotic χ^2 approximation

Table 11.2 Parameters in a few commonly used site models (models of variable ω ratios among sites)

Model	p	Parameters
M0 (one-ratio)	1	ω
M1a (neutral)	2	p_0 ($p_1 = 1 - p_0$), $\omega_0 < 1$ ($\omega_1 = 1$)
M2a (selection)	4	p_0, p_1 ($p_2 = 1 - p_0 - p_1$), $\omega_0 < 1, \omega_1 = 1, \omega_2 > 1$
M3 (discrete)	5	p_0, p_1 ($p_2 = 1 - p_0 - p_1$), $\omega_0, \omega_1, \omega_2$
M7 (beta)	2	p, q
M8 (beta&ω)	4	p_0 ($p_1 = 1 - p_0$), $p, q, \omega_s > 1$

Note: p is the number of parameters in the ω distribution.

are not met, and the correct null distribution is unknown. First, M1a is equivalent to M2a by fixing $p_2 = 0$, which is at the boundary of the parameter space. Second, when $p_2 = 0$, ω_2 is not identifiable. It thus appears that the difference between M1a and M2a is not as large as two free parameters and the use of χ_2^2 may be too conservative.

The second pair of models consists of the null model M7 (beta), which assumes a beta distribution for ω, with density

$$f(\omega; p, q) = \frac{1}{B(p, q)} \omega^{p-1}(1-\omega)^{q-1}, 0 < \omega < 1, p > 0, q > 0. \tag{11.8}$$

This has the mean $p/(p+q)$ and variance $pq/[(p+q)^2(p+q+1)]$. The beta distribution takes values in the interval $(0, 1)$ only and thus serves as a null model. It is also very flexible as it can take a variety of shapes depending on the two parameters p and q (Figure 11.3). The alternative model, M8 (beta&ω), adds another class of sites with $\omega_s > 1$ for positive selection. M8 has two more parameters than M7, so that χ_2^2 may be used to conduct the

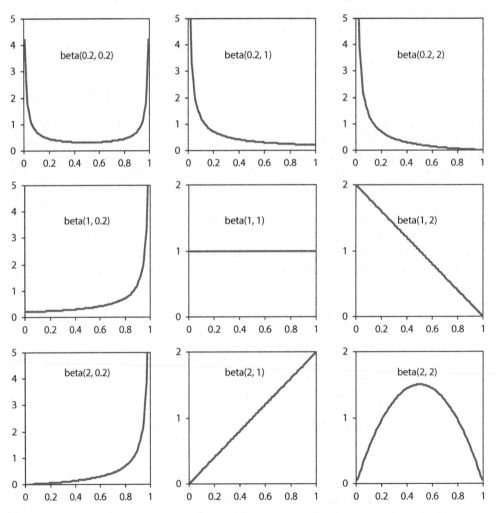

Fig. 11.3 Density of beta distribution: beta(p, q). The x-axis is the ω ratio, while the y-axis is proportional to the number of sites with that ω ratio.

LRT. As in the comparison between M1a and M2a, use of χ_2^2 appears to make the test conservative.

Another model, called M3 (discrete), may be mentioned here as well. This assumes a general discrete model, with the frequencies and the ω ratios (p_k and ω_k in equation (11.7)) for K site classes being estimated as free parameters. All models discussed here may be considered special cases of this general finite-mixture model. As is typical in such mixture models, often only a few classes can be fitted to real datasets. Model M3 may be compared with model M0 (one-ratio) to construct an LRT. However, this is more a test of variability in selective pressure among sites and should not be considered a reliable test of positive selection.

All the models discussed above assume that the synonymous rate is constant across the whole gene while the nonsynonymous rate varies. Kosakovsky Pond and Muse (2005) and Rubinstein et al. (2011) extended site model M3 (discrete) to allow both d_S and d_N to vary among sites.

11.4.3 Identification of sites under positive selection

When the LRTs suggest the presence of sites under positive selection, a natural question to ask is where these sites are. The EB approach can be used to calculate the posterior probability that each site is from a particular site class, and sites with high posterior probabilities (say, with $P > 95\%$) coming from the class with $\omega > 1$ are most likely under positive selection:

$$f(\omega_k|\boldsymbol{x}_h) = \frac{p_k f(\boldsymbol{x}_h|\omega_k)}{f(\boldsymbol{x}_h)} = \frac{p_k f(\boldsymbol{x}_h|\omega_k)}{\sum_j p_j f(\boldsymbol{x}_h|\omega_j)}. \tag{11.9}$$

This is the same methodology that is used in estimating substitution rates under the gamma or discrete-gamma models of rates for sites (see equation (4.21)). The approach makes it possible to detect positive selection and identify sites under positive selection even if the average ω ratio over all sites is much less than 1.

Equation (11.9) requires a knowledge of model parameters, such as the proportions and ω ratios for the site classes, as well as the branch lengths on the tree. Nielsen and Yang (1998) and Yang et al. (2000) replaced those parameters by their MLEs. This approach, known as naïve empirical Bayes (NEB), ignores sampling errors in parameters. It may be unreliable in small datasets, which do not contain enough information to estimate the parameters reliably. This deficiency has been explored in several computer simulation studies (e.g. Anisimova et al. 2002; Wong et al. 2004; Massingham and Goldman 2005; Scheffler and Seoighe 2005). A more reliable approach is implemented by Yang et al. (2005), known as the Bayes empirical Bayes (BEB; Deely and Lindley 1981). BEB accommodates uncertainties in the MLEs of parameters in the ω distribution by integrating numerically over a prior for the parameters. Other parameters such as branch lengths are fixed at their MLEs, as these are expected to have much less effect on inference concerning ω.

A hierarchical (full) Bayesian approach to dealing with uncertainties in parameter estimates has been implemented by Huelsenbeck and Dyer (2004), using MCMC to average over tree topologies, branch lengths, as well as other substitution parameters in the model. This involves intensive computation but may produce more reliable inference in small noninformative datasets, where the MLEs of branch lengths may involve large sampling errors (Scheffler and Seoighe 2005). Guindon et al. (2006) discussed the use of the false discovery rate in detection of positively selected amino acid sites.

11.4.4 Positive selection at the human MHC

Here we use the random-site models to analyse a dataset of class I major histocompatability (MHC or HLA) genes from humans, compiled by Yang and Swanson (2002). The dataset consists of 192 alleles from the A, B, and C loci, with 270 codons in each sequence, after removal of alignment gaps. The apparent selective force acting upon the class I MHC is to recognize and bind a large number of foreign peptides. The ARS is the cleft that binds foreign antigens (Bjorkman et al. 1987a, 1987b). The identification of the ARS enabled previous researchers to partition the data into ARS and non-ARS sites and to demonstrate positive selection in the ARS, even though positive selection was not detected in pairwise comparisons averaging rates over the entire sequence (Hughes and Nei 1988; Hughes et al. 1990).

The random-site models do not make use of the structural information. The tree topology is estimated using the neighbour-joining method (Saitou and Nei 1987) based on MLEs of pairwise distances (Goldman and Yang 1994). To save computation, the branch lengths are estimated under model M0 (one-ratio) and then fixed when other models are fitted to the data. Equilibrium codon frequencies are calculated using the base frequencies at the three codon positions (the F3×4 model), with the frequencies observed in the data used as estimates. The MLEs of parameters and the log likelihood values are given in Table 11.3 for several random-site models. For example, the MLEs under M2a suggest that about 8.4% of sites are under positive selection with $\omega = 5.4$. The LRT statistic comparing models M1a and M2a is $2\Delta\ell = 518.68$, much greater than critical values from the χ_2^2. The test using models M7 and M8 leads to the same conclusion. Parameter estimates of Table 11.3 can be used to calculate posterior probabilities that each site is from the different site classes, using the NEB procedure (equation (11.9)). The results obtained under M2a are presented in Figure 11.4, while amino acid residues with posterior probability $P > 95\%$ of coming from the site class of positive selection are shown on the three-dimensional structure in Figure 11.5. Most of these sites are on the list of 57 amino acid residues making up the ARS. Three of them are not on the list but are nevertheless in the same region

Table 11.3 Log likelihood values and parameter estimates under models of variable ω ratios among sites for 192 MHC alleles

Model	p	ℓ	Estimates of parameters	Positively selected sites
M0 (one-ratio)	1	−8,225.15	$\hat{\omega} = 0.612$	None
M1a (neutral)	2	−7,490.99	$\hat{p}_0 = 0.830, (\hat{p}_1 = 0.170),$ $\hat{\omega}_0 = 0.041, (\omega_1 = 1)$	Not allowed
M2a (selection)	4	−7,231.15	$\hat{p}_0 = 0.776, \hat{p}_1 = 0.140$ $(\hat{p}_2 = 0.084)$ $\hat{\omega}_0 = 0.058 (\omega_1 = 1),$ $\hat{\omega}_2 = 5.389$	**9F** 24A **45M** 62G **63E 67V** **70H 71S 77D 80T 81L** 82R 94T **95V 97R** 99Y 113Y **114H 116Y** **151H 152V 156L 163T** 167W
M7 (beta)	2	−7,502.79	$\hat{p} = 0.087, \hat{q} = 0.360$	Not allowed
M8 (beta&ω)	4	−7,238.01	$\hat{p}_0 = 0.915 (\hat{p}_1 = 0.085),$ $\hat{p} = 0.167, \hat{q} = 0.717,$ $\hat{\omega}_s = 5.079$	**9F** 24A **45M 63E 67V** 69A **70H** **71S 77D 80T 81L** 82R 94T **95V 97R** 99Y 113Y **114H 116Y** **151H 152V 156L 163T**

Note: p is the number of parameters in the ω distribution. Branch lengths are fixed at their MLEs under M0 (one-ratio). Estimates of the transition/transversion rate ratio κ range from 1.5 to 1.8 among models. Positively selected sites are inferred at the cutoff posterior probability $P \geq 95\%$ with those reaching 99% shown in bold. Amino acid residues are from the reference sequence in the PDB structure file 1AKJ. (Adapted from Yang and Swanson 2002; Yang et al. 2005)

11.4 AMINO ACID SITES UNDERGOING ADAPTIVE EVOLUTION

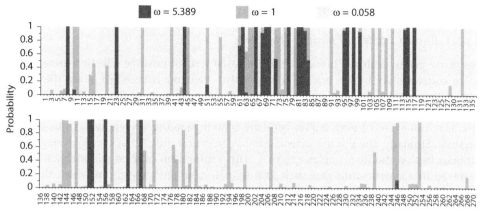

Fig. 11.4 Posterior probabilities that each site is from the three site classes under the M2a (selection) model, calculated using the NEB procedure for the dataset of 192 MHC alleles. The MLEs of parameters under the model suggest three site classes with $\omega_0 = 0.058$, $\omega_1 = 1$, and $\omega_2 = 5.389$ in proportions $p_0 = 0.776$, $p_1 = 0.140$, and $p_2 = 0.084$ (Table 11.3). These proportions are the prior probabilities that any site belongs to the three classes. The data at a site (codon configurations in different sequences) alter the prior probabilities dramatically, so that the posterior probabilities can be very different from the prior probabilities. For example, at site 4, the posterior probabilities are 0.925, 0.075, and 0.000, and thus the site is likely to be under strong purifying selection. At site 9, the posterior probabilities are 0.000, 0.000, and 1.000, and thus the site is almost certainly under positive selection. Sites with posterior probability $P > 0.95$ for site class 2 are listed in Table 11.3 and also shown on the protein structure in figure 11.5. The BEB procedure produced virtually identical posterior probabilities in this large dataset.

Fig. 11.5 The structure of the class I MHC allele H-2Db (Protein Data Bank file 1AKJ, chain A), with a bound antigen shown in stick and ball format. Amino acid residues identified to be under positive selection under the random-site model M2a are shown in spacefill, and all fall in the ARS domain. See also Table 11.3 and Figure 11.4. From Yang and Swanson (2002).

of the structure. It may be noted that the inferred sites are scattered along the primary sequence (Figure 11.4) but are clustered on the structure (Figure 11.5). This dataset is large, so that the parameters are estimated reliably. The NEB and BEB procedures produce almost identical posterior probabilities and lists of positively selected sites (see Yang et al. 2005). Furthermore, models M2a and M8 produce highly similar results (Table 11.3).

11.5 Adaptive evolution affecting particular sites and lineages

11.5.1 Branch-site test of positive selection

For many genes, both the branch- and site-based tests of positive selection are expected to be conservative. In the branch test, positive selection is detected along the branch only if the ω ratio averaged over all sites is significantly greater than 1. If most sites in a protein are under purifying selection with ω close to 0, the ω ratio averaged over all sites may not exceed 1 even if a few sites are evolving rapidly, driven by positive selective pressure. Similarly, the site test detects positive selection only if the ω ratio averaged over all branches on the tree is greater than 1. This assumption may be reasonable for genes involved in a genetic arms race such as host–pathogen antagonism, in which the genes may be under constant pressure of diversifying selection. This might explain the fact that the site test has been more successful than the branch test in identifying genes under positive selection. For most other genes, one might expect positive selection to affect only a few amino acid residues along particular lineages. The *branch-site* models (Yang and Nielsen 2002) attempt to detect signals of such local episodic natural selection.

In the branch-site models (Yang and Nielsen 2002), branches in the tree are divided *a priori* into foreground and background categories, and an LRT is constructed to compare an alternative model that allows for some sites under positive selection on the foreground branches with a null model that does not. However, a simulation study (Zhang 2004) has found that the test is sensitive to the underlying model assumptions, and when the assumptions are violated, it may generate excessive false positives. A slight modification was introduced and found to have much better performance (Yang et al. 2005; Zhang et al. 2005). Here we describe the modified test.

Table 11.4 summarizes the model, called branch-site model A. Along the background lineages, there are two classes of sites: conserved sites with $0 < \omega_0 < 1$ and neutral sites with $\omega_1 = 1$. Along the foreground lineages, a proportion $(1 - p_0 - p_1)$ of sites become under positive selection with $\omega_2 \geq 1$. Likelihood calculation under this model can be easily adapted from that for the site models. As we do not know *a priori* which site class each site is from, the probability of data at a site is an average over the four site classes. Let $I_h = 0, 1, 2a, 2b$ be the site class that site h is from. We have

$$f(\mathbf{x}_h) = \sum_{I_h} p_{I_h} f(\mathbf{x}_h | I_h). \tag{11.10}$$

The conditional probability $f(\mathbf{x}_h|I_h)$ of observing data $\mathbf{x}_h$ at site h, given that the site comes from site class I_h, is easy to calculate, because the site evolves under the one-ratio model if $I_h = 0$ or 1, and under the branch model if $I_h = 2a$ or 2b.

Table 11.4 The ω ratios assumed in branch-site model A

Site class	Proportion	Background ω	Foreground ω
0	p_0	$0 < \omega_0 < 1$	$0 < \omega_0 < 1$
1	p_1	$\omega_1 = 1$	$\omega_1 = 1$
2a	$(1-p_0-p_1)p_0/(p_0+p_1)$	$0 < \omega_0 < 1$	$\omega_2 \geq 1$
2b	$(1-p_0-p_1)p_1/(p_0+p_1)$	$\omega_1 = 1$	$\omega_2 \geq 1$

Note: Shown in the table is the alternative hypothesis in the branch-site test of positive selection, which involves four free parameters: p_0, p_1, ω_0, and ω_2. The null hypothesis of the test is the same model but with $\omega_2 = 1$ fixed, involving three free parameters.

In the branch-site test of positive selection, model A is the alternative hypothesis, while the null hypothesis is the same model A but with $\omega_2 = 1$ fixed (Table 11.4). The null model has one fewer parameter but since $\omega_2 = 1$ is fixed at the boundary of the parameter space of the alternative model, the null distribution should be a 1:1 mixture of point mass 0 and χ_1^2 (Self and Liang 1987). The critical values are 2.71 and 5.41 at the 5% and 1% levels, respectively. One may also use χ_1^2 (with critical values 3.84 at 5% and 5.99 at 1%) to be conservative and to guide against violations of model assumptions. In computer simulations, the branch-site test was found to have acceptable false positive rates (Yang and dos Reis 2011), and also more power than the branch test (Zhang et al. 2005). However, the model allows only two kinds of branches, the foreground and background, which may be unrealistic for some datasets.

The NEB and BEB procedures were implemented for the branch-site model as well, allowing identification of amino acid sites that are under positive selection along the foreground lineages, as in the site-based analysis (Yang et al. 2005). However, the posterior probabilities often do not reach values as high as 95%, indicating a lack of power in such analysis.

Similar to the branch test, the branch-site test requires the foreground branches to be specified *a priori*. This may be easy if a well-formulated biological hypothesis exists, for example, if we want to test adaptive evolution driving functional divergences after gene duplication. The test may be difficult to apply if no *a priori* hypothesis is available. To apply the test to several or all branches on the tree, one has to correct for multiple testing (Anisimova and Yang 2007).

11.5.2 Other similar models

Several other models of codon substitution have been implemented that allow the selective pressure indicated by the ω ratio to vary both among lineages and among sites. Forsberg and Christiansen (2003) and Bielawski and Yang (2004) implemented the clade models. Branches on the phylogeny are *a priori* divided into two or more clades, and an LRT is used to test for differences in selective pressure among the clades, with selective pressure measured by the ω ratio. The clade models are similar to the models of functional divergence of Gu (2001) and Knudsen and Miyamoto (2001), in which the amino acid substitution rate is used as an indicator of functional constraint.

Clade model C, implemented by Bielawski and Yang (2004), is summarized in Table 11.5. This assumes three site classes. Class 0 includes conserved sites with $0 < \omega_0 < 1$, while class 1 includes neutral sites with $\omega_1 = 1$; both apply to all lineages. Class 2 includes sites that are under different selective pressures in different clades, with ω_2 for clade 1, ω_3 for clade 2, and so on. There may not be any sites under positive selection with $\omega > 1$. With two clades, the model involves five parameters in the ω distribution:

Table 11.5 The ω ratios assumed in clade model C

Site class	Proportion	Clade 1	Clade 2
0	p_0	$0 < \omega_0 < 1$	$0 < \omega_0 < 1$
1	p_1	$\omega_1 = 1$	$\omega_1 = 1$
2	$p_2 = 1 - p_0 - p_1$	ω_2	ω_3

Note: With two clades, the model involves five parameters: $p_0, p_1, \omega_0, \omega_2,$ and ω_3. If there are more than two clades, site class 2 will have ratios $\omega_2, \omega_3, \omega_4,$ and so on.

p_0, p_1, ω_0, ω_2, and ω_3. The initial implementation allowed only two clades but this was extended to several clades (Yoshida et al. 2011). Bielawski and Yang (2004) constructed an LRT by comparing model C with the site model M1a (neutral), which assumes two site classes with two free parameters: p_0 and ω_0. Weadick and Chang (2012) point out that a more appropriate null model is a variation of M2a, with three site classes with $\omega_0 < 1$, $\omega_1 = 1$, and $0 < \omega_2 < \infty$. The degree of freedom should be $c - 1$ if c clades are in the model.

A switching model of codon substitution is implemented by Guindon et al. (2004), which allows the ω ratio at any site to switch among three different values $\omega_1 < \omega_2 < \omega_3$. Besides the Markov-process model of codon substitution, a hidden Markov chain runs over time and describes the switches of any site between different selective regimes (i.e. the three ω values). The model is similar in structure to the covarion model for variable substitution rates, in which every site switches between high and low rates (Tuffley and Steel 1998; Galtier 2001; Huelsenbeck 2002). The switching model is an extension of the site models (Nielsen and Yang 1998; Yang et al. 2000), which had a fixed ω for every site. An LRT test can thus be used to compare them. Guindon et al. (2004) found that the switching model fitted a dataset of the HIV-1 *env* genes much better than the site models. An EB procedure can be used to identify lineages and sites with high ω ratios. Compared with the branch-site model, the switching model does not require *a priori* partition of branches into foreground and background categories and allows a more flexible characterization of selective regimes on the background branches. It may thus be very useful for exploratory analysis, to generate hypotheses for experimental verification. However, if a biological theory is available to predict which branches may be affected by positive selection so that the foreground branches can be identified *a priori*, the branch-site test tends to have more power than tests based on the switching models (Shan et al. 2009).

11.5.3 Adaptive evolution in angiosperm phytochromes

Here we apply the branch and branch-site tests to detect positive selection in the evolution of the phytochrome gene *phy* subfamily. Phytochromes are the best characterized plant photosensors. They are chromoproteins that regulate the expression of a large number of light-responsive genes and many events in plant development, including seed germination, flowering, fruit ripening, stem elongation, and chloroplast development (Alba et al. 2000). All angiosperms characterized to date contain a small number of PHY apoproteins encoded by a small *phy* gene family. The data analysed here are from Alba et al. (2000). We use the alignment of the 16 sequences from the A and C/F subfamilies. The phylogenetic tree is shown in Figure 11.6. We test whether the gene was under positive selection along the branch separating the A and the C/F subfamilies, which represents a gene duplication.

The one-ratio model (M0) gives the estimate $\hat{\omega} = 0.089$ (Table 11.6a). This is an average over all branches and sites, and the small value reflects the dominating role of purifying selection in the evolution of the *phy* gene family. In the branch model (Table 11.6b), two ω ratios are assigned for the background (ω_0) and foreground (ω_1) branches. Although $\hat{\omega}_1 > \hat{\omega}_0$, $\hat{\omega}_1$ is not greater than 1. Furthermore, the null hypothesis $\omega_0 = \omega_1$ is not rejected by the data; the LRT statistic for comparing the one-ratio and two-ratios models is $2\Delta\ell = 2 \times 0.64 = 1.28$, with $p = 0.26$ at df = 1. Thus the branch test provides no evidence for positive selection.

We then conducted the branch-site test of positive selection. The test statistic is calculated to be $2\Delta\ell = 2 \times 9.94 = 19.88$, with $p = 8.2 \times 10^{-6}$ if we use χ_1^2 or

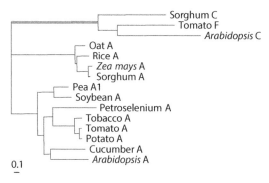

Fig. 11.6 Phylogeny for the phytochrome *phy* gene family in angiosperms. The branch separating the A and the C/F subfamilies is postulated to be under positive selection after gene duplication. This is the foreground branch in the branch and branch-site tests, while all other branches are the background branches (Table 11.6).

Table 11.6 Log likelihood values and parameter estimates under various models for the phytochrome gene *phy* AC and F subfamilies

Model	p	ℓ	Estimates of parameters
(a) One-ratio (M0)	1	−29,984.12	$\hat{\omega} = 0.089$
(b) Branch model (2-ratios)	2	−29,983.48	$\hat{\omega}_0 = 0.090$, $\hat{\omega}_1 = 0.016$
(c) Branch-site model A, with $\omega_2 = 1$ fixed	3	−29,704.74	$\hat{p}_0 = 0.774$, $\hat{p}_1 = 0.073$ ($\hat{p}_2 + \hat{p}_3 = 0.153$), $\hat{\omega}_0 = 0.078$, $\omega_1 = \omega_2 = 1$
(d) Branch-site model A, with $\omega_2 > 1$	4	−29,694.80	$\hat{p}_0 = 0.813$, $\hat{p}_1 = 0.075$ ($\hat{p}_2 + \hat{p}_3 = 0.111$), $\hat{\omega}_0 = 0.080$, $\hat{\omega}_2 = 131.1$

Note: p is the number of free parameters in the ω distribution. Estimates of κ are around 2.1. Estimates of branch lengths are not shown (see Figure 11.6). Codon frequencies are calculated using the base frequencies at the three codon positions under the F3×4 model.

$p = 8.2 \times 10^{-6}/2 = 4.1 \times 10^{-6}$ if we use the 1:1 mixture of 0 and χ_1^2. Thus the test provides strong evidence for positive selection. The parameter estimates under branch-site model A suggests about 11% of sites to be under positive selection along the foreground lineage with very high ω_2. Application of the BEB procedure under the model identified 27 amino acid sites potentially under positive selection along the foreground lineage at the cutoff posterior probability 95%. These are 55R, 102T, 105S, 117P, 130T, 147S, 171T, 216E, 227F, 252I, 304I, 305D, 321L, 440L, 517A, 552T, 560Y, 650T, 655S, 700A, 736K, 787N, 802V, 940T, 986M, 988Q, and 1087D (the amino acids refer to the *Zea mays* sequence). These are a subset of sites predicted to have changed along the branch using ancestral sequence reconstruction.

11.6 Assumptions, limitations, and comparisons

In this section we discuss some of the limitations of the codon substitution methods used for detecting positive selection. We also attempt to compare them with the tests of neutrality developed in population genetics, reviewed in §11.2.

11.6.1 Assumptions and limitations of current methods

Codon substitution models developed for detecting signals of natural selection in protein-coding genes involve a number of assumptions. First, like the many models of nucleotide substitution (see Chapter 1), the codon models describe substitutions and do not model mutation and selection explicitly as in a population genetics model. This formulation is both a drawback and an advantage. The drawback is that those codon models may not be useful for detecting natural selection at silent sites (Akashi 1995). If the silent substitution rate is high, the model cannot tell whether it is due to a high mutation rate, or a relaxed selective constraint, or even positive selection. The advantage is that the model, when used to detect selection acting on the protein, is not as sensitive to whether mutational bias or natural selection drives evolution at silent sites. In this regard, a number of authors have suggested that the use of the ω ratio to detect selection in protein-coding genes requires the assumption of neutral evolution at silent sites. This appears to be a widespread misconception. The ω ratio contrasts substitution rates before and after selection on the protein (see §2.5.2.1 and §2.5.4.2). Whether mutation or selection drives evolution at the silent sites should not affect this interpretation of the ω ratio or its use to detect positive selection at the protein level. This point is particularly clear when one considers the mutation–selection model (Yang and Nielsen 2008), which uses mutation bias and codon-fitness parameters to describe the frequencies of codons in the alignment, while ω is used to accommodate natural selection at the protein level (see §2.4.2). To quote Benner (2001), the ω ratio measures 'the difference between how proteins have divergently evolved in their past, and how they would have evolved had they been formless, functionless molecules'.

Second, even when considered as models of *substitution* and not of mutation and selection, the codon models may be highly unrealistic. For example, the same ω ratio is assumed for substitutions between different amino acids, whereas it is known that the relative substitution rates between amino acids are strongly influenced by their chemical properties (e.g. Zuckerkandl and Pauling 1962; Dayhoff et al. 1978). Incorporating chemical properties into codon models does improve the model's fit to data (Goldman and Yang 1994; Yang et al. 1998), but the improvement is not extraordinary. This appears partly due to our poor understanding of how chemical properties affect amino acid substitution rates. It is also unclear how to define positive selection in a model incorporating chemical properties. Some authors distinguish between radical and conservative amino acid replacements (i.e. between amino acids with very different and very similar chemical properties, respectively), and suggest that a higher radical than conservative rate is evidence for positive selection (Hughes et al. 1990; Rand et al. 2000; Zhang 2000). However, this criterion appears less convincing than the ω ratio and is also much more sensitive to model assumptions, such as unequal transition and transversion rates and unequal amino acid compositions (Dagan et al. 2002).

Third, tests based on the ω ratio are most likely to be conservative. The branch models average the ω ratio over all sites while the site models average it over all lineages, so that both may be expected to have low power. The models appear effective in detecting recurrent diversifying selection driving fixation of nonsynonymous mutations, but may lack power in detecting one-off directional selection that drives a new advantageous mutation quickly to fixation. In this regard, it is remarkable that the tests have successfully detected adaptive evolution in a number of genes and organisms where positive selection was not suspected before. By allowing the selective pressure to vary both among sites and among lineages, the branch-site test appears to have more power. Intuitively it is sensible to focus on a short time period and a few amino acid sites so that the signal of adaptive evolution will not be overwhelmed by the almost ubiquitous purifying selection. However, a short time period and a few amino acid sites may not offer enough opportunities for evolutionary changes to occur to generate a signal detectable by statistical tests.

Fourth, the site models discussed here assume that the synonymous rate is constant along the gene. Kosakovsky Pond and Muse (2005) and Rubinstein et al. (2011) extended site model M3 (discrete) to allow both d_S and d_N to vary among sites. While LRTs conducted by the authors suggest significant synonymous rate variation among sites in most datasets, its impact on detection of positively selected sites is less clear and appears to be overstated. The authors implemented the NEB approach, which ignores sampling errors in parameter estimates and does not perform well in simulations, and the site model they used for comparison was M3 (discrete). Simulations suggest that BEB is more reliable than NEB, and site models M2a (selection) and M8 (beta&ω) are more robust than M3; in particular, the use of a site class with $\omega_1 = 1$ in M2a appears to be effective in reducing false positives in detecting positively selected sites (e.g. Anisimova et al. 2001; Wong et al. 2004).

Fifth, current codon models assume that the same phylogenetic tree applies to all sites in the sequence. If intragenic recombination is frequent, as in some viral datasets, the LRTs may be misled into falsely claiming positive selection (Anisimova et al. 2003; Shriner et al. 2003). Unfortunately, most recombination detection methods (e.g. Hudson 2001; McVean et al. 2002) assume strict neutrality and may mistake selection as evidence of recombination. Methods that deal with both selection and recombination (e.g. Wilson and McVean 2006) may thus be preferable.

For practical data analysis, and in particular in large-scale genome-wide analysis, a common problem in applying codon-based models of positive selection appears to be wrong levels of sequence divergence. Codon-based analysis contrasts the rates of synonymous and nonsynonymous substitutions and relies on information about both types of changes. If the compared species are too close or the sequences are too similar, the sequence data will contain too little variation and thus too little information. On the other hand, if the sequences are too divergent, silent changes may have reached saturation and the data will contain too much noise. In computer simulations, the methods are quite tolerant of high divergences. However in real data, high divergences are often associated with a host of other problems, such as difficulties in alignment and different codon usage patterns in different sequences. Both simulation studies and empirical analyses have found that the LRTs based on codon models such as the site and branch-site tests are sensitive to errors in sequence alignment (Schneider et al. 2009; Fletcher and Yang 2010; Mallick et al. 2010; Jordan and Goldman 2012). If the alignment program misplaces nonsynonymous codons in the same column (site), that site will appear to have experienced excessive nonsynonymous substitutions and is very likely to be identified as being under positive selection by the LRTs. The quality of the sequence data and sequence alignment is thus important to detection of positive selection by those methods.

11.6.2 *Comparison of methods for detecting positive selection*

It is interesting to compare phylogenetic tests of positive selection based on the ω ratio and population genetics tests of neutrality, to understand what kind of selection each class of methods are likely to detect and how powerful the tests are (Nielsen 2001; Fay and Wu 2003). First, the two kinds of tests are designed for different data. The phylogenetic tests rely on an excess of amino acid replacements to detect positive selection and thus require the sequences to be quite divergent. They have virtually no power when applied to population data (Anisimova et al. 2002). In contrast, neutrality tests are designed for DNA samples taken in a population or from closely related species. Most tests also assume the infinite-site model, which may break down when sequences from different species are compared. Second, a test based on the ω ratio may provide more convincing evidence for positive selection as positive selective pressure is often the only convincing explanation for an excess of replacement substitutions. In comparison, neutrality tests are typically open to multiple interpretations. The strict neutral model in a neutrality test is a composite model, with a number of assumptions such as neutral evolution, constant population size, no population subdivision, no linkage to selected loci, and so on. Rejection of such a composite model may be caused by violations of any of the assumptions, besides selection at the locus, and often it is difficult to distinguish between them. For example, patterns of sequence variation at a neutral locus undergoing genetic hitchhiking are similar to those in an expanding population (Simonsen et al. 1995; Fay and Wu 2000). Similarly, patterns of sequence variation at a locus linked to a locus under balancing selection may be similar to sequence variations sampled from subdivided populations (Hudson 1990). Third, the tests may have different statistical power. Zhai et al. (2009) used computer simulation to investigate the power of several neutrality tests based on patterns of genetic diversity within and between species. They found that in the presence of repeated selective sweeps on relatively neutral background, tests based on the ω ratio using species data almost always have more power to detect selection than tests based on population genetic data, even if the overall level of divergence is low, because of the transient nature of positive selection.

11.7 Adaptively evolving genes

Several reviews have summarized empirical findings of positive selection detection. Hughes (1999) discussed several case studies in great detail. Yokoyama (2002) reviewed studies on visual pigment evolution in vertebrates. Vallender and Lahn (2004) provide a comprehensive summary of genes under positive selection, although with an anthropocentric bias. Yang (2006, Table 8.7) provides a sample of genes inferred to be undergoing adaptive evolution based on comparisons of synonymous and nonsynonymous rates. See also the Adaptive Evolution Database, compiled by Roth et al. (2005). Here we provide a summary of the empirical findings.

Most genes detected to be under positive selection based on the ω ratio fall into the following three categories. The first includes host genes involved in defence or immunity against viral, bacterial, fungal, or parasitic attacks, as well as viral or pathogen genes involved in evading host defence. The former includes, for example, the MHC (Hughes and Nei 1988, see also §11.4.4), lymphocyte protein CD45 (Filip and Mundy 2004), plant R-genes involved in pathogen recognition (Lehmann 2002), and the retroviral inhibitor *TRIM5α* in primates (Sawyer et al. 2005). The latter includes viral surface or capsid proteins (Haydon et al. 2001; Shackelton et al. 2005), *Plasmodium* membrane antigens (Polley and Conway 2001), and polygalacturonases produced by plant enemies, such as bacteria, fungi, oomycetes, nematodes, and insects (Götesson et al. 2002). One may expect the pathogen gene to be under selective pressure to evolve into new forms unrecognizable by the host defence system, while the host has to adapt and recognize the pathogen. Thus an evolutionary arms race ensues, driving new replacement mutations to fixation in both the host and the pathogen (Dawkins and Krebs 1979). This Red Queen hypothesis is also supported by direct experimental evidence in phage-bacteria co-evolution systems, in which the rate of molecular evolution in the phage was far higher when the phage coevolved with bacteria than when the phage evolved against a constant host genotype (Paterson et al. 2010). Toxins in snake or scorpion venoms are used to subdue prey and often evolve at fast rates under similar selective pressures (Duda and Palumbi 2000).

The second main category includes proteins or pheromones involved in sexual reproduction. In most species, the male produces numerous sperm using apparently very little resource. The egg is in contrast a big investment. The conflict of interests between the two sexes espouses a genetic battle: it is best for the sperm to recognize and fertilize the egg as soon as possible, while it is best for the egg to delay sperm recognition to avoid fertilization by multiple sperm (polyspermy), which causes loss of the egg. A number of studies have detected rapid evolution of proteins involved in sperm–egg recognition (Palumbi 1994) or in other aspects of male or female reproduction (Tsaur and Wu 1997; Hellberg and Vacquier 2000; Wyckoff et al. 2000; Swanson et al. 2001a, 2001b). It is also possible that natural selection on some of these genes may have accelerated or contributed to the origination of new species. See Swanson and Vacquier (2002a, 2002b) and Clark et al. (2006) for reviews.

The third category overlaps with the previous two and includes proteins that acquired new functions after gene duplication. Gene duplication is one of the primary driving forces in the evolution of genes, genomes, and genetic systems, and is believed to play a leading role in evolution of novel gene functions (Ohno 1970; Kimura 1983). The fate of duplicated genes depends on whether they bring a selective advantage to the organism. Most of duplicated genes are deleted or degrade into pseudogenes, disabled by deleterious mutations. Sometimes the new copies acquire novel functions under the driving force of adaptive evolution, due to different functional requirements from those of the parental genes (Walsh 1995; Lynch and Conery 2000; Zhang et al. 2001; Prince and Pickett 2002). Many genes were detected to experience accelerated protein evolution following gene duplication, including the DAZ gene family in primates (Bielawski and Yang 2001), chorionic gonadotropin in primates (Maston and Ruvolo 2002), the pancreatic ribonuclease genes in leaf-eating monkeys (Zhang et al. 1998, 2002) and xanthine dehydrogenase genes (Rodriguez-Trelles et al. 2003).

Many other proteins have been detected to be under positive selection as well, although they are not as numerous as those involved in evolutionary arms race such as the host–pathogen antagonism or reproduction. This pattern appears partly to be due to the limitations of the detection methods based on the ω ratio, which may miss one-off adaptive evolution in which an advantageous mutation arose and spread in the population quickly, followed by purifying selection.

Needless to say, a statistical test cannot prove that the gene is undergoing adaptive evolution. A convincing case may be built through experimental verification and functional assays, establishing a direct link between the observed amino acid changes with changes in protein folding, and with phenotypic differences such as different efficiencies in catalysing chemical reactions. In this regard, the statistical methods discussed in this chapter are useful for generating biological hypotheses for verification in the laboratory, and can significantly narrow down the possibilities to be tested. Numerous studies demonstrate the power of an approach that combines comparative analysis with carefully designed experiments. For example, Ivarsson et al. (2002) inferred positively selected amino acid residues in glutathione transferase, multifunctional enzymes that provide cellular defence against toxic electrophiles. They then used site-directed mutagenesis to confirm that these mutations were capable of driving functional diversification in substrate specificities. The evolutionary comparison provided a novel approach to designing new proteins, reducing the need for extensive mutagenesis or structural knowledge. Bielawski et al. (2004) detected amino acid sites under positive selection in proteorhodopsin, a retinal-binding membrane protein in marine bacteria that functions as a light-driven proton pump. Site-directed mutagenesis and functional assay demonstrated that those sites were responsible for fine-tuning the light absorption sensitivity of the protein to different light intensities in the ocean. Sawyer et al. (2005) inferred amino acid residues under positive selection in *TRIM5α*, a protein in the cellular antiviral defence system in primates that can restrict retroviruses such as HIV-1 and SIV in a species-specific manner. In particular a 13 amino acid 'patch' had a concentration of positively selected sites, implicating it as an antiviral interface. By creating chimeric *TRIM5α* genes, Sawyer et al. demonstrated that this patch is responsible for most of the species-specific antiretroviral activity. Finally Georgelis et al. (2009) analysed adaptive evolution of ADP-glucose pyrophosphorylase (AGPase), an enzyme that catalyses a rate-limiting step in glycogen synthesis in bacteria and starch synthesis in plants. The large subunit of AGPase has duplicated multiple times giving rise to groups with different kinetic and allosteric properties. Use of the branch-site test detected positive selection along lineages leading to those duplicated groups, and site-directed mutagenesis confirmed the importance of a number of inferred amino acid residues to the kinetic and allosteric properties of the enzyme. Georgelis et al. concluded that evolutionary comparisons have greatly facilitated enzyme structure-function analyses.

Relevant to this discussion of which genes are undergoing adaptive evolution, an interesting question is whether molecular adaptation is driven mainly by changes in regulatory genes or in structural genes (Hoekstra and Coyne 2007; Carroll 2008). Developmental biologists tend to emphasize the importance of regulatory genes, sometimes nullifying the significance of structural genes in evolution. This view derives support from the observation that dramatic shifts in organismal structure can arise from mutations at key regulatory loci. For example, mutations in the homeotic (*hox*) gene cluster are responsible for diversification of body plans in major animal phyla (Akam 1995; Carroll 1995). Nevertheless, examples like those discussed here demonstrate the adaptive significance of amino acid changes in structural genes. Whether regulatory or structural genes are of greater importance to molecular adaptation, it is a common observation that a few changes in the key genes may bring about large changes in morphology and behaviour, while many molecular substitutions appear to have little impact (e.g. Stewart et al. 1987). As a result, rates of morphological evolution are in general not well correlated with rates of molecular evolution. This pattern is in particularly conspicuous in species groups that have undergone recent adaptive radiations, such as the cichlid fish species in the lakes of East Africa (Meyer et al. 1990) or the plant species in the Hawaiian silversword alliance (Barrier et al. 2001). As the divergence time among the species is very short, there is very little genetic differentiation at most loci, while the dramatic adaptation in morphology and behaviour must be due to a few key changes at the molecular level. Identifying these key changes should provide exciting opportunities for understanding speciation and adaptation (Kocher 2004).

11.8 Problems

11.1 Re-analyse the MHC data of §11.4.4 under the site models M1a, M2a, M7, and M8, but using the mutation–selection model of Yang and Nielsen (2008) to accommodate codon frequencies instead of F3×4 used in §11.4.4. First run model M0 (one-ratio) to obtain MLEs of branch lengths. Then run the site models with the branch lengths fixed at the M0 estimates. Compare results with those of Table 11.3 and Figure 11.4 (see also Yang and Swanson 2002).

11.2 Use the branch and branch-site models to analyse the lysozyme genes from 24 primates of Messier and Steward (1997) to detect positive selection on the branch ancestral to the colobine monkeys. Use the F3×4 model of codon usage. (See results for the branch model in Yang 1998a.)

11.3 Use nucleotide and codon models to reconstruct ancestral sequences using the lysozyme genes from 24 primates to identify the likely amino acid changes on the branch ancestral to the colobine monkeys. Compare the results with the amino acids identified on the same branch by the branch-site test of Problem 11.2. For the nucleotide-based analysis, use the HKY85 + C model of Table 4.3 in §4.3.3, which accounts for different substitution rates, different transition/transversion rate ratios and different base compositions at the three codon positions.

11.4 (Yang and dos Reis 2011) Conduct a computer simulation to examine the false positive rate of the branch-site test, and examine how long the sequence should be for the asymptotic null distribution to be reliable. Note that according to theory, the test statistic $2\Delta\ell$ should be 0 in half of the datasets and χ_1^2 distributed in the other half if many datasets are simulated under the null model (see §11.5.1). Use the EVOLVER program in the PAML package to simulate 1,000 replicate datasets, on the unrooted tree of Figure 11.7, with all branch lengths equal to 0.5 substitutions per codon, and with branch a as the foreground branch. Assume $\kappa = 2$ for the transition/transversion rate ratio and equal codon frequencies ($\pi_j = 1/61$ for codon j). Generate data under the null model of the branch-site test, with $p_0 = 0.5$, $p_1 = 0.3$, $\omega_0 = 0.5$, and $\omega_1 = \omega_2 = 1$ (see Table 11.4). Then analyse each dataset under the null and alternative models to calculate the test statistic $2\Delta\ell$. Calculate the proportion of datasets in which $2\Delta\ell = 0$ and the proportion of datasets in which $2\Delta\ell > 2.71$. (Note that the latter proportion is the false positive rate of the test at the 5% level when the null distribution is the 1:1 mixture of 0 and χ_1^2.) Use different sequence lengths, such as $N = 50$, 100, 200, 500, and 1000.

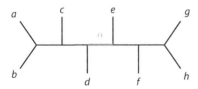

Fig. 11.7 A tree of eight species for simulating data under the branch-site model (Problem 11.4). The foreground branch a is highlighted.

CHAPTER 12

Simulating molecular evolution

12.1 Introduction

Computer simulation, also known as *stochastic simulation* or *Monte Carlo simulation*, is a virtual experiment in which we mimic a physical/biological process on a computer to study its properties. It is particularly useful for studying complex systems that are intractable analytically. Use of random numbers is a major feature of the method. Some authors use the term Monte Carlo simulation when the answer is deterministic, as in calculation of an integral by Monte Carlo integration, and stochastic simulation or computer simulation when the answer involves random variations, as in studies of biases and variances of parameter estimates. We do not make this distinction here.

Simulation is a useful approach for validating a theory or program implementation when the method of analysis is complex. Indeed, as it is easy to simulate even under a complex model, there is no need to make overly simplistic assumptions. Simulation is commonly used to compare different analytical methods, especially to study their robustness when the underlying assumptions are violated, in which case theoretical results are often unavailable. Simulation also plays a useful role in education. By simulating under a model and observing its behaviour, one gains intuitions about the system. Lastly, simulation forms the basis of many modern computation-intensive statistical methods; such as *bootstrapping, importance sampling*, and *Markov chain Monte Carlo* (see Chapter 7).

As a consequence of the ease of simulation and the availability of cheap and fast computers, simulation is widely used. One should nevertheless bear in mind that simulation is experimentation, and the same care should be exercised in the design and analysis of a simulation experiment as an ordinary experiment. A major limitation of simulation is that one can examine only a small portion of the parameter space, and the behaviour of the system may be different in other unexplored regions of this space. One should thus resist the temptation of over-generalization. Analytical results are in general superior as they apply to all parameter values.

This chapter provides an introduction to simulation techniques. The reader is invited to implement the algorithms discussed here. Any programming language, such as C/C++, JAVA, PERL, or R, will serve the purpose well. The reader may refer to a textbook on simulation for more extensive discussions, such as Ripley (1987) or Ross (1997).

12.2 Random number generator

Random variables from the uniform distribution $U(0, 1)$ are called *random numbers*. They are fundamentally important in computer simulation. First, they are used to simulate any random event that occurs with a given probability. For example, to simulate an event that

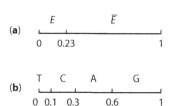

Fig. 12.1 Sampling from a discrete distribution. (**a**) To simulate an event that occurs with probability 0.23, draw a random number $u \sim U(0, 1)$. If it falls into the interval (0, 0.23), the event occurs (E). Otherwise the event does not occur ($\bar{E}$). (**b**) Similarly, to sample a nucleotide with probabilities 0.1, 0.2, 0.3, and 0.4, for T, C, A, and G, respectively, break the interval (0, 1) into four segments of lengths 0.1, 0.2, 0.3, and 0.4. Generate a random point u in the interval (0, 1) and choose the nucleotide depending on which segment u falls into.

occurs with probability 0.23, say, we draw a random number $u \sim U(0, 1)$. If $u < 0.23$, we decide that the event occurs; otherwise it does not (Figure 12.1). It is obvious that the probability that $u < 0.23$ is 0.23. Second, the $U(0, 1)$ random numbers are the basis for generating random variables from other distributions.

Hardware random number generators. Random numbers may be generated from 'unpredictable' physical processes, such as coin flipping, dice, or lottery machines. The outcome of a coin flip is unpredictable because it is sensitive to the minute details of the initial condition of the flip, even though the physical process is deterministic under Newtonian mechanics. More sophisticated hardware random number generators make use of a certain low-level microscopic process of random noise, such as photonic emission in semiconductors. Some aspect of this process is converted into an electrical signal, then into a digital signal of 0 or 1. By recording repeatedly the randomly varying signal, a series of random numbers can be generated. However, a process like this is too slow to be useful in computer simulation, so instead hardware random number generators are used to create cryptographic keys to encrypt data and to generate a seed to initiate a pseudo-random number generator.

Pseudo-random number generator. A pseudo-random number generator is a mathematical algorithm that produces a sequence of numbers between 0 and 1 that appear random. The algorithm is entirely deterministic and will produce the same fixed sequence of numbers every time it is run. The generated numbers are thus not random, and are called *pseudo-random numbers* but often simply random numbers for short.

One such algorithm is the *multiplication-congruential method*, given by the following formulae:

$$A_i = cA_{i-1} \bmod M, \tag{12.1}$$

$$u_i = A_i/M. \tag{12.2}$$

Here, A_i, c, and M are all positive integers: c is called the multiplier and M the modulus, with ($a \bmod M$) to be the remainder when a is divided by M. Note that A_i and cA_{i-1} have the same remainder when divided by M and are said to be congruent. A_0 is the initial value, known as the *seed*. $A_1, A_2, \ldots$, will be a sequence of integers between 0 and $M-1$, and $u_1, u_2, \ldots$, will be a sequence of (pseudo-)random numbers between 0 and 1.

As an example, imagine a computer using the familiar decimal system, with the register (storage) having only four digits. This computer can thus represent any integer in the range (0000, 9999). Let $M = 10^4, c = 13$, and $A_0 = 123$. Equation (12.1) then generates the A_i sequence $1599, 6677, 1271, 6333, 8959, \ldots$, corresponding to the u_i sequence $0.1599, 0.6677, 0.1271, 0.6333, 0.8959, \ldots$. Note that if the product cA_{i-1} is greater than 9999, we simply ignore the high digits to get the remainder A_i; most computer systems allow such 'overflow' without any warning. The sequence u_i does not show an obvious trend and appears random. However, as this computer can represent only 10,000 integers and many of them will never occur in the sequence, very soon a certain number A_i will reoccur and then the sequence will repeat itself. A real computer, however, has more digits, and by intelligent choices of M and c (and, to a lesser extent, A_0 as well), the period of the sequence can be made very long. Also real computers use the binary system. Thus the natural choice of M is 2^d, often with d to be the number of bits in an integer (say, 31, 32, or 64). Such a choice of M eliminates the need for the division in equation (12.1) for the modulus calculation and also makes the division in equation (12.2) a trivial calculation.

How do we decide whether the random numbers generated are random enough? The requirement is that they should be indistinguishable from random draws from the uniform distribution $U(0, 1)$. This can be examined using statistical tests, examining different departures from the theoretical expectation. For example, the random numbers should have the correct mean $\left(\frac{1}{2}\right)$ and variance $\left(\frac{1}{12}\right)$, and should not be autocorrelated. A large simulation study may require billions of random numbers, so that a reliable and computationally efficient generator may be very important. Some generators may produce 0 or 1, which can cause problems if one's code assumes that u is inside the interval $(0, 1)$ but not at the boundary. It is not advisable to design our own generators. Instead one should use algorithms that have been well tested (Ripley 1987; Knuth 1997). Ripley (1987) discussed various tests and commented that random number generators supplied by computer manufacturers or compiler writers were not trustworthy. For the exercises of this chapter, it is acceptable to use a generator provided by the programming language, such as rand() in C or PERL.

Random number seed and running simulations on computer clusters. A random number generator needs a seed (such as A_0 in equation (12.1)) to get started. As the algorithm is deterministic, running the program multiple times using the same seed produces identical results. This may be useful for debugging a simulation program, but is disastrous for a serious simulation study. In C and PERL, the function srand() can be used to specify a seed for the corresponding generator rand(). One approach is to use the wall-clock time in the computer to construct a seed. For example, the C function time() typically returns the number of seconds passed since 00:00:00, 1 January 1970. However, this approach may be unsafe if one uses modern multi-core multi-processor computers to generate and analyse many replicate datasets in parallel. Suppose we want to simulate and analyse 1,000 replicate datasets for a given parameter setting. One can prepare shell or PERL scripts to simulate and analyse those datasets in different folders. However, if one starts all 1,000 jobs on different computers/cores at the same time, hundreds of them may be started within the same second and will use exactly the same random number. If data generation is much faster than data analysis, the problem may be avoided by using one core to simulate all the datasets and then farming out the analysis onto different cores. Alternatively, modern UNIX systems provide a facility (a text file) called /dev/urandom, which contains random data generated by collecting environmental noise from device drivers and other sources so that the numbers are nearly truly random. One can open this file to read a few bytes of random data to 'seed' the random number generator.

12.3 Generation of discrete random variables

12.3.1 *Inversion method for sampling from a general discrete distribution*

A discrete random variable X can take only a set of possible values: $x_1, x_2, \ldots$, which may correspond to different outcomes of an experiment. The probabilities for the possible values

$$\Pr\{X = x_i\} = p_i, i = 1, 2, \ldots, \quad (12.3)$$

with $\sum_i p_i = 1$, are called the probability mass function. For convenience, suppose the values are ordered: $x_1 \leq x_2 \leq \ldots$. The cumulative distribution function (CDF) is then defined as

$$F_i = \Pr\{X \leq x_i\} = p_1 + p_2 + \cdots + p_i. \quad (12.4)$$

We illustrate how to generate a value of X from this distribution using an example. Suppose we want to sample a nucleotide at random with probabilities $0.1, 0.2, 0.3, 0.4$ for T, C, A, G, respectively.

Category (x_i)	1 (T)	2 (C)	3 (A)	4 (G)
Probability (p_i)	0.1	0.2	0.3	0.4
CDF (F_i)	0.1	0.3	0.6	1.0

We break the line segment $(0, 1)$ into four pieces, of lengths $0.1, 0.2, 0.3$, and 0.4, corresponding to the four nucleotides. We then pick a random point on the line segment and choose the nucleotide depending on where the point lands (Figure 12.1b). Specifically, draw a random number $u \sim U(0, 1)$. Compare u with 0.1, and choose T if $u < 0.1$. Otherwise compare u with 0.3 and choose C if $u < 0.3$. Otherwise compare u with 0.6 and choose A if $u < 0.6$. Otherwise choose G. Note that $0.1, 0.3, 0.6$, and 1.0 are the cumulative probabilities.

We can sample from any discrete distribution in this way. We generate the random number u, and then obtain the value of x such that $F(x) = u$ or $x = F^{-1}(u)$; in other words, given $F_{i-1} < u \leq F_i$, we set $X = x_i$. The algorithm thus inverts the CDF function and is known as the *inversion method*.

One may rearrange the high-probability categories before the low-probability categories to reduce the number of comparisons. In the above example, it takes 1, 2, 3, or 3 comparisons to sample T, C, A, or G, respectively, with $0.1 \times 1 + 0.2 \times 2 + 0.3 \times 3 + 0.4 \times 3 = 2.6$ comparisons to sample a nucleotide on average. If we order the nucleotides as G, A, C, T, it takes on average $0.4 \times 1 + 0.3 \times 2 + 0.2 \times 3 + 0.1 \times 3 = 1.9$ comparisons. If there are many categories, one may use a cruder classification to find the rough location of the sampled category before doing the finer comparisons.

12.3.2 The alias method for sampling from a discrete distribution

If there are many categories and if many samples are needed, the inversion algorithm may involve too many comparisons and may be very inefficient. A clever algorithm for sampling from a general discrete distribution with a fixed number of categories is the *alias method* (Walker 1974; Kronmal and Peterson 1979). This requires only one comparison to generate a discrete random variable, no matter how many categories there are in the distribution. The algorithm is based on the fact that any discrete distribution with n categories can be expressed as an equiprobable mixture of n two-point distributions. In other words, we can always find n distributions $q^{(m)} = \{q_i^{(m)}\}, m = 1, 2, \ldots, n$, such that $q_i^{(m)}$ (for $i = 1, 2, \ldots, n$) is nonzero for at most two values of i, and that

$$p_i = \frac{1}{n} \sum_{m=1}^{n} q_i^{(m)}, \text{ for all } i. \quad (12.5)$$

Table 12.1 Component distributions $q_i^{(m)}$ for the alias method of sampling from the discrete distribution (0.1, 0.2, 0.3, 0.4)

i	1 (T)	2 (C)	3 (A)	4 (G)	Sum
p	0.1	0.2	0.3	0.4	1
$q^{(1)}$	0.4		0.6		1
$q^{(2)}$		0.8		0.2	1
$q^{(3)}$			0.6	0.4	1
$q^{(4)}$				1.0	1

Algorithms for setting up the distributions $q_i^{(m)}$ are described by Walker (1974) and Kronmal and Peterson (1979); for a more accessible account, see Yang (2006, §9.4.6).

For the example discussed above, one way of dividing up the distribution $(p_1, p_2, p_3, p_4) = (0.1, 0.2, 0.3, 0.4)$ into four equiprobable mixture of component distributions $q^{(1)}, q^{(2)}, q^{(3)}$, and $q^{(4)}$ is shown in Table 12.1. Each component distribution $q^{(i)}$ has nonzero probabilities on two cells, and cells with zero probabilities are left blank. For example, $q^{(1)}$ assigns probability 0.4 to 1 (T), and 0.6 to 3 (A). After this table is set up, it is an easy matter to simulate according to equation (12.5). Generate a random number u and set $m = \lceil nu \rceil$. Then sample from the distribution $q^{(m)}$, which requires one comparison.

The alias method is efficient for generating many random variables from the same discrete distribution after the distributions $q^{(m)}$ have been set up. For example, it is useful for sampling from the multinomial distribution. It requires only one comparison to generate one variable, irrespective of n. This contrasts with the inversion method, which requires more comparisons for larger n.

12.3.3 Discrete uniform distribution

A random variable that takes n possible values, each with equal probability, is called a *discrete uniform distribution*. Let the possible values be 1, 2, ..., n. While the inversion method for simulating from a general discrete distribution can be used to sample from the discrete uniform, one can do much better. Generate $u \sim U(0, 1)$ and set $x = 1 + \lfloor nu \rfloor$, where $\lfloor a \rfloor$ means the integer part of a. Note that $\lfloor nu \rfloor$ takes values $0, 1, \ldots, n-1$, each with probability $1/n$.

As an example, the JC69 and K80 models predict equal proportions of the four nucleotides. We can use the discrete uniform distribution to generate a sequence for the root of the tree under any of those models. Let 1, 2, 3, and 4 represent T, C, A, and G, respectively. To sample the nucleotide at a site, generate $u \sim U(0, 1)$ and set $x = 1 + \lfloor 4u \rfloor$. Repeat l times to generate a sequence of l sites. Another use of discrete uniform distribution is nonparametric bootstrap, in which one samples sites in the sequence alignment at random with replacement (see §5.4.1). Generate $u \sim U(0, 1)$, set $x = 1 + \lfloor lu \rfloor$, and then the xth site is a random draw from among the l sites. Repeat l times to generate a bootstrap sample of the same size as the original dataset.

To sample a random pair (i and j) from the list $1, 2, \ldots, n$, one can sample i first, i.e. $i = 1 + \lfloor nu_1 \rfloor$, and then sample j from the remaining members of the list. Generate $j = 1 + \lfloor (n-1)u_2 \rfloor$, and if $j \leq i$, set $j = j + 1$. Here, u_1 and u_2 are two independent random numbers, and $\lfloor a \rfloor$ is the integer part of a. Another method is to sample from the $n(n-1)$ combinations, requiring only one random number. Let $y = \lfloor n(n-1)u \rfloor$, where u

is a random number. Then set $i = 1 + \lfloor y/(n-1) \rfloor$, and $j = 1 + y \bmod (n-1)$, and if $j \geq i$, set $j = j + 1$. Here ($a \bmod b$) is the remainder when a is divided by b.

To generate a random permutation of $1, 2, \ldots, n$, so that all $n!$ possible permutations are equally likely, one can proceed as follows. Start with any initial permutation, for example, with the number i in location i). Choose one out of the n locations at random and swap the number there with the number at location n. Then choose one of the first $n-1$ locations and swap the number there with the number at position $n-1$. Continue until only one location (the first) is left.

12.3.4 Binomial distribution

Suppose the probability of 'success' in a trial is p. Then the number of successes in n independent trials has a binomial distribution: $x \sim \text{bino}(n, p)$. To sample from the binomial distribution, one may simulate n trials and count the number of successes. For each trial, generate $u \sim U(0, 1)$; if $u < p$, we count a 'success' and otherwise we count a failure.

An alternative approach is to calculate the probability of x successes:

$$p_x = \binom{n}{x} p^x (1-p)^{n-x}, \tag{12.6}$$

for $x = 0, 1, \ldots, n$. One can then sample from the discrete distribution with $n + 1$ categories: $(p_0, p_1, \ldots, p_n)$. This involves the overhead of calculating p_x's but may be more efficient if many samples are to be taken from the same binomial distribution; see above about sampling from a general discrete distribution.

12.3.5 The multinomial distribution

In a binomial distribution, every trial has two outcomes: 'success' or 'failure'. If every trial has k possible outcomes, the distribution is called the *multinomial distribution*, represented as $\text{MN}(n, p_1, p_2, \ldots, p_k)$, where n is the number of trials. The multinomial variables are the counts of the k different outcomes $n_1, n_2, \ldots, n_k$, with $n_1 + n_2 + \cdots + n_k = n$. One can generate multinomial variables by sampling n times from the discrete distribution $(p_1, p_2, \ldots, p_k)$, and counting the number of times that each of the k outcomes is observed. If both n and k are large, it may be necessary to use an efficient algorithm to sample from the discrete distribution, such as the alias method mentioned above.

Under most of the substitution models discussed in this book, the sequence data follow a multinomial distribution. The sequence length is the number of trials n, and the possible site patterns correspond to the categories. For nucleotide data, there are $k = 4^s$ categories for s species or sequences. One can thus generate a sequence dataset by sampling from the multinomial distribution. See §12.6.1 for details.

12.3.6 The Poisson distribution

The random variable X is Poisson distributed with mean λ if

$$p_x = \Pr(X = x) = \frac{e^{-\lambda} \lambda^x}{x!}, x = 0, 1, \ldots. \tag{12.7}$$

Both the mean and variance are λ: $E(X) = V(X) = \lambda$. One can sample from the Poisson distribution using the algorithm for the general discrete distribution as follows.

Algorithm 12.1. Inversion method for generating Poisson variates with rate λ.

Step 1. Generate a random number u.
Step 2. Set $x = 0, F = p = e^{-\lambda}$.
Step 3. If $u < F$, set $X \leftarrow x$, and stop.
Step 4. Set $p \leftarrow p\lambda/(x+1)$, $F \leftarrow F + p$, and $x \leftarrow x + 1$. (F is now F_{x+1}.)
Step 5. Go to step 3. □

Here F records the CDF

$$F_x = \Pr\{X \leq x\} = p_0 + p_1 + \cdots + p_x, \tag{12.8}$$

and step 4 makes use of the fact that $p_{x+1} = p_x \lambda/(x+1)$ (see equation (12.7)) and $F_{x+1} = F_x + p_{x+1}$. The algorithm first calculates $F = e^{-\lambda} = p_0$, and checks whether the random number $u < F$. If this is the case, it sets $X = 0$. Otherwise it calculates $F = p_0 + p_1$, and checks whether $u < F$. If so, it sets $X = 1$. The values $0, 1, 2, \ldots$, are checked successively, with the CDF calculated on the way. The number of comparisons is thus $1 + x$, and the expected number of comparisons for generating one Poisson variable is $1 + \lambda$. When λ is large, it may be more efficient to reorder the Poisson values to reduce the number of comparisons. For example, values close to λ may have larger probabilities than values far away from λ since X has expectation λ. Furthermore, if multiple variables are needed from the same Poisson distribution, it may be more efficient to calculate the CDF first before generating the Poisson variables.

12.3.7 The composition method for mixture distributions

Suppose a random variable has a mixture distribution

$$f = \sum_{i=1}^{m} p_i f_i, \tag{12.9}$$

where f_i represents a discrete or continuous distribution, and $p_1, p_2, \ldots, p_m$ are the mixing proportions, which sum to one. Then f is said to be a *mixture* or *compound* distribution. To generate a random variable with distribution f, first sample the component (let it be i) from the discrete distribution $(p_1, p_2, \ldots, p_m)$, and then sample from the component distribution f_i. This method is known as the *composition method*.

For example, the 'I + Γ' model of variable rates for sites (see §4.3.1) assumes that a proportion p_0 of sites in the sequence are invariable with rate zero, while all other sites have rates drawn from the gamma distribution. To generate the rate for a site, first sample from the distribution $(p_0, 1 - p_0)$ to decide whether the rate is zero or is from the gamma distribution. Generate a random number $u \sim U(0, 1)$. If $u < p_0$, let the rate be 0; otherwise sample the rate from the gamma distribution. Mayrose et al. (2005) discussed a mixture of several gamma distributions with different parameters; rates can be sampled in a similar way from that model. Also the codon substitution model M8 (beta&ω) discussed in §11.4.2 is a mixture distribution, which assumes that a proportion p_0 of sites have the ω ratio drawn from the beta distribution while all other sites (with proportion $1 - p_0$) have the constant $\omega_s > 1$.

12.4 Generation of continuous random variables

Here we discuss two general methods for sampling from a continuous distribution: the transformation method, which includes the inversion method as a special case, and the rejection method.

12.4.1 The inversion method

Suppose random variable x has a CDF $F(x)$. Consider $u = F(x)$ as a function of x so that u is itself a random variable. It is known that u has a uniform distribution: $u = F(x) \sim U(0, 1)$. To see this, note that $F(x)$ is a monotonic function of x, so that $x_1 < x_2$ if and only if $F(x_1) < F(x_2)$. Thus $\Pr\{u_1 < u_2\} = \Pr\{F(x_1) < F(x_2)\} = \Pr\{x_1 < x_2\} = F(x_2) - F(x_1) = u_2 - u_1$, so that u has the CDF for a uniform random number. (An intuitive appreciation of this result, in terms of using a variable transform to stretch and squeeze the probability density $f(x)$, is provided in Appendix A immediately after Theorem 1.) In any case, if the inverse transform $x = F^{-1}(u)$ is available analytically, one can use it to sample x. Generate $u \sim U(0, 1)$, and set $x = F^{-1}(u)$. Here $F^{-1}(u)$ is the value of x such that $F(x) = u$. This is called the *inversion* method, and it applies only when the CDF is explicitly available and can be inverted easily. The method thus covers only a small number of cases.

1. *Uniform distribution.* To generate a random variable from the uniform distribution $x \sim U(a, b)$, note that the CDF $F(x) = (x - a)/(b - a)$, so that $u = (x - a)/(b - a)$ is a $U(0, 1)$ random number. Thus we generate $u \sim U(0, 1)$ and obtain x by inverse transform: $x = a + u(b - a)$.
2. *Exponential distribution.* The exponential distribution with mean θ has density function $f(x) = \frac{1}{\theta} e^{-x/\theta}$ and CDF $F(x) = \int_0^x \frac{1}{\theta} e^{-y/\theta} dy = 1 - e^{-x/\theta}$. Let $u = F(x)$. Then we have $x = F^{-1}(u) = -\theta \log(1 - u)$. Thus we generate $u \sim U(0, 1)$ and then $x = -\theta \log(1 - u)$ will have the desired exponential distribution. Since both u and $1 - u$ are $U(0, 1)$ variables, it suffices to use $x = -\theta \log(u)$. In particular $x = -\log(u)$ is an exponential random variable with mean 1 and density $f(x) = e^{-x}$.

12.4.2 The transformation method

A function of a random variable is itself a random variable. Thus if we want to simulate a random variable x, and if it is a function of another random variable y that is easier to simulate, with $x = h(y)$, then we can simulate y and obtain x as $x = h(y)$. The inversion method is a special case of the general transformation method, with $h(y) = F^{-1}(y)$ and $y \sim U(0, 1)$, where F^{-1} is the inverse CDF function for x.

If we can sample from the standard normal distribution $N(0, 1)$, we can easily sample from the normal distribution $N(\mu, \sigma^2)$ using a variable transform: if $z \sim N(0, 1)$, then $x = \mu + z\sigma \sim N(\mu, \sigma^2)$. If x has a gamma distribution $G(\alpha, \beta)$, then $y = 1/x$ has the inverse gamma distribution invG(α, β).

Similarly the exponential distribution can be used to generate some other random variables. Suppose y has the exponential distribution with mean 1, $y \sim \exp(1)$, then

$$x = \frac{1}{\beta} \sum_{j=1}^{n} y_i \sim G(n, \beta), \text{ for } n \text{ to be an integer.} \tag{12.10}$$

Thus to generate $x \sim G(n, \beta)$, we generate n i.i.d. exponential variables y_i, and use equation (12.10).

12.4.3 The rejection method

Suppose the random variable x has density $f(x)$. If we sample a random point (x, y) under the density curve (e.g. a random point in the shaded region in Figure 12.2a), the x coordinate will be a random draw from $f(x)$. To see this, note that the total area under the

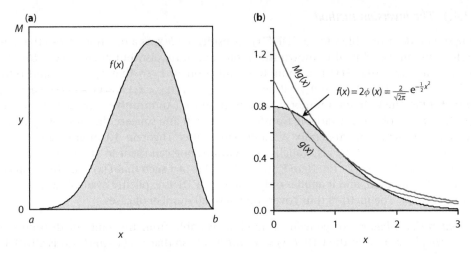

Fig. 12.2 Rejection method for generating a continuous random variable. In (a), the variable x, with $0 < x < 1$, has a beta distribution beta(5, 3) with mean 5/8. One generates a random point (x, y) in the rectangle, and accepts the point (the x value) if it is in the shaded region, below the density curve $f(x)$. In (b), the exponential distribution $g(x) = e^{-x}$ is used to sample the variable x from the folded normal distribution with density $f(x)$. A random point (x, y) under the curve $Mg(x)$ is generated, and it is accepted if it falls into the shaded area under $f(x)$. The accepted x value is then a random draw from $f(x)$.

curve is 1, and that the probability that x falls into any small bin of fixed width Δx is proportional to the height of the curve: $\Pr\{x < X < x + \Delta x\} \approx f(x)\Delta x$. In other words, x has density $f(x)$. Now if we can construct a rectangle to fully enclose the region under the density curve (the shaded region, Figure 12.2a), we can sample points in the rectangle at random, and remove (reject) those that are outside the shaded region. The remaining points will be distributed uniformly within the shaded region, and their x coordinates will be random variables from $f(x)$. The algorithm is as follows.

Algorithm 12.2. Rejection sampling from a uniform distribution to generate x from $f(x)$.

  Step 1. Generate a random point (x, y) in the rectangle (Figure 12.2a). In other words, generate $x \sim U(a, b)$ and $y \sim U(0, M)$.
  Step 2. If $y < f(x)$, accept x. Otherwise go back to step 1. □

What we have described is a particular form of *rejection sampling*, in which the sampling distribution is uniform (hence its density is a rectangle). There is no need for the sampling region to be rectangular; it can be arbitrary as long as it fully encloses the target distribution. In general, we can use rejection sampling if we can identify a sampling distribution $g(x)$ and a constant M such that

$$\frac{f(x)}{g(x)} \leq M \text{ for all } x. \tag{12.11}$$

Then the region below the curve $M \cdot g(x)$ fully encloses the shaded region below $f(x)$ (Figure 12.2b). If we generate a random point in the region below the curve $M \cdot g(x)$, and reject those outside the shaded area, the remaining points will be uniformly distributed within the shaded area below $f(x)$. The algorithm is as follows.

12.4 GENERATION OF CONTINUOUS RANDOM VARIABLES

Algorithm 12.3. Rejection sampling of x from f(x)..

Step 1. Generate a random point (x, y) under the curve $M \cdot g(x)$ (Figure 12.2b). In other words, generate x from $g(x)$, $u \sim U(0, 1)$, and set $y = uM \cdot g(x)$.

Step 2. If $y < f(x)$, accept x. Otherwise go back to step 1. □

Figure 12.2b shows an example of this algorithm. We sample from the region under the curve $Mg(x)$ and then reject all those points that are outside the shaded region under the target distribution $f(x)$. The function $Mg(x)$ is called the *envelope function* and it is important that it completely encloses the target density: $Mg(x) \geq f(x)$ for all x. To sample a variable that has no finite bounds (i.e. with $x < \infty$), $g(x)$ should be heavier-tailed than $f(x)$, but not vice versa. In other words, $g(x)$ should go to 0 more slowly than $f(x)$ when $x \to \infty$. For example, we can use the exponential distribution $g(x) = e^{-x}$ as the envelope to generate normal variables. The normal density goes to 0 at the rate of $e^{-\frac{1}{2}x^2}$, while the exponential density at the rate of e^{-x}, with $f(x)/g(x) \to 0$ when $x \to \infty$.

Since both $f(x)$ and $g(x)$ integrate to 1, equation (12.11) means that we must have $M \geq 1$, with $M = 1$ only if the two distributions are the same. We would like the sampled values to be accepted in step 2, since rejected values are a waste of effort. The acceptance probability in the algorithm is exactly $1/M$. To see this, note that for a given x, the acceptance probability is $\Pr\{y < f(x)\} = \Pr\{u < \frac{f(x)}{Mg(x)}\} = \frac{f(x)}{Mg(x)}$. By averaging over different x values generated from $g(x)$, we have

$$P_{accept} = \int \frac{f(x)}{Mg(x)} \cdot g(x) \, dx = \frac{1}{M} \int f(x) \, dx = \frac{1}{M}. \quad (12.12)$$

Thus it takes on average M iterations to generate one x variable. M should be chosen as close to 1 as possible: it is desirable for the envelope function to enclose the target density tightly.

Here we use rejection sampling to generate a variable from the standard normal distribution $N(0, 1)$. We do this by generating its absolute value and then assign a sign at random. Note that the absolute value of a standard normal variable has the density

$$f(x) = \frac{2}{\sqrt{2\pi}} e^{-\frac{1}{2}x^2}, \quad 0 \leq x < \infty. \quad (12.13)$$

We use as the sampling distribution the exponential distribution with mean 1:

$$g(x) = e^{-x}, \quad 0 \leq x < \infty. \quad (12.14)$$

Then

$$\frac{f(x)}{g(x)} = \sqrt{\frac{2}{\pi}} e^{x - \frac{1}{2}x^2} = \sqrt{\frac{2}{\pi}} e^{-\frac{1}{2}(x^2 - 2x + 1)} e^{\frac{1}{2}} = \sqrt{\frac{2e}{\pi}} e^{-\frac{1}{2}(x-1)^2} \leq \sqrt{\frac{2e}{\pi}}.$$

Thus if we let $M = \sqrt{2e/\pi}$, we have $\frac{f(x)}{g(x)} \leq M$ for all x. Then

$$\frac{f(x)}{M \cdot g(x)} = e^{-\frac{1}{2}(x-1)^2}. \quad (12.15)$$

See Figure 12.2b for plots of $f(x)$, $g(x)$, and $Mg(x)$. Thus we have the following algorithm.

Algorithm 12.4. Generate a standard normal variate: $x \sim N(0, 1)$.

Step 1. Generate a random point (x, y) under the curve $M \cdot g(x)$ (Figure 12.2b). Generate $u_1 \sim U(0, 1)$ and set $x = -\log u_1$. (Note that x is an exponential variable with mean 1.) Generate $u_2 \sim U(0, 1)$ and set $y = u_2 M \cdot g(x)$.

Step 2. If $y < f(x)$, accept x. Otherwise go back to step 1.
Step 3. Generate $u_3 \sim U(0, 1)$. If $u_3 < 0.5$, set $x = -x$. □

At the end of step 2, x is the absolute value of an $N(0, 1)$ variable. Step 3 changes its sign to negative with probability $\frac{1}{2}$. As $M = \sqrt{2e/\pi} = 1.3155$, it takes on average 1.3 iterations to generate one x.

We may wonder whether an exponential distribution with the mean θ different from 1,

$$g(x) = \frac{1}{\theta}e^{-x/\theta}, \tag{12.16}$$

instead of equation (12.14), may lead to a smaller M and thus be more efficient. When θ is given,

$$\frac{f(x)}{g(x)} = \theta\sqrt{\frac{2}{\pi}} \cdot e^{\frac{x}{\theta}-\frac{x^2}{2}} = \sqrt{\frac{2}{\pi}} \cdot \theta e^{-\frac{1}{2}\left(x-\frac{1}{\theta}\right)^2} \cdot e^{\frac{1}{2\theta^2}} \leq \sqrt{\frac{2}{\pi}} \cdot \theta e^{\frac{1}{2\theta^2}} = M(\theta), \tag{12.17}$$

with the upper bound $M(\theta)$ achieved at $x = 1/\theta$. We want $M(\theta)$ to be as small as possible. Setting $dM/d\theta = 0$ and solving for θ, we see that $M(\theta)$ achieves the minimum at $\theta = 1$. Thus our use of the exponential sampling distribution with mean 1 is indeed optimal.

12.4.4 Generation of a standard normal variate using the polar method

The CDF of the standard normal distribution is not analytically tractable, so the inversion method is not directly applicable. However, Box and Muller (1958) described a method for generating a pair of independent standard normal variables x and y using variable transform, based on the following observation. Let $x, y \sim N(0, 1)$ and consider (x, y) to be a point on the x–y plane (Figure 12.3a). The polar coordinates of the point, (r, θ), are given as

$$\begin{aligned} r^2 &= x^2 + y^2, \\ \theta &= \tan^{-1}\left(\frac{y}{x}\right). \end{aligned} \tag{12.18}$$

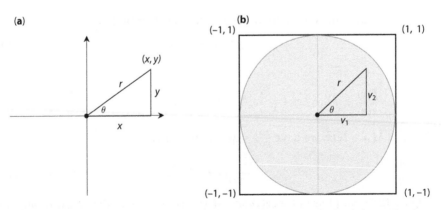

Fig. 12.3 Polar method for generating two normal variates x and y, both from $N(0, 1)$. (a) If $x, y \sim N(0, 1)$, the point (x, y) has a random angle $\theta \sim U(0, 2\pi)$, and $r^2 = x^2 + y^2$ is an exponential variable with mean 2. (b) Sample a random point inside the unit circle. Then $\theta \sim U(0, 2\pi)$, and $s = r^2 = v_1^2 + v_2^2 \sim U(0, 1)$.

12.4 GENERATION OF CONTINUOUS RANDOM VARIABLES

From the symmetry of the joint density of x and y, it is obvious that the angle θ is random: $\theta \sim U(0, 2\pi)$, and independent of the radius r. Also $r^2 = x^2 + y^2$ is a sum of two standard normal variables and has a χ^2 with df = 2, which is also the exponential distribution with mean 2. Formally it is easy to confirm that the joint PDF of $w = r^2$ and θ is

$$f(w, \theta) = \frac{1}{2} e^{-w/2} \cdot \frac{1}{2\pi}, \quad 0 \leq w < \infty, 0 \leq \theta < 2\pi \qquad (12.19)$$

(see Example A5 in Appendix A). As the exponential variable w (or r^2) and the uniform variable θ are easy to generate, we can use transformation to obtain x and y.

Algorithm 12.5. Box–Muller transform to generate normal variates $x, y \sim N(0, 1)$.

Step 1. Generate random numbers $u_1, u_2 \sim U(0, 1)$.
Step 2. Let $r^2 = -2\log(u_1)$, and $\theta = 2\pi u_2$. (Note that r^2 is exponential with mean 2 and $\theta \sim U(0, 2\pi)$.)
Step 3. Set

$$x = r\cos(\theta) = \sqrt{-2\log(u_1)} \times \sin(2\pi u_2), \qquad (12.20)$$

$$y = r\sin(\theta) = \sqrt{-2\log(u_1)} \times \cos(2\pi u_2). \qquad \square$$

This algorithm requires calculation of expensive sine and cosine functions of a random angle θ. To avoid this, we can generate a random point inside a circle (such as a unit circle of radius 1, Figure 12.3b) and then calculate these functions using the sides of a right triangle. Suppose the point is (v_1, v_2) with polar coordinates (r, θ). Then

$$\sin \theta = \frac{v_2}{r} = \frac{v_2}{\sqrt{v_1^2 + v_2^2}},$$

$$\cos \theta = \frac{v_1}{r} = \frac{v_1}{\sqrt{v_1^2 + v_2^2}}. \qquad (12.21)$$

Thus we can replace $\sin\theta$ and $\cos\theta$ in equation (12.20) by the functions in equation (12.21). However, the random number u_1 of equation (12.20) can be saved as well. If (v_1, v_2) is a random point inside the unit circle, $s = v_1^2 + v_2^2 \sim U(0, 1)$, independent of θ (see Example A6 in Appendix A). We can use s to replace u_1 in equation (12.20), saving one random number. The modified algorithm is as follows.

Algorithm 12.6. Polar method to generate normal variates $x, y \sim N(0, 1)$.

Step 1. Generate a random point (v_1, v_2) within the unit circle.
  a. Generate $u_1, u_2 \sim U(0, 1)$.
  b. Set $v_1 \leftarrow 2u_1 - 1, v_2 \leftarrow 2u_2 - 1$. (Note that (v_1, v_2) is a random point in the square of Figure 12.3b.)
  c. If $s = v_1^2 + v_2^2 > 1$, go back to step 1a.
Step 2. Set $x = \sqrt{-2\log(s)} \times v_1/\sqrt{s} = \sqrt{-2\log(s)/s} \times v_1$ and $y = \sqrt{-2\log(s)/s} \times v_2$. $\square$

Note that at the end of step 1c, the point (v_1, v_2) will be a random point inside the unit circle (Figure 12.3b). Since the area of the square is 4 and the area of the circle is π, the probability that a random point in the square falls within the circle is $\pi/4 = 0.785$, and it takes on average $4/\pi = 1.273$ iterations in step 1 to sample a random point in the unit circle.

12.4.5 Gamma, beta, and Dirichlet variables

The gamma distribution with shape parameter α and rate parameter β, $x \sim G(\alpha, \beta)$, has the density function:

$$f(x; \alpha, \beta) = \frac{\beta^\alpha}{\Gamma(\alpha)} e^{-\beta x} x^{\alpha-1}, \alpha > 0, \beta > 0, x > 0, \quad (12.22)$$

where $\Gamma(\alpha)$ is the gamma function. This has mean α/β and variance α/β^2. One can sample from the gamma distribution using specialized algorithms such as the Ziggurat algorithm of Marsaglia and Wang (2000).

The beta distribution $x \sim \text{beta}(p, q)$ has the density

$$f(x; p, q) = \frac{1}{B(p, q)} x^{p-1} (1-x)^{q-1}, 0 < x < 1, p > 0, q > 0, \quad (12.23)$$

where $B(p, q) = \frac{\Gamma(p)\Gamma(q)}{\Gamma(p+q)}$ is the beta function. This has mean $p/(p+q)$ and variance $pq/[(p+q)^2(p+q+1)]$. The beta variable can be generated using the fact that if y_1 and y_2 are gamma variables with the same rate parameter β (which can be set to 1), $y_1 \sim G(p, \beta)$ and $y_2 \sim G(q, \beta)$, then

$$x = \frac{y_1}{y_1 + y_2} \quad (12.24)$$

has the beta distribution beta(p, q). Thus we generate two gamma variables y_1 and y_2 and use equation (12.24) to generate x.

The Dirichlet distribution $\text{Dir}(\alpha_1, \alpha_2, \ldots, \alpha_K)$ is a multivariate extension to the beta distribution. Instead of two proportions x and $1-x$, there are K proportions, $\mathbf{x} = (x_1, x_2, \ldots, x_K)$, with $x_i > 0$, and $x_1 + x_2 + \cdots + x_K = 1$. The density is

$$f(\mathbf{x}; a_1, a_2, \ldots, a_K) = \frac{\Gamma(\alpha)}{\prod_{i=1}^{K} \Gamma(\alpha_i)} \prod_{i=1}^{K} x_i^{\alpha_i - 1}, a_i > 0, \quad (12.25)$$

where $\alpha = \sum_i \alpha_i$. Note that Dir(p, q) for two variables x and $1-x$ reduces to beta(p, q). To sample from the Dirichlet, one can generate K gamma variables $y_i \sim G(\alpha_i, 1)$, for $i = 1, \ldots, K$, and then set $x_i = y_i / \sum_j y_j$.

In R, the functions `runif`, `rnorm`, `rgamma`, and `rbeta` generate random variables from the uniform, normal, gamma, and beta distributions, respectively.

12.5 Simulation of Markov processes

12.5.1 Simulation of the Poisson process

Suppose that a certain event occurs at random at a constant rate λ and let $N(t)$ denote the number of events that occur in the time interval $[0, t]$, with $N(0) = 0$. Then $\{N(t), t \geq 0\}$ is said to constitute a Poisson process if:

a. The numbers of events occurring in disjoint time intervals are independent.
b. The probability that exactly one event occurs in a small time interval Δt is $\approx \lambda \Delta t$, the probability that no event occurs in the interval is $\approx 1 - \lambda \Delta t$, and the probability that two or more events occur in the interval is very small and can be ignored.

Then $N(t_0)$ has a Poisson distribution with mean λt_0:

$$\Pr\{N(t_0) = k\} = \frac{e^{-\lambda t_0}(\lambda t_0)^k}{k!}. \tag{12.26}$$

The Poisson process has the feature that the waiting times between successive events are independent exponential variables with rate λ or mean $1/\lambda$. Furthermore, it has the *memory-less property*: given that the event has not occurred by time s, the extra waiting time until the event occurs is still an exponential variable with rate λ (Problem 12.2).

The fact that the waiting times are exponential can be used to simulate the Poisson process.

Algorithm 12.7. Simulate the Poisson process of rate λ until time t_0

Step 1. Initialize $t = 0$, $N = 0$.
Step 2. Generate a random number u, and set $s \leftarrow -\frac{1}{\lambda} \log u$. (Note that s is an exponential variable with mean $1/\lambda$.)
Step 3. Set $t \leftarrow t + s$. If $t > t_0$, stop.
Step 4. Set $N \leftarrow N + 1$, $s_N \leftarrow t$.
Step 5. Go to step 2. □

The final value of N will be a realized value of $N(t_0)$, the number of events that occur by time t_0, and the time points $s_1, s_2, \ldots, s_N$ will be the event times. This algorithm may be slow if the rate λ is high and the time t_0 is long, as then there will be many events.

An alternative approach to simulating the Poisson process over time interval $[0, t_0]$ is to generate the total number of events $N(t_0)$ from the Poisson distribution with mean λt_0 (equation 12.26), for example, by using Algorithm 12.1, and then generating the $N(t_0)$ event times by sampling from $U(0, N(t_0))$. This is based on the fact that given the total number of events $N(t_0) = k$, the k event times are distributed independently and uniformly over $(0, t_0)$. This may be faster as it avoids the logarithm in generating the exponential variables in step 2 of Algorithm 12.7.

12.5.2 Simulation of the nonhomogeneous Poisson process

The nonhomogeneous (or variable-rate) Poisson process differs from the (homogeneous) Poisson process in that the rate λ varies deterministically over time. Thus we have a *rate or intensity function* $\lambda(t)$ (see §9.3.1). The probability that exactly one event occurs in a small time interval $(t, t + \Delta t)$ is $\lambda(t)\Delta t$, the probability that no event occurs in the interval is $1 - \lambda(t)\Delta t$, and the probability that two or more events occur in the interval is negligibly small. The number, $N(t_0)$, of events that occur in the time interval $[0, t_0]$ has the Poisson distribution with mean

$$\int_0^{t_0} \lambda(s) ds. \tag{12.27}$$

If $\lambda(t) = \lambda$ for all t, this reduces to λt_0.

One characterization of the nonhomogeneous Poisson process is as follows. Suppose events are occurring according to a Poisson process with rate λ_U, and suppose an event that occurs at time t is recorded with probability $q(t)$. Then the process of recorded events constitutes a nonhomogeneous Poisson process with rate function $\lambda(t) = \lambda_U \cdot q(t)$. This formulation also provides a strategy for simulating the process, called *thinning* or *sampling*. If there exists an upper bound λ_U such that

$$\lambda(t) \leq \lambda_U, \text{ for all } 0 < t \leq t_0, \tag{12.28}$$

we can simulate a Poisson process with rate λ_U but record (sample) an event that has occurred at time t with probability $q(t)$ to produce a nonhomogeneous Poisson process of recorded events.

Algorithm 12.8. Thinning algorithm to simulate the nonhomogeneous Poisson process with rate function $\lambda(t)$ until time t_0.

Step 1. Initialize $t = 0$, $N = 0$.
Step 2. Generate a random number u_1, and set $s \leftarrow -\frac{1}{\lambda_U} \log u_1$ and $t \leftarrow t + s$. If $t > t_0$, stop.
Step 3. Generate a random number u_2. If $u_2 \leq \lambda(t)/\lambda_U$, set $N \leftarrow N + 1$ and $s_N \leftarrow t$.
Step 4. Go to step 2. □

Note that in step 2, s is an exponential variable with mean $1/\lambda_U$, and in step 3 the event is accepted (recorded) with probability $q(t) = \lambda(t)/\lambda_U$. The final value of N will be a realized value of $N(t_0)$, the number of events that occur by time t_0, and $s_1, s_2, \ldots, s_N$ will be the event times.

If the rate function has a lower bound, with $\lambda(t) > \lambda_L$ for $0 \leq t \leq t_0$, one may simulate a constant-rate Poisson process with rate λ_L and another variable-rate Poisson process with rate function $\lambda(t) - \lambda_L$ and with the new upper bound $\lambda_U - \lambda_L$. Then the merging (or superposition) of the two processes yields the desired variable-rate Poisson process with rate function $\lambda(t)$.

Algorithm 12.8 may be inefficient if $\lambda(t)$ is much less than λ_U, as most of the initially generated events will be rejected. To make $\lambda(t)$ near λ_U, one may break the time interval $[0, t_0]$ into subintervals, with the upper bound λ_{Ui} for subinterval i. Simulation proceeds successively though the subintervals. Simulation for subinterval i ends when the last event time carries one beyond the desired boundary (see step 2 of Algorithm 12.8), at which point one should start the simulation for subinterval $i + 1$ afresh with rate $\lambda_{U,i+1}$. The last exponential waiting time for subinterval i is thus wasted (apart from the fact that it overshoots subinterval i). However, this waste can be avoided by using the memory-less property of the exponential distribution. Let s be the amount of overshoot for subinterval i. Then s is an exponential variable with rate λ_{Ui} (Problem 12.2), and $s\lambda_{Ui}/\lambda_{U,i+1}$ will be an exponential variable with rate $\lambda_{U,i+1}$ and can be used for simulation for subinterval $i + 1$.

This algorithm works well if the rate function $\lambda(t)$ changes continuously or at least is piece-wise continuous. Examples of such models may include deterministic change of molecular evolutionary rate, and exponential population size growth in a coalescent model.

If the rate function $\lambda(t)$ has a very simple form, one may be able to use the inversion method to simulate successively the event times $s_1, s_2, \ldots$, directly. Suppose an event occurs at time s. Then the additional time x until the next event has the CDF

$$F_s(x) = 1 - \Pr\{\text{no event in } (s, s + x) \,|\, \text{event at } s\} = 1 - \exp\left\{-\int_s^{s+x} \lambda(y) \, dy\right\}. \tag{12.29}$$

If the CDF and its inverse are analytically tractable, one can use the inversion method to sample from $F_s(x)$. Generate the first waiting time $x_1 \sim F_0(x)$ and set the first event time to $s_1 = x_1$. Generate the next waiting time $x_2 \sim F_{s_1}(x)$ and set the second event time to $s_2 = s_1 + x_2$, and so on. The simulation stops when the next event time would take us beyond the desired boundary t_0. The following example is from Ross (1997, pp. 80–81).

Example 12.1. Simulate the event times for a variable-rate Poisson process with the rate function

$$\lambda(t) = \frac{1}{t+a}, t \geq 0, a > 0. \tag{12.30}$$

We have the CDF for the waiting time x to be

$$F_s(x) = 1 - \exp\left\{-\int_s^{s+x} \frac{1}{y+a} dy\right\} = 1 - \exp\left\{-\log \frac{x+s+a}{s+a}\right\} = \frac{x}{x+s+a}. \tag{12.31}$$

This CDF can be easily inverted. If $u = F_s(x)$, we have

$$x = F_s^{-1}(u) = \frac{(s+a)u}{1-u}. \tag{12.32}$$

Thus we generate random numbers $u_1, u_2, \ldots$, and set

$$s_1 = x_1 = \frac{au_1}{1-u_1},$$

$$s_2 = s_1 + x_2 = s_1 + (s_1 + a)\frac{u_2}{1-u_2} = \frac{s_1 + au_2}{1-u_2},$$

and, in general,

$$s_j = s_{j-1} + x_j = s_{j-1} + (s_{j-1} + a)\frac{u_j}{1-u_j} = \frac{s_{j-1} + au_j}{1-u_j}, j \geq 2.$$

Then $s_1, s_2, \ldots$, etc., will constitute the event times. □

In general, if the CDF is easy to calculate but the inverse CDF is not, one may use a numerical root-finding algorithm to find the inverse CDF, i.e., the value of x such that $F_s(x) = u$. This is the case with the exponential population size growth model discussed in §9.3.2.

12.5.3 Simulation of discrete-time Markov chains

Suppose a discrete-time Markov chain has S states: $1, 2, \ldots, S$. Let $p_{ij} = \Pr\{X_{t+1} = j | X_t = i\}$ be the *transition probability*, the probability that the chain will be in state j in the next step given that it is currently in state i. The *transition matrix* is then given as

$$P = \{p_{ij}\} = \begin{bmatrix} p_{11} & p_{12} & \cdots & p_{1S} \\ p_{21} & p_{22} & \cdots & p_{2S} \\ \vdots & \vdots & \ddots & \vdots \\ p_{S1} & p_{S2} & \cdots & p_{SS} \end{bmatrix}. \tag{12.33}$$

Note that the probabilities $\{p_{i1}, p_{i2}, \ldots, p_{iS}\}$ on the ith row of the matrix specify the distribution of the state in the next step given the current state i. Thus, given the initial state $X_0 = i$, the state at next step X_1 can be generated from the discrete distribution $\{p_{i1}, p_{i2}, \ldots, p_{iS}\}$, for example, by using the inversion method. Suppose $X_1 = j$. Then X_2 can be generated using the probabilities in the jth row of P, and so on. One repeats the process t times to simulate the process over t steps. However when t is large, this simulation algorithm is inefficient.

Alternatively note that the m-step transition matrix is given as

$$P^{(m)} = \left\{p_{ij}^{(m)}\right\} = P^m, \tag{12.34}$$

where $p_{ij}^{(m)} = \Pr\{X_{t+m} = j | X_t = i\}$ is the m-step transition probability or the probability that the chain will be in state j after m steps given that it is currently in state i. If we can diagonalize P, i.e. if we can find a non-singular matrix U and a diagonal matrix $\Lambda = \text{diag}\{\lambda_1, \lambda_2, \ldots, \lambda_S\}$ such that

$$P = U\Lambda U^{-1}, \tag{12.35}$$

then

$$P^m = U\Lambda^m U^{-1}, \tag{12.36}$$

with $\Lambda^m = \{\lambda_1^m, \lambda_2^m, \ldots, \lambda_S^m\}$. One can then simulate the Markov chain over m steps in one step using P^m.

Example 12.2 Discrete-time simulation of nucleotide substitution under the K80 model. The evolution of a nucleotide site in a DNA sequence is described by a discrete-time Markov chain, with the transition probabilities between the four nucleotides T, C, A, and G given as

$$P = \begin{bmatrix} 1-(\alpha+2\beta) & \alpha & \beta & \beta \\ \alpha & 1-(\alpha+2\beta) & \beta & \beta \\ \beta & \beta & 1-(\alpha+2\beta) & \alpha \\ \beta & \beta & \alpha & 1-(\alpha+2\beta) \end{bmatrix}, \tag{12.37}$$

where $\alpha, \beta \ll 1$ are the probabilities of change over one step (e.g. one year). To be definite, let $\alpha = 10^{-8}$ and $\beta = 10^{-9}$ per site per year, based roughly on rate estimates for the mammalian mitochondrial genomes, in which the rate is $\alpha + 2\beta \approx 10^{-8}$ per site per year and $\kappa = \alpha/\beta \approx 10$. Simulate two sequences that separated 5 MY ago (with a total evolution time of 10 MY) under this discrete-time K80 model.

One can generate the first sequence by sampling the four nucleotides with equal probabilities and then 'evolve' every site in it over 10 MY. The process of evolution can be simulated in different ways. The first approach is to use the transition matrix P (equation (12.37)) over $m = 10^7$ steps (years). To speed up the simulation, a second approach is to take 1 MY as the time unit, redefine rates as $\alpha = 0.01$ and $\beta = 0.001$, and simulate the process over $m = 10$ steps. Obviously if the step is too large, the approximation may be too crude. A third approach is to use the m-step transition matrix P^m. Equation (12.37) can be rewritten as

$$P = \begin{bmatrix} 1 & 1 & 0 & 1 \\ 1 & 1 & 0 & -1 \\ 1 & -1 & 1 & 0 \\ 1 & -1 & -1 & 0 \end{bmatrix} \begin{bmatrix} 1 & 0 & 0 & 0 \\ 0 & 1-4\beta & 0 & 0 \\ 0 & 0 & 1-2(\alpha+\beta) & 0 \\ 0 & 0 & 0 & 1-2(\alpha+\beta) \end{bmatrix} \begin{bmatrix} \frac{1}{4} & \frac{1}{4} & \frac{1}{4} & \frac{1}{4} \\ \frac{1}{4} & \frac{1}{4} & -\frac{1}{4} & -\frac{1}{4} \\ 0 & 0 & \frac{1}{2} & -\frac{1}{2} \\ \frac{1}{2} & -\frac{1}{2} & 0 & 0 \end{bmatrix}. \tag{12.38}$$

Then the m-step transition matrix is (from equation (12.36))

$$P^m = \begin{bmatrix} p_0 & p_1 & p_2 & p_2 \\ p_1 & p_0 & p_2 & p_2 \\ p_2 & p_2 & p_0 & p_1 \\ p_2 & p_2 & p_1 & p_0 \end{bmatrix}, \tag{12.39}$$

where

$$p_0 = \frac{1}{4} + \frac{1}{4}(1-4\beta)^m + \frac{1}{2}(1-2\alpha-2\beta)^m,$$

$$p_1 = \frac{1}{4} + \frac{1}{4}(1-4\beta)^m - \frac{1}{2}(1-2\alpha-2\beta)^m,$$

$$p_2 = \frac{1}{4} - \frac{1}{4}(1-4\beta)^m.$$

Then one can use P^m to sample sites in the second sequence in one step. Note that $(1-4\beta)^m \approx e^{-4\beta m}$ and $(1-2\alpha-2\beta)^m \approx e^{-2(\alpha+\beta)m}$, so that sampling using P^m of equation (12.39) for the discrete Markov chain is equivalent to sampling from the transition probability matrix $P(t)$ of equation (1.10) for the continuous-time Markov chain. □

12.5.4 Simulation of continuous-time Markov chains

The continuous-time Markov chain can be simulated in multiple ways. One of the commonly used approaches is to simulate the exponential waiting times between events. Suppose the generator (rate matrix) is

$$Q = \{q_{ij}\} = \begin{bmatrix} q_{11} & q_{12} & \cdots & q_{1S} \\ q_{21} & q_{22} & \cdots & q_{2S} \\ \vdots & \vdots & \ddots & \vdots \\ q_{S1} & q_{S2} & \cdots & q_{SS} \end{bmatrix}. \tag{12.40}$$

Then $q_i = -q_{ii} = \sum_{j \neq i} q_{ij}$ is the rate of transition (change) when the chain is in state i. The waiting time until a jump or change of state has an exponential distribution with rate (intensity) q_i or with mean $1/q_i$. When a jump occurs, the chain moves to the alternative states $j \neq i$ with probability equal to q_{ij}/q_i (Figure 12.4). In other words, given a move, the target state is sampled from the following (discrete-time) Markov chain:

$$M = \begin{bmatrix} 0 & \frac{q_{12}}{q_1} & \cdots & \frac{q_{1S}}{q_1} \\ \frac{q_{21}}{q_2} & 0 & \cdots & \frac{q_{2S}}{q_2} \\ \vdots & \vdots & \ddots & \vdots \\ \frac{q_{S1}}{q_S} & \frac{q_{S2}}{q_S} & \cdots & 0 \end{bmatrix}. \tag{12.41}$$

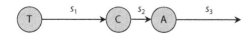

Fig. 12.4 Characterization of the Markov process as exponential waiting times and a jump chain. The waiting times (s_1, s_2, s_3, etc.) until the next jump or nucleotide substitution are independent random variables from the exponential distributions with means $1/q_T$, $1/q_C$, and $1/q_A$, etc., where q_T, q_C, and q_A are the substitution rates of nucleotides T, C, and A, respectively. If the waiting times are ignored, the visited states T, C, and A, etc., will constitute a (discrete-time) Markov chain, known as the jump chain.

Note that each row of M sums to 1. The discrete-time chain specified by M is called the *embedded Markov chain* or the *jump chain* of the continuous-time Markov chain specified by Q. The jump chain records the states visited by the continuous-time chain, with the waiting times ignored.

The following algorithm simulates the continuous-time Markov chain until time t_0. The final state of the process at time t_0 is recorded in i.

Algorithm 12.9. Simulating a Markov chain over time t_0 by using the exponential waiting times.

Step 1. Set $t = 0$ and the initial state $i = X(0)$. According to the model, $X(0)$ may be generated by sampling from the stationary distribution of the chain.

Step 2. Generate a random number u, and set $s \leftarrow -\frac{1}{q_i} \log u$. (Note that s is an exponential variable with mean $1/q_i$.)

Step 3. Set $t \leftarrow t + s$. If $t > t_0$, stop.

Step 4. Sample the new state j from the discrete distribution specified by the ith row of M (equation (12.41)). Set $i \leftarrow j$.

Step 5. Go to step 2.

Note that steps 2 and 3 are very similar to Algorithm 12.7 for simulating the Poisson process, although here the process is not Poisson as the rate $\lambda = q_i$ depends on the current state i. □

12.6 Simulating molecular evolution

12.6.1 Simulation of sequences on a fixed tree

Here we consider the generation of a nucleotide sequence alignment when the tree topology and branch lengths are given. The basic model assumes the same substitution process at all sites and along all branches, but we will also consider more complex models in which the evolutionary process may vary across sites or branches. We focus on nucleotide models; amino acid or codon sequences can be simulated using the same principles. Several approaches can be used and produce equivalent results.

12.6.1.1 Method 1: Sampling sites from the multinomial distribution of site patterns

If the model assumes independent evolution at different sites in the sequence and all sites evolve according to the same model, data at different sites will have independent and identical distributions; they are said to be i.i.d. Note that this class of models include the mixture models such as the gamma model for variable rates among sites (§4.3.1–2). The sequence dataset will then follow a multinomial distribution, with every site being a sample point, and every site pattern being a category (cell) of the multinomial. For a tree of s species, there are 4^s, 20^s, or 61^s possible site patterns for nucleotide, amino acid, or codon sequences, respectively. Calculation of the probability for every site pattern is explained in §4.2–4.3, which describes the calculation of the likelihood function under the model. Sampling from this multinomial distribution generates the site pattern counts, which are equivalent to and can be converted to a sequence alignment. If a pattern, say TTTC (for four species), is observed 50 times, one simply writes out 50 sites with the same data TTTC, either with or without randomizing the sites. Most phylogenetic inference programs, especially the likelihood and Bayesian programs, collapse sites into patterns to save computation, since the probabilities of observing sites with the same data are identical (§4.2.2.2). Some programs can also read the site pattern counts directly.

The multinomial sampling approach is not feasible for large trees because the number of categories becomes too large. However, it is very efficient for small trees with only four or five species, especially when combined with an efficient algorithm for sampling from the multinomial, such as the alias method.

12.6.1.2 *Method 2: Evolving sequences along the tree*

This approach 'evolves' sequences along branches on the given tree, and is the algorithm used in programs such as SEQ-GEN (Rambaut and Grassly 1997) and EVOLVER (Yang 2007b). First, we generate a sequence for the root of the tree, by sampling nucleotides according to their equilibrium distribution under the model: $\pi_T, \pi_C, \pi_A,$ and π_G. If the base frequencies are all equal, one can use the more efficient algorithm for the discrete uniform distribution. The sequence for the root is then allowed to evolve to produce sequences at the daughter nodes of the root. The procedure is repeated for every branch on the tree, generating the sequence at a node only after the sequence at its mother node has been generated. Finally sequences at the tips of the tree constitute the data, while sequences for ancestral nodes are discarded.

To simulate the evolution of a sequence along a branch of length t, calculate the transition probability matrix $P(t) = \{p_{ij}(t)\}$, and then simulate nucleotide substitutions at every site independently. For example, if a site is occupied by C in the source sequence, the nucleotide in the target sequence will be a random draw from the discrete distribution $\{p_{CT}(t), p_{CC}(t), p_{CA}(t), p_{CG}(t)\}$. This process is repeated to generate all sites in the target sequence. As the transition probabilities apply to all sites in the sequence, one has to calculate them only once for the whole sequence for that branch.

12.6.1.3 *Method 3: Simulating the waiting times of a Markov chain*

This is a variation of method 2, and simulates sequence evolution along each branch by generating the exponential waiting times and using the jump chain (Algorithm 12.9). As in method 2, we first generate a sequence for the root. Then we simulate evolution along each branch. Suppose the branch length is t, and the nucleotide at a site at the beginning of the branch is i. We generate a random waiting time s from the exponential distribution with mean $1/q_i$. If $s > t$, no substitution occurs before the end of the branch so that the target nucleotide at the site is i. Otherwise a substitution occurs and we sample the new nucleotide j with probability q_{ij}/q_i (Figure 12.4). The remaining time for the branch becomes $t - s$. We then draw the new waiting time from the exponential distribution with mean $1/q_j$. Continue the process until the branch length is exhausted.

This algorithm of simulating exponential waiting times and the jump chain may be applied to the whole sequence instead of one site. The total rate of substitution is the sum of the rates across sites, and the waiting time until a substitution occurs at any site in the whole sequence has an exponential distribution with the mean being equal to the reciprocal of the total rate. If a substitution occurs, it is assigned to sites with probabilities proportional to the rates at the sites.

An advantage of method 3 is that it does not require calculation of the transition probability matrix $P(t)$ over branch length t, as both the waiting times and the transition matrix for the jump chain are fully specified by the instantaneous rates. As a result, this procedure can be used to simulate sequences under more complex models of sequence evolution involving insertions, deletions, and genome-rearrangement events. One simply calculates the total rate for all events (including substitutions, insertions, and deletions) for the whole sequence, and simulates the exponential waiting time until the next event. If an event occurs before the end of the branch, one assigns the event to a site and to an event type (a substitution, insertion or deletion) with probabilities in proportion to the rates. This is the approach taken by simulation programs such as DAWG (Cartwright 2005), INDELIBLE (Fletcher and Yang 2009), and ALF (Dalquen et al. 2012). Note that most other programs for simulating sequence data under a model of insertions and deletions do not work correctly.

12.6.1.4 Simulation under JC69 and K80

Under the JC69 and K80 models of nucleotide substitution (Jukes and Cantor 1969; Kimura 1980), the four nucleotides have the same substitution rate, and the substitution process can be described by a Poisson process. Instead of simulating exponential waiting times for every site and every branch, a more efficient approach is to generate the number of substitutions over a branch for the entire sequence, which is a Poisson variable with mean $l\lambda t$ (if the sequence has l sites, the branch has length t, and the rate per site is λ) and then to assign every substitution to a site chosen at random. Under JC69, the substitution rates from any nucleotide to any other are equal, so that when a substitution event occurs at a site, one chooses any of the three other nucleotides with equal probability ($\frac{1}{3}$). Under the K80 model we need take into account the transition/transversion rate ratio κ: when a substitution event occurs, we choose the nucleotide with a transitional difference from the current nucleotide with probability $\kappa/(\kappa + 2)$ and the other two nucleotides with transversional differences with probability $1/(\kappa + 2)$.

Note that under models such as F81 (Felsenstein 1981), HKY85 (Hasegawa et al. 1985), etc., different nucleotides have different rates, so that the substitution process is not Poisson and the above algorithm cannot be used directly.

12.6.1.5 Simulation under more complex models

The three methods discussed in §12.6.1.1–3 can be modified to simulate more complex models, for example, models that allow different substitution parameters among branches (such as different transition/transversion rate ratio κ, different base frequencies, or different ω ratios for codon models). The multinomial sampling approach (method 1) applies as long as the site pattern probabilities are calculated correctly under the model. The approaches of evolving sequences along branches on the tree, either by calculating the transition probabilities (method 2) or by simulating the waiting times and the jump chain (method 3) are also straightforward; one simply use the appropriate model and parameters for the branch when simulating the evolutionary process along that branch.

One may also simulate under models that allow heterogeneity among sites. We will consider as an example variable substitution rates, but the approach applies to other kinds of among-site heterogeneity. There are two kinds of models that incorporate rate variation among sites (see §4.3.1–3). The first is the so-called *fixed-sites* or *partition* model, under which every site in the sequence belongs to a predetermined site partition. For example, one may simulate five genes evolving on the same tree but at different rates r_1–r_5. The transition probability for gene k is $p_{ij}(tr_k)$. Sites in different genes do not have the same distribution, but within each gene, the sites are i.i.d. Thus one can use any of the three methods discussed above to simulate the data for each gene separately and then merge them into one dataset.

A second kind of heterogeneous-site model is the so-called *random-site* or *mixture* model. Examples include the gamma models of variable rates for sites (Yang 1993, 1994a), the codon models of variable ω ratios among sites (Nielsen and Yang 1998; Yang et al. 2000), and the covarion-like models (Galtier 2001; Huelsenbeck 2002; Guindon et al. 2004). The rates (or some other features of the substitution process) are assumed to be random variables drawn from a common statistical distribution, and we do not know *a priori* which sites have high rates and which have low rates. Under such models, data at different sites are i.i.d. The approach of multinomial sampling can be used directly, although the site pattern probabilities have to be calculated under the heterogeneous-site model. One may also sample the rate for each site and apply the method of simulating evolution along branches. If a continuous distribution (such as the gamma) is used to model rate variation among sites, every site h will have a distinct rate r_h. Then one has to calculate the transition probability matrix $P(tr_h)$ for every site and every branch. This may be expensive. If a discrete distribution or a few site classes are used (as in the discrete-gamma model), one may sample rates for sites first, and then simulate data for the different rate classes separately, perhaps followed by a randomization of the sites. Within each site class, the sites have the same rate and their evolution on the branch can be simulated by using the same transition probability matrix. Note that under the mixture model, the number of sites in any site class varies among simulated replicates, whereas in the partition model, the number is fixed.

12.6.2 *Simulation of random trees*

Several models can be used to generate random trees with branch lengths, such as the standard coalescent model, the Yule branching model, and the birth–death process model either with or without species sampling. All those models assign equal probabilities to all possible labelled histories (see §3.1.1.5). A labelled history is a rooted tree with the internal nodes ranked according to their ages and can be generated by starting with the tips of the tree and joining nodes at random until there is only one lineage left.

In these models, the ages of the nodes are independent of the labelled history and can be attached to the tree afterwards. Under the coalescent model, the waiting time until the next coalescent event has an exponential distribution (see equation (9.8)). Under the birth–death sampling process model of Yang and Rannala (1997), the node ages on the labelled history are order statistics from a kernel density (see equation (10.11)) and can be simulated easily. Trees generated in this way have branch lengths conforming to a molecular clock. To generate trees in which the clock is violated, one may simulate a stochastic process of rate drift (see §10.4.3).

One may also generate random rooted or unrooted trees without assuming any biological model, by sampling at random from all possible trees. Branch lengths may also be sampled from arbitrary distributions such as the exponential or the gamma.

12.7 Validation of the simulation program

To verify that the simulation program does not have any bugs, one can use any of the 'standard' techniques for debugging computer programs. For example, one should try to debug the program in modules and isolate the problem. Often in simple cases (for example, when certain variables in the simulation program take special values), the answer can be worked out by hand for comparison with the simulated output. Printing out the values of certain variables as the program runs, and examining their correctness is a very effective approach.

Simulation may be the easiest thing one can do with a model (it is much harder to make inferences from the same model), yet it offers the uninitiated ample chances for mistakes. In molecular phylogenetics, errors in simulation programs and simulation studies are unfortunately common, as the programmer does not always respect the rules of probability calculus. Sometimes the behaviour of the simulation model is forcefully changed, under the illusion that the changes might make the simulation model more realistic biologically. For example, one common mistake is to fix the sequence at the root to be a real observed sequence when evolving sequences on the tree, while correctly it should be generated from the equilibrium distribution of the nucleotides or amino acids and should vary among replicate datasets.

12.8 Problems

12.1 If $x_0 = 11, x_n = (37x_{n-1} \bmod 1000)$, find $x_1, x_2, x_3, \ldots, x_{10}$ (see equation (12.1)).

12.2* *Memory-less property of exponential waiting time.* Suppose random variable X has the exponential distribution with rate λ or mean $1/\lambda$. Show that $\Pr\{X > a + x \mid X > a\} = \Pr\{X > x\}$. This result may be paraphrased as follows. Suppose the waiting time until the bus arrives is an exponential variable with mean $1/\lambda = 10$ minutes. Then, given that we have waited for $a = 9$ minutes, the extra time we have to wait for the bus to arrive is still an exponential variable with a mean of 10 minutes. [Hint. Note that $\Pr\{X > x\} = e^{-\lambda x}$.]

12.3 *Monte Carlo integration* (§6.4.5). Write a small program to calculate the integral $f(x)$ in the Bayesian estimation of sequence distance under the JC69 model, discussed in Example 6.4. The data are $x = 90$ differences out of $n = 948$ sites. Use the exponential prior with mean 0.2 for the sequence distance θ. Generate $N = 10^6$ or 10^8 random variables from the exponential prior: $\theta_1, \theta_2, \ldots, \theta_N$, and calculate

$$f(x) = \int_0^\infty f(\theta) f(x|\theta) \, d\theta \simeq \frac{1}{N} \sum_{i=1}^N f(x|\theta_i). \quad (12.42)$$

Note that the likelihood $f(x|\theta_i)$ may be too small to represent in the computer, so scaling may be needed. One way to do this is as follows. Compute the maximum log likelihood $\ell_m = \log\{f(x|\hat\theta)\}$, where $\hat\theta = 0.1015$ is the maximum likelihood estimate (MLE). Then multiply $f(x|\theta_i)$ in equation (12.42) by $e^{-\ell_m}$ before taking the sum:

$$e^{-\ell_m} \times \sum_{i=1}^N f(x|\theta_i) = \sum_{i=1}^N \exp\left(\log\{f(x|\theta_i)\} - \ell_m\right). \quad (12.43)$$

12.4 What is the optimal sequence divergence when a pair of sequences are used to estimate the transition/transversion rate ratio κ under the K80 model? Intuitively very similar sequences will have little information while very divergent sequences will have too much noise, so the optimum sequence divergence should be intermediate. Write a small simulation program to study the optimal sequence divergence. Each dataset consists of a pair of sequences, which can be generated using any of the three approaches discussed in §12.6.1. Alternatively you can use a simulation program such as SEQ-GEN or EVOLVER to simulate the datasets. Assume $\kappa = 2$ and use a sequence length of 500 sites. Consider several sequence distances, say, $d = 0.01, 0.02, \ldots, 2$. For each d, simulate 1,000 replicate datasets under K80 and analyse them under the same model to estimate d and κ using equation (1.12). Calculate the

---

* indicates a more difficult or technical problem.

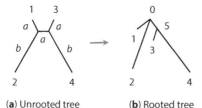

Fig. 12.5 (**a**) A tree of four species, with three short branches (of length a) and two long branches (with length b) for simulating data to demonstrate long-branch attraction. (**b**) The same tree rerooted at an ancestral node for simulation.

mean and variance of the estimate $\hat{\kappa}$ across replicate datasets. Calculate the standard deviation of $\hat{\kappa}$ and plot it against t.

12.5 Use a computer simulation to examine the null distribution of the likelihood ratio test (LRT) statistic for comparing JC69 and K80 using a pair of sequences. Simulate 10,000 replicate datasets as in Problem 12.4 under the JC69 model, and, for each of them, calculate the log likelihood values under the two models (ℓ_0 and ℓ_1) and $2\Delta\ell = 2(\ell_1 - \ell_0)$. See equations (1.47) and (1.52). Then construct a histogram (for example, using the R function hist). Compare it with the χ^2 distribution with one degree of freedom. Use different sequence distances ($d = 0.1, 0.5$, and 1, say) and sequence lengths ($l = 50, 100, 200$, and 500, say) to examine their impact.

12.6 *Long-branch attraction by parsimony*. Use the JC69 model to simulate datasets on a tree of four species (Figure 12.5a), with two different branch lengths $a = 0.1$ and $b = 0.5$ (in expected number of substitutions per site). Simulate 1,000 replicate datasets. For each dataset, count the sites with the three site patterns $xxyy$, $xyxy$, and $xyyx$, and determine the most parsimonious tree. To simulate a dataset, reroot the tree at an interior node as in Figure 12.5b, say. Generate a sequence for the root (node 0) by random sampling of the four nucleotides, and then evolve the sequence along the five branches of the tree. You may also use a program such as SEQ-GEN or EVOLVER. Consider a few sequence lengths, such as 100, 1,000, and 10,000 sites. Calculate the proportion of datasets in which parsimony recovers the true tree.

12.7 A useful test of a new and complex likelihood program is to generate a few very large datasets under the model and then analyse them under the same model, to confirm that the MLEs are close to the true values used in the simulation. As MLEs are consistent, they should approach the true values when the sample size (sequence length) becomes larger and larger. Use the program written for Problem 12.6 (or SEQ-GEN or EVOLVER) to generate one or two datasets of 10^6, 10^7, or 10^8 sites under JC69 and analyse them using a likelihood program (such as PAML, PHYML, or RAxML) under the same model, to see whether the MLEs of branch lengths are close to the true values. Beware that some programs may demand a lot of resources to process large datasets.

Appendices

Appendix A. Functions of random variables

This section includes two theorems concerning functions of random variables. Theorem 1 specifies the probability density of functions of random variables. Theorem 2 gives the proposal ratio in a Markov chain Monte Carlo (MCMC) algorithm when the Markov chain is formulated using variables x, but the proposal is made by changing variables y, which are functions of x.

Theorem 1. (**a**) Suppose X is a random variable with density $f_X(x)$, and $Y = h(X)$ and $X = h^{-1}(Y)$ constitute a one-to-one mapping between X and Y. Then the random variable Y has density

$$f_Y(y) = f_X(h^{-1}(y)) \times \left| \frac{dx}{dy} \right|. \tag{A.1}$$

Here we use the capital letters X and Y for the random variables and the small letters x and y for their realized values. Note that $\left|\frac{dx}{dy}\right| = 1 / \left|\frac{dy}{dx}\right|$ if $\left|\frac{dy}{dx}\right|$ is easier to calculate.

(**b**) The multivariate version is similar. Suppose random vectors $\boldsymbol{X} = \{X_1, X_2, \ldots, X_m\}$ and $\boldsymbol{Y} = \{Y_1, Y_2, \ldots, Y_m\}$ constitute a one-to-one mapping through $Y_i = h_i(\boldsymbol{X})$ and $X_i = h_i^{-1}(\boldsymbol{Y})$, $i = 1, 2, \ldots, m$, and that $\boldsymbol{X}$ has density $f_{\boldsymbol{X}}(\boldsymbol{x})$. Then $\boldsymbol{Y}$ has density

$$f_{\boldsymbol{Y}}(\boldsymbol{y}) = f_{\boldsymbol{X}}(\boldsymbol{h}^{-1}(\boldsymbol{y})) \times |J(\boldsymbol{y})|, \tag{A.2}$$

where $|J(\boldsymbol{y})|$ is the absolute value of the Jacobian determinant of the transform

$$J(\boldsymbol{y}) = \left| \frac{\partial \boldsymbol{x}}{\partial \boldsymbol{y}} \right| = \begin{vmatrix} \frac{\partial x_1}{\partial y_1} & \frac{\partial x_1}{\partial y_2} & \cdots & \frac{\partial x_1}{\partial y_m} \\ \frac{\partial x_2}{\partial y_1} & \frac{\partial x_2}{\partial y_2} & \cdots & \frac{\partial x_2}{\partial y_m} \\ \vdots & \vdots & \ddots & \vdots \\ \frac{\partial x_m}{\partial y_1} & \frac{\partial x_m}{\partial y_2} & \cdots & \frac{\partial x_m}{\partial y_m} \end{vmatrix}. \tag{A.3}$$

Due to the importance of this theorem, we will provide a heuristic proof. Consider the univariate case and assume that h is a monotonically increasing function. Note that the probability density function (PDF) is a smoothed histogram, scaled so that the area under the curve is 1. Suppose we construct a histogram for X using a large number of X values. The transform $Y = h(X)$ gives the same number of Y values. The proportion of X values in the bin $(x, x + \Delta x)$ is $f_X(x) \Delta x$. Those points will be in the bin $(y, y + \Delta y)$ in the histogram for Y, with $\Delta y = \Delta x \cdot \frac{dy}{dx}$, and constitute the same proportion of the whole sample. In other words, $f_X(x) \Delta x = f_Y(y) \Delta y$, which means $f_Y(y) = f_X(x) \cdot \frac{\Delta x}{\Delta y} \to f_X(x) \cdot \frac{dx}{dy}$ as Δx or $\Delta y \to 0$. If h is monotonically decreasing, we have $f_Y(y) = f_X(x) \cdot \left|\frac{dx}{dy}\right|$ since only the widths of the bins matter in the argument.

Molecular Evolution: A Statistical Approach. Ziheng Yang. © Ziheng Yang 2014.
Published 2014 by Oxford University Press.

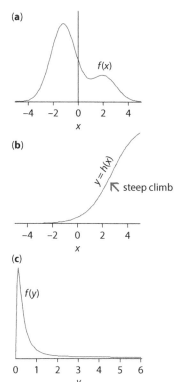

Fig. A1 A variable transform $y = h(x)$ can convert a two-moded distribution $f(x)$ into a single-moded distribution $f(y)$. The function $h(x)$ is steep, with large $h'(x)$, around the second mode for $f(x)$ and thus expands that part of the density.

From the above reasoning, we see that the density for X is stretched wherever $\left|\frac{dy}{dx}\right|$ is large and the $h(x)$ curve is steep, and that it is squeezed wherever $\left|\frac{dy}{dx}\right|$ is small and $h(x)$ is nearly flat. For example, the density of X in Figure A1a has two peaks. By applying a transform that is steep around the second peak (Figure A1b), that part of the density is stretched so that the peak disappears in the resulting density for Y, which has a single peak (Figure A1c). In particular, if $Y = F(X)$, where F is the cumulative distribution function (CDF) for X, the resulting distribution is uniform: $Y \sim U(0, 1)$. This is because around the mode for $f(X)$, $F(X)$ will be steep and the density will be stretched, and wherever $f(X)$ is low, $F(X)$ will be nearly flat and the density will be squeezed. There is just the right amount of stretching and squeezing so that the resulting density for Y is perfectly flat.

Example A1. Normal distribution. Suppose z has a standard normal distribution with density

$$\phi(z) = \frac{1}{\sqrt{2\pi}}e^{-\frac{1}{2}z^2}. \tag{A.4}$$

Let $x = \mu + \sigma z$, so that $z = (x - \mu)/\sigma$ and $dz/dx = 1/\sigma$. Then x has the density

$$f(x) = \phi\left(\frac{x-\mu}{\sigma}\right)\bigg/\sigma = \frac{1}{\sqrt{2\pi\sigma^2}}\exp\left\{-\frac{1}{2\sigma^2}(x-\mu)^2\right\}. \tag{A.5}$$

Thus x has a normal distribution with mean μ and variance σ^2. □

Example A2. Multivariate normal distribution. Suppose $z_1, z_2, \ldots, z_p$ are independent standard normal variables, so that the column vector $\mathbf{z} = (z_1, z_2, \ldots, z_p)^T$ has a standard p-variate normal density:

$$\phi_p(\mathbf{z}) = \frac{1}{(2\pi)^{p/2}} \exp\left\{-\frac{1}{2}\left(z_1^2 + z_2^2 + \cdots + z_p^2\right)\right\} == \frac{1}{(2\pi)^{p/2}} \exp\left\{-\frac{1}{2}\mathbf{z}^T\mathbf{z}\right\}. \quad (A.6)$$

Let $\mathbf{x} = \boldsymbol{\mu} + A\mathbf{z}$, where A is a $p \times p$ non-singular matrix, so that $\mathbf{z} = A^{-1}(\mathbf{x} - \boldsymbol{\mu})$ and $\partial \mathbf{z}/\partial \mathbf{x} = A^{-1}$. Let $\Sigma = AA^T$. Then $\mathbf{x}$ has the density

$$f_p(\mathbf{x}) = \phi_p(\mathbf{z}) \cdot \left|\frac{\partial \mathbf{z}}{\partial \mathbf{x}}\right| = \frac{1}{(2\pi)^{p/2}|\Sigma|^{1/2}} \exp\left\{-\frac{1}{2}(\mathbf{x}-\boldsymbol{\mu})^T \Sigma^{-1}(\mathbf{x}-\boldsymbol{\mu})\right\} \quad (A.7)$$

Note that $(A^{-1})^T A^{-1} = (AA^T)^{-1} = \Sigma^{-1}$ and $|\Sigma| = |AA^T| = |A| \cdot |A^T| = |A|^2$. Equation (A.7) is the density for a multivariate normal distribution with the mean vector $\boldsymbol{\mu}$ and variance–covariance matrix Σ. □

Example A3. Beta distribution. Suppose X has the (standard) beta distribution with density

$$f_X(x) = \frac{1}{B(p,q)} x^{p-1}(1-x)^{q-1},\ 0 \le x \le 1. \quad (A.8)$$

Let $Y = a + X(b-a)$ so that $a \le Y \le b$. Then $dx/dy = 1/(b-a)$, and Y has the density

$$f_Y(y) = \frac{1}{B(p,q)}\left(\frac{y-a}{b-a}\right)^{p-1}\left(\frac{b-y}{b-a}\right)^{q-1} \cdot \frac{1}{b-a} = \frac{1}{B(p,q)} \frac{(y-a)^{p-1}(b-y)^{q-1}}{(b-a)^{p+q-1}}, \quad (A.9)$$

with $a \le y \le b$. Y is said to have the beta distribution. □

Example A4. Inverse gamma distribution. Suppose X has the gamma distribution with density

$$f_X(x) = \frac{\beta^\alpha}{\Gamma(\alpha)} x^{\alpha-1} e^{-\beta x},\ x > 0. \quad (A.10)$$

Let $Y = 1/X$. We have $|dx/dy| = 1/y^2$, and Y has the density

$$f_Y(y) = \frac{\beta^\alpha}{\Gamma(\alpha)}\left(\frac{1}{y}\right)^{\alpha-1} e^{-\beta/y} \cdot \frac{1}{y^2} = \frac{\beta^\alpha}{\Gamma(\alpha)} y^{-\alpha-1} e^{-y/\beta},\ y > 0. \quad (A.11)$$

Y is known to have the inverse gamma distribution. □

Example A5. Polar transform of normal variables. Suppose X and Y are independent $N(0, 1)$ variables. Consider (x, y) as a point on the x–y plane and let (r, θ) be its polar coordinates, with

$$s = r^2 = x^2 + y^2,$$
$$\theta = \tan^{-1}\left(\frac{y}{x}\right). \quad (A.12)$$

Derive the PDF of (s, θ).

Note that the PDF of x and y is

$$f(x, y) = \frac{1}{2\pi} e^{-(x^2+y^2)/2},\ -\infty < x, y < \infty. \quad (A.13)$$

To calculate the Jacobi of the transform, note that $\frac{d\tan^{-1}(x)}{dx} = \frac{1}{1+x^2}$ and $\left|\frac{\partial(s,\theta)}{\partial(x,y)}\right| = \begin{vmatrix} 2x & 2y \\ -\frac{y}{x^2+y^2} & \frac{x}{x^2+y^2} \end{vmatrix} = 2$. Thus

$$f(s, \theta) = \frac{1}{2\pi} e^{-s/2} \times \frac{1}{2} = \frac{1}{2} e^{-s/2} \times \frac{1}{2\pi},\ 0 \le s < \infty, 0 \le \theta \le 2\pi. \quad (A.14)$$

This means that s is an exponential variable with mean 2, $\theta \sim U(0, 2\pi)$, and s and θ are independent. This transform is used to simulate $N(0, 1)$ variables. □

Example A6. Polar transform of uniform variables on the unit circle. Suppose x and y are uniform on the unit circle, so that the joint PDF is

$$f(x, y) = \frac{1}{\pi}, x^2 + y^2 \leq 1. \tag{A.15}$$

Find the PDF of r^2 and θ where (r, θ) is the polar coordinates, with

$$\begin{aligned} s &= r^2 = x^2 + y^2, \\ \theta &= \tan^{-1}\left(\frac{y}{x}\right). \end{aligned} \tag{A.16}$$

Derive the PDF of (s, θ).

As in Example A5, the Jacobi of the transform is 1/2. Thus

$$f(s, \theta) = \frac{1}{\pi} \times \frac{1}{2} = \frac{1}{2\pi}, 0 \leq s \leq 1, 0 \leq \theta \leq 2\pi. \tag{A.17}$$

Thus s and θ are independent, with $s \sim U(0, 1)$ and $\theta \sim U(0, 2\pi)$. □

Example A7. Prior distribution of sequence distance under the JC69 model. Suppose that the probability of different sites p between two sequences is assigned a uniform prior $f(p) = 4/3, 0 \leq p < 3/4$. What is the corresponding prior distribution of the sequence distance θ? Note that θ and p are related as

$$p = \frac{3}{4}\left(1 - e^{-\frac{4}{3}\theta}\right). \tag{A.18}$$

Thus $dp/d\theta = e^{-\frac{4}{3}\theta}$, so that the density of θ is $f(\theta) = \frac{4}{3}e^{-\frac{4}{3}\theta}, 0 \leq \theta < \infty$. Thus θ has an exponential distribution with mean $3/4$. Alternatively, suppose we assign a uniform prior on θ: $f(\theta) = 1/A, 0 \leq \theta \leq A$, where the upper bound A is a big number (say, 10 changes per site). Then $\frac{d\theta}{dp} = 1 / \left(\frac{dp}{d\theta}\right) = 1 / \left(1 - \frac{4p}{3}\right)$ and

$$f(p) = \frac{1}{A} / \left(1 - \frac{4}{3}p\right), 0 \leq p \leq \frac{3}{4}\left(1 - e^{-\frac{4}{3}A}\right). \tag{A.19}$$

Because p and θ do not have a linear relationship, it is impossible for them to have uniform distributions at the same time. Equivalently, we say that the prior is not invariant to reparametrization. □

Next we describe Theorem 2. In an MCMC algorithm, it is sometimes more convenient to propose changes to the state of the Markov chain using certain transformed variables instead of the original variables. Theorem 2 gives the proposal ratio for such a proposal.

Theorem 2. Suppose the original variables are $\boldsymbol{x} = \{x_1, x_2, \ldots, x_m\}$, but the proposal ratio is easier to calculate using transformed variables $\boldsymbol{y} = \{y_1, y_2, \ldots, y_m\}$, where $\boldsymbol{x}$ and $\boldsymbol{y}$ constitute a one-to-one mapping through $y_i = y_i(\boldsymbol{x})$ and $x_i = x_i(\boldsymbol{y}), i = 1, 2, \ldots, m$. Let $\boldsymbol{x}$ (or $\boldsymbol{y}$) be the current state of the Markov chain, and $\boldsymbol{x}^*$ (or $\boldsymbol{y}^*$) be the proposed state. Then

$$\frac{q(\boldsymbol{x}|\boldsymbol{x}^*)}{q(\boldsymbol{x}^*|\boldsymbol{x})} = \frac{q(\boldsymbol{y}|\boldsymbol{y}^*)}{q(\boldsymbol{y}^*|\boldsymbol{y})} \times \frac{|J(\boldsymbol{y}^*)|}{|J(\boldsymbol{y})|}. \tag{A.20}$$

The proposal ratio in the original variables $\boldsymbol{x}$ is the product of the proposal ratio in the transformed variables $\boldsymbol{y}$ and the ratio of the *Jacobian* determinants. To see this, note that

$$q_{\boldsymbol{Y}^*|\boldsymbol{Y}}(\boldsymbol{y}^*|\boldsymbol{y}) = q_{\boldsymbol{Y}^*|\boldsymbol{X}}(\boldsymbol{y}^*|\boldsymbol{x}) = q_{\boldsymbol{X}^*|\boldsymbol{X}}(\boldsymbol{x}^*|\boldsymbol{x}) \times |J(\boldsymbol{y}^*)|. \tag{A.21}$$

The first equation is because conditioning on y is equivalent to conditioning on x due to the one-to-one mapping. The second equation applies Theorem 1 to derive the density of y^* as a function of x^*, where $J(y^*) = \left|\frac{\partial x^*}{\partial y^*}\right|$ is the Jacobi determinant of the transform. Similarly, $q_{Y|Y^*}(y|y^*) = q_{X|X^*}(x|x^*) \times |J(y)|$. Equation (A.20) then follows.

Examples of applications of this theorem can be found in §7.2.

Appendix B. The delta technique

The delta technique is a general approach to deriving the means, variances, and covariances of functions of random variables. Suppose we are interested in the mean and variance of a function $g(x)$, where the random variable x has mean μ and variance σ^2. Note that if g is not a linear function of x, the mean value of a function is not equal to the function of the mean: $E(g(x)) \neq g(E(x))$. The Taylor expansion of g around the mean μ is

$$g = g(x) = g(\mu) + \frac{dg(\mu)}{dx} \cdot (x - \mu) + \frac{1}{2!} \cdot \frac{d^2 g(\mu)}{dx^2} \cdot (x - \mu)^2 + \cdots, \tag{A.22}$$

where the function $g(\cdot)$ and the derivatives are all evaluated at $x = \mu$; for example, $\frac{dg(\mu)}{dx} \equiv \frac{dg(x)}{dx}\big|_{x=\mu}$.

If we take the expectation on both sides and ignore terms of order 3 or higher, we get

$$E(g) \approx g(\mu) + \frac{1}{2} \frac{d^2 g(\mu)}{dx^2} \sigma^2, \tag{A.23}$$

because $E(x - \mu) = 0$ and $E(x - \mu)^2 = \sigma^2$. The derivatives are evaluated at $x = \mu$ and are thus constants when we take expectations over x. Similarly, the approximate variance of g is

$$\mathrm{var}(g) \approx E(g - E(g))^2 \approx \sigma^2 \cdot \left[\frac{dg(\mu)}{dx}\right]^2. \tag{A.24}$$

In statistical data analysis, x may be a parameter estimate, and g its function. Then μ and σ^2 may be replaced by their estimates from the dataset.

The multivariate version is similarly derived using a multivariate Taylor expansion. Suppose x is a random vector of n variables, and $y = h(x)$ is a function of x, with m elements. Then the approximate variance–covariance matrix of y is given as

$$\mathrm{var}(y) \approx J \cdot \mathrm{var}(x) \cdot J^\mathrm{T}, \tag{A.25}$$

where J is the $m \times n$ Jacobi matrix of the transform

$$J = \begin{pmatrix} \frac{\partial y_1}{\partial x_1} & \frac{\partial y_1}{\partial x_2} & \cdots & \frac{\partial y_1}{\partial x_n} \\ \frac{\partial y_2}{\partial x_1} & \frac{\partial y_2}{\partial x_2} & \cdots & \frac{\partial y_2}{\partial x_n} \\ \vdots & \vdots & \ddots & \vdots \\ \frac{\partial y_m}{\partial x_1} & \frac{\partial y_m}{\partial x_2} & \cdots & \frac{\partial y_m}{\partial x_n} \end{pmatrix}, \tag{A.26}$$

APPENDIX B. THE DELTA TECHNIQUE 447

and J^T is its transpose. In particular, the variance of a single-valued function $g(x)$ of x is approximately

$$\text{var}(g) \approx \sum_{i=1}^{n} \sum_{j=1}^{n} \text{cov}(x_i, x_j) \left(\frac{\partial g}{\partial x_i}\right) \left(\frac{\partial g}{\partial x_j}\right), \tag{A.27}$$

where $\text{cov}(x_i, x_j)$ is the covariance of x_i and x_j if $i \neq j$ or the variance of x_i if $i = j$. The mean of $g(\mathbf{x})$ is approximately

$$E(g) \approx g(\mu_1, \mu_2, \ldots, \mu_n) + \frac{1}{2} \sum_{i=1}^{n} \sum_{j=1}^{n} \text{cov}(x_i, x_j) \frac{\partial^2 g}{\partial x_i \partial x_j}, \tag{A.28}$$

where $\mu_i = E(x_i)$.

Example B1. The variance of the JC69 distance. Suppose x out of n sites are different between two sequences. Under JC69 the sequence distance is given as $\hat{d} = -\frac{3}{4} \log(1 - \frac{4}{3}\hat{p})$, where $\hat{p} = x/n$ is the proportion of different sites. This is a binomial proportion, with variance $\text{var}(\hat{p}) = \hat{p}(1-\hat{p})/n$. Consider $\hat{d}$ as a function of $\hat{p}$, and note that $d\hat{d}/d\hat{p} = \frac{1}{(1-4\hat{p}/3)}$. Thus the approximate variance of $\hat{d}$ is $\frac{\hat{p}(1-\hat{p})}{n(1-4\hat{p}/3)^2}$, as given in equation (1.8). □

Example B2. The expectation and variance of the ratio of two random variables. Suppose x and y are two random variables. Let $\mu_x = E(x)$, $\mu_y = E(y)$, $\sigma_x^2 = \text{var}(x)$, $\sigma_y^2 = \text{var}(y)$, and $\sigma_{xy} = \text{cov}(x,y)$. One can then use equations (A.28) and (A.27) to work out the approximate mean and variance of the ratio x/y as

$$E\left(\frac{x}{y}\right) \approx \frac{\mu_x}{\mu_y} - \frac{\sigma_{xy}}{\mu_y^2} + \frac{\mu_x \sigma_y^2}{\mu_y^3}, \tag{A.29}$$

$$\text{var}\left(\frac{x}{y}\right) \approx \frac{\sigma_x^2}{\mu_y^2} - \frac{2\mu_x \sigma_{xy}}{\mu_y^3} + \frac{\mu_x^2 \sigma_y^2}{\mu_y^4}. \tag{A.30}$$

These equations may be used to derive the mean and variance of the transition/transversion ratio κ under the K80 model. □

Example B3. The variance of the sequence distance under the K80 model. The maximum likelihood estimates (MLEs) of the sequence distance d and the transition/transversion rate ratio κ are given in equation (1.12), as functions of the proportions of sites with transitional and tranversional differences between two sequences (S and V). Note that S and V are multinomial proportions, with variance–covariance matrix

$$\text{var}\begin{pmatrix} S \\ V \end{pmatrix} = \begin{pmatrix} S(1-S)/n & -SV/n \\ -SV/n & V(1-V)/n \end{pmatrix}, \tag{A.31}$$

where n is the number of sites in the sequence. Now consider the estimates $\hat{d}$ and $\hat{\kappa}$ as functions of S and V, so that

$$\text{var}\begin{pmatrix} \hat{d} \\ \hat{\kappa} \end{pmatrix} = J \cdot \text{var}\begin{pmatrix} S \\ V \end{pmatrix} \cdot J^T, \tag{A.32}$$

where J is the Jacobi matrix for the transform

$$J = \begin{pmatrix} \dfrac{\partial \hat{d}}{\partial S} & \dfrac{\partial \hat{d}}{\partial V} \\ \dfrac{\partial \hat{\kappa}}{\partial S} & \dfrac{\partial \hat{\kappa}}{\partial V} \end{pmatrix}$$

$$= \begin{pmatrix} \dfrac{1}{1-2S-V} & \dfrac{1}{2(1-2V)} + \dfrac{1}{2(1-2S-V)} \\ -\dfrac{4}{(1-2S-V)\log(1-2V)} & -\dfrac{2}{(1-2S-V)\log(1-2V)} + \dfrac{4\log(1-2S-V)}{(1-2V)(\log(1-2V))^2} \end{pmatrix}.$$
(A.33)

In particular, the variance of $\hat{d}$ can be derived using equation (A.27), as

$$\mathrm{var}(\hat{d}) = \left(\dfrac{\partial \hat{d}}{\partial S}\right)^2 \mathrm{var}(S) + 2 \cdot \dfrac{\partial \hat{d}}{\partial S} \cdot \dfrac{\partial \hat{d}}{\partial V} \cdot \mathrm{cov}(S, V) + \left(\dfrac{\partial \hat{d}}{\partial V}\right)^2 \mathrm{var}(V) \quad \text{(A.34)}$$
$$= [a^2 S + b^2 V - (aS + bV)^2]/n,$$

as given in equation (1.14). □

Example B4. The variances of MLEs of d_S and d_N under a codon model (§2.5.2). Suppose the MLEs of t, κ, and ω have been obtained. Their variance–covariance matrix is given by the Hessian matrix, which can be calculated using the difference method or generated in the numerical optimization algorithm (see §4.5.2). We then view $\hat{d}_S$ and $\hat{d}_N$ as functions of $\hat{t}, \hat{\kappa}$, and $\hat{\omega}$. Let the Jacobi matrix for the transform be

$$J = \begin{pmatrix} \dfrac{\partial d_S}{\partial t} & \dfrac{\partial d_S}{\partial \kappa} & \dfrac{\partial d_S}{\partial \omega} \\ \dfrac{\partial d_N}{\partial t} & \dfrac{\partial d_N}{\partial \kappa} & \dfrac{\partial d_N}{\partial \omega} \end{pmatrix}, \quad \text{(A.35)}$$

evaluated at the MLEs. Then

$$\mathrm{var}\begin{pmatrix} \hat{d}_S \\ \hat{d}_N \end{pmatrix} = J \cdot \mathrm{var}\begin{pmatrix} \hat{t} \\ \hat{\kappa} \\ \hat{\omega} \end{pmatrix} \cdot J^{\mathrm{T}}. \quad \text{(A.36)}$$

The Jacobi matrix can be easily calculated using difference approximation. □

Appendix C. Phylogenetic software

Here we give an overview of several widely used programs or software packages in molecular phylogenetics. An almost exhaustive list has been compiled by Joseph Felsenstein, at http://evolution.gs.washington.edu/phylip/software.html. The easiest way of finding these programs is to type their names into a search engine.

CLUSTAL is a program for progressive alignment of multiple sequences (Thompson et al. 1994; Sievers et al. 2011). It does pairwise alignments using the Needleman–Wunsch algorithm (Needleman and Wunsch 1970) to calculate pairwise distances, which are used to reconstruct an NJ tree. It then uses the NJ tree as a guide to progressively align multiple sequences. Clustal is available at http://www.clustal.org/.

PRANK (Löytynoja and Goldman 2005, 2008), available at http://www.ebi.ac.uk/goldman-srv/prank/, is a probabilistic multiple alignment program for DNA, codon, and amino acid sequences. It treats insertions differently from other programs (arguably correctly) and avoids over-estimation of the number of deletion events. As a result, PRANK can produce very different alignments from other programs.

PHYLIP (Phylogeny Inference Package) is a package of about 30 C programs for parsimony, distance, and likelihood methods of phylogeny reconstruction, developed by Joseph Felsenstein. It is available at http://evolution.gs.washington.edu/phylip.html.

PAUP* 4 (Phylogenetic Analysis Using Parsimony* and other methods) is a program for phylogenetic analysis of molecular and morphological data using distance, parsimony, and likelihood methods, written by David Swofford and distributed by Sinauer Associates. It is available through the program web site at http://paup.csit.fsu.edu/.

MEGA (for Molecular Evolutionary Genetic Analysis) is a Windows program for downloading sequences from GenBank, and conducting alignment and phylogenetic analysis using distance, parsimony, and likelihood methods (Tamura et al. 2011). It has a graphical user interface. It is available at http://www.megasoftware.net/.

PAML (Phylogenetic Analysis by Maximum Likelihood) is my package for likelihood analysis of nucleotide, amino acid, and codon sequences (Yang 2007b). It is not good for making trees, but implements many substitution models, and can be used to reconstruct ancestral sequences, detect positive selection, and estimate species divergence times under relaxed molecular clock models. It is available at http://abacus.gene.ucl.ac.uk/software/paml.html.

PHYML is a fast ML tree search program (Guindon and Gascuel 2003). Executables for various platforms are available at http://code.google.com/p/phyml/.

RAxML is an efficient ML tree search program (Stamatakis et al. 2012). The parallel versions implemented using MPI and Pthreads are especially powerful and can be used to analyse large datasets. It is available at http://sco.h-its.org/exelixis/software.html.

MrBayes is a Bayesian MCMC program for phylogenetic inference using nucleotide, amino acid, and codon sequences. It is available at http://mrbayes.sourceforge.net/ (Ronquist et al. 2012b). It can analyse multiple heterogeneous datasets while accommodating their differences.

BEAST is a Bayesian MCMC program for phylogenetic analysis of molecular sequences (Drummond et al. 2006; Drummond and Rambaut 2007). It implements strict or relaxed molecular clock models and works on rooted trees only. It is available at http://beast.bio.ed.ac.uk/.

PhyloBayes is a Bayesian MCMC program for phylogenetic reconstruction. It includes the CAT model and its variants (Lartillot and Philippe 2004; Lartillot et al. 2009), which may be important for inferring deep phylogenies.

FigTree is a graphical program for viewing and printing trees, written by Andrew Rambaut. It is available at http://tree.bio.ed.ac.uk/software/figtree/.

References

Ababneh, F., L. S. Jermiin, C. Ma, and J. Robinson. 2006. Matched-pairs tests of homogeneity with applications to homologous nucleotide sequences. *Bioinformatics* **22**, 1225–1231.

Abascal, F., D. Posada, and R. Zardoya. 2007. MtArt: a new model of amino acid replacement for Arthropoda. *Mol. Biol. Evol.* **24**, 1–5.

Abramowitz, M., and I. A. Stegun. 1972. *Handbook of Mathematical Functions*. Dover, New York.

Adachi, J., and M. Hasegawa. 1996a. MolPhy Version 2.3: programs for molecular phylogenetics based on maximum likelihood. *Computer Science Monographs* **28**, 1–150.

Adachi, J., and M. Hasegawa. 1996b. Model of amino acid substitution in proteins encoded by mitochondrial DNA. *J. Mol. Evol.* **42**, 459–468.

Adachi, J., P. J. Waddell, W. Martin, and M. Hasegawa. 2000. Plastid genome phylogeny and a model of amino acid substitution for proteins encoded by chloroplast DNA. *J. Mol. Evol.* **50**, 348–358.

Akaike, H. 1974. A new look at the statistical model identification. *IEEE Trans. Autom. Contr.* AC **19**, 716–723.

Akam, M. 1995. Hox genes and the evolution of diverse body plans. *Philos. Trans. R. Soc. Lond. B. Biol. Sci.* **349**, 313–319.

Akashi, H. 1994. Synonymous codon usage in Drosophila melanogaster: natural selection and translational accuracy. *Genetics* **136**, 927–935.

Akashi, H. 1995. Inferring weak selection from patterns of polymorphism and divergence at 'silent' sites in Drosophila DNA. *Genetics* **139**, 1067–1076.

Akashi, H. 1999a. Within- and between-species DNA sequence variation and the 'footprint' of natural selection. *Gene* **238**, 39–51.

Akashi, H. 1999b. Inferring the fitness effects of DNA mutations from polymorphism and divergence data: statistical power to detect directional selection under stationarity and free recombination. *Genetics* **151**, 221–238.

Akashi, H., P. Goel, and A. John. 2007. Ancestral state inference and the study of codon bias evolution: implications for molecular evolutionary analysis of the *Drosophila melanogaster* subgroup. *PloS One* **2**, e1065.

Akerborg, O., B. Sennblad, and J. Lagergren. 2008. Birth-death prior on phylogeny and speed dating. *BMC Evol. Biol.* **8**, 77.

Alba, R., P. M. Kelmenson, M.-M. Cordonnier-Pratt, and L. H. Pratt. 2000. The phytochrome gene family in tomato and the rapid differential evolution of this family in angiosperms. *Mol. Biol. Evol.* **17**, 362–373.

Albert, V. A. 2005. *Parsimony, Phylogeny, and Genomics*. Oxford University Press, Oxford, UK.

Aldous, D. J. 2001. Stochastic models and descriptive statistics for phylogenetic trees. *Stat. Sci.* **16**, 23–34.

Alfaro, M. E., and M. T. Holder. 2006. The posterior and the prior in Bayesian phylogenetics. *Ann. Rev. Ecol. Syst.* **37**, 19–42.

Allman, E. S., and J. A. Rhodes. 2006. The identifiability of tree topology for phylogenetic models, including covarion and mixture models. *J. Comput. Biol.* **13**, 1101–1113.

Allman, E. S., C. Ane, and J. A. Rhodes. 2008. Identifiability of a Markovian model of molecular evolution with gamma-distributed rates. *Adv. Appl. Prob.* **40**, 228–249.

Altekar, G., S. Dwarkadas, J. P. Huelsenbeck, and F. Ronquist. 2004. Parallel Metropolis coupled Markov chain Monte Carlo for Bayesian phylogenetic inference. *Bioinformatics* **20**, 407–415.

Andersen, L. N., T. Mailund, and A. Hobolth. 2014. Efficient computation in the IM model. *J. Math. Biol.* in press.

Andolfatto, P. 2005. Adaptive evolution of non-coding DNA in Drosophila. *Nature* **437**, 1149–1152.

Ané, C., B. Larget, D. A. Baum et al. 2007. Bayesian estimation of concordance among gene trees. *Mol. Biol. Evol.* **24**, 412–426.

Anisimova, M. 2012. Parametric models of codon substitution. Pp. 12–33 *in* G. Cannarozzi, and A. Schneider, eds. *Codon Evolution: Mechanisms and Models*. Oxford University Press, New York.

Anisimova, M., and O. Gascuel. 2006. Approximate likelihood ratio test for branches: a fast, accurate and powerful alternative. *Syst. Biol.* **55**, 539–552.

Anisimova, M., and C. Kosiol. 2009. Investigating protein-coding sequence evolution with probabilistic codon substitution models. *Mol. Biol. Evol.* **26**, 255–271.

Anisimova, M., and C. Kosiol. 2012. Selection on the protein-coding genome. Pp. 113–140 *in* M. Anisimova, ed. *Evolutionary Genomics: Statistical and Computational Methods, Volume 2*. Springer, New York.

Anisimova, M., and D. A. Liberles. 2007. The quest for natural selection in the age of comparative genomics. *Heredity*. **99**: 567–579.

Anisimova, M., and Z. Yang. 2007. Multiple hypothesis testing to tetect adaptive protein evolution affecting individual branches and sites. *Mol. Biol. Evol.* **24**, 1219–1228.

Anisimova, M., J. P. Bielawski, and Z. Yang. 2001. The accuracy and power of likelihood ratio tests to detect positive selection at amino acid sites. *Mol. Biol. Evol.* **18**, 1585–1592.

Anisimova, M., J. P. Bielawski, and Z. Yang. 2002. Accuracy and power of Bayes prediction of amino acid sites under positive selection. *Mol. Biol. Evol.* **19**, 950–958.

Anisimova, M., R. Nielsen, and Z. Yang. 2003. Effect of recombination on the accuracy of the likelihood method for detecting positive selection at amino acid sites. *Genetics* **164**, 1229–1236.

Antoniak, C. E. 1974. Mixtures of Dirichlet processes with applications to Bayesian nonparametric problems. *Ann. Stat.* **2**, 1152–1174.

Aris-Brosou, S. 2007. Dating phylogenies with hybrid local molecular clocks. *PLOS One* **2**, e879.

Atchadé, Y. F., G. O. Roberts, and J. S. Rosenthal. 2011. Towards optimal scaling of metropolis-coupled Markov chain Monte Carlo. *Stat. Comput.* **21**, 555–568.

Atkinson, A. C. 1970. A method of discriminating between models. *J. R. Stat. Soc. B*. **32**, 323–353.

Avise, J. C. 2000. *Phylogeography: The History and Formation of Species*. Harvard University Press, Cambridge, Massachusetts.

Baele, G., P. Lemey, T. Bedford et al. 2012. Improving the accuracy of demographic and molecular clock model comparison while accommodating phylogenetic uncertainty. *Mol. Biol. Evol.* **29**, 2157–2167.

Bahlo, M., and R. C. Griffiths. 2000. Inference from gene trees in a subdivided population. *Theor. Popul. Biol.* **57**, 79–95.

Barker, D. 2004. LVB: parsimony and simulated annealing in the search for phylogenetic trees. *Bioinformatics* **20**, 274–275.

Barrier, M., R. H. Robichaux, and M. D. Purugganan. 2001. Accelerated regulatory gene evolution in an adaptive radiation. *Proc. Natl. Acad. Sci. U.S.A.* **98**, 10208–10213.

Barry, D., and J. A. Hartigan. 1987a. Statistical analysis of hominoid molecular evolution. *Stat. Sci.* **2**, 191–210.

Barry, D., and J. A. Hartigan. 1987b. Asynchronous distance between homologous DNA sequences. *Biometrics* **43**, 261–276.

Baudry, E., and F. Depaulis. 2003. Effect of misoriented sites on neutrality tests with outgroup. *Genetics* **165**, 1619–1622.

Bauer, A. M., J. F. Parham, R. M. Brown et al. 2011. Availability of new Bayesian-delimited gecko names and the importance of character-based species descriptions. *Proc. R. Soc. Lond. B. Biol. Sci.* **278**, 490–492.

Baum, D. A. 1992. Phylogenetic species concepts. *Trends Ecol. Evol.* **7**, 1–2.

Baum, D. A. 2007. Concordance trees, concordance factors, and the exploration of reticulate genealogy. *Taxon* **56**, 417–426.

Baum, D. A., and K. L. Shaw. 1995. Genealogical perspectives on the species problem. Pp. 289–303 *in* P. C. Hoch, and A. G. Stephenson, eds. *Molecular and Experimental Approaches to Plant Biosystematics*. Missouri Botanical Garden, St. Louis.

Bayes, T. 1763. An essay towards solving a problem in the doctrine of chance, with an Introduction and an Appendix by Richard Price. *Philos. Trans. R. Soc. Lond.* **53**, 370–418.

Beaumont, M. A. 1999. Detecting population expansion and decline using microsatellites. *Genetics* **153**, 2013–2029.

Beerli, P. 2004. Effect of unsampled populations on the estimation of population sizes and migration rates between sampled populations. *Mol. Ecol.* **13**, 827–836.

Beerli, P. 2006. Comparison of Bayesian and maximum-likelihood inference of population genetic parameters. *Bioinformatics* **22**, 341–345.

Beerli, P., and J. Felsenstein. 1999. Maximum-likelihood estimation of migration rates and effective population numbers in two populations using a coalescent approach. *Genetics* **152**, 763–773.

Beerli, P., and J. Felsenstein. 2001. Maximum likelihood estimation of a migration matrix and effective population sizes in n subpopulations by using a coalescent approach. *Proc. Natl. Acad. Sci. U.S.A.* **98**, 4563–4568.

Benner, S. A. 2001. Natural progression. *Nature* **409**, 459.

Benner, S. A. 2002. The past as the key to the present: resurrection of ancient proteins from eosinophils. *Proc. Natl. Acad. Sci. U.S.A.* **99**, 4760–4761.

Benton, M. J., P. C. J. Donoghue, and R. J. Asher. 2009. Calibrating and constraining molecular clocks. Pp. 35–86 *in* B. S. Hedges, and S. Kumar, eds. *The Timetree of Life*. Oxford University Press, Oxford, UK.

Berg, B. A., and T. Neuhaus. 1991. Multicanonical algorithms for 1st order phase-transitions. *Phys. Lett. B* **267**, 249–253.

Berger, J. O., and J. M. Bernardo. 1992. On the development of reference priors (with Discussion). Pp. 35–60 *in* J. M. Bernardo, J. O. Berger, D. V. Lindley, and A. F. M. Smith, eds. *Bayesian Statistics 4*. Oxford University Press, Oxford, UK.

Berger, J. O., J. M. Bernardo, and D. Sun. 2009. The formal definition of reference priors. *Ann. Stat.* **37**, 905–938.

Bernardo, J. M. 1979. Reference posterior distributions for Bayesian inference. *J. R. Stat. Soc. B* **41**, 113–147.

Bernardo, J. M. 2005. Reference analysis. *Handb. Stat.* **25**, 17–90.

Berry, I. M., R. Ribeiro, M. Kothari et al. 2007. Unequal evolutionary rates in the human immunodeficiency virus type 1 (HIV-1) pandemic: the evolutionary rate of HIV-1 slows down when the epidemic rate increases. *J. Virol.* **81**, 10625–10635.

Berry, V., and O. Gascuel. 1996. On the interpretation of bootstrap trees: appropriate threshold of clade selection and induced gain. *Mol. Biol. Evol.* **13**, 999–1011.

Besag, J., and P. J. Green. 1993. Spatial statistics and Bayesian computation. *J. R. Stat. Soc. B* **55**, 25–37.

Bielawski, J. P., and J. R. Gold. 2002. Mutation patterns of mitochondrial H- and L-strand DNA in closely related Cyprinid fishes. *Genetics* **161**, 1589–1597.

Bielawski, J. P., and Z. Yang. 2001. Positive and negative selection in the DAZ gene family. *Mol. Biol. Evol.* **18**, 523–529.

Bielawski, J. P., and Z. Yang. 2004. A maximum likelihood method for detecting functional divergence at individual codon sites, with application to gene family evolution. *J. Mol. Evol.* **59**, 121–132.

Bielawski, J. P., K. Dunn, and Z. Yang. 2000. Rates of nucleotide substitution and mammalian nuclear gene evolution: approximate and maximum-likelihood methods lead to different conclusions. *Genetics* **156**, 1299–1308.

Bielawski, J. P., K. A. Dunn, G. Sabehi, and O. Beja. 2004. Darwinian adaptation of proteorhodopsin to different light intensities in the marine environment. *Proc. Natl. Acad. Sci. U.S.A.* **101**, 14824–14829.

Bierne, N., and A. Eyre-Walker. 2003. The problem of counting sites in the estimation of the synonymous and nonsynonymous substitution rates: implications for the correlation between the synonymous substitution rate and codon usage bias. *Genetics* **165**, 1587–1597.

Bininda-Emonds, O. R. P. 2004. *Phylogenetic Supertrees: Combining Information to Reveal the Tree of Life*. Kluwer Academic, Dordrecht, the Netherlands.

Bishop, M. J., and A. E. Friday. 1985. Evolutionary trees from nucleic acid and protein sequences. *Proc. R. Soc. Lond. B. Biol. Sci.* **226**, 271–302.

Bishop, M. J., and A. E. Friday. 1987. Tetropad relationships: the molecular evidence. Pp. 123–139 in C. Patterson, ed. *Molecules and Morphology in Evolution: Conflict or Compromise?* Cambridge University Press, Cambridge, UK.

Bishop, M. J., and E. A. Thompson. 1986. Maximum likelihood alignment of DNA sequences. *J. Mol. Biol.* **190**, 159–165.

Bjorklund, M. 1999. Are third positions really that bad? A test using vertebrate cytochrome *b*. *Cladistics* **15**, 191–197.

Bjorkman, P. J., S. A. Saper, B. Samraoui et al. 1987a. Structure of the class I histocompatibility antigen, HLA-A2. *Nature* **329**, 506–512.

Bjorkman, P. J., S. A. Saper, B. Samraoui et al. 1987b. The foreign antigen binding site and T cell recognition regions of class I histocompatibility antigens. *Nature* **329**, 512–518.

Blanquart, S., and N. Lartillot. 2006. A Bayesian compound stochastic process for modeling nonstationary and nonhomogeneous sequence evolution. *Mol. Biol. Evol.* **23**, 2058–2071.

Blanquart, S., and N. Lartillot. 2008. A site- and time-heterogeneous model of amino acid replacement. *Mol. Biol. Evol.* **25**, 842–858.

Bollback, J. P. 2006. SIMMAP: stochastic character mapping of discrete traits on phylogenies. *BMC Bioinformatics* **7**, 88.

Bouchard-Coté, A., S. Sankararaman, and M. I. Jordan. 2012. Phylogenetic inference via sequential Monte Carlo. *Syst. Biol.* **61**, 579–593.

Bourlat, S. J., T. Juliusdottir, C. J. Lowe et al. 2006. Deuterostome phylogeny reveals monophyletic chordates and the new phylum Xenoturbellida. *Nature* **444**, 85–88.

Boussau, B., and M. Gouy. 2006. Efficient likelihood computations with nonreversible models of evolution. *Syst. Biol.* **55**, 756–768.

Box, G. E. P. 1979. Robustness in the strategy of scientific model building. Pp. 202 in R. L. Launer, and G. N. Wilkinson, eds. *Robustness in Statistics*. Academic Press, New York.

Box, G. E. P., and M. E. Muller. 1958. A note on the generation of random normal deviates. *Ann. Math. Stat.* **29**, 610–611.

Brandley, M. C., A. D. Leqach, D. L. Warren, and J. A. McGruire. 2006. Are unequal clade priors problematic for Bayesian phylogenetics? *Syst. Biol.* **55**, 158–146.

Braverman, J. M., R. R. Hudson, N. L. Kaplan et al. 1995. The hitchhiking effect on the site frequency spectrum of DNA polymorphisms. *Genetics* **140**, 783–796.

Bremer, K. 1988. The limits of amino acid sequence data in angiosperm phylogenetic reconstruction. *Evolution* **42**, 795–803.

Brenner, S. 1988. The molecular evolution of genes and proteins: a tale of two serines. *Nature* **334**, 528–530.

Brent, R. P. 1973. *Algorithms for Minimization without Derivatives*. Prentice-Hall Inc., Englewood Cliffs, New Jersey.

Brinkmann, H., M. van der Giezen, Y. Zhou et al. 2005. An empirical assessment of long-branch attraction artefacts in deep eukaryotic phylogenomics. *Syst. Biol.* **54**, 743–757.

Britten, R. J. 1986. Rates of DNA sequence evolution differ between taxonomic groups. *Science* **231**, 1393–1398.

Britton, T. 2005. Estimating divergence times in phylogenetic trees without a molecular clock. *Syst. Biol.* **54**, 500–507.

Bromham, L. 2011. The genome as a life-history character: why rate of molecular evolution varies between mammal species. *Phil. Trans. R. Soc. B: Biol. Sci.* **366**, 2503–2513.

Bromham, L., and D. Penny. 2003. The modern molecular clock. *Nat. Rev. Genet.* **4**, 216–224.

Bromham, L., D. Penny, A. Rambaut, and M. D. Hendy. 2000. The power of relative-rates tests depends on the data. *J. Mol. Evol.* **50**, 296–301.

Brooks, S. P., P. Giudici, and G. O. Roberts. 2003. Efficient construction of reversible jump Markov chain Monte Carlo proposal distributions. *J. R. Stat. Soc. B.* **65**, 3–39.

Brown, J. M., and A. R. Lemmon. 2007. The importance of data partitioning and the utility of Bayes factors in Bayesian phylogenetics. *Syst. Biol.* **56**, 643–655.

Brown, J. M., S. M. Hedtke, A. R. Lemmon, and E. M. Lemmon. 2010. When trees grow too long: investigating the causes of highly inaccurate Bayesian branch-length estimates. *Syst. Biol.* **59**, 145–161.

Brown, W. M., E. M. Prager, A. Wang, and A. C. Wilson. 1982. Mitochondrial DNA sequences of primates: tempo and mode of evolution. *J. Mol. Evol.* **18**, 225–239.

Brunet, M., F. Guy, D. Pilbeam et al. 2002. A new hominid from the upper Miocene of Chad, central Africa. *Nature* **418**, 145–151.

Bruno, W. J. 1996. Modeling residue usage in aligned protein sequences via maximum likelihood. *Mol. Biol. Evol.* **13**, 1368–1374.

Bruno, W. J., and A. L. Halpern. 1999. Topological bias and inconsistency of maximum likelihood using wrong models. *Mol. Biol. Evol.* **16**, 564–566.

Bruno, W. J., N. D. Socci, and A. L. Halpern. 2000. Weighted neighbor joining: a likelihood-based approach to distance-based phylogeny reconstruction. *Mol. Biol. Evol.* **17**, 189–197.

Bryant, D. 2003. A classification of consensus methods for phylogenetics. Pp. 163–184 *in* M. Janowitz, F.-J. Lapointe, F. R. McMorris, B. Mirkin, and F. S. Roberts, eds. *BioConsensus, DIMACS Series in Discrete Mathematics and Theoretical Computer Science*. American Mathematical Society, Providence, Rhode Island.

Bryant, D., and P. J. Waddell. 1998. Rapid evaluation of least-squares and minimum-evolution criteria on phylogenetic trees. *Mol. Biol. Evol.* **15**, 1346–1359.

Bryant, D., R. Bouckaert, J. Felsenstein et al. 2012. Inferring species trees directly from biallelic genetic markers: bypassing gene trees in a full coalescent analysis. *Mol. Biol. Evol.* **29**, 1917–1932.

Bulmer, M. G. 1990. Estimating the variability of substitution rates. *Genetics* **123**, 615–619.

Burgess, R., and Z. Yang. 2008. Estimation of hominoid ancestral population sizes under Bayesian coalescent models incorporating mutation rate variation and sequencing errors. *Mol. Biol. Evol.* **25**, 1979–1994.

Burridge, C. P., D. Craw, and J. M. Waters. 2006. River capture, range expansion, and cladogenesis: the genetic signature of freshwater vicariance. *Evolution* **60**, 1038–1049.

Bustamante, C. D., R. Nielsen, and D. L. Hartl. 2003. Maximum likelihood and Bayesian methods for estimating the distribution of selective effects among classes of mutations using DNA polymorphism data. *Theor. Popul. Biol.* **63**, 91–103.

Bustamante, C. D., J. Wakeley, S. Sawyer, and D. L. Hartl. 2001. Directional selection and the site-frequency spectrum. *Genetics* **159**, 1779–1788.

Bustamante, C. D., R. Nielsen, S. A. Sawyer et al. 2002. The cost of inbreeding in Arabidopsis. *Nature* **416**, 531–534.

Butler, G., M. D. Rasmussen, M. F. Lin et al. 2009. Evolution of pathogenicity and sexual reproduction in eight Candida genomes. *Nature* **459**, 657–662.

Calderhead, B., and M. Girolami. 2009. Estimating Bayes factors via thermodynamic integration and population MCMC. *Comput. Stat. Data Analysis* **48**, 4028–4045.

Camin, J. H., and R. R. Sokal. 1965. A method for deducing branching sequences in phylogeny. *Evolution* **19**, 311–326.

Cannarozzi, G., and A. Schneider. 2012. *Codon Evolution: Mechanisms and Models*. Oxford University Press, New York.

Cao, Y., K. S. Kim, J. H. Ha, and M. Hasegawa. 1999. Model dependence of the phylogenetic inference: relationship among Carnivores, Perissodactyls and Cetartiodactyls as inferred from mitochondrial genome sequences. *Genes Genet. Syst.* **74**, 211–217.

Cao, Y., J. Adachi, A. Janke et al. 1994. Phylogenetic relationships among eutherian orders estimated from inferred sequences of mitochondrial proteins: instability of a tree based on a single gene. *J. Mol. Evol.* **39**, 519–527.

Cao, Y., A. Janke, P. J. Waddell et al. 1998. Conflict among individual mitochondrial proteins in resolving the phylogeny of eutherian orders. *J. Mol. Evol.* **47**, 307–322.

Carlin, B. P., and S. Chib. 1995. Bayesian model choice through Markov chain Monte Carlo. *J. R. Stat. Soc. B* **57**, 473–483.

Carlin, B. P., and T. A. Louis. 2000. *Bayes and Empirical Bayes Methods for Data Analysis*. Chapman and Hall, London.

Carroll, S. B. 1995. Homeotic genes and the evolution of the arthropods and chordates. *Nature* **376**, 479–485.

Carroll, S. B. 2008. Evo-devo and an expanding evolutionary synthesis: a genetic theory of morphological evolution. *Cell* **134**, 25–36.

Carstens, B. C., and T. A. Dewey. 2010. Species delimitation using a combined coalescent and information-theoretic approach: an example from North American Myotis bats. *Syst. Biol.* **59**, 400–414.

Cartwright, R. A. 2005. DNA assembly with gaps (Dawg): simulating sequence evolution. *Bioinformatics* **21**, iii31–38.

Cavalli-Sforza, L. L., and A. W. F. Edwards. 1966. Estimation procedures for evolutionary branching processes. *Bull. Int. Stat. Inst.* **21**, 803–808.

Cavalli-Sforza, L. L., and A. W. F. Edwards. 1967. Phylogenetic analysis: models and estimation procedures. *Evolution* **21**, 550–570.

Cavender, J. A. 1978. Taxonomy with confidence. *Math. Biosci.* **40**, 271–280.

Chang, B. S., and M. J. Donoghue. 2000. Recreating ancestral proteins. *Trends Ecol. Evol.* **15**, 109–114.

Chang, J. T. 1996a. Full reconstruction of Markov models on evolutionary trees: identifiability and consistency. *Math. Biosci.* **137**, 51–73.

Chang, J. T. 1996b. Inconsistency of evolutionary tree topology reconstruction methods when substitution rates vary across characters. *Math. Biosci.* **134**, 189–215.

Charleston, M. A. 1995. Toward a characterization of landscapes of combinatorial optimization problems, with special attention to the phylogeny problem. *J. Comput. Biol.* **2**, 439–450.

Chen, F.-C., and W.-H. Li. 2001. Genomic divergences between humans and other Hominoids and the effective population size of the common ancestor of humans and chimpanzees. *Am. J. Hum. Genet.* **68**, 444–456.

Chen, M.-H., and Q.-M. Shao. 1999. Monte Carlo estimation of Bayesian credible and HPD intervals. *J. Comput. Graph. Stat.* **8**, 69–92.

Chen, M.-H., L. Kuo, and P. Lewis. 2014. *Bayesian Phylogenetics: Methods, Algorithms, and Applications*. Chapman & Hall/CRC, London. in press.

Cheon, S., and F. Liang. 2009. Bayesian phylogeny analysis via stochastic approximation Monte Carlo. *Mol. Phylogenet. Evol.* **53**, 394–403.

Chernoff, H. 1954. On the distribution of the likelihood ratio. *Ann. Math. Stat.* **25**, 573–578.

Chor, B., and S. Snir. 2004. Molecular clock fork phylogenies: closed form analytic maximum likelihood solutions. *Syst. Biol.* **53**, 963–967.

Chor, B., B. R. Holland, D. Penny, and M. D. Hendy. 2000. Multiple maxima of likelihood in phylogenetic trees: an analytic approach. *Mol. Biol. Evol.* **17**, 1529–1541.

Chung, Y., and C. Ané. 2011. Comparing two Bayesian methods for gene tree/species tree reconstruction: simulations with incomplete lineage sorting and horizontal gene transfer. *Syst. Biol.* **60**, 261–275.

Clark, B. 1970. Selective constraints on amino-acid substitutions during the evolution of proteins. *Nature* **228**, 159–160.

Clark, N. L., J. E. Aagaard, and W. J. Swanson. 2006. Evolution of reproductive proteins from animals and plants. *Reproduction* **131**, 11–22.

Collins, T. M., P. H. Wimberger, and G. J. P. Naylor. 1994. Compositional bias, character-state bias, and character-state reconstruction using parsimony. *Syst. Biol.* **43**, 482–496.

Comeron, J. M. 1995. A method for estimating the numbers of synonymous and nonsynonymous substitutions per site. *J. Mol. Evol.* **41**, 1152–1159.

Cooper, A., and R. Fortey. 1998. Evolutionary explosions and the phylogenetic fuse. *Trends Ecol. Evol.* **13**, 151–156.

Cox, D. R. 1961. Tests of separate families of hypotheses. *Proc. 4th Berkeley Symp. Math. Stat. Prob.* **1**, 105–123.

Cox, D. R. 1962. Further results on tests of separate families of hypotheses. *J. R. Stat. Soc. B.* **24**, 406–424.

Cox, D. R., and D. V. Hinkley. 1974. *Theoretical Statistics*. Chapman and Hall, London.

Coyne, J. A., and H. A. Orr. 2004. *Speciation*. Sinauer Assoc., Sunderland, Massachusetts.

Cranston, K. A., B. Hurwitz, D. Ware et al. 2009. Species trees from highly incongruent gene trees in rice. *Syst. Biol.* **58**, 489–500.

Crawford, N. G., B. C. Faircloth, J. E. McCormack et al. 2012. More than 1000 ultraconserved elements provide evidence that turtles are the sister group of archosaurs. *Biol. Lett.* **8**, 783–786.

Cummings, M. P., S. P. Otto, and J. Wakeley. 1995. Sampling properties of DNA sequence data in phylogenetic analysis. *Mol. Biol. Evol.* **12**, 814–822.

Cummings, M. P., S. A. Handley, D. S. Myers et al. 2003. Comparing bootstrap and posterior probability values in the four-taxon case. *Syst. Biol.* **52**, 477–487.

Cutler, D. J. 2000. Understanding the overdispersed molecular clock. *Genetics* **154**, 1403–1417.

Dagan, T., Y. Talmor, and D. Graur. 2002. Ratios of radical to conservative amino acid replacement are affected by mutational and compositional factors and may not be indicative of positive Darwinian selection. *Mol. Biol. Evol.* **19**, 1022–1025.

Dalquen, D. A., M. Anisimova, G. H. Gonnet, and C. Dessimoz. 2012. ALF: a simulation framework for genome evolution. *Mol. Biol. Evol.* **29**, 1115–1123.

Dasmahapatra, K. K., G. Iamas, F. Simpson, and J. Mallet. 2010. The anatomy of a 'suture zone' in Amazonian butterflies: a coalescent-based test for vicariant geographic divergence and speciation. *Mol. Ecol.* **19**, 4283–4301.

Datta, G. S., and M. Ghosh. 1996. On the invariance of noninformative priors. *Ann. Stat.* **24**, 141–159.

Davison, A. C., and D. V. Hinkley. 1997. *Bootstrap Methods and their Application*. Cambridge University Press, , Cambridge, UK.

Dawid, A. P. 1992. Prequential analysis, stochastic complexity and Bayesian inference (with Discussion). Pp. 109–125 *in* J. M. Bernardo, J. O. Berger, A. P. Dawid, and A. F. M. Smith, eds. *Bayesian Statistics*. Clarendon Press, Oxford, UK.

Dawkins, R., and J. R. Krebs. 1979. Arms races between and within species. *Proc. R. Soc. Lond. B. Biol. Sci.* **205**, 489–511.

Dayhoff, M. O., R. V. Eck, and C. M. Park. 1972. Evolution of a complex system: the immunoglobulins. Pp. 31–40. *Atlas of Protein Sequence and Structure*. National Biomedical Research Foundation, Maryland.

Dayhoff, M. O., R. M. Schwartz, and B. C. Orcutt. 1978. A model of evolutionary change in proteins. Pp. 345–352. *Atlas of Protein Sequence and Structure, Volume 5, Suppl. 3*. National Biomedical Research Foundation, Washington DC.

De Queiroz, K. 2007. Species concepts and species delimitation. *Syst. Biol.* **56**, 879–886.

Dean, A. M., and J. W. Thornton. 2007. Mechanistic approaches to the study of evolution: the functional synthesis. *Nat. Rev. Genet.* **8**, 675–688.

DeBry, R. 2001. Improving interpretation of the decay index for DNA sequences. *Syst. Biol.* **50**, 742–752.

DeBry, R. W. 1992. The consistency of several phylogeny-inference methods under varying evolutionary rates. *Mol. Biol. Evol.* **9**, 537–551.

Deely, J. J., and D. V. Lindley. 1981. Bayes empirical Bayes. *J. Am. Stat. Assoc.* **76**, 833–841.

Degnan, J. H., and N. A. Rosenberg. 2006. Discordance of species trees with their most likely gene trees. *PLoS Genet.* **2**, e68.

Degnan, J. H., and N. A. Rosenberg. 2009. Gene tree discordance, phylogenetic inference and the multispecies coalescent. *Trends Ecol. Evol.* **24**, 332–340.

Degnan, J. H., and L. A. Salter. 2005. Gene tree distributions under the coalescent process. *Evolution* **59**, 24–37.

Degnan, J. H., N. A. Rosenberg, and T. Stadler. 2012. The probability distribution of ranked gene trees on a species tree. *Math. Biosci.* **235**, 45–55.

DeGroot, M. H., and M. J. Schervish. 2002. *Probability and Statistics*. Addison-Wesley, Boston, Massachusetts.

Delson, E., I. Tattersall, J. A. Van Couvering, and A. S. Brooks. 2000. Pp. 166–171 *in* E. Delson, I. Tattersall, J. A. Van Couvering, and A. S. Brooks, eds. *Encyclopedia of Human Evolution and Prehistory*. Garland, New York.

Desper, R., and O. Gascuel. 2002. Fast and accurate phylogeny reconstruction algorithms based on the minimum-evolution principle. *J. Comput. Biol.* **9**, 687–705.

Desper, R., and O. Gascuel. 2004. Theoretical foundation of the balanced minimum evolution method of phylogenetic inference and its relationship to weighted least-squares tree fitting. *Mol. Biol. Evol.* **21**, 587–598.

Desper, R., and O. Gascuel. 2005. The minimum-evolution distance-based approach to phylogenetic inference. Pp. 1–32 *in* O. Gascuel, ed. *Mathematics of Evolution and Phylogeny*. Oxford University Press, Oxford, UK.

Dieckmann, U., and M. Doebeli. 1999. On the origin of species by sympatric speciation. *Nature* **400**, 354–357.

Diggle, P. J. 1990. *Time Series: A Biostatistical Introduction*. Oxford University Press, Oxford, UK.

Dimmic, M. W., J. S. Rest, D. P. Mindell, and R. A. Goldstein. 2002. rtREV: an amino acid substitution matrix for inference of retrovirus and reverse transcriptase phylogeny. *J. Mol. Evol.* **55**, 65–73.

Dobzhansky, T. G. 1937. *Genetics and the Origin of Species*. Columbia University, New York.

Donnelly, P., and S. Tavaré. 1997. *Progress in Population Genetics and Human Evolution*. Springer-Verlag, New York.

Donnelly, P., and S. Tavaré. 2005. Coalescents and genealogical structure under neutrality. *Ann. Rev. Genet.* **29**, 401–421.

Doolittle, F. W. 1998. You are what you eat: a gene transfer ratchet could account for bacterial genes in eukaryotic nuclear genomes. *Trends Genet.* **14**, 307–311.

Doolittle, R. F., and B. Blomback. 1964. Amino-acid sequence investigations of fibrinopeptides from various mammals: evolutionary implications. *Nature* **202**, 147–152.

Dornburg, A., F. Santini, and M. E. Alfaro. 2008. The influence of model averaging on clade posteriors: an example using the triggerfishes (Family Balistidae). *Syst. Biol.* **57**, 905–919.

Doron-Faigenboim, A., and T. Pupko. 2007. A combined empirical and mechanistic codon model. *Mol. Biol. Evol.* **24**, 388–397.

dos Reis, M., and Z. Yang. 2011. Approximate likelihood calculation for Bayesian estimation of divergence times. *Mol. Biol. Evol.* **28**, 2161–2172.

dos Reis, M., and Z. Yang. 2013a. The unbearable uncertainty of Bayesian divergence time estimation. *J. Syst. Evol.* **51**, 30–43.

dos Reis, M., and Z. Yang. 2013b. Why do more divergent sequences produce smaller nonsynonymous/synonymous rate ratios in pairwise sequence comparisons? *Genetics* **195**, 195–204.

dos Reis, M., J. Inoue, M. Hasegawa et al. 2012. Phylogenomic data sets provide both precision and accuracy in estimating the timescale of placental mammal evolution. *Proc. R. Soc. Lond. B. Biol. Sci.* **279**, 3491–3500.

Douady, C. J., F. Delsuc, Y. Boucher et al. 2003. Comparison of Bayesian and maximum likelihood bootstrap measures of phylogenetic reliability. *Mol. Biol. Evol.* **20**, 248–254.

Drummond, A. J., and A. Rambaut. 2007. BEAST: Bayesian evolutionary analysis by sampling trees. *BMC Evol. Biol.* **7**, 214.

Drummond, A. J., S. Y. W. Ho, M. J. Phillips, and A. Rambaut. 2006. Relaxed phylogenetics and dating with confidence. *PLoS Biol.* **4**, e88.

Drummond, A. J., G. K. Nicholls, A. G. Rodrigo, and W. Solomon. 2002. Estimating mutation parameters, population history and genealogy simultaneously from temporally spaced sequence data. *Genetics* **161**, 1307–1320.

Drummond, A. J., A. Rambaut, B. Shapiro, and O. G. Pybus. 2005. Bayesian coalescent inference of past population dynamics from molecular sequences. *Mol. Biol. Evol.* **22**, 1185–1192.

Duda, T. F., and S. R. Palumbi. 2000. Evolutionary diversification of multigene families: allelic selection of toxins in predatory cone snails. *Mol. Biol. Evol.* **17**, 1286–1293.

Duret, L. 2002. Evolution of synonymous codon usage in metazoans. *Curr. Opin. Genet. Dev.* **12**, 640–649.

Duret, L., M. Semon, G. Piganeau et al. 2002. Vanishing GC-rich isochores in mammalian genomes. *Genetics* **162**, 1837–1847.

Dutheil, J., T. Pupko, A. Jean-Marie, and N. Galtier. 2005. A model-based approach for detecting coevolving positions in a molecule. *Mol. Biol. Evol.* **22**, 1919–1928.

Eck, R. V., and M. O. Dayhoff. 1966. Inference from protein sequence comparisons. *in* M. O. Dayhoff, ed. *Atlas of Protein Sequence and Structure*. National Biomedical Research Foundation, Maryland.

Edgeworth, F. Y. 1885. Observations and statistics. *Trans. Cam. Phil. Soc.* **14**, 138–169.

Edwards, A. W. F. 1970. Estimation of the branch points of a branching diffusion process (with discussion). *J. R. Stat. Soc. B.* **32**, 155–174.

Edwards, A. W. F. 1974. A problem in the doctrine of chances. Pp. 43–60 *in* O. Barndorff-Nielsen, P. Balaesild, and G. Schou, eds. *Proceedings of the Conference on Foundational Questions in Statistical Inference*. Institute of Mathematics, University of Aarhus, Denmark.

Edwards, A. W. F. 1992. *Likelihood*. John Hopkins University Press, London.

Edwards, A. W. F. 1996. The origin and early development of the method of minimum evolution for the reconstruction of phylogenetic trees. *Syst. Biol.* **45**, 79–91.

Edwards, A. W. F. 2009a. Statistical methods for evolutionary trees. *Genetics* **183**, 5–12.

Edwards, A. W. F., and L. L. Cavalli-Sforza. 1963a. A method for cluster analysis (Abstract). *The 5th International Biometrics Conference*, Cambridge, UK.

Edwards, A. W. F., and L. L. Cavalli-Sforza. 1963b. The reconstruction of evolution (Abstract). *Ann. Hum. Genet.* **27**, 105.

Edwards, A. W. F., and L. L. Cavalli-Sforza. 1964. Reconstruction of evolutionary trees. *Phenetic. Phylogenet. Classificat. Syst. Assoc. Publ.* **6**, 67–76.

Edwards, S. V. 2009b. Is a new and general theory of molecular systematics emerging? *Evolution* **63**, 1–19.

Edwards, S. V., and P. Beerli. 2000. Gene divergence, population divergence, and the variance in coalescence time in phylogeographic studies. *Evolution* **54**, 1839–1854.

Edwards, S. V., W. B. Jennings, and A. M. Shedlock. 2005. Phylogenetics of modern birds in the era of genomics. *Proc. R. Soc. B.* **272**, 979–992.

Efron, B. 1979. Bootstrap methods: another look at the jackknife. *Ann. Stat.* **7**, 1–26.

Efron, B. 1986. Why isn't everyone a Bayesian? (with discussion). *Am. J. Stat. Assoc.* **40**, 1–11.

Efron, B., and D. V. Hinkley. 1978. Assessing the accuracy of the maximum likelihood estimator: observed and expected information. *Biometrika* **65**, 457–487.

Efron, B., and R. J. Tibshirani. 1993. *An Introduction to the Bootstrap*. Chapman and Hall, London.

Efron, B., and R. J. Tibshirani. 1998. The problem of regions. *Ann. Stat.* **26**, 1687–1718.

Efron, B., E. Halloran, and S. Holmes. 1996. Bootstrap confidence levels for phylogenetic trees. *Proc. Natl. Acad. Sci. U.S.A.* **93**, 13429–13434 [corrected and republished article originally printed in Proc. Natl. Acad. Sci. U.S.A. 1996, **93**, 7085–7090].

Ence, D. D., and B. C. Carstens. 2011. SpedeSTEM: a rapid and accurate method for species delimitation. *Mol. Ecol. Resour.* **11**, 473–480.

Erixon, P., B. Svennblad, T. Britton, and B. Oxelman. 2003. Reliability of Bayesian posterior probabilities and bootstrap frequencies in phylogenetics. *Syst. Biol.* **52**, 665–673.

Everitt, B. S., S. Landau, and M. Leese. 2001. *Cluster Analysis*. Arnold, London.

Ewens, W. J. 1990. Population genetics theory – the past and the future. Pp. 177–227 *in* S. Lessard, ed. *Mathematical and Statistical Developments of Evolutionary Theory*. Kluwer Academic, Amsterdam.

Excoffier, L., and Z. Yang. 1999. Substitution rate variation among sites in the mitochondrial hypervariable region I of humans and chimpanzees. *Mol. Biol. Evol.* **16**, 1357–1368.

Eyre-Walker, A. 1998. Problems with parsimony in sequences of biased base composition. *J. Mol. Evol.* **47**, 686–690.

Farris, J. S. 1969. A successive approximation approach to character weighting. *Syst. Zool.* **18**, 374–385.

Farris, J. S. 1973. A probability model for inferring evolutionary trees. *Syst. Zool.* **22**, 250–256.

Farris, J. S. 1977. Phylogenetic analysis under Dollo's law. *Syst. Zool.* **26**, 77–88.

Farris, J. S. 1983. The logical basis of phylogenetic analysis. Pp. 7–26 *in* N. Platnick, and V. Funk, eds. *Advances in Cladistics*. Columbia University Press, New York.

Farris, J. S. 1989. The retention index and the rescaled consistency index. *Cladistics* **5**, 417–419.

Fay, J. C., and C.-I. Wu. 2001. The neutral theory in the genomic era. *Curr. Opinion Genet. Dev.* **11**, 642–646.

Fay, J. C., and C. I. Wu. 2000. Hitchhiking under positive Darwinian selection. *Genetics* **155**, 1405–1413.

Fay, J. C., and C. I. Wu. 2003. Sequence divergence, functional constraint, and selection in protein evolution. *Ann. Rev. Genomics Hum. Genet.* **4**, 213–235.

Feder, J. L., S. P. Egan, and P. Nosil. 2012. The genomics of speciation-with-gene-flow. *Trends Genet.* **28**, 342–350.

Felsenstein, J. 1973a. Maximum-likelihood estimation of evolutionary trees from continuous characters. *Am. J. Hum. Genet.* **25**, 471–492.

Felsenstein, J. 1973b. Maximum likelihood and minimum-steps methods for estimating evolutionary trees from data on discrete characters. *Syst. Zool.* **22**, 240–249.

Felsenstein, J. 1978a. Cases in which parsimony and compatibility methods will be positively misleading. *Syst. Zool.* **27**, 401–410.
Felsenstein, J. 1978b. The number of evolutionary trees. *Syst. Zool.* **27**, 27–33.
Felsenstein, J. 1981. Evolutionary trees from DNA sequences: a maximum likelihood approach. *J. Mol. Evol.* **17**, 368–376.
Felsenstein, J. 1983. Statistical inference of phylogenies. *J. R. Stat. Soc. A.* **146**, 246–272.
Felsenstein, J. 1985a. Phylogenies and the comparative method. *Am. Nat.* **125**, 1–15.
Felsenstein, J. 1985b. Confidence limits on phylogenies with a molecular clock. *Evolution* **34**, 152–161.
Felsenstein, J. 1985c. Confidence limits on phylogenies: an approach using the bootstrap. *Evolution* **39**, 783–791.
Felsenstein, J. 1988. Phylogenies from molecular sequences: inference and reliability. *Ann. Rev. Genet.* **22**, 521–565.
Felsenstein, J. 1992. Estimating effective population size from samples of sequences: inefficiency of pairwise and segregating sites as compared to phylogenetic estimates. *Genet. Res.* **59**, 139–147.
Felsenstein, J. 2001a. Taking variation of evolutionary rates between sites into account in inferring phylogenies. *J. Mol. Evol.* **53**, 447–455.
Felsenstein, J. 2001b. The troubled growth of statistical phylogenetics. *Syst. Biol.* **50**, 465–467.
Felsenstein, J. 2004. *Inferring Phylogenies*. Sinauer Associates, Sunderland, Massachusetts.
Felsenstein, J., and G. A. Churchill. 1996. A hidden Markov model approach to variation among sites in rate of evolution. *Mol. Biol. Evol.* **13**, 93–104.
Felsenstein, J., and H. Kishino. 1993. Is there something wrong with the bootstrap on phylogenies? A reply to Hillis and Bull. *Syst. Biol.* **42**, 193–200.
Felsenstein, J., and E. Sober. 1986. Parsimony and likelihood: an exchange. *Syst. Zool.* **35**, 617–626.
Ferguson, T. 1973. Bayesian analysis of some nonparametric problems. *Ann. Stat.* **1**, 209–230.
Ferreira, M. A. R., and M. A. Suchard. 2008. Bayesian analysis of elapsed times in continuous-time Markov chains. *Can. J. Stat.* **36**, 355–368.
Filip, L. C., and N. I. Mundy. 2004. Rapid evolution by positive Darwinian selection in the extracellular domain of the abundant lymphocyte protein CD45 in primates. *Mol. Biol. Evol.* **21**, 1504–1511.
Fisher, R. 1930a. The distribution of gene ratios for rare mutations. *Proc. R. Soc. Edin.* **50**, 205–220.
Fisher, R. 1930b. *The Genetic Theory of Natural Selection*. Clarendon Press, Oxford, UK.
Fisher, R. A. 1970. *Statistical Methods for Research Workers*. Oliver and Boyd, Edinburgh.
Fitch, W. M. 1970. Distinguishing homologous from analogous proteins. *Syst. Zool.* **19**, 99–113.
Fitch, W. M. 1971a. Toward defining the course of evolution: minimum change for a specific tree topology. *Syst. Zool.* **20**, 406–416.
Fitch, W. M. 1971b. Rate of change of concomitantly variable codons. *J. Mol. Evol.* **1**, 84–96.
Fitch, W. M. 1976. Molecular evolutionary clocks. Pp. 160–178 *in* F. J. Ayala, ed. *Molecular Evolution*. Sinauer Associates, Sunderland, Massachusetts.
Fitch, W. M., and E. Margoliash. 1967. Construction of phylogenetic trees. *Science* **155**, 279–284.
Fitch, W. M., R. M. Bush, C. A. Bender, and N. J. Cox. 1997. Long term trends in the evolution of H(3) HA1 human influenza type A. *Proc. Natl. Acad. Sci. U.S.A.* **94**, 7712–7718.
Fleissner, R., D. Metzler, and A. von Haeseler. 2005. Simultaneous statistical multiple alignment and phylogeny reconstruction. *Syst. Biol.* **54**, 548–561.
Fletcher, R. 1987. *Practical Methods of Optimization*. Wiley, New York.
Fletcher, W., and Z. Yang. 2009. INDELible: a flexible simulator of biological sequence evolution. *Mol. Biol. Evol.* **26**, 1879–1888.
Fletcher, W., and Z. Yang. 2010. The effect of insertions, deletions and alignment errors on the branch-site test of positive selection. *Mol. Biol. Evol.* **27**, 2257–2267.
Foote, M., J. P. Hunter, C. M. Janis, and J. J. Sepkoski. 1999. Evolutionary and preservational constraints on origins of biologic groups: divergence times of eutherian mammals. *Science* **283**, 1310–1314.
Forsberg, R., and F. B. Christiansen. 2003. A codon-based model of host-specific selection in parasites, with an application to the influenza A virus. *Mol. Biol. Evol.* **20**, 1252–1259.
Foster, P. G. 2004. Modeling compositional heterogeneity. *Syst. Biol.* **53**, 485–495.
Freeland, S. J., and L. D. Hurst. 1998. The genetic code is one in a million. *J. Mol. Evol.* **47**, 238–248.

Friel, N., and A. N. Pettitt. 2008. Marginal likelihood estimation via power posteriors. *J. Roy. Stat. Soc. B* **70**, 589–607.

Frigessi, A., C. R. Hwang, and L. Younes. 1992. Optimal spectral structure of reversible stochastic matrices, Monte Carlo methods and the simulation of Markov random fields. *Ann. Appl. Prob.* **2**, 610–628.

Fu, Y.-X. 1997. Statistical tests of neutrality of mutations against population growth, hitchhiking and backgroud selection. *Genetics* **147**, 915–925.

Fu, Y. 1994. Estimating effective population size or mutation rate using the frequencies of mutations of various classes in a sample of DNA sequences. *Genetics* **138**, 1375–1386.

Fu, Y. X., and W. H. Li. 1993. Statistical tests of neutrality of mutations. *Genetics* **133**, 693–709.

Fujisawa, T., and T. G. Barraclough. 2013. Delimiting species using single-locus data and the generalized mixed yule coalescent approach: a revised method and evaluation on simulated data sets. *Syst. Biol.* **62**, 707–724.

Fujita, M. K., and A. D. Leaché. 2011. A coalescent perspective on delimiting and naming species: a reply to Bauer et al. *Proc. R. Soc. Lond. B. Biol. Sci.* **278**, 493–495.

Fujita, M. K., A. D. Leaché, F. T. Burbrink et al. 2012. Coalescent-based species delimitation in an integrative taxonomy. *Trends Ecol. Evol.* **27**, 480–488.

Fukami-Kobayashi, K., and Y. Tateno. 1991. Robustness of maximum likelihood tree estimation against different patterns of base substitutions. *J. Mol. Evol.* **32**, 79–91.

Fukami, K., and Y. Tateno. 1989. On the maximum likelihood method for estimating molecular trees: uniqueness of the likelihood point. *J. Mol. Evol.* **28**, 460–464.

Gadagkar, S. R., and S. Kumar. 2005. Maximum likelihood outperforms maximum parsimony even when evolutionary rates are heterotachous. *Mol. Biol. Evol.* **22**, 2139–2141.

Galtier, N. 2001. Maximum-likelihood phylogenetic analysis under a covarion-like model. *Mol. Biol. Evol.* **18**, 866–873.

Galtier, N., and M. Gouy. 1998. Inferring pattern and process: maximum-likelihood implementation of a nonhomogeneous model of DNA sequence evolution for phylogenetic analysis. *Mol. Biol. Evol.* **15**, 871–879.

Galtier, N., N. Tourasse, and M. Gouy. 1999. A nonhyperthermophilic common ancestor to extant life forms. *Science* **283**, 220–221.

Gascuel, O. 1994. A note on Sattath and Tversky's, Saitou and Nei's, and Studier and Keppler's algorithms for inferring phylogenies from evolutionary distances. *Mol. Biol. Evol.* **11**, 961–963.

Gascuel, O. 1997. BIONJ: an improved version of the NJ algorithm based on a simple model of sequence data. *Mol. Biol. Evol.* **14**, 685–695.

Gascuel, O. 2000. On the optimization principle in phylogenetic analysis and the minimum-evolution criterion. *Mol. Biol. Evol.* **17**, 401–405.

Gascuel, O., and M. Steel. 2006. Neighbor-joining revealed. *Mol. Biol. Evol.* **23**, 1997–2000.

Gascuel, O., D. Bryant, and F. Denis. 2001. Strengths and limitations of the minimum evolution principle. *Syst. Biol.* **50**, 621–627.

Gaucher, E. A., and M. M. Miyamoto. 2005. A call for likelihood phylogenetics even when the process of sequence evolution is heterogeneous. *Mol. Phylogenet. Evol.* **37**, 928–931.

Gaucher, E. A., S. Govindarajan, and O. K. Ganesh. 2008. Palaeotemperature trend for Precambrian life inferred from resurrected proteins. *Nature* **451**, 704–707.

Gaut, B. S. 1998. Molecular clocks and nucleotide substitution rates in higher plants. *Evol. Biol.* **30**, 93–120.

Gaut, B. S., and P. O. Lewis. 1995. Success of maximum likelihood phylogeny inference in the four-taxon case. *Mol. Biol. Evol.* **12**, 152–162.

Gelfand, A. E., and A. F. M. Smith. 1990. Sampling-based approaches to calculating marginal densities. *J. Am. Stat. Assoc.* **85**, 398–409.

Gelman, A., and X. L. Meng. 1998. Simulating normalizing constants: from importance sampling to bridge sampling to path sampling. *Stat. Sci.* **13**, 163–185.

Gelman, A., and D. B. Rubin. 1992. Inference from iterative simulation using multiple sequences (with discussion). *Stat. Sci.* **7**, 457–511.

Gelman, A., G. O. Roberts, and W. R. Gilks. 1996. Efficient Metropolis jumping rules. Pp. 599–607 in J. M. Bernardo, J. O. Berger, A. P. Dawid, and A. F. M. Smith, eds. *Bayesian Statistics 5*. Oxford University Press, Oxford, UK.

Gelman, S., and G. D. Gelman. 1984. Stochastic relaxation, Gibbs distributions and the Bayes restoration of images. *IEEE Trans. Pattn. Anal. Mach. Intel.* **6**, 721–741.

Georgelis, N., J. R. Shaw, and L. C. Hannah. 2009. Phylogenetic analysis of ADP-glucose pyrophosphorylase subunits reveals a role of subunit interfaces in the allosteric properties of the enzyme. *Plant Physiol.* **151**, 67–77.

Geyer, C. J. 1991. Markov chain Monte Carlo maximum likelihood. Pp. 156–163 *in* E. M. Keramidas, ed. *Computing Science and Statistics: Proc. 23rd Symp. Interface*. Interface Foundation, Fairfax Station.

Geyer, C. J. 1992. Practical Markov chain Monte Carlo. *Stat. Sci.* **7**, 473–511.

Gilks, W. R., S. Richardson, and D. J. Spielgelhalter. 1996. *Markov Chain Monte Carlo in Practice*. Chapman and Hall, London.

Gill, P. E., W. Murray, and M. H. Wright. 1981. *Practical Optimization*. Academic Press, London.

Gillespie, J. H. 1984. The molecular clock may be an episodic clock. *Proc. Natl. Acad. Sci. U.S.A.* **81**, 8009–8013.

Gillespie, J. H. 1986a. Natural selection and the molecular clock. *Mol. Biol. Evol.* **3**, 138–155.

Gillespie, J. H. 1986b. Rates of molecular evolution. *Ann. Rev. Ecol. Syst.* **17**, 637–665.

Gillespie, J. H. 1991. *The Causes of Molecular Evolution*. Oxford University Press, Oxford, UK.

Gillespie, J. H. 1998. *Population Genetics: a Concise Guide*. John Hopkins University Press, Baltimore, Maryland.

Godsill, S. J. 2001. On the relationship between Markov chain Monte Carlo methods for model uncertainty. *J. Comput. Graph. Stat.* **10**, 230–248.

Gogarten, J. P., H. Kibak, P. Dittrich et al. 1989. Evolution of the vacuolar H^+-ATPase: implications for the origin of eukaryotes. *Proc. Natl. Acad. Sci. U.S.A.* **86**, 6661–6665.

Gojobori, T. 1983. Codon substitution in evolution and the 'saturation' of synonymous changes. *Genetics* **105**, 1011–1027.

Gojobori, T., W. H. Li, and D. Graur. 1982. Patterns of nucleotide substitution in pseudogenes and functional genes. *J. Mol. Evol.* **18**, 360–369.

Golding, G. B. 1983. Estimates of DNA and protein sequence divergence: an examination of some assumptions. *Mol. Biol. Evol.* **1**, 125–142.

Golding, G. B., and A. M. Dean. 1998. The structural basis of molecular adaptation. *Mol. Biol. Evol.* **15**, 355–369.

Goldman, N. 1990. Maximum likelihood inference of phylogenetic trees, with special reference to a Poisson process model of DNA substitution and to parsimony analysis. *Syst. Zool.* **39**, 345–361.

Goldman, N. 1993a. Simple diagnostic statistical tests of models for DNA substitution. *J. Mol. Evol.* **37**, 650–661.

Goldman, N. 1993b. Statistical tests of models of DNA substitution. *J. Mol. Evol.* **36**, 182–198.

Goldman, N. 1994. Variance to mean ratio, R(t), for Poisson processes on phylogenetic trees. *Mol. Phylogenet. Evol.* **3**, 230–239.

Goldman, N. 1998. Phylogenetic information and experimental design in molecular systematics. *Proc. R. Soc. Lond. B Biol. Sci.* **265**, 1779–1786.

Goldman, N., and Z. Yang. 1994. A codon-based model of nucleotide substitution for protein-coding DNA sequences. *Mol. Biol. Evol.* **11**, 725–736.

Goldman, N., J. P. Anderson, and A. G. Rodrigo. 2000. Likelihood-based tests of topologies in phylogenetics. *Syst. Biol.* **49**, 652–670.

Goldman, N., J. L. Thorne, and D. T. Jones. 1998. Assessing the impact of secondary structure and solvent accessibility on protein evolution. *Genetics* **149**, 445–458.

Goldstein, D. B., and D. D. Pollock. 1994. Least squares estimation of molecular distance–noise abatement in phylogenetic reconstruction. *Theor. Popul. Biol.* **45**, 219–226.

Goldstein, R. A., and D. D. Pollock. 2006. Observations of amino acid gain and loss during protein evolution are explained by statistical bias. *Mol. Biol. Evol.* **23**, 1444–1449.

Goloboff, P. A. 1999. Analyzing large data sets in reasonable times: solutions for composite optima. *Cladistics* **15**, 415–428.

Goloboff, P. A., and D. Pol. 2005. Parsimony and Bayesian phylogenetics. Pp. 148–159 *in* V. A. Albert, ed. *Parsimony, Phylogeny, and Genomics*. Oxford University Press, Oxford, UK.

Golub, G. H., and C. F. Van Loan. 1996. *Matrix Computations*. Johns Hopkins University Press, Baltimore, Maryland.

Gonnet, G. H., M. A. Cohen, and S. A. Benner. 1992. Exhaustive matching of the entire protein sequence database. *Science* **256**, 1443–1445.

Goswami, A., and P. Upchurch. 2010. The dating game: a reply to Heads (2010). *Zool. Scr.* **39**, 406–409.

Götesson, A., J. S. Marshall, D. A. Jones, and A. R. Hardham. 2002. Characterization and evolutionary analysis of a large polygalacturonase gene family in the oomycete pathogen Phytophthora cinnamomi. *Mol. Plant Microbe Interact.* **15**, 907–921.

Grantham, R. 1974. Amino acid difference formula to help explain protein evolution. *Science* **185**, 862–864.

Graur, D., and W.-H. Li. 2000. *Fundamentals of Molecular Evolution*. Sinauer Associates, Massachusetts.

Graur, D., and W. Martin. 2004. Reading the entrails of chickens: molecular timescales of evolution and the illusion of precision. *Trends Genet.* **20**, 80–86.

Green, P. J. 1995. Reversible jump Markov chain Monte Carlo computation and Bayesian model determination. *Biometrika* **82**, 711–732.

Green, P. J. 2003. Trans-dimensional Markov chain Monte Carlo. Pp. 179–196 *in* P. J. Green, N. L. Hjort, and S. Richardson, eds. *Highly Structured Stochastic Systems*. Oxford University Press, Oxford, UK.

Green, P. J., and X. L. Han. 1992. Metropolis methods, Gaussian proposals and antithetic variables. Pp. 142–164 *in* P. Barone, A. Frigessi, and M. Piccioni, eds. *Stochastic Models, Statistical Methods & Algorithms in Image Analysis*. Springer, New York.

Green, P. J., and S. Richardson. 2001. Modelling heterogeneity with and without the Dirichlet process. *Scand. J. Stat.* **28**, 355–375.

Griffiths, R. C., and S. Tavaré. 1994. Sampling theory for neutral alleles in a varying environment. *Philos. Trans. R. Soc. Lond. B. Biol. Sci.* **344**, 403–410.

Grimmett, G. R., and D. R. Stirzaker. 1992. *Probability and Random Processes*. Clarendon Press, Oxford.

Gronau, I., M. J. Hubisz, B. Gulko et al. 2011. Bayesian inference of ancient human demography from individual genome sequences. *Nat. Genet.* **43**, 1031–1034.

Gu, X. 2001. Maximum-likelihood approach for gene family evolution under functional divergence. *Mol. Biol. Evol.* **18**, 453–464.

Gu, X., and W.-H. Li. 1996. A general additive distance with time-reversibility and rate variation among nucleotide sites. *Proc. Natl. Acad. Sci. U.S.A.* **93**, 4671–4676.

Gu, X., Y. X. Fu, and W. H. Li. 1995. Maximum likelihood estimation of the heterogeneity of substitution rate among nucleotide sites. *Mol. Biol. Evol.* **12**, 546–557.

Guindon, S. 2013. From trajectories to averages: an improved description of the heterogeneity of substitution rates along lineages. *Syst. Biol.* **62**, 22–34.

Guindon, S., and O. Gascuel. 2003. A simple, fast, and accurate algorithm to estimate large phylogenies by maximum likelihood. *Syst. Biol.* **52**, 696–704.

Guindon, S., M. Black, and A. Rodrigo. 2006. Control of the false discovery rate applied to the detection of positively selected amino acid sites. *Mol. Biol. Evol.* **23**, 919–926.

Guindon, S., A. G. Rodrigo, K. A. Dyer, and J. P. Huelsenbeck. 2004. Modeling the site-specific variation of selection patterns along lineages. *Proc. Natl. Acad. Sci. U.S.A.* **101**, 12957–12962.

Hacking, I. 1965. *Logic of Scientific Inference*. Cambridge University Press, Cambridge, UK.

Haldane, J. B. S. 1931. A note on inverse probability. *Proc. Cam. Phil. Soc.* **28**, 55–61.

Haldane, J. B. S. 1932. *The Causes of Evolution*. Longmans Green & Co., London.

Halpern, A. L., and W. J. Bruno. 1998. Evolutionary distances for protein-coding sequences: modeling site-specific residue frequencies. *Mol. Biol. Evol.* **15**, 910–917.

Hanson-Smith, V., B. Kolaczkowski, and J. W. Thornton. 2010. Robustness of ancestral sequence reconstruction to phylogenetic uncertainty. *Mol. Biol. Evol.* **27**, 1988–1999.

Harris, H. 1966. Enzyme polymorphism in man. *Proc. R. Soc. Lond. B. Biol. Sci.* **164**, 298–310.

Hartigan, J. A. 1973. Minimum evolution fits to a given tree. *Biometrics* **29**, 53–65.

Hartl, D. L., and A. G. Clark. 1997. *Principles of Population Genetics*. Sinauer Associates, Sunderland, Massachusetts.

Hartl, D. L., E. N. Moriyama, and S. A. Sawyer. 1994. Selection intensity for codon bias. *Genetics* **138**, 227–234.

Harvey, P. H., and M. Pagel. 1991. *The Comparative Method in Evlutionary Biology*. Oxford University Press, Oxford, UK.

Harvey, P. H., and A. Purvis. 1991. Comparative methods for explaining adaptations. *Nature* **351**, 619–624.

Hasegawa, M., and M. Fujiwara. 1993. Relative efficiencies of the maximum likelihood, maximum parsimony, and neihbor joining methods for estimating protein phylogeny. *Mol. Phylogenet. Evol.* **2**, 1–5.

Hasegawa, M., and H. Kishino. 1989. Confidence limits on the maximum-likelihood estimate of the Hominoid tree from mitochondrial DNA sequences. *Evolution* **43**, 672–677.

Hasegawa, M., and H. Kishino. 1994. Accuracies of the simple methods for estimating the bootstrap probability of a maximum likelihood tree. *Mol. Biol. Evol.* **11**, 142–145.

Hasegawa, M., J. Adachi, and M. C. Milinkovitch. 1997. Novel phylogeny of whales supported by total molecular evidence. *J. Mol. Evol.* **44**, S117–S120.

Hasegawa, M., Y. Cao, and Z. Yang. 1998. Preponderance of slightly deleterious polymorphism in mitochondrial DNA: replacement/synonymous rate ratio is much higher within species than between species. *Mol. Biol. Evol.* **15**, 1499–1505.

Hasegawa, M., H. Kishino, and N. Saitou. 1991. On the maximum likelihood method in molecular phylogenetics. *J. Mol. Evol.* **32**, 443–445.

Hasegawa, M., H. Kishino, and T. Yano. 1985. Dating the human-ape splitting by a molecular clock of mitochondrial DNA. *J. Mol. Evol.* **22**, 160–174.

Hasegawa, M., J. L. Thorne, and H. Kishino. 2003. Time scale of eutherian evolution estimated without assuming a constant rate of molecular evolution. *Genes Genet. Syst.* **78**, 267–283.

Hasegawa, M., T. Yano, and H. Kishino. 1984. A new molecular clock of mitochondrial DNA and the evolution of Hominoids. *Proc. Japan Acad. B.* **60**, 95–98.

Hastings, W. K. 1970. Monte Carlo sampling methods using Markov chains and their application. *Biometrika* **57**, 97–109.

Haydon, D. T., A. D. Bastos, N. J. Knowles, and A. R. Samuel. 2001. Evidence for positive selection in foot-and-mouth-disease virus capsid genes from field isolates. *Genetics* **157**, 7–15.

Heads, M. 2005. Dating nodes on molecular phylogenies: a critique of molecular biogeography. *Cladistics* **21**, 62–78.

Heath, T. A., M. T. Holder, and J. P. Huelsenbeck. 2012. A Dirichlet process prior for estimating lineage-specific substitution rates. *Mol. Biol. Evol.* **29**, 939–955.

Hebert, P. D., M. Y. Stoeckle, T. S. Zemlak, and C. M. Francis. 2004. Identification of birds through DNA barcodes. *PLoS Biol.* **2**, 1657–1663.

Hedges, S. B., and S. Kumar. 2004. Precision of molecular time estimates. *Trends Genet.* **20**, 242–247.

Hedges, S. B., P. H. Parker, C. G. Sibley, and S. Kumar. 1996. Continental breakup and the ordinal diversification of birds and mammals. *Nature* **381**, 226–229.

Hein, J., J. L. Jensen, and C. N. Pedersen. 2003. Recursions for statistical multiple alignment. *Proc. Natl. Acad. Sci. U.S.A.* **100**, 14960–14965.

Hein, J., M. H. Schierup, and C. Wiuf. 2005. *Gene Genealogies, Variation and Evolution: A Primer in Coalescent Theory*. Oxford University Press, Oxford, UK.

Hein, J., C. Wiuf, B. Knudsen et al. 2000. Statistical alignment: computational properties, homology testing and goodness-of-fit. *J. Mol. Biol.* **302**, 265–279.

Heled, J., and A. J. Drummond. 2008. Bayesian inference of population size history from multiple loci. *BMC Evol. Biol.* **8**, 289.

Heled, J., and A. J. Drummond. 2010. Bayesian inference of species trees from multilocus data. *Mol. Biol. Evol.* **27**, 570–580.

Heled, J., and A. J. Drummond. 2012. Calibrated tree priors for relaxed phylogenetics and divergence time estimation. *Syst. Biol.* **61**, 138–149.

Heled, J., D. Bryant, and A. J. Drummond. 2013. Simulating gene trees under the multispecies coalescent and time-dependent migration. *BMC Evol. Biol.* **13**, 44.

Hellberg, M. E., and V. D. Vacquier. 2000. Positive selection and propeptide repeats promote rapid interspecific divergence of a gastropod sperm protein. *Mol. Biol. Evol.* **17**, 458–466.

Hendy, M. D. 2005. Hadamard conjugation: an analytical tool for phylogenetics. Pp. 143–177 *in* O. Gascuel, ed. *Mathematics of Evolution and Phylogeny*. Oxford University Press, Oxford, UK.

Hendy, M. D., and D. Penny. 1982. Branch and bound algorithms ro determine minimum-evolution trees. *Math. Biosci.* **60**, 133–142.

Hendy, M. D., and D. Penny. 1989. A framework for the quantitative study of evolutionary trees. *Syst. Zool.* **38**, 297–309.

Henikoff, S., and J. Henikoff. 1992. Amino acid substitution matrices from protein blocks. *Proc. Natl. Acad. Sci. U.S.A.* **89**, 10915–10919.

Hernandez, R. D., S. H. Williamson, and C. D. Bustamante. 2007. Context dependence, ancestral misidentification, and spurious signatures of natural selection. *Mol. Biol. Evol.* **24**, 1792–1800.

Hey, J. 2010. Isolation with migration models for more than two populations. *Mol. Biol. Evol.* **27**, 905–920.

Hey, J., and R. Nielsen. 2004. Multilocus methods for estimating population sizes, migration rates and divergence time, with applications to the divergence of *Drosophila pseudoobscura* and *D. persimilis*. *Genetics* **167**, 747–760.

Hickerson, M. J., C. P. Meyer, and C. Moritz. 2006. DNA barcoding will often fail to discover new animal species over broad parameter space. *Syst. Biol.* **55**, 729–739.

Hillis, D. M., and J. J. Bull. 1993. An empirical test of bootstrapping as a method for assessing confidence in phylogenetic analysis. *Syst. Biol.* **42**, 182–192.

Hillis, D. M., J. J. Bull, M. E. White et al. 1992. Experimental phylogenetics: generation of a known phylogeny. *Science* **255**, 589–592.

Ho, S. Y. W., and M. J. Phillips. 2009. Accounting for calibration uncertainty in phylogenetic estimation of evolutionary divergence times. *Syst. Biol.* **58**, 367–380.

Hobolth, A., L. N. Andersen, and T. Mailund. 2011. On computing the coalescence time density in an isolation-with-migration model with few samples. *Genetics* **187**, 1241–1243.

Hodgkinson, A., and A. Eyre-Walker. 2011. Variation in the mutation rate across mammalian genomes. *Nat. Rev. Genet.* **12**, 756–766.

Hoekstra, H. E., and J. A. Coyne. 2007. The locus of evolution: evo devo and the genetics of adaptation. *Evolution* **61**, 995–1016.

Höhna, S., and A. J. Drummond. 2012. Guided tree topology proposals for Bayesian phylogenetic inference. *Syst. Biol.* **61**, 1–11.

Höhna, S., M. Defoin-Platel, and A. J. Drummond. 2008. Clock-constrained tree proposal operators in Bayesian phylogenetic inference. *8th IEEE International Conference on BioInformatics and BioEngineering. Athens (Greece): BIBE*:7.

Holder, M., and P. O. Lewis. 2003. Phylogeny estimation: traditional and Bayesian approaches. *Nat. Rev. Genet.* **4**, 275–284.

Holder, M. T., P. O. Lewis, D. L. Swofford, and B. Larget. 2005. Hastings ratio of the LOCAL proposal used in Bayesian phylogenetics. *Syst. Biol.* **54**, 961–965.

Holmes, I. 2005. Using evolutionary expectation maximization to estimate indel rates. *Bioinformatics* **21**, 2294–2300.

Holmes, S. 2003. Bootstrapping phylogenetic trees: theory and methods. *Stat. Sci.* **18**, 241–255.

Horai, S., K. Hayasaka, R. Kondo et al. 1995. Recent African origin of modern humans revealed by complete sequences of hominoid mitochondrial DNAs. *Proc. Natl. Acad. Sci. U.S.A.* **92**, 532–536.

Huang, H., and L. L. Knowles. 2009. What is the danger of the anomaly zone for empirical phylogenetics? *Syst. Biol.* **58**, 527–536.

Hudson, R. R. 1983a. Testing the constant-rate neutral alele model with protein sequence data. *Evolution* **37**, 203–217.

Hudson, R. R. 1983b. Properties of a neutral allele model with intragenic recombination. *Theor. Popul. Biol.* **23**, 183–201.

Hudson, R. R. 1990. Gene genealogies and the coalescent process. Pp. 1–44 *in* D. J. Futuyma, and J. D. Antonovics, eds. *Oxford Surveys in Evolutionary Biology*. Oxford University Press, New York.

Hudson, R. R. 2001. Two-locus sampling distributions and their application. *Genetics* **159**, 1805–1817.

Hudson, R. R., and J. A. Coyne. 2002. Mathematical consequences of the genealogical species concept. *Evolution* **56**, 1557–1565.

Hudson, R. R., M. Kreitman, and M. Aguade. 1987. A test of neutral molecular evolution based on nucleotide data. *Genetics* **116**, 153–159.

Huelsenbeck, J. P. 1995a. The performance of phylogenetic methods in simulation. *Syst. Biol.* **44**, 17–48.

Huelsenbeck, J. P. 1995b. The robustness of two phylogenetic methods: four-taxon simulations reveal a slight superiority of maximum likelihood over neighbor joining. *Mol. Biol. Evol.* **12**, 843–849.

Huelsenbeck, J. P. 1998. Systematic bias in phylogenetic analysis: is the Strepsiptera problem solved? *Syst. Biol.* **47**, 519–537.

Huelsenbeck, J. P. 2002. Testing a covariotide model of DNA substitution. *Mol. Biol. Evol.* **19**, 698–707.

Huelsenbeck, J. P., and P. Andolfatto. 2007. Inference of population structure under a Dirichlet process model. *Genetics* **175**, 1787–1802.

Huelsenbeck, J. P., and J. P. Bollback. 2001. Empirical and hierarchical Bayesian estimation of ancestral states. *Syst. Biol.* **50**, 351–366.

Huelsenbeck, J. P., and K. A. Dyer. 2004. Bayesian estimation of positively selected sites. *J. Mol. Evol.* **58**, 661–672.

Huelsenbeck, J. P., and K. M. Lander. 2003. Frequent inconsistency of parsimony under a simple model of cladogenesis. *Syst. Biol.* **52**, 641–648.

Huelsenbeck, J. P., and R. Nielsen. 1999. Variation in the pattern of nucleotide substitution across sites. *J. Mol. Evol.* **48**, 86–93.

Huelsenbeck, J. P., and B. Rannala. 2004. Frequentist properties of Bayesian posterior probabilities of phylogenetic trees under simple and complex substitution models. *Syst. Biol.* **53**, 904–913.

Huelsenbeck, J. P., and F. Ronquist. 2001. MrBayes: Bayesian inference of phylogenetic trees. *Bioinformatics* **17**, 754–755.

Huelsenbeck, J. P., and M. A. Suchard. 2007. A nonparametric method for accommodating and testing across-site rate variation. *Syst. Biol.* **56**, 975–987.

Huelsenbeck, J. P., M. E. Alfaro, and M. A. Suchard. 2011. Biologically inspired phylogenetic models strongly outperform the no common mechanism model. *Syst. Biol.* **60**, 225–232.

Huelsenbeck, J. P., J. J. Bull, and C. W. Cunningham. 1996. Combining data in phylogenetic analysis. *Trends Ecol. Evol.* **11**, 152–158.

Huelsenbeck, J. P., B. Larget, and M. E. Alfaro. 2004. Bayesian phylogenetic model selection using reversible jump Markov chain Monte Carlo. *Mol. Biol. Evol.* **21**, 1123–1133.

Huelsenbeck, J. P., B. Larget, and D. Swofford. 2000a. A compound Poisson process for relaxing the molecular clock. *Genetics* **154**, 1879–1892.

Huelsenbeck, J. P., R. Nielsen, and J. P. Bollback. 2003. Stochastic mapping of morphological characters. *Syst. Biol.* **52**, 131–158.

Huelsenbeck, J. P., B. Rannala, and B. Larget. 2000b. A Bayesian framework for the analysis of cospeciation. *Evolution* **54**, 352–364.

Huelsenbeck, J. P., F. Ronquist, R. Nielsen, and J. P. Bollback. 2001. Bayesian inference of phylogeny and its impact on evolutionary biology. *Science* **294**, 2310–2314.

Hughes, A. L. 1999. *Adaptive Evolution of Genes and Genomes*. Oxford University Press, Oxford, UK.

Hughes, A. L., and M. Nei. 1988. Pattern of nucleotide substitution at major histocompatibility complex class I loci reveals overdominant selection. *Nature* **335**, 167–170.

Hughes, A. L., T. Ota, and M. Nei. 1990. Positive Darwinian selection promotes charge profile diversity in the antigen-binding cleft of class I major-histocompatibility-complex molecules. *Mol. Biol. Evol.* **7**, 515–524.

Hurley, I. A., R. L. Mueller, K. A. Dunn et al. 2007. A new time-scale for ray-finned fish evolution. *Proc. R. Soc. Lond. B. Biol. Sci.* **274**, 489–498.

Hurvich, C. M., and C.-L. Tsai. 1989. Regression and time series model selection in small samples. *Biometrika* **76**, 297–307.

Ina, Y. 1995. New methods for estimating the numbers of synonymous and nonsynonymous substitutions. *J. Mol. Evol.* **40**, 190–226.

Inoue, J., P. C. H. Donoghue, and Z. Yang. 2010. The impact of the representation of fossil calibrations on Bayesian estimation of species divergence times. *Syst. Biol.* **59**, 74–89.

Issac, N. J. B., J. Mallet, and G. M. Mace. 2004. Taxonomic inflation: its influence on macroecology and conservation. *Trends Ecol. Evol.* **19**, 464–469.

Ivarsson, Y., A. J. Mackey, M. Edalat et al. 2002. Identification of residues in glutathione transferase capable of driving functional diversification in evolution: a novel approach to protein design. *J. Biol. Chem.* **278**, 8733–8738.

Iwabe, N., K. Kuma, M. Hasegawa et al. 1989. Evolutionary relationship of archaebacteria, eubacteria, and eukaryotes inferred from phylogenetic trees of duplicated genes. *Proc. Natl. Acad. Sci. U.S.A.* **86**, 9355–9359.

Iwabe, N., Y. Hara, Y. Kumazawa et al. 2005. Sister group relationship of turtles to the bird-crocodilian clade revealed by nuclear DNA-coded proteins 10.1093/molbev/msi075. *Mol. Biol. Evol.* **22**, 810–813.

Jasra, A., C. C. Holmes, and D. A. Stephens. 2005. Markov chain Monte Carlo methods and the label switching problem in Bayesian mixture modeling. *Stat. Sci.* **1**, 50–67.

Jeffreys, H. 1935. Some tests of significance, treated by the theory of probability. *Proc. Cam. Phil. Soc.* **31**, 203–222.

Jeffreys, H. 1961. *Theory of Probability*. Oxford University Press, Oxford, UK.

Jennings, W. B., and S. V. Edwards. 2005. Speciational history of Australian grass finches (Poephila) inferred from thirty gene trees. *Evolution* **59**, 2033–2047.

Jensen, J. D., A. Wong, and C. F. Aquadro. 2007. Approaches for identifying targets of positive selection. *Trends Genet.* **23**, 568–577.

Jermann, T. M., J. G. Opitz, J. Stackhouse, and S. A. Benner. 1995. Reconstructing the evolutionary history of the artiodactyl ribonuclease superfamily. *Nature* **374**, 57–59.

Jermiin, L. S., V. Jayaswal, F. Ababneh, and J. Robinson. 2008. Phylogenetic model evaluation (Chapter 16) *in* J. M. Keith, ed. *Bioinformatics, Volume I: Data, Sequence Analysis, and Evolution*. Humana Press (Springer), Totowa, New Jersey.

Jin, L., and M. Nei. 1990. Limitations of the evolutionary parsimony method of phylogenetic analysis [Erratum in Mol. Biol. Evol. 1990 7, 201]. *Mol. Biol. Evol.* **7**, 82–102.

Jones, D. T., W. R. Taylor, and J. M. Thornton. 1992. The rapid generation of mutation data matrices from protein sequences. *CABIOS* **8**, 275–282.

Jordan, G., and N. Goldman. 2012. The effects of alignment error and alignment filtering on the sitewise detection of positive selection. *Mol. Biol. Evol.* **29**, 1125–1139.

Jordan, I. K., F. A. Kondrashov, I. A. Adzhubei et al. 2005. A universal trend of amino acid gain and loss in protein evolution. *Nature* **433**, 633–638.

Jukes, T. H. 1987. Transitions, transversions, and the molecular evolutionary clock. *J. Mol. Evol.* **26**, 87–98.

Jukes, T. H., and C. R. Cantor. 1969. Evolution of protein molecules. Pp. 21–123 *in* H. N. Munro, ed. *Mammalian Protein Metabolism*. Academic Press, New York.

Jukes, T. H., and J. L. King. 1979. Evolutionary nucleotide replacements in DNA. *Nature* **281**, 605–606.

Kafatos, F. C., A. Efstratiadis, B. G. Forget, and S. M. Weissman. 1977. Molecular evolution of human and rabbit ß-globin mRNAs. *Proc. Natl. Acad. Sci. U.S.A.* **74**, 5618–5622.

Kalbfleisch, J. G. 1985. *Probability and Statistical Inference, Volume 2, Statistical Inference*. Springer-Verlag, New York.

Kalbfleisch, J. G., and D. A. Sprott. 1970. Application of likelihood methods to models involving large numbers of parameters (with discussions). *J. R. Stat. Soc. B.* **32**, 175–208.

Kao, E. P. C. 1997. *An Introduction to Stochastic Processes*. ITP, Belmont, California.

Karlin, S., and H. M. Taylor. 1975. *A First Course in Stochastic Processes*. Academic Press, San Diego, California.

Kass, R. E., and A. E. Raftery. 1995. Bayes factors. *J. Am. Stat. Assoc.* **90**, 773–795.

Katoh, K., K. Kuma, and T. Miyata. 2001. Genetic algorithm-based maximum-likelihood analysis for molecular phylogeny. *J. Mol. Evol.* **53**, 477–484.

Keilson, J. 1979. *Markov Chain Models: Rarity and Exponentiality*. Springer-Verlag, New York.

Keller, A., F. Förster, T. Müller et al. 2010. Including RNA secondary structures improves accuracy and robustness in reconstruction of phylogenetic trees. *Biol. Direct.* **5**, 4.

Kelly, C., and J. Rice. 1996. Modeling nucleotide evolution: a heterogeneous rate analysis. *Math. Biosci.* **133**, 85–109.

Kemeny, J. G., and J. L. Snell. 1960. *Finite Markov Chains*. Van Nostrand, Princeton, New Jersey.

Kendall, D. G. 1948. On the generalized birth-and-death process. *Ann. Math. Stat.* **19**, 1–15.

Kidd, K. K., and L. A. Sgaramella-Zonta. 1971. Phylogenetic analysis: concepts and methods. *Am. J. Hum. Genet.* **23**, 235–252.

Kim, J. 1996. General inconsistency conditions for maximum parsimony: effects of branch lengths and increasing numbers of taxa. *Syst. Biol.* **45**, 363–374.

Kimura, M. 1957. Some problems of stochastic processes in genetics. *Ann. Math. Stat.* **28**, 882–901.

Kimura, M. 1968. Evolutionary rate at the molecular level. *Nature* **217**, 624–626.

Kimura, M. 1977. Prepondence of synonymous changes as evidence for the neutral theory of molecular evolution. *Nature* **267**, 275–276.

Kimura, M. 1980. A simple method for estimating evolutionary rate of base substitution through comparative studies of nucleotide sequences. *J. Mol. Evol.* **16**, 111–120.

Kimura, M. 1981. Estimation of evolutionary distances between homologous nucleotide sequences. *Proc. Natl. Acad. Sci. U.S.A.* **78**, 454–458.

Kimura, M. 1983. *The Neutral Theory of Molecular Evolution*. Cambridge University Press, Cambridge, UK.

Kimura, M. 1987. Molecular evolutionary clock and the neutral theory. *J. Mol. Evol.* **26**, 24–33.

Kimura, M., and T. Ohta. 1971a. *Theoretical Topics in Population Genetics*. Princeton University Press, Princeton, New Jersey.

Kimura, M., and T. Ohta. 1971b. Protein polymorphism as a phase of molecular evolution. *Nature* **229**, 467–469.

Kimura, M., and T. Ohta. 1972. On the stochastic model for estimation of mutational distance between homologous proteins. *J. Mol. Evol.* **2**, 87–90.

King, C. E., and T. H. Jukes. 1969. Non-Darwinian evolution. *Science* **164**, 788–798.

Kingman, J. F. C. 1982a. On the genealogy of large populations. *J. Appl. Prob.* **19A**:27–43.

Kingman, J. F. C. 1982b. The coalescent. *Stochastic Process Appl.* **13**, 235–248.

Kingman, J. F. C. 2000. Origins of the coalescent: 1974–1982. *Genetics* **156**, 1461–1463.

Kirkpatrick, S., C. D. Gelatt, and M. P. Vecchi. 1983. Optimization by simulated annealing. *Science* **220**, 671–680.

Kishino, H., and M. Hasegawa. 1989. Evaluation of the maximum likelihood estimate of the evolutionary tree topologies from DNA sequence data, and the branching order in hominoidea. *J. Mol. Evol.* **29**, 170–179.

Kishino, H., and M. Hasegawa. 1990. Converting distance to time: application to human evolution. *Methods Enzymol.* **183**, 550–570.

Kishino, H., T. Miyata, and M. Hasegawa. 1990. Maximum likelihood inference of protein phylogeny and the origin of chloroplasts. *J. Mol. Evol.* **31**, 151–160.

Kishino, H., J. L. Thorne, and W. J. Bruno. 2001. Performance of a divergence time estimation method under a probabilistic model of rate evolution. *Mol. Biol. Evol.* **18**, 352–361.

Kluge, A. G., and J. S. Farris. 1969. Quantitateive phyletics and the evolution of anurans. *Syst. Zool.* **18**, 1–32.

Knoll, A. H., and S. B. Carroll. 1999. Early animal evolution: emerging views from comparative biology and geology. *Science* **284**, 2129–2137.

Knowles, L. L. 2009. Statistical phylogeography. *Ann. Rev. Ecol. Syst.* **40**, 593–612.

Knowles, L. L., and B. C. Carstens. 2007. Delimiting species without monophyletic gene trees. *Syst. Biol.* **56**, 887–895.

Knudsen, B., and M. M. Miyamoto. 2001. A likelihood ratio test for evolutionary rate shifts and functional divergence among proteins. *Proc. Natl. Acad. Sci. U.S.A.* **98**, 14512–14517.

Knuth, D. E. 1997. *The Art of Computer Programming: Fundamental Algorithms*. Addison-Wesley, Reading, Massachusetts.

Kocher, T. D. 2004. Adaptive evolution and explosive speciation: the cichlid fish model. *Nat. Rev. Genet.* **5**, 288–298.

Kolaczkowski, B., and J. W. Thornton. 2004. Performance of maximum parsimony and likelihood phylogenetics when evolution is heterogeneous. *Nature* **431**, 980–984.

Kolaczkowski, B., and J. W. Thornton. 2008. A mixed branch length model of heterotachy improves phylogenetic accuracy. *Mol. Biol. Evol.* **25**, 1054–1066.

Kondrashov, A. S., and F. A. Kondrashov. 1999. Interactions among quantitative traits in the course of sympatric speciation. *Nature* **400**, 351–354.

Kong, A., M. L. Frigge, G. Masson et al. 2012. Rate of de novo mutations and the importance of father's age to disease risk. *Nature* **488**, 471–475.

Kosakovsky Pond, S. L., and S. D. W. Frost. 2005a. Not so different after all: a comparison of methods for detecting amino acid sites under selection. *Mol. Biol. Evol.* **22**, 1208–1222.

Kosakovsky Pond, S. L., and S. D. W. Frost. 2005b. A genetic algorithm approach to detecting lineage-specific variation in selection pressure. *Mol. Biol. Evol.* **22**, 478–485.

Kosakovsky Pond, S. L., and S. V. Muse. 2004. Column sorting: rapid calculation of the phylogenetic likelihood function. *Syst. Biol.* **53**, 685–692.

Kosakovsky Pond, S. L., and S. V. Muse. 2005. Site-to-site variation of synonymous substitution rates. *Mol. Biol. Evol.* **22**, 2375–2385.

Koshi, J. M., and R. A. Goldstein. 1996a. Probabilistic reconstruction of ancestral protein sequences. *J. Mol. Evol.* **42**, 313–320.

Koshi, J. M., and R. A. Goldstein. 1996b. Correlating structure-dependent mutation matrices with physical- chemical properties. *Pac. Symp. Biocomput.*:488–499.

Koshi, J. M., D. P. Mindell, and R. A. Goldstein. 1999. Using physical-chemistry-based substitution models in phylogenetic analyses of HIV-1 subtypes. *Mol. Biol. Evol.* **16**, 173–179.

Kosiol, C., and N. Goldman. 2005. Different versions of the Dayhoff rate matrix. *Mol. Biol. Evol.* **22**, 193–199.

Kosiol, C., I. Holmes, and N. Goldman. 2007. An empirical codon model for protein sequence evolution. *Mol. Biol. Evol.* **24**, 1464–1479.

Kou, S. C., Q. Zhou, and W. H. Wong. 2006. Equi-energy sampler with applications in statistical inference and statistical mechanics. *Ann. Stat.* **34**, 1581–1619.

Kreitman, M. 2000. Methods to detect selection in populations with applications to the human. *Ann. Rev. Genomics Hum. Genet.* **1**, 539–559.

Kreitman, M., and H. Akashi. 1995. Molecular evidence for natural selection. *Ann. Rev. Ecol. Syst.* **26**, 403–422.

Kronmal, R. A., and A. V. Peterson. 1979. On the alias method for generating random variables from a discrete distribution. *Am. Stat.* **33**, 214–218.

Kryazhimskiy, S., and J. B. Plotkin. 2008. The population genetics of dN/dS. *PLoS Genet.* **4**, e1000304.

Kubatko, L. S., and J. H. Degnan. 2007. Inconsistency of phylogenetic estimates from concatenated data under coalescence. *Syst. Biol.* **56**, 17–24.

Kubatko, L. S., B. C. Carstens, and L. L. Knowles. 2009. STEM: species tree estimation using maximum likelihood for gene trees under coalescence. *Bioinformatics* **25**, 971–973.

Kuhner, M. K., and J. Felsenstein. 1994. A simulation comparison of phylogeny algorithms under equal and unequal evolutionary rates (Erratum in Mol. Biol. Evol. 1995; **12**, 525). *Mol. Biol. Evol.* **11**, 459–468.

Kuhner, M. K., J. Yamato, and J. Felsenstein. 1995. Estimating effective population size and mutation rate from sequence data using Metropolis-Hastings sampling. *Genetics* **140**, 1421–1430.

Kumar, S. 2005. Molecular clocks: four decades of evolution. *Nat. Rev. Genet.* **6**, 654–662.

Kumar, S., and S. B. Hedges. 1998. A molecular timescale for vertebrate evolution. *Nature* **392**, 917–920.

Kumar, S., and S. Subramanian. 2002. Mutation rate in mammalian genomes. *Proc. Natl. Acad. Sci. U.S.A.* **99**, 803–808.

Kumar, S., K. Tamura, and M. Nei. 2005a. MEGA3: integrated software for molecular evolutionary genetics analysis and sequence alignment. *Brief Bioinform.* **5**, 150–163.

Kumar, S., A. Filipski, V. Swarna et al. 2005b. Placing confidence limits on the molecular age of the human-chimpanzee divergence. *Proc. Natl. Acad. Sci. U.S.A.* **102**, 18842–18847.

Laird, C. D., B. L. McConaughy, and B. J. McCarthy. 1969. Rate of fixation of nucleotide substitutions in evolution. *Nature* **224**, 149–154.

Lake, J. A. 1994. Reconstructing evolutionary trees from DNA and protein sequences: paralinear distances. *Proc. Natl. Acad. Sci. U.S.A.* **91**, 1455–1459.

Lakner, C., P. van der Mark, J. P. Huelsenbeck et al. 2008. Efficiency of Markov chain Monte Carlo tree proposals in Bayesian phylogenetics. *Syst. Biol.* **57**, 86–103.

Lang, S. 1987. *Linear Algebra*. Springer-Verlag, New York.

Langley, C. H., and W. M. Fitch. 1974. An examination of the constancy of the rate of molecular evolution. *J. Mol. Evol.* **3**, 161–177.

Larget, B., and D. L. Simon. 1999. Markov chain Monte Carlo algorithms for the Bayesian analysis of phylogenetic trees. *Mol. Biol. Evol.* **16**, 750–759.

Lartillot, N. 2006. Conjugate Gibbs sampling for Bayesian phylogenetic models. *J. Comput. Biol.* **13**, 1701–1722.

Lartillot, N., and H. Philippe. 2004. A Bayesian mixture model for across-site heterogeneities in the amino-acid replacement process. *Mol. Biol. Evol.* **21**, 1095–1109.

Lartillot, N., and H. Philippe. 2006. Computing Bayes factors using thermodynamic integration. *Syst. Biol.* **55**, 195–207.

Lartillot, N., and H. Philippe. 2008. Improvement of molecular phylogenetic inference and the phylogeny of Bilateria. *Philos. Trans. R. Soc. Lond. B. Biol. Sci.* **363**, 1463–1472.

Lartillot, N., H. Brinkmann, and H. Philippe. 2007. Suppression of long-branch attraction artefacts in the animal phylogeny using a site-heterogeneous model. *BMC Evol. Biol.* **7 Suppl 1**, S4.

Lartillot, N., T. Lepage, and S. Blanquart. 2009. PhyloBayes 3, a Bayesian software package for phylogenetic reconstruction and molecular dating. *Bioinformatics* **25**, 2286–2288.

Le, S. Q., and O. Gascuel. 2008. An improved general amino acid replacement matrix. *Mol. Biol. Evol.* **25**, 1307–1320.

Le, S. Q., N. Lartillot, and O. Gascuel. 2008. Phylogenetic mixture models for proteins. *Philos. Trans. R. Soc. Lond. B. Biol. Sci.* **363**, 3965–3976.

Leaché, A. D. 2009. Species tree discordance traces to phylogeographic clade boundaries in North American fence lizards (sceloporus). *Syst. Biol.* **58**, 547–559.

Leaché, A. D., and M. K. Fujita. 2010. Bayesian species delimitation in West African forest geckos (*Hemidactylus fasciatus*). *Proc. R. Soc. Lond. B. Biol. Sci.* **277**, 3071–3077.

Leaché, A. D., and D. G. Mulcahy. 2007. Phylogeny, divergence times and species limits of spiny lizards (Sceloporus magister species group) in western North American deserts and Baja California. *Mol. Ecol.* **16**, 5216–5233.

Leaché, A. D., and B. Rannala. 2011. The accuracy of species tree estimation under simulation: a comparison of methods. *Syst. Biol.* **60**, 126–137.

Leaché, A. D., and T. W. Reeder. 2002. Molecular systematics of the Eastern Fence Lizard (Sceloporus undulatus): a comparison of Parsimony, Likelihood, and Bayesian approaches. *Syst. Biol.* **51**, 44–68.

Leaché, A. D., R. B. Harris, B. Rannala, and Z. Yang. 2014. The influence of gene flow on Bayesian species tree estimation: a simulation study. *Syst. Biol.* **63**, 17–30.

Leaché, A. D., M. S. Koo, C. L. Spencer et al. 2009. Quantifying ecological, morphological, and genetic variation to delimit species in the coast horned lizard species complex (*Phrynosoma*). *Proc. Natl .Acad. Sci. U.S.A.* **106**, 12418–12423.

LeCam, L. 1953. On some asymptotic properties of maximum likelihood estimates and related Bayes estimates. *Univ. Calf. Publ. Stat.* **1**, 277–330.

Lee, M. S. Y. 2000. Tree robustness and clade significance. *Syst. Biol.* **49**, 829–836.

Lee, Y., and J. A. Nelder. 1996. Hierarchical generalized linear models. *J. R. Stat. Soc. B.* **58**, 619–678.

Lehmann, P. 2002. Structure and evolution of plant disease resistance genes. *J. Appl. Genet.* **43**, 403–414.

Leigh, J. W., E. Susko, M. Baumgartner, and A. J. Roger. 2008. Testing congruence in phylogenomic analysis. *Syst. Biol.* **57**, 104–115.

Lemmon, A. R., and M. C. Milinkovitch. 2002. The metapopulation genetic algorithm: an efficient solution for the problem of large phylogeny estimation. *Proc. Natl. Acad. Sci. U.S.A.* **99**, 10516–10521.

Lemmon, A. R., and E. C. Moriarty. 2004. The importance of proper model assumption in Bayesian phylogenetics. *Syst. Biol.* **53**, 265–277.

Leonard, T., and J. S. J. Hsu. 1999. *Bayesian Methods*. Cambridge University Press, Cambridge, UK

Lepage, T., D. Bryant, H. Philippe, and N. Lartillot. 2007. A general comparison of relaxed molecular clock models. *Mol. Biol. Evol.* **24**, 2669–2680.

Letsch, H., and K. Kjer. 2011. Potential pitfalls of modelling ribosomal RNA data in phylogenetic tree reconstruction: evidence from case studies in the Metazoa. *BMC Evol. Biol.* **11**, 146.

Lewis, P. O. 1998. A genetic algorithm for maximum-likelihood phylogeny inference using nucleotide sequence data. *Mol. Biol. Evol.* **15**, 277–283.

Lewis, P. O. 2001. A likelihood approach to estimating phylogeny from discrete morphological character data. *Syst. Biol.* **50**, 913–925.

Lewis, P. O., M. T. Holder, and K. E. Holsinger. 2005. Polytomies and Bayesian phylogenetic inference. *Syst. Biol.* **54**, 241–253.

Lewontin, R. 1989. Inferring the number of evolutionary events from DNA coding sequence differences. *Mol. Biol. Evol.* **6**, 15–32.

Lewontin, R. C., and J. L. Hubby. 1966. A molecular approach to the study of genic heterozygosity in natural populations. II. Amount of variation and degree of heterozygosity in natural populations of Drosophila pseudoobscura. *Genetics* **54**, 595–609.

Li, S., D. Pearl, and H. Doss. 2000. Phylogenetic tree reconstruction using Markov chain Monte Carlo. *J. Am. Stat. Assoc.* **95**, 493–508.

Li, W.-H. 1976. Distribution of nucleotide differences between two randomly chosen cistrons in a subdivided population: the finite island model. *Theor. Popul. Biol.* **10**, 303–308.

Li, W.-H. 1986. Evolutionary change of restriction cleavage sites and phylogenetic inference. *Genetics* **113**, 187–213.

Li, W.-H. 1989. A statistical test of phylogenies estimated from sequence data. *Mol. Biol. Evol.* **6**, 424–435.

Li, W.-H. 1993. Unbiased estimation of the rates of synonymous and nonsynonymous substitution. *J. Mol. Evol.* **36**, 96–99.

Li, W.-H. 1997. *Molecular Evolution*. Sinauer Associates, Massachusetts.

Li, W.-H., and M. Gouy. 1991. Statistical methods for testing molecular phylogenies. Pp. 249–277 in M. Miyamoto, and J. Cracraft, eds. *Phylogenetic Analysis of DNA Sequences*. Oxford University Press, Oxford, UK.

Li, W.-H., and M. Tanimura. 1987. The molecular clock runs more slowly in man than in apes and monkeys. *Nature* **326**, 93–96.

Li, W.-H., M. Tanimura, and P. M. Sharp. 1987. An evaluation of the molecular clock hypothesis using mammalian DNA sequences. *J. Mol. Evol.* **25**, 330–342.

Li, W.-H., C.-I. Wu, and C.-C. Luo. 1985. A new method for estimating synonymous and nonsynonymous rates of nucleotide substitutions considering the relative likelihood of nucleotide and codon changes. *Mol. Biol. Evol.* **2**, 150–174.

Li, W. L. S., and A. J. Drummond. 2012. Model averaging and Bayes factor calculation of relaxed molecular clocks in Bayesian phylogenetics. *Mol. Biol. Evol.* **29**, 751–761.

Liang, F. 2005. Generalized Wang–Landau algorithm for Monte Carlo computation. *J. Am. Stat. Assoc.* **100**, 1311–1327.

Liang, F., and W. H. Wong. 2001. Real parameter evolutionary Monte Carlo with applications in Bayesian mixture models. *J. Am. Stat. Assoc.* **96**, 653–666.

Liang, F., C. Liu, and R. J. Carroll. 2007. Stochastic approximation in Monte Carlo computation. *J. Am. Stat. Assoc.* **102**, 305–320.

Liang, F., C. Liu, and R. J. Carroll. 2010. *Advanced Markov chain Monte Carlo: Learning from Past Samples*. Wiley, New York.

Liberles, D. A. 2009. *Ancestral Sequence Reconstruction*. Oxford University Press, New York.

Libertini, G., and A. Di Donato. 1994. Reconstruction of ancestral sequences by the inferential method, a tool for protein engineering studies. *J. Mol. Evol.* **39**, 219–229.

Lindley, D. V. 1957. A statistical paradox. *Biometrika* **44**, 187–192.

Lindley, D. V. 1962. Discussion on 'Confidence sets for the mean of a multivariate normal distribution' by C. Stein. *J. R. Stat. Soc. B.* **24**, 265–296.

Lindley, D. V., and L. D. Phillips. 1976. Inference for a Bernoulli process (a Bayesian view). *Am. Stat.* **30**, 112–119.

Lindsey, J. K. 1974a. Construction and comparison of statistical models. *J. R. Stat. Soc. B.* **36**, 418–425.

Lindsey, J. K. 1974b. Comparison of probability distributions. *J. R. Stat. Soc. B.* **36**, 38–47.

Linhart, H. 1988. A test whether two AIC's differ significantly. *S. Afr. Stat. J.* **22**, 153–161.

Little, R. A. J., and D. B. Rubin. 1987. *Statistical Analysis with Missing Data*. Wiley, New York.

Liu, J. S. 2001. *Monte Carlo Strategies in Scientific Computing*. Springer, New York.

Liu, L. 2008. BEST: Bayesian estimation of species trees under the coalescent model. *Bioinformatics* **24**, 2542–2543.

Liu, L., and D. K. Pearl. 2007. Species trees from gene trees: reconstructing Bayesian posterior distributions of a species phylogeny using estimated gene tree distributions. *Syst. Biol.* **56**, 504–514.

Liu, L., L. Yu, and S. V. Edwards. 2010a. A maximum pseudo-likelihood approach for estimating species trees under the coalescent model. *BMC Evol. Biol.* **10**, 302.

Liu, L., L. Yu, and D. K. Pearl. 2010b. Maximum tree: a consistent estimator of the species tree. *J. Math. Biol.* **60**, 95–106.

Liu, L., L. Yu, D. K. Pearl, and S. V. Edwards. 2009a. Estimating species phylogenies using coalescence times among sequences. *Syst. Biol.* **58**, 468–477.

Liu, L., L. Yu, L. Kubatko et al. 2009b. Coalescent methods for estimating phylogenetic trees. *Mol. Phylogenet. Evol.* **53**, 320–328.

Lockhart, P., P. Novis, B. G. Milligan et al. 2006. Heterotachy and tree building: a case study with plastids and Eubacteria. *Mol. Biol. Evol.* **23**, 40–45.

Lockhart, P. J., M. A. Steel, M. D. Hendy, and D. Penny. 1994. Recovering evolutionary trees under a more realistic model of sequence evolution. *Mol. Biol. Evol.* **11**, 605–612.

Lohse, K., and N. H. Barton. 2011. A general method for calculating likelihoods under the coalescent process. *Genetics*. **189**, 977–987.

Löytynoja, A., and N. Goldman. 2005. An algorithm for progressive multiple alignment of sequences with insertions. *Proc. Natl. Acad. Sci. U.S.A.* **102**, 10557–10562.

Löytynoja, A., and N. Goldman. 2008. Phylogeny-aware gap placement prevents errors in sequence alignment and evolutionary analysis. *Science* **320**, 1632–1635.

Lunter, G., I. Miklos, A. Drummond et al. 2005. Bayesian coestimation of phylogeny and sequence alignment. *BMC Bioinformatics* **6**, 83.

Lynch, M. 2010. Rate, molecular spectrum, and consequences of human mutation. *Proc. Natl. Acad. Sci. U.S.A.* **107**, 961–968.

Lynch, M., and J. S. Conery. 2000. The evolutionary fate and consequences of duplicate genes. *Science* **290**, 1151–1155.

Maddison, D. 1991. The discovery and importance of multiple islands of most-parsimonious trees. *Syst. Zool.* **33**, 83–103.

Maddison, D. R., and W. P. Maddison. 2000. *MacClade 4, Analysis of Phylogeny and Character Evolution*. Sinauer Associates, Inc., Sunderland, Massachusettes.

Maddison, W. P. 1995. Calculating the probability distributions of ancestral states reconstructed by parsimony on phylogenetic trees. *Syst. Biol.* **44**, 474–481.

Maddison, W. P. 1997. Gene trees in species trees. *Syst. Biol.* **46**, 523–536.

Maddison, W. P., and L. L. Knowles. 2006. Inferring phylogeny despite incomplete lineage sorting. *Syst. Biol.* **55**, 21–30.

Maddison, W. P., and D. R. Maddison. 1982. *MacClade: Analysis of Phylogeny and Character Evolution*. Sinauer Associates, Inc., Sunderland, Massachusettes.

Magallon, S., K. W. Hilu, and D. Quandt. 2013. Land plant evolutionary timeline: gene effects are secondary to fossil constraints in relaxed clock estimation of age and substitution rates. *Am. J. Bot.* **100**, 556–573.

Makova, K. D., M. Ramsay, T. Jenkins, and W. H. Li. 2001. Human DNA sequence variation in a 6.6-kb region containing the melanocortin 1 receptor promoter. *Genetics* **158**, 1253–1268.

Malcolm, B. A., K. P. Wilson, B. W. Matthews et al. 1990. Ancestral lysozymes reconstructed, neutrality tested, and thermostability linked to hydrocarbon packing. *Nature* **345**, 86–89.

Mallet, J. 2013. Concepts of species. Pp. 679–691 *in* S. A. Levin, ed. *Encyclopedia of Biodiversity*. Academic Press, Massachusetts.

Mallick, S., S. Gnerre, P. Muller, and D. Reich. 2010. The difficulty of avoiding false positives in genome scans for natural selection. *Genome Res.* **19**, 922–933.

Margoliash, E. 1963. Primary structure and evolution of cytochrome c. *Proc. Natl. Acad. Sci. U.S.A.* **50**, 672–679.

Marinari, E., and G. Parisi. 1992. Simulated tempering: a new Monte Carlo scheme. *Europhys. Lett.* **19**, 451–458.

Maritz, J. S., and T. Lwin. 1989. *Empirical Bayes Methods*. Chapman and Hall, London.

Marko, P. B. 2002. Fossil calibration of molecular clocks and the divergence times of geminate species pairs separated by the Isthmus of Panama. *Mol. Biol. Evol.* **19**, 2005–2021.

Marsaglia, G., and W. W. Tsang. 2000. A simple method for generating gamma variables. *ACM Trans. Math. Soft.* **26**, 363–372.

Marshall, D. C. 2010. Cryptic failure of partitioned Bayesian phylogenetic analyses: lost in the land of long trees. *Syst. Biol.* **59**, 108–117.

Martin, A. P., and S. R. Palumbi. 1993. Body size, metabolic rate, generation time, and the molecular clock. *Proc. Natl. Acad. Sci. U.S.A.* **90**, 4087–4091.

Massingham, T., and N. Goldman. 2005. Detecting amino acid sites under positive selection and purifying selection. *Genetics* **169**, 1753–1762.

Massingham, T., and N. Goldman. 2007. Statistics of the log-det estimator. *Mol. Biol. Evol.* **24**, 2277–2285.

Maston, G. A., and M. Ruvolo. 2002. Chorionic gonadotropin has a recent origin within primates and an evolutionary history of selection. *Mol. Biol. Evol.* **19**, 320–335.

Mateiu, L. M., and B. Rannala. 2006. Inferring complex DNA substitution processes on phylogenies using uniformization and data augmentation. *Syst. Biol.* **55**, 259–269.

Mau, B., and M. A. Newton. 1997. Phylogenetic inference for binary data on dendrograms using Markov chain Monte Carlo. *J. Comput. Graph. Stat.* **6**, 122–131.

Mau, B., M. A. Newton, and B. Larget. 1999. Bayesian phylogenetic inference via Markov chain Monte Carlo methods. *Biometrics* **55**, 1–12.

Maynard Smith, J., and J. Haigh. 1974. The hitch-hiking effect of a favorable gene. *Genet. Res. (Camb.)* **23**, 23–35.

Mayr, E. 1942. *Systematics and the Origin of Species from the Viewpoint of a Zoologist*. Columbia University Press, New York.

Mayrose, I., N. Friedman, and T. Pupko. 2005. A gamma mixture model better accounts for among site rate heterogeneity. *Bioinformatics* **21**, 151–158.

McDonald, J. H., and M. Kreitman. 1991. Adaptive protein evolution at the *Adh* locus in Drosophila. *Nature* **351**, 652–654.

McGuire, G., M. C. Denham, and D. J. Balding. 2001. Models of sequence evolution for DNA sequences containing gaps. *Mol. Biol. Evol.* **18**, 481–490.

McVean, G. A., and B. Charlesworth. 2000. The effects of Hill-Robertson interference between weakly selected mutations on patterns of molecular evolution and variation. *Genetics* **155**, 929–944.

McVean, M., P. Awadalla, and P. Fearnhead. 2002. A coalescent-based method for detecting and estimating recombination from gene sequences. *Genetics* **160**, 1231–1241.

Mengersen, K. L., and R. L. Tweedie. 1996. Rates of convergence of the Hastings and Metropolis algorithms. *Ann. Stat.* **24**, 101–121.

Meredith, R. W., J. E. Janecka, J. Gatesy et al. 2011. Impacts of the Cretaceous terrestrial revolution and KPg extinction on mammal diversification. *Science* **334**, 521–524.

Messier, W., and C.-B. Stewart. 1997. Episodic adaptive evolution of primate lysozymes. *Nature* **385**, 151–154.

Metropolis, N., A. W. Rosenbluth, M. N. Rosenbluth et al. 1953. Equations of state calculations by fast computing machines. *J. Chem. Phys.* **21**, 1087–1092.

Meyer, A., T. D. Kocher, P. Basasibwaki, and A. C. Wilson. 1990. Monophyletic origin of Lake Victoria cichlid fishes suggested by mitochondrial DNA sequences. *Nature* **347**, 550–553.

Minin, V. N., and M. A. Suchard. 2008. Fast, accurate and simulation-free stochastic mapping. *Philos. Trans. R. Soc. Lond. B. Biol. Sci.* **363**, 3985–3995.

Minin, V. N., E. W. Bloomquist, and M. A. Suchard. 2008. Smooth skyride through a rough skyline: Bayesian coalescent-based inference of population dynamics. *Mol. Biol. Evol.* **25**, 1459–1471.

Mira, A. 2001. Ordering and improving the performance of Monte Carlo Markov chains. *Stat. Sci.* **16**, 340–350.

Miyata, T., and T. Yasunaga. 1980. Molecular evolution of mRNA: a method for estimating evolutionary rates of synonymous and amino acid substitutions from homologous nucleotide sequences and its applications. *J. Mol. Evol.* **16**, 23–36.

Miyata, T., S. Miyazawa, and T. Yasunaga. 1979. Two types of amino acid substitutions in protein evolution. *J. Mol. Evol.* **12**, 219–236.

Moler, C., and C. F. Van Loan. 2003. Nineteen dubious ways to compute the exponential of a matrix, twenty-five years later. *SIAM Review* **45**, 3–49.

Mooers, A. Ø., and D. Schluter. 1999. Reconstructing ancestor states with maximum likelihood: support for one- and two-rate models. *Syst. Biol.* **48**, 623–633.

Morgan, G. J. 1998. Emile Zuckerkandl, Linus Pauling, and the molecular evolutionary clock. *J. Hist. Biol.* **31**, 155–178.

Moriyama, E. N., and J. R. Powell. 1997. Synonymous substitution rates in *Drosophila*: mitochondrial versus nuclear genes. *J. Mol. Evol.* **45**, 378–391.

Mossel, E., and S. Roch. 2010. Incomplete lineage sorting: consistent phylogeny estimation from multiple loci. *IEEE/ACM Trans. Comput. Biol. Bioinform.* **7**, 166–171.

Mossel, E., and E. Vigoda. 2005. Phylogenetic MCMC algorithms are misleading on mixtures of trees. *Science* **309**, 2207–2209.

Mulcahy, D. G., B. P. Noonan, T. Moss et al. 2012. Estimating divergence dates and evaluating dating methods using phylogenomic and mitochondrial data in squamate reptiles. *Mol. Phylogenet. Evol.* **65**, 974–991.

Müller, T., and M. Vingron. 2000. Modeling amino acid replacement. *J. Comput. Biol.* **7**, 761–776.

Muse, S. V. 1996. Estimating synonymous and nonsynonymous substitution rates. *Mol. Biol. Evol.* **13**, 105–114.

Muse, S. V., and B. S. Gaut. 1994. A likelihood approach for comparing synonymous and nonsynonymous nucleotide substitution rates, with application to the chloroplast genome. *Mol. Biol. Evol.* **11**, 715–724.

Muse, S. V., and B. S. Gaut. 1997. Comparing patterns of nucleotide substitution rates among chloroplast loci using the relative ratio test. *Genetics* **146**, 393–399.

Muse, S. V., and B. S. Weir. 1992. Testing for equality of evolutionary rates. *Genetics* **132**, 269–276.

Nachman, M. W., S. Boyer, and C. F. Aquadro. 1996. Non-neutral evolution at the mitochondrial NADH dehydrogenase subunit 3 gene in mice. *Proc. Natl. Acad. Sci. U.S.A.* **91**, 6364–6368.

Nakamura, K., T. Oshima, T. Morimoto et al. 2011. Sequence-specific error profile of Illumina sequencers. *Nucl. Acids Res.* **39**, e90.

Nath, H. B., and R. C. Griffiths. 1993. The coalescent in two colonies with symmetric migration. *J. Math. Biol.* **31**, 841–852.

Neal, R. M. 2001. Markov chain sampling methods for Dirichlet process mixture models. *J. Comput. Graph. Stat.* **9**, 249–265.

Needleman, S. G., and C. D. Wunsch. 1970. A general method applicable to the search for similarities in the amino acid sequence of two proteins. *J. Mol. Biol.* **48**, 443–453.

Nei, M. 1977. Standard error of immunological dating of evolutionary time. *J. Mol. Evol.* **9**, 203–211.

Nei, M. 1987. *Molecular Evolutionary Genetics*. Columbia University Press, New York.

Nei, M. 1996. Phylogenetic analysis in molecular evolutionary genetics. *Ann. Rev. Genet.* **30**, 371–403.

Nei, M., and T. Gojobori. 1986. Simple methods for estimating the numbers of synonymous and nonsynonymous nucleotide substitutions. *Mol. Biol. Evol.* **3**, 418–426.

Nei, M., and L. Jin. 1989. Variances of the average numbers of nucleotide substitutions within and between populations. *Mol. Biol. Evol.* **6**, 290–300.

Nei, M., and S. Kumar. 2000. *Molecular evolution and phylogenetics*. Oxford University Press, Oxford, UK.

Nei, M., S. Kumar, and K. Takahashi. 1998. The optimization principle in phylogenetic analysis tends to give incorrect topologies when the number of nucleotides or amino acids used is small [In Process Citation]. *Proc. Natl. Acad. Sci. U.S.A.* **95**, 12390–12397.

Nei, M., J. C. Stephens, and N. Saitou. 1985. Methods for computing the standard errors of branching points in an evolutionary tree and their application to molecular data from humans and apes. *Mol. Biol. Evol.* **2**, 66–85.

Newton, M. A. 1996. Bootstrapping phylogenies: large deviations and dispersion effects. *Biometrika* **83**, 315–328.

Newton, M. A., and A. E. Raftery. 1994. Approximating Bayesian inference with the weigthed likelihood bootstrap. *J. R. Stat. Soc. B* **56**, 3–48.

Neyman, J. 1971. Molecular studies of evolution: a source of novel statistical problems. Pp. 1–27 *in* S. S. Gupta, and J. Yackeleds., *Statistical Decision Theory and Related Topics*. Academic Press, New York.

Nichols, R. 2001. Gene trees and species trees are not the same. *Trends Ecol. Evol.* **16**, 358–364.

Nielsen, R. 1997. Site-by-site estimation of the rate of substitution and the correlation of rates in mitochondrial DNA. *Syst. Biol.* **46**, 346–353.

Nielsen, R. 2001. Statistical tests of selective neutrality in the age of genomics. *Heredity* **86**, 641–647.

Nielsen, R. 2002. Mapping mutations on phylogenies. *Syst. Biol.* **51**, 729–739.

Nielsen, R., and J. Wakeley. 2001. Distinguishing migration from isolation: a Markov chain Monte Carlo approach. *Genetics* **158**, 885–896.

Nielsen, R., and Z. Yang. 1998. Likelihood models for detecting positively selected amino acid sites and applications to the HIV-1 envelope gene. *Genetics* **148**, 929–936.

Nixon, K. C. 1999. The parsimony ratchet, a new method for rapid parsimony analysis. *Cladistics* **15**, 407–414.

Nordborg, M. 2007. Coalescent theory. Pp. 843–877 *in* D. Balding, M. Bishop, and C. Cannings, eds. *Handbook of Statistical Genetics*. Wiley, New York..

Norris, J. R. 1997. *Markov Chains*. Cambridge University Press, Cambridge, UK.

Notohara, M. 1990. The coalescent and the genealogical process in geographically structured populations. *J. Math. Biol.* **29**, 59–75.

Nylander, J. A. A., F. Ronquist, J. P. Huelsenbeck, and J. L. Nieves-Aldrey. 2004. Bayesian phylogenetic analysis of combined data. *Syst. Biol.* **53**, 47–67.

O'Hagan, A., and J. Forster. 2004. *Kendall's Advanced Theory of Statistics: Bayesian Inference*. Arnold, London.

O'Meara, B. C. 2010. New heuristic methods for joint species delimitation and species tree inference. *Syst. Biol.* **59**, 59–73.

Ohno, S. 1970. *Evolution by Gene Duplication*. Springer-Verlag, New York.

Ohta, T. 1973. Slightly deleterious mutant substitutions in evolution. *Nature* **246**, 96–98.

Ohta, T. 1992. Theoretical study of near neutrality. II. Effect of subdivided population structure with local extinction and recolonization. *Genetics* **130**, 917–923.

Ohta, T. 1995. Synonymous and nonsynonymous substitutions in mammalian genes and the nearly neutral theory. *J. Mol. Evol.* **40**, 56–63.

Ohta, T. 2002. Near-neutrality in evolution of genes and gene regulation. *Proc. Natl. Acad. Sci. U.S.A.* **99**, 16134–16137.

Ohta, T., and J. H. Gillespie. 1996. Development of neutral and nearly neutral theories. *Theor. Popul. Biol.* **49**, 128–142.

Ohta, T., and M. Kimura. 1971. On the constancy of the evolutionary rate of cistrons. *J. Mol. Evol.* **1**, 18–25.

Ohta, T., and H. Tachida. 1990. Theoretical study of near neutrality. I. Heterozygosity and rate of mutant substitution. *Genetics* **126**, 219–229.

Oliver, J. C. 2013. Microevolutionary processes generate phylogenomic discordance at ancient divergences. *Evolution* **67**, 1823–1830.

Olsen, G. J., H. Matsuda, R. Hagstrom, and R. Overbeek. 1994. fastDNAML: a tool for construction of phylogenetic trees of DNA sequences using maximum likelihood. *Comput. Appl. Biosci.* **10**, 41–48.

Opgen-Rhein, R., L. Fahrmeir, and K. Strimmer. 2005. Inference of demographic history from genealogical trees using reversible jump Markov chain Monte Carlo. *BMC Evol. Biol.* **5**, 6.

Osawa, S., and T. H. Jukes. 1989. Codon reassignment (codon capture) in evolution. *J. Mol. Evol.* **28**, 271–278.

Page, R. D., and M. A. Charleston. 1997. From gene to organismal phylogeny: reconciled trees and the gene tree/species tree problem. *Mol. Phylogenet. Evol.* **7**, 231–240.

Pagel, M. 1994. Detecting correlated evolution on phylogenies: a general method for the comparative analysis of discrete characters. *Proc. R. Soc. Lond. B. Biol. Sci.* **255**, 37–45.

Pagel, M. 1999. The maximum likelihood approach to reconstructing ancestral character states of discrete characters on phylogenies. *Syst. Biol.* **48**, 612–622.

Pagel, M., and A. Meade. 2004. A phylogenetic mixture model for detecting pattern-heterogeneity in gene sequence or character-state data. *Syst. Biol.* **53**, 571–581.

Pagel, M., and A. Meade. 2008. Modelling heterotachy in phylogenetic inference by reversible-jump Markov chain Monte Carlo. *Philos. Trans. R. Soc. Lond. B. Biol. Sci.* **363**, 3955–3964.

Pagel, M., A. Meade, and D. Barker. 2004. Bayesian estimation of ancestral character states on phylogenies. *Syst. Biol.* **53**, 673–684.

Palumbi, S. R. 1994. Genetic divergence, reproductive isolation and marine speciation. *Ann. Rev. Ecol. Syst.* **25**, 547–572.

Pamilo, P., and N. O. Bianchi. 1993. Evolution of the *Zfx* and *Zfy* genes – rates and interdependence between the genes. *Mol. Biol. Evol.* **10**, 271–281.

Pamilo, P., and M. Nei. 1988. Relationships between gene trees and species trees. *Mol. Biol. Evol.* **5**, 568–583.

Papadopoulou, A., I. Anastasiou, and A. P. Vogler. 2010. Revisiting the insect mitochondrial molecular clock: the mid-Aegean trench calibration. *Mol. Biol. Evol.* **27**, 1659–1672.

Parham, J., P. Donoghue, C. Bell et al. 2012. Best practices for applying paleontological data to molecular divergence dating analyses. *Syst. Biol.* **61**, 346–359.

Paterson, S., T. Vogwill, A. Buckling et al. 2010. Antagonistic coevolution accelerates molecular evolution. *Nature* **464**, 275–278.

Pauling, L., and E. Zuckerkandl. 1963. Chemical paleogenetics: molecular "restoration studies" of extinct forms of life. *Acta Chem. Scand.* **17**, S9–S16.

Pauplin, Y. 2000. Direct calculation of a tree length using a distance matrix. *J. Mol. Evol.* **51**, 41–47.

Penny, D., and M. D. Hendy. 1985. The use of tree comparison metrics. *Syst. Zool.* **34**, 75–82.

Perler, F., A. Efstratiadis, P. Lomedica et al. 1980. The evolution of genes: the chicken preproinsulin gene. *Cell* **20**, 555–566.

Perna, N. T., and T. D. Kocher. 1995. Unequal base frequencies and the estimation of substitution rates. *Mol. Biol. Evol.* **12**, 359–361.

Peskun, P. H. 1973. Optimum Monte-Carlo sampling using Markov chains. *Biometrika* **60**, 607–612.

Philippe, H., Y. Zhou, H. Brinkmann et al. 2005. Heterotachy and long-branch attraction in phylogenetics. *BMC Evol. Biol.* **5**, 50.

Pickett, K. M., and C. P. Randle. 2005. Strange Bayes indeed: uniform topological priors imply non-uniform clade priors. *Mol. Phylogenet. Evol.* **34**, 203–211.

Polley, S. D., and D. J. Conway. 2001. Strong diversifying selection on domains of the *Plasmodium falciparum* apical membrane antigen 1 gene. *Genetics* **158**, 1505–1512.

Pons, J., T. G. Barraclough, J. Gomez-Zurita et al. 2006. Sequence-based species delimitation for the DNA taxonomy of undescribed insects. *Syst. Biol.* **55**, 595–609.

Posada, D. 2008. jModelTest: phylogenetic model averaging. *Mol. Biol. Evol.* **25**, 1253–1256.

Posada, D., and T. R. Buckley. 2004. Model selection and model averaging in phylogenetics: advantages of Akaike Informtaion Criterion and Bayesian approaches over likelihood ratio tests. *Syst. Biol.* **53**, 793–808.

Posada, D., and K. Crandall. 2001. Simple (wrong) models for complex trees: a case from retroviridae. *Mol. Biol. Evol.* **18**, 271–275.

Posada, D., and K. A. Crandall. 1998. MODELTEST: testing the model of DNA substitution. *Bioinformatics* **14**, 817–818.

Prince, V. E., and F. B. Pickett. 2002. Splitting pairs: the diverging fates of duplicated genes. *Nat. Rev. Genet.* **3**, 827–837.

Pritchard, J. K., M. Stephens, and P. Donnelly. 2000. Inference of population structure using multilocus genotype data. *Genetics* **155**, 945–959.

Pupko, T., I. Pe'er, R. Shamir, and D. Graur. 2000. A fast algorithm for joint reconstruction of ancestral amino acid sequences. *Mol. Biol. Evol.* **17**, 890–896.

Pupko, T., D. Huchon, Y. Cao et al. 2002a. Combining multiple data sets in a likelihood analysis: which models are the best? *Mol. Biol. Evol.* **19**, 2294–2307.

Pupko, T., I. Pe'er, M. Hasegawa et al. 2002b. A branch-and-bound algorithm for the inference of ancestral amino-acid sequences when the replacement rate varies among sites: application to the evolution of five gene families. *Bioinformatics* **18**, 1116–1123.

Pybus, O. G., A. Rambaut, and P. H. Harvey. 2000. An integrated framework for the inference of viral population history from reconstructed genealogies. *Genetics* **155**, 1429–1437.

Raaum, R. L., K. N. Sterner, C. M. Noviello et al. 2005. Catarrhine primate divergence dates estimated from complete mitochondrial genomes: concordance with fossil and nuclear DNA evidence. *J. Hum. Evol.* **48**, 237–257.

Rambaut, A. 2000. Estimating the rate of molecular evolution: incorporating non-comptemporaneous sequences into maximum likelihood phylogenetics. *Bioinformatics* **16**, 395–399.

Rambaut, A., and L. Bromham. 1998. Estimating divergence dates from molecular sequences. *Mol. Biol. Evol.* **15**, 442–448.

Rambaut, A., and N. C. Grassly. 1997. Seq-Gen: an application for the Monte Carlo simulation of DNA sequence evolution along phylogenetic trees. *CABIOS* **13**, 235–238.

Rand, D., M. Dorfsman, and L. Kann. 1994. Neutral and nonneutral evolution of *Drosophila* mitochondrial DNA. *Genetics* **138**, 741–756.

Rand, D. M., D. M. Weinreich, and B. O. Cezairliyan. 2000. Neutrality tests of conservative-radical amino acid changes in nuclear- and mitochondrially-encoded proteins. *Gene* **261**, 115–125.

Rannala, B. 2002. Identifiability of parameters in MCMC Bayesian inference of phylogeny. *Syst. Biol.* **51**, 754–760.

Rannala, B., and Z. Yang. 1996. Probability distribution of molecular evolutionary trees: a new method of phylogenetic inference. *J. Mol. Evol.* **43**, 304–311.

Rannala, B., and Z. Yang. 2003. Bayes estimation of species divergence times and ancestral population sizes using DNA sequences from multiple loci. *Genetics* **164**, 1645–1656.

Rannala, B., and Z. Yang. 2007. Inferring speciation times under an episodic molecular clock. *Syst. Biol.* **56**, 453–466.

Rannala, B., and Z. Yang. 2013. Improved reversible jump algorithms for Bayesian species delimitation. *Genetics* **194**, 245–253.

Rannala, B., T. Zhu, and Z. Yang. 2012. Tail paradox, partial identifiability and influential priors in Bayesian branch length inference. *Mol. Biol. Evol.* **29**, 325–335.

Ranwez, V., and O. Gascuel. 2002. Improvement of distance-based phylogenetic methods by a local maximum likelihood approach using triplets. *Mol. Biol. Evol.* **19**, 1952–1963.

Redelings, B. D., and M. A. Suchard. 2005. Joint Bayesian estimation of alignment and phylogeny. *Syst. Biol.* **54**, 401–418.

Reeves, J. H. 1992. Heterogeneity in the substitution process of amino acid sites of proteins coded for by mitochondrial DNA. *J. Mol. Evol.* **35**, 17–31.

Ren, F., H. Tanaka, and Z. Yang. 2005. An empirical examination of the utility of codon-substitution models in phylogeny reconstruction. *Syst. Biol.* **54**, 808–818.

Ren, F., H. Tanaka, and Z. Yang. 2009. A likelihood look at the supermatrix-supertree controversy. *Gene* **441**, 119–125.

Ripley, B. 1987. *Stochastic Simulation*. Wiley, New York.

Roach, J. C., G. Glusman, A. F. A. Smit et al. 2010. Analysis of genetic inheritance in a family quartet by whole-genome sequencing. *Science* **328**, 636–639.

Robbins, H. 1955. An empirical Bayes approach to statistics. *Proc. 3rd Berkeley Symp. Math. Stat. Prob.* **1**, 157–164.

Robbins, H. 1983. Some thoughts on empirical Bayes estimation. *Ann. Stat.* **1**, 713–723.

Robert, C. P., and G. Casella. 2004. *Monte Carlo Statistical Methods*. Springer-Verlag, New York.

Roberts, G. O., and R. L. Tweedie. 1996. Geometric convergence and central limit theorems for multidimensional Hastings and Metropolis algorithms. *Biometrika* **83**, 95–110.

Robinson, D. F., and L. R. Foulds. 1981. Comparison of phylogenetic trees. *Math. Biosci.* **53**, 131–147.

Rocha, E. P., J. M. Smith, L. D. Hurst et al. 2006. Comparisons of dN/dS are time dependent for closely related bacterial genomes. *J. Theor. Biol.* **239**, 226–235.

Rodrigo, A. G., and J. Felsenstein. 1999. Coalescent approaches to HIV population genetics. pp. 233–271 in K. Crandall, ed. *Molecular Evolution of HIV*. Johns Hopkins University Press, Baltimore, Maryland.

Rodrigue, N., H. Philippe, and N. Lartillot. 2008. Uniformization for sampling realizations of Markov processes: applications to Bayesian implementations of codon substitution models. *Bioinformatics* **24**, 56–62.

Rodrigue, N., H. Philippe, and N. Lartillot. 2010. Mutation-selection models of coding sequence evolution with site-heterogeneous amino acid fitness profiles. *Proc. Natl. Acad. Sci. U.S.A.* **107**, 4629–4634.

Rodriguez-Trelles, F., R. Tarrio, and F. J. Ayala. 2003. Convergent neofunctionalization by positive Darwinian selection after ancient recurrent duplications of the xanthine dehydrogenase gene. *Proc. Natl. Acad. Sci. U.S.A.* **100**, 13413–13417.

Rodriguez, F., J. F. Oliver, A. Marin, and J. R. Medina. 1990. The general stochastic model of nucleotide substitutions. *J. Theor. Biol.* **142**, 485–501.

Rogers, J. S. 1997. On the consistency of maximum likelihood estimation of phylogenetic trees from nucleotide sequences. *Syst. Biol.* **46**, 354–357.

Rogers, J. S., and D. L. Swofford. 1998. A fast method for approximating maximum likelihoods of phylogenetic trees from nucleotide sequences. *Syst. Biol.* **47**, 77–89.

Rogers, J. S., and D. L. Swofford. 1999. Multiple local maxima for likelihoods of phylogenetic trees: a simulation study. *Mol. Biol. Evol.* **16**, 1079–1085.

Rokas, A., D. Kruger, and S. B. Carroll. 2005. Animal evolution and the molecular signature of radiations compressed in time. *Science* **310**, 1933–1938.

Ronquist, F. 1998. Fast Fitch-parsimony algorithms for large data sets. *Cladistics* **14**, 387–400.

Ronquist, F., and J. P. Huelsenbeck. 2003. MrBayes 3, Bayesian phylogenetic inference under mixed models. *Bioinformatics* **19**, 1572–1574.

Ronquist, F., S. Klopfstein, L. Vilhelmsen et al. 2012a. A total-evidence approach to dating with fossils, applied to the early radiation of the Hymenoptera. *Syst. Biol.* **61**, 973–999.

Ronquist, F., M. Teslenko, P. van der Mark et al. 2012b. MrBayes 3.2, efficient Bayesian phylogenetic inference and model choice across a large model space. *Syst. Biol.* **61**, 539–542.

Rosenberg, N. A., and M. Nordborg. 2002. Genealogical trees, coalescent theory and the analysis of genetic polymorphisms. *Nat. Rev. Genet.* **3**, 380–390.

Ross, R. 1997. *Simulation*. Academic Press, London.

Ross, S. 1996. *Stochastic Processes*. Springer-Verlag, New York.

Rota-Stabelli, O., Z. Yang, and M. Telford. 2009. MtZoa: a general mitochondrial amino acid substitutions model for animal evolutionary studies. *Mol. Phylogenet. Evol.* **52**, 268–272.

Roth, C., M. J. Betts, P. Steffansson et al. 2005. The Adaptive Evolution Database (TAED): a phylogeny based tool for comparative genomics. *Nucl. Acids Res.* **33**, D495–D497.

Rubin, D. B., and N. Schenker. 1986. Efficiently simulating the coverage properties of interval estimates. *Appl. Stat.* **35**, 159–167.

Rubinstein, N. D., I. Mayrose, A. Doron-Faigenboim, and T. Pupko. 2011. Evolutionary models accounting for layers of selection in protein coding genes and their impact on the inference of positive selection. *Mol. Biol. Evol.* **28**, 3297–3308.

Russo, C. A., N. Takezaki, and M. Nei. 1996. Efficiencies of different genes and different tree-building methods in recovering a known vertebrate phylogeny. *Mol. Biol. Evol.* **13**, 525–536.

Rzhetsky, A. 1995. Estimating substitution rates in ribosomal RNA genes. *Genetics* **141**, 771–783.

Rzhetsky, A., and M. Nei. 1992. A simple method for estimating and testing minimum-evolution trees. *Mol. Biol. Evol.* **9**, 945–967.

Rzhetsky, A., and M. Nei. 1993. Theoretical foundation of the minimum-evolution method of phylogenetic inference. *Mol. Biol. Evol.* **10**, 1073–1095.

Rzhetsky, A., and M. Nei. 1994. Unbiased estimates of the number of nucleotide substitutions when substitution rate varies among different sites. *J. Mol. Evol.* **38**, 295–299.

Rzhetsky, A., and M. Nei. 1995. Tests of applicability of several substitution models for DNA sequence data. *Mol. Biol. Evol.* **12**, 131–151.

Rzhetsky, A., and T. Sitnikova. 1996. When is it safe to use an oversimplified substitution model in tree-making? *Mol. Biol. Evol.* **13**, 1255–1265.

Saitou, N. 1988. Property and efficiency of the maximum likelihood method for molecular phylogeny. *J. Mol. Evol.* **27**, 261–273.

Saitou, N., and M. Nei. 1986. The number of nucleotides required to determine the branching order of three species, with special reference to the human-chimpanzee-gorilla divergence. *J. Mol. Evol.* **24**, 189–204.

Saitou, N., and M. Nei. 1987. The neighbor-joining method: a new method for reconstructing phylogenetic trees. *Mol. Biol. Evol.* **4**, 406–425.

Salter, L. A. 2001. Complexity of the likelihood surface for a large DNA dataset. *Syst. Biol.* **50**, 970–978.

Salter, L. A., and D. K. Pearl. 2001. Stochastic search strategy for estimation of maximum likelihood phylogenetic trees. *Syst. Biol.* **50**, 7–17.

Sanderson, M. J. 1997. A nonparametric approach to estimating divergence times in the absence of rate constancy. *Mol. Biol. Evol.* **14**, 1218–1232.

Sanderson, M. J. 2002. Estimating absolute rates of molecular evolution and divergence times: a penalized likelihood approach. *Mol. Biol. Evol.* **19**, 101–109.

Sanderson, M. J., and J. Kim. 2000. Parametric phylogenetics? *Syst. Biol.* **49**, 817–829.

Sankoff, D. 1975. Minimal mutation trees of sequences. *SIAM J. Appl. Math.* **28**, 35–42.

Sarich, V. M., and A. C. Wilson. 1967. Rates of albumin evolution in primates. *Proc. Natl. Acad. Sci. U.S.A.* **58**, 142–148.

Sarich, V. M., and A. C. Wilson. 1973. Generation time and genomic evolution in primates. *Science* **179**, 1144–1147.

Satta, Y., M. Hickerson, H. Watanabe et al. 2004. Ancestral population sizes and species divergence times in the primate lineage on the basis of intron and BAC end sequences. *J. Mol. Evol.* **59**, 478–487.

Saunders, I. W., S. Tavaré, and G. A. Watterson. 1984. On the genealogy of nested subsamples from a haploid population. *Adv. Appl. Prob.* **16**, 471–491.

Savage, L. J. 1962. *The Foundations of Statistical Inference*. Metheun & Co., London.

Savill, N. J., D. C. Hoyle, and P. G. Higgs. 2001. RNA sequence evolution with secondary structure constraints: comparison of substitution rate models using maximum-likelihood methods. *Genetics* **157**, 399–411.

Sawyer, K. R. 1984. Multiple hypothesis testing. *J. R. Stat. Soc. B.* **46**, 419–424.

Sawyer, S. A., and D. L. Hartl. 1992. Population genetics of polymorphism and divergence. *Genetics* **132**, 1161–1176.

Sawyer, S. L., L. I. Wu, M. Emerman, and H. S. Malik. 2005. Positive selection of primate TRIM5a identifies a critical species-specific retroviral restriction domain. *Proc. Natl. Acad. Sci. U.S.A.* **102**, 2832–2837.

Scally, A., J. Y. Dutheil, L. W. Hillier et al. 2012. Insights into hominid evolution from the gorilla genome sequence. *Nature* **483**, 169–175.

Scheffler, K., and C. Seoighe. 2005. A Bayesian model comparison approach to inferring positive selection. *Mol. Biol. Evol.* **22**, 2531–2540.

Schluter, D. 1995. Uncertainty in ancient phylogenies. *Nature* **377**, 108–110.

Schluter, D. 2000. *The Ecology of Adaptive Radiation*. Oxford University Press, Oxford, UK.

Schmidt-Lebuhn, A. N., J. M. de Vos, B. Keller, and E. Conti. 2012. Phylogenetic analysis of Primula section Primula reveals rampant non-monophyly among morphologically distinct species. *Mol. Phylogenet. Evol.* **65**, 23–34.

Schmidt, H. A., K. Strimmer, M. Vingron, and A. von Haeseler. 2002. TREE-PUZZLE: maximum likelihood phylogenetic analysis using quartets and parallel computing. *Bioinformatics* **18**, 502–504.

Schneider, A., G. M. Cannarozzi, and G. Gonnet. 2005. Empirical codon substitution matrix. *BMC Bioinformatics* **6**, 134.

Schneider, A., A. Souvorov, N. Sabath et al. 2009. Estimates of positive Darwinian selection are inflated by errors in sequencing, annotation, and alignment. *Genome Biol. Evol.* **2009**, 114–118.

Schoeniger, M., and A. von Haeseler. 1994. A stochastic model for the evolution of autocorrelated DNA sequences. *Mol. Phylogenet. Evol.* **3**, 240–247.

Schott, J. R. 1997. *Matrix Analysis for Statistics*. Wiley, New York.

Schultz, T. R., and G. A. Churchill. 1999. The role of subjectivity in reconstructing ancestral character states: a Bayesian approach to unknown rates, states, and transformation asymmetries. *Syst. Biol.* **48**, 651–664.

Schwarz, G. 1978. Estimating the dimension of a model. *Ann. Stat.* **6**, 461–464.

Self, S. G., and K.-Y. Liang. 1987. Asymptotic properties of maximum likelihood estimators and likelihood ratio tests under nonstandard conditions. *J. Am. Stat. Assoc.* **82**, 605–610.

Semple, C., and M. Steel. 2003. *Phylogenetics*. Oxford University Press, New York.

Seo, T. K., H. Kishino, and J. L. Thorne. 2004. Estimating absolute rates of synonymous and non-synonymous nucleotide substitution in order to characterize natural selection and date species divergences. *Mol. Biol. Evol.* **21**, 1201–1213.

Shackelton, L. A., C. R. Parrish, U. Truyen, and E. C. Holmes. 2005. High rate of viral evolution associated with the emergence of carnivore parvovirus. *Proc. Natl. Acad. Sci. U.S.A.* **102**, 379–384.

Shan, H., L. Zahn, S. Guindon et al. 2009. Evolution of plant MADS box transcription factors: evidence for shifts in selection associated with early angiosperm diversification and concerted gene duplications. *Mol. Biol. Evol.* **26**, 2229–2244.

Shapiro, B., A. Rambaut, and A. J. Drummond. 2006. Choosing appropriate substitution models for the phylogenetic analysis of protein-coding sequences. *Mol. Biol. Evol.* **23**, 7–9.

Sharp, P. M. 1997. In search of molecular Darwinism. *Nature* **385**, 111–112.

Shaw, K. L. 1998. Species and the diversity of natural groups. pp. 44–56 *in* D. J. Howard, and S. J. Berlocher, eds. *Endless Forms: Species and Speciation*. Oxford University Press, Oxford, U.K.

Shimodaira, H. 2002. An approximately unbiased test of phylogenetic tree selection. *Syst. Biol.* **51**, 492–508.

Shimodaira, H., and M. Hasegawa. 1999. Multiple comparisons of log-likelihoods with applications to phylogenetic inference. *Mol. Biol. Evol.* **16**, 1114–1116.

Shimodaira, H., and M. Hasegawa. 2001. CONSEL: for assessing the confidence of phylogenetic tree selection. *Bioinformatics* **17**, 1246–1247.

Shindyalov, I. N., N. A. Kolchanov, and C. Sander. 1994. Can three-dimensional contacts in protein structures be predicted by analysis of correlated mutations? *Protein Eng.* **7**, 349–358.

Shriner, D., D. C. Nickle, M. A. Jensen, and J. I. Mullins. 2003. Potential impact of recombination on sitewise approaches for detecting positive natural selection. *Genet. Res.* **81**, 115–121.

Siddall, M. E. 1998. Success of parsimony in the four-taxon case: long branch repulsion by likelihood in the Farris zone. *Cladistics* **14**, 209–220.

Siepel, A., and D. Haussler. 2004. Phylogenetic estimation of context-dependent substitution rates by maximum likelihood. *Mol. Biol. Evol.* **21**, 468–488.

Sievers, F., A. Wilm, D. Dineen et al. 2011. Fast, scalable generation of high-quality protein multiple sequence alignments using Clustal Omega. *Mol. Syst. Biol.* **7**, 539.

Silverman, B. W. 1986. *Density Estimation for Statistics and Data Analysis*. Chapman and Hall, London.

Simes, R. J. 1986. An improved Bonferroni procedure for multiple tests of significance. *Biometrika* **73**, 751–754.

Simmons, M. P., K. M. Pickett, and M. Miya. 2004. How meaningful are Bayesian support values? *Mol. Biol. Evol.* **21**, 188–199.

Simonsen, K. L., G. A. Churchill, and C. F. Aquadro. 1995. Properties of statistical tests of neutrality for DNA polymorphism data. *Genetics* **141**, 413–429.

Singh, N. D., P. F. Arndt, A. G. Clark, and C. F. Aquadro. 2009. Strong evidence for lineage and sequence specificity of substitution rates and patterns in Drosophila. *Mol. Biol. Evol.* **26**, 1591–1605.

Sites, J. W., and J. C. Marshall. 2004. Delimiting species: a renaissance issue in systematic biology. *Trends Ecol. Evol.* **18**, 462–470.

Sitnikova, T., A. Rzhetsky, and M. Nei. 1995. Interior-branch and bootstrap tests of phylogenetic trees. *Mol. Biol. Evol.* **12**, 319–333.

Slatkin, M. 1991. Inbreeding coefficients and coalescence times. *Genet. Res.* **58**, 167–175.

Slatkin, M., and R. R. Hudson. 1991. Pairwise comparisons of mitochondrial DNA sequences in stable and exponentially growing populations. *Genetics* **129**, 555–562.

Sneath, P. H. A. 1962. The construction of taxonomic groups. Pp. 289–332 *in* G. C. Ainsworth, and P. H. A. Sneath, eds. *Microbial Classification*. Cambridge University Press, Cambridge, UK.

Sober, E. 1988. *Reconstructing the Past: Parsimony, Evolution, and Inference*. MIT Press, Cambridge, Massachusetts.

Sober, E. 2004. The contest between parsimony and likelihood. *Syst. Biol.* **53**, 644–653.

Sokal, A. D. 1989. *Monte Carlo Methods in Statistical Mechanics: Foundations and New Algorithms*. Lecture Notes for the Cours de Troisieme Cycle de la Physique en Suisse Romande, Lausanne, Switzerland (June 1989).

Sourdis, J., and M. Nei. 1988. Relative efficiencies of the maximum parsimony and distance-matrix methods in obtaining the correct phylogenetic tree. *Mol. Biol. Evol.* **5**, 298–311.

Spencer, M., E. Susko, and A. J. Roger. 2005. Likelihood, parsimony, and heterogeneous evolution. *Mol. Biol. Evol.* **22**, 1161–1164.

Springer, M. S., W. J. Murphy, E. Eizirik, and S. J. O'Brien. 2003. Placental mammal diversification and the Cretaceous-Tertiary boundary. *Proc. Natl. Acad. Sci. U.S.A.* **100**, 1056–1061.

Stackhouse, J., S. R. Presnell, G. M. McGeehan et al. 1990. The ribonuclease from an ancient bovid ruminant. *FEBS Lett.* **262**, 104–106.

Stadler, T. 2010. Sampling-through-time in birth-death trees. *J. Theor. Biol.* **267**, 396–404.

Stadler, T. 2013. How can we improve accuracy of macroevolutionary rate estimates. *Syst. Biol.* **62**, 321–329.

Stadler, T., and Z. Yang. 2013. Dating phylogenies with sequentially sampled tips. *Syst. Biol.* **62**, 674–688.

Stadler, T., R. Kouyos, V. von Wyl et al. 2012. Estimating the basic reproductive number from viral sequence data. *Mol. Biol. Evol.* **29**, 347–357.

Stamatakis, A. 2006. RAxML-VI-HPC: maximum likelihood-based phylogenetic analyses with thousands of taxa and mixed models. *Bioinformatics* **22**, 2688–2690.

Stamatakis, A., A. J. Aberer, C. Goll et al. 2012. RAxML-Light: a tool for computing terabyte phylogenies. *Bioinformatics* **28**, 2064–2066.

Steel, M. 2011. Can we avoid 'SIN' in the house of 'no common mechanism'? *Mol. Biol. Evol.* **60**, 96–109.

Steel, M., and K. M. Pickett. 2006. On the impossibility of uniform priors on clades. *Mol. Phylogenet. Evol.* **39**, 585–586.

Steel, M. A. 1994a. The maximum likelihood point for a phylogenetic tree is not unique. *Syst. Biol.* **43**, 560–564.

Steel, M. A. 1994b. Recovering a tree from the leaf colourations it generates under a Markov model. *Appl. Math. Lett.* **7**, 19–24.

Steel, M. A. 2005. Should phylogenetic models be trying to 'fit an elephant'? *Trends Genet.* **21**, 307–309.

Steel, M. A., and D. Penny. 2000. Parsimony, likelihood, and the role of models in molecular phylogenetics. *Mol. Biol. Evol.* **17**, 839–850.

Stein, C. 1956. Inadmissibility of the usual estimator for the mean of a multivariate normal distribution. *Proc. 3rd Berkeley Symp. Math. Stat. Prob.* **1**, 197–206.

Stein, C. 1964. Inadmissibility of the usual estimator for the variance of a multivariate normal distribution with unknown mean. *Ann. Inst. Math.* **16**, 155–160.

Steiper, M. E., N. M. Young, and T. Y. Sukarna. 2004. Genomic data support the hominoid slow-down and an Early Oligocene estimate for the hominoid-cercopithecoid divergence. *Proc. Natl. Acad. Sci. U.S.A.* **101**, 17021–17026.

Stephens, M. 2000. Dealing with label switching in mixture models. *J. R. Stat. Soc. B.* **62**, 795–809.

Stephens, M., and P. Donnelly. 2000. Inference in molecular population genetics (with discussions). *J. R. Stat. Soc. B.* **62**, 605–655.

Stewart, C.-B., J. W. Schilling, and A. C. Wilson. 1987. Adaptive evolution in the stomach lysozymes of foregut fermenters. *Nature* **330**, 401–404.

Stigler, S. M. 1982. Thomas Bayes's Bayesian inference. *J. R. Stat. Soc. A.* **145**, 250–258.

Strimmer, K., and O. G. Pybus. 2001. Exploring the demographic history of DNA sequences using the generalized skyline plot. *Mol. Biol. Evol.* **18**, 2298–2305.

Strimmer, K., and A. von Haeseler. 1996. Quartet puzzling: a quartet maximum-likelihood method for reconstructing tree topologies. *Mol. Biol. Evol.* **13**, 964–969.

Strobeck, K. 1987. Average number of nucleotide differences in a sample from a single subpopulation: a test for population subdivision. *Genetics* **117**, 149–153.

Stuart, A., K. Ord, and S. Arnold. 1999. *Kendall's Advanced Theory of Statistics*. Arnold, London.

Studier, J. A., and K. J. Keppler. 1988. A note on the neighbor-joining algorithm of Saitou and Nei. *Mol. Biol. Evol.* **5**, 729–731.

Suchard, M., and A. Rambaut. 2009. Many-core algorithms for statistical phylogenetics. *Bioinformatics* **25**, 1370–1376.

Suchard, M. A., R. E. Weiss, and J. S. Sinsheimer. 2001. Bayesian selection of continuous-time Markov chain evolutionary models. *Mol. Biol. Evol.* **18**, 1001–1013.

Suchard, M. A., C. M. Kitchen, J. S. Sinsheimer, and R. E. Weiss. 2003. Hierarchical phylogenetic models for analyzing multipartite sequence data. *Syst. Biol.* **52**, 649–664.

Sueoka, N. 1995. Intrastrand parity rules of DNA base composition of cyprinid fishes in subgenus Notropis inferred from nucleotide and usage biases of synonymous codons. *J. Mol. Evol.* **40**, 318–325.

Sugiura, N. 1978. Further analysis of the data by Akaike's information criterion and the finite corrections. *Commun. Stat. A – Theory Methods* **7**, 13–26.

Sullivan, J., and D. L. Swofford. 2001. Should we use model-based methods for phylogenetic inference when we know that assumptions about among-site rate variation and nucleotide substitution pattern are violated? *Syst. Biol.* **50**, 723–729.

Sullivan, J., K. E. Holsinger, and C. Simon. 1995. Among-site rate variation and phylogenetic analysis of 12S rRNA in sigmodontine rodents. *Mol. Biol. Evol.* **12**, 988–1001.

Sullivan, J., D. L. Swofford, and G. J. P. Naylor. 1999. The effect of taxon-sampling on estimating rate heterogeneity parameters on maximum-likelihood models. *Mol. Biol. Evol.* **16**, 1347–1356.

Susko, E. 2008. On the distributions of bootstrap support and posterior distributions for a star tree. *Syst. Biol.* **57**, 602–612.

Susko, E. 2009. Bootstrap support is not first-order correct. *Syst. Biol.* **58**, 211–223.

Susko, E. 2010. First-order correct bootstrap support adjustments for splits that allow hypothesis testing when using maximum likelihood estimation. *Mol. Biol. Evol.* **27**, 1621–1629.

Suzuki, Y. 2004. New methods for detecting positive selection at single amino acid sites. *J. Mol. Evol.* **59**, 11–19.

Suzuki, Y., and T. Gojobori. 1999. A method for detecting positive selection at single amino acid sites. *Mol. Biol. Evol.* **16**, 1315–1328.

Suzuki, Y., G. V. Glazko, and M. Nei. 2002. Overcredibility of molecular phylogenies obtained by Bayesian phylogenetics. *Proc. Natl. Acad. Sci. U.S.A.* **99**, 16138–16143.

Swanson, W. J., and V. D. Vacquier. 2002a. Reproductive protein evolution. *Ann. Rev. Ecol. Syst.* **33**, 161–179.

Swanson, W. J., and V. D. Vacquier. 2002b. The rapid evolution of reproductive proteins. *Nat. Rev. Genet.* **3**, 137–144.

Swanson, W. J., Z. Yang, M. F. Wolfner, and C. F. Aquadro. 2001a. Positive Darwinian selection in the evolution of mammalian female reproductive proteins. *Proc. Natl. Acad. Sci. U.S.A.* **98**, 2509–2514.

Swanson, W. J., A. G. Clark, H. M. Waldrip-Dail et al. 2001b. Evolutionary EST analysis identifies rapidly evolving male reproductive proteins in Drosophila. *Proc. Natl. Acad. Sci. U.S.A.* **98**, 7375–7379.

Swofford, D. L. 2000. *PAUP*: Phylogenetic Analysis by Parsimony*, Version 4. Sinauer Associates, Sanderland, Massachusetts.

Swofford, D. L., and G. J. Olsen. 1990. Phylogeny reconstruction. Pp. 411–501 *in* D. M. Hillis, and C. Moritz, eds. *Molecular Systematics*. Sinauer Associates, Sunderland, Massachusetts.

Swofford, D. L., G. J. Olsen, P. J. Waddell, and D. M. Hillis. 1996. Phylogeny inference. Pp. 407–514 *in* D. M. Hillis, C. Moritz, and B. K. Mable, eds. *Molecular Systematics*. Sinauer Associates, Sunderland, Massachusetts.

Swofford, D. L., P. J. Waddell, J. P. Huelsenbeck et al. 2001. Bias in phylogenetic estimation and its relevance to the choice between parsimony and likelihood methods. *Syst. Biol.* **50**, 525–539.

Tajima, F. 1983. Evolutionary relationship of DNA sequences in finite populations. *Genetics* **105**, 437–460.

Tajima, F. 1989. Statistical method for testing the neutral mutation hypothesis by DNA polymorphism. *Genetics* **123**, 585–595.

Tajima, F. 1993. Simple methods for testing the molecular evolutionary clock hypothesis. *Genetics* **135**, 599–607.

Tajima, F., and M. Nei. 1982. Biases of the estimates of DNA divergence obtained by the restriction enzyme technique. *J. Mol. Evol.* **18**, 115–120.

Tajima, F., and N. Takezaki. 1994. Estimation of evolutionary distance for reconstructing molecular phylogenetic trees. *Mol. Biol. Evol.* **11**, 278–286.

Takahata, N. 1983. Gene identity and genetic differentiation of populations in the finite island model. *Genetics* **104**, 497–512.

Takahata, N. 1988. The coalescent in two partially isolated diffusion populations. *Genet. Res. (Camb.)* **52**, 213–222.

Takahata, N. 1989. Gene genealogy in three related populations: consistency probability between gene and population trees. *Genetics* **122**, 957–966.

Takahata, N., and M. Nei. 1985. Gene genealogy and variance of interpopulational nucleotide differences. *Genetics* **110**, 325–344.

Takahata, N., Y. Satta, and J. Klein. 1995. Divergence time and population size in the lineage leading to modern humans. *Theor. Popul. Biol.* **48**, 198–221.

Takezaki, N., and T. Gojobori. 1999. Correct and incorrect vertebrate phylogenies obtained by the entire mitochondrial DNA sequences. *Mol. Biol. Evol.* **16**, 590–601.

Takezaki, N., and M. Nei. 1994. Inconsistency of the maximum parsimony method when the rate of nucleotide substitution is constant. *J. Mol. Evol.* **39**, 210–218.

Takezaki, N., A. Rzhetsky, and M. Nei. 1995. Phylogenetic test of the molecular clock and linearized trees. *Mol. Biol. Evol.* **12**, 823–833.

Tamura, K. 1992. Estimation of the number of nucleotide substitutions when there are strong transition-transversion and G+C content biases. *Mol. Biol. Evol.* **9**, 678–687.

Tamura, K., and M. Nei. 1993. Estimation of the number of nucleotide substitutions in the control region of mitochondrial DNA in humans and chimpanzees. *Mol. Biol. Evol.* **10**, 512–526.

Tamura, K., D. Peterson, N. Peterson et al. 2011. MEGA5, molecular evolutionary genetics analysis using maximum likelihood, evolutionary distance, and maximum parsimony methods. *Mol. Biol. Evol.* **28**, 2731–2739.

Tanner, M. A., and W. H. Wong. 2000. From EM to data augmentation: the emergence of MCMC Bayesian computation in the 1980s. *Stat. Sci.* **25**, 506–516.

Tateno, Y., N. Takezaki, and M. Nei. 1994. Relative efficiencies of the maximum-likelihood, neighbor-joining, and maximum-parsimony methods when substitution rate varies with site. *Mol. Biol. Evol.* **11**, 261–277.

Tautz, D., P. Arctander, A. Minelli et al. 2003. A plea for DNA taxonomy. *Trends Ecol. Evol.* **18**, 70–74.

Tavaré, S. 1986. Some probabilistic and statistical problems on the analysis of DNA sequences. *Lect. Math. Life Sci.* **17**, 57–86.

Tavaré, S., C. R. Marshall, O. Will et al. 2002. Using the fossil record to estimate the age of the last common ancestor of extant primates. *Nature* **416**, 726–729.

Telford, M. J., M. J. Wise, and V. Gowri-Shankar. 2005. Consideration of RNA secondary structure significantly improves likelihood-based estimates of phylogeny: examples from the bilateria. *Mol. Biol. Evol.* **22**, 1129–1136.

Templeton, A. R. 1983. Phylogenetic inference from restriction endonuclease cleavage site maps with particular reference to the evolution of man and the apes. *Evolution* **37**, 221–224.

Thompson, E. A. 1975. *Human Evolutionary Trees*. Cambridge University Press, Cambridge, UK.

Thompson, J. D., D. G. Higgins, and T. J. Gibson. 1994. CLUSTAL W: improving the sensitivity of progressive multiple sequence alignment through sequence weighting, position-specific gap penalties and weight matrix choice. *Nucl. Acids Res.* **22**, 4673–4680.

Thorne, J. L., and H. Kishino. 1992. Freeing phylogenies from artifacts of alignment. *Mol. Biol. Evol.* **9**, 1148–1162.

Thorne, J. L., and H. Kishino. 2005. Estimation of divergence times from molecular sequence data. Pp. 233–256 *in* R. Nielsen, ed. *Statistical Methods in Molecular Evolution*. Springer-Verlag, New York.

Thorne, J. L., N. Goldman, and D. T. Jones. 1996. Combining protein evolution and secondary structure. *Mol. Biol. Evol.* **13**, 666–673.

Thorne, J. L., H. Kishino, and J. Felsenstein. 1991. An evolutionary model for maximum likelihood alignment of DNA sequences [Erratum in *J. Mol. Evol.* 1992, 34, 91]. *J. Mol. Evol.* **33**, 114–124.

Thorne, J. L., H. Kishino, and J. Felsenstein. 1992. Inching toward reality: an improved likelihood model of sequence evolution. *J. Mol. Evol.* **34**, 3–16.

Thorne, J. L., H. Kishino, and I. S. Painter. 1998. Estimating the rate of evolution of the rate of molecular evolution. *Mol. Biol. Evol.* **15**, 1647–1657.

Thornton, J. 2004. Resurrecting ancient genes: experimental analysis of extinct molecules. *Nat. Rev. Genet.* **5**, 366–375.

Thornton, J. W., E. Need, and D. Crews. 2003. Resurrecting the ancestral steroid receptor: ancient origin of estrogen signaling. *Science* **301**, 1714–1717.

Tillier, E. R., and R. A. Collins. 1998. High apparent rate of simultaneous compensatory base-pair substitutions in ribosomal RNA. *Genetics* **148**, 1993–2002.

Tillier, E. R. M. 1994. Maximum likelihood with multiparameter models of substitution. *J. Mol. Evol.* **39**, 409–417.

Tsaur, S. C., and C.-I. Wu. 1997. Positive selection and the molecular evolution of a gene of male reproduction, *Acp26Aa* of *Drosophila*. *Mol. Biol. Evol.* **14**, 544–549.

Tucker, A. 1995. *Applied Combinatorics*. Wiley, New York.

Tuff, P., and P. Darlu. 2000. Exploring a phylogenetic approach for the detection of correlated substitutions in proteins. *Mol. Biol. Evol.* **17**, 1753–1759.

Tuffley, C., and M. Steel. 1997. Links between maximum likelihood and maximum parsimony under a simple model of site substitution. *Bull. Math. Biol.* **59**, 581–607.

Tuffley, C., and M. Steel. 1998. Modeling the covarion hypothesis of nucleotide substitution. *Math. Biosci.* **147**, 63–91.

Tzeng, Y. H., R. Pan, and W. H. Li. 2004. Comparison of three methods for estimating rates of synonymous and nonsynonymous nucleotide substitutions. *Mol. Biol. Evol.* **21**, 2290–2298.

Ugalde, J. A., B. S. W. Chang, and M. V. Matz. 2004. Evolution of coral pigments recreated. *Science* **305**, 1433.

Vallender, E. J., and B. T. Lahn. 2004. Positive selection on the human genome. *Hum. Mol. Genet.* **13**, R245–R254.

Vinh, L. S., and A. von Haeseler. 2005. Shortest triplet clustering: reconstructing large phylogenies using representative sets. *BMC Bioinformatics* **6**, 92.

Vuong, Q. H. 1989. Likelihood ratio tests for model selection and non-nested hypotheses. *Econometrica* **57**, 307–333.

Waddell, P. J., and M. A. Steel. 1997. General time-reversible distances with unequal rates across sites: mixing gamma and inverse Gaussian distributions with invariant sites. *Mol. Phylogenet. Evol.* **8**, 398–414.

Waddell, P. J., D. Penny, and T. Moore. 1997. Hadamard conjugations and modeling sequence evolution with unequal rates across sites. *Mol. Phylogenet. Evol.* **8**, 33–50. [Erratum in *Mol. Phylogenet. Evol.* 1997, **8**, 446]

Wakeley, J. 1994. Substitution-rate variation among sites and the estimation of transition bias. *Mol. Biol. Evol.* **11**, 436–442.

Wakeley, J. 2009. *Coalescent Theory: An Introduction*. Roberts & Co. Publishers, Greenwood Village, Colorado.

Wakeley, J., and N. Aliacar. 2001. Gene genealogies in a metapopulation. *Genetics* **159**, 893–905.

Wald, A. 1949. Note on the consistency of the maximum likelihood estimate. *Ann. Math. Stat.* **20**, 595–601.

Walker, A. J. 1974. New fast method for generating discrete random numbers with arbitrary frequency distributions. *Electron. Let.* **10**, 127–128.

Walsh, J. B. 1995. How often do duplicated genes evolve new functions? *Genetics* **139**, 421–428.

Wang, F., and D. P. Landau. 2001. Efficient, multiple-range random-walk algorithm to calculate the density of states. *Phys. Rev. Lett.* **86**, 2050–2053.

Wang, Y., and J. Hey. 2010. Estimating divergence parameters with small samples from a large number of loci. *Genetics* **184**, 363–379.

Waterston, R. H. Lindblad-TohK. E. Birney et al. 2002. Initial sequencing and comparative analysis of the mouse genome. *Nature* **420**, 520–562.

Watterson, G. A. 1975. On the number of segregating sites in genetical models without recombination. *Theor. Popul. Biol.* **7**, 256–276.

Weadick, C. J., and B. S. Chang. 2012. An improved likelihood ratio test for detecting site-specific functional divergence among clades of protein-coding genes. *Mol. Biol. Evol.* **29**, 1297–1300.

Whelan, S., and N. Goldman. 2000. Statistical tests of gamma-distributed rate heterogeneity in models of sequence evolution in phylogenetics. *Mol. Biol. Evol.* **17**, 975–978.

Whelan, S., and N. Goldman. 2001. A general empirical model of protein evolution derived from multiple protein families using a maximum likelihood approach. *Mol. Biol. Evol.* **18**, 691–699.

Whelan, S., P. Liò, and N. Goldman. 2001. Molecular phylogenetics: state of the art methods for looking into the past. *Trends Genet.* **17**, 262–272.

Wiens, J. J. 2007. Species delimitation: new approaches for discovering diversity. *Syst. Biol.* **56**, 875–878.

Wiley, E. O. 1981. *Phylogenetics: The Theory and Practice of Phylogenetic Systematics*. Wiley, New York.

Wilkinson-Herbots, H. M. 1998. Genealogy and subpopulation differentiation under various models of population structure. *J. Math. Biol.* **37**, 535–585.

Wilkinson-Herbots, H. M. 2008. The distribution of the coalescence time and the number of pairwise nucleotide differences in the 'isolation with migration' model. *Theor. Popul. Biol.* **73**, 277–288.

Wilkinson-Herbots, H. M. 2012. The distribution of the coalescence time and the number of pairwise nucleotide differences in a model of population divergence or speciation with an initial period of gene flow. *Theor. Popul. Biol.* **82**, 92–108.

Wilkinson, M., F.-J. Lapointe, and D. J. Gower. 2003. Branch lengths and support. *Syst. Biol.* **52**, 127–130.

Wilkinson, M., D. Pisani, J. A. Cotton, and I. Corfe. 2005. Measuring support and finding unsupported relationships in supertrees. *Syst. Biol.* **54**, 823–831.

Wilkinson, M., J. O. McInerney, R. P. Hirt et al. 2007. Of clades and clans: terms for phylogenetic relationships in unrooted trees. *Trends Ecol. Evol.* **22**, 114–115.

Wilkinson, R. D., M. E. Steiper, C. Soligo et al. 2011. Dating primate divergences through an integrated analysis of palaeontological and molecular data. *Syst. Biol.* **60**, 16–31.

Williams, P. D., D. D. Pollock, B. P. Blackburne, and R. A. Goldstein. 2006. Assessing the accuracy of ancestral protein reconstruction methods. *PLoS Comput. Biol.* **2**, e69.

Williamson, S., and M. E. Orive. 2002. The genealogy of a sequence subject to purifying selection at multiple sites. *Mol. Biol. Evol.* **19**, 1376–1384.

Williamson, S. H., M. J. Hubisz, A. G. Clark et al. 2007. Localizing recent adaptive evolution in the human genome. *PLoS Genet.* **3**, e90.

Wilson, A. C., S. S. Carlson, and T. J. White. 1977. Biochemical evolution. *Ann. Rev. Biochem.* **46**, 573–639.

Wilson, D. J., and G. McVean. 2006. Estimating diversifying selection and functional constraint in the presence of recombination. *Genetics* **172**, 1411–1425.

Wilson, I. J., and D. J. Balding. 1998. Genealogical inference from microsatellite data. *Genetics* **150**, 499–510.

Wilson, I. J., M. E. Weal, and D. J. Balding. 2003. Inference from DNA data: population histories, evolutionary processes and forensic match probabilities. *J. R. Stat. Soc. A* **166**, 155–201.

Wong, W. H., and F. Liang. 1997. Dynamic weighting in Monte Carlo and optimization. *Proc. Natl. Acad. Sci. U.S.A.* **94**, 14220–14224.

Wong, W. S., and R. Nielsen. 2004. Detecting selection in noncoding regions of nucleotide sequences. *Genetics* **167**, 949–958.

Wong, W. S. W., Z. Yang, N. Goldman, and R. Nielsen. 2004. Accuracy and power of statistical methods for detecting adaptive evolution in protein coding sequences and for identifying positively selected sites. *Genetics* **168**, 1041–1051.

Wray, G. A., J. S. Levinton, and L. H. Shapiro. 1996. Molecular evidence for deep Precambrian divergences. *Science* **274**, 568–573.

Wright, F. 1990. The 'effective number of codons' used in a gene. *Gene* **87**, 23–29.

Wright, S. 1931. Evolution in Mendelian populations. *Genetics* **16**, 97–159.

Wright, S. 1943. Isolation by distance. *Genetics* **28**, 114–138.

Wu, C.-I., and W.-H. Li. 1985. Evidence for higher rates of nucleotide substitution in rodents than in man. *Proc. Natl. Acad. Sci. U.S.A.* **82**, 1741–1745.

Wu, C. I., and C. T. Ting. 2004. Genes and speciation. *Nat. Rev. Genet.* **5**, 114–122.

Wyckoff, G. J., W. Wang, and C.-I. Wu. 2000. Rapid evolution of male reproductive genes in the descent of man. *Nature* **403**, 304–309.

Xia, X. 1998. How optimized is the translational machinery in Escherichia coli, Salmonella typhimurium and Saccharomyces cerevisiae? *Genetics* **149**, 37–44.

Xie, W., P. O. Lewis, Y. Fan et al. 2011. Improving marginal likelihood estimation for Bayesian phylogenetic model selection. *Syst. Biol.* **60**, 150–160.

Yang, Z. 1993. Maximum-likelihood estimation of phylogeny from DNA sequences when substitution rates differ over sites. *Mol. Biol. Evol.* **10**, 1396–1401.

Yang, Z. 1994a. Statistical properties of the maximum likelihood method of phylogenetic estimation and comparison with distance matrix methods. *Syst. Biol.* **43**, 329–342.

Yang, Z. 1994b. Estimating the pattern of nucleotide substitution. *J. Mol. Evol.* **39**, 105–111.

Yang, Z. 1994c. Maximum likelihood phylogenetic estimation from DNA sequences with variable rates over sites: approximate methods. *J. Mol. Evol.* **39**, 306–314.

Yang, Z. 1995a. Evaluation of several methods for estimating phylogenetic trees when substitution rates differ over nucleotide sites. *J. Mol. Evol.* **40**, 689–697.

Yang, Z. 1995b. A space-time process model for the evolution of DNA sequences. *Genetics* **139**, 993–1005.

Yang, Z. 1995c. On the general reversible Markov-process model of nucleotide substitution: a reply to Saccone et al. *J. Mol. Evol.* **41**, 254–255.

Yang, Z. 1996a. Maximum-likelihood models for combined analyses of multiple sequence data. *J. Mol. Evol.* **42**, 587–596.

Yang, Z. 1996b. Among-site rate variation and its impact on phylogenetic analyses. *Trends Ecol. Evol.* **11**, 367–372.

Yang, Z. 1996c. Phylogenetic analysis using parsimony and likelihood methods. *J. Mol. Evol.* **42**, 294–307.

Yang, Z. 1997a. How often do wrong models produce better phylogenies? *Mol. Biol. Evol.* **14**, 105–108.

Yang, Z. 1997b. PAML: a program package for phylogenetic analysis by maximum likelihood. *Comput. Appl. Biosci.* **13**, 555–556.

Yang, Z. 1998a. Likelihood ratio tests for detecting positive selection and application to primate lysozyme evolution. *Mol. Biol. Evol.* **15**, 568–573.

Yang, Z. 1998b. On the best evolutionary rate for phylogenetic analysis. *Syst. Biol.* **47**, 125–133.

Yang, Z. 2000a. Maximum likelihood estimation on large phylogenies and analysis of adaptive evolution in human influenza virus A. *J. Mol. Evol.* **51**, 423–432.

Yang, Z. 2000b. Complexity of the simplest phylogenetic estimation problem. *Proc. R. Soc. B: Biol. Sci.* **267**, 109–116.

Yang, Z. 2002a. Likelihood and Bayes estimation of ancestral population sizes in Hominoids using data from multiple loci. *Genetics* **162**, 1811–1823.

Yang, Z. 2002b. Inference of selection from multiple species alignments. *Curr. Opinion Genet. Devel.* **12**, 688–694.

Yang, Z. 2004. A heuristic rate smoothing procedure for maximum likelihood estimation of species divergence times. *Acta Zool. Sinica* **50**, 645–656.

Yang, Z. 2005. Bayesian inference in molecular phylogenetics. Pp. 63–90 *in* O. Gascuel, ed. *Mathematics of Evolution and Phylogeny*. Oxford University Press, Oxford, UK.

Yang, Z. 2006. *Computational Molecular Evolution*. Oxford University Press, Oxford, UK.

Yang, Z. 2007a. PAML 4, Phylogenetic analysis by maximum likelihood. *Mol. Biol. Evol.* **24**, 1586–1591.

Yang, Z. 2007b. Fair-balance paradox, star-tree paradox and Bayesian phylogenetics. *Mol. Biol. Evol.* **24**, 1639–1655.

Yang, Z. 2008. Empirical evaluation of a prior for Bayesian phylogenetic inference. *Phil. Trans. R. Soc. Lond. B.* **363**, 4031–4039.

Yang, Z. 2010. A likelihood ratio test of speciation with gene flow using genomic sequence data. *Genome Biol. Evol.* **2**, 200–211.

Yang, Z., and J. P. Bielawski. 2000. Statistical methods for detecting molecular adaptation. *Trends Ecol. Evol.* **15**, 496–503.

Yang, Z., and M. dos Reis. 2011. Statistical properties of the branch-site test of positive selection. *Mol. Biol. Evol.* **28**, 1217–1228.

Yang, Z., and S. Kumar. 1996. Approximate methods for estimating the pattern of nucleotide substitution and the variation of substitution rates among sites. *Mol. Biol. Evol.* **13**, 650–659.

Yang, Z., and R. Nielsen. 1998. Synonymous and nonsynonymous rate variation in nuclear genes of mammals. *J. Mol. Evol.* **46**, 409–418.

Yang, Z., and R. Nielsen. 2000. Estimating synonymous and nonsynonymous substitution rates under realistic evolutionary models. *Mol. Biol. Evol.* **17**, 32–43.

Yang, Z., and R. Nielsen. 2002. Codon-substitution models for detecting molecular adaptation at individual sites along specific lineages. *Mol. Biol. Evol.* **19**, 908–917.

Yang, Z., and R. Nielsen. 2008. Mutation-selection models of codon substitution and their use to estimate selective strengths on codon usage. *Mol. Biol. Evol.* **25**, 568–579.

Yang, Z., and B. Rannala. 1997. Bayesian phylogenetic inference using DNA sequences: a Markov chain Monte Carlo Method. *Mol. Biol. Evol.* **14**, 717–724.

Yang, Z., and B. Rannala. 2005. Branch-length prior influences Bayesian posterior probability of phylogeny. *Syst. Biol.* **54**, 455–470.

Yang, Z., and B. Rannala. 2006. Bayesian estimation of species divergence times under a molecular clock using multiple fossil calibrations with soft bounds. *Mol. Biol. Evol.* **23**, 212–226.

Yang, Z., and B. Rannala. 2010. Bayesian species delimitation using multilocus sequence data. *Proc. Natl. Acad. Sci. U.S.A.* **107**, 9264–9269.

Yang, Z., and B. Rannala. 2012. Molecular phylogenetics: principles and practice. *Nat. Rev. Genet.* **13**, 303–314.

Yang, Z., and D. Roberts. 1995. On the use of nucleic acid sequences to infer early branchings in the tree of life. *Mol. Biol. Evol.* **12**, 451–458.

Yang, Z., and C. E. Rodríguez. 2013. Searching for efficient Markov chain Monte Carlo proposal kernels. *Proc. Natl. Acad. Sci. U.S.A.* **110**, 19307–19312.

Yang, Z., and W. J. Swanson. 2002. Codon-substitution models to detect adaptive evolution that account for heterogeneous selective pressures among site classes. *Mol. Biol. Evol.* **19**, 49–57.

Yang, Z., and T. Wang. 1995. Mixed model analysis of DNA sequence evolution. *Biometrics* **51**, 552–561.

Yang, Z., and A. D. Yoder. 2003. Comparison of likelihood and Bayesian methods for estimating divergence times using multiple gene loci and calibration points, with application to a radiation of cute-looking mouse lemur species. *Syst. Biol.* **52**, 705–716.

Yang, Z., N. Goldman, and A. Friday. 1994. Comparison of models for nucleotide substitution used in maximum-likelihood phylogenetic estimation. *Mol. Biol. Evol.* **11**, 316–324.

Yang, Z., N. Goldman, and A. E. Friday. 1995a. Maximum likelihood trees from DNA sequences: a peculiar statistical estimation problem. *Syst. Biol.* **44**, 384–399.

Yang, Z., S. Kumar, and M. Nei. 1995b. A new method of inference of ancestral nucleotide and amino acid sequences. *Genetics* **141**, 1641–1650.

Yang, Z., I. J. Lauder, and H. J. Lin. 1995c. Molecular evolution of the hepatitis B virus genome. *J. Mol. Evol.* **41**, 587–596.

Yang, Z., R. Nielsen, and M. Hasegawa. 1998. Models of amino acid substitution and applications to mitochondrial protein evolution. *Mol. Biol. Evol.* **15**, 1600–1611.

Yang, Z., W. S. W. Wong, and R. Nielsen. 2005. Bayes empirical Bayes inference of amino acid sites under positive selection. *Mol. Biol. Evol.* **22**, 1107–1118.

Yang, Z., R. Nielsen, N. Goldman, and A.-M. K. Pedersen. 2000. Codon-substitution models for heterogeneous selection pressure at amino acid sites. *Genetics* **155**, 431–449.

Yap, V. B., H. Lindsay, S. Easteal, and G. Huttley. 2010. Estimates of the effect of natural selection on protein-coding content. *Mol. Biol. Evol.* **27**, 726–734.

Yoder, A. D., and Z. Yang. 2000. Estimation of primate speciation dates using local molecular clocks. *Mol. Biol. Evol.* **17**, 1081–1090.

Yokoyama, S. 2002. Molecular evolution of color vision in vertebrates. *Gene* **300**, 69–78.

Yoshida, I., W. Sugiura, J. Shibata et al. 2011. Change of positive selection pressure on HIV-1 envelope gene inferred by early and recent samples. *PLOS One* **6**, e18630.

Yu, N., Z. Zhao, Y. X. Fu et al. 2001. Global patterns of human DNA sequence variation in a 10-kb region on chromosome 1. *Mol. Biol. Evol.* **18**, 214–222.

Zang, L.-L., X.-H. Zou, F.-M. Zhang et al. 2011. Phylogeny and species delimitation of the C-genome diploid species in Oryza. *J. Syst. Evol.* **49**, 386–395.

Zardoya, R., and A. Meyer. 1996. Phylogenetic performance of mitochondrial protein-coding genes in resolving relationships among vertebrates. *Mol. Biol. Evol.* **13**, 933–942.

Zhai, W., R. Nielsen, and M. Slatkin. 2009. An investigation of the statistical power of neutrality tests based on comparative and population genetic data. *Mol. Biol. Evol.* **26**, 273–283.

Zhang, C., B. Rannala, and Z. Yang. 2012. Robustness of compound Dirichlet priors for Bayesian inference of branch lengths. *Syst. Biol.* **61**, 779–784.

Zhang, C., D.-X. Zhang, T. Zhu, and Z. Yang. 2011. Evaluation of a Bayesian coalescent method of species delimitation. *Syst. Biol.* **60**, 747–761.

Zhang, J. 2000. Rates of conservative and radical nonsynonymous nucleotide substitutions in mammalian nuclear genes. *J. Mol. Evol.* **50**, 56–68.

Zhang, J. 2003. Evolution of the human ASPM gene, a major determinant of brain size. *Genetics* **165**, 2063–2070.

Zhang, J. 2004. Frequent false detection of positive selection by the likelihood method with branch-site models. *Mol. Biol. Evol.* **21**, 1332–1339.

Zhang, J., and M. Nei. 1997. Accuracies of ancestral amino acid sequences inferred by the parsimony, likelihood, and distance methods. *J. Mol. Evol.* **44**, S139–146.

Zhang, J., S. Kumar, and M. Nei. 1997. Small-sample tests of episodic adaptive evolution: a case study of primate lysozymes. *Mol. Biol. Evol.* **14**, 1335–1338.

Zhang, J., R. Nielsen, and Z. Yang. 2005. Evaluation of an improved branch-site likelihood method for detecting positive selection at the molecular level. *Mol. Biol. Evol.* **22**, 2472–2479.

Zhang, J., H. F. Rosenberg, and M. Nei. 1998. Positive Darwinian selection after gene duplication in primate ribonuclease genes. *Proc. Natl. Acad. Sci. U.S.A.* **95**, 3708–3713.

Zhang, J., Y. P. Zhang, and H. F. Rosenberg. 2002. Adaptive evolution of a duplicated pancreatic ribonuclease gene in a leaf-eating monkey. *Nat. Genet.* **30**, 411–415.

Zhang, L., B. S. Gaut, and T. J. Vision. 2001. Gene duplication and evolution. *Science* **293**, 1551.

Zhang, R., Z. Peng, G. Li et al. 2013. Ongoing speciation in the tibetan plateau gymnocypris species complex. *PLoS One* **8**, e71331.

Zhao, Z., L. Jin, Y. X. Fu et al. 2000. Worldwide DNA sequence variation in a 10-kilobase noncoding region on human chromosome 22. *Proc. Natl. Acad. Sci. U.S.A.* **97**, 11354–11358.

Zharkikh, A. 1994. Estimation of evolutionary distances between nucleotide sequences. *J. Mol. Evol.* **39**, 315–329.

Zharkikh, A., and W.-H. Li. 1993. Inconsistency of the maximum parsimony method: the case of five taxa with a molecular clock. *Syst. Biol.* **42**, 113–125.

Zharkikh, A., and W.-H. Li. 1995. Estimation of confidence in phylogeny: the complete-and-partial bootstrap technique. *Mol. Phylogenet. Evol.* **4**, 44–63.

Zhou, Y., H. Brinkmann, N. Rodrigue et al. 2010. A Dirichlet process covarion mixture model and its assessments using posterior predictive discrepancy tests. *Mol. Biol. Evol.* **27**, 371–384.

Zhu, L., and C. D. Bustamante. 2005. A composite likelihood approach for detecting directional selection from DNA sequence data. *Genetics* **170**, 1411–1421.

Zhu, T., and Z. Yang. 2012. Maximum likelihood implementation of an isolation-with-migration model with three species for testing speciation with gene flow. *Mol. Biol. Evol.* **29**, 3131–3142.

Zierke, S., and J. Bakos. 2010. FPGA acceleration of the phylogenetic likelihood function for Bayesian MCMC inference methods. *BMC Bioinformatics* **11**, 184.

Zoller, S., and A. Schneider. 2010. Empirical analysis of the most relevant parameters of codon substitution models. *J. Mol. Evol.* **70**, 605–612.

Zuckerkandl, E. 1964. Further principles of chemical paleogenetics as applied to the evolution of hemoglobin. Pp. 102–109 *in* P. H., ed. *Peptides of the Biological Fluids*. Elsevier, Amsterdam.

Zuckerkandl, E., and L. Pauling. 1962. Molecular disease, evolution, and genetic heterogeneity. Pp. 189–225 *in* M. Kasha, and B. Pullman, eds. *Horizons in Biochemistry*. Academic Press, New York.

Zuckerkandl, E., and L. Pauling. 1965. Evolutionary divergence and convergence in proteins. Pp. 97–166 *in* V. Bryson, and H. J. Vogel, eds. *Evolving Genes and Proteins*. Academic Press, New York.

Zwickl, D. 2006. Genetic algorithm approaches for the phylogenetic analysis of large biological sequence datasets under the maximum likelihood criterion. *Ph.D. Thesis*: University of Texas at Austin.

Zwickl, D. J., and M. T. Holder. 2004. Model parameterization, prior distributions, and the general time-reversible model in Bayesian phylogenetics. *Syst. Biol.* **53**, 877–888.

Index

Note: Locators in bold refer to important concepts.

12s RNA 7, 9, 12, 17, 21–5, 28, 30, 34, 115–19, 172, 188–9, 200, 216, 230–1, 260, 269, 365, 389

A

acceptance proportion (P_{jump}) 217, **218–19**, 224, 229, **232–3**, 234–7, 241, 252–5, 265
   maximum for cross-model moves 322–3
   optimum 236–8
   step length 217–18, 236–8
acceptance rate 37, 46, 60
acceptance ratio 215, **218**, 248, 253
adaptive evolution 390–1, 358
   examples 414–16
additive-tree method 89
agglomerative method 82–3
Akaike Information Criterion (AIC) 148–50
alias method 421
alignment error 110, 413
alignment gap 7, 42, 50, 109, **111–13**, 146
ambiguous nucleotides 111
amino acid exchangeability 36–8, 46, 271
amino acid model 35–9, 110
ancestral polymorphism 80–1, 297, 335, 351–2, 358
ancestral state
   reconstruction 125–33
   bias 131–3
   detection of selection 398, 401
   empirical Bayes 127–9
   hierarchical Bayesian 129–30
   joint 127–9
   likelihood 127–9
   marginal 127–9
   morphological characters 130–1, 96–8
   parsimony 81, 95–9, 129
angiosperm 410–11
anomalous gene tree 333
antigen-recognition site (ARS) 401, 406–7
ape 108–9, 119, 149, 306, 330

AU test 178
autocorrelation time 214
auto-discrete-gamma model 121

B

Bayes, Thomas 182, 187–8, 197
Bayes factor, (see also marginal likelihood) 196, 256–60
Bayes's theorem 183–6
Bayesian information criterion (BIC) 149–50
Bayesian method
   ancestral reconstruction 127–30
   coalescent model 319–20
   empirical 120, 127, 197
   hierarchical 197
   molecular clock dating 375–89
   objective 184
   subjective 184
   θ (theta) 319–20
   species delimitation 349–59
   species tree 335–43
   tree reconstruction 263–4
Bayesian simulation 242
BEAST 264, 276, 325, 340–2, 375, 377, 382, 385, 449
BEST 338, 340–2
beta distribution **186–7**, 198, 201–2, 270–1, 307, 394, 403–4, 406, 426, 430, 444
BFGS 138
BIC, see Bayesian information criterion
bifurcating (binary) tree 73
binomial distribution 7, 18, **34**, 154, **186–8**, 191–3, 198, 200, 307, 432
biological species concept (BSC) 349
bipartition, see split
birth-death process 74, 197, 277, 340, 379–82, 385, 387, 439
body size 362
bootstrap 172–7, 372
   confidence interval 172
   interpretation 175–7
   parametric 146–7
   standard error 172
bound
   hard 379

   optimization 138
   P_{jump} 233
   reflection 222, 225–6
   soft 375, 379–80
   variance 155
Box-Muller transformation 428
BPP 319–20, 329–30, 341–4, 355–8, 360
branch support, see decay index
branch swapping 84, 86
Bremer support, see decay index
BUCKy 340
burn-in 217, 243

C

Cambrian explosion 362
Chapman-Kolmogorov theorem 6, 30, 107
character length 95
chi-square distribution 21, 23, 68, 145–6, 148, 417
   mixture 119, 145, 149, 177, 411, 417
CI, see confidence interval and credibility interval
clade 78
cluster method 82–92
coalescent 309–15
   demography 320–5
   history 308
   multispecies 325–31
   one population 309–15
   simulation 315–16
codon model 42–7
   branch 398–400
   branch-site 408–9
   clade 409–10
   covarion 410
   empirical 46
   site 400–7
   switching 410
codon usage 43–5, 48, 51, 57–8, 61, 65, 396, 399, 413
combined analysis 123–5
compensatory mutation 46
compound distribution, see mixture model
confidence interval (CI) 7, 20–1, 172–3, 175, 190–1
consensus tree 77–8

majority-rule 77, 129, 174, 296
   strict 77
consistency index 180
consistency 19, 24, 154, **158**, 170, 441
counting method 47–50, 54
covarion model 121–2, 151, 167, 438
coverage probability 190, 242
convergence rate
   optimization 135, 137
   Markov chain 10
   MCMC 226–30, 238
credibility interval (CI) 185, 243–4
   equal tail 185, 243–4
   highest probability density (HPD) 185–6, 189, 243–4
cross-validation 369–70

D

Dayhoff 36–9
decay index 180
deletion 113, 437–8
delta technique 7–8, 56, 446–8
detailed balance 11, 29, 36
Dirichlet distribution 176, **267–70**, 276, 280, 379, 430
Dirichlet process 167, **272–5**, 340, 370
discrete distribution 114, 272, 403, 405, 419–23
discrete uniform distribution 422
discrete-gamma model 116–19
distance methods 88–95
divisive method 92
d_N/d_S 47–58, 61–2
   adaptive evolution 390–1
   counting method 47–50, 54
   definition 43
   estimation 47–58
   estimates 54, 62
   interpretation 45, 61, 412
   maximum likelihood 55–7
dynamic programming
   algorithm 96–9, 104, 128

E

effective number of codons (ENC) 65
effective sample size (ESS) 214
effective population size (N_e) 310
efficiency 155, 159
eigenvalue 10, 12, 17, 66, 117, 232, 234, 238, 245
eigenvector 10, 66
empirical Bayes method 120, 197
   Bayes 405, 407–8
   naive 405, 407–8
empirical model 35, 40–1

ENC, *see* effective number of codons
equal-tail credibility interval 185, 243–4
error
   random 154
   systematic 154
ESS, *see* effective sample size
exhaustive search 82
expected information 20, 155, 200
exponential distribution 315, 440

F

false positive rate 175–6, 183–4, 409, 417
finite-sites model 318
Fisher information, *see* expected information
Fisher, Ronald A. 45, 155, 182, 188, 308–9, 396
Fisher-Wright model 309–12
fixation probability 45
fixed-sites model 123, 438
fossil calibration 378–82
   hard bound 375, 379
   soft bound 375, 379–80
   uncertainty 370–2, 378–82
four-fold degenerate site 60
Frequentist statistics 183, 189–93
Fu and Li's D 394–5

G

Galton, Francis 183
gamma model 15–17, 39–40, 115–19
   shape parameter 15, 40
   simulation 430
gene tree 79–81, 297, 308
general time reversible (GTR) model 3, 9, 14, 26–30, 36, 41, 45, 47, 66, 139, 151, 270–2, 316
generation time 312, 320, 362, 366, 378, 392
generator 26
genetic arms race 408, 414–15
genetic algorithm 88, 140, 143, 400
genetic code 38
genetic hitchhiking 395, 414
geometric Brownian motion 370, 325, 369, 375, 377–8
Gibbs sampler 221, 274
GLASS 338
Golden section search 134
gradient 135–7
GTR, *see* general time-reversible model

H

Hadamard conjugation 107
Hastings ratio, *see* proposal ratio
Hessian matrix 20
heterotachy 164
hidden Markov model (HMM) 121
highest probability density (HPD) 185, 243–4
HKA test, *see* Hudson-Kreitman-Aquade test
HKY85 model 2–3, 9, 12–14
   codon model 45
   divergence time estimation 387
   nonhomogeneous 125
   rate matrix 2–3, 9
   sequence error 112
   LRT 144–7
   transition probability 10
homoplasy 180
Horner's rule 103
HPD, *see* highest posterior density
Hudson-Kreitman-Aquade (HKA) test 397–8
hypothesis testing 191–3, 195

I

i.i.d. model 120–2, 124, 151, 158, 173, 210–12, 266–9, 276, 436
identifiability 28, 154, 336, 368, 371–2, 382
ILS, *see* incomplete lineage sorting
IM model, *see* isolation-with-migration model
IMa 344
importance sampling 210–12
incomplete lineage sorting (ILS) 80–1, 297, 335, 351–2, 358
inconsistency
   likelihood 166
   parsimony 99, 152, 170
index of dispersion 366
infinite-sites model 316, 393, 396
infinite-sites plot 383, 388
infinite-sites theory 382–4
information
   expected 20, 137, 155, 200
   observed 137, 200
ingroup 71, 365
insertion 113, 437–8
integration
   Laplacian approximation 203
   Monte Carlo 210, 440
   numerical 204–5
   Gaussian quadrature 205
   mid-point 204–5
   trapezoid 204–5
interior branch test 177

invariable-sites model 15, 115, 119–20, 272, 424
invariance 19
inversion method 420, 425
isolation by distance 357
isolation-with-migration (IM) model 343–6
  simulation 349

J

JC69 model 3–7, 253, 318, 438, 445
  d_N/d_S estimation 47–9
  partial site pattern 106
  rate matrix 2–4
  rjMCMC (vs. K80), 253–4
  sequence error 112
  simulation 438
  super-efficiency 162–3
  test 22–3, 144–5
  test of adequacy 146–7
  transition probability 4–5
JC69 distance
  Bayesian 188–9
  gamma 16
  local peak 231
  maximum likelihood 18–21
  marginal likelihood 206–9
  MCMC 216–17
  prior 147, 200–1, 445
  time and rate 229
  variance 7, 447
JTT 36
jump chain 435–6
jump probability (P_{jump}), see acceptance proportion
jumping kernel, see proposal density

K

K80 model 2–3, 7–8, 434–5
  ancestral reconstruction 127–9
  d_N/d_S estimation 51–4, 68
  likelihood calculation 103–6
  LRT 22–3, 145, 441
  partial site pattern 106
  rate matrix 2–3, 7–8
  simulation 438
  transition probability 8, 434–5
  transition/transversion rate ratio 13–14, 440
K80 distance 8
  gamma 16–17
  maximum likelihood 22
  MCMC 260
  prior 270
  profile likelihood 24–6
  integrated likelihood 24–6
  rjMCMC (vs. JC69) 253–4
  variance 8, 447–8

Kishino-Hasegawa (K-H) test 178
K-L divergence, see Kullback-Leibler divergence
K-Pg boundary 363
Kullback-Leibler (K-L) divergence 141, 159, **202**, 248

L

labelled history 74–5, 313, 439
Laplace, Pierre-Simon 182–3, 185, 187, 197, 203–4, 261
Laplacian approximation 203
lateral gene transfer (LGT) 80
law of total probability 6, 110
least squares 89–95
  generalized 91
  ordinary 89–91
  weighted 91
likelihood
  integrated 24–6, 168
  marginal 24–6, 185, 187, 195–7, 200, 203
  penalized 369–70
  profile 24–6
likelihood equation 133
likelihood function 18, 102–3
  singularity 338–9, 360
likelihood interval 20–1, 24–6, 172
likelihood ratio test (LRT) 22, 144–6
  molecular clock 364–5
  positive selection 398–410
  substitution model 22–3, 144–6
limiting distribution 6
Lindley's paradox 196
line search 134, 137
lizard 79, 269–70, 306, 341–3
local peaks 86–8, 139–40, 241, 244–7
LogDet 31
log-normal distribution 119, 369, 371–3
long-branch attraction 99, 152, 441
LRT, see likelihood ratio test
LS, see least squares

M

major histocompatibility complex (MHC) 406–7
mammals 80, 108, 386–8
MAP tree 82, 168, 282, **295**
marginal likelihood 24, 185, 187, 195–6, 203
  Gaussian quadrature 205–6
  numerical integration 204–9
  Monte Carlo integration 210–12

Laplacian approximation 203
  stepping-stones 258
  path sampling 259
Markov chain 26–7
  amino acid substitution 35–9
  codon substitution 42–4
  IM model 344–8
  irreducible 232
  MCMC 219
  nucleotide substitution 1–13
  reversible 11, 29, 36
  simulation 433–6
Markov chain Monte Carlo (MCMC) 214–19
  convergence 226–30
  mixing efficiency 214, 230–8
  monitoring 241–2
  transmodel 247–52
  transdimensional (rjMCMC) 252–4
Markov chain Monte Carlo proposal
  Bactrian 223
  multiplier 225–6, 248–9
  single-component 220
  sliding window 222, 224
Markov process, see Markov chains
Markovian property 1
maximum likelihood estimate (MLE) 17
  consistency 441
maximum likelihood method (ML)
  ancestral reconstruction 127–9
  computation, see optimisation
  distance estimation 17–22, 40–2
  divergence times 366–8
  d_S and d_N 55–6
  rate matrix 29, 36
  species tree 336–9
  theta (θ) 316–18
  tree reconstruction 81, 102–25
maximum parsimony 81, 95–100
  ancestral state reconstruction 81, 95–9, 129
  inconsistency 99, **170**
  long-branch attraction 99, 152, 441
  philosophy 165–71
maximum tree 336–8
McDonald-Kreitman test 395–7
MCMC, see Markov chain Monte Carlo
MCMCMC 244–7
mechanistic model 36, 39
metabolic rate 362
Metropolis-Hastings (M-H) algorithm, see Markov chain Monte Carlo

MHC, see major histocompatibility complex
mid-point rooting 70
MIGRATE 343–4
migration 343–9
  rate 343–4
minimum-evolution 91, 95
missing data 110–12
mitochondrial protein 109, 119, 149–50, 179, 306
mixing efficiency 214, 232–8
  acceptance proportion (P_{jump}) 232–3
  step length 217–18, 236–40
mixture model 114–23
  branch-length prior 267–8
  d_N/d_S for sites 400–7
  heterotachy 164, 167
  rates for sites 114–23
  simulation 424, 438
ML, see maximum likelihood
MLE, see maximum likelihood estimate
model adequacy 146–8, 150–1, 163
model averaging 255–6
model robustness 150–1, 155, 163
model selection 144–50, 196
molecular clock 361–3
  controversies 361–3
  global 366
  local 367–8
  neutral theory 361–2
  rooting 70
  test 363–6
molecular clock dating
  Bayesian 375–88
  likelihood 366–8
  rate smoothing 368–72
monophyly 78, 350–1
Monte Carlo integration 210, 440
Monte Carlo simulation, see simulation
MRBAYES 230, 244, 263, 266, 269, 272, 276, 278, 297–8, 306–7, 340, 375, 449
MRCA (most recent common ancestor) 309
multifurcating tree 73
multinomial distribution 8, 12, 19, 28, 149, 158, 299, 422–3, 436
multiple substitutions (multiple hits) 1–2, 49
multiplier 225–6, 248–9
multiplication-congruent method 419
multispecies coalescent 325–30
  species tree estimation 335–43
  species delimitation 349–59
mutation-selection model 44

N
N_e, see effective population size
nearest neighbour interchange (NNI) 85, 284–6, 292–3
nearly neutral theory 362, 391–3
negative binomial distribution 34, 191–2
neighbour joining 91–5
neutral evolution 43, 61, 309, 358–9
neutral theory 391–3
  claims 392
  tests 393–8
Newick format 72
Newton's method (Newton-Raphson method) 135–7
Neyman, Jerzy, vii, 182, 191
NG86 48–52
NJ, see neighbour joining
NNI, see nearest neighbour interchange
no-common mechanism model 167
node slider 294–5
nonhomogeneous model 125, 275
noninformative prior, see prior distribution
nonlinear programming, see optimization
nonstationary model 125
nonsynonymous site 48
nonsynonymous substitution 35, 43
nonsynonymous/synonymous rate ratio, see d_N/d_S
normal distribution 20, 126–9, 145, 190, 195, 199, 203–4, 210–12, 219, 223–4, 227, 236–7, 245, 249, 299, 339, 425–9, 443–5
nuisance parameter 24

O
objective function 81, 88, 134
Ockham's Razor, see principle of parsimony
optimization 133–4
  bound 138
  multivariate 136–8
  univariate 134–6
outgroup rooting 70–1

P
p distance 1
p53 gene 42
PAM 36, 39
PAML 34, 54, 64, 66, 68, 152, 374, 385, 417, 441, 449
paradox
  fair coin 307
  fair balance 301–3
  Lindley's 196–7
  star tree 298–300
  tail 227
parametric bootstrap 146–7
parsimony method 95–100
  ancestral reconstruction 95–9
  positive selection 398, 401
  tree reconstruction 95–100
partition distance 75
partition model 123–4, 128, 276, 438
PAUP 338, 449
Pearson, Egon 182, 191
Pearson, Karl 170, 182
phylogram 74
PHYLIP 12, 181, 449
PHYLOBAYES 167, 264, 375, 449
PHYML 122, 143, 152, 278, 264, 441, 449
phytochrome gene 410–11
P_{jump}, see acceptance proportion
Poisson distribution 198, 366, 369, 423–4
Poisson model 40–1, 110
Poisson process 275, 315–16, 321, 327, 430–1, 438
  simulation 430–3
  variable-rate 321–2, 328, 336, 345–6, 349, 385, 431–3
Poisson random field 393, 397
polymorphism 309, 312, 392, 395–7
polytomy 73
posterior split probability 277–8, 296–8, 309
posterior distribution 182–6
post-order tree traversal 105
post-trial evaluation 191
Potential scale reduction statistic 242–3
pre-order tree traversal 105
pre-trial betting 191
primate 373–5, 385–6
principle of indifference 187
principle of parsimony 169
prior distribution 182–6
  branch lengths 266–9
  conjugate 198–9
  controversy 183, 187
  diffuse 197
  hyper 197, 267
  ignorance 188, 193
  Jeffreys 200–2
  non-informative 188, 193
  reference 202
  sensitivity 194, 196–7, 212, 259–60, 297, 356, 381
  tree topology 276–9
  vague 192

proposal, *see* MCMC proposal
proposal density 218
proposal ratio 218, 220, 445
pruning algorithm 103–6
pseudo-random number, *see* random number
pulley principle 107–8
purine 7
pyrimidine 7
θ (theta) **312**, 316–20

R

random variable generation
   inversion 420, 425
   rejection 425–8
   transformation 425
random number 410–20
   seed 420
random-sites model 402–5
rate smoothing 368–72
RAxML 143, 264, 441, 449
rbcL gene 50, 53–4, 56–7, 59–60, 63–4, 144–7, 152
reciprocal monophyly 350–1
recombination 88, 308, 310, 317–18, 326, 329, 340, 393, 397, 413
reflection 222
regulatory gene 416
relative rate test 363–4
RELL bootstrap 174–5
reproductive isolation 356
retention index 180
REV, *see* general time-reversible model
reversibility 11, 30, 66, 216
   root of tree 28, 30, 107
   test 148
rjMCMC 252–5, 352–5
   JC69–K80 253–4
root 70
rooted tree 70, 74
rooting 70
rule of succession 187

S

safe-guided Newton algorithm 136
saturation 33, 110
segregating sites 317–18, 393–7
selection 35–6, 43, 51, 56, 61–2, 309, 390–8
   negative 43, 391
   positive 43, 391
   purifying 43, 391
   codon usage 45
selection bias 176, 178
sequence error 110–13
sequential addition, *see* stepwise addition

Shimodaira-Hasegawa (S-H) test 178
simulated annealing 88
simulation 419
   Bayesian 242
site configuration, *see* site pattern
site length, *see* character length
site pattern 106
site-frequency spectrum 394–7
site-specific model 124
sliding-window proposal
   uniform 222
   normal 223–4
   Bactrian 223
   optimal step length 218, 236–8
   bounds and reflection 222
speciation 78, 80–1, 272, 326, 331–5, 345
   allopatric 350, 358–9
   sympatric 350, 359
species concept 349–50
   biological species concept 349
   phylogenetic species concept 350
species tree 79–81, 325–43
species tree-gene tree conflict 331–5
spectral decomposition 10, 66
split 75, 79, 174–7, 180, 283–4
   prior probability 277–9, 283
   posterior probability 277–8, 296–8, 309
SPR, *see* subtree pruning and regrafting
star decomposition 83
stationary (steady-state) distribution 6, 27, 33, 219, 226–7, 235, 238, 436
steepest ascent 136
steepest descent 136
STEM 338–9, 341, 351
step matrix 96, 100
stepwise addition 73–4, 82–4
stochastic simulation, *see* simulation
substitution rate matrix 3–4, 7, 9, 26, 29–31, 36, 43, 66, 269, 274, 403, 435
subtree pruning and regrafting (SPR) 85, 287–9, 293–4
subtree swapping 291–2
super-efficiency 162–3
super-matrix 124
super-tree 124
switching model 121
synonymous site 48
synonymous substitution 35, 43

T

Tajima's D 393–4
Taylor expansion 66, 376
TN93 3, 9–13

TBR, *see* tree bisection and reconnection
thinning 243
transition 7
transition kernel 219
transition probability 4–6, 10, 219, 433–4
transition probability matrix 4, 10, 26, 65–7, 230, 239, 347
   computation 65–7
transition/transversion rate ratio 8, 13–14, 51–3
transversion 7
transversion parsimony 96
tree bisection and reconnection (TBR) 85, 289–91
tree height (t_{MRCA}) 313
tree length 91, 109, 180, 266–70, 286, 313–14, 317
tree rearrangement, *see* branch swapping
tree reconstruction method
   Bayesian 81, 263
   cluster 81
   distance 88
   likelihood 102
   minimum-evolution 91
   optimality-based 81
   parsimony 95
tree search 140
   branch swapping 84, 86
   exhaustive 82
   heuristic 82
   stochastic 88
tree space 86, 140–3
   local peaks 86–8, 140–3

U

unbiased estimate 19, 155, 214, 393
uniform distribution 187, 193, 222, 418, 421–2, 425
unrooted tree 70–1
UPGMA 81

V

viral divergence 372–3, 384–5

W

ω, *see* d_N/d_S
WAG 36–7, 42, 272
weighted parsimony 96
winning-sites test 180
Wright-Fisher model, *see* Fisher-Wright model

Y

YN00 53–4, 61
Yule process 74, 197, 263, 277, 340, 350, 381–2, 439

Printed and bound by CPI Group (UK) Ltd, Croydon, CR0 4YY